CMOS Front-End Electronics for Radiation Sensors

Devices, Circuits, and Systems

Series Editor
Krzysztof Iniewski
CMOS Emerging Technologies Research Inc.,
Vancouver, British Columbia, Canada

PUBLISHED TITLES:

Atomic Nanoscale Technology in the Nuclear Industry
Taeho Woo

Biological and Medical Sensor Technologies
Krzysztof Iniewski

Building Sensor Networks: From Design to Applications
Ioanis Nikolaidis and Krzysztof Iniewski

Circuits at the Nanoscale: Communications, Imaging, and Sensing
Krzysztof Iniewski

CMOS: Front-End Electronics for Radiation Sensors
Angelo Rivetti

Design of 3D Integrated Circuits and Systems
Rohit Sharma

Electrical Solitons: Theory, Design, and Applications
David Ricketts and Donhee Ham

Electronics for Radiation Detection
Krzysztof Iniewski

Electrostatic Discharge Protection of Semiconductor Devices and Integrated Circuits
Juin J. Liou

Embedded and Networking Systems: Design, Software, and Implementation
Gul N. Khan and Krzysztof Iniewski

Energy Harvesting with Functional Materials and Microsystems
Madhu Bhaskaran, Sharath Sriram, and Krzysztof Iniewski

Graphene, Carbon Nanotubes, and Nanostuctures: Techniques and Applications
James E. Morris and Krzysztof Iniewski

High-Speed Devices and Circuits with THz Applications
Jung Han Choi

High-Speed Photonics Interconnects
Lukas Chrostowski and Krzysztof Iniewski

PUBLISHED TITLES:

High Frequency Communication and Sensing: Traveling-Wave Techniques
Ahmet Tekin and Ahmed Emira

Integrated Microsystems: Electronics, Photonics, and Biotechnology
Krzysztof Iniewski

Integrated Power Devices and TCAD Simulation
Yue Fu, Zhanming Li, Wai Tung Ng, and Johnny K.O. Sin

Internet Networks: Wired, Wireless, and Optical Technologies
Krzysztof Iniewski

Labs on Chip: Principles, Design, and Technology
Eugenio Iannone

Laser-Based Optical Detection of Explosives
Paul M. Pellegrino, Ellen L. Holthoff, and Mikella E. Farrell

Low Power Emerging Wireless Technologies
Reza Mahmoudi and Krzysztof Iniewski

Medical Imaging: Technology and Applications
Troy Farncombe and Krzysztof Iniewski

Metallic Spintronic Devices
Xiaobin Wang

MEMS: Fundamental Technology and Applications
Vikas Choudhary and Krzysztof Iniewski

Micro- and Nanoelectronics: Emerging Device Challenges and Solutions
Tomasz Brozek

Microfluidics and Nanotechnology: Biosensing to the Single Molecule Limit
Eric Lagally

MIMO Power Line Communications: Narrow and Broadband Standards, EMC, and Advanced Processing
Lars Torsten Berger, Andreas Schwager, Pascal Pagani, and Daniel Schneider

Mobile Point-of-Care Monitors and Diagnostic Device Design
Walter Karlen

Multisensor Data Fusion: From Algorithm and Architecture Design to Applications
Hassen Fourati

Nano-Semiconductors: Devices and Technology
Krzysztof Iniewski

Nanoelectronic Device Applications Handbook
James E. Morris and Krzysztof Iniewski

Nanopatterning and Nanoscale Devices for Biological Applications
Šeila Selimović

PUBLISHED TITLES:

Nanoplasmonics: Advanced Device Applications
James W. M. Chon and Krzysztof Iniewski

Nanoscale Semiconductor Memories: Technology and Applications
Santosh K. Kurinec and Krzysztof Iniewski

Novel Advances in Microsystems Technologies and Their Applications
Laurent A. Francis and Krzysztof Iniewski

Optical, Acoustic, Magnetic, and Mechanical Sensor Technologies
Krzysztof Iniewski

Optical Fiber Sensors: Advanced Techniques and Applications
Ginu Rajan

Optical Imaging Devices: New Technologies and Applications
Dongsoo Kim and Ajit Khosla

Organic Solar Cells: Materials, Devices, Interfaces, and Modeling
Qiquan Qiao

Radiation Detectors for Medical Imaging
Jan S. Iwanczyk

Radiation Effects in Semiconductors
Krzysztof Iniewski

Semiconductor Radiation Detection Systems
Krzysztof Iniewski

Smart Grids: Clouds, Communications, Open Source, and Automation
David Bakken

Smart Sensors for Industrial Applications
Krzysztof Iniewski

Soft Errors: From Particles to Circuits
Jean-Luc Autran and Daniela Munteanu

Solid-State Radiation Detectors: Technology and Applications
Salah Awadalla

Technologies for Smart Sensors and Sensor Fusion
Kevin Yallup and Krzysztof Iniewski

Telecommunication Networks
Eugenio Iannone

Testing for Small-Delay Defects in Nanoscale CMOS Integrated Circuits
Sandeep K. Goel and Krishnendu Chakrabarty

VLSI: Circuits for Emerging Applications
Tomasz Wojcicki

Wireless Technologies: Circuits, Systems, and Devices
Krzysztof Iniewski

Wireless Transceiver Circuits: System Perspectives and Design Aspects
Woogeun Rhee

FORTHCOMING TITLES:

Advances in Imaging and Sensing
Shuo Tang, Dileepan Joseph, and Krzysztof Iniewski

Analog Electronics for Radiation Detection
Renato Turchetta

Cell and Material Interface: Advances in Tissue Engineering, Biosensor, Implant, and Imaging Technologies
Nihal Engin Vrana

Circuits and Systems for Security and Privacy
Farhana Sheikh and Leonel Sousa

CMOS Time-Mode Circuits and Systems: Fundamentals and Applications
Fei Yuan

Gallium Nitride (GaN): Physics, Devices, and Technology
Farid Medjdoub and Krzysztof Iniewski

Ionizing Radiation Effects in Electronics: From Memories to Imagers
Marta Bagatin and Simone Gerardin

Mixed-Signal Circuits
Thomas Noulis and Mani Soma

Magnetic Sensors: Technologies and Applications
Simone Gambini and Kirill Poletkin

MRI: Physics, Image Reconstruction, and Analysis
Angshul Majumdar and Rabab Ward

Nanoelectronics: Devices, Circuits, and Systems
Nikos Konofaos

Nanomaterials: A Guide to Fabrication and Applications
Gordon Harling, Krzysztof Iniewski, and Sivashankar Krishnamoorthy

Physical Design for 3D Integrated Circuits
Aida Todri-Sanial and Chuan Seng Tan

Power Management Integrated Circuits and Technologies
Mona M. Hella and Patrick Mercier

Radio Frequency Integrated Circuit Design
Sebastian Magierowski

Reconfigurable Logic: Architecture, Tools, and Applications
Pierre-Emmanuel Gaillardon

Silicon on Insulator System Design
Bastien Giraud

Structural Health Monitoring of Composite Structures Using Fiber Optic Methods
Ginu Rajan and Gangadhara Prusty

FORTHCOMING TITLES:

Terahertz Sensing and Imaging: Technology and Devices
Daryoosh Saeedkia and Wojciech Knap

Tunable RF Components and Circuits: Applications in Mobile Handsets
Jeffrey L. Hilbert

Wireless Medical Systems and Algorithms: Design and Applications
Pietro Salvo and Miguel Hernandez-Silveira

ABOUT THE SERIES EDITOR:

Krzysztof (Kris) Iniewski is managing R&D at Redlen Technologies Inc., a start-up company in Vancouver, Canada. Redlen's revolutionary production process for advanced semiconductor materials enables a new generation of more accurate, all-digital, radiation-based imaging solutions. Kris is also a President of CMOS Emerging Technologies Research Inc (www.cmosetr.com), an organization of high-tech events covering Communications, Microsystems, Optoelectronics, and Sensors. In his career Dr. Iniewski has held numerous faculty and management positions at University of Toronto, University of Alberta, SFU, and PMC-Sierra Inc. He has published over 100 research papers in international journals and conferences. He holds 18 international patents granted in USA, Canada, France, Germany, and Japan. He is a frequent invited speaker and has consulted for multiple organizations internationally. He has written and edited several books for CRC Press, Cambridge University Press, IEEE Press, Wiley, McGraw-Hill, Artech House, and Springer. His personal goal is to contribute to healthy living and sustainability through innovative engineering solutions. In his leisure time Kris can be found hiking, sailing, skiing or biking in beautiful British Columbia. He can be reached at kris.iniewski@gmail.com.

CMOS Front-End Electronics for Radiation Sensors

Angelo Rivetti

ISTITUTO NAZIONALE DI FISICA NUCLEARE, TORINO, ITALY

CRC Press is an imprint of the
Taylor & Francis Group, an **informa** business

CRC Press
Taylor & Francis Group
6000 Broken Sound Parkway NW, Suite 300
Boca Raton, FL 33487-2742

First issued in paperback 2017

ISBN-13: 978-1-4665-6310-0 (hbk)
ISBN-13: 978-1-138-82738-7 (pbk)

Visit the Taylor & Francis Web site at
http://www.taylorandfrancis.com

and the CRC Press Web site at
http://www.crcpress.com

Dedication

To my family and my friends

Contents

Preface

Radiation detectors are indispensable tools in many areas of science. They provide the eyes through which nuclear, particle physics and astrophysics unravel the secrets of Nature at the most fundamental levels, and they find use in several applied fields, such as medical imaging, material analysis and detection of hazardous substances, to name just a few. Modern radiation detectors are complex systems, composed of different devices that perform specific functions. A sensor converts part of the energy of the impinging radiation into an electrical signal, which is pre-processed by the front-end electronics before being further treated by a digital readout circuitry and eventually archived on disk for later inspection. Sophisticated computer programs are necessary to control the instrument operation and to analyze the data. In many applications, the design of the detector housing and of its cooling system is not trivial and poses interesting challenges also to the mechanical engineer.

This book focuses on the front-end electronics of those sensors that allow the detection of individual charged particles or photons with spatial accuracy down to a few micrometers, good energy resolution, high timing precision and low power consumption and are able to capture up to billions of events per second in a surface of the order of a square centimeter. Such performance can only be achieved by segmenting the sensitive area into many independent channels and processing the resulting signals with highly parallelized electronics. The readout and data acquisition tasks are often implemented with Field Programmable Gate Arrays (FPGA), commercially available chips which deploy an impressive amount of gates and even full microprocessors and can be programmed to execute very complex algorithms. The large number of channels and the wide spectrum of possible requirements impose instead in many cases that the very front-end electronics, located immediately after the sensor, is realized in the form of custom integrated circuits fabricated through modern Very Large Scale of Integration (VLSI) technologies.

Historically, the use of custom integrated front-ends coupled to highly segmented radiation detectors was introduced in the mid-eighties of the last century, when the first silicon microstrip sensors were used to study the decays of short living particles in high energy physics experiments. The need of building large detectors to equip huge particle accelerators such as the Tevatron and the RHIC in the United States, the LEP and the LHC at CERN boosted the field, leading to the creation of dedicated groups at the major national and international laboratories and in several Universities, where electronics engineers and physicists work full time at the design of complex front-end chips. The performance improvements made possible by these highly integrated systems were soon recognized in other domains beyond particle physics. Highly segmented radiations sensors coupled to custom designed front-ends started to be employed, for example, in space science, in medical imaging and in radiation dosimetry, while new applications are being added continuously to the list.

Modern front-ends are mixed-signal circuits in which the ultimate performance is set by the analog circuitry that receives the signal from the sensor and transforms it in a binary representation that is further processed either on chip or off-chip with digital techniques. For this reason, the book concentrates on the implementation of the most critical analog building blocks encountered in such systems. While the design of analog integrated circuits is a topic widely covered in University courses and many excellent textbooks are available on the subject, the optimization of the very same circuits in a radiation sensor front-end is a body of knowledge mostly found in papers published in specialized journals. This provided the main motivation for writing this book. Although bipolar transistors have been used in some designs, the large majority of integrated front-ends for radiation sensors is fabricated in CMOS technologies, which allows the realization of complex mixed-signal ASICs at an affordable cost. This consideration led to the choice of restricting the presented material only to CMOS circuits. In compiling a specialized text, sometimes it becomes necessary to recall more general notions, and avoiding the mere repetition of concepts better described by other sources is always a challenge. When I review topics such as the MOS transistor or the design of basic CMOS amplifiers, I therefore try to outline those aspects which are particularly relevant in front-end design and as such not treated in detail in general purpose textbooks.

This book is organized in ten chapters and two appendices.

Chapter 1 describes the typical specifications of a front-end system and presents the most common architectures employed in integrated front-ends. The main purpose of the chapter is to provide an introduction to the electronic engineer that for the first time is involved in the design of a front-end chip for a radiation detector.

Chapter 2 is a review of the basic properties of the MOS transistor. The quadratic I-V relationship, appropriate only for strong inversion, is still widely adopted in standard analog design courses to analyze circuits. Transistors in front-end amplifiers however typically work in weak or moderate inversion, and these operating regions are discussed with some detail in the second part of the chapter. The equations based on the inversion coefficient are also introduced, because they are extensively used in the rest of the book, in particular in the practical examples.

Chapter 3 introduces the design of the very input stages of a front-end. The circuits are built with a step-by-step approach, to give the reader also the opportunity of refreshing basic notions on the design of CMOS amplifiers and current mirrors.

Chapter 4 extends the study of the same circuits to the frequency domain, including stability issues, and addresses the basic properties of Charge Sensitive Amplifiers.

Chapter 5 focuses on noise, because its minimization is one of the key steps in a front-end design. Understanding basic noise properties and mastering the techniques of noise calculation in circuits is thus of utmost importance. Basic noise concepts are discussed here and the most common noise sources are introduced. The necessity of filtering the signal after the very input stage to optimize the signal-to-noise ratio, performing what in the radiation detection community is commonly known as "pulse shaping" , is also presented.

Chapter 6 and Chapter 7 treat respectively time invariant and time variant shapers. In these two chapters, the most common types of shapers are analyzed, and their noise and signal processing properties are calculated. The material is elaborated using circuit macromodels rather than transistor-level representation, starting from highly idealized descriptions and introducing progressively the limitations unavoidably found in real-life circuits. Chapter 7 presents also noise calculation in the time domain and the concept of weighing function, which is an indispensable tool in studying the noise behavior of time variant systems.

Chapter 8 discusses the transistor-level implementation of front-end amplifiers. Noise-aware device sizing, practical implementation of feedback networks, transistor level design of pulse shapers and baseline holders are the key topics addressed in this chapter.

Chapter 9 concentrates on the design of comparators or, as they are often called in the context of radiation detection, discriminators. These blocks are of particular relevance as they are needed to separate small signals from the noise, transforming a small analogue signal in a clean digital pulse, thus allowing the proper identification of hits. An increasing number of applications emphasize the capture of the time of arrival of the impinging particle or photon with a resolution well below the nanosecond. The design of these "timing discriminators" is treated with particular emphasis.

Chapter 10 deals with Data Converters, and in particular with Analog-to-Digital and Time-to-Digital Converters. Data Conversion is an extremely wide and complex field, hence the attention here is limited to those topologies that are of particular relevance in front-end design.

Appendix 1 provides a quick reference to the key properties of CMOS differential and operational amplifiers.

Appendix 2 reviews the critical design steps and gives basic insights into a few recurrent issues encountered in front-end implementation: the reduction of interference between analog and digital circuits integrated on the same chip, the biasing of multi-channel systems and the problem of radiation tolerance.

Acknowledgments

My first thanks go to the CRC editorial and technical staff: Kris Iniewski for encouraging me to write this book, Nora Konopka, Jennifer Ahringer and Sashi Kumar for their technical assistance. I am particularly in debt with Nora for her constant support and her endless patience in accepting a few delays as the number of pages was growing beyond my original intention.

Many persons have helped me in the preparation of the manuscript. Gian Franco Dalla Betta (University of Trento) and Gianluigi De Geronimo (Brookhaven National Laboratory) reviewed some not-so-easy concepts presented in the text. Thanushan Kugathasan (CERN) took the burden of cross-checking many cumbersome mathematical derivations.

My co-workers and PhD students at INFN in Torino: Sara Garbolino, Serena Panati, Valentino Di Pietro, Ennio Monteil, Jonhatan Olave, Luca Pacher, Alberto Riccardi and Manuel Rolo were keen in proof-reading the different chapters, and they were very effective in spotting mistakes and inconsistencies. Since I performed the final revision, the responsibility of the remaining ones is entirely mine. I owe Sara and Manuel additional thanks for helping me so much in important projects when the writing kept me particularly busy.

Barbara Pini arranged the cover layout, relieving me from a task that, given my artistic skills, would have been the most difficult part of the book. The suggestions and comments initially given by Grzegorz Deptuch (Fermilab) and Alberto Aloisio (University of Naples) on the content selection were also very helpful. I must thank also my colleagues Gianni Mazza and Giulio Dellacasa, with whom I have shared so many years of work in chip design, and all the other colleagues at INFN for providing such a wonderful professional environment.

Last, but not least, I extend my deepest gratitude to my family and my friends for their constant support and their patience in tolerating the countless hours, in the evenings and during the weekends, I could not spend with them in the last couple of years.

List of Tables

About the Author

Angelo Rivetti received his degree in Physics from the University of Torino, Italy in 1995 and the Ph.D. in Electrical Engineering from the Polytechnic Institute of the same town in 2000. From 1998 to 2000 he worked at CERN on the implementation of radiation tolerant integrated circuits in commercial deep submicron CMOS technologies. From 2000 to 2001 he was assistant professor with the Faculty of Physics of the University of Torino. In December 2001 he joined the Italian National Institute for Nuclear Physics (INFN), where he developed VLSI front-end circuits now in use in the ALICE and COMPASS experiments at CERN. He is currently a senior member of the research and technology staff of INFN in Torino. His research interests are in the design of mixed signal front-end electronics for hybrid and monolithic radiation detectors employed in high energy physics, medical imaging and industrial applications.

1 Front-end Specifications and Architecture Overview

Radiations detectors convert part of the energy lost by an incident particle into an electrical signal, which is then processed by suitable electronic techniques. The first component of the readout chain is often an integrated circuit optimized to serve a specific purpose (Application Specific Integrated Circuit, ASIC). In many cases, the front-end chip must perform different tasks (amplification, filtering, analog-to-digital conversion, high speed data transmission), making necessary the integration of complex analog and digital circuits on the same silicon substrate. The physics of the process under study, the sensor characteristics and other additional constraints such as the power consumption, the number of channels and the space to fit the system in determine the choice of the front-end electronics parameters.

Before the transistor-level implementation of a front-end chip begins, a few preliminary steps are necessary. First, the application requirements and the sensor properties need to be understood with adequate detail. The essential sensor features must also be captured in a compact electrical network that can be simulated with the standard computer programs employed for circuit design. The first section of the chapter therefore provides a brief introduction to the sensor characteristics that are most relevant for the front-end electronics performance and to the equivalent circuits commonly adopted to model a radiation detector. To make the discussion more concrete, a few examples of widely used sensors are also discussed.

The next step is the analysis of the circuit specifications and their trade-offs. This is a critical point, because often engineers or physicists defining the requirements are not ASIC designers, so they can easily miss important aspects which are peculiar of integrated circuit design and manufacturing. Electronic engineers, on the other hand, may lack a comprehensive view of the final application the chip is intended for. The situation is further complicated because electronics for radiation detectors uses a specific terminology and sometimes even a jargon which is different from the one found in other fields of electronics. Unsatisfactory performance often stems from poorly defined and/or not adequately understood specifications rather than from actual design mistakes. The second section of the chapter thus focuses on the key front-end electronics parameters and on their interplay.

Once the specifications have been properly worked-out, a suitable chip architecture must be identified. The architecture definition is an "open problem" because the same set of requirements can be fulfilled through different circuit implementations. Before the transistor or gate level implementation takes place, high level simulations are hence necessary to validate the fundamental choices. The last part of the chapter briefly examines the most common front-end architectures which have emerged along the years.

In the book, we will make extensive use of basic electronics concepts and methods. Mastering in particular feedback amplifiers, linear network analysis and Laplace transforms is important to take full profit of the presented material. By reading the following pages, the reader can also cross-check her/his familiarity with those notions. In case it is felt that this is not solid enough, it is advised that such topics are reviewed before proceeding further with the reading. Without an adequate knowledge of those basic notions it is still possible to grasp the general ideas, but important details can easily be missed.

1.1 BASIC FEATURES AND ELECTRICAL MODELING OF RADIATION SENSORS

Many different kinds of radiation sensors have been developed along the years: proportional counters, gas electron multipliers, silicon strips, pixels and drift detectors, DEPFETs, CCDs, active pixel sensors, vacuum tube photomultipliers, avalanche photodiodes, silicon photomultipliers provide an already long, but not yet exhaustive list. A detailed description of these devices and of their working principles requires a full book in itself and it is well beyond the scope of our treatment. For that, the reader is referred to one of the many textbooks available on the subject [1–9]. In the following, we will review only the basic concepts which are essential to understand the specifications of front-end circuits and to appreciate how sensors are modeled with simple electrical networks. It is in fact important to realize that an optimized readout can not be designed without an adequate knowledge of the sensor and of the physical measurement the system is intended for.

1.1.1 SIGNAL FORMATION IN DETECTORS

Fig. 1.1 a) shows the working principle of the ionization chamber, formed by enclosing a suitable material between two conductive plates. These are kept at different potentials, in order to establish an electric field inside the detecting volume. For many years, only gasses have been employed as the sensing medium but, starting from the sixties of the last century, the use of semiconductors became more and more common [10]. Insulators are not used because due to a phenomenon known as polarization, their signal quickly deteriorates with the device operation [3]. A charged particle crossing the sensor interacts with its atoms, creating ion-electron pairs in a gas and hole-electron pairs in a semiconductor. A sufficiently energetic photon can eject from an atomic orbital an electron, which then loses the acquired energy, ionizing the medium. In sensors without internal amplification, only the primary charges directly set free by the impinging radiation contribute to the signal formation. This is the case of many semiconductor detectors. In other devices, like gaseous detectors, the primary charge is too small to be efficiently detected and it is thus first transported into a region of higher electric field, where the carriers gain enough energy to create secondary ionization. The carriers originating from this avalanche process induce then the signal

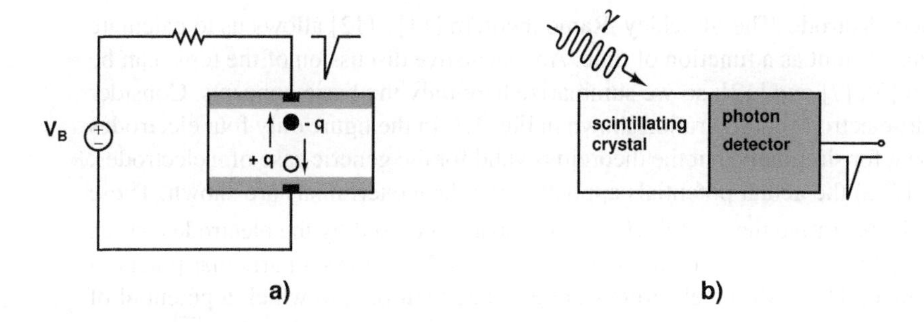

FIGURE 1.1 a) Principle of the ionization chamber. b) Sketch of a scintillator-based detector.

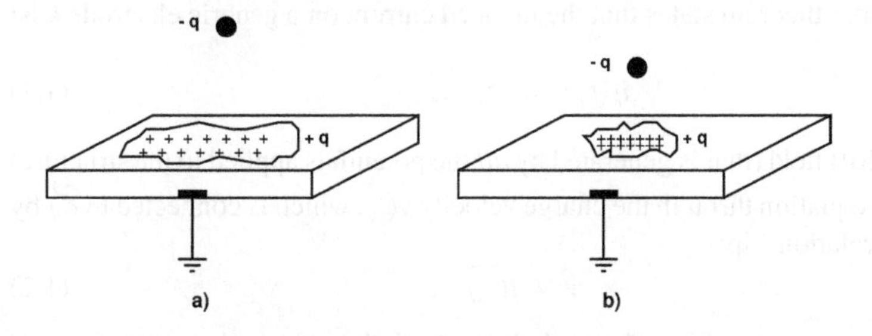

FIGURE 1.2 Charge induction on a grounded electrode.

which is measured by the front-end electronics[1]. Fig. 1.1 b) shows an example of detection involving multiple devices. Here, a gamma ray is first absorbed by a crystal, which re-emits part of the absorbed energy in the form of visible photons. These are detected through photo-electric effect by a photon detector that produces an electrical output pulse.

With radiation sensors, the electrical signal observed by the front-end electronics originates from the *induction* of charge carriers that move inside a system of electrodes. To better understand this point, let's first consider the situation of Fig. 1.2, where a negative charge $-q$ "floats" on top of an ideal grounded conductor. From basic electrostatics, it is well known that this alters the charge distribution at the conductor surface, inducing on it a charge $+q$ of equal magnitude and opposite sign. If the charge $-q$ is now put closer to the conductor, the total induced charge is always the same, but its distribution changes, becoming more peaked around the imaginary line connecting $-q$ to the conductor surface. By dynamically changing from the situation in Fig. 1.2 a) to the one in Fig. 1.2 b), we alter the charge distribution at the conductor surface, causing a motion of charges and thus an electric current. This signal is due by induction and it vanishes if the charge $-q$ stops moving or if it "lands" on the

[1]Gaseous detectors without internal amplification are not used to detect single quanta of radiation. Due to their particularly simple and cheap construction, they are however employed as integrating detectors to measure the electrical current formed by the superposition of many ionizing events.

grounded electrode. The Shockley–Ramo theorem [11], [12] allows us to calculate the signal current as a function of time. An exhaustive discussion of the topic can be found in [5], [7] and [13], so we summarize here only the basic concepts. Consider the multi-electrode configuration shown in Fig. 1.3. In the figure only four electrodes are shown for simplicity, but the theorem is valid for the generic case of n electrodes. In Fig. 1.3 a) the actual potentials applied to the device terminals are shown. These potentials determine the *drift field* $\overrightarrow{\mathscr{E}_d}$ in the space enclosed by the electrodes which is responsible for the movement of charge.[2] Fig. 1.3 b) shows a particular potential arrangement. Here, all the electrodes are grounded but one, to which a potential of 1 V is applied. This configuration is used to calculate the *weighting field* $\overrightarrow{\mathscr{E}_w}$, which determines how a moving charge couples to the specific terminal which has been kept at 1 V. The Ramo theorem states that the induced current on a generic electrode k is:

$$i_k(t) = q\overrightarrow{v(t)} \cdot \overrightarrow{\mathscr{E}_w} \tag{1.1}$$

Note that the drift field (that is generated by *all* the potentials applied to the structure) enters into the equation through the charge velocity $\overrightarrow{v(t)}$, which is connected to $\overrightarrow{\mathscr{E}_d}$ by the following relationship:

$$\vec{v} = \mu\overrightarrow{\mathscr{E}_d} \tag{1.2}$$

where μ is the carrier mobility. The analytical calculation of $i_k(t)$ is straightforward for a simple sensor like the one of Fig. 1.1 a). Suppose in fact that L is the distance between the two electrodes. If a bias voltage V_B is applied, the drift field inside the volume is uniform and equal to V_B/L. The weighting field is obtained by applying on the electrode of interest (usually, the one connected to the front-end electronics) a unit potential, so it is simply given by $1/L$. Notice that in this case the drift and weighting field differ only by a scale factor, but in more general cases the maps of the two fields inside the sensor are *different*. Assume now that a charge pair is generated in the middle of the device. As soon as the carriers are set free, they start moving towards the terminals under the effect of the drift field. Note that electrons and holes/ions move in opposite directions , but the signals they induce on a given electrode add up, as their charges are also opposite. Using the Ramo theorem, we can write the total induced current as:

$$i = q\left(\mu_n + \mu_p\right)\frac{V_B}{L}\frac{1}{L} = q\left(\mu_n + \mu_p\right)\frac{V_B}{L^2} \tag{1.3}$$

Consider first the case in which the carriers have the same mobility, $\mu_n = \mu_p = \mu$, hence (1.3) reads:

$$i = q2\mu\frac{V_B}{L^2} \tag{1.4}$$

The two carriers start from the same point and move at identical speed, so they take the same time to arrive at their respective terminals. The time t_c that a carrier needs

---

[2] If the detector volume contains space charge, this must also be included in the field calculation.

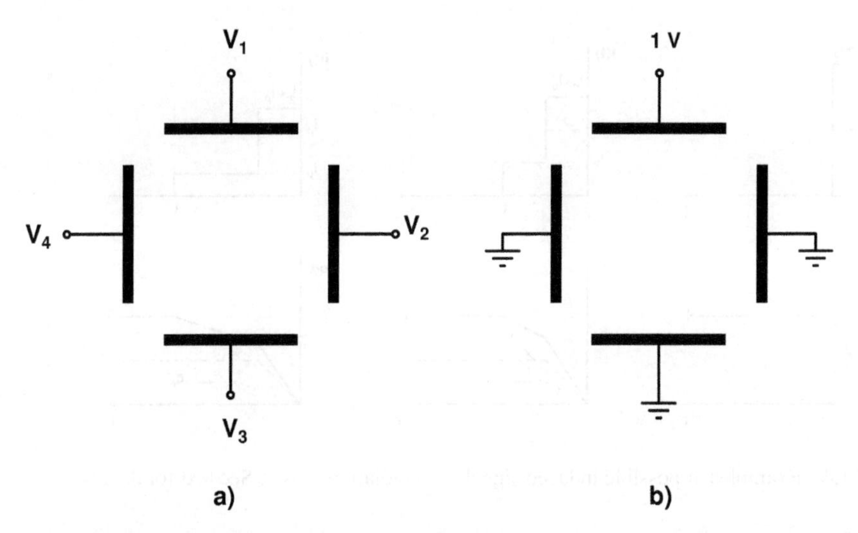

FIGURE 1.3 Potential configurations in a four-electrode system to calculate the drift (left) and the weighting field (right).

to reach the electrode is called the *collection time*. The current is induced for a time t_c and can be expressed as:

$$t_c = \frac{L}{2v} = \frac{L}{2\mu\frac{V_B}{L}} = \frac{L^2}{2\mu V_B} \tag{1.5}$$

Notice that in this particular case both the positive and negative charge induces the same current for the same time. The total charge induced on the electrode is obtained by integrating the induced current signal for a time equal to t_c and, in the case under study, is given by:

$$Q = 2q\mu\frac{V_B}{L^2}\frac{L^2}{2\mu V_B} = q \tag{1.6}$$

If the integration is done for a time smaller than t_c, only a fraction of q is observed. The induced currents and charge as a function of time are shown in the two plots in the leftmost part of Fig. 1.4. The current rises abruptly to the peak value, given by the sum of the currents induced by the two carriers, and stops immediately as the charges reach their electrodes. Observe that the *total* induced charge is q and *not* $2q$.

Consider now the more general case in which the mobility of the charges is different. Suppose, for instance, that the positive carrier is slower and that $\mu_p = \mu_n/2$. The time to collect the faster carrier is:

$$t_{cn} = \frac{L}{2v_n} = \frac{L}{2\mu_n\frac{V_B}{L}} = \frac{L^2}{2\mu_n V_B} \tag{1.7}$$

For the slower positive carrier, the time is instead:

$$t_{cp} = \frac{L}{2v_p} = \frac{L}{2\mu_p\frac{V_B}{L}} = \frac{L^2}{\mu_n V_B} \tag{1.8}$$

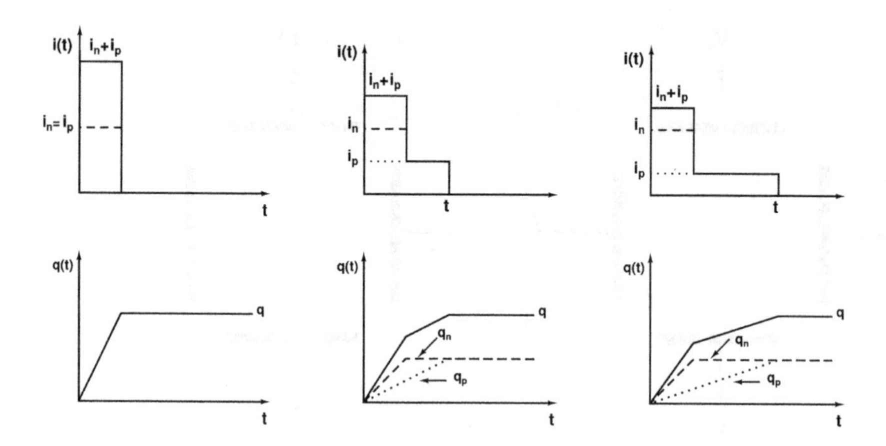

FIGURE 1.4 Examples of possible induced signals in a radiation sensor. See text for details.

The charges have to travel the same distance, but now the positive carrier takes twice as much the time than the negative one. During time between 0 and t_{cn}, both carriers are moving and induce upon the electrode a current given by:

$$i_n + i_p = q(\mu_n + \mu_p)\frac{V_B}{L^2} = q\frac{3}{2}\mu_n\frac{V_B}{L^2} \tag{1.9}$$

At $t = t_{cn}$ the negative carrier arrives on its terminal and stops contributing to the signal. During the time t_{cn} the total induced charge is:

$$q_1 = (i_n + i_p)t_{cn} = q\frac{3}{2}\mu_n\frac{V_B}{L^2}\frac{L^2}{2\mu_n V_B} = \frac{3}{4}q \tag{1.10}$$

Notice that in this first phase, *both* the negative and the positive carriers contribute to the signal. At $t = t_{cn}$ the current sharply drops to:

$$i_p = q\mu_p\frac{V_B}{L^2} = q\frac{1}{2}\mu_n\frac{V_B}{L^2} \tag{1.11}$$

Only the positive carrier contributes to the signal between t_{cn} and t_{cp} and when it arrives upon its terminal, the induction completely stops and the signal ends. The charge induced by i_2 is:

$$q_2 = i_2(t_{cp} - t_{cn}) = \frac{1}{2}q\mu_n\frac{V_B}{L^2}\left(\frac{L^2}{\mu_n V_B} - \frac{L^2}{2\mu_n V_B}\right) = \frac{q}{4} \tag{1.12}$$

The total charge is given by $q_1 + q_2$ and it is again equal to q. This second case is shown in the central part of Fig. 1.4. Observe that if we stop the integration of the signal after t_{cn} we measure only 75% of the total induced charge. Finally, the rightmost part of the figure shows the case where $\mu_p = 1/3\mu_n$. The calculation is similar to the ones just discussed and it is left as an exercise to the reader.

In radiation sensors, the negative and positive charge carriers have in general different mobility. The total charge is measured only after all the carriers have been collected and if the integration time in the front-end electronics is smaller than the signal collection time only one part of the charge is measured. In gaseous detectors, the mobility of ions can be three order of magnitude lower than the one of electrons [4]. The ions thus give rise to very slow signal components. Such long tails are undesirable, as signals due to different interactions will overlap. Some gas detectors are designed so that the ions do not induce signals on the electrodes connected to the front-end electronics. If this is not the case, the detector equivalent circuit must take properly into account the presence of the slow ion tail in the signal and the front-end electronics must be designed to be insensitive to it.

Finally, consider the case of a pair created at a generic position x inside the sensor. If the negative carrier has to travel for a distance x, then the positive one must travel for $L-x$. To reach its electrode, the negative carrier takes a time t_{cn} given by:

$$t_{cn} = \frac{x}{v_n} = \frac{x}{\mu_n \frac{V_B}{L}} = \frac{x}{\mu_n}\frac{L}{V_B} \tag{1.13}$$

During this time, it induces the current:

$$i_n = q\mu_n \frac{V_B}{L^2} \tag{1.14}$$

The total charge induced by the negative carrier is thus:

$$q_n = i_n t_{cn} = q\mu_n \frac{V_B}{L^2}\frac{xL}{\mu_n V_B} = q\frac{x}{L} \tag{1.15}$$

For the positive carrier the induced current is instead:

$$i_p = q\mu_p \frac{V_B}{L^2} \tag{1.16}$$

while its travel time can be written as:

$$t_{cp} = \frac{(L-x)}{\mu_p \frac{V_B}{L}} \tag{1.17}$$

The induced charge is thus given by:

$$q_p = i_p t_{cp} = q\mu_p \frac{V_B}{L^2}\frac{(L-x)L}{\mu_p V_B} = q\left(1-\frac{x}{L}\right) \tag{1.18}$$

Observe that the total induced charge, obtained by summing (1.15) and (1.18) is always q. Notice also that the current induced on the two electrodes has equal magnitude and opposite polarity.

The analytical derivation of induced currents become quickly cumbersome as more realistic arrangements are considered. As we will see, at least one collection electrode is usually segmented to allow for adequate space resolution, the field inside the detector might not be uniform, mechanisms may exist that trap the moving charges, and so on. A numerical approach becomes hence necessary and a sensor is optimized with the help of extensive computer simulations, that allow us to obtain a more accurate description of the induced signal.

1.1.2 SIGNAL POLARITY

We can now examine the less idealized configuration depicted in Fig. 1.5. Here, one plate of the detector is connected to a high voltage supply, which is filtered to reduce noise. The other arm is not grounded, but it is tied to the input of a front-end amplifier, obtained by connecting a suitable network in the feedback path of a high gain voltage amplifier. The amplifier output voltage must be within the power supply rails and if it is small enough that the circuit does not saturate, the input voltage experiences very little variations, giving rise to the "virtual ground" condition. In case of Fig. 1.5, the high voltage is negative with respect to the common ground reference. Hence, a positive charge moves towards the high voltage while a negative one moves "upstream" towards the amplifier input. While the charges are moving, they induce a current signal on the two plates. We stress once more that the observed current pulse is generated by the charge carriers when they move inside the sensor and not when they arrive at the sensor terminals. As soon as a carrier reaches the electrode, it stops contributing to the signal and when all the charges are collected, the signal is fully over. For the configuration of Fig. 1.5, the negative charge moves towards the upper electrode, repelling other negative charges from it towards the amplifier input. This determines a negative bounce at the input node, which translates into a signal swinging upwards at the output. In making an electrical model of the sensor, one has to remember that in electric circuits current is represented as a flow of positive charges. Therefore, electrons flowing towards the amplifier input are equivalent to positive charges moving away from it, and the representation of Fig. 1.5 b) is obtained. The complementary situation, in which one detector arm is connected to a power supply much more positive than the amplifier input is shown in Fig. 1.6, and results in reversed signal polarities. We notice that the device sketched in Fig. 1.1 resembles a parallel plate capacitor, with the two electrodes acting as the capacitor arms and the sensing material playing the role of the dielectric. This justifies the representation of a detector as a current source with a capacitor in parallel, indicated as C_d in Fig. 1.5 and Fig. 1.6.

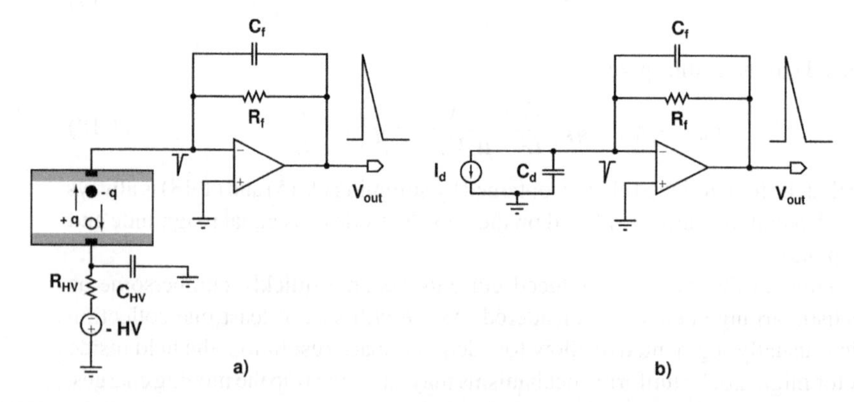

FIGURE 1.5 Signal polarity in a simple parallel plate detector in case the high voltage power supply is negative.

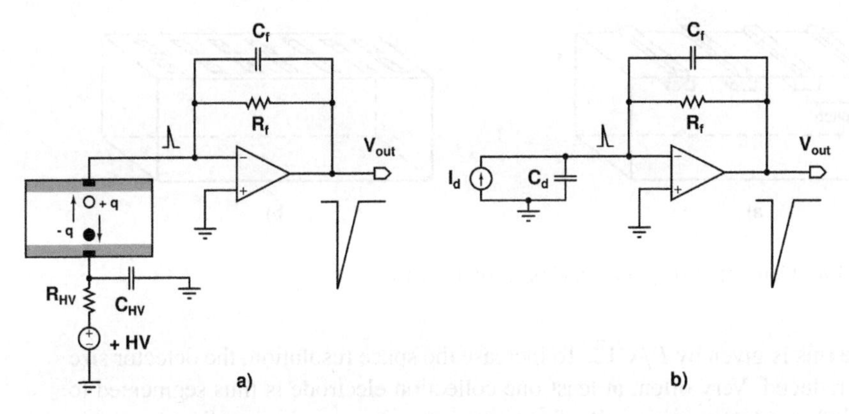

FIGURE 1.6 Signal polarity in a simple parallel plate detector in case the high voltage power supply is positive.

1.1.3 SPACE RESOLUTION AND DETECTOR SEGMENTATION

Consider a simple two-terminal detector in which the electrode width is P. As shown in Fig. 1.7, we can define a coordinate system with the origin in the middle of the device. A particle can cross the detector at any point between $-P/2$ and $+P/2$, so each point has the same probability $1/P$ of being hit. Any time an event is detected, the coordinate of the central point is conventionally assigned. When the particle crosses exactly in the center, the error made in measuring the space coordinate is thus zero, while if it passes close to the borders, the error is maximum and equal either to $-P/2$ or $+P/2$. The variance of a uniform distribution with zero average can be calculated as:

$$\sigma_P^2 = \int_{-P/2}^{+P/2} \frac{x^2}{P} dx = \frac{P^2}{12} \tag{1.19}$$

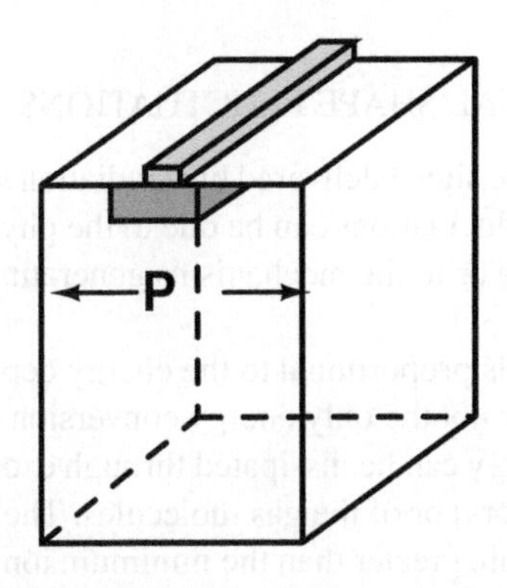

FIGURE 1.7 The space resolution of a single channel detector in a given direction is $P/\sqrt{12}$, where P is the cell size in the considered direction.

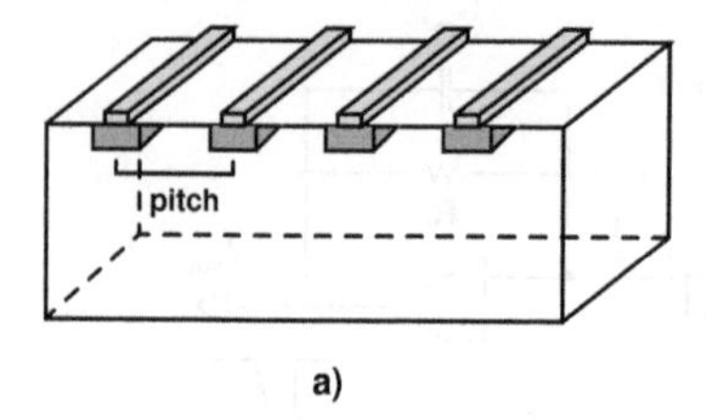

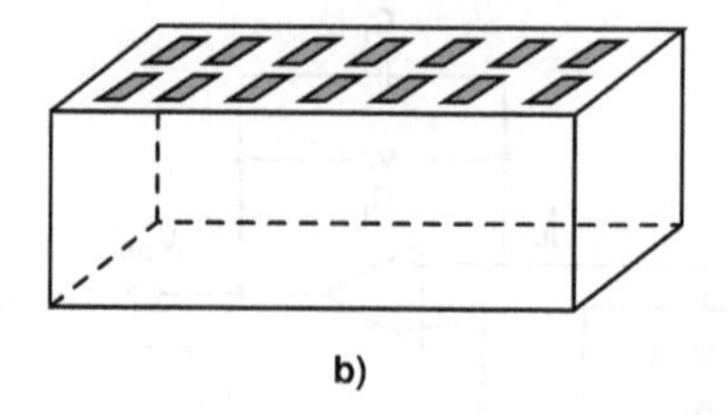

FIGURE 1.8 Concept of strip (in a) and pixel (in b) sensor.

while the rms is given by $P/\sqrt{12}$. To increase the space resolution, the detector size must be reduced. Very often, at least one collection electrode is thus segmented to provide independent sensing cells. Segmentation may occur in one direction (strip sensors) or in both (pixels), as shown in Fig. 1.8. The distance between the centers of two neighboring electrodes is called the *pitch*. In a segmented sensor, the capacitance of each cell has two components: one between the collecting node and the backplane and another one between the electrode and its neighbors, which is often the dominant one. In a multi-electrode system, the charge can be collected by one or more electrodes. In the former case, the space resolution is still given by $P/\sqrt{12}$. In the latter, if the amplitude of the signals is recorded, the coordinate of the impact point can be deduced by calculating the center of gravity of the resulting charge distribution. In this case, a resolution significantly better than the pure geometrical one can be obtained. Although the improved space resolution is often the primary reason to increase granularity, sensor segmentation has additional benefits. In fact, for a given particle flux, the event rate per channel is reduced, allowing the use of slower and less power-demanding front-end circuits. Also the capacitance of the collecting electrode becomes smaller and this is beneficial for the noise performance. Despite the number of front-end channels increases, the power per channel can be reduced. Highly segmented detectors may therefore offer overall better system performance. The high granularity reached by modern detectors is thus the main motivation beyond the design of highly integrated front-end chips.

1.1.4 AMPLITUDE AND SIGNAL SHAPE FLUCTUATIONS

The amplitude and the shape of the signal delivered by a radiation sensor may change from one event to the next. Such fluctuations can be due to the physics of the process that produces the radiation quanta or to the mechanisms generating the signal inside the sensor.

In linear detectors, the charge is proportional to the energy deposited by the radiation pulse. Ionization is however not the only energy conversion mechanism acting within the device. Part of the energy can be dissipated through excitation of the crystal lattice (in semiconductor sensors) or of the gas molecules. The average energy to create a charge pair, E_P, is therefore greater than the minimum ionization energy, E_I. As an example, in silicon E_I is 1.2 electron-Volt (eV), while E_P is 3.6 eV. In gases, E_I is typically between 10 eV and 20 eV, while E_P ranges from 20 eV to 50 eV [4].

If E is the energy lost by the impinging quantum, the average number of generated pairs is thus given by:

$$N_P = \frac{E}{E_P} \tag{1.20}$$

The energy is lost in the sensor through different mechanisms, therefore the number of produced pairs is regulated by probabilistic laws and fluctuates from one event to the next even in case the energy released is the same. From the physics point of view, such fluctuations are undesirable, because they limit the sensor accuracy in measuring the radiation energy. Applying Poisson statistics, one would expect that the standard deviation of N_P is given by $\sqrt{N_P}$, but this prediction is in general not correct. In fact, if only events releasing the same energy are considered, the total energy lost in the sensor is constrained and the number of possible energy dissipation modes, although high, is not infinite. The standard deviation of N_P can thus be expressed as:

$$\sigma_{NP} = \sqrt{FN_P} \tag{1.21}$$

where F is known as the Fano factor [14]. For many materials (including silicon and most gases) F is smaller than one, so the energy resolution is actually better than what is predicted by elementary statistics. For charged particles, the average energy loss per unit path length, indicated as dE/dx, depends on the particle type, particle energy and characteristics of the material and can be calculated with the Bethe-Block formula [4]. Highly energetic charged particles can traverse completely a material and lose only a small fraction of their original energy. In thin absorbers, dE/dx shows large fluctuations, which are described by the Landau distribution [4], an example of which is shown in Fig. 1.9. The plot reports the specific loss of a beam of monoenergetic particles traversing a 100 μm thick silicon sensor. The x axis represents the amplitude of the signal in arbitrary units while the y axis gives the number of entries per each amplitude. The distribution has a peak at about 28 amplitude units, but a few signals yield values significantly bigger than the most probable one. For the same deposited energy, sensor signals fluctuate also in shape. The electric field inside the device is in fact not perfectly uniform and the field lines bend at the borders of the cell. Therefore, the charge collection is faster if the primary particle traverses the sensor in the middle and slower if the crossing occurs in proximity of the sensor boundaries. For the same deposited charge, the collection time may thus change from event to event, resulting in fluctuations of the sensor signal shape.

1.1.5 SENSOR CAPACITANCE

The sensor capacitance plays a central role in defining the system performance. In the first front-end circuits, developed in the thirties of the last century and based on vacuum tubes, the detector capacitance was used as the conversion element between the charge and the voltage domain and the resulting voltage signal, given by Q/C_d, was amplified with high-gain voltage amplifiers. In modern front-ends, the charge-to-voltage conversion is usually done by a circuit component placed in the feedback path of a high gain amplifier, but the sensor capacitance limits nevertheless the speed and

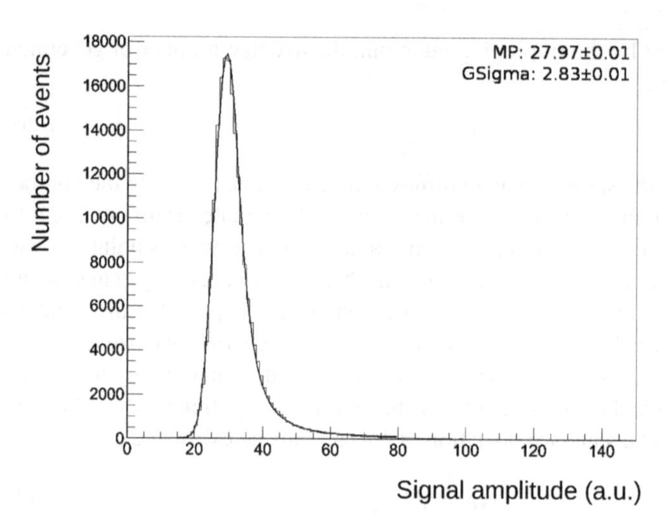

FIGURE 1.9 Example of a Landau distribution. The plot was obtained from a beam of 2.75 GeV/c protons traversing a hybrid silicon detector 100 μm thick. Courtesy of the PANDA-MVD collaboration.

noise figures of these systems. The Q/C_d ratio i.e. the ratio between the average charge yield and the capacitance in a given sensor, still provides a useful metric of the "effort" that is needed to obtain a good signal-to-noise ratio in the front-end electronics. The value of sensor capacitance may range from a few femto-Farads typical of monolithic CMOS sensors to the nano-Farads found in large area photodiodes.

1.1.6 LEAKAGE CURRENT

Some sensors drive also a DC current. This is particularly important in semiconductor detectors, where the current originates from electron-hole pairs thermally generated in the depletion region and swept by the electric field. In semiconductors, leakage current is a sensitive function of temperature and in silicon it doubles approximately for every 8° C of temperature increase. A reference value for a good silicon sensor is 1 nA/cm^2 at room temperature. For Germanium detectors, the leakage current is much higher and these devices need to be cooled with liquid Nitrogen to be of practical use. Leakage increases significantly if the device bulk is damaged by non-ionizing radiation. The leakage current has two effects on the front-end electronics. First, by flowing in the input circuitry, it is amplified and it can shift significantly the DC operating points of the front-end amplifier. This is addressed either by AC coupling the sensor to the front-end or by implementing on chip suitable compensating techniques. Second, leakage current is formed by discrete carriers crossing a junction and is subjected to random fluctuations. It therefore generates also noise, which, being a broad-band signal, can not be suppressed by the AC coupling.

1.1.7 SENSOR EQUIVALENT CIRCUIT

A front-end circuit can not be designed without a realistic electrical model of the sensor. From the point of view of the signal processing chain, radiation sensors can be described by fairly simple linear networks made of basic elements such as current sources, capacitors, resistors and inductors. In most cases, the circuit shown in Fig. 1.10 is adequate. The model consists of a current source, I_d, representing the sensor signal as a function of time. Connected in parallel to the source, C_d models the capacitive load the sensor presents to the front-end electronics. The DC source I_L takes into account the device leakage current, while R_d models the resistive component of the sensor output impedance (usually negligible) or physical resistors that might be used in the sensor biasing network. Inductors L_t and resistor R_t describe respectively the parasitic inductance and resistance of the connections. In many circumstances these can be omitted but care must be taken to properly handle them in high speed and very low noise applications. The direction of the current source must be chosen to properly reflect the signal polarity, as discussed in the previous section. While the model of Fig. 1.10 can be used for many detectors, the values of the parameters depend on the particular device and may change by orders of magnitudes even between sensors relying on the same working principles, but optimized for different measurements and operating under different conditions (bias voltage, temperature, radiation exposure, etc.). Circuit simulation programs offer the possibility of feeding input stimuli to the circuit under study with different shapes. The most common functions, such as square waves, rectangular pulses, sinusoidal and exponential waveforms are found in the form of built-in generators that can be easily parametrized. In modeling the current source I_d different approaches can hence be taken. In the simplest one, all the details of the signal shape are neglected and the current pulse is modeled as a Dirac-delta, represented by a narrow rectangular pulse conveying the input charge Q_{in}. This model is very often employed and it is useful as a first step, because it allows us to study the impulse response of the front-end electronics. However, it is adequate for a final validation only if the impulse response of the front-end is long compared to the sensor signal duration (the "collection time"). If this is not the case, a more accurate description of the detector signal becomes necessary. Some signal features can be introduced by using other type of sources. For instance, the tail due the slow ion motion in a gas detector can be modeled with an exponential waveform

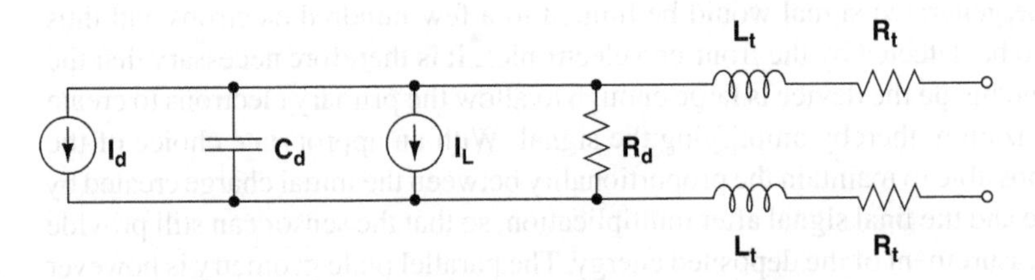

FIGURE 1.10 Equivalent small signal circuit that can be used to model many radiation sensors.

with appropriate decay time or with a combination of exponentials. The most accurate results are obtained by giving input signals generated by a dedicated software that simulate the particle interactions within the sensor. In this case, the information can be fed to the circuit simulator as a set of data points. To make a realistic study of the front-end response, many iterations must be run, each corresponding to a particular input signal. Professional programs allow us to do such detailed studies in a reasonable time because, with suitable scripting languages, the CAD can be instructed to do automatically all the key operations (input file reading, netlist generation, simulation, extraction and plotting of the relevant features).

1.1.8 MODELING OF COMPOSITE SYSTEMS

In many applications, the electrical signal which is fed to the front-end electronics is the result of a multi-step process involving different devices. For instance, in the fields of nuclear spectroscopy, medical imaging and detection of hazardous materials, photon detectors are used in conjunction with scintillators to detect highly penetrating radiation such as gamma-rays [4]. The gamma-ray is absorbed by the scintillator, which re-emits part of the excitation energy as visible or near UV photons, detected by a photosensor (in most cases, a vacuum tube photomultiplier). In this case, the capacitance presented to the front-end electronics is still defined by the photosensor. However, the signal shape is given by the convolution of the scintillator and the photomultiplier response and it is primarily dominated by the decay time of the former. When detection is achieved by a chain of elements, the device connected to the front-end electronics determines the equivalent circuit to be used. However, while some values of the model parameters are defined only by the last component (such as the capacitance), others, like the signal shape, depend on the behavior of the full chain. The complete system must thus be studied and modeled with adequate care before starting the design of its front-end electronics.

1.2 EXAMPLE OF RADIATION SENSORS

To further understand the applicability and the limits of the model in Fig. 1.10, we now briefly consider a few examples of sensors. Gas-based detectors are still widely used as they allow us to cover large areas at a fairly low cost. In a gas detector, the simple parallel plate geometry shown in Fig. 1.1 is not adequate to detect single particles because the generated signal would be limited to a few hundred electrons and thus too weak to be detected by the front-end electronics. It is therefore necessary that the electric field inside the device is large enough to allow the primary electrons to create further ionization, thereby amplifying the signal. With an appropriate choice of the field, it is possible to maintain the proportionality between the initial charge created by the particle and the final signal after multiplication, so that the sensor can still provide a linear measurement of the deposited energy. The parallel plate geometry is however not suitable to this purpose [15], because the length of the avalanche depends on where the primary charge is created inside the sensitive volume. The same charge released in different points would thus result in signals of different amplitude, thus preventing

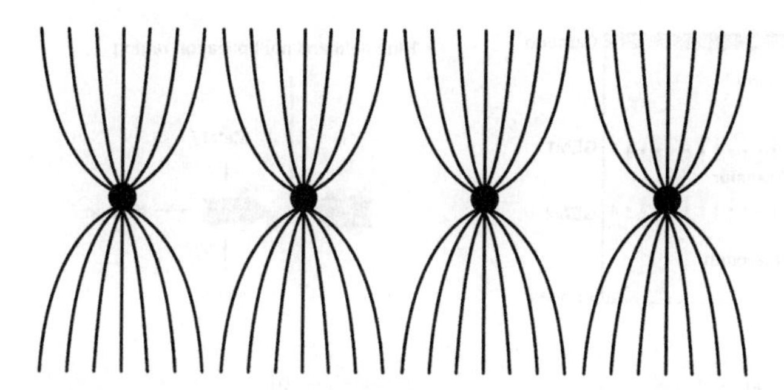

FIGURE 1.11 Field lines in a typical Multi-Wire Proportional Chamber (MWPC). Electric field is stronger in proximity of the wires, where the field lines are more dense.

the sensor to work in proportional mode. The avalanche multiplication must thus be confined to a small and well localized region. This is typically achieved by using cylindrical geometries, with the avalanche occurring in proximity of a thin anode wire. More wires can be located next to each other in the same gas volume, giving rise to a multi-wire proportional chamber (MWPC) [16]. Invented by G. Charpak and coworkers in the late sixties, this detector is of great historical importance as it paved the way to the use of electronic detectors in the world of high resolution particle tracking. Fig. 1.11 shows the typical field line arrangement inside the device. In most part of the volume, the field is too weak to create an avalanche and it is just used to transport by drift the primary charge in proximity of the anodes, where the electric field becomes strong enough to cause a localized avalanche that brings the signal to a detectable level. Typical multiplication factors are $10^4 - 10^5$. To improve the space resolution and rate limitation of traditional gas detectors, new concepts have been proposed, among which the Gas Electron Multiplier (GEM) has been particularly successful [17–19]. The principle consists in cladding a Kapton foil with Copper and drilling holes through the structure. A high voltage difference is applied between the two sides of the foil, so that in the holes a region of high electric field is formed. Fig. 1.12 a) shows an arrangement in which two GEM foils are cascaded to reach high gain. The primary particle creates ionization in the drift region. Avalanche multiplication occurs when the primary charge traverses the first GEM foil. The resulting charge is transferred to the second GEM, when multiplication occurs again. Signal induction on the readout electrodes starts only when the charges emerge from the second GEM. By keeping the induction region short, a fast signal can be obtained, with typical duration in the order of few tens of nanoseconds. In GEM the induced signal is entirely due to electrons, so the long ion tail is suppressed. Fig. 1.12 b) shows the detail of a hole, which is typically manufactured with a diameter of $50 - 70\ \mu$m and a pitch of $150 - 200\ \mu$m. Notice that multiplication occurs only when the charges pass through the holes, while no amplification takes place in the transfer region. The capacitance presented to the front-end electronics depends on the size of the landing pad, but it rarely exceeds 50 pF. A large variety of different gas-based detectors are available. For a complete overview, see for instance [4] and [7].

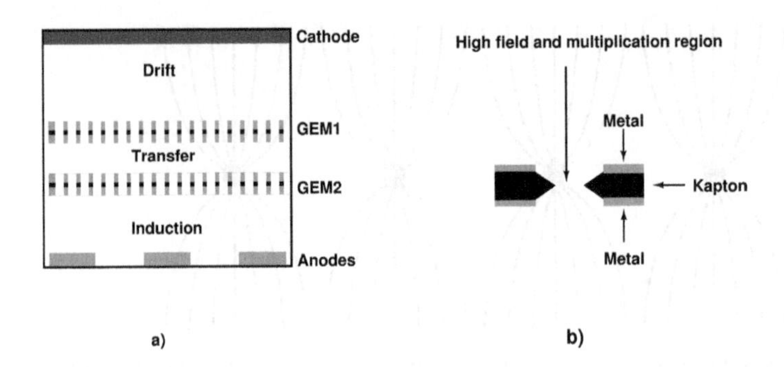

FIGURE 1.12 Example of GEM detector. In b) the sketch of a single hole is shown.

Semiconducting materials are also largely used to detect ionizing radiation. Semiconductor detectors often exploit the rectifying junction which forms at the boundary between two oppositely doped regions, as shown in Fig. 1.13. In case of reverse biasing, a depleted zone extends around the junction. This part of the device is depleted of mobile carriers and contains only ions fixed in the crystal lattice that give rise to a space charge volume. The undepleted regions are populated by free carriers and can be thus conductive. Therefore, there is no electric field here and the reverse bias voltage is entirely applied across the depleted zone. Charges set free by radiation in the undepleted region move only by diffusion and can eventually recombine. The portion of charge that reaches the depleted part is swept towards the collection electrode by the electric field. To enhance the signal, semiconductor sensors are often fully depleted by applying a sufficiently strong reverse bias voltage. The *pn* junction will be discussed in detail in chapter 2. Here we just remind that the depletion region extends more on the less-doped side of the junction and the depletion voltage is larger if the

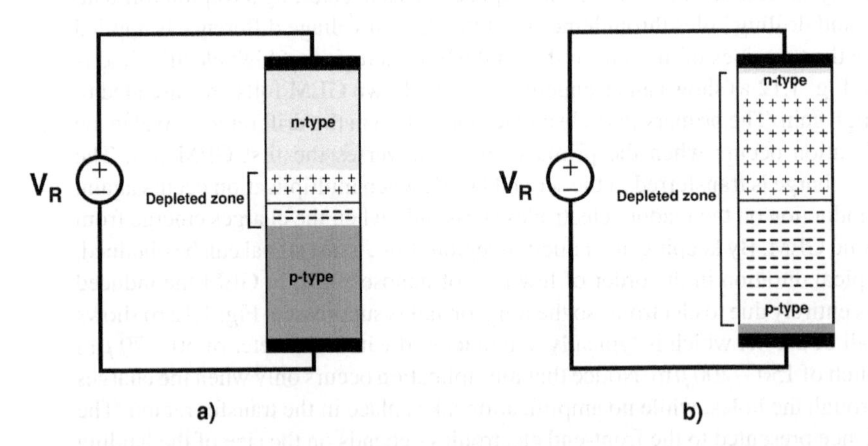

FIGURE 1.13 A reverse-biased semiconductor junction can be used as a radiation detector. In a) a partially depleted junction is shown, in b) a fully depleted one.

doping is higher. Therefore, radiation detectors exploiting *pn* junctions are fabricated on high resistivity substrates. The impressive developments of the microelectronics industry have favored the widespread use of silicon sensors, which are well suited for the detection of charged particles. Microstrip and pixel silicon detectors have thus become the workhorse for the construction of high resolution particle trackers and vertex detectors in high energy physics experiments [20–23].

Semiconductor detectors are also of primary interest for X and gamma-ray spectroscopy, because they offer much better energy resolution than gas or scintillator-based materials. Thick sensors are required to provide an adequate stopping power that allows for an efficient detection of highly penetrating radiation. To reach the necessary sensitive volume, a resistivity as close as possible to the one of the intrinsic semiconductor is needed, which imposes the use of very high purity crystals. In Silicon, this is achieved by drifting Lithium ions into the silicon substrate. The ions compensate for the impurities present in the bulk, allowing us to obtain a material that behaves as extremely pure Silicon, with a resistivity high enough to allow depletion up to several millimeters [1]. The low atomic number of Silicon ($Z = 14$) makes it not suitable to detect efficiently photons of energy above 30 keV. For higher energy photons, Germanium-based detectors are employed. For Germanium detectors, cooling with liquid Nitrogen is mandatory, while for Si(Li) it is desirable to optimize the performance. This has stimulated the research of alternative solutions that can combine good detection efficiency for high energy photons with room temperature operation. Compound semiconductor materials are used to this end [24], with Cadmium Zinc Telluride (CdZnTe) having emerged as the one offering the best trade-offs between energy resolution, ease of operation and possibility of fabricating fairly large crystals, albeit the single crystal size is not comparable with the one achieved with Si and Ge. The research on compound materials for room temperature X and gamma-ray detectors however continues [25]. If the resistivity of the semiconductor is high enough, it is not necessary to deplete the sensor through a junction, because when the drift field is applied, the current is low enough to not impair the performance. The sensor can thus be operated in resistive mode, with an ohmic region providing the electrical contact to the sensitive volume. Due to their high resistivity, CdZnTe sensors can be used in this way [4].

Fast detection of visible photons is another field that shows the variety of technologies that have been developed to measure the same type of radiation. The most common sensor in this field is still the vacuum tube photomultiplier, whose basic structure is sketched in Fig. 1.14. A photocathode coated with a suitable material emits an electron when a visible or near UV photon impinges upon it. Several electrodes, called dinodes, are found between the photocathode and the anode. Going towards the anode, the bias voltage of the dinodes become progressively more positive. The potential difference applied between any two electrodes is sufficient to cause secondary emission. The photoelectron arriving on the first dinode thus extract other electrons from it, starting an exponential multiplication. For one initial photoelectron, a number as high as one million electrons can be found in the output signal at the end of the chain. When the electrons in this avalanche emerge from the last dinode, the

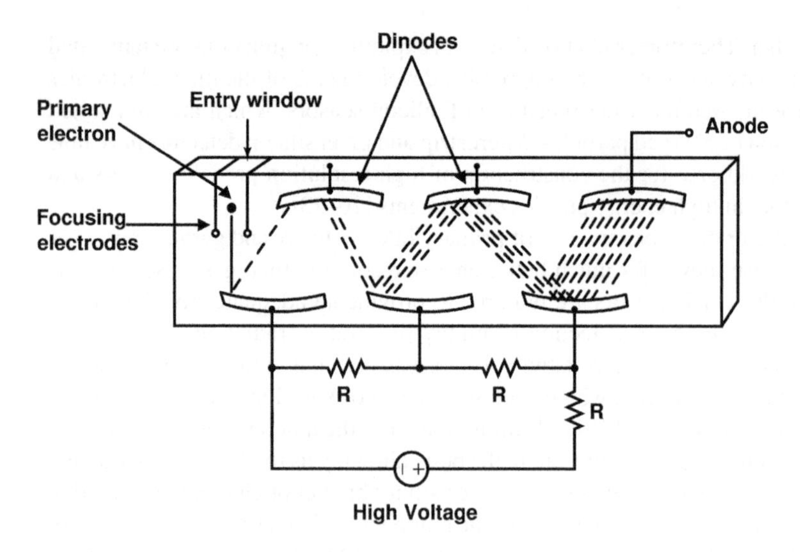

FIGURE 1.14 Concept of a vacuum tube photomultiplier.

signal current starts being induced on the anode. The device can be again represented with the model of Fig. 1.10. The capacitance in this case originates from the capacitance between the last dinode and the anode and by the stray capacitance due to the interconnections.

Prompt detection of low-level light can also be achieved with silicon sensors. Fig. 1.15 shows the schematic view of the avalanche photodiode (APD). In this device, the photons generate electron-hole pairs which are guided by the electric field towards the avalanche region. Here, the field rises and the primary carriers acquire enough energy to extract additional electrons from the lattice atoms (impact ionization). In this process, electrons are more efficient than holes by typically one order of magnitude. In APDs, the gain is between 100 and 1000, so they are adequate to detect small numbers of photons, but not single ones. Since every photoelectron is multiplied by the same gain, the output current pulse is proportional to the number of incident photons. The sensor can be again modeled with the circuit of Fig. 1.10. For an APD, the bias voltage is of several hundreds Volts and may exceed 2000 V in some designs. This favors also the generation of leakage current, which is found in the $1 \div 10$ nA range for a typical device. Depending on the sensor area, the capacitance of such a device may range from 0.5 pF to 300 pF. The APD is thus a good example of how sensor parameters can change by order of magnitudes even within the same device family. An avalanche diode can be designed so that the field in the avalanche region is strong enough to allow a single photoelectron to trigger the junction breakdown. The sensor gives hence a constant output independent of the number of impinging photons. To stop the avalanche, a quenching resistor is put in series with the diode. If there is no current, the voltage at the cathode is equal to the bias voltage V_B. When the current increases the voltage drop across the resistor reduces the cathode voltage

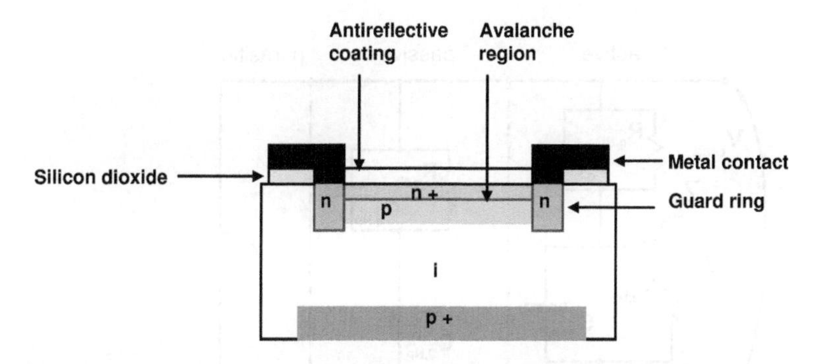

FIGURE 1.15 Schematic view of an Avalanche Photo-Diode (APD).

below the breakdown point and the avalanche is stopped. Such a device is known as Single Photon Avalanche Diode (SPAD) or Geiger-mode Avalanche Photodiode (GAPD). To restore proportionality between the output signal and the number of photons arriving on the sensor, the device is segmented into small independent diodes, called microcells. In analog Silicon PhotoMultipliers (SiPMs) [26–28] the microcells are then put in parallel and their signals are summed at the output. The concept is depicted in Fig. 1.16. The area of each cell is kept sufficiently small (in the order of 50 μm$\times$50 μm), so that the probability of having more than one photon arriving simultaneously on the same diode is small. The number of photons per event is thus recovered from the number of firing microcells. SiPMs are developed as replacement of standard photomultiplier tubes for applications requiring high magnetic field and high space resolution. Combining high granularity with very good time resolution, they offer the possibility of building 3D imaging cameras, with the third coordinate

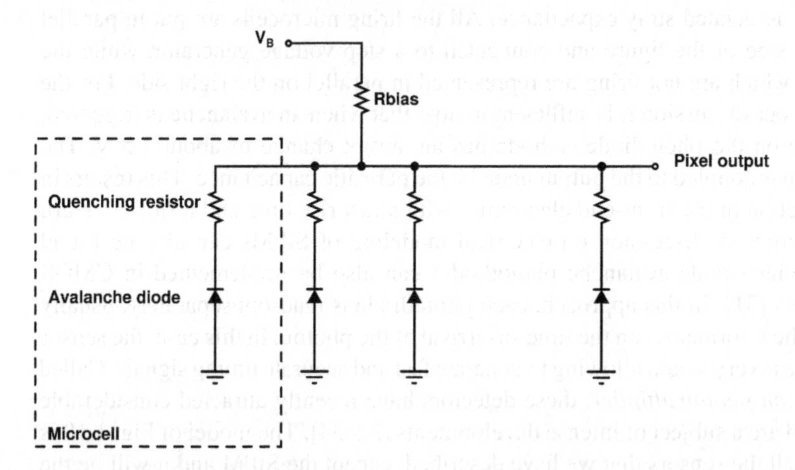

FIGURE 1.16 Concept of Silicon Photomultiplier. Each pixel is formed by an array of microcells connected in parallel. Each microcell consists of Geiger-mode avalanche photodiode plus its quenching resistor.

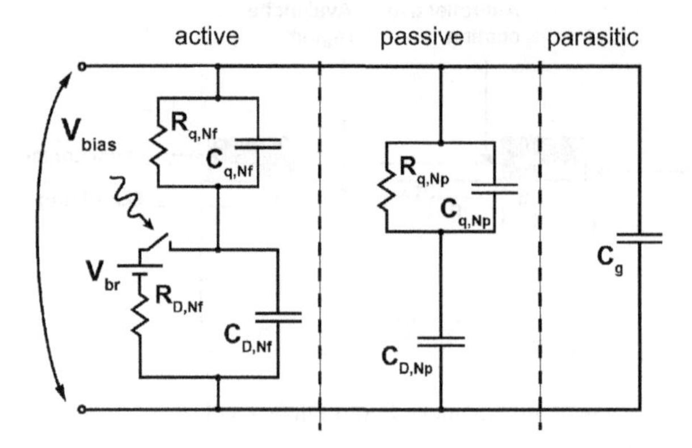

FIGURE 1.17 Electrical model of a Silicon Photomultiplier. © [2009] IEEE. From [29], reprinted with permission.

obtained by measuring the time of arrival of the photons. When many microcells are put in parallel, the sensor capacitance increases. Depending on the final pixel size (a pixel in this context is a group of microcells sharing the same output), the capacitance may range from 30 pF to 300 pF. For a SiPM the circuit model of Fig. 1.10 is not appropriate and may lead to overestimation of the signal rise time. A more accurate model is shown in Fig. 1.17 and is described in detail in [29]. In general, a SiPM contains a total of N microcells. In any particular event, only a fraction N_f of microcells is hit by photons, while the other $N - N_f$ are inactive. In this model, C_d is the capacitance of the reverse biased diode associated to the fired microcell, R_d is the resistance in the microplasma of the avalanche, R_q the quenching resistor and C_q the associated stray capacitance. All the firing microcells are put in parallel on the left side of the figure and connected to a step voltage generator, while the microcells which are not firing are represented in parallel on the right side. For the purpose of our discussion it is sufficient to note that when an avalanche is triggered, the voltage on the photodiode cathode has an abrupt change of about 1.5 V. The voltage step is coupled to the output node by the parasitic capacitance. This results in signal injection in the front-end electronics with a fast rise time and a slower return to the baseline. A discussion on electrical modeling of SiPMs can also be found in [30]. Geiger-mode avalanche photodiodes can also be implemented in CMOS technologies [31]. In this approach, each photodiode is read-out separately, usually encoding the information on the time-of-arrival of the photon. In this case, the sensor capacitance is very small, allowing to generate fast and accurate timing signals. Called *digital silicon photomultiplier*, these detectors have recently attracted considerable interest and are a subject of intense developments [32–34]. The model of Fig. 1.10 is usable for all the sensors that we have described, except the SiPM and it will be the reference in this book. However, the SiPM case shows that the model is not always applicable and that attention must always be paid in deriving and using the appropriate sensor model in circuit simulations.

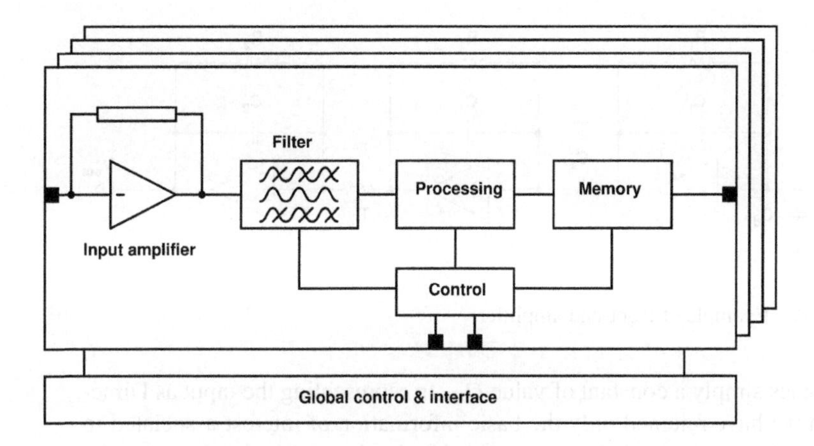

FIGURE 1.18 Generic architecture for a front-end ASIC.

1.3 KEY PARAMETERS IN FRONT-END ELECTRONICS

Fig. 1.18 shows a very general model of a front-end. The ASIC consists of an array of identical channels performing in parallel the same functions. Each channel contains an input amplifier, followed by a filter that optimizes the signal-to-noise ratio. Further signal processing can then be carried-out and a memory finally saves the data while they are waiting for readout. Depending on the application, some of the blocks may take a very simple form or can even be omitted. In a front-end, several key parameters are defined by the input amplifier and the filter. For the sake of clarity, we illustrate them with the help of a simple circuit.

1.3.1 A FIRST FRONT-END AMPLIFIER

Fig. 1.19 shows the block scheme of a typical front-end amplifier. It must be pointed-out that detector signals are very fast and preserving their shape with high fidelity would require circuits with very large bandwidth and thus high power consumption. The key information associated to an event can however be extracted without keeping the original sensor signal shape, thus the frequency content of the input can be altered. A front-end amplifier is usually obtained by cascading a few stages. The first one, connected directly to sensor, is the preamplifier, while the following stages are band-limited and determine the frequency spectrum of the output pulse and hence its shape, forming the "pulse shaper". Front-end amplifiers will be treated in detail in chapter 6 and chapter 7. Here, we consider a few general aspects of pulse processing by inspecting the transfer function of the circuit of Fig. 1.19 under the following hypotheses:

- The detector signal can be approximated with a Dirac-delta. In the time domain, the input can thus be written as $I_{in}(t) = Q_{in}\delta(t)$, where Q_{in} is the total charge contained in the sensor pulse, while in the Laplace domain it

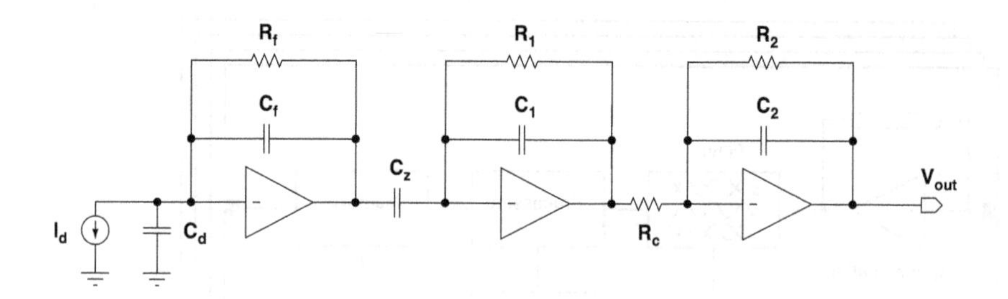

FIGURE 1.19 Example of front-end amplifier.

becomes simply a constant of value Q_{in}. In representing the input as Dirac-delta we have retained only the basic information of interest associated to the amplitude: the total charge contained in the signal, which is linked to the energy released in the sensor by the impinging quantum of radiation.

- The core amplifiers are ideal voltage amplifiers, with infinite gain and bandwidth.
- The preamplifier works as an ideal integrator. This implies that its feedback resistance R_f has only the purpose of establishing a DC feedback that allows a proper biasing of the stage and it is chosen large enough that its contribution to the signal processing can be ignored.
- The time constants in the feedback loops of the second and the third stage are matched, i.e. $R_1C_1 = R_2C_2 = \tau$.

With these assumptions, the circuit response to a Dirac-delta input in the Laplace domain can be written as:

$$V_{out}(s) = \frac{Q_{in}}{sC_f} sC_z \frac{R_1R_2}{R_c} \frac{1}{(1+s\tau)^2} = Q_{in} \frac{C_z}{C_f} \frac{R_1R_2}{R_c} \frac{1}{(1+s\tau)^2} \quad (1.22)$$

Taking the Inverse Laplace Transform of the above equation yields the signal representation as a function of time:

$$V_{out}(t) = Q_{in} \frac{C_z}{C_f} \frac{1}{\tau} R_1 \frac{R_2}{R_c} \frac{t}{\tau} e^{-\frac{t}{\tau}} \quad (1.23)$$

Notice that this is the impulse response of the network. We can now insert in equation 1.23 the value of $\tau = R_1C_1$ to further simplify the expression[3]

$$V_{out}(t) = \frac{Q_{in}}{C_f} \frac{C_z}{C_1} \frac{R_2}{R_c} \frac{t}{\tau} e^{-\frac{t}{\tau}} \quad (1.24)$$

Fig. 1.20 reports the waveform represented by equation 1.24, plotted with the following values of the input signal and of the circuit components:

[3]The choice of $\tau = R_2C_2$ is of course also legitimate and leads to an equivalent formulation. Observe that, following the standard notation used in the field, the substitution $\tau = R_1C_1$ is only made in the constant term $1/\tau$, but not in the function $\frac{t}{\tau}e^{-\frac{t}{\tau}}$.

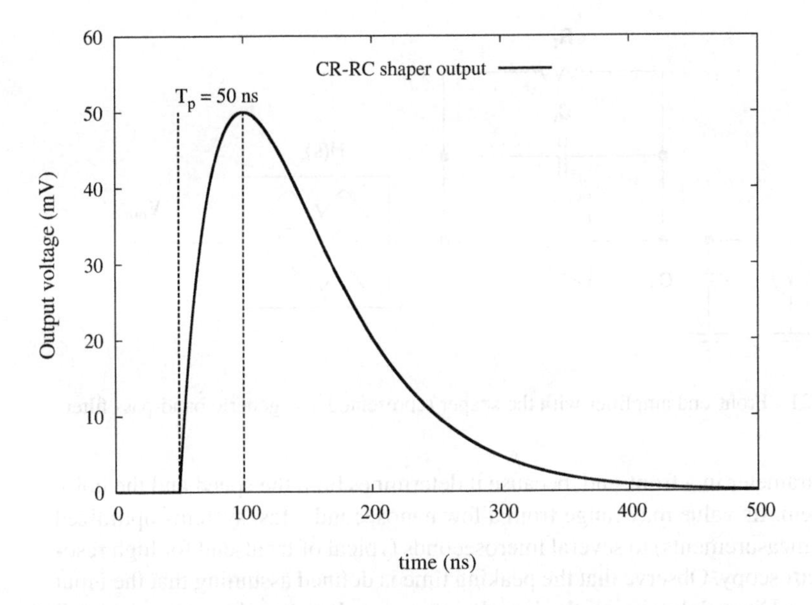

FIGURE 1.20 Impulse response of the front-end discussed in the text.

- $Q_{in} = 1 \cdot 10^{-15}\, C$
- $C_f = 500$ fF
- $C_z = 12.5$ pF
- $C_1 = C_2 = 0.5$ pF
- $R_2 = R_1 = 100$ kΩ
- $R_c = 36.8$ kΩ

The strategy for sizing resistors and capacitors in the amplifier is discussed in chapter 8, while in the following we concentrate on the main system parameters. The front-end described so far is just one of the many possible implementations. In particular, many different architectures are available for the pulse shaper. To indicate a generic front-end, we will thus use the scheme reported in Fig. 1.21, where the pulse shaper is represented as a generic band-pass filter with transfer function $H(s)$.

1.3.2 PEAKING TIME

Consider again the impulse response of the circuit shown in Fig. 1.20. Notice that to make the plot more readable, the input stimulus is applied at $t = 50$ ns and the quiescent point of the output is assumed to be 0 V. In practical cases, the output signal is superimposed on some DC level or baseline, which is needed for the proper biasing of the transistors forming the core amplifiers. The time required for the signal to swing from the baseline to its maximum is called the peaking time[4] and is a

[4]Note that the peaking time is measured from the baseline to the peak, so from 0% to 100% of the signal, while the rise time is taken between the 10% and 90% points.

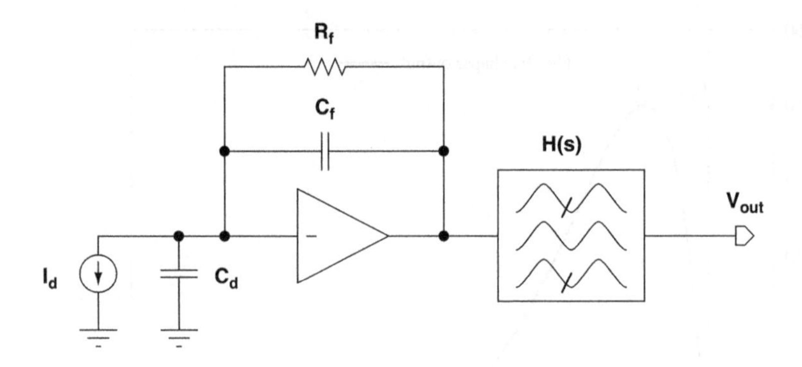

FIGURE 1.21 Front-end amplifier with the shaper represented as a generic band-pass filter.

primary parameter in a front-end, because it determines both the speed and the noise of the system. Its value may range from a few nanoseconds (fast systems optimized for timing measurements) to several microseconds typical of front-end for high resolution spectroscopy. Observe that the peaking time is defined assuming that the input stimulus is a Dirac-delta, i.e on the impulse response. If the peaking time is much longer than the sensor signal collection time, the impulse response describes with good approximation also the front-end response to the actual detector signal. If this is not the case, the output is obtained from the convolution between the detector signal and the front-end impulse response and, for the same total charge, exhibits a smaller peak than the one observed with a Dirac-delta input. This amplitude loss is called ballistic deficit.

1.3.3 GAIN AND SIGNAL POLARITY

The gain of a front-end is usually given as the ratio between the peak of the output voltage and the input charge. In the example of Fig. 1.20, we have fed our system with a charge of 1 fC, obtaining and output peak of 50 mV, hence the gain is 50 mV/fC. In some cases, the gain is also expressed in Volt per electron, although this terminology is less common in the radiation detector community. Knowing that 1 fC is the charge of 6250 electrons, we can thus also say that the gain is 8 μV/electron. In a linear system, the gain should be chosen so that the maximum signal of interest brings the amplifier at the onset of saturation. As an example, the output stage of an amplifier working with 2.5 V power supply may saturate if the output signal is below 0.2 V or above 2.3 V, resulting in a maximum linear output range of 2.1 V. The ratio between the maximum output voltage for which the front-end amplifier still preserves the proportionality between the input and the output signal and the rms noise level at the output is called the output linear dynamic range and, divided by the gain, yields the input linear dynamic range. In our example the amplifier was designed for a gain of 50 mV/fC. If the front-end output stage has a linear range of 2.1 V, the maximum signal that can be treated by the circuit without saturating is 42 fC. The magnitude of the signal of interest is often defined in terms of average amplitudes. For instance, a system might be specified to treat input signals from 1 fC to 10 fC, providing a linear response in this range. If the amplifier has a maximum output range of 2.1 V, one could thus choose a gain of 210 mV/fC. However, as discussed above, the charge generation process in the sensor is affected by statistical fluctuations, so the maximum signal to be treated may well exceed the value of 10 fC. Conversely, in a segmented detector the charge can be collected by more electrodes, which reduces the minimum signal to be measured. In this example, the actual range might rather span from 0.3 fC up to $30 \div 40$ fC and the appropriate gain would hence be around 50 mV/fC. When choosing the gain, one must always cross-check that the expected fluctuations of the input signals have been properly incorporated in the specifications. In Section 1.1 we have seen that different detectors may deliver signals of different polarity. The polarity of the full front-end output depends on how many stages are employed and on their nature. For instance, the output signal of the circuit in Fig. 1.19, which contains three inverting stages, swings upwards if the model of Fig. 1.5 b) is presented to its input and downwards in case the model of Fig. 1.6 b) is used. To maximize the linear dynamic range, the baseline in each stage must thus be appropriately chosen. If the front-end of Fig. 1.19 is used to readout a sensor collecting electrons, the DC level of the first stage should be close to the negative rail, the one of the second stage close to the positive power supply and the baseline of the third stage is again set close to the negative rail. If the polarity is instead as shown in Fig. 1.6 b), the DC levels of the first stage must be set close to the positive rail, the one of the second stage close to ground, while the output must seat close to the positive supply. The gain of each stage should be also chosen so that the maximum signal brings each core amplifier at the edge of its saturation region. Once a given detector is selected, the signal polarity is also defined. Nevertheless, developing a chip that can handle either polarity adds a modest overhead to the design and it is usually a good policy because this increases the chance of re-using the ASIC with different detectors. Observe also that an easy way to make a system compatible with signals of either polarity is to fix the DC bias voltage at the output of each stage half-way between the supply rails, leaving equal headroom for signals swinging upwards and downwards. However, in this case one uses for each polarity only one half of the possible dynamic range. Since, for a given detector, the signal polarity does not change during operation, circuits in which the baseline of the stages can be programmed to a value near the appropriate rail are thus preferable.

1.3.4 NOISE

Fig. 1.22 a) shows the typical output which is observed when the same input signal is repetitively applied. In this example, an input charge of 1 fC was sent to the front-end described in Section1.3.1, with a gain set to 50 mV/fC and 50 ns peaking time. As one can see, the different output waveforms do not overlap, but they lay within a band. This is the effect of noise. In this context, the term "noise" refers to disturbances generated within the sensor and the front-end amplifier. Such noise is intrinsic to the system and can not be eliminated just by shielding it from the outside world.

Physically, noise has its roots in the fact that the mobile charges in electronic devices are in finite numbers and move at finite speed. Any fluctuation in the number

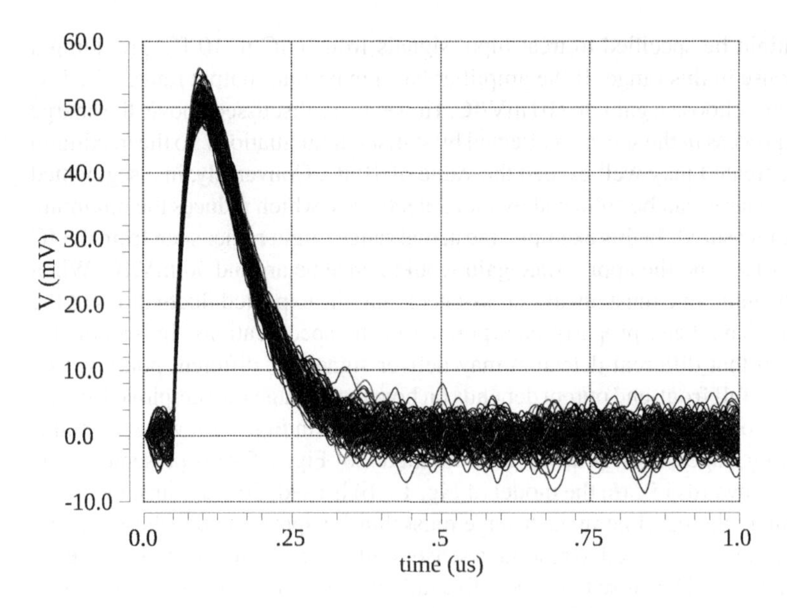

FIGURE 1.22 Output of the front-end amplifier of Fig. 1.19 including the effect of noise.

of carriers (such as the one induced by trapping and detrapping phenomena) or in their speed results in variations of currents or voltages inside the circuit. If we sample the peak of the waveform in Fig. 1.22 we see that it fluctuates around its average value of 50 mV. To estimate the noise strength we need to employ quantities related to the square of the noisy waveform. In measurements, the sample standard deviation is used. From elementary statistics, we know that this is defined as:

$$V_{std} = \sqrt{\frac{1}{N-1}\sum_{k=1}^{k=N}(V_k - V_{av})^2} \qquad (1.25)$$

where V_k is the amplitude of the kth pulse and V_{av} is the average value. The quantity thus obtained is called the rms output noise. In front-end terminology, the noise is usually given as Equivalent Noise Charge (ENC) referred to the amplifier input. In fact, the total charge delivered by the sensor is one of the primary quantity of interest, so quoting the noise in terms of ENC allows for an immediate comparison between the noise floor and the signal to be measured. As an example, suppose that a system with a gain of 50 mV/fC has an rms output noise of 2.2 mV. The ENC is obtained dividing the noise by the gain, so:

$$\frac{2.2\ \text{mV}}{50\ \text{mV/fC}} = 0.044\ \text{fC} \qquad (1.26)$$

which corresponds to a charge of 275 electrons. The noise requirement may vary from a few electrons in high resolution spectroscopy to thousands of electrons, allowed in the readout of sensors that combine large capacitance with strong signals, such as Silicon Photomultipliers.

Noise is generated by the devices forming the front-end amplifier, by the sensor leakage current and by the detector bias resistors. In circuit analysis, the effect of noise is calculated using the concept of equivalent noise sources, through which the noise of a given device is referred to the system input. Depending on the position in the network of the considered device, its noise, when seen from the circuit input, can be described as a voltage or as a current contribution. Voltage noise sources are connected in series with the amplifier input (series noise), while current noise sources are connected in parallel (parallel noise), as shown in Fig. 1.23. The quantity associated to these sources is the "noise spectral density" and represents the power delivered by the noise source into a resistor of 1 Ω and in a bandwidth of 1 Hertz. Therefore, the units are V^2/Hz or A^2/Hz. As a function of frequency, the spectral density may be flat (white noise) or may display some dependency (colored noise). Due to the thermal agitation, the speed of the charge carriers in conductors or semiconductors fluctuates and this originates a noise (thermal noise) which has a white spectrum, at least up to the frequencies presently of interest in front-end electronics. Note that even when there is no net current flowing, the charge carriers still move randomly because of the kinetic energy associated to the temperature and noise is generated. The most common form of colored noise is the flicker or 1/f noise, for which the spectral density takes the form:

$$v_{n,f}^2 = \frac{K_f}{f} \tag{1.27}$$

where K_f is constant for a given device. The noise introduced by the amplifier itself is usually dominated by the series noise and it has both a white and a $1/f$ component. Typically, the input-referred noise spectral density of an amplifier is in the range of a few nV/$\sqrt{\text{Hz}}$ and in well designed circuits, the major noise contribution is given by the input transistor. Thermal noise can be decreased by burning more power in the input device.

Noise is also generated by the sensor and its biasing network. For instance, in semiconductor sensors, the leakage current associated to the reverse bias junctions is often an important noise source, which can be modeled by a current source in parallel to the front-end input. It can be proven that the shot noise due to the leakage current

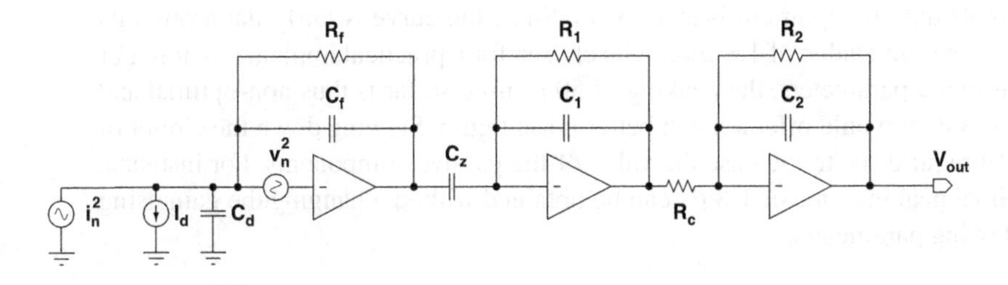

FIGURE 1.23 Front-end amplifier with input-referred parallel and serial noise sources.

has a white spectral density given by:

$$i_n^2 = 2qI_L \tag{1.28}$$

where q is the electron charge and I_L is the leakage current.

Detailed calculations, that will be discussed in the following chapters, reveal that in a continuous time front-end the equivalent noise charge can be expressed as:

$$ENC^2 = (C_d + C_{in})^2 \left(A_w v_n^2 \frac{1}{T_p} + A_f K_f \right) + A_p i_n^2 T_p \tag{1.29}$$

In this equation, v_n^2 and i_n^2 are respectively the input-referred voltage and current spectral noise density, C_d is the sensor capacitance and C_{in} is the sum of all the other capacitances appearing in parallel to the input, including the amplifier input capacitance. T_p is the peaking time of the front-end amplifier, while A_w, A_f and A_p are constant coefficients that depend on the front-end transfer function. Taking the square root of (1.29) we get the equivalent noise charge in Coulomb and dividing this by the electron charge we arrive at the ENC expressed in number of electrons. A few important points must be noted about (1.29):

- The effect of series noise (both thermal and 1/f) is directly proportional to the input capacitance, therefore sensors presenting a small capacitance to the front-end amplifier are preferable.
- The thermal series noise and the parallel noise are weighted in opposite ways by the peaking time. Long peaking times reduce the contribution of series noise and emphasize the one of parallel noise, and viceversa.
- The peaking time does not affect 1/f noise.

For given series and parallel noise spectral densities, an optimal value of peaking time can be found that optimizes the noise performance. Fig. 1.24 shows an example of this trade-off. Here, it has been assumed that thermal noise v_n is 1 nV/$\sqrt{\text{Hz}}$ and the parallel noise i_n is 5.6 fA/$\sqrt{\text{Hz}}$, corresponding to a leakage current of 100 pA. The input capacitance has been fixed to 10 pF. For simplicity, the frequency-dependent noise has been neglected and the assumption $A_w = A_p = 1$ has been made. The figure shows that the optimal peaking time is between 1.5 and 2 μs and a detail calculation reveals that the optimum is at 1.78 μs. Since the curve is fairly flat around its minimum, a round value of 1.8 μs can be chosen for a practical implementation. For the considered parameters, the peaking of 50 ns used so far is thus non-optimal and a slower system would offer a much better noise figure. Slowing down the circuit of Fig. 1.19 requires us to increase the value of the passive components. For instance, the optimal peaking time of 1.8 μs can be obtained without changing the gain using the following parameters:

- $C_f = 0.5$ pF
- $C_z = 75$ pF
- $C_1 = C_2 = 3$ pF

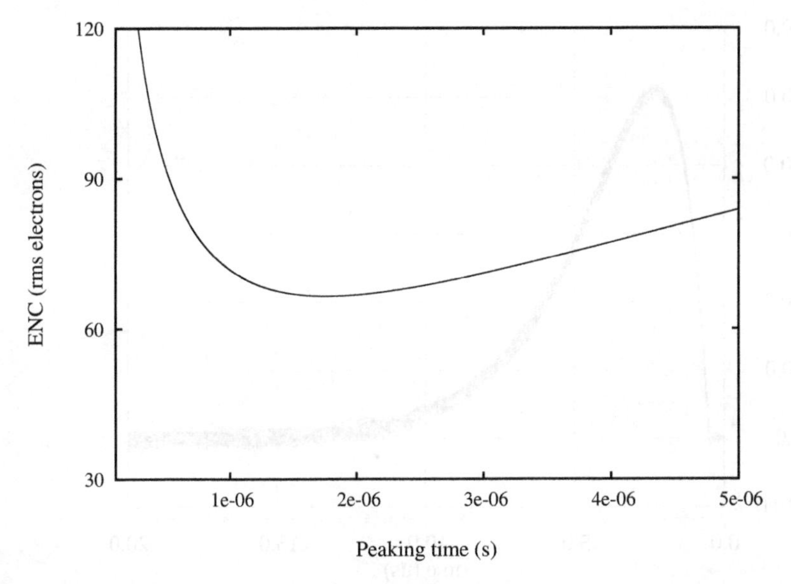

FIGURE 1.24 ENC versus peaking time for a leakage current of 100 pA. See text for discussion.

- $R_1 = R_2 = 600\ \text{k}\Omega$
- $R_c = 220\ \text{k}\Omega$

Observe that the increased values of resistors and capacitors lead to a circuit occupying a much larger silicon area. Fig. 1.25 shows the response of the amplifier with 1.8 μs peaking time including noise. As one can see, the spread between the waveforms is now substantially smaller, because with the optimization the noise has been reduced to 67 electrons, which are expected to produce an output fluctuation of 0.5 mV rms.

Fig. 1.26 shows the change in noise when the leakage current is increased to 1 nA. We can notice that the optimal peaking time has moved to 400 ns and that the minimum achievable noise has increased by a factor of two. To reduce the noise below this level, one needs to reduce the series noise term, which implies increasing the power in the front-end. The example shows that even small values of leakage current have a relevant impact on the noise performance of a front-end. Furthermore, the common-sense notion that noise can be reduced by reducing the bandwidth or, in other words, by slowing down the system does not always apply.

1.3.5 TIME RESOLUTION

The noise optimization discussed so far assumes that the information of interest is the energy and the quantity to be maximized is the ratio between the peak of the output signal and the noise floor. It this case, it is advantageous to work with a peaking time as long as permitted by parallel noise. The peaking time optimization changes in case the primary variable of interest is the time of occurrence of the hit. Events must be

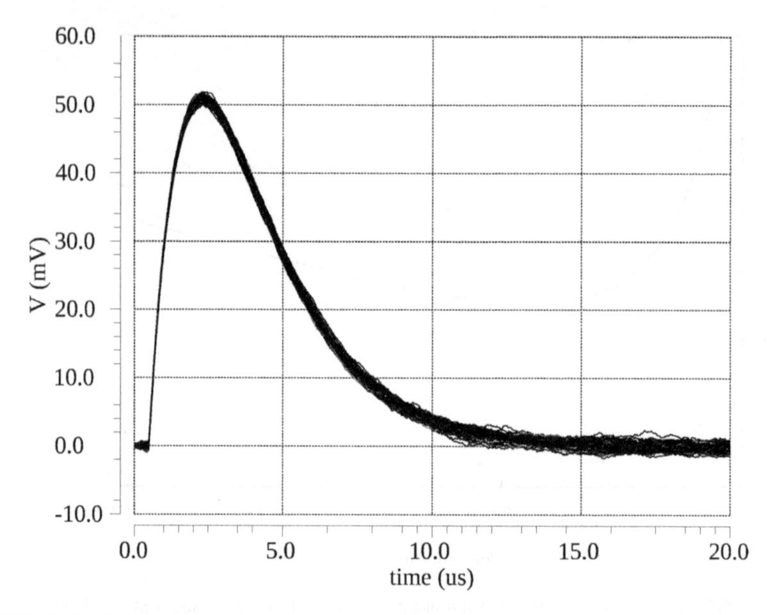

FIGURE 1.25 Output noise waveforms including noise of the circuit of Fig. 1.19. The value of the passive components has been changed in order to keep the gain unaffected while increasing the peaking time to its optimal value. Observe the significant reduction in noise with respect to the case of 50 ns peaking time.

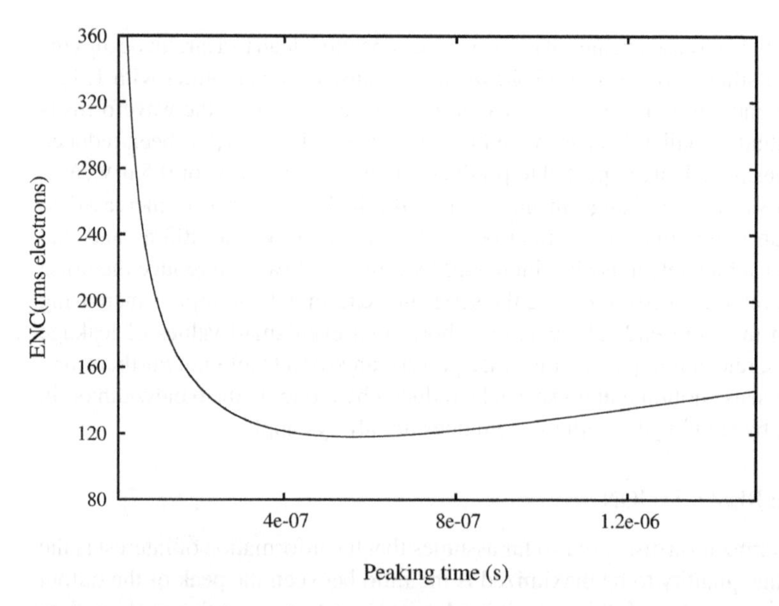

FIGURE 1.26 ENC versus peaking time of the circuit discussed in the text for a leakage current of 1 nA.

ordered in time, with an accuracy that strongly depends on the application. In Time of Flight measurements (ToF) the time taken by a particle to fly over a known distance is used to identify the particle itself. In other cases, time measurements are used to derive a space coordinate. In these applications, the required time resolution can be as low as 100 ps. On the other hand, in integrating systems, in which many events are accumulated in a given time-frame, timing accuracy is not a concern. Fig. 1.27 shows a simple implementation of a time measuring system. A counter is used to count clock pulses and the resulting word is fan-out to the individual channels. A comparator is connected to the output of the front-end amplifier. When the input signal crosses a predefined threshold, the comparator fires and the status of the counter is recorded into local registers in the channel. If we now repeat the exercise of sending identical pulses to the system and we observe the comparator output, we see that the transition point moves back and forth in time around its average value. Called timing jitter, this phenomenon is due to the uncertainty in the threshold crossing time caused by the noise present at the amplifier output. To understand this point, let's consider the situation shown in Fig. 1.28, which reports a zoom of the rising edge of the signal. Around this point, the amplifier output can be approximated with a first order Taylor expansion. If t_0 is the nominal time at which the signal crosses the threshold we can write:

$$V_{out}(t) = V_{out}(t_0) + \left.\frac{dV}{dt}\right|_{t=t_0}(t-t_0) \tag{1.30}$$

Observe that the coefficient $(dV/dt)_{t=t_0}$ represents the slope of the signal presented to the comparator. Suppose now that when the signal is just about to cross the threshold, a negative noise pulse brings it down by an amount ΔV. A time given by the amplitude of the noise pulse divided by the amplifier signal slope is now necessary to reach again the threshold and make the comparator flip. The comparator will therefore switch later than it would do with a noiseless input. Symmetrically, when the amplifier output is still below the threshold, a positive noise pulse can anticipate the crossing time.

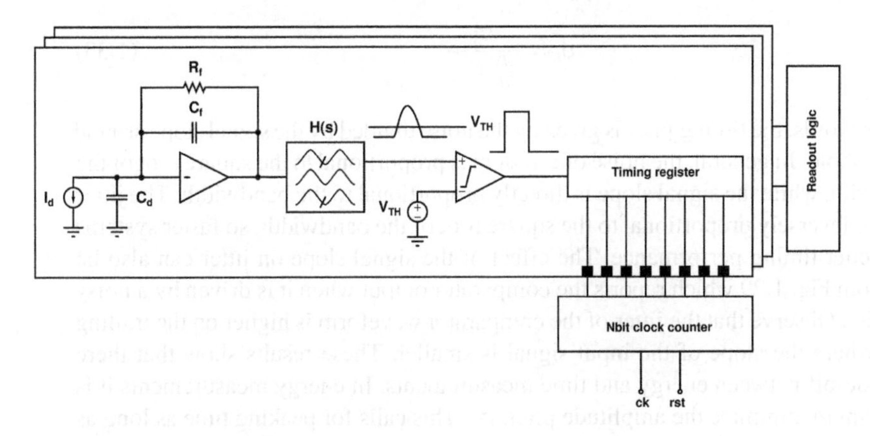

FIGURE 1.27 Example of front-end for time measurement.

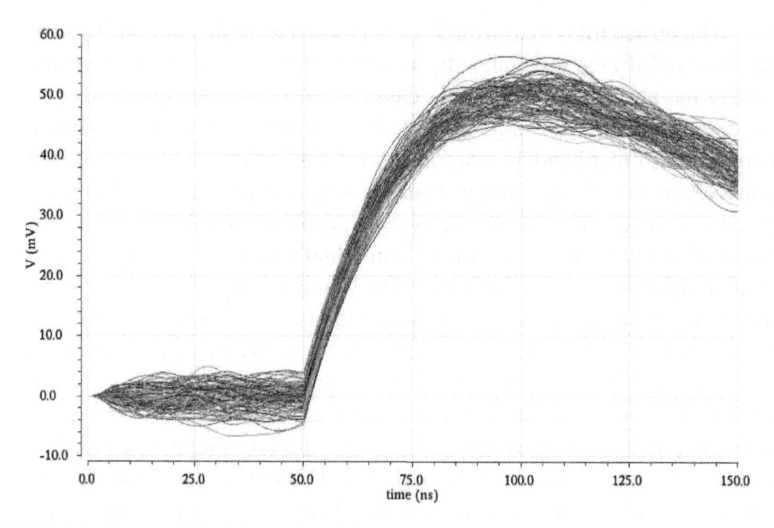

FIGURE 1.28 Zoom of the rising edge of the front-end amplifier, including noise. The peaking time is 50 ns.

Using (1.30), we can write a relationship between the voltage and time uncertainty:

$$\Delta V_{out} = \left.\frac{dV}{dt}\right|_{t=t_0} \Delta t \tag{1.31}$$

The noise standard deviation at the amplifier output can be assumed as a measure of the typical ΔV, thus we have:

$$\sigma_V = \left.\frac{dV}{dt}\right|_{t=t_0} \sigma_t \tag{1.32}$$

which allows to express the timing jitter as:

$$\sigma_t = \frac{\sigma_V}{\left.\frac{dV}{dt}\right|_{t=t_0}} \tag{1.33}$$

In other words, the timing jitter is given by the noise divided by the signal slope around the threshold. In general, the noise of a system is proportional to the square root of the bandwidth, while the signal slope is directly proportional to the bandwidth. The jitter is hence inversely proportional to the square root of the bandwidth, so faster systems have better timing performance. The effect of the signal slope on jitter can also be seen from Fig. 1.29 which reports the comparator output when it is driven by a noisy amplifier. Observe that the jitter of the comparator waveform is higher on the trailing edge, where the slope of the input signal is smaller. These results show that there is a trade-off between energy and time measurements. In energy measurements it is important to minimize the amplitude jitter, σ_v. This calls for peaking time as long as permitted by rate and/or parallel noise considerations. In timing applications instead it is important to minimize the slope-to-noise ratio, which demands fast peaking times.

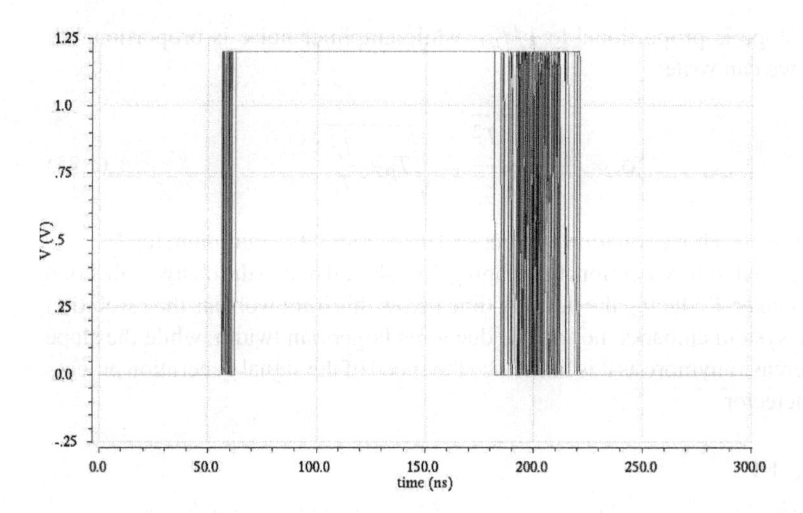

FIGURE 1.29 Comparator output with jitter.

This is what motivates the dual path topology recurring in many systems that have to measure with good accuracy both the energy and the time of arrival of the event. The principle is shown in Fig. 1.30. After the very input stage, the signal is split in two independent branches. In the timing branch, a fast shaper optimizes the slope to noise ratio to achieve good timing resolution, while in the energy branch a slow shaper minimizes the amplitude variations to allow for good energy resolution. Finally, it must be noted that in timing measurement the ultimate limit on the shaping time is given by the speed of the sensor signal. Suppose in fact that the sensor signal collection time is T_c. The total rise time at the amplifier output can be estimated as:

$$T_{tot} = \sqrt{T_p^2 + T_c^2} \tag{1.34}$$

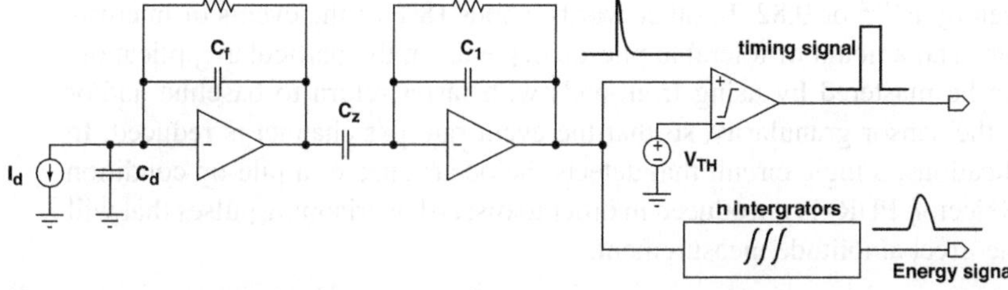

FIGURE 1.30 Front-end with a branch optimized for timing and one for energy measurement.

The signal slope is proportional to $1/T_{tot}$, while amplifier noise is proportional to $1/\sqrt{T_p}$, so we can write:

$$\sigma_t \propto \frac{\sqrt{T_p^2 + T_c^2}}{\sqrt{T_p}} = \sqrt{T_p + \frac{T_c^2}{T_p}} \tag{1.35}$$

Differentiating the above equation with respect to T_p, one gets a minimum for $T_p = T_c$, i.e. for an optimal time resolution, the shaping time should match the charge collection time in the sensor. Reducing the peaking time below this limit worsens the resolution because the system embarks more noise due to its larger bandwidth, while the slope does not increase anymore as it is limited by the speed of the signal generation process inside the detector.

1.3.6 PILE-UP

Noise considerations are not the only ones affecting the choice of the peaking time, as this must be compatible also with rate requirements. In fact, the amplifier output must return to the baseline before a new pulse can be processed, otherwise the two signals will pile-up. The time of arrivals of the events on a radiation detector usually follows a Poisson distribution:

$$P(n) = \mu^n \frac{e^{-\mu}}{n!} \tag{1.36}$$

The distribution allows us to calculate the probability of observing n events in a process which has a mean value of μ. Suppose that we have a front-end that whenever it is hit responds with a pulse lasting 1 μs. Note that in this calculation the relevant parameter is the total signal duration, which can be significantly longer than the peaking time. If the event rate is 200 kHz, the average interval between two events is 5 μs. To avoid pile-up, we must require that during the busy time no other event arrives on the occupied channel. The probability for this to happen is obtained by (1.36) by putting $n = 0$ (we require that no event arrives in the considered time) and $\mu = 0.2$, because, with a 200 kHz rate, we have on average 0.2 events in the 1 μs interval we are interested in. The probability of not having an event piling-up on the previous one is thus given by $e^{-0.2}$ or 0.82. In other words, about 18% of the events of interests will overlap. The amount of tolerable pile-up depends on the particular application. Pile-up can be mastered by using front-ends with faster return to baseline and/or increasing the sensor granularity, so that the event rate per channel is reduced. In some applications, a logic circuit that detects the occurrence of a pile-up condition (Pile-Up Rejector, PUR) is introduced in order to discard overlapping pulses that will result in incorrect amplitude measurement.

1.3.7 DETECTION EFFICIENCY AND DERANDOMIZATION

Consider now a slightly different situation. Suppose that in the system of Fig. 1.18 a very fast front-end is used, so that the channel dead-time is dominated by the time needed to read the data out. Assume also that the events at the input follow a Poisson distribution with an average rate of 100 kHz. We ask now the following question: what is the maximum dead-time that we can tolerate for a given detection efficiency? If the incoming hits were uniformely distributed, a dead-time just below the interval between two events would be sufficient to avoid any loss. For instance, if the channel needs 8 μs to process one event, it is already free 2 μs before the next hit is presented to the input. Unfortunately, the random distribution of the arrival times complicates the picture and if the channel is blocked for 8 μs each time an event is detected a significant event loss occurs. With a rate of 100 kHz, we have on average 0.8 events in 8 μs. The probability that an event arrives and has to be rejected because the channel is already occupied can be calculated again with (1.36), using $\mu = 0.8$ and it is:

$$P_{loss} = 1 - P(0) = 1 - e^{-0.8} \approx 0.55 \qquad (1.37)$$

Therefore, 55% of the events will be lost. The straightforward way to reduce the event loss is to speed-up the system. For instance, to keep the inefficiency below 1%, we must require that:

$$P_{loss} = 1 - P(0) = 1 - e^{-\mu} \approx 0.01 \qquad (1.38)$$

The condition is fulfilled if $\mu \approx 0.01$ or, equivalently, if the dead time is reduced by a factor of 80, shrinking it from 8 μs to 100 ns. Faster systems demand however more complex and power hungry circuits. Suppose instead that a memory is added, so that the data can be stored locally while waiting for read-out, making again the channel sensitive to a possible new hit. In this case, it is necessary that at least two events arrive in the considered time window of 8 μs for one to be lost. Therefore, the loss probability becomes:

$$P_{loss} = 1 - P(0) - P(1) \approx 0.19 \qquad (1.39)$$

From the above equation, we see that now 19% of the hits are undetected. Adding more buffers we can further reduce the data loss. If we can store n events in the channel, the probability of losing one hit becomes:

$$P_{loss} = 1 - \sum_{0}^{n} \frac{\mu^n e^{-\mu}}{n!} \qquad (1.40)$$

Applying (1.40), we see that if we use four buffers, the probability to miss one hit drops to 0.5%. This means that, after the buffers, the speed of the system can be tuned on the average arrival rate rather than on the maximum one. The random input data stream has thus been regularized or, in other words "derandomized". In our discussion we have supposed that the time needed to extract the data from the channel was the dominant source of inefficiency. However, the bottleneck can be in any of the blocks shown in Fig. 1.18. The derandomization technique can in principle be applied to all of them, using more elements in a time-interleaved configuration, so that when a unit is busy, there is statistically at least another identical one free which can take-over. The approach implies however switching between different circuit elements. In analog circuits with high gain, such as the input amplifiers, this is more problematic because the noise associated to the switching procedure can distrupt the system performance. Derandomization is thus applied on already amplified signals, i.e. after the preamplifier and the pulse shaper.

1.4 FRONT-END ARCHITECTURES

The diversity of applications and system constraints has motivated the development of many different front-end architectures. In some applications, in fact, it is sufficient to register if a particular detector channel has been hit. In this case, the signal processing unit of Fig. 1.18 takes the simple form of a threshold comparator followed by a few registers. A common example is provided by tracking in High Energy Physics, where the trajectory of a charged particle is sampled by multiple layers of sensors, usually located in a magnetic field to allow also the measurement of the particle momentum. In nuclear spectroscopy, where energy measurement is the primary objective, the signal amplitude must instead be recorded with very good resolution, making analog circuits with large dynamic range and Analog to Digital Converters necessary. In other fields, like Positron Emission Tomography (PET), both the time of occurrence of the event and the signal amplitude need to be extracted. In the following, the most recurrent front-end topologies are introduced in order of increasing complexity. The section ends with a short discussion of the problem of data readout and transmission.

1.4.1 BINARY FRONT-END

In a binary system, a comparator detects if the front-end output goes above a preset threshold, generating a digital pulse. No direct recording of the analog amplitude is provided [35–43]. Fig. 1.31 shows a possible implementation. Before the measurement starts, a reset is issued, so that the two flip-flops are in a known state. The clock input of the first flip-flop is driven by the comparator, which triggers the storage of a "one" when the front-end output crosses the threshold. The *clear, ck, reset* and *sel* signals are common to all channels. To read out the circuit, a clock pulse is first sent to copy the stored bit from FF1 into FF2. A clear is issued to FF1, that becomes again ready to capture a new event. The state of the multiplexer in then changed with the *sel* signal, so that the output of FF2 in a channel becomes the input of the corresponding flip flop in the next one, forming a shift register. The clock pulses are again applied, shifting the bits from one channel to the next till the output is reached. Observe that only a minimum of digital circuitry is required beyond the front-end, making it possible to fit each channel in a small silicon area. Binary systems are hence well suited for those applications in which the hit position in space is the primary information of interest.

In a binary system, any signal crossing the threshold is processed as a good event. Therefore, it is important to minimize the number of hits due to noise, which requires a proper setting of the comparator threshold. It can be proven that the frequency of noisy hits can be estimated as [5]:

$$f_n = \frac{1}{4\sqrt{3}\tau} e^{-\frac{A_{th}^2}{2A_n^2}} \tag{1.41}$$

where f_n is the number of noisy hits per second, while A_{th} is the applied threshold voltage measured from the front-end amplifier baseline and A_n is the rms noise voltage measured at the front-end amplifier output. The quantities A_{th} and A_n can be voltages

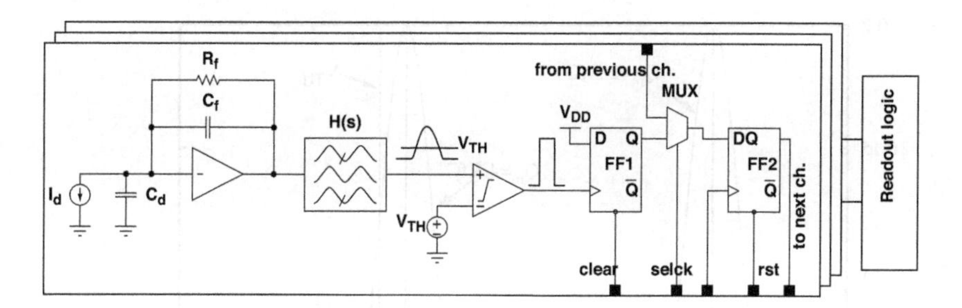

FIGURE 1.31 Example of binary front-end.

or currents. The ratio between threshold and noise can also directly be expressed in terms of the corresponding charges:

$$f_n = \frac{1}{4\sqrt{3}\tau} e^{-\frac{Q_{th}^2}{2Q_n^2}} \tag{1.42}$$

The above equations are strictly correct only if the amplifier behaves as a first order low pass filter, with the bandwidth defined by a single time constant $\tau = RC$. However, they can be used to provide a first estimate of the threshold necessary to achieve a given noise suppression. For a time constant τ of 50 ns and a threshold-to-noise ratio of four, equations (1.41)-(1.42) tell us that there are 968 noisy hits per second in one channel. This number drops to $5.5 \cdot 10^{-16}$ if the threshold-to-noise ratio is augmented to ten. Increasing the threshold strongly reduces the spurious hits, but it also affects the detection efficiency on good signals. As a rule of thumb, a signal-to-noise ratio of between 10 and 15 on the minimum signal of interest allows us to work with thresholds high enough to suppress almost completely the noise while preserving a good efficiency. As an example, suppose that an amplifier has an output baseline of 500 mV, responds with a positive output signal and the minimum amplitude of interest is 30 mV. If the output noise is 2 mV rms, a threshold of 515 mV (i.e. 15 mV above the baseline) yields a threshold-to-noise ratio of 7.5. Assuming again a time constant of 50 ns, a noise hit rate per channel of $1.8 \cdot 10^{-6}$ hits per second results. In other words, in a system containing one million channels, on average only two channels per second would fire on noise. Lowering the thresholds increases the efficiency on small signals, but also the fake hit rate.

A binary system is usually tested with the method of the “S-curve”. To understand how this technique works, let’s consider a practical example. Suppose we have a front-end amplifier with a gain of 10 mV/fC, in which the baseline is set at 450 mV. The front-end output is presented to a discriminator with a threshold fixed to 500 mV. An input signal of a fixed amplitude is repeatedly sent to the circuit. If we can observe the analog output with an oscilloscope and make an histogram of the peak voltage, we notice that, due to noise, the samples are distributed around a mean value, usually following a Gaussian distribution. Fig. 1.32 shows the situations for two cases. The curve on the left of the graph corresponds to an input signal of 2.5 fC, for which a value of 25 mV above the baseline is expected. The average total peak voltage is thus

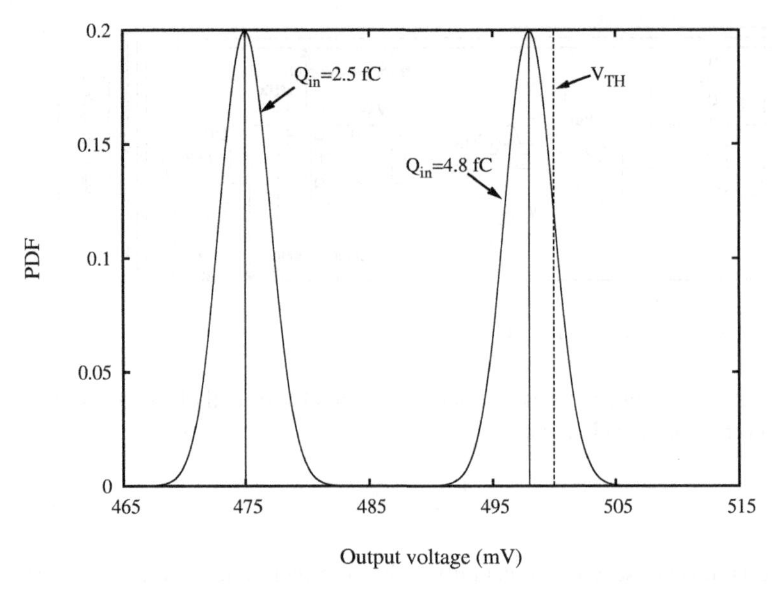

FIGURE 1.32 Noise effect for signals with average value smaller than the discriminator threshold.

475 mV. It has been assumed that the output noise voltage is purely Gaussian, with a rms of 2 mV. Since the gain is assumed to be 10 mV/fC, this corresponds to an ENC of 1250 electrons. The curve on the right is for an input signal of 4.8 fC, which yields an output of 48 mV above the baseline, bringing the average peak value to 498 mV. The plot has the voltage in mV on the x axis and the probability density function (PDF) of the Gaussian on the y axis. The threshold value is indicated with a vertical line. The probability that a signal with a given average amplitude exceeds the threshold can be calculated as follows:

$$P(V > V_{TH}) = \frac{1}{\sqrt{2\pi}\sigma} \int_{V_{TH}}^{\infty} e^{-\frac{(V-\mu)^2}{2\sigma^2}} dV \tag{1.43}$$

We see from Fig. 1.32 that the distribution corresponding to an input of 2.5 fC is well below the threshold and the probability that a value extracted from this curve exceeds the threshold is extremely small. Therefore, if the same signal is sent many times to the circuit, the comparator would almost never fire. The distribution associated to an input of 4.8 fC intersects instead the threshold, hence the probability that the same value falls beyond V_{TH}, making the comparator switch, is significant. Fig. 1.33 shows two other interesting cases. The first one, on the left, is the one in which the input signal results in an output amplitude distribution with an average equal to threshold. In this case, if many identical signals are sent, half of them will make the comparator flip, causing a detection. The curve on the right corresponds to a signal generating an average peak well above the threshold, which is thus always detected. Based on these considerations, one can thus build an efficiency curve as follows. The value of

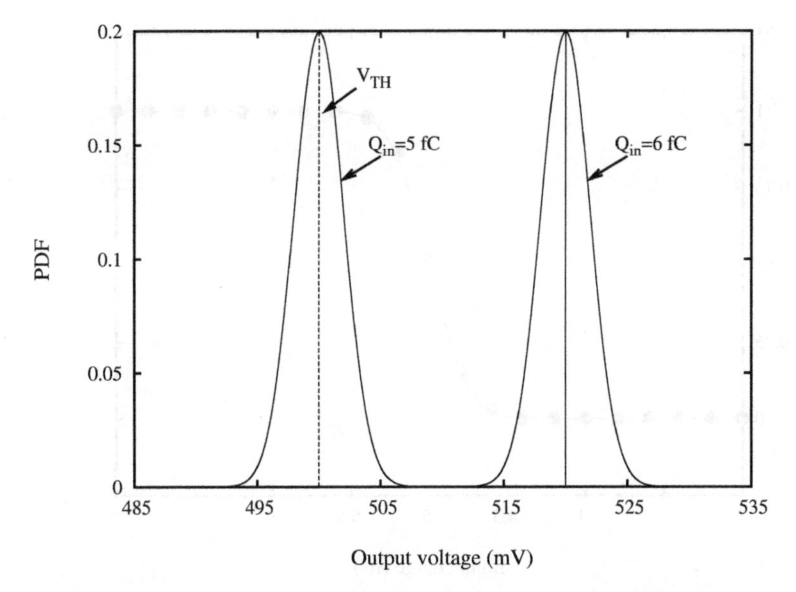

FIGURE 1.33 Noise effect for signals with average value equal or greater than the discriminator threshold.

the threshold voltage is fixed and many signals with the same amplitude are sent to the front-end. The number of times the comparator fires is counted and the efficiency η is calculated as:

$$\eta = \frac{\text{Number of answers}}{\text{Number of sent signals}} \tag{1.44}$$

The amplitude of the input signal is then increased and the procedure is repeated. The points obtained for the circuit of the example with a threshold set to 500 mV are shown in Fig. 1.34. Notice that the known variables of the problem are the nominal injected input charge, the nominal threshold voltage and the measured efficiency. Therefore, η is expressed directly as a function of the input charge and the data can be fitted with a sigmoid function:

$$\eta(Q) = \frac{1}{2}\left[erf\left(\frac{Q-\mu}{\sqrt{2}\sigma}\right)+1\right] \tag{1.45}$$

where the parameters μ and σ need to be extracted from the fit. The point $\eta =$ 0.5 corresponds to input signals generating an output whose average is identical to the threshold. Differentiating (1.45), the Gaussian distribution representing the probability density function of the considered output voltage is obtained and the resulting σ yields a measurement of the noise at the discriminator input. Fig. 1.35 shows as an example a similar plot obtained from a system with the same gain, but a noise twice as big. Observe that the point $\eta = 0.5$ occurs always for the same value of the input charge, but transition between $\eta = 0$ and $\eta = 1$ is now broader due to the increased noise. By repeating the measurement with different threshold voltages, one can also obtain a curve $V_{out} = f(Q_{in})$, where V_{out} is the inferred analog output voltage

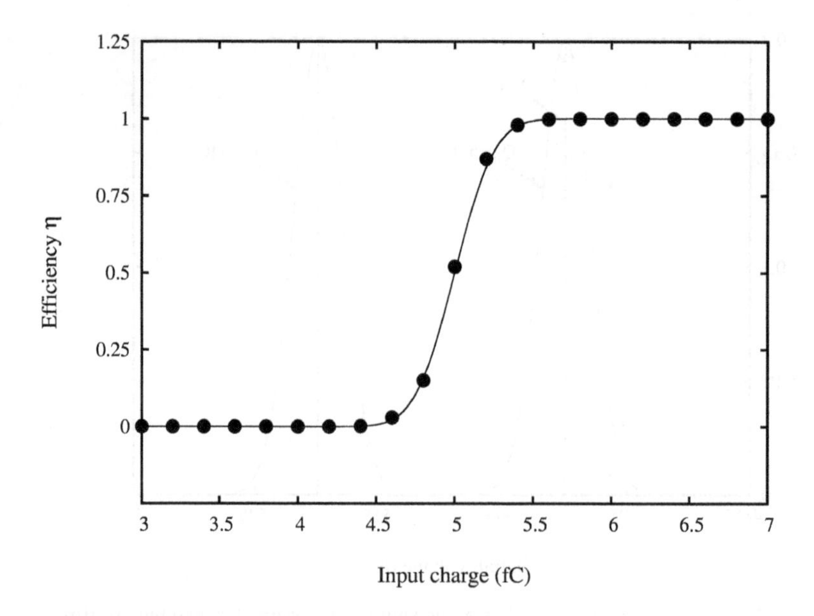

FIGURE 1.34 Example of "S-curve" for a binary system.

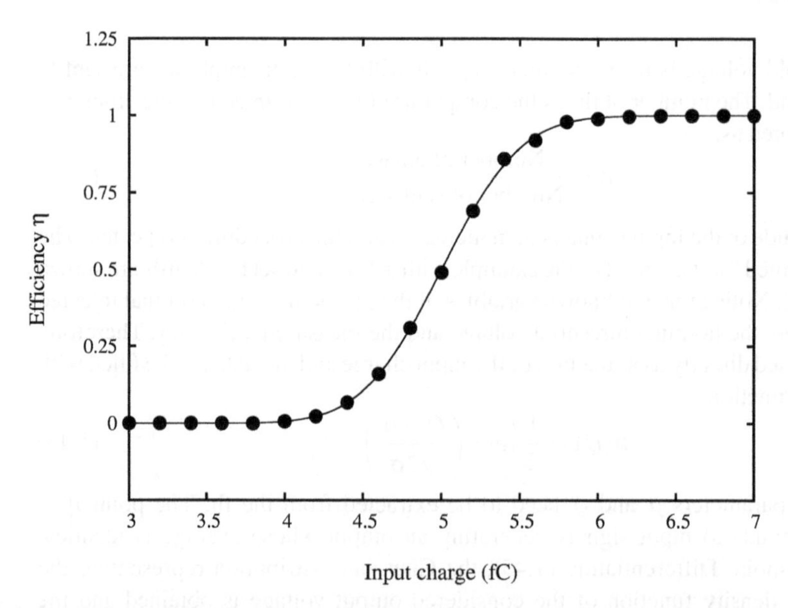

FIGURE 1.35 Example of "S-curve" with greater noise. All the other system parameters are the same.

of the front-end. In this way it is possible to extract the gain of the analog front-end preceding the discriminator. As an alternative, the input signal can be kept fixed, while the threshold is swept. The procedure usually begins with a threshold low enough so

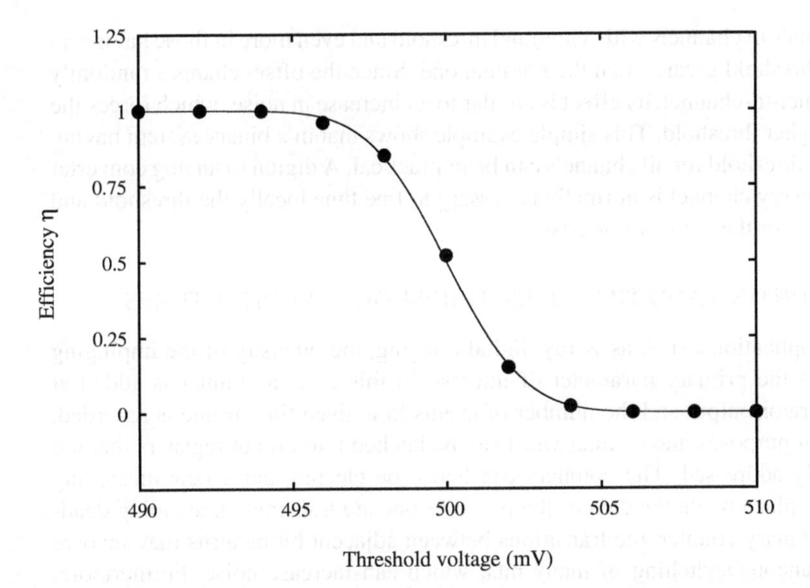

FIGURE 1.36 Example of "S-curve" obtained by scanning the threshold voltage at fixed signal amplitude.

that all the signals sent can be recorded. The threshold is then incremented till the efficiency starts decreasing and eventually no hits are detected anymore. The resulting data set and its fitting function are shown in Fig. 1.36. In this example, the threshold voltage is directly expressed in mV, but it could be equivalently expressed in unit of charge. Notice that in this case the fitting sigmoid is associated to the complementary error function:

$$\eta(V) = 1 - \frac{1}{2}\left[erf\left(\frac{V-500}{\sqrt{2}\sigma}\right)+1\right] \tag{1.46}$$

In the above example, the noise figure allows in principle to set a threshold significantly lower than the considered value of 500 mV. In fact, with a baseline of 450 mV and rms noise of 2 mV, a threshold of 462 mV is already six sigma above the noise, reducing the spurious hits to a usually negligible level. Random deviations in the properties of the transistors forming the comparator however generate offsets, i.e. random shifts of the effective threshold from the nominal one. The offset can thus be represented as a voltage source connected in series to one of the comparator inputs that applies an effective threshold different from the nominal and desired one. To visualize the effect, we can use again the parameters of the above example and consider a chip with 128 identical channels. Suppose that, due to the offsets, the effective threshold voltage is randomly distributed with a sigma of 5 mV. If one of the comparators has a negative shift of two sigmas, its effective threshold would be 452 mV, close enough to the baseline to cause a significant amount of spurious hits. To have an effective threshold of 12 mV above noise for this channel, one needs to apply an external threshold of 472 mV. If the threshold is common to the whole chip, this may cut the signals in channels with a nominal threshold and even more in those having an effective threshold greater than the nominal one. Since the offset changes randomly from channel-to-channel, its effect is similar to an increase in noise, which forces the use of a higher threshold. This simple example shows that in a binary system having a common threshold for all channels can be impractical. A digital to analog converter (DAC) in every channel is normally necessary to fine tune locally the threshold and compensate for the comparator offset.

1.4.2 COUNTING AND TIME-OVER-THRESHOLD ARCHITECTURES

In some applications, such as X-ray digital imaging, the intensity of the impinging radiation is the primary parameter of interest. In this case, a counter is added at the comparator output and the number of events in a given time frame is recorded. For readout purposes, the counter word can be latched into output registers that are sequentially addressed. The counters can hence be cleared and a new measuring cycle takes place while the data of the previous one are transmitted, avoiding dead-time. In a binary counter, the transitions between adjacent bit patterns may involve the simultaneous switching of many bits, which can increase noise. Furthermore, when the counter value is sampled, an error in latching one bit can determine huge information loss. For both aspects, the most critical transition is the one from $011\cdots11$ to $100\cdots00$, occurring at the middle of the counter dynamic range, which involves the simultaneous switching of all bits. If in this transition the most significant bit is misinterpreted as 0, the word $000\cdots00$ is stored and an error equal to half the dynamic range of the counter is made. The issue becomes less severe as the weight of the bits decreases. For these reasons, pseudo-random counters that reduce the number of simultaneous commutations are used. In this respect, the Gray codes, in which two subsequent words always differ by only one bit, is particularly useful. In a counting front-end, some energy discrimination is sometimes provided by using a few comparators working with different thresholds and connected to different counters. Examples of counting front-ends can be found in [44–47]. With few modifications, a binary system can be transformed into an amplitude measuring device. In fact, the time spent over threshold by the amplifier output is in general proportional to the magnitude of the input signal. Therefore, by measuring the duration of the comparator response, one can extract the information about the input charge [48–51]. A particular interesting approach is shown in Fig. 1.37. Here the amplifier is a simple integrator obtained by connecting a capacitor C_f in the feedback loop of a high-gain amplifier. The sensor current pulse is integrated on C_f, which is discharged by a constant feedback current I_{FEED}. The time needed to remove the charge is given by:

$$T = \frac{Q_{in}}{I_{FEED}} \tag{1.47}$$

The output signal has a triangular shape and the *Time over Threshold* (ToT) is linearly proportional to the input charge. Notice that the ToT is independent on the value of C_f. The ratio between I_{FEED} and C_f defines the slope of the triangular signal when it returns to the baseline. The peak amplitude is given by Q_{in}/C_f, therefore if C_f is

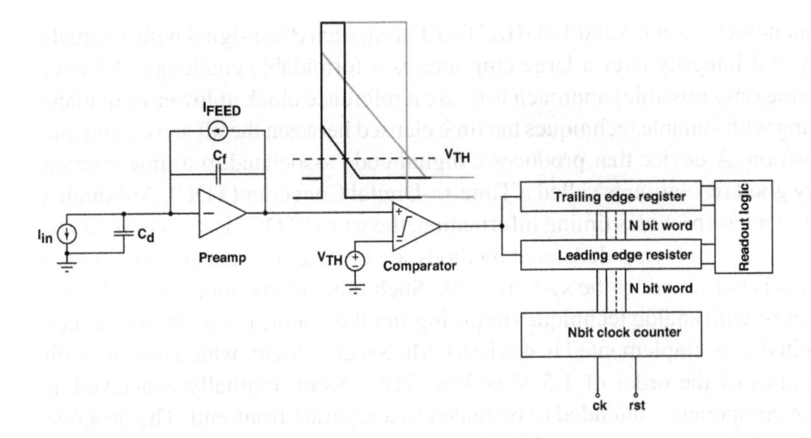

FIGURE 1.37 Front-end with linear Time-over-Threshold measurement.

halved, both the slope and the peak amplitude are doubled, and the time needed to return below threshold is unchanged. Another interesting feature of ToT systems is that the core amplifier can saturate without automatically compromising the linearity of the measurement. In fact, when the amplifier saturates, its DC gain drops. The virtual ground approximation is not valid anymore and the extra charge is integrated on the amplifier input node. This charge is however removed by the feedback current, so (1.47) still holds. The linearity of the measurement is affected if the voltage swing at the input is high enough to push the transistors that are used in the constant current generator out of their desired working region. It must be pointed out that the swing of the input node can be problematic when there is a significant capacitance between adjacent channels of the sensor, as this increases significantly the cross-talk. The ToT time can be measured in different ways. A clock counter, enabled when the comparator output is high, can be put in every channel. Alternatively, the clock counter is centralized and its output is fan-out to the whole chip. The counter state is latched into local registers when the leading and trailing edge transitions are detected and the ToT is obtained by difference. ToT systems use the same principle of single slope ADC, so they can be intrinsically slow. As an example, if a 200 MHz clock is used and the required dynamic range is 8 bits, the maximum measuring time is 1.28 μs. To speed up the conversion, the use of local ring oscillators, which can allow a counting frequency in the GHz range, has also been proposed [52].

1.4.3 TIME PICK-OFF SYSTEMS

Range finding, particle identification in High Energy Physics and photon time of flight measurement in Positron Emission Tomography (PET) are examples of applications in which the time of arrival of the event must be detected with very good accuracy. In this case, a binary front-end can be used, but the firing time of the discriminator must be captured with a precision often better than 100 ps rms. The simple method of counting clock pulses just described becomes impractical, because it would require clock frequencies between 5 and 10 GHz. Distributing such a fast signal with adequate uniformity and integrity over a large chip area is a formidable challenge. A better (and often the only possible) approach is to use a reference clock at lower frequency, interpolating with suitable techniques the time elapsed between the hit arrival and one clock transition. A device that produces a digital code associated to a time interval with a very good resolution is called a Time-to-Digital Converter (TDC). Although a simple counter also provides timing information, the word "TDC" is usually reserved to those systems employing dedicated methods to enhance the time resolution well beyond what is permitted by the system clock. Such time interpolators are built with digital gates or with analog techniques requiring small dynamic range, therefore they are well suited to be implemented in modern CMOS technologies which operate with power supplies of the order of 1.5 V or less. TDCs were originally conceived as stand-alone components, intended to be mated to a separate front-end. The progress in microelectronics technology and design techniques makes it possible today to integrate them together with the front-end amplifier and the discriminator on a per-channel basis [53–57]. This offers the opportunity of building highly integrated timing systems. Furthermore, using a TDC, ToT measurement can be done with high accuracy while keeping the pulse duration short, thus achieving high rate capability. In this way, fast systems which can measure both the time of arrival and the charge associated to the event can be built with low power and using mainly digital components.

1.4.4 SAMPLE AND HOLD AND PEAK DETECTORS

The conceptual scheme of a sample and hold system is shown in Fig. 1.38. During acquisition, switch S_1 is closed while S_2 and S_3 are open. The hold capacitor C_h is thus connected to the front-end output. When a peak is detected, S_1 is open, storing the peak value in C_h. For read-out, S_2 and S_3 are closed, and the source follower acts as a buffer for the sampling capacitor. The analog samples can be digitized on chip, but in simpler implementations it is common to send analog data off-chip. Fig. 1.39 shows an alternative implementation. Here also the front-end is a discrete-time circuit. Before

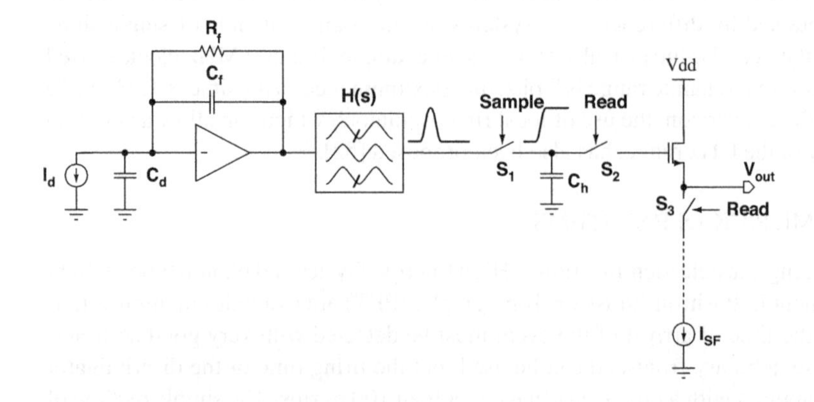

FIGURE 1.38 Front-end with sample and hold.

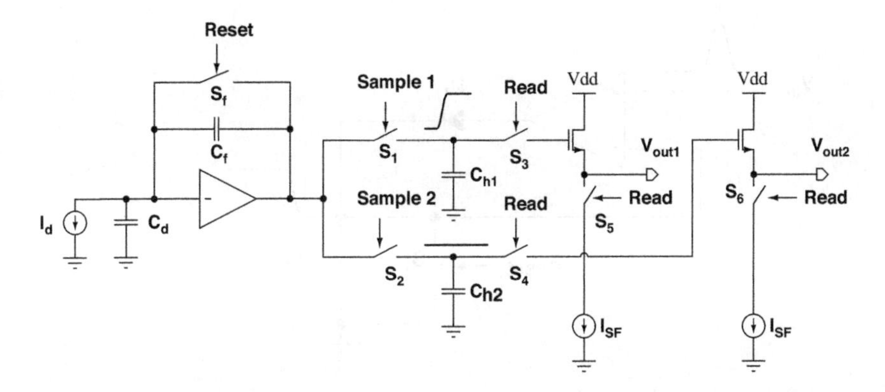

FIGURE 1.39 Front-end with sample and hold and Correlated-Double Sampling (CDS).

the acquisition starts, the feedback switch S_f is closed to reset the feedback capacitor and establishes a suitable DC bias point for the core amplifier. When a signal arrives, the detector current is integrated on C_f providing an output step voltage. For readout, the front-end output is first sampled on C_{h1}. Then, S_f is closed and the value after reset is captured on C_{h2}. The two voltages are then readout. The useful information is contained in the difference between the voltage stored in C_{h1} and the one sampled in C_{h2}. Called *Correlated Double Sampling* (CDS), this approach allows us to subtract offsets that are common to both signal and baseline and to suppress slow varying noise components which exhibit little changes between the times the two samples are taken. Interestingly, sample and hold circuits with active reset were the standard technique employed in the first full custom ASICs designed for silicon microstrip detectors [58]. Due to the additional charge injection and noise introduced by the feedback reset switch, the method was later superseded by continuous time approaches that offered a better noise figure [59]. Modern deep submicron technologies provide however the opportunity of building very small CMOS switches with minimal charge injection, making again discrete-time systems competitive [60]. An issue is sample and hold circuits is the proper choice of the sampling time. For a discrete-time front-end, the problem is less severe, because once the detector signal has been integrated on C_f, the front-end output voltage stays constant for a fairly long time. The only discharge path for the feedback capacitor is in fact provided by the leakage current of the reset switch. It is therefore sufficient to sample and read-out the information before the next pulse arrives or before the charge stored on C_f is washed away by the leakage current, a process which usually requires several hundreds of microseconds. For a front-end with a continuous-time transfer function, like the one of Fig. 1.38, it is instead desirable to capture the peak, where the SNR is the highest. The peaking and the sampling times must thus be properly synchronized. Peak detectors provide an interesting solution to the problem. Fig. 1.40 shows the principle. A unidirectional element, represented in the figure as a diode, is inserted in the feedback path of an operational amplifier which is configured as a voltage follower. By closing the reset switch S_1 the hold capacitor is discharged to ground. The front-end output is connected to the "plus" terminal of the

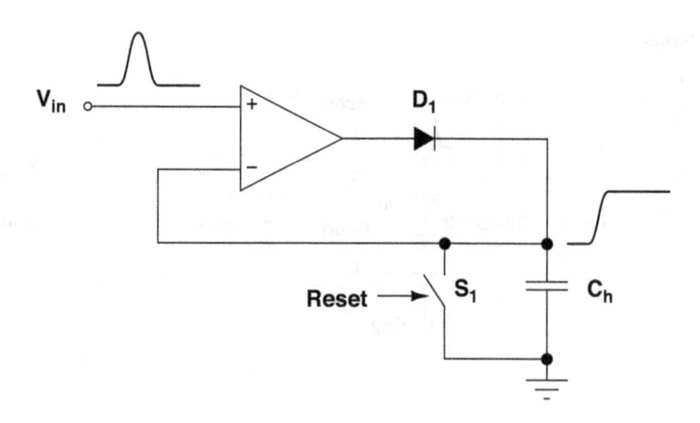

FIGURE 1.40 Peak detector concept.

op-amp. If there is no signal ($V_{in} \leq 0$), the diode is off. When the front-end output starts swinging upwards, the op-amp amplifies the difference between the input signal and the voltage on its negative terminal (which is zero when the signal appears), turning the diode abruptly on. At this point, the voltage on the hold capacitor starts tracking the one at the amplifier input and the circuit behaves as a voltage follower. After reaching the peak, the front-end output begins its return to the baseline. To follow the input, the hold capacitor should be discharged, which requires to sink current from the node. This can not happen because the diode allows the flow of current only in the forward direction. Therefore, the voltage on the "plus" terminal of the op-amp becomes quickly smaller than the one on C_h, pulling the op-amp output down and turning the diode abruptly off. The peak voltage is thus captured and held on C_h. The information is then read-out by connecting C_h to a voltage buffer. In CMOS technologies floating diodes are usually not available and the unidirectional component must hence be implemented with a transistor. Fig. 1.41 shows a possible scheme using a PMOS transistor. The signal at the op-amp output is also reported. Notice that when the signal arrives, the gate of M_1 is quickly pulled down, so that M_1 is switched on and V_{in} can be tracked on C_h. As soon as the peak is over, the op-amp output swings upwards, shutting M_1 off and preventing C_h from being discharged. Observe that the hold node is protected with a source follower because it has a high impedance and thus a very poor driving capability. In the tracking mode, peak detectors are closed-loop systems and their frequency stability must be carefully checked. The design of these circuits entails several interesting issues, which are discussed in [61, 62], while full front-ends based on this principle can be found in [63–65]. Sample and hold and peak detectors are often complemented with one discriminator per channel to flag the hit detection. In this way, only channels containing data above noise are addressed for readout, reducing dead time. Notice that in peak detectors, the op-amp output makes a sharp transition when the peak is found. In time-invariant front-ends, the peak occurs ideally always at the same time, hence connecting a comparator to the op-amp output one can also capture the time of arrival of the hit with a fairly good accuracy. In the circuits described so far a new event can not be acquired before the previous one has

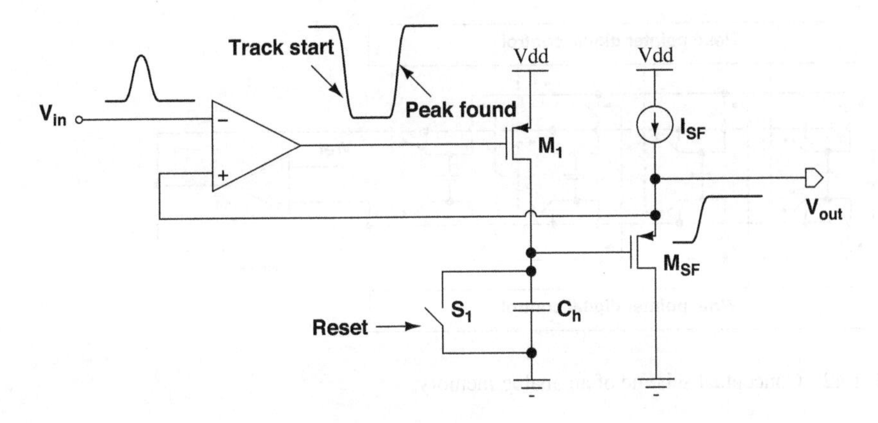

FIGURE 1.41 Peak detector using only MOS transistors.

been readout. This causes dead-time, which in many applications can not be tolerated. The problem is circumvented by implementing more storage cells per channel, so that if a given cell is full and waiting for readout, another one can be made available to capture the next hit. Notice that in a system based on peak detection, only the hold capacitor and the switches need to be replicated.

1.4.5 ANALOG MEMORIES

In an analog memory an array of capacitors is available in every channel, as shown in Fig. 1.42. In the figure, only four cells are drawn for simplicity. In practical cases, the number of cells per channel ranges between 16 and 256, but it can reach several thousand units in some applications [66]. Each cell has two sets of CMOS switches, one for writing and the other for reading back the stored information. Through the write switches, any given cell is connected for a fixed amount of time to the output of the front-end amplifier. The value of the signal is sampled by opening the switches associated to the interested capacitor. The switches of the cell coming next in the array are then closed, in order to prepare the system for sampling the new value. When a cell must be read, it is put in feedback to an operational amplifier through the read switches. After readout, the cell is usually cleared by closing the reset switch. This switch serves also to connect the read amplifier as a voltage follower when the readout does not occur, thus keeping always the op-amp in a well defined closed-loop feedback condition and precharging the read bus to the reference voltage. In some applications, analog memories are used as derandomizing buffers. In this case, the event is usually confined in a small number of cells, and more events can be captured simultaneously in the memory. The write and read operations can thus be concurrent and two different logic blocks are necessary. In other cases, the analog memory stores only one event at a time and when it is readout a new event can not be acquired. In this configuration, one logic unit is sufficient to control both the write and readout phases and the design of the overall circuit becomes simpler. An interesting feature of analog

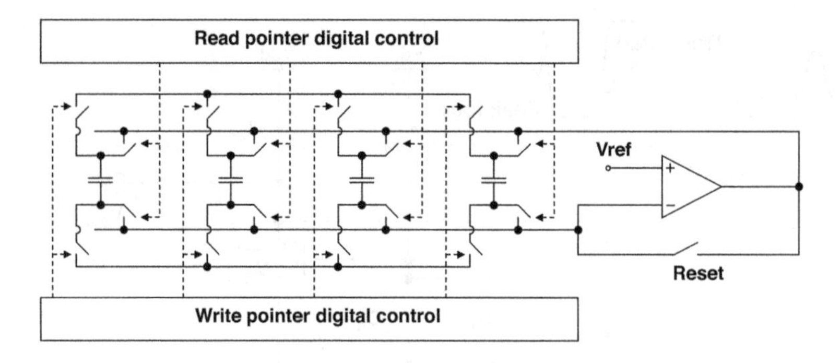

FIGURE 1.42 Conceptual scheme of an analog memory.

memories is that they allow to capture the full waveform generated by the front-end amplifier, provided of course that the sampling frequency is high enough. Having the full waveform available can help in identifying undesirable conditions, such as the occurrence of pile-up. Sampling an analog value involves only switches and capacitors and can be accomplished with low power consumption, while fast analog to digital conversion is more power demanding. Analog memories are therefore particularly interesting to detect fast transients which occur rarely in time. In this case, one can sample the signal at high rate and then read and digitize the analogue values at a lower speed only if an interesting event is found. The ADC can be put either on-chip [67–70] or off-chip [71, 72].

In designing analogue memories, two key choices need to be made: the selection of the sampling capacitors and the design of the sequencer that drives the sampling switches. Modern CMOS technologies offer two types of capacitors: metal-metal and MOS capacitor. Metal-metal capacitors are very linear and very fast, but they offer a relatively small density (typically around 1 fF/μm^2 or less). They are therefore mostly suitable for very high speed applications. MOS capacitors exploit the thin oxide employed for the gate of MOS transistors, suffer from poor linearity and are slower, but their capacitance density easily exceeds 5 fF/μm^2. They are thus preferable when many sampling cells per channel are necessary. When linearity is an issue, it is important that the same quantity is stored and read back from the sampling capacitor, that must not be used to transfer the signal between different domains (e.g. converting a current into a voltage by integration). A discussion on the usage of MOS capacitors in analog memories can be found in [73]. For the sequencer, two major topologies can be employed. The first one is based on the use of a shift register to create a pointer which addresses the cell to be written. A clock signal determines the shift of the pointer between one location and the next. The sampling time is thus well defined by the clock period and samples are taken at very uniform intervals. The other alternative is to use a digital delay line. Fig. 1.43 shows a possible implementation. Digital buffers (each one made of two inverters) are cascaded and a pulse is sent through the line. The time difference between two consecutive samples is given by the buffer propagation delay. In modern CMOS technologies, this can be less than 100 ps, making thus

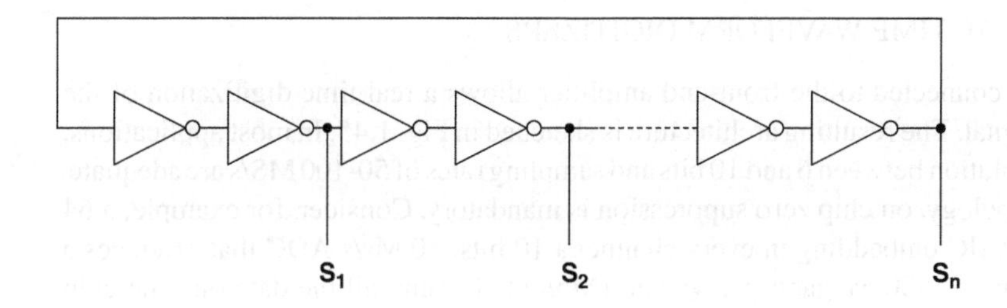

FIGURE 1.43 Buffer chain employed in fast sampling analog memories.

possible very fast sampling. The buffers can be arranged in a ring, so that the samples are overwritten unless an interesting event is detected. The circuitry to freeze the memory can be put either on-chip or off-chip. The buffer delay can be controlled by limiting the maximum current that can be driven by the digital gates. One issue is the control of the delay value, which can change with process variations and with temperature. The problem is often addressed by enclosing the delay line in a feedback loop that constrains the sum of all the delays in the chain to be locked to a reference value. Shown in Fig. 1.44, the technique works by comparing the signal before and after the delay line, imposing that delay is equal to one period of a reference clock. The total delay is then subdivided among the buffers in the chain. Mismatches between the individual taps however lead to sampling non-uniformity. Fast waveform recorders achieved a sampling rate of 700 MHz already in technologies with a minimum feature size as coarse as 2 μm [74]. In submicron CMOS technologies sampling rates greater than 1 GS/s have been reported [66, 75–77], with most recent designs reaching up to 15GS/s [70]. From the sample data, the time of arrival of a hit can be reconstructed with a precision significantly better than the peaking time of the input signal. Fast analog memories can thus be used as an alternative to TDCs also in those applications demanding very high time resolution [78, 79].

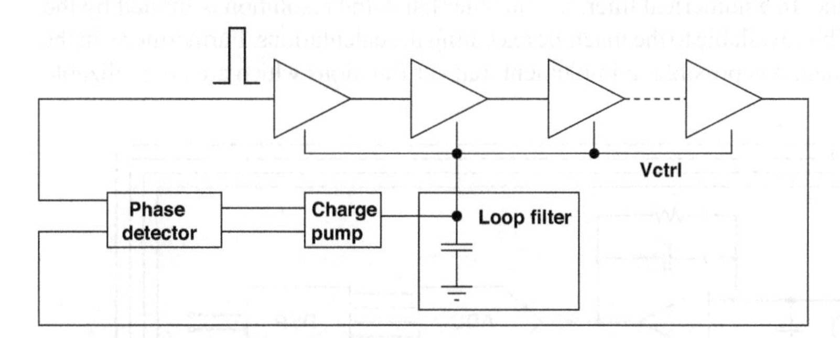

FIGURE 1.44 A Delay Locked Loop (DLL) can be used to generate fast sampling signals in a high speed analog memory.

1.4.6 REAL TIME WAVEFORM DIGITIZERS

An ADC connected to the front-end amplifier allows a real time digitization of the output signal. The resulting architecture is sketched in Fig. 1.45. In most applications, ADC resolution between 6 and 10 bits and sampling rates of 50-100 MS/s are adequate. In this topology, on chip zero suppression is mandatory. Consider, for example, a 64 channel ASIC embedding in every channel a 10 bits, 50 Ms/s ADC that produces a data stream of 500 Megabit per second (Mbit/s). To send all the data out, the chip needs an output bandwidth of 32 Gbit/s, which is challenging to achieve. In addition, the system would dissipate a large amount power, most of which is spent to ship off-chip uninteresting noise samples. The data selection is fully performed in the digital domain. In the Digital Signal Processor (DSP), gain and offset errors can be corrected and long term drifts can be identified and suppressed [80, 81]. The clean data are then compared against a numerical threshold and only those above it are preserved and stored into local memories. To properly order the events in time, a time-stamp, usually derived by counting clock pulses, is attached to each sample. Samples belonging to the same event are normally grouped into clusters, so that the common information (time stamp, channel address, etc.) is not repeated beyond necessity. The stored data are transmitted with fast serial links, to avoid overloading the system with a huge numbers of cables.

The architecture of Fig. 1.45 offers several advantages. First, it preserves the maximum of information on the signal. Second, it minimizes the amount of analog circuitry and it is therefore suitable for modern deep-submicron CMOS technologies, which are mainly intended for digital applications. Third, data selection can be done more accurately in the digital than in the analog domain. In the analog domain, in fact, the accuracy of a given operation is limited by the unavoidable tolerances affecting the components that form the signal processing network. For instance, in RC filters, the time constants change if either the resistor or the capacitor deviates from the nominal values. In high precision applications the filter components are programmable, so they can be tuned to reach the desired performance, but the calibration procedure is cumbersome. In a numerical filter, on the other hand, the resolution is limited by the number of bits available to the machine executing the calculations. Furthermore, in the digital domain it is possible to implement transfer functions which are not realizable

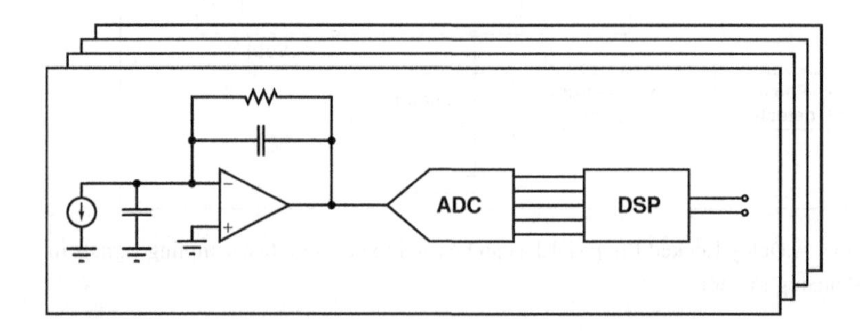

FIGURE 1.45 Front-end with embedded digital signal processing.

with linear networks. The ultimate performance of the system of Fig. 1.45 are thus set by the front-end amplifier and the ADC. This approach is hence very attractive and it is in fact used in those applications where the limited number of channels and the space and power constraints allow to implement it with off-the-shelf components. The main obstacle to the use of a real time digitizing front-end in the read-out of high granularity detectors has been the power consumption of ADCs with high resolution and adequate sampling rates. Recently, the progress in ADC design, combined with the scaling of CMOS technologies, have lead to the implementation of converters with 8-10 bit resolution and less than 1 mW of power consumption for sampling rates as high as 50 Ms/s. With such performance, it is envisageable to integrate one converter per channel and keep the power of the full front-end below 5 mW per channel. Deep submicron technologies are also ideally suited to implement complex digital functions. So one can expect that low power digitizing front-end embedding a significant amount of digital signal processing will become more and more common in the future.

1.4.7 DATA READOUT AND TRANSMISSION

A variety of techniques has been deployed to readout multi-channel front-end ASICs. A simple approach consists in sending a token that enables the receiving channel to put its data on a common output line or set of lines. The same amount of time is allocated to every channel, regardless of its information content. A counter is thus sufficient to keep track of the channel which is being addressed. The data are further processed in the chip periphery and those that match the established selection criteria are put in a buffer together with the appropriate channel address and finally queued for off-chip transmission. The method minimizes the complexity of the in-channel logic, but it is not efficient if only few channels in a given readout cycle contain useful information. In this case, it is preferable to address only those channels that have a potentially interesting content, performing a sparsified readout. Additional in-channel circuitry allows to decide locally if there are data and issue a readout request. A counter in the periphery is not sufficient to generate the addresses and more elaborated schemes need to be applied. A first possibility is to define for each channel a hard-wired address which is transmitted together with the data. Another method is to use priority encoders, awarding to each channel a rank in a hierarchy. A combinatorial circuit generates the address of the channel that contains data and has the highest priority. Once its data have been acquired, the channel is reset, so that the next busy channel in the priority hierarchy can take control of the common data bus, and so on. In priority-based readout there is the risk of creating a bias in favor of the channels that have higher priority. Therefore, a given channel is usually allowed to talk only once during a readout cycle.

The selection of the information to be transmitted can be done fully on-chip or can be supported by external control signals. In the first case, one speaks of a data-push, data-driven or self-triggered readout. In the second, one speaks of a triggered architecture. Triggered systems are common in those applications where a rare signal must be extracted from a strong physics background. A common example is found in high energy and nuclear physics experiments, where the interaction producing a specific particle must be identified among many others that yield uninteresting results. In this case, a set of dedicated detectors look for prompt signatures of the event under study and generates a trigger signal that is propagated back to the full system to initiate the readout. The time elapsing from the instant in which the event actually takes place to the one in which the trigger is received is called the *trigger latency* and is often found in the range of a few microseconds. The front-end electronics must thus have sufficient memory capability to store the data during the latency time. When a trigger is received, a trigger matching operation takes place to identify the data that belong to the selected events. Only those fulfilling the matching criteria are then transmitted. A simple example helps to clarify this concept. Consider a particle accelerator in which two counter-rotating beams are made to collide every 100 ns. If a potentially interesting condition is detected, a trigger is sent back to the detector with a latency of 2 μs. The machine clock is used to count the interaction number, producing on-chip a time-stamp which is attached to any signal going above the noise threshold. Suppose now that an interesting event occurs at the fourth interaction. The associated trigger will arrive only after the latency, when the time-stamp counter is already at the number 24. A second counter is thus started with a delay equal to the latency, i.e. 20 clock cycles in this case. When the trigger arrives, the value of the second counter is latched, producing the correct time-stamp (4 in our example) of the interaction of interest. The data whose time-stamp matches the selected interaction are thus readout. The use of a trigger allows us to cut significantly the amount of data to be shipped off-chip. In recent years, however, the advent of deep-submicron technologies has allowed a remarkable improvement in the data transmission speed and in the computational power of programmable logic. There is therefore a trend in sending out of the front-end chip as many data as possible, performing event selections on devices such as Field Programmable Gate Arrays (FPGA) in which the trigger condition can be easily redefined by changing the firmware.

In the past, it was common to send off-chip analog data. In some cases to avoid high speed clock on-chip, also digital addresses were sent in analog form, transmitting simultaneously more bits over the same line by driving it with a Digital-to-Analog Converter [82]. Today, it is preferred to send off-chip only digital data, using high speed digital differential links such as LVDS or SLVS or, in some cases, Current-Mode Logic (CML) line drivers.

REFERENCES

1. G.F. Knoll. *Radiation Detection and Measurement*. 4th ed., Wiley, 2010.
2. W.R. Leo. *Techniques for Nuclear and Particle Physics Experiments*. 2nd ed., Springer, 1994.
3. C. Grupen and B. Shwartz. *Particle Detectors*. Cambridge University Press, 2011.
4. C. Grupen and I. Buvat. *Handbook of Particle Detection and Imaging*. Springer, 2012.
5. H. Spieler. *Semiconductor Detector Systems*. Springer, 2005.

6. D. Green. *The Physics of Particle Detectors*. Cambridge University Press, 2005.
7. W. Blum, W. Riegler, and L. Rolandi. *Particle Detection with Drift Chambers*. 2nd ed., Springer, 2008.
8. S. Tavernier. *Experimental Techniques in Nuclear and Particle Physics*. Springer, 2010.
9. G. Lutz. *Semiconductor Radiation Detectors: Device Physics*. Springer, 2007.
10. J. McKenzie. Development of the semiconductor radiation detector. *Nucl. Instr. and Meth.*, 162:49–73, 1979.
11. W. Shockley. Currents to conductors induced by a moving point charge. *Journal of Applied Physics*, 9:635–636, 1938.
12. S. Ramo. Currents induced by electron motion. *Proceedings of The Institute of Radio Engineers*, 27:584–585, 1939.
13. Z. He. Review of the Shockley–Ramo theorem and its application in semiconductor gamma-ray detectors. *Nucl. Instr. Meth.*, A463:250–267, 2001.
14. U. Fano. Ionization yield of radiations. II. The fluctuations of the number of ions. *Phys. Rev.*, 72:26–29, Jul 1947.
15. F. Sauli. Principles of operation of multiwire proportional and drift chambers. *Phys. Rev.*, CERN 77-09:92, May 1977.
16. G. Charpak, R. Bouclier, T. Bressani, J. Favier, and C. Zupancic. The use of multiwire proportional counters to select and localize charged particles. *Nucl. Instr. Meth.*, 62:262–268, 1968.
17. F. Sauli. GEM: A new concept for electron amplification in gas detectors. *Nucl. Instr. Meth.*, A386:531–534, 1997.
18. R. Bouclier et al. The gas electron multiplier (GEM). *IEEE Trans. Nucl. Sci.*, 44:646–650, 1997.
19. S. Bachmann, A. Bressan, L. Ropelewski, F. Sauli, A. Sharma, and D. Mörmann. Charge amplification and transfer processes in the gas electron multiplier. *Nucl. Instr. Meth.*, A438:376–408, 1999.
20. J. M. Heuser. Experience with the construction, operation and maintenance of vertex detectors at LEP. *Nucl. Instr. Meth.*, A418:1–8, 1998.
21. P. Weilhammer. Overview: silicon vertex detectors and trackers. *Nucl. Instr. Meth.*, A453:60–70, 2000.
22. N. Wermes. Pixel detectors for charged particles. *Nucl. Instr. Meth.*, A604:370–379, 2009.
23. A. Nomerotski. Silicon detectors for tracking and vertexing. *Nucl. Instr. Meth.*, A598:33–40, 2009.
24. A. Owens and A. Peacock. Compound semiconductor radiation detectors. *Nucl. Instr. Meth.*, A531:18–37, 2004.
25. D. Kahler et al. Performance of novel materials for radiation detection: Tl_3AsSe_3, $TlGaSe_2$, and Tl_4HgI_6. *Nucl. Instr. Meth.*, A652:183–185, 2011.
26. P. Buzhan et al. Silicon photomultiplier and its possible applications. *Nucl. Instr. Meth.*, A504:48–52, 2003.
27. V. D. Kovaltchouk, G. J. Lolos, Z. Papandreou, and K. Wolbaum. Comparison of a silicon photomultiplier to a traditional vacuum photomultiplier. *Nucl. Instr. Meth.*, A538:408–415, 2005.

28. B. Dolgoshein et al. Status report on silicon photomultiplier development and its applications. *Nucl. Instr. Meth.*, A563:368–376, 2006.
29. S. Seifert et al. Simulation of silicon photomultiplier signals. *IEEE Trans. Nucl. Sci.*, 56:3276–3733, 2009.
30. F. Corsi et al. Modelling a silicon photomultiplier (SiPM) as a signal source for optimum front-end design. *Nucl. Instr. Meth.*, A572:416–418, 2007.
31. T. Frach, G. Prescher, C. Degenhardt, R. de Gruyter, A. Schmitz, and R. Ballizany. The digital silicon photomultiplier — principle of operation and intrinsic detector performance. In *IEEE NSS-MIC Conference Records*, pages 1959–1965, 2009.
32. S. Mandai and E. Charbon. A 128-channel, 9 ps column-parallel two-stage TDC based on time difference amplification for time-resolved imaging. In *Solid-State Circuits European Conference*, pages 119–122, 2011.
33. T. Frach. Optimization of the digital silicon photomultiplier for Cherenkov light detection. *JINST*, 7:C01112, 2012.
34. S. Mandai, M. W. Fishburn, Y. Maruyama, and E. Charbon. A wide spectral range single-photon avalanche diode fabricated in an advanced 180 nm CMOS technology. *Journal of Chemical and Engineering Data*, 2012.
35. W. Dabrowski et al. Design and performance of the ABCD chip for the binary readout of silicon strip detectors in the ATLAS semiconductor tracker. *IEEE Trans. Nucl. Sci.*, 47:1843–1850, 2000.
36. R. Dinapoli et al. A front-end for silicon pixel detectors in ALICE and LHCb. *Nucl. Instr. Meth.*, A461:492–495, 2001.
37. W. Bonivento, P. Jarron, D. Moraes, W. Riegler, and F. dos Santos. Development of the CARIOCA front-end chip for the LHCb muon detector. *Nucl. Instr. Meth.*, A491:233–243, 2002.
38. F. Anghinolfi, P. Jarron, F. Krummenacher, E. Usenko, and M. C. S. Williams. NINO: an ultrafast low-power front-end amplifier discriminator for the time-of-flight detector in the ALICE experiment. *IEEE Trans. Nucl. Sci.*, 51:1974–1978, 2004.
39. J. Kaplon and W. Dabrowski. Fast CMOS binary front end for silicon strip detectors at LHC experiments. *IEEE Trans. Nucl. Sci.*, 52:2713–2720, 2005.
40. F. Campabadal et al. Design and performance of the ABCD3TA ASIC for readout of silicon strip detectors in the ATLAS semiconductor tracker. *Nucl. Instr. Meth.*, A552:292–328, 2005.
41. M. Chiosso, O. Cobanoglu, G. Mazza, D. Panzieri, and A. Rivetti. A fast binary front-end ASIC for the RICH detector of the COMPASS experiment at CERN. In *IEEE NSS-MIC Conference Records*, pages 1495–1500, 2008.
42. J. Kaplon and M. Noy. Front end electronics for silicon strip detectors in 90 nm CMOS technology: advantages and challenges. *JINST*, 5:C11024, 2010.
43. W. Ferguson et al. The CBC microstrip readout chip for CMS at the High Luminosity LHC. *JINST*, 7:C08006, 2012.
44. P. Grybos et al. RX64DTH: a fully integrated 64-channel ASIC for a digital X-ray imaging system with energy window selection. *IEEE Trans. Nucl. Sci.*, 52:839–846, 2005.

45. R. Szczygiel, P. Grybos, P. Maj, A. Tsukiyama, K. Matsushita, and T. Taguchi. RG64—High count rate low noise multichannel ASIC with energy window selection and continuous readout mode. *IEEE Trans. Nucl. Sci.*, 56:487–495, 2009.
46. R. Szczygiel, P. Grybos, and P. Maj. A prototype pixel readout IC for high count rate X-ray imaging systems in 90 nm CMOS technology. *IEEE Trans. Nucl. Sci.*, 57:1664–1674, 2010.
47. R. Steadman, C. Herrmann, O. Mülhens, and D. G. Maeding. ChromAIX: Fast photon-counting ASIC for spectral computed tomography. *Nucl. Instr. Meth.*, A648, 2011.
48. I. Kipnis et al. A time-over-threshold machine: the readout integrated circuit for the BABAR Silicon Vertex Tracker. *IEEE Trans. Nucl. Sci.*, 44:289–297, 1997.
49. Ivan Peric et al. The FEI3 readout chip for the ATLAS pixel detector. *Nucl. Instr. Meth.*, A565:178–187, 2006.
50. T. Kugathasan, G. Mazza, A. Rivetti, and L. Toscano. A 15 μW, 12-bit dynamic range charge measuring front-end in 0.13 μm CMOS. In *IEEE NSS-MIC Conference Records*, pages 1667–1673, 2010.
51. K. Kasinski, R. Szczygiel, and P. Grybos. TOT01, a time-over-threshold based readout chip in 180 nm CMOS technology for silicon strip detectors. *JINST*, 6:C01026, 2011.
52. F. Zappon, M. van Beuzekom, V. Gromov, R. Kluit, X. Fang, and A. Kruth. GOSSIPO-4: an array of high resolution TDCs with a PLL control. *JINST*, 7:C01081, 2012.
53. S. Martoiu et al. A pixel front-end ASIC in 0.13 μm CMOS for the NA62 experiment with on pixel 100 ps Time-to-Digital Converter. In *IEEE NSS-MIC Conference Records*, pages 55–60, 2009.
54. P. Fischer, I. Peric, M. Ritzert, and M. Koniczek. Fast self triggered multi channel readout ASIC for time and energy measurement. *IEEE Trans. Nucl. Sci.*, 56:1153–1158, 2009.
55. X. C. Fang, W. Gao, Ch. Hu-Guo, D. Brasse, B. Humbert, and Y. Hu. Development of a low-noise front-end readout chip integrated with a high-resolution TDC for APD-based small-animal PET. *IEEE Trans. Nucl. Sci.*, 58:370–377, 2011.
56. X. Fang et al. IMOTEPAD: A mixed-signal 64-channel front-end ASIC for small-animal PET imaging. *Nucl. Instr. Meth.*, A634:106–112, 2011.
57. M. D. Rolo, L. N. Alves, E. V. Martins, A. Rivetti, M. B. Santos, and J. Varela. A low-noise CMOS front-end for TOF-PET. *JINST*, 6:P09003, 2011.
58. J. Walker, S. Parker, B. Hyams, and S. Shapiro. Development of high density readout for silicon strip detectors. *Nucl. Instr. Meth.*, A226:200–203, 1984.
59. E. Beuville et al. Amplex, a low-noise, low-power analog CMOS signal processor for multi-element silicon particle detectors. *Nucl. Instr. Meth.*, A288:157–167, 1990.
60. M. Havranek et al. Pixel front-end development in 65 nm CMOS technology. *JINST*, 9:C01003, 2014.
61. G. De Geronimo, P. O'Connor, and A. Kandasamy. Analog CMOS peak detect and hold circuits. Part 1. Analysis of the classical configuration. *Nucl. Instr. Meth.*, A484:533–543, 2002.

62. G. De Geronimo, P. O'Connor, and A. Kandasamy. Analog CMOS peak detect and hold circuits. Part 2. The two-phase offset-free and derandomizing configuration. *Nucl. Instr. Meth.*, A484:544–556, 2002.
63. G. De Geronimo, A. Kandasamy, and P. O'Connor. Analog peak detector and derandomizer for high-rate spectroscopy. *IEEE Trans. Nucl. Sci.*, 49:1769–1773, 2002.
64. G. De Geronimo et al. Front-end ASIC for a GEM based time projection chamber. *IEEE Trans. Nucl. Sci.*, 51:1312–1317, 2004.
65. G. De Geronimo et al. Front-end ASIC for a liquid Argon TPC. *IEEE Trans. Nucl. Sci.*, 58:1376–1385, 2011.
66. S. Ritt, R. Dinapoli, and U. Hartmann. Application of the DRS chip for fast waveform digitizing. *Nucl. Instr. Meth.*, A623:486–488, 2010.
67. G. Mazza, A. Rivetti, G. Anelli, F. Anghinolfi, M. I. Martinez, and F. Rotondo. A 32-channel, 0.25 μm CMOS ASIC for the readout of the silicon drift detectors of the ALICE experiment. *IEEE Trans. Nucl. Sci.*, 51:1942–1947, 2004.
68. A. Rivetti et al. The front-end system of the silicon drift detectors of ALICE. *Nucl. Instr. Meth.*, A541:267–273, 2005.
69. S. Tedja, J. Van der Speigel, and H. H. Williams. A CMOS low-noise and low-power charge sampling integrated circuit for capacitive detector/sensor interfaces. *IEEE J. of Solid-state Circ.*, 30:110–119, 1995.
70. E. Oberla, J.-F. Genat, H. Grabas, H. Frisch, K. Nishimura, and G. Varner. A 15 GSa/s, 1.5 GHz bandwidth waveform digitizing ASIC. *Nucl. Instr. Meth.*, A735:452–461, 2014.
71. M. J. French et al. Design and results from the APV25, a deep sub-micron CMOS front-end chip for the CMS tracker. *Nucl. Instr. Meth.*, A466:359–365, 2001.
72. E. Albuquerque et al. Experimental characterization of the 192 channel Clear-PEM frontend ASIC coupled to a multi-pixel APD readout of LYSO:Ce crystals. *Nucl. Instr. Meth.*, A598:802–814, 2009.
73. G. Anelli, F. Anghinolfi, and A. Rivetti. A large dynamic range radiation-tolerant analog memory in a quarter-micron CMOS technology. *IEEE Trans. Nucl. Sci.*, 48:435–439, 2001.
74. G. M. Haller and B. A. Wooley. A 700 MHz switched-capacitor analog waveform sampling circuit. *IEEE Journal of Solid-state Circuits*, 29:500–508, 1994.
75. E. Delagnes, Y. Degerli, P. Goret, P. Nayman, F. Toussenel, and P. Vincent. SAM: A new GHz sampling ASIC for the H.E.S.S.-II front-end electronics. *Nucl. Instr. Meth.*, A567:21–26, 2006.
76. G. S. Varner et al. The large analog bandwidth recorder and digitizer with ordered readout (LABRADOR) ASIC. *Nucl. Instr. Meth.*, A583:447–460, 2007.
77. S. Kleinfelder. Gigahertz waveform sampling and digitization circuit design and implementation. *IEEE Trans. Nucl. Sci.*, 50:955–962, 2003.
78. J. F. Genat, G. Varner, F. Tang, and H. Frisch. Signal processing for picosecond resolution timing measurements . *Nucl. Instr. Meth.*, A607:387–393, 2009.
79. L. L. Ruckman and G. S. Varner. Sub-10 ps monolithic and low-power photodetector readout. *Nucl. Instr. Meth.*, A602:438–445, 2009.

80. R. Esteve Bosch, A. Jimenez de Parga, B. Mota, and L. Musa. The ALTRO chip: a 16-channel A/D converter and digital processor for gas detectors. *IEEE Trans. Nucl. Sci.*, 50:2460–2469, 2003.
81. P. Aspell, M. De Gaspari, H. Franca, E. G. Garcia, and L. Musa. Super-ALTRO 16: A front-end system on chip for DSP based readout of gaseous detectors. *IEEE Trans. Nucl. Sci.*, 60:1289–1295, 2013.
82. W. Erdmann. The 0.25 μm front-end for the CMS pixel detector. *Nucl. Instr. Meth.*, A549:153–156, 2005.

2 MOS Transistor Properties

The Metal Oxide Semiconductor Field Effect Transistor ("MOSFET", or, in short "MOS") is the key building block of all the circuits that will be studied in the book. In this chapter, we provide a concise introduction to the device.

To make the material accessible also to readers with little back-ground in electronics, we start reviewing a few concepts of semiconductor physics that are essential to understand the working principles of the MOSFET. A broader treatment of these notions can be found in standard textbooks on semiconductor devices, such as [1] and [2]. In the central part of the chapter we examine the most important terms contributing to the transistor threshold voltage and we derive the key relationships between the device current and the voltages applied to its terminals. We then study the parasitic capacitance affecting the MOSFET behavior at high frequency and we introduce the small signal equivalent circuits commonly employed in hand circuit analysis. In this treatment, we make use of a very simplified description of the MOSFET physics, allowing to derive analytical equations with a minimum of mathematical complexity. We are thus forced to neglect fundamental effects that play a major role in devices fabricated in modern deep submicron CMOS technologies. These aspects are dealt with in the last part of the chapter.

An exhaustive treatment of the MOS transistor is provided in [3]. Further details on the topics discussed in the last part of the chapter can be found in standard textbooks on CMOS microelectronics such as [4] and [5].

2.1 SILICON PROPERTIES

Silicon is the most widely used semiconductor in the microelectronics industry. Its success stems from several reasons, which are briefly summarized hereafter:

- Silicon is an abundant element as it makes about 26% of the Earth crust.
- It is relatively easy and cheap to grow it in large crystals of very high purity.
- The electrical properties of silicon can be easily modified by introducing in the crystal lattice atoms of other appropriate chemical elements, in a process called doping.
- Devices fabricated in silicon can operate over a wide range of temperatures (conventional range spans from -55°C to 125°C).
- Silicon easily forms an oxide, SiO_2, which behaves as an almost ideal insulator. Furthermore, SiO_2 is not soluble in water.
- Due to its excellent thermal and mechanical properties, silicon can be easily patterned into integrated circuits through photo-lithographic techniques.

The main properties of silicon are summarized in table 2.1.

TABLE 2.1
Fundamental properties of silicon.

Atomic number	14
Atomic weight	28.0855
Electron distribution per shell	2, 8, 4
Density	2.33 g cm^{-3}
Melting point	1414°C
Bandgap energy at 300 K	1.12 eV
Intrinsic carrier concentration, n_i	$1.45(1.08)\cdot 10^{10}$ cm^{-3}
Electron mobility	1500 $cm^2/V\cdot s$
Hole mobility	500 $cm^2/V\cdot s$

2.1.1 SILICON BAND STRUCTURE

Chemically, silicon is a tetravalent metalloid. Each atom contains four electrons in the outer shell that can be shared with other atoms to form chemical bonds. Silicon can exist in a crystal form, with a diamond cubic lattice in which a pattern of 8 atoms repeats itself along the crystal. The Pauli exclusion principle prevents two electrons from occupying the same state. When atoms are closely packed together in a solid, the energy levels therefore split, giving rise to bands of possible states. The semiconductor properties of silicon originate from its bands structure, which is schematically shown in Fig. 2.1. Electrons involved in bonds between atoms occupy states in the valence band and they are localized in proximity of the atoms. Electrons in the conduction band are not "attached" to any atom in particular and they can move easily through the crystal under the action of an electric field. The two bands are separated by an energy gap of 1.12 eV, in which no accessible states are present.

At 0 K (Fig. 2.1 left) all the states in the valence band are occupied and all the states in the conduction band are empty. No free carriers are available for conduction and the material behaves as an insulator. As the temperature rises above the absolute zero, electrons acquire thermal energy. At some point, a few of them may have enough

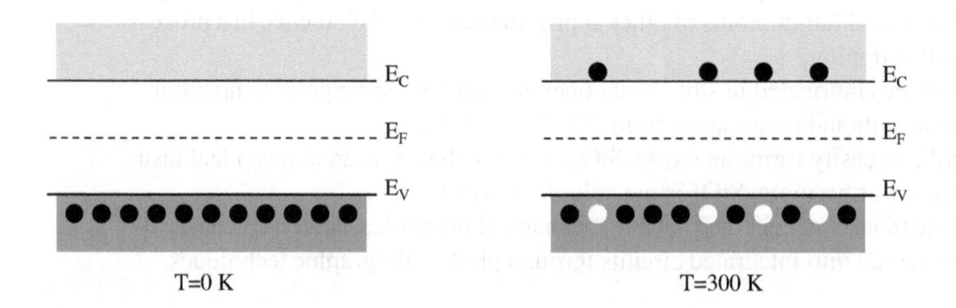

FIGURE 2.1 Band structure of silicon with electron population at 0 K and 300 K.

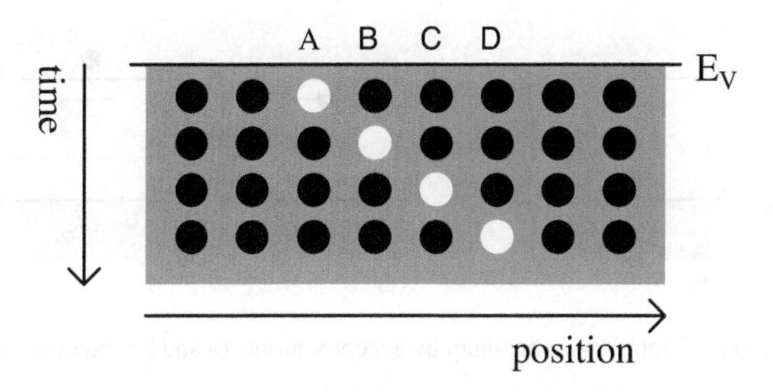

FIGURE 2.2 Transport of holes in the valence band.

energy to make a transition from the valence into the conduction band. When this happens, the electron leaves behind in the valence band an empty location, called a hole. Electrons and holes can also recombine. At thermal equilibrium the generation and the recombination processes balance and yield a constant average number of electrons-hole pairs. The number of carriers per unit volume in pure (also called "intrinsic") silicon is indicated with $n_i(T)$. This means that at given temperature, $n_i(T)$ electrons per unit volume are found in the conduction band. Since these electrons can only originate from the valence band, the concentration of holes is also equal to $n_i(T)$. The traditionally accepted value for $n_i(T)$ at room temperature is $1.45 \cdot 10^{10}$ cm^{-3}. Another proposed value is $1.08 \cdot 10^{10}$ cm^{-3} [6, 7]. An empirical fit that can be used to evaluate the carrier concentration as a function of temperature is given by:

$$n_i = 9.38 \cdot 10^{19} \left(\frac{T}{300} \right)^2 e^{-\frac{6884}{T}} \tag{2.1}$$

where T is the absolute temperature.

The presence of holes allows the transport of charge within the valence band. This is illustrated in Fig.2.2, where holes and electrons are designed respectively as white and black circles. The hole is originally in position A, as shown in the first row of the figure. An electron in B moves to the left, filling the hole in A and leaving behind a new one in B, thus creating the pattern shown in the second row. Another electron can now move from C to B and so on. As time progresses, the hole moves from left to right. From the point of view of charge transport, the movement of an electron from right to left is equivalent to the movement of a positive charge in the opposite direction. Holes can therefore be treated as fictitious particles with positive charge $+q$, where q is the absolute value of the electron charge (1.6×10^{-19} Coulomb).

2.1.2 DOPING

Doping allows changing the concentration of electrons and holes in silicon, altering substantially its ability to conduct electrical current. The process consists in introducing into the silicon lattice foreign atoms which have either three or five electrons in

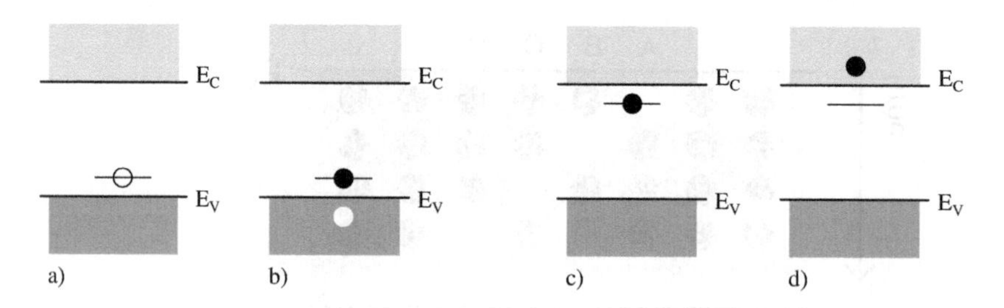

FIGURE 2.3 Creations of states in the bandgap by acceptor atoms (a and b) and donor atoms (c and d).

their outer shells. If the atom has only three valence electrons, it will form three bonds with its neighbors and it will have a free location that can easily take up an electron from a nearby silicon atom. Fig. 2.3 illustrates the phenomenon. The fact that the acceptor atom has only three valence electrons results in the creation of additional states in the bandgap. These states are localized in proximity of the acceptor atom and their energy is very close to the valence band, typically 50 meV above it (see Fig. 2.3a). It is therefore very easy to find electrons in the valence band with enough energy to reach and occupy those states (Fig. 2.3b). When this happens, the acceptor atom acquires an extra electron and becomes negatively charged. At room temperature, practically all the states associated to acceptors atoms are filled. A corresponding number of holes, left by the electrons which have filled the acceptor states, are present in the valence band. If an electric field is applied, the holes allow the transport of charge in the valence band with the mechanism illustrated in Fig. 2.2. Ionized acceptors, on the other hand, are bound to fix positions in the lattice and they can not contribute to the current flow. Silicon doped with acceptor atoms is called p-type. The acceptor concentration is usually indicated as N_A. A common material used in p-type doping is Boron (B).

Fig. 2.3 c) and d) show the effect of doping with donor atoms. A donor has five electrons in its outer shell. Four electrons participate in bonds with the tetravalent silicon neighbors, while the extra electron is loosely bound to the parent dopant atom. This electron resides in a state in the bandgap very close to the conduction band. At room temperature, all the electrons in these conditions have enough thermal energy to jump into the conduction band, living behind an empty state in the bandgap. Once in the conduction band, the electrons can move freely along the crystal. Since the donor has lost one electron, it becomes positively charged. As for acceptors, also ionized donors, being fix in the lattice, can not contribute to the current flow. Silicon doped with donors is called n-type and the donor concentration is indicated as N_D. Common materials employed in n-type doping are Phosphorus (P), Arsenic (As) and Antimony (Sb).

In p-type silicon there is an excess of holes. The probability for an electron to recombine with a hole is thus increased and the electron concentration is reduced. Similarly, in n-type silicon the hole concentration is less than the one found in the intrinsic material. Therefore, in doped material one type of carrier is prevalent with respect to the other and is called the "majority" carrier. Electrons are majority carriers in n-type silicon and minority carriers in p-type silicon. The viceversa is true for holes. It can be proven that the product of electron and hole concentration is constant (law of mass action):

$$np = n_i^2 \tag{2.2}$$

where n and p are respectively the electron and the hole concentration, while n_i is the carrier concentration in intrinsic silicon. At room temperature, all the donor or acceptor atoms are ionized. Therefore, in n-type silicon the electron concentration n is equal to the donor concentration N_D and the hole concentration is $p = n_i^2/N_D$. In p-type silicon the hole concentration p is equal to the acceptor atom concentration N_A and the electron concentration is given by $n = n_i^2/N_A$. The law of mass action is very useful in studying the behavior of doped semiconductors. However, it is not valid in a few peculiar situations:

- At very low temperature, when it is not true that all the dopant atoms are ionized.
- At very high temperature, when the condition $N_{dopant} \gg n_i$ might not be met (remember that n_i increases with temperature, so at high temperature the dopants might not outnumber the intrinsic carriers).
- When the doping concentration becomes very high (greater than 10^{18} cm^{-3}). In this case the semiconductor is called "degenerate".

Note that while the first two conditions are not commonly encountered in ordinary electronics, the third is. Very high doping concentrations are in fact routinely used in standard electronics devices, for example when non-rectifying contacts must be established between the semiconductor and a metal.

2.2 CHARGE TRANSPORT IN SILICON

Two independent mechanisms allow the transport of charge in silicon: diffusion and drift, that may or may not act simultaneously within a given electronic device.

2.2.1 CHARGE TRANSPORT BY DRIFT

Charge transport by drift originates from the collective movement of the charge carriers under the effect of an electric field. Let us consider the slab of silicon shown in Fig. 2.4. The slab has a length L, a section A and is connected to a battery with the polarity shown in the figure. Under the effect of the electric field the electrons move from right to left, generating a current flow in the opposite direction, as indicated by the arrow. Remember that, conventionally, we think of electrical currents as if the particles involved in the charge transport were positively charged. This is not true for metal and semiconductors, because the mobile carriers are electrons. These actually move in the opposite direction a positive charge would follow under the action of the same electric field. However, from the point of view of charge transport a negative charge moving to the left is equivalent to a positive charge going to the right. To min-

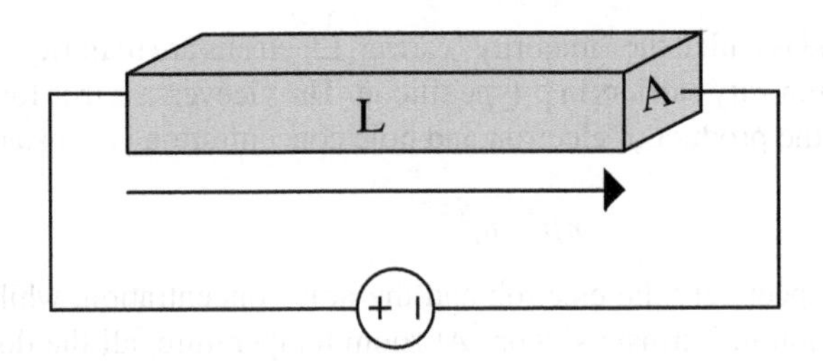

FIGURE 2.4 Silicon bar connected to a voltage power supply.

imize the mathematical complexity, we study the problem only in one dimension. In the slab we have n electrons per unit volume, therefore the total number of electrons is given by $N = nAL$. It takes a time $t = L/v$ before all those electrons leave the slab and are replaced by an equal number provided by the battery. Therefore, the current and the current density are given by:

$$I = Q/t = qnAv; \quad J = qnv \tag{2.3}$$

where q is the electron charge. Any electron is accelerated by a force $F = -q\mathscr{E}$. If the electrons were moving in free space they would be subjected to a constant acceleration. However, the electrons move in a medium (the crystal) and from time to time they experience collisions which may reduce their velocity to zero. Therefore, in between two collisions, the electron velocity changes uniformly from zero to the maximum value. To calculate the maximum velocity reached by the electrons we note that:

$$\frac{dp_n}{dt} = -q\mathscr{E} \tag{2.4}$$

where p_n is the electron momentum. Therefore, integrating we have:

$$\int_0^{p_{max}} dp = -\int_0^{\tau_c} q\mathscr{E}\, dt \tag{2.5}$$

The second member of equation (2.5) is integrated from time 0 (when the acceleration starts) to τ_c, which is the mean free time between two collisions. The maximum drift velocity is hence obtained from:

$$p_{max} = mv_{nd,max} = -q\mathscr{E}\tau_c \tag{2.6}$$

The drift velocity is thus uniformly distributed between zero and $v_{nd,max}$. Since the average of a uniform distribution is one half of its maximum value, the average drift velocity attained by the electrons, indicated simply as v_{nd}, is given by:

$$v_{nd} = -\frac{q\mathscr{E}\tau_c}{2m} \tag{2.7}$$

The interaction between the electrons and the crystal is much more complex than what we have described in our simplistic model and rigorous treatment is possible only in the framework of quantum mechanics. Fortuitously, in many situations the behavior of the electron can be described through the concept of a "quasi particle", i.e. a particle with charge $-q$ and a mass m_n, called effective mass, which is different from the one of the free electrons. In other words, all the complexity of the phenomenon is "hidden" just by treating the electron as classical particle with the same charge and a different mass. Equation (2.7) can thus be rewritten as:

$$v_{nd} = -\frac{q\tau_c}{m_n}\mathscr{E} = -\mu_n\mathscr{E} \tag{2.8}$$

where m_n is the electron effective mass. We see that the electron velocity is proportional to the electric field. The constant of proportionality, μ_n is called mobility and it is one of the primary parameters that characterize the behavior of a semiconductor device. The sign in equation 2.8 means that the actual direction of the electrons (from right to left in Fig. 2.4) is opposite to the one of the electric field. The current density associated to the electron movement is given by:

$$J_n = n(-q)(-\mu_n\mathscr{E}) = nq\mu_n\mathscr{E} \tag{2.9}$$

The same arguments applied to the electrons can be used also for the holes. Hence, the hole drift velocity is:

$$v_{pd} = \frac{q\tau_c}{m_p}\mathscr{E} \tag{2.10}$$

where m_p is the effective mass for holes. The hole mobility and the hole current density are respectively given by:

$$\mu_p = \frac{q\tau_c}{m_p} \tag{2.11}$$

$$J_p = pq\mu_p\mathscr{E} \tag{2.12}$$

The total current density flowing through the slab is due to the sum of the hole and electron contributions:

$$J = J_n + J_p = (nq\mu_n + pq\mu_p)\,\mathscr{E} = \sigma\mathscr{E} \tag{2.13}$$

The quantity $\sigma = (nq\mu_n + pq\mu_p)$ is the conductivity, while its reciprocal is the resistivity:

$$\rho = \frac{1}{nq\mu_n + pq\mu_p} \tag{2.14}$$

Consider now the case in which the potential applied across the slab is uniform, so the field is constant. We can write the current as:

$$I = JA = \sigma A\mathscr{E} = \sigma A\frac{V}{L} \tag{2.15}$$

Therefore,

$$V = \frac{1}{\sigma}\frac{L}{A}I = \rho\frac{L}{A}I = RI \tag{2.16}$$

which is the well known Ohm's law. Let's now practice these concepts with a simple example.

Example 2.1: Resistance of intrinsic and doped silicon.

A bar of silicon has a length of 10 μm, a width of 1 μm and a thickness of 0.2 μm. Calculate the potential necessary to support a DC current of 1 nA in the following conditions:

1. The bar is made of intrinsic silicon;
2. The bar is made of n-type silicon with $N_D = 10^{16}$ cm^{-3};
3. The bar is made of p-type silicon with $N_A = 10^{16}$ cm^{-3}.

* * *

We can use equations (2.14) and (2.16). Assuming the traditionally accepted value of $n_i = 1.45 \cdot 10^{10}$ cm^{-3} for the carrier concentration in intrinsic silicon, $\mu_n = 1500$ cm^2/V·s and $\mu_p = 500$ cm^2/V·s, we have:

$$\rho = \frac{1}{n_i q (\mu_n + \mu_p)} = 2155\ \Omega \cdot \mathrm{m} = 215500\ \Omega \cdot \mathrm{cm} \tag{2.17}$$

The resistance of the bar is thus given by:

$$R = \rho \frac{L}{A} = 2155 \times \frac{1 \cdot 10^{-5}}{1 \cdot 10^{-6} \times 2 \cdot 10^{-7}} = 1.078 \cdot 10^{11}\ \Omega \tag{2.18}$$

To support a current of 1 nA across the slab we need to apply a voltage given by $V = RI = 107.8$ V. This shows that the resistivity of intrinsic silicon is quite high, and the material is closer to an insulator than to a good conductor.

In a doped material we can assume that at room temperature all the dopant atoms are ionized. Therefore, if we have $N_D = 10^{16}$ cm^{-3}, we have 10^{16} electrons per cubic centimeter. The hole concentration is given by the law of mass action as:

$$p = \frac{n_i^2}{N_D} = \frac{2.1 \cdot 10^{20}}{1 \cdot 10^{16}} = 21000\ \mathrm{cm}^{-3} \tag{2.19}$$

The electrons thus outnumber the holes and the latter can be neglected in the calculation. The resistivity of the n-type material is:

$$\rho \approx \frac{1}{N_D q \mu_n} = 4.17 \cdot 10^{-3}\ \Omega \cdot \mathrm{m} = 0.417\ \Omega \cdot \mathrm{cm} \tag{2.20}$$

The resistance of the n-doped bar is hence 208 kΩ and the voltage necessary to have a current of 1 nA is 208 μV. Although the doping is the same, the p-type doped slab has a higher resistance because the mobility of holes is lower. Assuming that for the p-type $p = N_A$ and neglecting the residual electrons, one gets a resistance of 624 kΩ and a voltage of 624 μV.

2.2.2 MOBILITY IN DOPED SILICON

The value of resistance we have found above for doped silicon is underestimated because we have assumed that the electrons mobility in the doped material is equal to the one of intrinsic silicon. Several mechanisms however affect the carrier mobility. The two primary ones are phonon scattering and impurity scattering. Phonons are the result of the thermal excitation of atoms in the lattice, which oscillate around their equilibrium point. Impurities, such as the ones used in doping, are ionized and they act as scattering centers, usually being the key players in mobility degradation. In a doped material there is a high number of impurities and all of them are active and thus ionized at room temperature. As a result, it must be expected that the higher the doping concentration is, the lower the mobility becomes. Figure 2.5 shows as an example the behavior of mobility as a function of doping concentration for an n-type silicon doped with arsenic. Interpolating the curve of Fig. 2.5 one can see that for a doping of 10^{16} cm^{-3} the mobility is about 1180 cm^2/V·s. Introducing this value in our previous exercise we get a voltage drop of 264 μV instead of 208 μV.

2.2.3 CHARGE TRANSPORT BY DIFFUSION

We now turn our attention to a different conduction mechanism which comes into play if there is a concentration difference between two points. Consider a semiconductor slab with two regions separated by an imaginary plane *A*, as shown in Fig. 2.6. To the left of *A* the electron concentration is bigger than to the right. If thermal equilibrium is assumed, all electrons have the same average velocity and their direction of movement is random. Since there are more electrons in the left part on the slab, statistically the

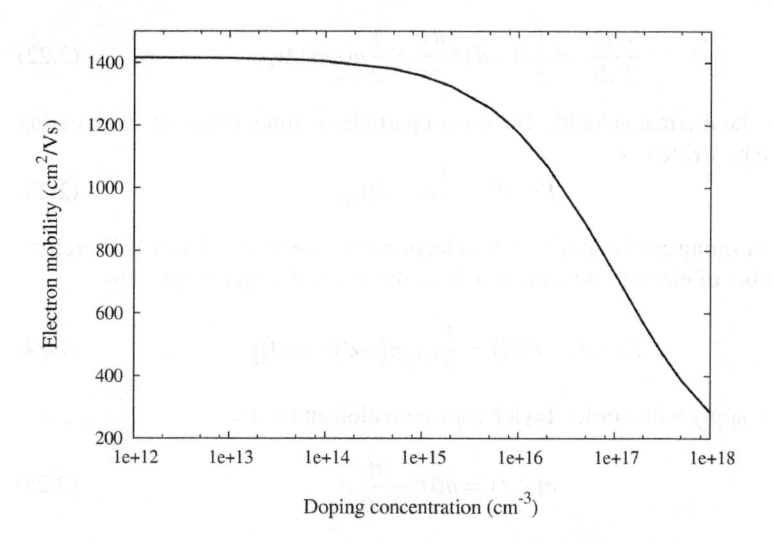

FIGURE 2.5 Electron mobility versus doping concentration.

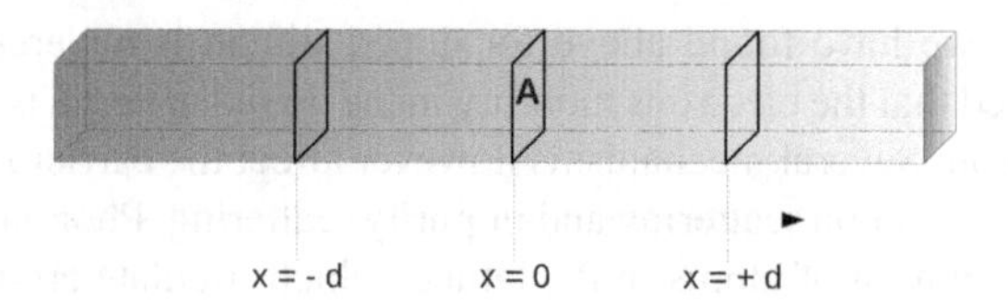

FIGURE 2.6 A concentration gradient originates motion by diffusion.

number of electrons that cross A from left to right is higher than the number of electrons crossing A from right to left. To derive a simple mathematical description of the phenomenon, we make the assumption that the difference in concentration is present only in the x direction. Therefore, a net displacement of particles occurs only along x. We fix the origin of the x axis in correspondence of A and we call $n(-d)$ and $n(d)$ the concentrations respectively at a distance $-d$ to the left and $+d$ to the right. In an infinitesimal length dx around $-d$, the concentration can be considered constant and the number of electrons in the infinitesimal volume is given by:

$$dN = n(-d)Adx \tag{2.21}$$

If we take d equal to the mean free path, electrons in $-d$ will have equal probability of moving towards 0 or in the opposite direction. Electrons that move towards 0 reach it after a time τ_c. The total number of particles that starting from $-d$ reach A in the unit time is given by:

$$\frac{1}{2}\frac{dN}{dt} = \frac{1}{2}n(-d)A\frac{dx}{dt} = \frac{1}{2}n(-d)Av_{th} \tag{2.22}$$

where v_{th} is the thermal velocity. The flux of particles coming from $-d$ and crossing A can hence be written as:

$$F(-d) = \frac{1}{2}n(-d)v_{th} \tag{2.23}$$

The same reasoning can be applied to the electrons at a distance $+d$ from A. Therefore, the net number of electrons that cross A from the left to the right is given by:

$$F(-d) - F(d) = \frac{1}{2}v_{th}\left[n(-d) - n(d)\right] \tag{2.24}$$

We can now apply a first order Taylor approximation and write:

$$n(-d) = n(0) - \frac{dn}{dx}d \tag{2.25}$$

$$n(d) = n(0) + \frac{dn}{dx}d \tag{2.26}$$

Inserting (2.25) and (2.26) in (2.24) we therefore have:

$$F = \frac{1}{2}v_{th}\left[n(0) - \frac{dn}{dx}d - n(0) - \frac{dn}{dx}d\right] = -v_{th}d\frac{dn}{dx} = -D_n\frac{dn}{dx} \tag{2.27}$$

In (2.27) we have put $v_{th}d = D_n$. D_n is called the diffusivity of electrons. Since electrons are negatively charged particles, their net displacement along the gradient, i.e. from left to right, implies an electrical current in the opposite direction. The current density due to electron diffusion can hence be written as:

$$J_n = qD_n\frac{dn}{dx} \tag{2.28}$$

The same concepts can be applied in case we are in the presence of a gradient in the hole concentration. A net displacement of holes results and the hole current density due to diffusion is

$$J_p = -qD_p\frac{dp}{dx} \tag{2.29}$$

The minus sign in equation (2.29) stems from the fact that holes are positively charged, so the current will be directed in the same sense of the gradient. If we suppose that the concentration is decreasing towards increasing x, the hole current is directed in the same sense, but dp/dx is negative, so we need the minus sign to have a net current with the same direction of the hole gradient. If diffusion and drift act simultaneously, the total current densities for electrons and holes are respectively given by:

$$J_n = nq\mu_n\mathscr{E} + qD_n\frac{dn}{dx} \tag{2.30}$$

$$J_p = pq\mu_p\mathscr{E} - qD_p\frac{dp}{dx} \tag{2.31}$$

2.2.4 EINSTEIN RELATIONSHIP

We have introduced the diffusivity for electrons and holes, which can be written in the form $D = v_{th}d$, where v_{th} is the thermal velocity and d is the mean free path. Since we have supposed that we have a concentration gradient only in one direction, we are interested in the component of the velocity vector only along that direction, x in our case. Writing the equipartition theorem in one dimension we have:

$$E = \frac{1}{2}m{v_{th}}^2 = \frac{1}{2}k_BT \tag{2.32}$$

where m is the effective mass of electrons or holes[1]. The thermal velocity can also be expressed as:

$$v_{th} = \frac{d}{\tau_c} \tag{2.33}$$

[1]Remember that we considered here problems in one dimension only, thus the cinetic energy is $E = (1/2)k_BT$.

where d and τ_c are respectively the mean free path and the mean free time. Therefore, the diffusivity can be written as:

$$D = {v_{th}}^2 \tau_c = \frac{k_B T \tau_c}{m} \tag{2.34}$$

where k_B is the Boltzmann constant (k_B=1.38·10^{-23} J/K) and T is the absolute temperature. Remember now that the mobility for electrons and holes is defined as:

$$\mu_n = \frac{q\tau_c}{m_n} \tag{2.35}$$

$$\mu_p = \frac{q\tau_c}{m_p} \tag{2.36}$$

Replacing (2.35) and (2.36) in (2.34) we therefore have:

$$D_n = \frac{k_B T \mu_n}{q} \tag{2.37}$$

$$D_p = \frac{k_B T \mu_p}{q} \tag{2.38}$$

Equations (2.37) and (2.38) can be rewritten in the following form:

$$\frac{D_n}{\mu_n} = \frac{D_p}{\mu_p} = \frac{k_B T}{q} \tag{2.39}$$

Called Einstein's relationship, equation (2.39) relates the diffusivity of the charge carriers to their mobility and it proves very useful in studying the effects of concentration gradients at thermal equilibrium.

2.2.5 GRADED DOPING

Consider now the situation in Fig. 2.7. Here a slab of silicon with a gradient in the electron doping concentration is considered. The electron concentration $n(x)$ decreases from left to right, so in point 1 we have more electrons than in point 2. Since there is an excess of electrons on the left, we must expect a diffusion current towards the right of the slab. However, as electrons move from the region of higher concentration they leave behind ionized donor atoms. These have positive electrical charge and are bound in the crystal lattice. Therefore, a potential and an electric field exist that tend to attract the electrons back towards the left. At the equilibrium, no net current can be present in the slab. This equilibrium is however the result of two components that cancel each other: a diffusion current, that move electrons from left to right, and a drift current, that moves them in the opposite direction. We can then calculate the potential difference between region 1 and region 2 by using (2.30) and the Einstein's relationship. In fact:

$$J_n = nq\mu_n \mathscr{E} + qD_n \frac{dn}{dx} = 0 \tag{2.40}$$

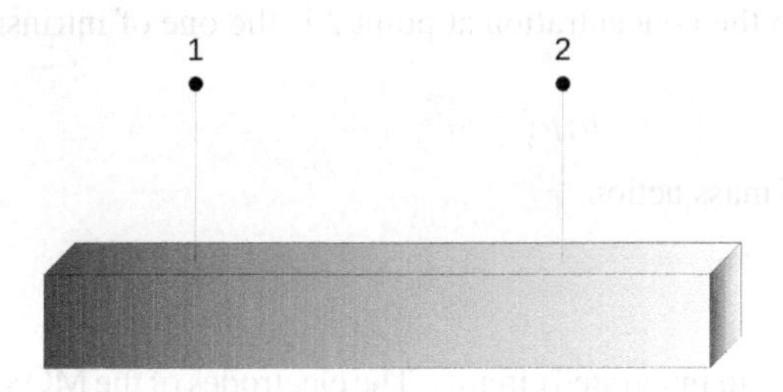

FIGURE 2.7 Silicon bar with non-uniform doping.

$$n\mathscr{E} = -\frac{D_n}{\mu_n}\frac{dn}{dx} \tag{2.41}$$

$$n\left(-\frac{dV}{dx}\right) = -\frac{k_B T}{q}\frac{dn}{dx} \tag{2.42}$$

$$dV = \phi_T \frac{dn}{n} \tag{2.43}$$

In (2.43) we have made the substitution $\frac{k_B T}{q} = \phi_T$, where ϕ_T is called the thermal voltage and is equal to 25.9 mV at room temperature. Integrating (2.43) we have:

$$\int_{V_1}^{V_2} dV = \phi_T \int_{n_1}^{n_2} \frac{dn}{n} \Rightarrow V_2 - V_1 = \phi_T \ln\frac{n_2}{n_1} \Rightarrow V_1 - V_2 = \phi_T \ln\frac{n_1}{n_2} \tag{2.44}$$

Therefore, the potential in region 1 is more positive than in region 2. Equation (2.44) relates the concentrations of the carriers in two zones of a semiconductor to the potential difference which is present between the same zones and is only valid at the thermal equilibrium. Let's now consider the situation from the point of view of the holes. If p_1 and p_2 are the hole concentrations in point 1 and point 2 of the slab, we can write:

$$J_p = pq\mu_p\mathscr{E} - qD_p\frac{dp}{dx} = 0 \tag{2.45}$$

$$p\left(-\frac{dV}{dx}\right) = \frac{D_p}{\mu_p}\frac{dp}{dx} \Rightarrow -dV = \phi_T \frac{dp}{dx} \tag{2.46}$$

Integrating (2.46) we obtain:

$$V_1 - V_2 = \phi_T \ln\frac{p_2}{p_1} \tag{2.47}$$

Since the potential difference between the two regions must be the same, we have:

$$\ln\frac{n_1}{n_2} = \ln\frac{p_2}{p_1} \Rightarrow \frac{n_1}{n_2} = \frac{p_2}{p_1} \Rightarrow n_1 p_1 = n_2 p_2 \tag{2.48}$$

The two points were arbitrarily chosen, so we see from (2.48) that the product of the electron and hole concentration at any point in the slab is constant. If we consider a particular case in which the concentration at point 2 is the one of intrinsic silicon, equation (2.48) reads:

$$n_1 p_1 = n_i^2 \tag{2.49}$$

which is again the law of mass action.

2.3 PN JUNCTIONS

Pn junctions are ubiquitous in integrated circuits. The electrodes of the MOS transistor are insulated from the device substrate by reverse-biased pn junctions. Junctions are also employed to protect the Input/Output (I/O) ports of an integrated circuit against electrostatic discharge. In CMOS technologies, pn junctions are normally used reversed-biased, while forward biased diodes are mainly found in bandgap reference voltage generators. An adequate background on this fundamental structure is thus necessary to understand the behavior of the MOS transistor.

2.3.1 BUILT-IN VOLTAGE

Consider now the structure reported in Fig. 2.8. The left part of the slab has a uniform p-type doping. The hole concentration is N_A and the electron concentration is $n_p = n_i^2/N_A$. The right part has a uniform n-type doping, so the electron concentration is N_D and the hole concentration is $p_n = n_i^2/N_D$. When the junction is formed, holes migrate towards the n-side leaving behind negatively ionized acceptor atoms, shown with the "-" sign in the figure. Electrons migrate from the n-side towards the p-side, leaving behind positively charged donor atoms, represented with the "+" sign. When the equilibrium is reached, the electric field due the ionized atoms counteracts the diffusion and the net current at the junction is zero. We can use the results of equation (2.44) to write the potential difference between the n-side and the p-side. We have:

$$\phi_n - \phi_p = \phi_T \ln \frac{n_n}{n_p} \tag{2.50}$$

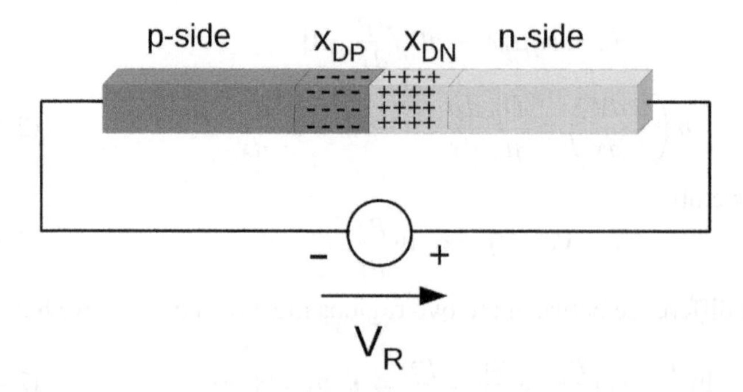

FIGURE 2.8 pn junction.

where n_n is the electron concentration in the n-side and n_p is the electron concentration in the p-side. Since $n_n = N_D$ and $n_p = n_i{}^2/N_A$ we have:

$$\phi_n - \phi_p = \psi_{BI} = \phi_T \ln \frac{N_A N_D}{n_i{}^2} \tag{2.51}$$

Called built-in potential, the voltage given in (2.51) depends only on the doping concentration in the two sides of the junction and on the absolute temperature.

Example 2.2: Calculation of the built-in potential.

Calculate the built-potential of a junction with $N_A = 10^{16}$ cm^{-3} and $N_D = 10^{17}$ cm^{-3} at room temperature and at 100° C.

* * *

At room temperature (300 K) the carrier concentration of intrinsic silicon is $n_i = 1.45 \cdot 10^{10}$ cm^{-3}. The thermal voltage $k_B T/q$ is 25.9 mV. Therefore, we have:

$$\psi_{BI} = 25.9 \ln \frac{10^{16} \times 10^{17}}{2.1 \cdot 10^{20}} = 0.756 \text{ V} \tag{2.52}$$

Using (2.1) we can estimate the intrinsic carrier concentration at 100° C, i.e. 373 K. Replacing $T = 373$ in (2.1) one obtains $n_i(373) = 1.4 \cdot 10^{12}$ cm^{-3}. The thermal voltage at 373 K is 32 mV. Inserting the new numbers in (2.52) one obtains for ψ_{BI} at 373 K the value of 0.642 V.

2.3.2 DEPLETION REGION

We must note that at the junction there is in either side a region which does not contain mobile charge carriers. This zone is called depletion region and it plays an important role in the physics of the MOS transistor. In the derivation of the built in potential, we have assumed that the battery shown in Fig. 2.8 does not impose any voltage. We note that to preserve the charge neutrality, the number of ionized acceptor atoms on the p-side must be equal to one of the donor atoms on the n-side. Referring to Fig. 2.8 we call A the area of the junction, x_{dp} the extension of the depletion region in the p-side, and x_{dn} the extension of the depletion region on the n-side. The volume of the depleted region on the p-side is hence given by Ax_{dp} while the charge density per unit volume is $-qN_A$. Similarly, on the n-side, the depleted volume is Ax_{dn} and the charge density is qN_D. If we fix the origin of the x axis at the junction we can write:

$$(-qN_A)\left(-x_{dp}A\right) = qN_d x_{dn} A \tag{2.53}$$

which simplifies to:

$$N_A x_{dp} = N_D x_{dn} \Rightarrow \frac{N_A}{N_D} = \frac{x_{dn}}{x_{dp}} \tag{2.54}$$

From (2.54) we see that the ratio between the extensions of the depletion region is inversely proportional to the one of the doping concentrations in the respective zones. That is, the depletion region is wider in the less doped side of the junction.

The electric field and the potential in the depletion region can be found by solving the Poisson equation. For the p-side we can write:

$$\frac{d^2V}{dx^2} = -\frac{\rho}{\varepsilon_{Si}} = -\left(-q\frac{N_A}{\varepsilon_{Si}}\right) = q\frac{N_A}{\varepsilon_{Si}} \tag{2.55}$$

In (2.55) ε_{Si} is dielectric constant of silicon. Integrating (2.55) we have:

$$\frac{dV}{dx} = q\frac{N_A}{\varepsilon_{Si}}x + A \tag{2.56}$$

No electric field must exist in the neutral part of the p-side, so:

$$\frac{dV}{dx} = 0 \text{ for } x = -x_{dp} \tag{2.57}$$

Hence:

$$-q\frac{N_A}{\varepsilon_{Si}}x_{dp} + A = 0 \Rightarrow A = q\frac{N_A}{\varepsilon_{Si}}x_{dp} \tag{2.58}$$

and

$$\frac{dV}{dx} = q\frac{N_A}{\varepsilon_{Si}}x + q\frac{N_A}{\varepsilon_{Si}}x_{dp} \tag{2.59}$$

Integrating now (2.59) we obtain:

$$V(x) = \frac{1}{2}\frac{qN_A}{\varepsilon_{Si}}x^2 + \frac{qN_A}{\varepsilon_{Si}}xx_{dp} + B \tag{2.60}$$

To fix the integration constant B we can chose arbitrarily $V(x) = 0$ in $x = x_{dp}$, which yields:

$$\frac{1}{2}\frac{qN_A}{\varepsilon_{Si}}{x_{dp}}^2 - \frac{qN_A}{\varepsilon_{Si}}{x_{dp}}^2 + B = 0 \Rightarrow B = \frac{1}{2}\frac{qN_A}{\varepsilon_{Si}}{x_{dp}}^2 \tag{2.61}$$

Therefore, combining (2.60) and (2.61) we get:

$$V(x) = \frac{1}{2}\frac{qN_A}{\varepsilon_{Si}}x^2 + \frac{qN_A}{\varepsilon_{Si}}xx_{dp} + \frac{1}{2}\frac{qN_A}{\varepsilon_{Si}}{x_{dp}}^2; \quad -x_{dp} \le x \le 0 \tag{2.62}$$

We can now calculate the potential for the n-side of the junction. Here we have:

$$\frac{d^2V}{dx^2} = -\frac{\rho}{\varepsilon_{Si}} = -q\frac{N_D}{\varepsilon_{Si}} \tag{2.63}$$

which after a first integration becomes:

$$\frac{dV}{dx} = -q\frac{N_D}{\varepsilon_{Si}}x + D \tag{2.64}$$

We can impose again the condition that we do not have any electrical field outside the depletion region, so:

$$\frac{dV}{dx} = 0 \text{ for } x = x_{dn} \Rightarrow -\frac{qN_D}{\varepsilon_{Si}}x_{dn} + D = 0 \Rightarrow D = \frac{qN_D}{\varepsilon_{Si}}x_{dn} \tag{2.65}$$

Substituting the expression of D in (2.64) and integrating again we have:

$$V(x) = -\frac{1}{2}\frac{qN_D}{\varepsilon_{Si}}x^2 + \frac{qN_D}{\varepsilon_{Si}}xx_{dn} + E \quad 0 \le x \le x_{dn} \tag{2.66}$$

The potential must be continuous in $x = 0$. Therefore, equating (2.62) and (2.66) in $x = 0$ we have:

$$E = \frac{1}{2}\frac{qN_A}{\varepsilon_{Si}}{x_{dp}}^2 \tag{2.67}$$

and

$$V(x) = -\frac{1}{2}\frac{qN_D}{\varepsilon_{Si}}x^2 + \frac{qN_D}{\varepsilon_{Si}}xx_{dn} + \frac{1}{2}\frac{qN_A}{\varepsilon_{Si}}{x_{dp}}^2; \quad 0 \le x \le x_{dn} \tag{2.68}$$

The electric field in the depleted p-side is:

$$\mathscr{E}_P(x) = -\frac{dV}{dx} = -\frac{qN_A}{\varepsilon_{Si}}x - \frac{qN_A}{\varepsilon_{Si}}x_{dp} \tag{2.69}$$

while the electric field in the depleted n-side is:

$$\mathscr{E}_N(x) = -\frac{dV}{dx} = \frac{qN_D}{\varepsilon_{Si}}x - \frac{qN_D}{\varepsilon_{Si}}x_{dn} \tag{2.70}$$

At the junction, the field must be continuous, which implies:

$$\mathscr{E}_P(0) = \mathscr{E}_N(0) \Rightarrow \frac{qN_A}{\varepsilon_{Si}}x_{dp} = \frac{qN_D}{\varepsilon_{Si}}x_{dn} \tag{2.71}$$

Note that by imposing the continuity of the field we find again the condition $N_A x_{dp} = N_D x_{dn}$. If we assume that the external voltage imposed by the battery is zero, the potential difference between the two extremes of the depletion region must be equal to the junction built-in potential. Hence, we can write:

$$V(x_{dn}) - V(x_{dp}) = \psi_{BI} = \frac{1}{2}\frac{q}{\varepsilon_{Si}}\left(N_A{x_{dp}}^2 + N_D{x_{dn}}^2\right) \tag{2.72}$$

Noting that:

$$x_{dp} = \frac{N_D}{N_A}x_{dn} \tag{2.73}$$

and replacing (2.73) in (2.72) we obtain:

$$\psi_{BI} = \frac{1}{2}\frac{q}{\varepsilon_{Si}} N_D {x_{dn}}^2 \left(\frac{N_A + N_D}{N_A}\right) \tag{2.74}$$

Solving for x_{dn} we finally have:

$$x_{dn} = \sqrt{\frac{2\varepsilon_{Si}\psi_{BI}}{qN_D}\frac{N_A}{N_A + N_D}} \tag{2.75}$$

Similarly, we can write:

$$x_{dn} = \frac{N_A}{N_D} x_{dp} \tag{2.76}$$

Replacing (2.76) in (2.72) and proceeding as before, we get:

$$x_{dp} = \sqrt{\frac{2\varepsilon_{Si}\psi_{BI}}{qN_A}\frac{N_D}{N_D + N_A}} \tag{2.77}$$

Note that, mnemonically, equations (2.77) and (2.75) can be obtained one from the other by exchanging N_A and N_D. Suppose that we now apply a positive voltage to the n-side and a negative voltage to the p-side. No field can exist in the neutral p and n regions which are conductive. Therefore, the additional voltage imposed by the battery is all applied across the depletion region. Since we have made the n-side more positive, more electrons are attracted away from the junction and the depletion region expands. The same is true for the holes in the p-side. The external voltage is superimposed on the built-in potential and the two have the same polarity. We can hence rewrite the expressions for x_{dn} and x_{dp} as follows:

$$x_{dn} = \sqrt{\frac{2\varepsilon_{Si}(\psi_{BI} + V_R)}{qN_D}\frac{N_A}{N_A + N_D}} \tag{2.78}$$

$$x_{dp} = \sqrt{\frac{2\varepsilon_{Si}(\psi_{BI} + V_R)}{qN_A}\frac{N_D}{N_D + N_A}} \tag{2.79}$$

We now calculate the total junction width as a function of the doping concentrations. Noting that:

$$(x_{dn} + x_{dp})^2 = {x_{dn}}^2 + {x_{dp}}^2 + 2x_{dn}x_{dp} \tag{2.80}$$

we first evaluate the last term of equation (2.80):

$$2x_{dn}x_{dp} = 2\sqrt{\frac{2\varepsilon_{Si}(\psi_{BI} + V_R)}{qN_D}\left(\frac{N_A}{N_A + N_D}\right)\frac{2\varepsilon_{Si}(\psi_{BI} + V_R)}{qN_A}\left(\frac{N_D}{N_A + N_D}\right)} \tag{2.81}$$

Simplifying equation (2.81) we obtain:

$$2x_{dn}x_{dp} = 4\frac{\varepsilon_{Si}(\psi_{BI} + V_R)}{q(N_A + N_D)} \tag{2.82}$$

Combining together (2.80),(2.82),(2.78) and (2.79) we can now write:

$$\left(x_{dn}+x_{dp}\right)^2 = 2\varepsilon_{Si}\left(\psi_{BI}+V_R\right)\frac{N_A{}^2+N_D{}^2+2N_AN_D}{qN_AN_D\left(N_A+N_D\right)} \quad (2.83)$$

Noting that:

$$N_A{}^2+N_A{}^2+2N_AN_D=(N_A+N_D)^2 \quad (2.84)$$

we have:

$$\left(x_{dn}+x_{dp}\right)^2 = 2\varepsilon_{Si}\left(\psi_{BI}+V_R\right)\frac{(N_A+N_D)}{qN_AN_D} \Rightarrow x_{dn}+x_{dp}=\sqrt{\frac{2\varepsilon_{Si}\left(\psi_{BI}+V_R\right)\left(N_A+N_D\right)}{qN_AN_D}} \quad (2.85)$$

Equation (2.85) allows us to calculate the total width of a pn junction as a function of the doping levels and of the reverse bias voltage applied to the structure. Let's clarify this with the following example.

Example 2.3: Calculation of the depletion width.

Calculate the width of the depletion region for a junction with $N_A = 10^{15}$ cm^{-3} and $N_D = 10^{17}$ cm^{-3}. A reverse bias of 3 V is applied to the junction. Assume that the junction is kept at room temperature.

* * *

We must first calculate the built-in potential of the junction. Employing (2.51) and replacing the given doping levels we obtain:

$$\psi_{BI} = 25.9\ln\frac{10^{15}\times 10^{17}}{2.1\cdot 10^{20}} = 0.716\ V \quad (2.86)$$

We can now use equation (2.85). Putting in the numbers we have:

$$x_{dn}+x_{dp}=\sqrt{\frac{2\times(0.716+3)\times 11.9\times 8.85\cdot 10^{-14}\times\left(10^{15}+10^{17}\right)}{1.6\cdot 10^{-19}\times 10^{15}\times 10^{17}}} = 2.22\ \mu\text{m} \quad (2.87)$$

In equation (2.86) we have used the fact that the dielectric constant of silicon is

$$\varepsilon_{Si} = \varepsilon_{sr}\varepsilon_0 \quad (2.88)$$

where the relative dielectric constant of silicon ε_{sr} is 11.9 and the vacuum dielectric constant ε_0 is 8.85·10^{-14} F/cm. Inserting the values of the doping levels and of the potential in (2.78) and (2.79) we can now evaluate the depth of the depletion region in the n and in the p sides, respectively. We obtain 20 nm for x_{dn} and 2.2 μm for x_{dp}. As expected, the depletion region extends mostly in the less doped side of the junction, the ratio between x_{dp} and x_{dn} being defined by (2.54).

2.3.3 BREAKDOWN VOLTAGE

The electric field reaches its maximum at the metallurgical junction, i.e., with respect to Fig. 2.8, in $x = 0$. Using equation (2.71), we have:

$$|\mathscr{E}_{max}| = \frac{qN_D}{\varepsilon_{Si}} x_{dn} \tag{2.89}$$

We can now replace in (2.89) the expression of x_{dn} given by (2.78):

$$|\mathscr{E}_{max}| = \frac{qN_D}{\varepsilon_{Si}} \sqrt{\frac{2\varepsilon_{Si}}{qN_D} \frac{N_A}{N_D + N_A} (\psi_{BI} + V_R)} = \sqrt{\frac{2q}{\varepsilon_{Si}} \frac{N_A N_D}{N_A + N_D} (\psi_{BI} + V_R)} \tag{2.90}$$

When the electric field in silicon is larger than $3{\cdot}10^5$ V/cm, breakdown occurs and the current rises abruptly. The minimum value of $\mathscr{E}$ sufficient to cause breakdown is called critical electric field and it is indicated as $\mathscr{E}_{crit}$. Two mechanisms are involved in breakdown: avalanche multiplication and Zener breakdown. In the avalanche process the carriers acquire enough energy to extract electrons from the silicon atoms. These electrons have themselves enough energy to cause further ionization. Zener breakdown consists of a direct extraction of valence electrons by the high electric field. Putting $|\mathscr{E}_{max}| = |\mathscr{E}_{crit}|$ in equation (2.90) and solving for V_R we obtain the maximum reverse voltage that can be applied to a junction before causing breakdown:

$$V_{R,max} = \frac{\varepsilon_{Si}(N_A + N_D)}{2qN_A N_D} \mathscr{E}_{crit}^{\,2} - \psi_{BI} \tag{2.91}$$

Example 2.4: Evaluation of breakdown voltage.

A symmetric pn junction operating at room temperature has doping concentrations of $N_A = N_D = 1 \cdot 10^{17}$ cm^{-3}. Calculate the maximum reverse bias voltage that the junction can sustain. Repeat the calculation in case the doping on the p side in decreased by two orders of magnitude.

* * *

Using (2.51) we first calculate the built-in voltage of the junction:

$$\psi_{BI} = 25.9 \ln \frac{1 \times 10^{34}}{1.16 \times 10^{20}} = 0.831 \text{ V} \tag{2.92}$$

The maximum applicable reverse voltage is given by (2.91), so we have:

$$V_{R,max} = \frac{11.9 \times 8.85 \cdot 10^{-14} \times 2 \cdot 10^{17}}{2 \times 1.6 \cdot 10^{-19} \times 10^{34}} \times \left(3 \cdot 10^5\right)^2 - 0.831 \text{ V} = 5.093 \text{ V} \tag{2.93}$$

If we now repeat the calculation for a doping concentration on the p side of $N_A = 10^{15}$ cm^{-3} we obtain $V_{R,max} = 298.3$ V. We see that decreasing the doping concentration in one or both sides of the junction increases the breakdown voltage. In fact, as we have seen above, the reverse bias voltage appears all across the depletion region. If the doping concentration is decreased, the depletion region gets wider and the intensity of the electric field is reduced.

2.3.4 JUNCTION CAPACITANCE

We observe now that the depletion layer of a pn junction stores an amount of charge which depends on the voltage applied to the junction. Therefore, there is a capacitance associated to the junction as any change in the applied bias voltage will determine a change in the stored charge. The junction capacitance can be studied assimilating it to a parallel plate capacitor. This might seem inappropriate at the first sight, since in the typical parallel plate capacitor the charge is stored on the surface of two conductive arms which are separated by a dielectric. In a pn junction, the charge is instead found in the entire depleted volume. However, when the bias voltage is varied, the charge is added or removed only at the two borders of the depletion region. Therefore, the relationship between the change in the bias voltage and the one in the stored charge can be calculated by considering a parallel plate capacitor in which the arms have the area of the junction and the depleted volume acts as the dielectric. We are hence interested in calculating the capacitance associated to the junction and defined as:

$$C_j = \frac{dQ}{dV_R} = \frac{dQ}{dx_{dn}}\frac{dx_{dn}}{dV_R} \tag{2.94}$$

Note that we are calculating the change in the charge on one arm of the capacitor (at the n side of the junction, in this case) since, as in any capacitor, the change on the other arm will be equal in magnitude and opposite in sign. We have also that:

$$Q_D = AN_D x_{dn} \Rightarrow dQ = AN_D dx_{dn} \Rightarrow \frac{dQ}{dx_{dn}} = AN_D \tag{2.95}$$

Since x_{dn} depends explicitly on V_R as shown in equation (2.78), differentiating (2.78) we have:

$$\frac{dx_{dn}}{dV_R} = \frac{1}{2}\left(\frac{\frac{2\varepsilon_{Si}}{qN_D}\frac{N_A}{(N_A+N_D)}}{\sqrt{\frac{2\varepsilon_{Si}}{qN_D}\frac{N_A}{(N_A+N_D)}(\psi_{BI}+V_R)}}\right) = \sqrt{\frac{\varepsilon_{Si}}{2qN_D}\frac{N_A}{(N_A+N_D)}}\frac{1}{\sqrt{\psi_{BI}+V_R}} \tag{2.96}$$

Now we can insert (2.95) and (2.96) in (2.94) to obtain:

$$C_j = A\sqrt{\frac{\varepsilon_{Si}}{2q}\frac{N_A N_D}{(N_A+N_D)}}\frac{1}{\sqrt{\psi_{BI}}}\frac{1}{\sqrt{1+\frac{V_R}{\psi_{BI}}}} = \frac{C_{j0}}{\left(1+\frac{V_R}{\psi_{BI}}\right)^{1/2}} \tag{2.97}$$

Equation (2.97) expresses the junction capacitance as the capacitance in case the reverse bias voltage is zero (C_{j0}) divided by a term which depends on the applied

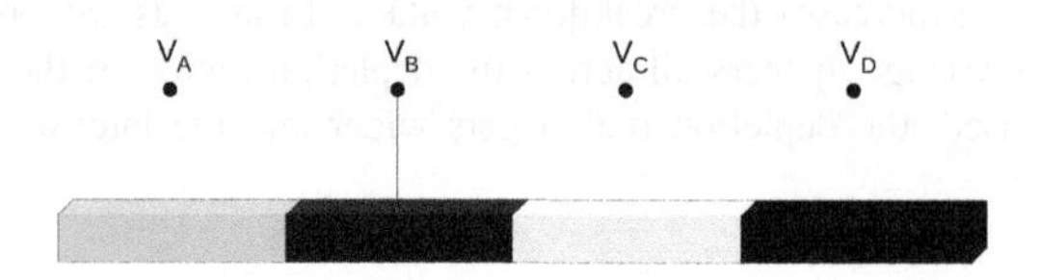

FIGURE 2.9 Contact potentials in a chain of materials.

bias voltage. We note that increasing the reverse bias voltage the capacitance is decreased. This is a consequence of the widening of the depletion layer with the reverse bias voltage, which increases the thickness of the dielectric of our capacitor. Equation (2.97) is valid even if the junction gets forward biased. Therefore, it is customary to write (2.97) as:

$$C_j = \frac{C_{j0}}{\left(1 - \frac{V_D}{\psi_{BI}}\right)^{1/2}} \tag{2.98}$$

Here V_D is positive when the junction is forward biased and negative when it is reverse biased. Equation (2.98) predicts that the capacitance increases as the junctions becomes more and more forward biased (which is true) and that the capacitance tends to infinity as V_D approaches ψ_{BI}, which is indeed false. In fact, when a significant forward current starts flowing the assumption that the depletion region is void of mobile carriers does not hold and (2.98) is not applicable anymore. Equation (2.98) has been derived assuming an abrupt junction. In practical cases, the transition between the p and the n sides will not be perfectly sharp. In this case, it can be proven that (2.98) still holds, but with a different exponent at the denominator:

$$C_j = \frac{C_{j0}}{\left(1 - \frac{V_D}{\psi_{BI}}\right)^{m}} \tag{2.99}$$

where $0.2 \leq m \leq 0.5$.

2.3.5 CONTACT POTENTIALS

In 2.3.1 we derived an expression for the electrostatic potential that develops when p-type and n-type semiconductors are brought in contact and the junction is kept at thermal equilibrium. The concepts elaborated there can be applied to the general case where any two materials in which the electrons have different energies are brought in contact and the thermal equilibrium is maintained. Consider the case shown in Figure 2.9 where a chain of four different materials is formed. The contact potential between the extremes of the chain can be written as:

$$(V_A - V_B) + (V_B - V_C) + (V_C - V_D) = V_A - V_D \tag{2.100}$$

so it depends only on the first and last material in the chain. We can define V_A, V_B, etc. by taking the contact potential that the considered material develops when put in contact

TABLE 2.2
Contact potentials of relevant materials to intrinsic silicon.

Material	Contact potential (V)
Aluminum	+0.6
Copper	+0.0
Nickel	+0.15
Gold	−0.3
Degenerate, n-type doped polysilicon	−0.56
Degenerate, p-type doped polysilicon	+0.56

with a reference one. In our context, it makes sense to take intrinsic silicon as the reference material. Consider now the particular case of a junction formed by intrinsic silicon and p-type silicon doped with N_A acceptors/cm^3. Applying equation (2.51) we have:

$$\phi_p - \phi_i = -\phi_T \ln \frac{N_A n_i}{n_i{}^2} = \phi_T \ln \frac{n_i}{N_A} \equiv -\phi_{Fp} \tag{2.101}$$

Note that the potential is more positive in the side of the intrinsic silicon. Therefore, the contact potential of p-type silicon with respect to intrinsic silicon is negative. The opposite of the contact potential between p-type and intrinsic silicon is called the Fermi potential of the p-type silicon. Similarly, we can define the contact potential of n-type silicon doped with N_D donors/cm^3 with respect to intrinsic silicon:

$$\phi_n - \phi_i = \phi_T \ln \frac{N_D n_i}{n_i{}^2} = \phi_T \ln \frac{N_D}{n_i} \equiv -\phi_{Fn} \tag{2.102}$$

The concept of contact potential is very practical as it allows the study of many phenomena without using the more complex energy band descriptions [3]. Table 2.2 reports the contact potentials of some materials which are relevant in the treatment of MOS transistors.

Example 2.5: Evaluation of contact potentials

Calculate the contact potentials and the Fermi voltages of p-type and n-type silicon with $N_A = 10^{18}$ cm^{-3} and $N_D = 10^{17}$ cm^{-3}. Evaluate the contact potential between the n-type and the p-type silicon.

* * *

Using equation (2.101) and (2.102) we have:

$$\phi_p = -25.9 \ln \frac{10^{16}}{1.08 \cdot 10^{10}} = -0.356 \text{ V} \tag{2.103}$$

$$\phi_n = 25.9 \ln \frac{10^{18}}{1.08 \cdot 10^{10}} = 0.475 \text{ V} \tag{2.104}$$

The Fermi potentials are just the opposite of the above quantities, so we have $\phi_{Fn} = -0.475$ V and $\phi_{Fp} = +0.356$ V. The contact potential between the n-type silicon and the p-type silicon is:

$$\phi_n - \phi_p = 0.475 - (-0.356) = 0.831 \text{ V} \tag{2.105}$$

2.4 A FIRST LOOK AT MOS TRANSISTORS

The original idea of the field effect transistor is credited to J. Lilienfeld [3], who patented it in 1925 (US patent 1745175, describing a MESFET-like device) and in 1928 (US patent 1900018, describing a thin film MOSFET). Other patents describing MOS-like structures were obtained by Oskar Heil in the middle thirties. Despite these early findings, it took almost three decades before MOS transistors could be routinely fabricated, because producing a reliable gate oxide turned out to be a major challenge. The basic idea beyond the MOS transistor is that one electrode, the gate, controls through capacitive coupling the conductivity between two other electrodes (called drain and source). The control can be sharp, i.e. the channel is either formed, allowing current to flow between drain and source, or blocked. In other words, the transistor behaves like a switch that establishes or prevents an electrical contact between two points in a circuit. The use of the MOS transistor as a switch is the basis for digital circuits, but switches are extensively used also in analog applications. The charge in the channel can also be smoothly modulated by the gate to produce signals which vary continuously. This is the basis for the design of linear analog circuits. CMOS technologies offer two types of transistors, NMOS and PMOS, which are both essential to implement high performance analog and digital circuits.

2.4.1 THE NMOS TRANSISTORS

Fig. 2.10 shows the structure of an NMOS transistor. The device consists of a p-doped region, called substrate or body, in which two n-type electrodes are implanted. In principle, the electrodes are identical and their role is defined by the potentials applied to them during transistor operation. This point will be further clarified when we will discuss in more detail the device physics and its electrical characteristics. However, for simplicity, we identify already now one electrode as the "source" and the other one as the "drain". The "n^+" indicates that the doping concentration in the electrodes is higher than in the substrate. Typical values are $N_D = 10^{19}$ cm^{-3} for source and drain and $N_A = 10^{16}$ cm^{-3} for the body. The electrodes form with the substrate pn junctions that must always be reverse biased. The region between drain and source is the transistor channel. Over this area, a very thin layer of high quality silicon dioxide, SiO_2, (represented in Fig. 2.10 by the white area shaded in gray), is

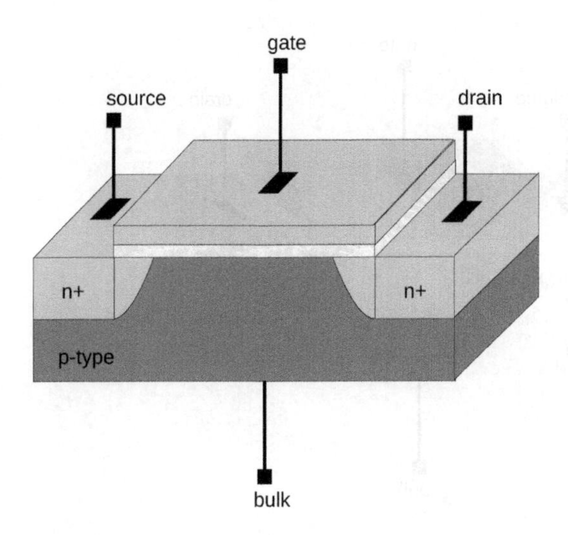

FIGURE 2.10 NMOS transistor.

grown. Typical values of gate dielectric thickness range today from 1.2 nm to 7 nm, depending on the technology and on the particular type of device. On top of the silicon dioxide the gate, which is the control electrode, is deposited. The gate is usually made of silicon that, being grown over an oxide, can not form a single crystal. Therefore the gate material consists of several grains of silicon with different crystal orientations. For this reason it is called poly-crystalline silicon or, in short, polysilicon or poly. For reasons that will become clear when we will discuss the threshold voltage, the gate of an NMOS transistor is usually doped n^+. The distance between drain and source is the channel length, indicated as L. The channel dimension orthogonal to the length is the channel width, indicated as W. During device fabrication, the source and drain regions slightly diffuse underneath the gate. The lateral diffusion, L_D, is the distance by which source and drain protrude inside the channel. The effective channel length, L_{eff}, is hence smaller than the designed channel length by $2L_D$ and we can write:

$$L_{eff} = L - 2L_D \tag{2.106}$$

However, it is customary in the MOS literature to call L the effective channel length. In this book, we will stick to this convention unless otherwise specified. The minimum achievable gate length is taken as a metric of the integration density offered by a given process. Different technology generations, called also technology nodes, are classified on the basis of the minimum gate length they allow to fabricate. At the time of this writing, the most advanced integrated circuits are produced in 14 nm technologies. Front-end electronics for radiation detectors is currently implemented exploiting many different technology nodes, from 0.35 μm down to 45 nm.

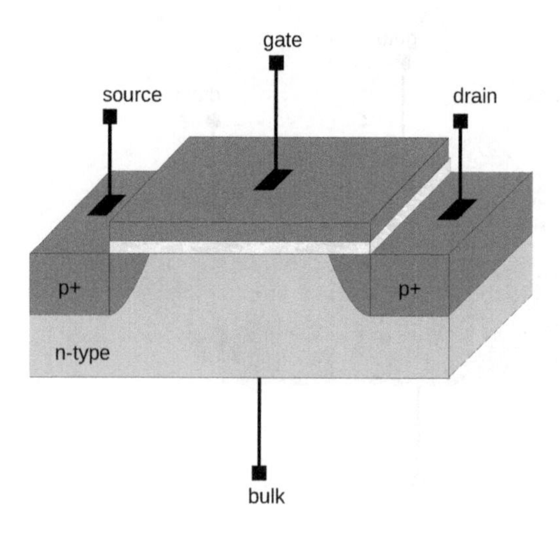

FIGURE 2.11 PMOS transistor.

2.4.2 THE PMOS TRANSISTORS

Fig. 2.11 shows a PMOS transistor. The device has the same structure of the NMOS, but the doping types are exchanged. The bulk of a PMOS transistor is n-type, while source and drain are p^+ doped, as well as the gate. All the definitions given above for the NMOS applies also to the PMOS.

2.4.3 TRANSISTOR REPRESENTATIONS

Transistors are tridimensional structures. However the parameters related to the depth inside the wafer, such as the thickness of the oxide and of the various metal films, as well as the doping level in the different areas, are defined by the silicon foundry. Therefore, they are not in the hand of the circuit designer, who can basically choose only the width and the length of the transistors. From the design point of view, a two-dimensional representation is hence sufficient and this is what is provided by the CAD tools employed to design the layout of integrated circuits. The layout is the circuit representation closer to the silicon reality. As such, it is the source of the fabrication database used by the foundry to produce the masks necessary for patterning the chip on silicon. Fig. 2.12 shows a view of an NMOS and PMOS transistor similar to the one displayed by a layout drawing tool. The black squares are the contacts. In the figure, the p^+ contact to the substrate and the n^+ contact to the nwell are shown, respectively to the left of the NMOS and to the right of the PMOS. These contacts are necessary to define properly the potential of the transistor substrate. The layout is the final step in the design of an integrated circuit. A lot of its properties can be studied using a higher level representation, in which devices are drawn as idealized,

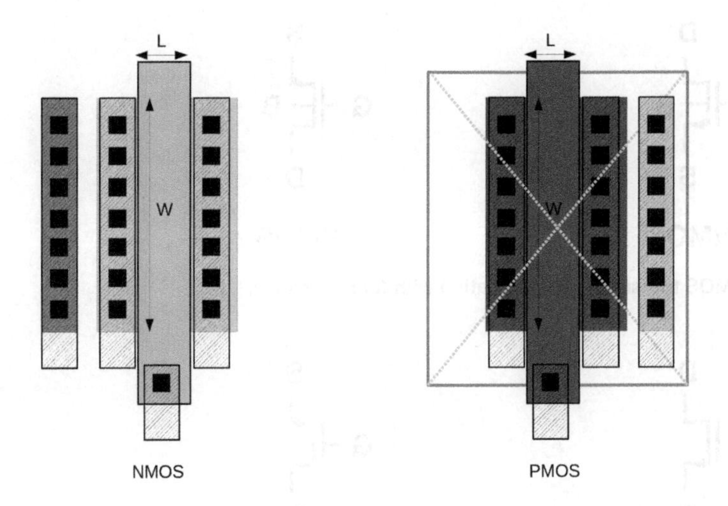

FIGURE 2.12 Top view of an NMOS (left) and a PMOS (right) transistor.

multi-terminal objects. Associating to these objects a model file which contains the device parameters, simulates the circuit through a computer program. A network in which transistors are represented with more idealized symbols is called a schematic. Unfortunately, there is not a unique symbolic representation for MOS transistors. Fig. 2.13 shows four frequently used set of symbols. In Fig. 2.13 a), the device is distinguished by an arrow on one terminal. This arrow indicates the conventional sense of current flow, which enters in the source of a PMOS and exits from the source of an NMOS. These will be the symbols used in this book. In the representation of Fig. 2.13 b), the bulk terminal is omitted. If this is the case, it is understood that the substrate is connected to the most available negative potential for NMOS transistors and to the most positive potential for PMOS ones. In Fig. 2.13 c) the transistor is identified by an arrow on the bulk. The arrow shows the direction in which the bulk-electrode diodes would get forward biased. Finally, Figure 2.13 d) reports symbols frequently used in digital design. The PMOS is distinguished from the NMOS by the circle on the gate, suggesting the logic operation NOT.

2.5 CMOS TECHNOLOGIES

The success of CMOS technologies stems from their ability to fabricate on the same substrate both NMOS and PMOS transistors. This is indispensable to design digital gates dissipating very small static power. As we will discuss in the next chapter, having both devices simultaneously available is also the key to building high performance amplifiers. Fig. 2.14 shows the simplified cross section of a standard bulk CMOS process and allows us to grasp some important features of these technologies. The transistors are built over a common substrate, that in the example is p-type. In CMOS fabrication, p-type substrates tend to be preferred because p-doped wafers are

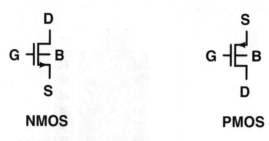

a) MOS transistor representation with four terminals.

b) MOS transistor representation with three terminals.

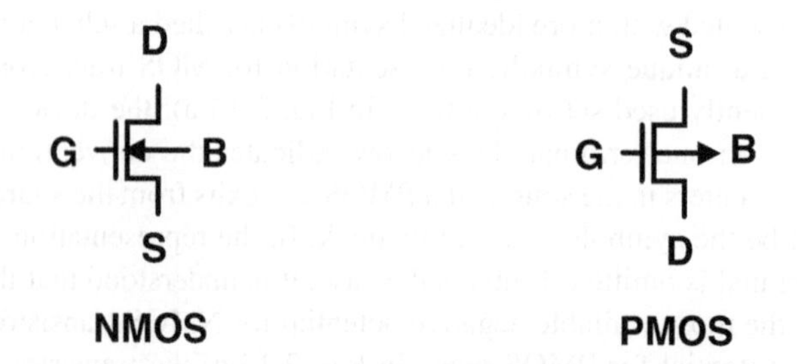

c) MOS transistor representation with bulk polarity indication. The arrows indicates the direction in which the bulk-electrode junction would be forward biased.

d) MOS transistor representation with three terminals used in digital design.

FIGURE 2.13 Symbols used in circuit drawings.

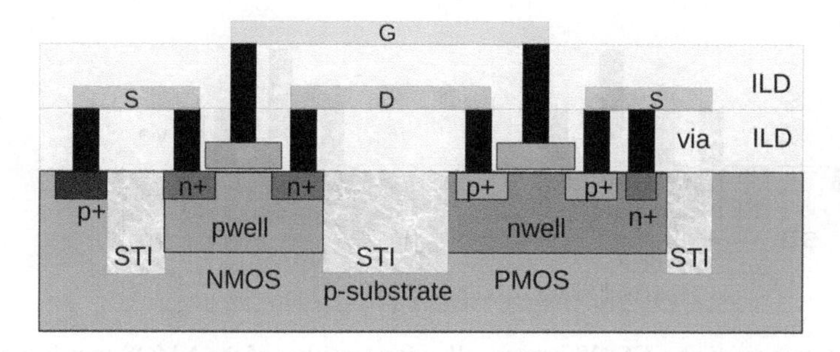

FIGURE 2.14 Simplified cross view of a standard twin-well CMOS process.

easier and cheaper to produce. Furthermore, since the hole mobility is lower, for the same doping concentration the resistivity is higher, which can be useful in limiting the propagation of noise through the substrate in complex ICs. To create the body for PMOS transistors the wafer is counter-doped in selected areas, called nwells. Note that if we start from a material doped with N_A acceptors and we introduce the same concentration of donors (i.e. $N_D = N_A$) the material behaves again like intrinsic silicon, since the additional electrons and holes, being present in the same amount, will completely recombine. In this case the material is said to be compensated. If we further increase the donor concentration, so that $N_D > N_A$, an excess of electrons equal to $N_{eff} = N_D - N_A$, will be present. These extra electrons give the material n-type properties. A careful adjustment of the substrate doping is required also in regions where NMOS devices are fabricated. Hence, a pwell with an optimized doping profile is defined in correspondence of the NMOS transistors. Modern CMOS technologies are therefore "twin-well" or "twin-tub" processes. However, since the substrate is p-type, the local adjustment made with the pwells does not lead to the electrical isolation between the pwells and the global p substrate. In a circuit, the latter must be connected to the most negative potential to avoid the risk of forward biasing the junctions between the electrodes and the bulk in NMOS transistors. An nwell, on the other hand, is isolated from the substrate and from other nwells by reverse-biased junctions. Hence it can be tied to a private voltage line, provided that the junctions between the electrodes and the transistor body never become forward biased. In modern technologies, it is standard to have the possibility of putting also NMOS transistors in fully isolated pwells. Called *triple well transistor*, the device is illustrated in Fig. 2.15. First, an additional nwell is created which is deeper than the one used for PMOS devices and thus called a *deep nwell*. Inside the deep nwell, a pwell is implanted to provide the bulk for the NMOS transistor, which is hence isolated from the main substrate. A p^+ doped area provides an ohmic contact to the pwell and an n^+ one allows contact to the deep nwell. To ensure reverse biasing of the junctions, the deep nwell is usually tied to the positive rail, whereas the p well can be also connected to the source electrode. Due to the use of two oppositely doped areas one enclosed in the other and to the necessity of providing room for the well contacts, triple well transistors occupy more area than standard ones and are typi-

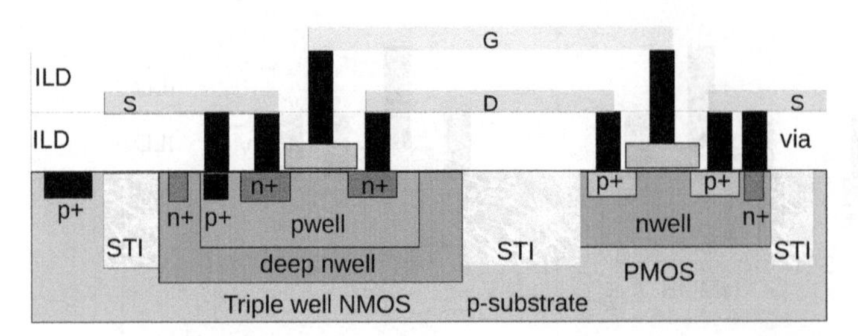

FIGURE 2.15 Cross view of a CMOS process allowing isolation of the NMOS transistors from the main wafer substrate.

cally used only in noise-sensitive circuits that need to be shielded from the common substrate. Depending on the process, it can be possible to put more transistors in the same deep nwell to realize higher density circuits. Note that the triple well option is an *addition* to a standard twin-tub process, thus triple-well and standard bulk NMOS coexist on the same wafer. In addition to the NMOS and PMOS transistors, a bulk CMOS process contains also parasitic bipolar transistors, as shown in Fig. 2.16 a). In fact, the p^+ electrode of the PMOS transistor, its nwell and the substrate form respectively the emitter, base and collector of a *vertical pnp* transistor. The $n+$ electrode of the NMOS device, the substrate and the nwell containing the PMOS provide instead the emitter, base and collector of a *lateral npn* transistor. The substrate has a non-negligible resistance, therefore the collector of the vertical pnp is connected to ground through a parasitic resistor, called R_S in Fig. 2.16. Similarly, the collector of the npn is connected to V_{DD} via the nwell, which has an equivalent resistance given by R_W. Fig. 2.16 b) redraws the circuit to make more clear the connections in this parasitic structure. Each bipolar forms with its collector resistance a common emitter amplifier and the two amplifiers are connected in a positive feedback loop. Injecting, for instance, some current in R_S increases the base voltage of Q_1, which starts drawing current from V_{DD}, lowering its collector voltage and thus the base voltage of Q_2. This in turn increases the current in Q_2 and in R_S further increasing the base voltage of Q_1. In this way, the system can quickly diverge, drawing a large current and practically shorting the power supply rails. Called *latch-up*, the phenomenon, once triggered, can only be stopped by cutting the power supply to the circuit and if it is not detected and interrupted in time it can cause fatal damage to the chip. To prevent latch-up, the nwell and substrate resistors must be as low as possible. This requires that an adequate number of low impedance substrate and nwell contacts are used. In addition, the current gain of the parasitic bipolars must be reduced, which is obtained by making their base wider. This in practice implies that the distance between the nwell and the NMOS device can not be too small. The rules on the number of contacts and nwell spacing are provided by the silicon foundry and must be tightly followed to ensure that the design is adequately immune against latch-up.

Bulk CMOS technologies may employ two types of wafers. In epitaxial wafers shown in Fig. 2.17 a), the substrate for device fabrication is provided by a lowly doped layer which is grown by epitaxy on a highly doped wafer. Typical resistivity is

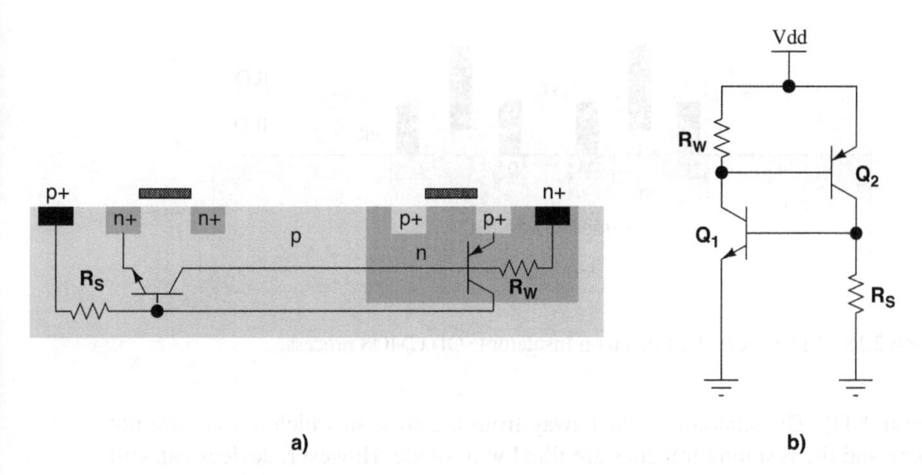

FIGURE 2.16 In bulk CMOS processes parasitic bipolar transistors, combined with the parasitic resistance of the substrate and the nwells, are at the origin of the latch-up phenomenon.

20 Ω·cm for the epitaxial layer and 20-50 mΩ·cm for the bulk material. The highly conductive substrate provides a low impedance path that reduces the latch-up sensitivity, thus relaxing the requirements on contact density and NMOS to PMOS distance and favoring more compact designs. In mixed signal chips digital and analog circuits share the same bulk. The CMOS gates switch between the power supply rails and inject through the junction capacitance a significant amount of parasitic current into the substrate. This current can reach easily the analog circuits through the low resistivity substrate, creating a *substrate noise* that affects significantly their performance. To mitigate this problem, uniformly doped wafers with a resistivity of 10 ÷ 20 Ω·cm can be employed, as shown in Fig. 2.17 b). As shown in Fig. 2.14 and Fig. 2.15, in a bulk process device isolation is achieved with the use of shallow trenches (Shallow Trench

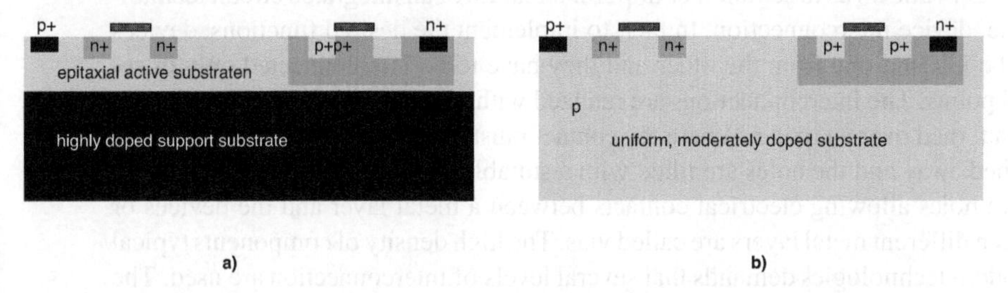

FIGURE 2.17 Substrate types in bulk CMOS: epitaxial layer on highly doped substrate (in a) and homogeneous, moderately doped substrate (in b). The design is not in scale. The active area containing the transistors is limited to a few microns from the wafer surface. Typical total wafer thickness is 700-800 μm to guarantee adequate mechanical stiffness during processing. After fabrication, wafers are usually thinned down to 100-200 μm and diced to extract the individual chips.

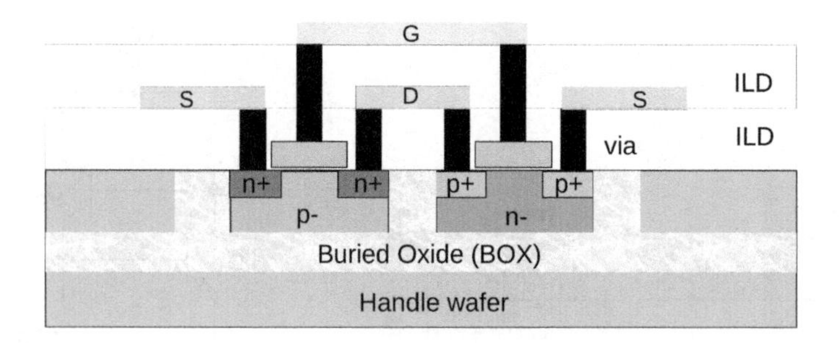

FIGURE 2.18 Cross view of a Silicon on Insulator (SOI) CMOS process.

Isolation, STI). The silicon is etched away from the areas in which devices are not foreseen and the resulting trenches are filled with oxide. However, devices can still interfere with each other through the common substrate. A radical way to increase the device isolation is the use of Silicon On Insulator (SOI) technologies, sketched in Fig. 2.18. In these processes, the transistors are fabricated on very thin silicon layers, located on top of a thick oxide, called *buried oxide*, or BOX. In this way, each device is fully isolated from the others, eliminating the latch-up issue and allowing for more compact circuits. In addition to preventing latch-up, the technique also reduces the parasitic capacitance seen from the transistor junctions, thus increasing the device speed. SOI wafers are more expensive to produce than standard bulk material, but the advantages they offer can offset the initial cost. Their use is in fact becoming more common in very deep submicron technologies. Bulk processes are however still prevalent on the market and in the following we will thus focus on them. At present, homogeneous wafers with moderate and uniform resistivity are more often used for general purpose devices and in particular in Radio-Frequency applications, whereas epitaxial wafers are employed when also sensors have to be embedded in the chip, as, for instance, in monolithic CMOS cameras.

Fig. 2.14 allows us to see another important feature of an integrated circuit technology, i.e. device interconnection. In fact, to implement the desired functions, devices must be isolated one from the other and they have to be interconnected only in selected points. The interconnections are realized with metal lines which are deposited and patterned over an oxide. Where the contact must take place, the protecting oxide is etched away and the holes are filled with a suitable conducting material. The conductive holes allowing electrical contacts between a metal layer and the devices or between different metal layers are called vias. The high density of components typical of modern technologies demands that several levels of interconnection are used. The different metals are insulated by oxide layers (Inter Level Dielectric, ILD) and interconnected where needed through vias. A good planarity of the ILD is fundamental to allow reliable fabrication of the vias. If the oxide is too thick or too thin, under-etching or over-etching will result and the contacts will fail. To achieve a good ILD planarity the structures which are underneath a given oxide layer must be distributed as uniformly as possible. This is at the origin of the design rules which impose that a given layer cover the silicon wafer within a given percentage (e.g., a common requirement is that metal levels fill between 25% and 70% of the chip area). Called *pattern density rules* these guidelines are given by the silicon foundry and must be tightly followed to avoid severe fabrication issues.

2.5.1 CMOS RADIATION SENSORS

Before continuing with our study of the MOS transistor, it is interesting to observe that CMOS technologies can be used also to implement radiation sensing devices. In the previous chapter we have seen that pn junctions can be exploited to collect the charges set free by the interaction between an impinging particle and the semiconducting material. Standard silicon sensors such as hybrid pixels or microstrips are produced on optimized substrates and then mated to separate front-end circuits usually fabricated in CMOS technologies. Embedding the sensor and its front-end electronics on the same CMOS wafer is therefore very attractive because the interconnection between the two is already realized during fabrication. Furthermore, CMOS technologies are very cost-effective for large volume production. Fig. 2.19 shows how a standard CMOS process can be used to fabricate a monolithic radiation detector. The junction between the wafer substrate and the nwell, that usually hosts PMOS transistors, provides the sensing device. A lightly doped epitaxial substrate is normally adopted to maximize the size of the depletion region around the collecting diode. The detecting elements are then arranged in a bidimensional matrix, where the pitch between diodes is usually in the 5-10 μm range. The main drawback of the approach is that only NMOS transistors can be used in the sensitive area, since the additional nwells necessary to implement PMOS transistors will drain the charges away from the collecting electrode, reducing the signal. This constrains the complexity of the front-end electronics that can be embedded close to the sensor, forcing to adopt simple readout schemes in which the information is acquired by periodically scanning the matrix, thereby limiting the acquisition rate. For applications requiring a modest frame rate (in the order of 1 kHz or less), this is not a major issue and monolithic CMOS sensors have thus become the standard choice for cheap digital cameras, such as the one instrumenting mobile telephones. The big market potential has driven the performance improvement of these sensors, making them today competitive with the more traditional Charge-Coupled Device (CCD). The use of standard

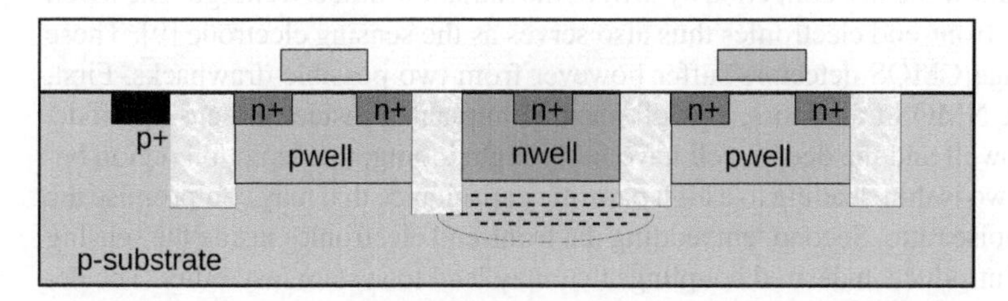

FIGURE 2.19 Radiation sensor implemented in a standard CMOS technology.

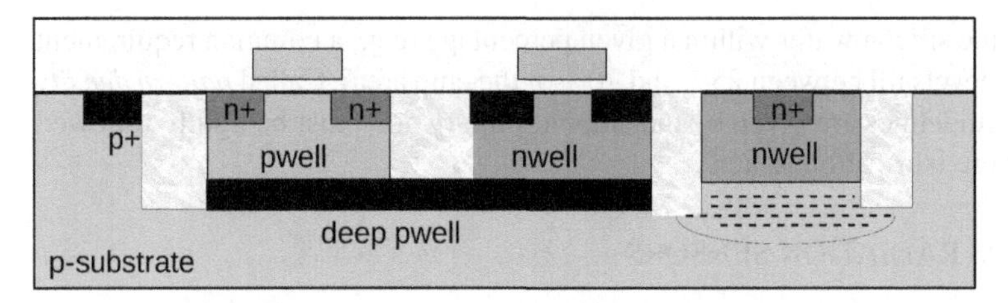

FIGURE 2.20 Quadruple well CMOS sensors. A deep pwell shields the nwells where PMOS transistors are fabricated from the substrate, allowing the implementation of complex CMOS circuits in close proximity of the detecting diode.

monolithic sensors to detect fast radiation pulses like the ones produced by highly energetic charged particles is however more problematic because the impossibility of using PMOS transistors prevents the design of more complex signal processing electronics which is often needed in fast applications. Furthermore, the charge is primarily collected by diffusion, because no external field is applied to enhance the depletion volume around the sensing electrode. The resulting long charge collection times (which can exceed 100 ns) lead to poor timing performance and increased sensitivity to radiation damage[2]. The ideal monolithic sensor should therefore allow the integration of both PMOS and NMOS transistors close to the sensing diode and the biasing of the substrate with a voltage high enough to grant a prompt collection of the signal charge. The first issue can be addressed as shown in Fig. 2.20. Here a deep pwell is implanted underneath the area where the front-end electronics is fabricated, whereas the sensing diode is embedded directly in the bulk. The pwell thus shields the nwells hosting the circuitry, and the charges are collected only by the exposed nwell [8]. Fig. 2.21 shows instead a technology that applies a significant reverse bias voltage to the substrate. For power management applications, it is in fact necessary to fabricate devices that can withstand bias voltages in the order of 50-60 V. These power transistors are then controlled by suitable electronics that need to be fast and it is thus implemented with standard MOS transistors with thin gate oxide. To isolate the signal processing devices from the high voltage, a deep nwell is used. The doping of the substrate is reduced to guarantee a sufficiently high breakdown voltage. As a consequence, a fairly deep depletion region is created and the applied field can be strong enough to allow the fast collection by drift of the radiation-induced charge. The nwell hosting the front-end electronics thus also serves as the sensing electrode [9]. These High Voltage CMOS detectors suffer however from two possible drawbacks. First, to fabricate NMOS transistors, a pwell must be implanted inside the sensing diode. Since the pwell and the deep nwell have fairly high doping, the depletion region between the two is thin, leading to a high parasitic capacitance that may compromise the signal-to-noise ratio. Second, embedding the front-end electronics inside the sensing diode can introduce undesired couplings that may lead to system instability. The detection of a signal triggers in fact voltage swings inside the front-end electronics that

[2]Radiation damage is discussed in more detail in Appendix 2.

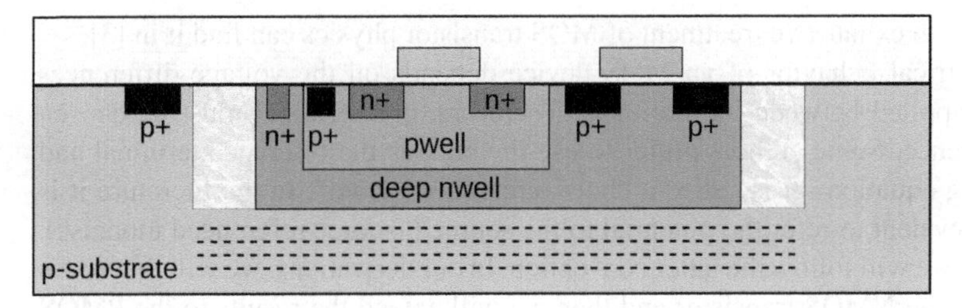

FIGURE 2.21 High Voltage CMOS sensors. A deep nwell fabricated on a lightly doped substrate can sustain a significant reverse bias voltage to allow charge collection by drift.

can couple through parasitic capacitance to the sensing nwell, thus inducing spurious signals that further stimulate the front-end electronics. If not properly mastered, the resulting positive feedback can severly compromise the sensor performance. These two side effects impose some restrinction on the design of the front-end electronics inside the sensing diode. The handling wafer in an SOI technology can also be chosen with appropriate doping and it can be patterned to produce a detector [10]. The buried oxide guarantees the isolation between the high voltage applied to the sensor and the front-end electronics located on top. Through the buried oxide, the high voltage and the front-end electronics, affecting the performance of the latter. Also in this case, a shielding deep pwell can be inserted underneath the transistors in the top layer. More important, the buried oxide is fairly thick and this increases the sensitivity to the damage caused by ionizing radiation. The research on fast monolithic pixels sensors is still ongoing. For the detection of fast radiation pulses, their front-end electronics is similar to that employed for hybrid systems and relies on the basic principles and circuit topologies described in the following chapters. The interested reader can find more information in [11] and references therein.

2.6 ELECTRICAL CHARACTERISTICS OF MOS TRANSISTORS

The physics of MOS devices is very complex and an exhaustive treatment is well beyond the scope of this chapter. Fortunately, essential properties can be studied with relatively simple mathematical derivations. The number of details that must be taken into account to accurately describe the behavior of a device increases as technologies with smaller and smaller feature size are considered. Therefore, the simplified equations derived in this section may lead to inaccurate results when they are applied to transistors implemented in modern CMOS processes. These equations have rather the purpose of making the reader familiar with key aspects of MOS devices without asking her/him to accept on "good faith" a set of mathematical formulas without any link to the device physics. To achieve this goal, many simplifications are necessary. Additional effects, not essential to get the basic picture, but fundamental for implementing in practice any serious design, will be dealt with in sections 2.8 and 2.9. However, an accurate description of the MOS behavior leads to very complex equations and reliable results can not be predicted without the help of computer simulations. Readers interested in an exhaustive treatment of MOS transistor physics can find it in [3].

The electrical behavior of an MOS device depends on the voltage differences which are applied between its terminals. Therefore, one terminal must be chosen as the reference. Some authors prefer to use the bulk as the reference terminal and the resulting equation set is called a "bulk referenced model". In the literature it is however prevalent to refer the potential to the source ("source referenced models"). In this book we will follow the latter convention. In our derivations, we will first focus attention on the NMOS transistor and then we will extend the results to the PMOS case.

2.6.1 THE THRESHOLD VOLTAGE

The channel in an MOS transistor is formed only when the gate-source voltage exceeds a critical limit, called the threshold voltage. To understand why threshold voltages arise in MOS structures we start by studying a simplified two-terminal device: the MOS capacitor. This is conceptually an MOS transistor without the source and drain electrodes. Note that since the gate and the bulk are conductors and they are separated by an insulator (the SiO_2) the MOS structure resembles a parallel plate capacitor. Let's now consider the two situations shown in Fig. 2.22. In Fig. 2.22 a) the gate, the substrate and the interconnects are all made of the same material. The gate is shorted to the substrate. Since everything is made of the same material, no contact potential arises. In this case there is no reason for mobile charges to pile-up in any particular region of the structure, which is uniformly neutral. Figure 2.22 b) shows a more realistic situation. Here the gate is made of n^+ doped silicon, while the substrate is p-doped. The gate and the substrate are connected to a battery by metal leads. We now ask ourselves the following question: which is the voltage that the battery must apply to attract a significant number of electrons underneath the gate? Answering this question means calculating the threshold voltage of the MOS structure. In this

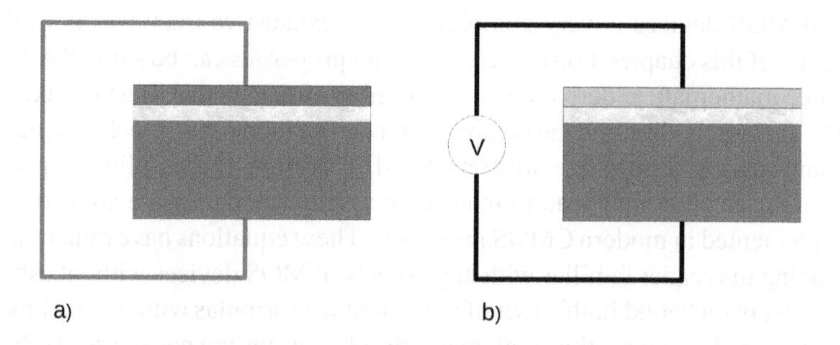

FIGURE 2.22 MOS capacitors. The figure on the left shows the idealized case in which only one type of material is employed. The figure on the right describes the more realistic case in which the top plate is made of polysilicon and the bottom plate by doped silicon. Interconnections are realized with metals.

process, we can end up in two situations. In some cases it will be straightforward to calculate the voltage which is needed to achieve a given charge distribution in the device. In other situations the calculations will lead us first to the charge which is stored in the gate capacitance. Then, by dividing the charge per unit area by the gate capacitance per unit area we will find the required voltage. Applying the formula of a parallel plate capacitor, we can write the gate capacitance per unit area as follows:

$$C_{ox} = \frac{\varepsilon_{ox}}{t_{ox}} \tag{2.107}$$

In the above equation t_{ox} is the oxide thickness and $\varepsilon_{ox} = \varepsilon_r \varepsilon_0$ is the dielectric constant of SiO_2. The relative dielectric constant of silicon dioxide is 3.9. To gain a feeling of the orders of magnitude, consider that a gate dielectric with 5 nm thickness has a capacitance per unit area of 6.9 fF/μm^2. We now come back to the structure of Fig. 2.22 b). Here we have created junctions between different materials. Therefore, we expect to find several contact potentials. In our sequence of materials, the two extremes are given by the gate and the substrate, which are also the two arms of our capacitor. From our discussion on contact potentials, we know that if we have a chain of materials, the difference between the contact potentials of the two extremes of the chain depends only on the properties of the first and last material, $\phi_G - \phi_B$, where ϕ_G and ϕ_B are the contact potentials of the gate and of the bulk measured with respect to intrinsic silicon. Let's clarify further this concept with an example.

Example 2.6: Calculation of gate to bulk contact potential.

Calculate the contact potential which develops between a degenerate n$^+$ gate with $N_D = 10^{19}$ cm^{-3} and a p-type substrate with $N_A = 10^{17}$ cm^{-3}.

* * *

Since the gate is doped with $N_D = 10^{19}$ cm^{-3} it can be considered degenerate. Its contact potential with respect to intrinsic silicon is given by $+0.56\ V$. The contact potential of the substrate can be calculated as:

$$\phi_B = -\frac{k_B T}{q} \ln\frac{N_A}{n_i} = -25.9 \ln\frac{1 \cdot 10^{17}}{1.08 \cdot 10^{10}} = -0.416\ V \tag{2.108}$$

Therfore we have:

$$\phi_G - \phi_B = 0.56 - (-0.416) = +0.976\ V \tag{2.109}$$

We note from the above example that the difference between the gate and bulk contact potential is positive. Since these are the arms of our MOS capacitors, we must expect that the MOS capacitor is naturally charged. In other words, the difference between the gate and bulk contact potentials alters the distribution of charges inside the structure: negative charges tend to pile-up at the interface between the bulk and the gate dielectric and positive charges tend to accumulate at the interface between the polysilicon and the SiO_2. Note that device as whole is always neutral, but uncompensated charges appear in particular zones. The fabrication process of the gate oxide may leave small quantities of positive charge trapped in the oxide. In modern fabrication processes the density per unit area of this charge is low, approximately 10^{10} cm^{-2}. Positive charge trapped in the oxide further attracts negative charges towards the bulk-SiO_2 interface. Therefore, it acts in the direction of charging up further the MOS capacitor. As discussed above, the voltage contribution of this charge is obtained by dividing the charge per unit area by the gate capacitance per unit area. If we use the gate capacitance corresponding to a 5 nm thick oxide, for a charge density of 10^{10} cm^{-2} we have a contribution of only 2.3 mV. The total voltage at which our capacitor is charged is obtained by summing the two contributions:

$$\phi_G - \phi_B + \frac{Q_{ox}}{C_{ox}} = 0.976\ \text{V} + 0.002\ \text{V} = 0.978\ \text{V}. \tag{2.110}$$

We can now make the MOS structure uniformly neutral (i.e. no pile-up of any charges in any point of the device) by inserting between gate and bulk an external battery with a voltage equal to the one given in (2.110) but with an opposite sign. In the energy band description, the non-uniform distribution of charges inside a structure is represented as a bending of the valence and conduction band. Therefore, the voltage imposed by the battery is the one needed to avoid the bending of the bands inside a structure and to guarantee a uniform distribution of charges over the whole sample. For this reason, the quantity:

$$V_{FB} = \phi_B - \phi_G - \frac{Q_{ox}}{C_{ox}} = \phi_{MS} - \frac{Q_{ox}}{C_{ox}} \tag{2.111}$$

is called the flat band voltage. In equation (2.111) we have made the substitution $\phi_B - \phi_G = \phi_{MS}$, as it is common practice in literature about MOS devices. Starting now from the flat band voltage we try to charge up our capacitor again. We then rise the potential on the gate side, to bring on the gate positive charges (which in practice means removing electrons, which are the mobile charge carriers). The positive charge on the gate repels the holes from the SiO_2-bulk interface underneath the gate. Removing the holes, we are left with a depletion region populated by negatively ionized acceptor atoms. Increasing further the gate potential, the depletion region in the substrate widens and also electrons start to pile up at the interface. In an MOS structure like the one we are considering these electrons are thermally generated in the depletion region, so the build up of a layer of mobile charges is quite a slow process. Note however that by acting on the gate potential we are inducing electrons at the surface of the bulk. In other words, we are populating the bulk surface close to the gate oxide with carriers of the opposite sign of the ones that would be naturally present in the semiconductor as a result of the doping it has received. With an external potential we are locally reversing the doping of the material. When this process will be fully completed, we say that the surface has been inverted. This field-induced doping is not a permanent effect, as it disappears once the external voltage is removed. In an MOS structure the following conditions must be satisfied:

$$V_{GB} = V_{FB} + \psi_{ox} + \psi_{sub} \tag{2.112}$$

$$Q_G + Q_{ox} + Q_{ch} = 0 \tag{2.113}$$

The first equation states the conservation of energy. The voltage imposed by the battery must be equal to the sum of the voltage drops inside the device. In this equation ψ_{ox} is the voltage across the gate oxide and ψ_{sub} is the potential difference between the bulk surface at the Si-SiO_2 interface and the neutral silicon bulk. The second equation states the conservation of charge and the overall neutrality of the capacitor. The application of an external voltage does not create charges, but just moves them from one place to another. If we apply now a change in the gate potential we can write:

$$\Delta V_{GB} = \Delta\psi_{ox} + \Delta\psi_{sub} \tag{2.114}$$

$$\Delta Q_G + \Delta Q_{ch} = 0 \tag{2.115}$$

In fact, both the flat band voltage and the positive charge stored in the oxide are constant. Therefore, any variation in the gate-bulk voltage implies a change in the oxide and in the surface potential. Additionally, any change in the charge accumulated on the gate must be compensated by a change of the opposite sign in the charge stored in the channel. Suppose that we have in the channel an electron concentration indicated by n_s. Using equation (2.44) we can relate the electron concentration at the surface of the Si, n_s and the one in the undisturbed bulk, n_0 to the potential difference between the two regions. We can write:

$$\psi_{sub} = \phi_T \ln\frac{n_s}{n_0} \tag{2.116}$$

Remember the ψ_{sub} is the potential difference between the surface just underneath the gate and the deep, uniformly neutral silicon bulk. Rearranging (2.116) we have:

$$n_s = n_0 e^{\frac{\psi_{sub}}{\phi_T}} \tag{2.117}$$

Using the law of mass action we can also write:

$$n_0 p_0 = n_i^2 \Rightarrow n_0 = \frac{n_i^2}{p_0} \tag{2.118}$$

Inserting (2.118) in (2.117) we obtain:

$$n_s = \frac{n_i^2}{p_0} e^{\frac{\psi_{sub}}{\phi_T}} \tag{2.119}$$

Making now some algebra we can put n_s in a particularly interesting form. We recall in fact that the Fermi potential of the bulk can be expressed as:

$$\phi_F = \phi_T \ln\frac{p_0}{n_i} \Rightarrow \frac{p_0}{n_i} = e^{\frac{\phi_F}{\phi_T}} \Rightarrow n_i = p_0 e^{-\frac{\phi_F}{\phi_T}} \tag{2.120}$$

Inserting the last part of equation (2.120) in (2.119) and recalling that the hole concentration p_0 is equal to the acceptor concentration N_A we can write:

$$n_s = N_A e^{\frac{(\psi_{sub} - 2\phi_F)}{\phi_T}} \tag{2.121}$$

The above equation is interesting because it relates the electron concentration at the surface to the Fermi potential of the bulk. Here we see that if $\psi_{sub} < 2\phi_F$ the field-induced electron concentration is small. If $\psi_{sub} = 2\phi_F$, the electron concentration at the substrate is exactly equal to the hole concentration in the undisturbed bulk. Conventionally, this point is considered the onset of strong inversion in the MOS structure. If $\psi_{sub} - 2\phi_F$ is greater than 3 times ϕ_T the electron concentration at the surface is already 20 times N_A. Therefore, small increments of ψ_{sub} determine large variation of the surface charge. The term $2\phi_F$ is the minimum voltage that must be applied to create an inversion layer and adds positively to the threshold voltage.

To complete the picture, we need to consider a last contribution. The fact that we have induced a n^+ layer in a p-type substrate implies that we have to support also a n^+p junction. This junction is not permanent, since it disappears when the voltage that supports the inversion layer is removed. Nevertheless, being a junction, when it is formed it has an associated depletion layer, which stores a well defined amount of charge. To calculate this charge we make the following assumptions:

1. When the surface is inverted, the concentration of the electrons in the inversion layer is much greater than the one of holes in the substrate, so the depletion region extends mostly in the latter.
2. To have large concentration of electrons it is sufficient that ψ_{sub} exceeds $2\phi_F$ by a few tens of millivolts. Therefore, as a first approximation, after the inversion layer is formed the substrate potential can be consider almost constant and approximated by $2\phi_F$.

Applying now the results on the extension of the depletion region in case of an asymmetric junction we have:

$$x_{dp} = \sqrt{\frac{2\varepsilon_{Si}}{qN_A} 2\phi_F} \tag{2.122}$$

The total charge stored in the depletion region is $Q_{BT} = qN_A W L x_{dp}$ and the charge per unit area is $Q_B = qN_A x_{dp}$. We can then write:

$$Q_B = qN_A x_{dp} = qN_A \sqrt{\frac{2\varepsilon_{Si}}{qN_A} 2\phi_F} = \sqrt{2\varepsilon_{Si} q N_A 2\phi_F} \tag{2.123}$$

To maintain the storage of this charge we need to apply a voltage on the gate equal to Q_B/C_{ox}. Summing all the contributions, the total voltage needed to support an inversion layer in the MOS structure is given by:

$$V_{TH0} = V_{FB} + 2\phi_F + \frac{1}{C_{ox}} \sqrt{2\varepsilon_{Si} q N_A 2\phi_F} \tag{2.124}$$

Finally, we must note that the bulk charge, Q_B, contributes to the total charge balance defined in equation (2.113). Therefore, the charge in the channel Q_C in (2.113) can be written as:

$$Q_C = Q_B + Q_I \tag{2.125}$$

We see from (2.125) that the charge on the channel that balances the one on the gate is made of two components:

1. Q_B due to ionized dopant atoms, which are not mobile and can not participate in the current flow
2. Q_I which is the inversion layer charge, made of mobile carriers.

2.6.2 REGIONS OF OPERATION OF THE MOS TRANSISTOR

Let's now turn back our attention to the full NMOS transistor shown in Fig. 2.10. A front view of the device is reported in Fig. 2.23. The charge in the depletion layers associated to each n^+p junction is also shown. The source and drain electrodes serve two purposes: first, they establish a contact to the inversion layer charge that piles-up in the channel under the gate. This applies a potential across the channel region, thereby allowing a current to flow. Second, the highly doped electrodes serve as source of carriers to form the inversion layer. We now study how the transistor responds to the voltages applied to its terminal. We start from a simple situation, in which source, drain and bulk are kept at zero and the gate is connected to a voltage supply. We can make a simple picture of the MOS behavior treating the gate and the channel as the two arms of a parallel plate capacitor. If we charge the gate terminal, we induce a charge equal in magnitude and opposite in sign in the channel (see also equation 2.115). If the gate is connected to a negative voltage, holes are attracted underneath the gate oxide. Shown in Fig. 2.24 a), this region of operation is called "accumulation". In this region no current can flow between drain and source and the device behaves simply as a capacitor with the gate and the bulk acting as the

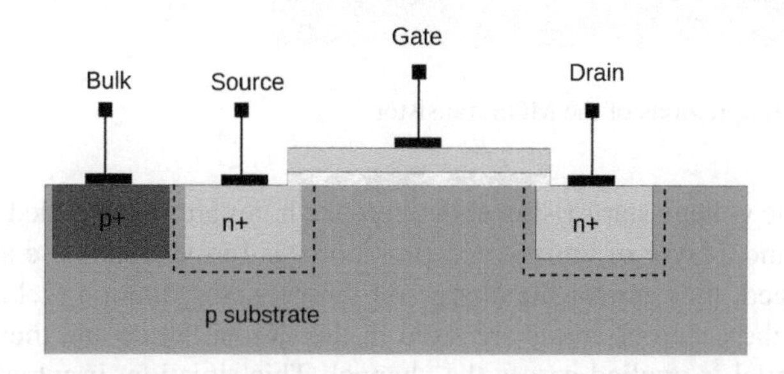

FIGURE 2.23 Cross section of an NMOS device. Associated to each n^+p junction there is a depletion region that extends mostly into the less doped substrate.

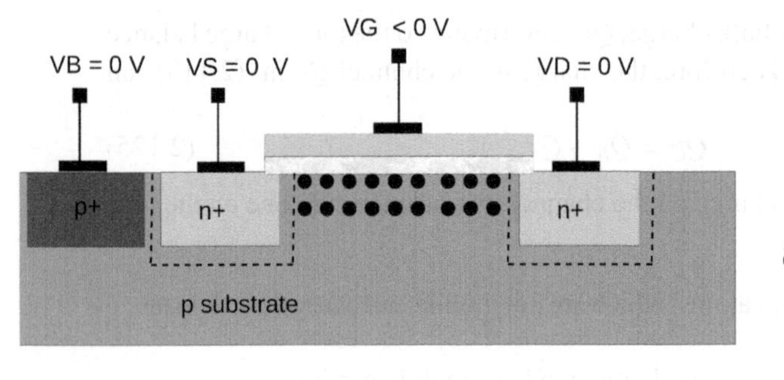

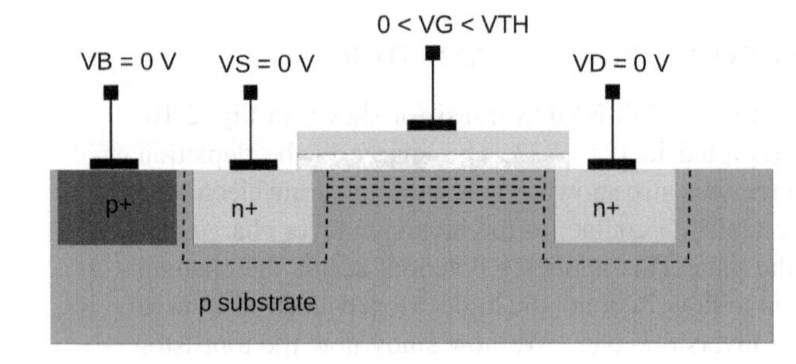

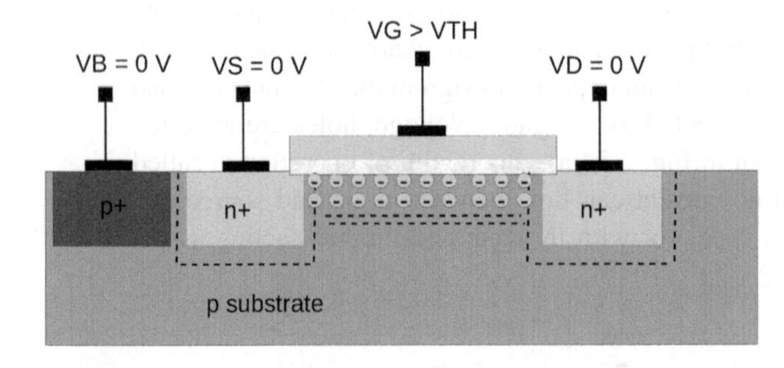

FIGURE 2.24 Operating regions of the MOS transistor.

terminals. As the gate voltage starts rising above zero the holes are first repelled from the channel region and a layer of ionized acceptor atoms is formed. As these atoms are negatively charged, they mirror the charge on the gate, so equation (2.115) is satisfied. However, the acceptor atoms are fixed in the silicon lattice and they can not move if a potential is applied across the channel. This situation, in which the channel region is deprived of any mobile charge is called depletion and is depicted in Fig. 2.24 b). If we further increase the voltage on the gate, electrons start appearing in the channel. The charge on the gate is mirrored partially by acceptors atoms and partially by electrons. We have seen that the electron density in the channel depends exponentially on the difference between the channel potential, ψ_S, and $2\phi_F$, where ϕ_F is the Fermi potential of the substrate. When the relationship $\psi_s > 2\phi_F$ is verified the electrons in the channel quickly outnumber the negative acceptor atoms and the electron density in the channel rapidly approaches the one in the source and drain regions. The mirror charge needed to satisfy equation (2.115) is now provided mostly by electrons. This condition, called inversion, is shown in Fig. 2.24 c).

In the remainder of our discussion, we will take the source potential as the reference potential and we will indicate the gate-source voltage as V_{GS} and the drain-source voltage as V_{DS}. In the classical MOS model, it is assumed that the electron density in the channel is relevant only when V_{GS} is above the threshold voltage V_{TH} and that the transistor turns on and off abruptly as the V_{GS} crosses the threshold. This is an extremely crude approximation for devices implemented in modern CMOS technologies. The subthreshold region plays in fact an important role in many analog circuits implemented in these processes. We stick for the moment to the classical model and we will discuss in section 2.8 the implication of subthreshold operation.

2.6.3 MOS CHARACTERISTICS IN THE LINEAR REGION

In strong inversion, all the positive charge accumulated on the gate after the threshold voltage has been reached is mirrored by electrons in the channel. Therefore, the total charge in the channel available for conduction is given by:

$$Q = C_{ox}WL(V_{GS} - V_{TH}) \tag{2.126}$$

where W is the channel width, L is the channel length and C_{ox} is the gate capacitance per unit area. Let's now rise also the potential voltage V_{DS}. If V_{DS} is small, we can assume that the electric field is constant along the channel. Therefore:

$$\mathscr{E} = \frac{V_{DS}}{L} \tag{2.127}$$

The current can be written as $I = Q/t$, where t is the time an electron needs to cross the channel. We can write:

$$t = \frac{L}{v} \tag{2.128}$$

$$v = \mu_n \mathscr{E} = \mu_n \frac{V_{DS}}{L} \tag{2.129}$$

where v is the electron velocity and μ_n is the electron mobility. Combining equations (2.128) and (2.129) we have:

$$\frac{1}{t} = \mu_n \frac{V_{DS}}{L^2} \tag{2.130}$$

The current I_{DS} can therefore be expressed as:

$$I_{DS} = \frac{Q}{t} = \mu_n C_{ox} \frac{W}{L} (V_{GS} - V_{TH}) V_{DS} \tag{2.131}$$

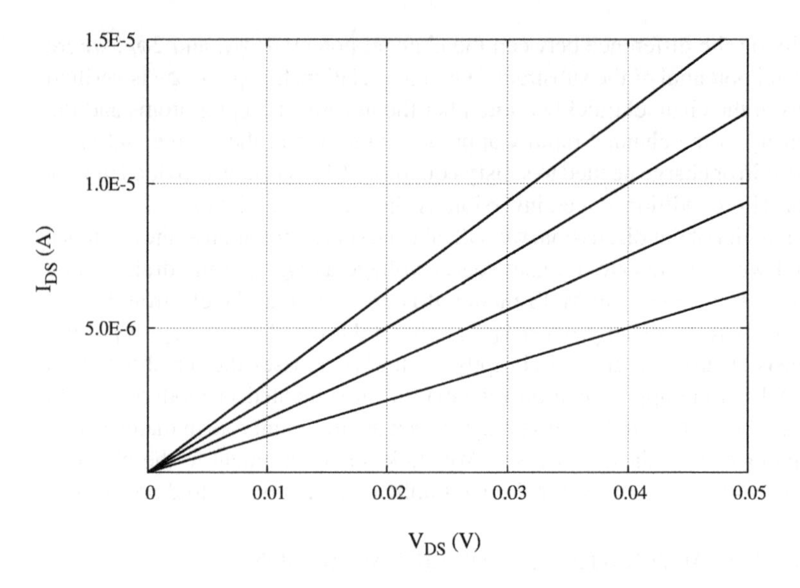

FIGURE 2.25 Drain current for small values of V_{DS} and for different gate voltages.

Under the above hypothesis, the relationship between the drain-source current and the drain-source voltage is linear. The transistor behaves as a voltage controlled linear resistor with a resistance value defined by:

$$R_{ON} = \frac{V_{DS}}{I_{DS}} = \frac{1}{\mu_n C_{ox} \frac{W}{L} (V_{GS} - V_{TH})} \tag{2.132}$$

Fig. 2.25 shows the drain-source current as a function of V_{DS} for different values of $V_{GS} - V_{TH}$. In this plot, the following parameters were assumed: $W/L = 1$, $\mu_n C_{ox} = 250\ \mu A/V^2$ and $V_{TH} = 0.5\ V$. The gate voltage was changed from 1 V to 1.75 V in steps of 0.25 V. The slope of the curves increases with $V_{GS} - V_{TH}$. Note that in Fig. 2.25 the sweep on V_{DS} is limited to low voltage values. In fact, as V_{DS} keeps on increasing, the charge density in the channel can no longer be considered uniform. To study this, we can divide our transistor in many sub-transistors connected in series, as shown in Fig. 2.26. Each sub-device has a channel width equal to the one of the global transistor and an infinitesimal channel length, dx. In each sub-transistor the potential in the channel can be considered constant. Let's now take a generic point x in the channel, as shown in Fig. 2.26. The source voltage of the associated infinitesimal transistor will be $V(x)$, therefore the effective gate voltage that supports the inversion of the local channel will be $(V_{GS} - V_{TH} - V(x))$. The charge in the local device will be $dQ = C_{ox} W dx (V_{GS} - V_{TH} - V(x))$. Noting that all the infinitesimal transistors are in series and have the same drain-source current I_{DS}, we can write:

$$I_{DS} = \frac{dQ}{dt} = C_{ox} W \frac{dx}{dt} (V_{GS} - V_{TH} - V(x)) = C_{ox} W (V_{GS} - V_{TH} - V(x))\, v \tag{2.133}$$

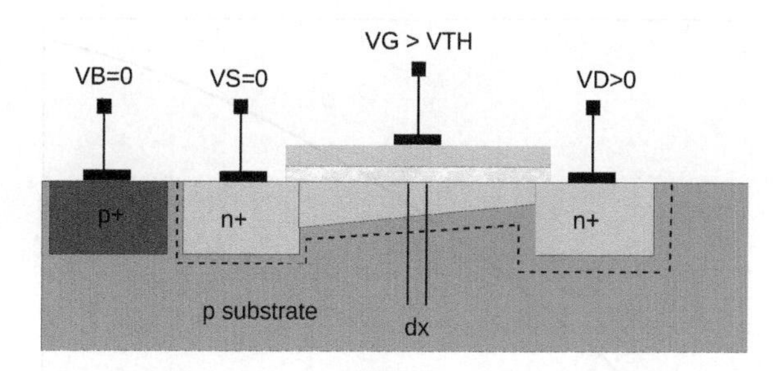

FIGURE 2.26 Transistor in linear region with non-uniform channel.

where $v = dx/dt$ is the electron velocity in the channel. Recall now that the relationship between velocity and drift field is given by:

$$v = \mu_n \mathscr{E} = \mu_n \frac{dV}{dx} \tag{2.134}$$

Introducing (2.134) in (2.133) we obtain:

$$I_{DS} = C_{ox} W \left(V_{GS} - V_{TH} - V(x)\right) \mu_n \frac{dV}{dx} \tag{2.135}$$

We can now rearrange (2.135) and integrate:

$$\int_0^L I_{DS} dx = \int_0^{V_{DS}} \mu_n C_{ox} W \left(V_{GS} - V_{TH} - V(x)\right) dV \tag{2.136}$$

The integration limits are justified by the fact that as we move in the channel from 0 (the source) to L (the drain) the voltage rises from 0 to V_{DS}. After integration we can write:

$$I_{DS} = \mu_n C_{ox} \frac{W}{L} \left[\left(V_{GS} - V_{TH}\right) V_{DS} - \frac{V_{DS}{}^2}{2} \right] \tag{2.137}$$

In (2.137) we see that the drain-source current has a quadratic dependence on V_{DS}. In other words, the transistor behaves in this region as voltage controlled non-linear resistor. Fig. 2.27 shows the I_{DS} versus V_{DS} plot for the same transistor considered before with $V_{GS} - V_{TH} = 0.75$ V. Note that in Fig. 2.26 the channel becomes thinner as we move from the source to the drain, indicating that the charge channel density progressively decreases in the direction of the drain. Fig. 2.27 shows the characteristics of the transistors when the drain-source voltage is high enough to make apparent the non-uniformity of the channel.

2.6.4 MOS CHARACTERISTICS IN SATURATION

Equation (2.137) predicts that as V_{DS} increases the current reaches a maximum and then inverts its polarity. Shown in Fig. 2.28, this behavior is not physical. In fact

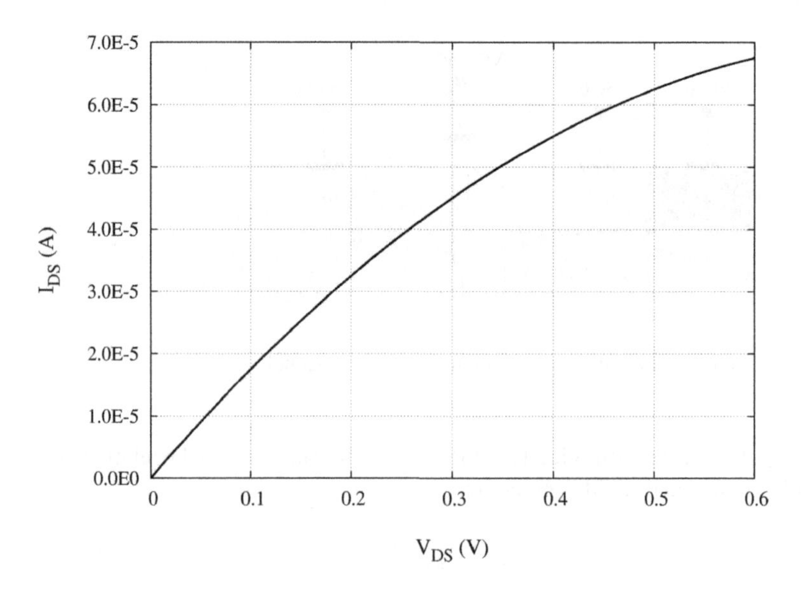

FIGURE 2.27 Drain current versus V_{DS} in the linear region for large values of V_{DS}.

to have carriers at a point x in the channel we must have $(V_{GS} - V_{TH} - V(x)) > 0$. Therefore, for the points in which $V(x) \geq (V_{GS} - V_{TH})$ the inversion layer does not form and this portion of the channel does not contribute to the current flow with its own carriers. When this condition is reached, the channel is said to be pinched-off. Electrons coming from the inverted zones of the channel can however traverse this region and reach the drain. The phenomenon is represented is Fig. 2.29. Here the channel becomes thinner as we move from the source to the drain. If $V_{DS} > V_{GS} - V_{TH}$ we must limit the integration in the right member of (2.136) to $V_{GS} - V_{TH}$. The current then becomes:

$$I_{DS} = \frac{1}{2}\mu_n C_{ox}\frac{W}{L}(V_{GS} - V_{TH})^2 \tag{2.138}$$

Equation (2.138) describes the transistor behavior in the saturation region, which is widely employed in analog circuit design. According to (2.138) the current does not depend anymore on V_{DS}. The transistor hence behaves as a voltage controlled current source. Remember now that an ideal current source constrains the current between the points of the circuit to which it is connected to a given value, regardless of the voltages that the rest of the circuit will develop to support the flow of that particular current. A transistor can work as a current source only if the drain-source voltage V_{DS} is greater than the *saturation voltage* $V_{GS} - V_{TH}$. In practice, the transition between linear and saturation region will be smooth, and adequate headroom must be guaranteed to keep the device saturated. How much margin must be left will depend on the particular application. However, as a rule of thumb, we can say that V_{DS} should always be greater than $V_{GS} - V_{TH} + \Delta V$ where ΔV typically lies between 0.1 and 0.2 V. Fig. 2.30 shows an example of characteristics in which the transistor reaches the saturation region. As we see from Fig. 2.30, the curve does not flatten out completely after the device

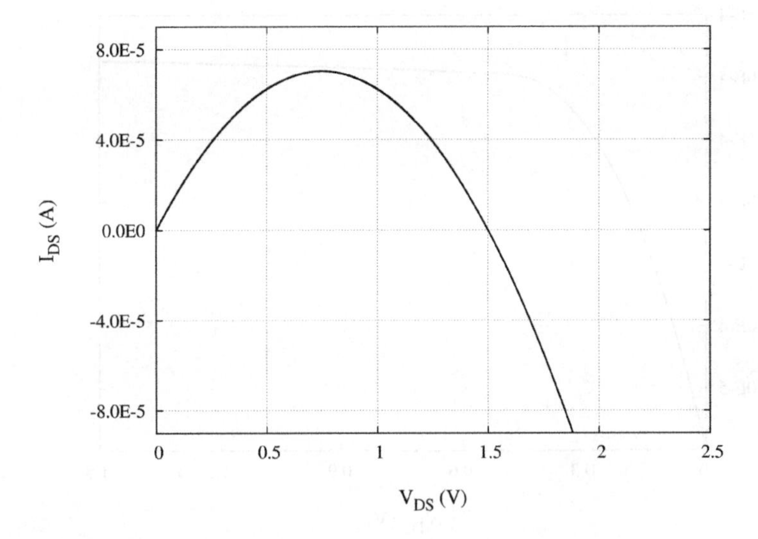

FIGURE 2.28 Behavior predicted by equation (2.137) for large values of V_{DS}.

enters in saturation. To explain this residual dependence of I_{DS} on V_{DS} we recall that the increase of V_{DS} also widens the depletion region around the drain. Once we are in saturation, this translates into a modulation of the pinch-off point. More precisely, as V_{DS} is increased, the pinch-off point moves towards the source and the effective channel length is decreased. Called channel length modulation, this effect translates into a residual control over the drain-source current by V_{DS} when the device works in saturation. To put this on a quantitative basis, we can write:

$$L_{eff} = L - \Delta L \tag{2.139}$$

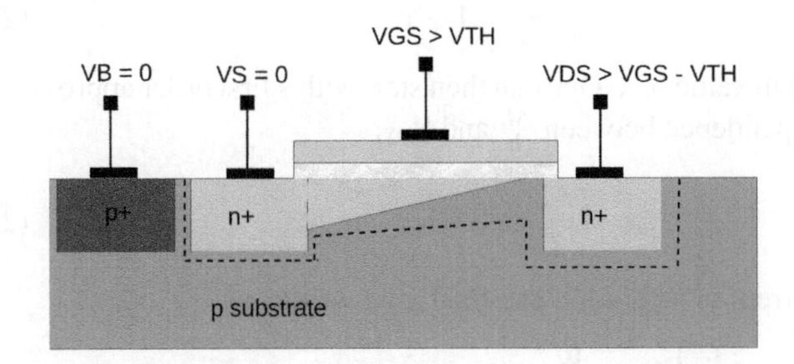

FIGURE 2.29 Illustration of channel pinch-off.

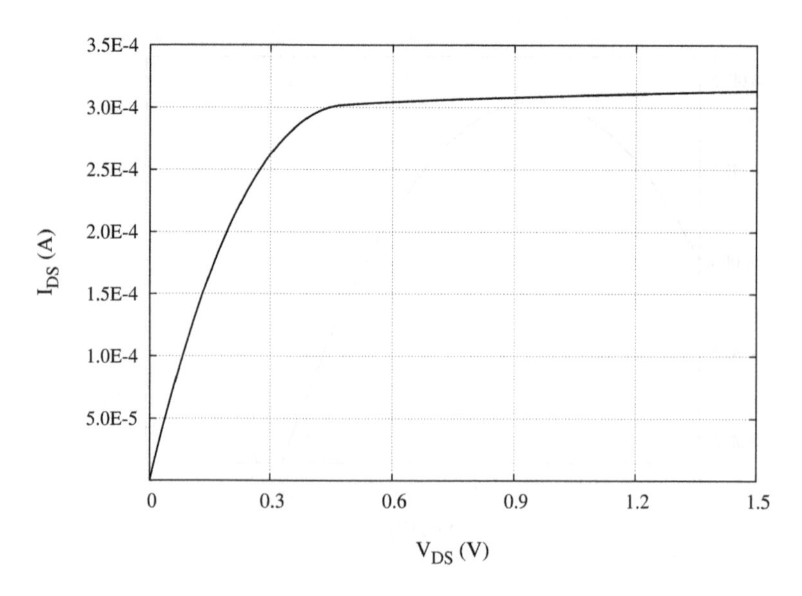

FIGURE 2.30 Characteristics of an NMOS transistor from the linear to the saturation region.

where in (2.139) ΔL is the variation of the channel length due to the displacement of the pinch-off point. Introducing the value of L_{eff} in equation (2.138) we obtain:

$$I_{DS} = \frac{1}{2}\mu_n C_{ox}\frac{W}{L-\Delta L}(V_{GS}-V_{TH})^2 \tag{2.140}$$

We can rearrange (2.140) in the following way:

$$I_{DS} = \frac{1}{2}\mu_n C_{ox}\frac{W}{L(1-\frac{\Delta L}{L})}(V_{GS}-V_{TH})^2 = \frac{1}{2}\mu_n C_{ox}\frac{W}{L}(V_{GS}-V_{TH})^2(1+\frac{\Delta L}{L}) \tag{2.141}$$

Since ΔL is small with respect to L, in the right part of equation (2.141) we have used the approximation:

$$\frac{1}{1-x} \approx 1+x \tag{2.142}$$

which is valid for small value of x. One can then start with a first order approximation, assuming a linear dipendence between $\frac{\Delta L}{L}$ and V_{DS}:

$$\frac{\Delta L}{L} = \lambda V_{DS} \tag{2.143}$$

The drain-source current in saturation can finally be written as:

$$I_{DS} = \frac{1}{2}\mu_n C_{ox}\frac{W}{L}(V_{GS}-V_{TH})^2(1+\lambda V_{DS}) \tag{2.144}$$

The "channel length modulation parameter", λ defines the sensitivity of the drain-source current to the drain-source voltage when the device works in saturation.

From (2.141) we see that $\Delta L/L$ is bigger for shorter channel length. Hence, the channel length modulation effect is more severe for devices with small gate length.This fact implies that λ can not be considered a technological parameter valid for all devices implemented in a given CMOS process. We will discuss further these aspects in section 2.7.

2.6.5 THE BODY EFFECT

In our considerations so far we have always kept the source of the transistor grounded. We now relax this condition by studying the example of Fig. 2.31. In Fig. 2.31 a) the source is grounded, the gate is at 1 V and the drain voltage is swept from 0 to 1.5 V. In Fig. 2.31 b) the source is kept at 1 V and the gate is risen to 2 V to maintain V_{GS} constant at 1 V. The drain voltage is swept from 1 (corresponding to $V_{DS} = 0$) to 2.5 V (corresponding to $V_{DS} = 1.5$). Fig. 2.32 shows the I_{DS} versus V_{DS} characteristics obtained for the two situations. We see that when the source is kept at 1 V, the current is smaller and the transistor enters the saturation region earlier. Since the onset of the saturation region is given by $V_{DS} = V_{GS} - V_{TH}$ and V_{GS} is not changed between the two plots, we deduce that the threshold voltage must have increased. Note that the only potential which has not been changed is the bulk potential, which in both cases is kept at zero. By doing so, when we move from the condition of Fig. 2.31 a) to the one of Fig. 2.31 b) we increase the reverse bias voltage between the electrode-bulk junctions

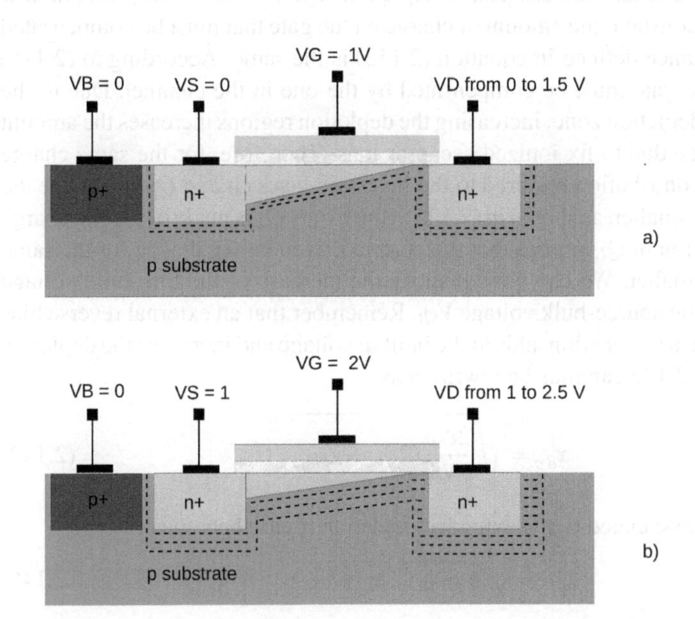

FIGURE 2.31 Illustration of the body effect.

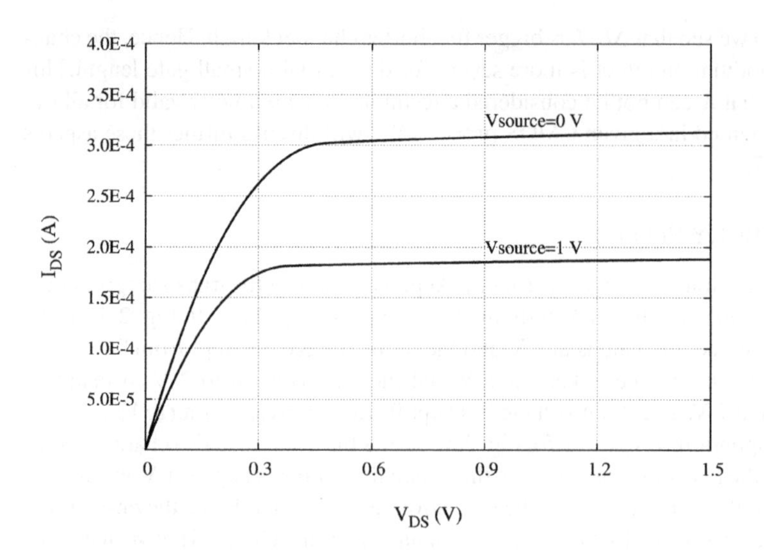

FIGURE 2.32 Characteristics of an NMOS transistor with the source grounded and kept at 1 V. In both cases the voltage difference between the gate, source and drain terminals is the same.

and the bulk. As a result, the associated depletion regions widen. We observe now that since V_{GS} is constant, the amount of charge on the gate that must be compensated in the charge balance defined in equation (2.113) is the same. According to (2.115) the charge on the gate must be compensated by the one in the channel. Due to the widening of the depletion zone, increasing the depletion regions increases the amount of negative charge due to fix ionized acceptor ions. Therefore, for the same charge on the gate, the contribution required to the inversion layer charge Q_I to achieve the charge balance is smaller, as shown in (2.125). Since only Q_I is made of mobile charge carriers, a reduction in Q_I implies that the current driven by the device for the same bias settings is smaller. We can now quantify the increase of the threshold voltage as a function of the source-bulk voltage V_{SB}. Remember that an external reverse bias voltage applied to a pn junction adds to the built in voltage and increases the depletion region. Equation 2.122 can thus be rewritten as:

$$x'_{dp} = \sqrt{\frac{2\varepsilon_{Si}}{qNA}\,(2\phi_F + V_{SB})} \tag{2.145}$$

The bulk charge associated to the extended depletion region becomes:

$$Q_B = \sqrt{2\varepsilon_{Si} N_A\,(2\phi_F + V_{SB})} \tag{2.146}$$

while the threshold voltage is:

$$V_{TH} = V_{FB} + 2\phi_F + \frac{Q_B}{C_{ox}} \tag{2.147}$$

The bulk charge Q_{B0} when the source-bulk voltage is zero is given by:

$$Q_{B0} = \sqrt{2\varepsilon_{Si} q N_A 2\phi_F} \tag{2.148}$$

We can now write:

$$Q_B = Q_B - Q_{B0} + Q_{B0} = \sqrt{2\varepsilon_{Si} q N_A}\left(\sqrt{2\phi_F + V_{SB}} - \sqrt{2\phi_F}\right) + \sqrt{2\varepsilon_{Si} q N_A 2\phi_F} \tag{2.149}$$

Replacing (2.149) in (2.147) we have:

$$V_{TH} = V_{FB} + 2\phi_F + \frac{1}{C_{ox}}\sqrt{2\varepsilon_{Si} q N_A 2\phi_F} + \frac{1}{C_{ox}}\sqrt{2\varepsilon_{Si} q N_A}\left(\sqrt{2\phi_F + V_{SB}} - \sqrt{2\phi_F}\right) \tag{2.150}$$

We note that the first three terms in (2.150) form the threshold voltage as defined in (2.124). Therefore we can compact (2.150) rewriting it as:

$$V_{TH} = V_{TH0} + \frac{1}{C_{ox}}\sqrt{2\varepsilon_{Si} q N_A}\left(\sqrt{2\phi_F + V_{SB}} - \sqrt{2\phi_F}\right) \tag{2.151}$$

The threshold voltage can be thus expressed by the sum of the threshold voltage when V_{SB} is zero plus a term that depends on the source-bulk voltage. Often, the following substitution is made:

$$\gamma = \frac{\sqrt{2\varepsilon_{Si} q N_A}}{C_{ox}} \tag{2.152}$$

so that (2.153) becomes:

$$V_{TH} = V_{TH0} + \gamma\left(\sqrt{2\phi_F + V_{SB}} - \sqrt{2\phi_F}\right) \tag{2.153}$$

Called body effect coefficient, the parameter γ is measured in units of $V^{1/2}$ and depends on the substrate doping and on the gate oxide capacitance. A typical value for γ is 0.5 $V^{1/2}$.

2.6.6 MOS CAPACITANCE

We consider now the main capacitances which are present in the structure of the MOS transistor. Such capacitances determine the device performance at high frequency and are shown in Fig. 2.33. The source and drain form with the bulk pn junctions, hence we expect to have a capacitance associated to the depletion layer, as discussed in 2.3.4. Since the electrodes are three-dimensional structures, for a more accurate modeling, the total capacitance between the source and drain and the substrate is split in two components. One component is given by the capacitance between the bottom of the electrodes and the body and is proportional to the source/drain area. This contribution is represented by C_{DBB} and C_{SBB} in Fig. 2.33. The other contribution stems from the depletion region which extends on the sides of the electrodes. Called sidewall capacitance, this is proportional to the perimeter of the source and drain and

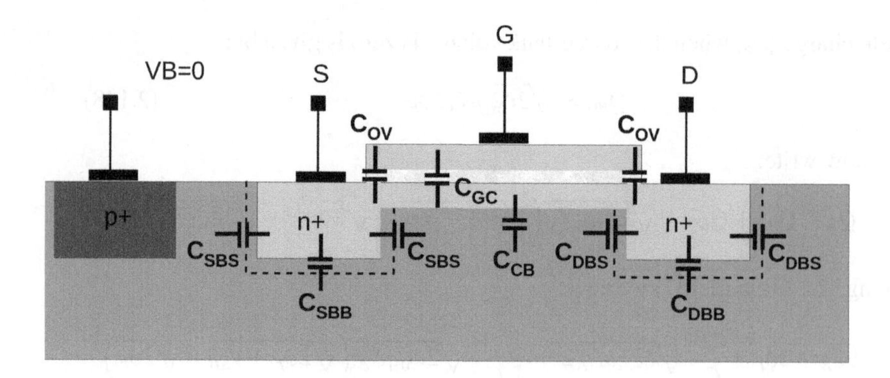

FIGURE 2.33 Transistor parasitic capacitances.

is represented by C_{DBS} and C_{SBS}. Therefore, summing up the two contributions, for the source-bulk capacitance we can write:

$$C_{SB} = \frac{CJ \times AS}{\left(1 - \frac{V_{BS}}{\psi_{BI}}\right)^{MJ}} + \frac{CJSW \times PS}{\left(1 - \frac{V_{BS}}{\psi_{BI}}\right)^{MJSW}} \tag{2.154}$$

For the drain bulk capacitance, the equation reads:

$$C_{DB} = \frac{CJ \times AD}{\left(1 - \frac{V_{BD}}{\psi_{BI}}\right)^{MJ}} + \frac{CJSW \times PD}{\left(1 - \frac{V_{BD}}{\psi_{BI}}\right)^{MJSW}} \tag{2.155}$$

In the above equations CJ and $CJSW$ are, respectively the capacitance per unit area and unit length measured when the external reverse bias voltage is zero, AS and AD are the source and drain areas, while PS and PD are the source and drain perimeters. The exponents MJ and $MJSW$ are the junction grading coefficients and keep into account the fact that the transition between the n-side and the p-side of the junction might be gradual. This is especially true for $MJSW$ which typically ranges from 0.3 to 0.4, while MJ is normally closer to the theoretical value (0.5) expected for an abrupt junction. Equations (2.154) and (2.155) are valid when the junctions are reverse-biased, which is the normal operating condition of an MOS transistor.

Several capacitances are formed between the gate and the other terminals of the transistors. The source and drain regions side-diffuse underneath the gate. Therefore, there is an overlap capacitance between gate and source and gate and drain. This capacitance can be written as:

$$C_{ov} = LDWC_{ox} \tag{2.156}$$

where LD is the amount of lateral diffusion, W is the transistor width and C_{ox} is the gate capacitance per unit area. Since LD and C_{ox} are constant for a given technology, we can incorporate them in a single parameter, CGO which expresses the gate overlap capacitance per unit length. This parameter is usually given separately for source and drain, so we have $CGSO$ and $CGDO$. Since the gate area is not infinite, there is a fringe

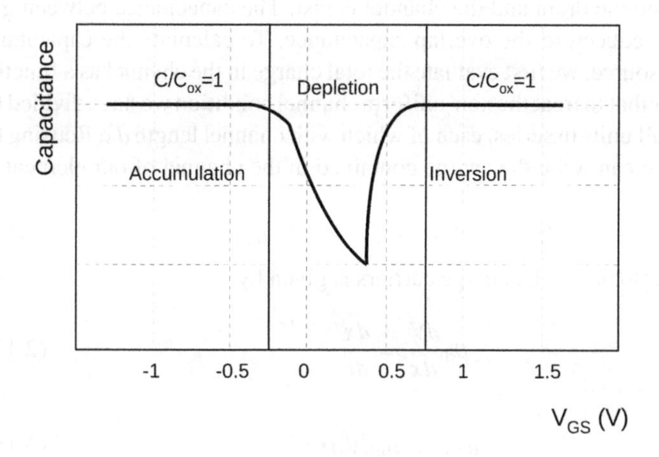

FIGURE 2.34 Qualitative trend of the gate capacitance versus gate voltage.

field at the border of the gate, with the field line starting from the gate and ending on the source or drain. Therefore, the simple overlap model is not fully accurate and the capacitance per unit length must be obtained from more elaborate calculations. The capacitance between the gate and the channel depends on the transistor operating region. If we are in accumulation, the bulk majority carriers are attracted towards the surface, when they form a highly conductive layer. The structure is well approximated by a parallel plate capacitor of value $C = C_{ox}WL$. If we rise the gate voltage to switch the transistor on, first we enter the depletion region and then we start attracting minority carriers into the channel. In this condition, the capacitance between the gate and the bulk can be viewed by the series of two capacitances: the gate-channel capacitance and the channel-bulk capacitance, due to the depletion layer. The channel-bulk capacitance decreases as we push the transistor more and more towards the inversion as the depletion layer widens. Finally, as the inversion layer is fully formed, it shields the gate from the bulk, as all the field lines start from the gate and terminate on the mobile charges in the channel. The parallel plate approximation holds again and the gate capacitance can be calculated as $C = C_{ox}WL$. Fig. 2.34 sketches the value of the gate capacitance as we move from accumulation to inversion. The capacitance first decreases reaching a minimum and then rises back to its maximum value in strong inversion. We examine now in more detail the gate capacitance in strong inversion. Note that the channel is contacted through the source and drain terminals. Therefore, the capacitance between the gate and the channel appears as a capacitance between the gate and the source and/or the gate and the drain. If the transistor is in the linear region the channel can be considered uniform. The channel capacitance in this condition is equally distributed between drain and source, so we have $C_{gs} = 1/2C_{ox}WL$, $C_{ds} = 1/2C_{ox}WL$. In saturation, the channel is pinched-off and the direct connection between the drain and the channel is lost. The capacitance between gate and drain is then reduced to the overlap capacitance. To calculate the capacitance between gate and source, we first evaluate the total charge in the channel as a function of V_{GS}. Remember that to treat the non-uniform channel condition we have divided the transistor into small units in series, each of which with channel length dx. Reusing the same approach, we can write the charge contained in the channel of our elementary transistor as:

$$dQ = C_{ox} W dx (V_{GS} - V - V_{TH}) \tag{2.157}$$

We note that the velocity of the charge carriers is given by:

$$\mu_n \frac{dV}{dx} = \frac{dx}{dt} \tag{2.158}$$

which yields:

$$(dx)^2 = \mu_n dV dt \tag{2.159}$$

Introducing (2.159) into (2.157) we have:

$$(dQ)^2 = W^2 {C_{ox}}^2 \mu_n dt (V_{GS} - V - V_{TH})^2 dV \tag{2.160}$$

The above equation can also be rewritten as:

$$dQ = W^2 {C_{ox}}^2 \mu_n \frac{dt}{dQ} (V_{GS} - V - V_{TH})^2 dV \tag{2.161}$$

Noting that $dt/dQ = 1/I_{DS}$ we have:

$$dQ = \frac{W^2 {C_{ox}}^2 \mu_n}{I_{DS}} (V_{GS} - V - V_{TH})^2 dV \tag{2.162}$$

We can now integrate equation (2.162). The left side of the equation is integrated from zero to the total channel charge, Q_T. The right side is integrated from the source voltage to the pinch-off voltage given by $V_{GS} - V_{TH}$. In fact, points in the channel with a potential higher than the pinch-off voltage are not inverted and do not contribute to the total channel charge. We thus have:

$$\int_0^{Q_T} dQ = \frac{\mu_n {C_{ox}}^2 W^2}{I_{DS}} \int_0^{V_{GS}-V_{TH}} (V_{GS} - V - V_{TH})^2 dV \tag{2.163}$$

which yields:

$$Q_T = \frac{\mu_n {C_{ox}}^2 W^2}{I_{DS}} \frac{1}{3} (V_{GS} - V_{TH})^3 \tag{2.164}$$

We can replace I_{DS} in the above equation with its expression in saturation given by (2.138) to obtain:

$$Q_T = \frac{2}{3} C_{ox} W L (V_{GS} - V_{TH}) \tag{2.165}$$

By differentiating with respect to V_{GS} we finally have:

$$\frac{dQ_T}{dV_{GS}} = C_{gs} = \frac{2}{3}C_{ox}WL \tag{2.166}$$

Equation (2.166) shows that in strong inversion and in saturation, the channel capacitance fully appears between gate and source, with a value which is two-thirds of the total gate capacitance.

2.6.7 THE PMOS TRANSISTOR

We can now extend the results to the PMOS transistor. A front view of the device is shown in Figure 2.35. As we have already discussed, in a PMOS the substrate in n-type and the electrodes are p-type. To invert the substrate we need to repel electrons and attract holes in the channel region. The gate must therefore be more negative than the source by at least one threshold voltage. In this sense, we say that the threshold voltage of a PMOS device is negative. The junctions between the electrodes and the bulk must always be reverse biased. Hence the bulk voltage must always be more positive than (or at least equal to) the bias voltage applied to the source. To avoid dealing with negative quantities we can invert the sense of the measurement of the potential differences with respect to the NMOS. So we speak of the source-gate voltage, V_{SG} and of the source-drain voltage V_{SD}, while taking the absolute value of the threshold voltage. The source-drain current in the linear region can be expressed as:

$$I_{SD} = \mu_p C_{ox}\frac{W}{L}\left[(V_{SG} - |V_{THP}|)\,V_{SD} - \frac{V_{SD}{}^2}{2}\right] \tag{2.167}$$

The channel is pinched-off when the source-drain voltage is more positive than the source-gate voltage minus the absolute value of the threshold, i.e.:

$$V_{SD} \geq (V_{SG} - |V_{THP}|) \tag{2.168}$$

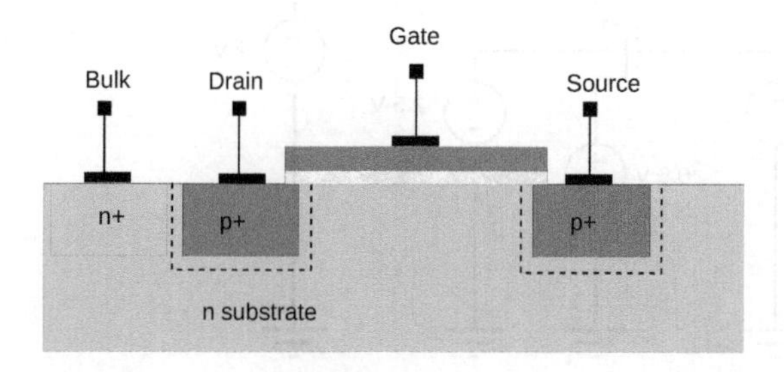

FIGURE 2.35 Cross section of a PMOS transistor.

The source-drain current in saturation, including the channel modulation effect, is:

$$I_{SD} = \frac{1}{2}\mu_p C_{ox}\frac{W}{L}(V_{SG} - |V_{THP}|)^2 (1 + \lambda V_{SD}) \tag{2.169}$$

For the capacitances, the same considerations made for the NMOS apply.

Example 2.7: Current calculation in a PMOS transistor.

Calculate the drain current for a PMOS transistor with $W = 20$ μm and $L = 1$ μm biased as shown in Figure 2.36. Use the following parameters: $\mu_p C_{ox} = 50$ μA/V^2, $V_{THP} = -0.5$ V, $\gamma = 0.5$ V$^{1/2}$, $\lambda_P = 0.05$ V^{-1}, $2\phi_F = 0.9$ V.

* * *

First, we note that the bulk and the source are not at the same potential. In a PMOS transistor, the source is the terminal at higher potential. To avoid the risk of forward biasing the junctions, the bulk potential must be at least equal to or higher than the source one. Due to the body effect the threshold voltage is increased and we can calculate it as:

$$|V_{THP}| = |V_{THO}| + \gamma\left(\sqrt{2|\phi_F| + V_{BS}} - \sqrt{2|\phi_F|}\right) \tag{2.170}$$

Putting in numbers, we have:

$$|V_{THP}| = 0.5 + 0.5 \times \left(\sqrt{1.4} - \sqrt{0.9}\right) = 0.62 \text{ V} \tag{2.171}$$

The source-gate voltage is 1 V. The source-drain saturation voltage is:

$$V_{SD,sat} = V_{SG} - V_{THP} = 1\ V - 0.62\ V = 0.38 \text{ V} \tag{2.172}$$

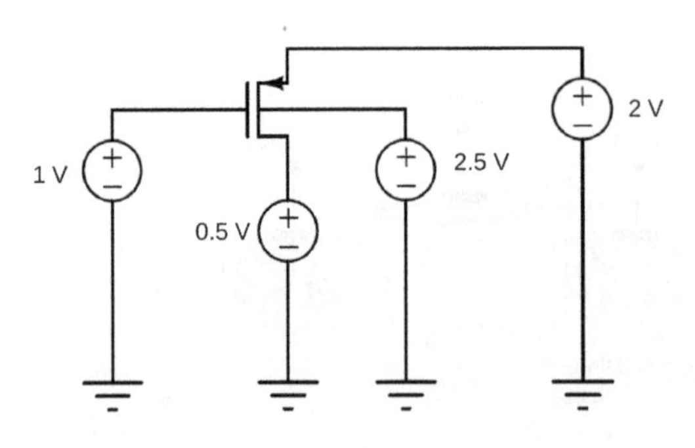

FIGURE 2.36 PMOS transistor biased with voltage sources.

The source-drain voltage is 1.5 V, so the device is in saturation. We need to use equation (2.169) and we obtain:

$$I_{SD} = \frac{1}{2} \times \frac{20}{1} \times 50 \times 10^{-6} \times 0.38^2 \times (1 + 0.05 \times 1.5) = 77.62\ \mu\text{A} \tag{2.173}$$

2.7 MOS SMALL SIGNAL PARAMETERS

The characteristics of the MOS transistor derived so far are non-linear. However, for the purpose of circuit analysis it is useful to have a model of the device that can be treated with the standard methods of linear networks. To do so, we recall that a function which is differentiable at a point x_0 can be approximated around that point with a Taylor's series. Here we are interested only to first order approximations, so we have:

$$f(x) = f(x_0) + \left(\frac{\partial f}{\partial x}\right)_{x=x_0} (x - x_0) \tag{2.174}$$

Consider now a variation of our function around x_0:

$$\Delta f = \left(\frac{\partial f}{\partial x}\right)_{x=x_0} \Delta x \tag{2.175}$$

Equation (2.175) establishes a linear relationship between *variations* that occur around an equilibrium point. Is it important to stress that the small signal parameters always relate *variations* of quantities in the vicinity of a point and not their values at that point. In case of a circuit, $f(x_0)$ can be interpreted as the value of a device voltage or current in a quiescent state, i.e. when no signal is applied. Equation (2.175) allows us to calculate the effect of a signal, provided that the value of $\partial f/\partial x$ in the point of interest is not noteably influenced by the signal itself. We start deriving the small signal parameters for the NMOS transistor and we then extend the results to the PMOS case.

2.7.1 GATE TRASCONDUCTANCE

A primary small signal parameter of the transistor is the transconductance, defined as the partial derivative of the drain-source current with respect to the gate-source voltage. The transconductance in hence the parameter that links *variations* of the device current I_{DS} to *variations* of the control voltage V_{GS}. Neglecting the channel length modulation, we have:

$$g_m = \frac{\partial I_{DS}}{\partial V_{GS}} = \mu_n C_{ox} \frac{W}{L} (V_{GS} - V_{TH}) \tag{2.176}$$

Using (2.138) we can express $(V_{GS} - V_{TH})$ as function of I_{DS}, i.e.:

$$V_{GS} - V_{TH} = \sqrt{\frac{2L}{\mu_n C_{ox} W} I_{DS}} \tag{2.177}$$

Inserting (2.177) in (2.176) we obtain an alternative formulation of the transconductance:

$$g_m = \sqrt{2\mu_n C_{ox} \frac{W}{L} I_{DS}} \tag{2.178}$$

The context will suggest which of the two expressions of g_m is more appropriate to use in a given circuit analysis. Since it is a ratio between a current and a voltage, the physical dimensions of g_m are Ω^{-1} or Siemens. Note also that the variation of the current is measured between drain and source, while the variation of the voltage is measured between gate and source. In other words, the terminals between which the cause (variation of V_{GS}) and the effect (variation of I_{DS}) are measured are not the same, hence the name *trans*conductance.

2.7.2 BULK TRANSCONDUCTANCE

The bulk-source voltage also modulates the drain-source current through the modulation of the threshold voltage (the so called back-gate effect). We can hence define the bulk transconductance in the following way:

$$g_{mb} = \frac{\partial I_{DS}}{\partial V_{BS}} = \frac{\partial I_{DS}}{\partial V_{TH}} \frac{\partial V_{TH}}{\partial V_{BS}} = -\frac{\partial I_{DS}}{\partial V_{TH}} \frac{\partial V_{TH}}{\partial V_{SB}} \tag{2.179}$$

Using the formulation of V_{TH} given by (2.151) we can write:

$$g_{mb} = \mu_n C_{ox} \frac{W}{L} (V_{GS} - V_{TH}) \frac{\gamma}{2\sqrt{2\phi_F + V_{SB}}} = \eta g_m \tag{2.180}$$

The bulk transconductance is hence proportional to g_m through the coefficient η that typically ranges from 0.2 to 0.3.

2.7.3 OUTPUT CONDUCTANCE

In saturation, the drain-source voltage can still influence the current through the channel length modulation. The output conductance of the transistor is defined as:

$$g_{ds} = \frac{\partial I_{DS}}{\partial V_{DS}} = \lambda \mu_n C_{ox} \frac{W}{L} (V_{GS} - V_{TH}) \tag{2.181}$$

Note that here both variations are measured between drain and source, so we speak of conductance and not of transconductance. The reciprocal of g_{ds} is the output resistance (usually indicated with r_0):

$$r_0 \approx \frac{1}{\lambda I_{DS}} \tag{2.182}$$

As we shall see, r_0 plays a fundamental role in defining the ultimate gain achievable in MOS amplifiers. The reciprocal of λ defines a voltage. In analogy with bipolar transistors, this voltage, changed in sign, is called the "Early voltage" (indicated here as V_A) and gives the point at which the extrapolation of the I_{DS} curve in saturation meets the V_{DS} axis, as sketched in Fig. 2.37. Note that the higher the slope of I_{DS}

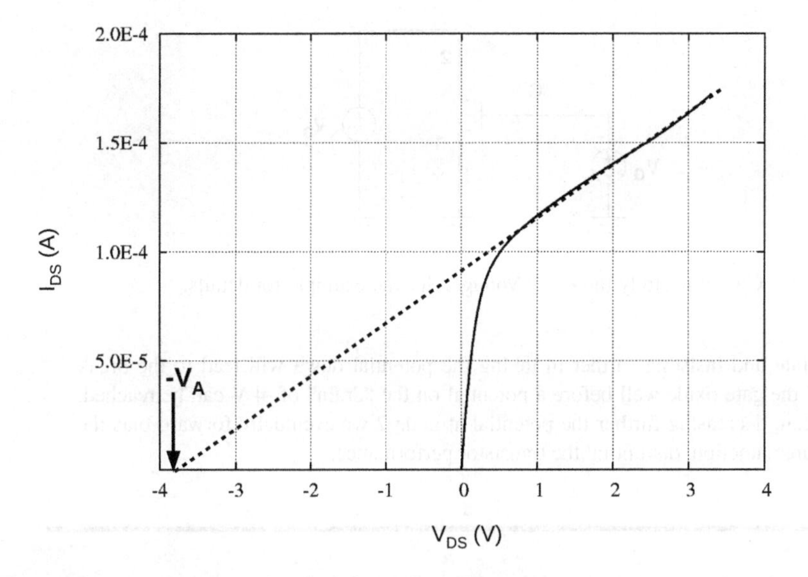

FIGURE 2.37 Illustration of the Early voltage for an NMOS transistor

versus V_{DS} is saturation, the smaller the Early voltage is. Conceptually, the Early voltage represents the V_{DS} voltage that would bring the transistor current to zero if channel length modulation was the *only* effect experienced by the device when the drain-source voltage is changed. This is clearly an abstraction, as shown in the following example.

Example 2.8: Reaching the Early Voltage.

Consider an NMOS transistor implemented in a 0.35 μm process that has a maximum power supply of 3.3 V. The device is biased with a gate voltage of 3 V and an initial drain voltage of 3 V. Explain what happens when the drain voltage is moved towards the Early voltage of the device, which is assumed to be -4 V.

* * *

We refer to the circuit of Fig. 2.38 where the source and drain have been labeled as "1" and "2", respectively and we indicate their potential difference as V_{21}. In the initial condition, node 1 is at zero and acts as the source, node 2 is at 3 V and acts as the drain, the device is on and in saturation, with the current flowing from node 2 towards node 1. We now start lowering the drain voltage. In doing so, we reduce V_{21}, driving eventually the device in the linear region. The current stops flowing when node 2 reaches ground, i.e. when $V_{21} = 0$. If now make V_{21} negative, node 2 starts acting as the source, and the current flows from node 1 towards node 2. When the potential on node 2 is more negative than -0.3 V, we exceed the maximum gate-source voltage at which the device

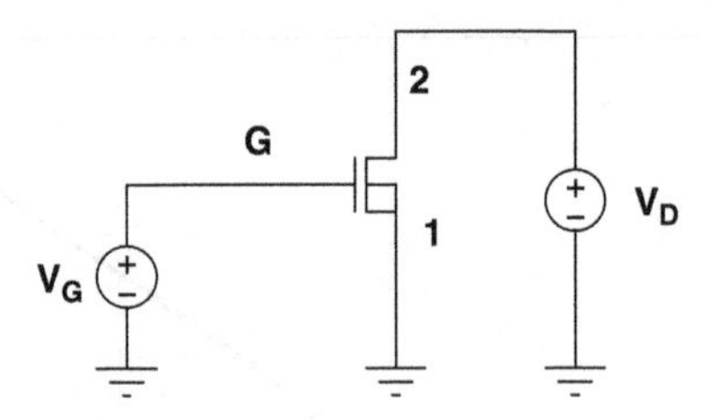

FIGURE 2.38 Circuit to study the Early Voltage. See the example for details.

can operate and insisting further in taking the potential down will lead to the break down of the gate oxide well before a potential on the "drain" of -4 V can be reached. In addition, decreasing further the potential at node 2 we eventually forward bias the bulk-source junction, disrupting the transistor performance.

We have already seen that the channel length modulation is more important for short channel devices. However, for the *same* device, g_{ds} depends also on the drain-source voltage at which the transistor operates. The detail study of the problem is very complex and the interested reader can find more details in [3] and [12]. To get a basic understanding of the issue, let's consider again equation (2.140) which is repeated here for convenience:

$$I_{DS} = \frac{1}{2}\mu_n C_{ox}\frac{W}{L-\Delta L}\left(V_{GS}-V_{TH}\right)^2 \tag{2.183}$$

We now write the output conductance putting in evidence the fact that ΔL changes with V_{DS}:

$$g_{ds} = \frac{\partial I_{DS}}{\partial V_{DS}} = \frac{\partial I_{DS}}{\partial \Delta L}\frac{\partial \Delta L}{\partial V_{DS}} \tag{2.184}$$

Using (2.183) we have:

$$g_{ds} = \frac{1}{2}\mu_n C_{ox}\frac{W}{(L-\Delta L)^2}\left(V_{GS}-V_{TH}\right)^2\frac{\partial \Delta L}{\partial V_{DS}} \tag{2.185}$$

We can now write:

$$\frac{1}{(L-\Delta L)^2} = \frac{1}{L^2\left(1-\frac{\Delta L}{L}\right)^2} \approx \frac{1}{L^2} \tag{2.186}$$

where we have used the fact that $\Delta L \ll L$. Note that this assumption might not be true for very short channel devices. We can then write the output conductance as:

$$g_{ds} = \frac{1}{2}\mu_n C_{ox}\frac{W}{L}\left(V_{GS}-V_{TH}\right)^2\frac{1}{L}\frac{\partial \Delta L}{\partial V_{DS}} = I_{DS0}\frac{1}{L}\frac{\partial \Delta L}{\partial V_{DS}} \tag{2.187}$$

where I_{DS0} is the drain-source current calculated without taking into account the channel length modulation. Note now that the ratio I_{DS0}/g_{ds} defines a voltage:

$$V_A = \frac{I_{DS0}}{g_{ds}} = L\left(\frac{\partial \Delta L}{\partial V_{DS}}\right)^{-1} \tag{2.188}$$

The reciprocal of the derivative of ΔL with respect to V_{DS} defines the Early voltage per unit of channel length. In the traditional approach, the dependence of ΔL on V_{DS} considers only the horizontal field due to V_{DS}. Unfortunately, the calculation of V_A for modern devices is much more complex, because vertical fields due to the gate and the drain have an impact on the channel length modulation as well. An expression that can be used is the following ([12], pp. 134-135):

$$V_A = L\frac{V_E}{\lambda_C L_C}\left(1+\frac{V_{DS}-V_{DSsat}}{V_E}\right) \tag{2.189}$$

where λ_C is a fit parameter, V_E is determined experimentally and L_C is a parameter typical of any given process and depends on the oxide thickness, t_{ox} and the junction depth, x_j: $L_C \approx \sqrt{3t_{ox}x_j}$ [3], [12]. The key point to take from (2.189) is that for the *same* device the Early voltage and hence the output conductance or, equivalently, the output resistance, are not constant but depend on the drain-source voltage and can be significantly degraded as the device approaches the linear region. Hence, biasing a transistor deeper in saturation (higher V_{DS}) results in greater output resistance, at the expense of the voltage headroom. Fig. 2.39 shows a simulation of the output conductance as a function of the drain-source voltage. Typical parameters of a 0.35 μm process have been assumed, the device length has been chosen to 0.7 μm, (twice the minimum allowed by the process) and the width has been fixed to 10 μm. The transistors needs a saturation voltage of 0.3 V, but note that when V_{DS} is only 100 mV above V_{DSsat} the output conductance rises sharply, degrading the device output resistance. A V_{DS} above 0.8 V is necessary to have an output resistance greater that 40 kΩ.

2.7.4 MOS SMALL SIGNAL MODEL.

Fig. 2.40 shows the equivalent circuit of the NMOS transistor at low frequency. The gate is left floating, to indicate that the resistance looking into the gate terminal is very high. SiO_2 is in fact an excellent insulator and the DC resistance between the gate and the other terminals may exceed 10^{12} Ω. The effect of the transconductance, g_m, is modeled with a voltage controlled current source. Current flows from drain to source and is controlled by the gate-source voltage. The current increases if V_{GS} increases and viceversa. The bulk transconductance, g_{mb} is modeled with a voltage controlled current source connected in parallel to the one representing g_m. The drain-source current increases if the bulk-source voltage increases. In practice, the bulk voltage will in general be constant and the source will move. If the variation of v_{bs} is negative (i.e. the source becomes more positive that the bulk) the current is reduced, and viceversa. Figure 2.41 shows the high-frequency equivalent model of the NMOS transistor including also the main capacitances discussed in 2.6.6. For a

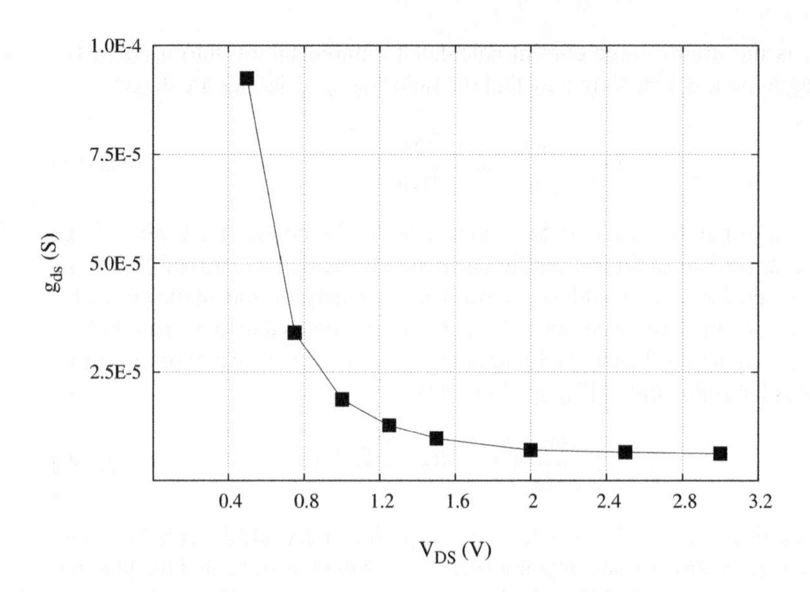

FIGURE 2.39 Effect of V_{DS} on the output conductance.

PMOS transistor, the small signal parameters are defined in the same way as for the NMOS and can be calculated starting from (2.169). Figure 2.42 reports the small signal equivalent PMOS model in saturation.

2.8 WEAK AND MODERATE INVERSION

The characteristics of MOS transistors derived in (2.28) and (2.137) predict that the current drops abruptly to zero as V_{GS} goes below the threshold. Such an approximation is not true and a more detailed analysis reveals that below threshold the drain-source current has an exponential dependence on the gate voltage. Called "weak inversion", this region has become increasingly important in the design of front-end electronics and analog circuits in general. Fig. 2.43 shows a more realistic trend of the drain-source current as a function of V_{DS} on a semi-log scale. Note that the device still drives a significant amount of current when the gate-source voltage is below V_{TH}.

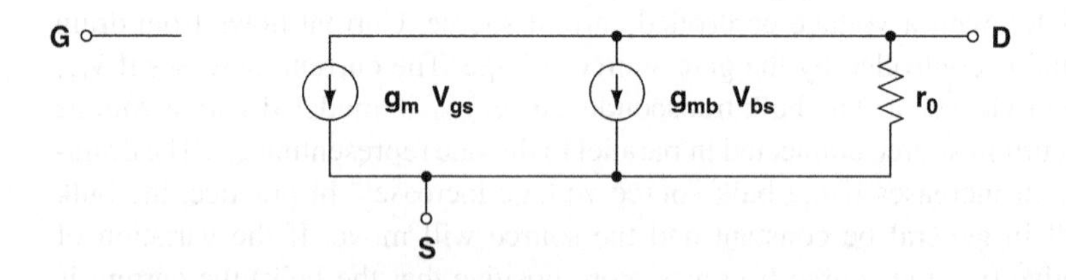

FIGURE 2.40 Linear small signal model of an NMOS transistor at low frequency.

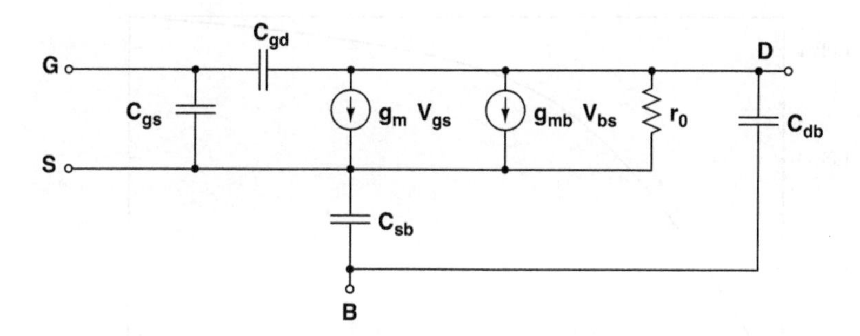

FIGURE 2.41 Linear small signal model of an NMOS transistor including the key capacitances limiting the device performance at high frequency.

The derivation of the MOS characteristics for weak inversion is more elaborate than the one for strong inversion. In fact, the simple assumption that the charge in the channel is given by $Q = C_{ox}WL(V_{GS} - V_{TH})$ does not hold anymore. An adequate derivation of I_{DS} in weak inversion can be found for instance, in [3]. Here we give a brief summary of the problem. As a first step, we must find the mobile charge in the channel. This can be obtained by solving the Poisson equation:

$$\frac{d^2\psi_s}{d^2y} = -\frac{\rho}{\varepsilon_{Si}} \tag{2.190}$$

where $\psi_s(y)$ is the surface potential and ρ is the charge density, given by:

$$\rho = -n_0 e^{\frac{\psi_s(y)}{\phi_T}} + p_0 e^{-\frac{\psi_s(y)}{\phi_T}} - N_A \tag{2.191}$$

Beyond depletion, the term due to the holes ($+p_0 e^{-\frac{\psi_s(y)}{\phi_T}}$) can be neglected. In this case, it can be proven that the solution of (2.190) yields the following result:

$$Q_{ch} = -\sqrt{2\varepsilon_{Si} q N_A}\sqrt{\psi_s + \phi_T e^{\frac{(\psi_s - 2\phi_F)}{\phi_T}}} \tag{2.192}$$

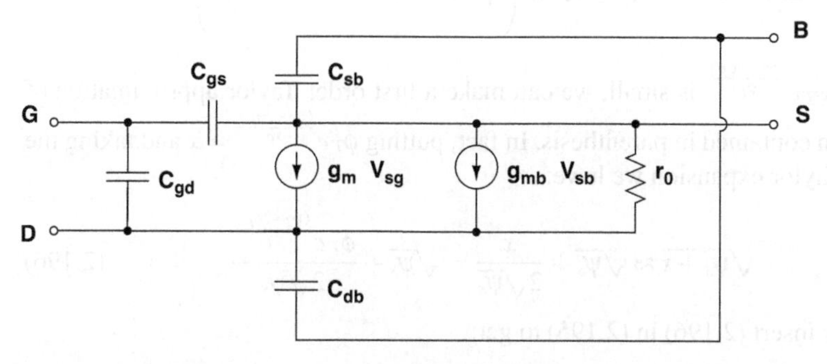

FIGURE 2.42 Linear small signal model of a PMOS transistor including the key capacitances limiting the device performance at high frequency.

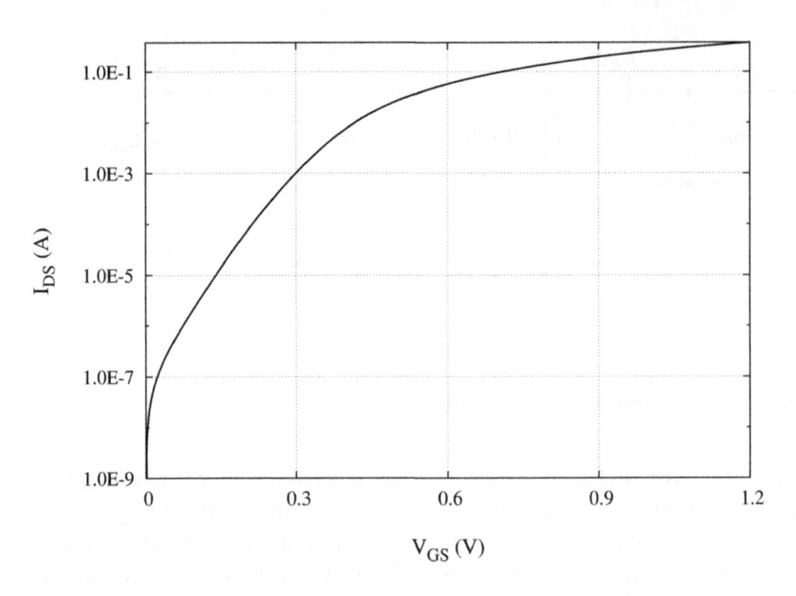

FIGURE 2.43 Drain-source current versus gate-source voltage on a semi-log scale.

where Q_{ch} is the *charge density per unit area* in the considered point in the channel. The total charge balance is given by the sum of the mobile charge Q_M due to electrons and the fixed charge Q_d due to ionized acceptor atoms. Therefore we have:

$$Q_{ch} = Q_M + Q_d \tag{2.193}$$

The charge density due to the depletion layer can be written as:

$$Q_d = -\sqrt{2\varepsilon_{Si} q N_A \psi_s} \tag{2.194}$$

where the minus sign indicates that the charge due to acceptor ions is negative. We insert now (2.192) and (2.194) in (2.193) to calculate Q_M, which yields:

$$Q_M = Q_{ch} - Q_d = -\sqrt{2\varepsilon_{Si} q N_A}\left(\sqrt{\psi_s + \phi_T e^{\frac{(\psi_s - 2\phi_F)}{\phi_T}}} - \sqrt{\psi_s}\right) \tag{2.195}$$

If the term $\phi_T e^{\frac{(\psi_s - 2\phi_F)}{\phi_T}}$ is small, we can make a first order Taylor approximation of the first term contained in parenthesis. In fact, putting $\phi_T e^{\frac{(\psi_s - 2\phi_F)}{\phi_T}} = x$ and taking the first order Taylor expansion we have:

$$\sqrt{\psi_s + x} \approx \sqrt{\psi_s} + \frac{x}{2\sqrt{\psi_s}} = \sqrt{\psi_s} + \frac{\phi_T e^{\frac{(\psi_s - 2\phi_F)}{\phi_T}}}{2\sqrt{\psi_s}} \tag{2.196}$$

We can now insert (2.196) in (2.195) to get:

$$Q_M = -\frac{\sqrt{2\varepsilon_{Si} q N_A}}{2\sqrt{\psi_s}}\phi_T e^{\frac{\psi_s - 2\phi_F}{\phi_T}} \tag{2.197}$$

As a further step, we can introduce the dependency on the channel bulk voltage, V_{CB}, replacing $2\phi_F$ with $2\phi_F + V_{CB}$ in equation (2.197):

$$Q_M = -\frac{\sqrt{2\varepsilon_{Si} q N_A}}{2\sqrt{\psi_s}} \phi_T e^{\frac{\psi_S - 2\phi_F - V_{CB}}{\phi_T}} = -\frac{\sqrt{2\varepsilon_{Si} q N_A}}{2\sqrt{\psi_s}} \phi_T e^{\frac{\psi_S - 2\phi_F}{\phi_T}} e^{-\frac{V_{CB}}{\phi_T}} \tag{2.198}$$

The electrons that cross the potential barrier at the source found themselves in a space where very few other electrons are present, thus they start diffusing towards the drain. Diffusion is hence the dominant mechanism supporting the flow of current in weak inversion. Assuming a linear gradient from source to drain, we can write the diffusion current as follows:

$$I_{DS} = -\mu_n \phi_T W \left(\frac{Q_S - Q_D}{L} \right) = -\mu_n \phi_T \frac{W}{L} Q_S \left(1 - \frac{Q_D}{Q_S} \right) \tag{2.199}$$

where Q_D and Q_S are respectively the charge per unit area at the source and drain and the Einstein relationship has been used to express the diffusion coefficient. Starting from (2.198) we have:

$$Q_S = -\frac{\sqrt{2\varepsilon_{Si} q N_A}}{2\sqrt{\psi_s}} \phi_T e^{\frac{\psi_S - 2\phi_F}{\phi_T}} e^{-\frac{V_{SB}}{\phi_T}} \tag{2.200}$$

$$Q_D = -\frac{\sqrt{2\varepsilon_{Si} q N_A}}{2\sqrt{\psi_s}} \phi_T e^{\frac{\psi_S - 2\phi_F}{\phi_T}} e^{-\frac{V_{DB}}{\phi_T}} \tag{2.201}$$

Therefore, the ratio Q_D/Q_S is simply:

$$\frac{Q_D}{Q_S} = e^{-\frac{V_{DS}}{\phi_T}} \tag{2.202}$$

We now consider the terms in Q_S. The quantity before the exponential has a weak dependence of ψ_s, as it appears under the square root. We can then approximate ψ_s with $2\phi_F + V_{SB}$, which is the value it has at the top of the weak inversion region. Remember in fact, that, conventionally, the channel is considered inverted when the surface potential ψ_s is equal to $2\phi_F$ (or $2\phi_F + V_{SB}$ in case the source and the bulk are not at the same potential). The depletion layer charge can thus be rewritten as:

$$Q_d = \sqrt{2\varepsilon_{Si} q N_A (2\phi_F + V_{SB})} \tag{2.203}$$

Differentiating the above expression with respect to V_{SB} one obtains the depletion layer capacitance:

$$C_d = \frac{dQ_d}{dV_{SB}} = \frac{\sqrt{2\varepsilon_{Si} q N_A}}{2\sqrt{2\phi_F + V_{SB}}} \tag{2.204}$$

Let's now introduce the slope factor (the name will become clear shortly) defined as following:

$$n = 1 + \frac{C_d}{C_{ox}} \rightarrow \frac{C_d}{C_{ox}} = n - 1 \tag{2.205}$$

Consider first the term that is in front of the exponential in the expression of Q_S given in (2.200):

$$Q_{S0} = -\frac{\sqrt{2\varepsilon_{Si} q N_A}}{2\sqrt{\psi_s}} \phi_T \tag{2.206}$$

Since in (2.206) ψ_s appears under a square-root, Q_{S0} is a weak function of ψ_s, which can thus be approximated with the fixed value: $\psi_s = 2\phi_F + V_{SB}$:

$$Q_{S0} = -\frac{\sqrt{2\varepsilon_{Si} q N_A}}{2\sqrt{2\phi_F + V_{SB}}} \phi_T \tag{2.207}$$

In we now multiply and divide the above expression by C_{ox} and we use (2.204) and (2.205), we can write Q_{S0} as:

$$Q_{S0} = -C_{ox}\phi_T(n-1) \tag{2.208}$$

Inserting (2.208) and (2.201) into (2.199), we get:

$$I_{DS} = \mu_n C_{ox}(n-1)\frac{W}{L}\phi_T^2 e^{\frac{\psi_s - 2\phi_F - V_{SB}}{\phi_T}}\left(1 - e^{-\frac{V_{DS}}{\phi_T}}\right) \tag{2.209}$$

Note that the approximation $\psi_s = 2\phi_F + V_{SB}$ has not been made in the exponential because it is instead a strong function of ψ_s To evaluate the argument of the exponential, consider that in the source referenced model it is the overdrive voltage $V_{GS} - V_{TH}$ which determines the charge in a strong inverted channel. In weak inversion, the charge in the channel is not sufficient to shield the bulk charges and from the gate one sees two capacitors in series, the gate oxide capacitance C_{ox} and the depletion layer capacitance C_d. The voltage which is effective in attracting charges in the channel is hence given by the capacitor divider rule and we can make the following approximation:

$$\psi_s - 2\phi_F - V_{SB} \approx (V_{GS} - V_{TH}(V_{SB}))\frac{C_{ox}}{C_d + C_{ox}} = \frac{(V_{GS} - V_{TH}(V_{SB}))}{n} \tag{2.210}$$

Combining all the above, the drain-source current in weak inversion can be written as:

$$I_{DS} = \mu_n C_{ox}(n-1)\frac{W}{L}{\phi_T}^2 e^{\frac{V_{GS} - V_{TH}}{n\phi_T}}\left(1 - e^{-\frac{V_{DS}}{\phi_T}}\right) \tag{2.211}$$

where for simplicity the explicit dependence of V_{TH} on V_{SB} has been omitted. Note also that the expression describing I_{DS} in weak inversion is completely different from the one describing it in strong inversion. This illustrates the difficulty of deriving equations that account for the MOS behavior across all its possible range of operation only from first principles. In fact, to build models that adequately take into account the complexity of real transistors, measurements are done and equations based on physics are fined tuned in order to reproduce the experimental data. The parameters in the equations (μC_{ox}, V_{TH}, n) are then extracted from the measurements and compiled into files employed in computer simulations. The advantage of having models based as much as possible on device physics with a limited number of parameters and particularly suited for analog design, drove the development of the EKV model [13]. The EKV model is born as a bulk referenced model. When referenced to the source, its equations describing respectively weak, strong and moderate inversion in absence of channel length modulation appear as follows [12]:

$$I_{DS} = 2n\mu C_{ox}\frac{W}{L}{\phi_T}^2 e^{\frac{V_{GS}-V_{TH}}{n\phi_T}} \tag{2.212}$$

$$I_{DS} = \frac{1}{2}\mu C_{ox}\frac{W}{nL}\left(V_{GS}-V_{TH}\right)^2 \tag{2.213}$$

$$I_{DS} = 2n\mu C_{ox}\frac{W}{L}{\phi_T}^2\left[\ln\left(1+e^{\frac{V_{GS}-V_{TH}}{2n\phi_T}}\right)\right]^2 \tag{2.214}$$

The above equations do not consider small geometry effects which will be illustrated in 2.9. We note that in equation (2.212) the parameter $2n$ replaces $n-1$ of (2.211). The difference can be compensated by an adjustment of the value of V_{TH}. Since V_{TH} appears in the exponential, very small tuning is sufficient. The technicalities associated to such procedures are discussed in [3]. The current in strong inversion found in (2.138) is divided by n in (2.213). In the derivation of I_{DS} with the gradual channel approximation, we have in fact neglected that as we move from the source to the drain, the rising of the channel potential increases also the source voltage of the elementary transistors in which we have segmented the main device. Therefore, as we move along the channel, our elementary units experience a progressively higher body effect, which increases the local value of the threshold voltage. This implies that equation (2.138) overestimates the current, while (2.213) gives a more realistic prediction ([12], [14]). Equation (2.214) is an interpolation between the two physically based equations describing strong and weak inversion. In fact, for $V_{GS} \ll V_{TH}$, the exponential is small and the approximation $\ln(1+x) \approx x$ can be used, which, inserted in (2.214), gives back (2.212). For $V_{GS} > V_{TH}$, the exponential becomes quickly dominant, and neglecting "1" in the argument of the logarithm yields (2.213).

Equation (2.211) and equation 2.212 can be used to derive two interesting properties of weak inversion operation. First, the influence of V_{DS} on I_{DS} become negligible if V_{DS} is above $4\phi_T$, which means that a voltage 0.1 V and independent of $V_{GS}-V_{TH}$ is sufficient to keep the transistor in saturation, favoring operation with low power supply voltages. Second, but more important, the transconductance in weak inversion is directly proportional to the device current. In fact, using (2.212) to calculate g_m, we obtain:

$$g_m = \frac{\partial I_{DS}}{\partial V_{DS}} = \frac{I_{DS}}{n\phi_T} \tag{2.215}$$

The ratio g_m/I_{DS} defines the efficiency of the device in using the bias current to generate transconductance. For an MOS transistor, this is constant in weak inversion and inversely proportional to $\sqrt{I_{DS}}$ in strong inversion. But how can we predict the region of operation of a given device? To do so, we introduce a boundary current between strong and weak inversion defined as the current for which the value of g_m calculated with (2.212) is equal to the one found employing (2.213). We therefore have:

$$\frac{I_{DS,wi-si}}{n\phi_T} = \sqrt{2\mu C_{ox}\frac{W}{nL}I_{DS,wi-si}} \rightarrow I_{DS,wi-si} = 2n\mu C_{ox}\frac{W}{L}{\phi_T}^2 \tag{2.216}$$

Based on the above relationship, we can introduce two important parameters: the technology boundary current and the inversion coefficient. The technology boundary current is the boundary current between strong and weak inversion divided by the aspect ratio of the device:

$$I_{DS,wi-si0} = 2n\mu C_{ox}{\phi_T}^2 \tag{2.217}$$

Note that the above quantity depends only on fundamental constants, on n and μC_{ox}, so it is a characteristics property of any given process. The inversion coefficient is defined as the ratio between the actual bias current at which the device operates and its boundary current, $I_{DS,wi-si}$:

$$I_C = \frac{I_{DS}}{2n\mu C_{ox}\frac{W}{L}{\phi_T}^2} \tag{2.218}$$

A device can be considered in weak inversion if $I_C < 0.1$ and in strong inversion if $I_C > 10$. The region where $0.1 < I_C < 10$, centered at $I_C = 1$, defines an intermediate situation, called moderate inversion, for which equation (2.214) should be used. Observe also that the boundary current between weak and strong inversion can be obtained from (2.212) by putting $V_{GS} = V_{TH}$ and neglecting the dependence on V_{DS}. If the transistor had only the strong inversion region, increasing W/L would provide a means of reaching arbitrarily high transconductance for a fixed bias current. However, equation (2.216) tells us that in any technology, for a given bias current there is a W/L ratio that drives the device in weak inversion. When this region is reached, the transconductance is "pinned" at the value $I_{DS}/n\phi_T$ and increasing the aspect ratio further leads only to more parasitic capacitance without any extra increase in g_m. In general, driving a device towards weak inversion for a given bias current implies increasing its aspect ratio. This results in a better power efficiency (more g_m for the same current) and less bandwidth, because the device parasitic capacitances are also increased. An inversion coefficient less than 0.1 or greater than 10 is attained only if very small or very large bias currents are used. As a consequence, the I_C is often found between 0.1 and 10 and the moderate inversion region is extensively used in the design of low power front-end circuits. In moderate inversion, the charge in the channel and thus the gate capacitance are not so straightforward to calculate as in strong inversion.

Once the inversion coefficient is introduced, the relevant MOS small signal parameters can be written as functions of I_C, in a formulation which is valid in all regions of inversion [12]. With this approach, the expression of the transconductance becomes:

$$g_m = \frac{I_D}{n\phi_T} \frac{1}{\sqrt{I_C + 0.5\sqrt{I_C} + 1}} \tag{2.219}$$

Another useful quantity is the inversion factor γ, which is given by:

$$\gamma = \frac{1}{2} + \frac{1}{6}\frac{I_C}{I_C + 1} \tag{2.220}$$

Note that γ is related to I_C, but not coincident with it. Its main use is in quantifying the effect of the degree of channel inversion on the thermal noise generated in the device channel. This aspect will be treated in more detail in chapter 5. The total gate capacitance can be expressed as:

$$C_g = C_{gs} + C_{gb} \tag{2.221}$$

where C_{gs} is the gate-source capacitance, C_{gb} is the gate-bulk capacitance. The sum of C_{gs} and C_{gb} can be written in the following form [12]:

$$C_g = C(x)C_{ox}WL \tag{2.222}$$

where the coefficient $C(x)$ is defined as:

$$C(x) = \frac{n - (1+x)/3}{n} \tag{2.223}$$

The variable x is in turn related to the inversion coefficient by the following expression:

$$x = \frac{\left(\sqrt{I_C + 0.25} + 0.5\right) + 1}{\left(\sqrt{I_C + 0.25} + 0.5\right)^2} \tag{2.224}$$

Consider now the decimal logarithm of the current in weak inversion:

$$\log I_{DS} = \log K + \log e^{\frac{V_{GS} - V_{TH}}{n\phi_T}} = \log K + \frac{V_{GS} - V_{TH}}{n\phi_T} \frac{1}{\ln 10} \approx 0.434 \frac{V_{GS} - V_{TH}}{n\phi_T} \tag{2.225}$$

If we take the partial derivative of the above equation with respect to V_{GS} we get:

$$\frac{\partial \log I_{DS}}{\partial V_{GS}} = 0.434 \frac{1}{n\phi_T} \tag{2.226}$$

This quantity is called the weak inversion slope. If we take $n = 1.3$, we have that $0.434 \cdot (1/n\phi_T)$ is about 12.5. This means that a change of 80 mV in V_{GS} determines a change of 1 in $\log I_{DS}$, implying (depending on the sign of the variation) a reduction/increase of the current by a factor of ten. For instance, if the current in a given device when $V_{GS} = V_{TH}$ is 10 μA, V_{GS} must go 240 mV below V_{TH} to reduce the device current to 10 nA. The reciprocal of the weak inversion slope is the weak inversion swing:

$$S = 2.3n\phi_T \tag{2.227}$$

and it is measured in mV/decade. A reduction of the weak inversion slope (hence an increase in the weak inversion swing) implies that more voltage is needed to change the transistor current in weak inversion. In other words, if the weak inversion slope is degraded, the transistor turns off more gradually and the off-state current (i.e. the residual current for $V_{GS} = 0$) is bigger. We must observe that even for the ideal case where n=1 at least 60 mV of V_{GS} are needed to change the device current by one order of magnitude. This prevents the use of too small threshold voltages, since the DC power consumption due to subthreshold current in digital gates will become prohibitive.

2.9 THE DEEP SUBMICRON MOS TRANSISTOR

In the paper "Cramming more components onto integrated circuits" [15], published in 1965 Gordon Moore, co-founder of Intel, observed that the number of components on integrated circuits had doubled every year and predicted that this trend would continue in the future, leading to integrated circuits with "as many as 65000 transistors in 1975". In practice, the transistor count per integrated circuit has doubled approximately every 18 months in a trend which has become known as "Moore's law". Today, state of the art Field Programmable Gate Arrays (FPGAs), produced with 28 nm CMOS technology, incorporate almost 7 billion transistors. This impressive evolution has been possible thanks to the reduction of the feature size in MOS devices. In principle, scaling is the process by which a device with given dimensions is replaced by a smaller one, described by the same characteristics. Unfortunately, the procedure is not that simple and shrinking the transistor affects significantly also its electrical behavior. Device scaling is optimized to have smaller and faster digital gates with reduced cost and power consumption per function and its impact on analog circuits is not always beneficial. Scaling is a fascinating and complex subject. In the following, we just cover a few key aspects to provide a general background. This will serve as a basis to discuss in later chapters the implications of using scaled CMOS processes to implement circuit topologies often recurring in front-end designs.

2.9.1 SCALING METHODS

The ideal scaling approach is the *constant field scaling*: all the lateral and vertical dimensions are shrunk by the scaling factor κ, with $\kappa > 1$. At the same time, all the operating voltages, including the threshold, are reduced by the same amount to keep constant the electric fields inside the device. The first interesting consequence of scaling is the *increase* of the gate capacitance per unit area, C_{ox}, due to the reduction of the gate oxide thickness by the scaling factor κ. Therefore, $C_{ox} \rightarrow \kappa C_{ox}$. In section 2.8 we have seen that the limit between strong and weak inversion is given by:

$$I_{DS,wi-si} = 2n\mu C_{ox}{\phi_T}^2 \frac{W}{L} \tag{2.228}$$

Hence, increasing the gate capacitance per unit area by κ, increases the boundary between weak and strong inversion by the same factor. Therefore, for a fixed bias current, a transistor of a given aspect ratio will operate closer to weak inversion when implemented in a scaled technology. The scaling of the geometrical dimensions apply also to the depletion regions associated to the junctions. In particular, it is important to reduce the depletion region at the drain junction to avoid excessive channel length modulation. The depletion region width can be estimated as:

$$d \approx \sqrt{\frac{2\varepsilon_{Si}}{qN_B} V_R} \tag{2.229}$$

where N_B is the doping of the bulk and V_R is the reverse bias voltage applied to the drain-bulk junction. If we increase the doping by κ and reduce the voltage by κ we have:

$$d' \approx \sqrt{\frac{2\varepsilon_{Si}}{q\kappa N_B}\frac{V_R}{\kappa}} = \frac{d}{\kappa} \tag{2.230}$$

In strong inversion, the current driven by the scaled device is:

$$I_{DS}' = \frac{1}{2}\mu\kappa C_{ox}\frac{\frac{W}{\kappa}}{\frac{L}{\kappa}}\left(\frac{V_{GS}}{\kappa} - \frac{V_{TH}}{\kappa}\right)^2 = \frac{I_{DS}}{\kappa} \tag{2.231}$$

so the current drive capability is reduced by the scaling factor. The transconductance in strong inversion does not scale, as one can see from the following equation:

$$g_m' = \kappa\mu C_{ox}\frac{\frac{W}{\kappa}}{\frac{L}{\kappa}}\frac{(V_{GS} - V_{TH})}{\kappa} = g_m \tag{2.232}$$

In fact, the reduction in the overdrive voltage (or, equivalently, in the bias current) is compensated by the increase in the density of the gate capacitance. Let's now apply the constant field scaling rules while moving from a 350 to a 65 nm process. The geometrical scaling factor is 5.4. Therefore, the power supply voltage should be scaled from 3.3 V (typical of a 350 nm node) to 0.6 V and the threshold voltage from 0.7 V to 0.13 V. Unfortunately, the subthreshold slope is defined by very fundamental physical quantities and *does not scale* with the shrinking of the feature size. With a threshold of 130 mV the current with $V_{GS} = 0$, when the transistor should be theoretically fully off, it is still 2% of the value it has for V_{GS}=V_{TH}. A device driving 10 μA at $V_{GS} = V_{TH}$ would still sink 200 nA when $V_{GS} = 0$. The off-leakage current, which defines the static power consumption of digital gates, would become obviously intolerable (7 billion transistors leaking 200 nA each imply 1.4 kA of static current per chip!). The constant field scaling has its fundamental limit in the fact the subthreshold swing prevents excessive reduction of the threshold voltage. The opposite scaling method would be the *constant voltage scaling*: the physical dimensions are shrunk by the scaling factor, the threshold and power supply voltages are kept constant and the electric fields inside the device are increased by κ. Unfortunately, also this method does not work, because it would lead to breakdown in the gate oxide and in the junctions. Therefore, a hybrid scaling has been in practice applied. The power supply and threshold voltages have been reduced from one technology node to the next, but less than what is required to keep constant the electric fields, which thereby increase at each technology generation. Furthermore, different properties can be scaled by different factors to optimize as much as possible the device performance (*selective scaling*).

2.9.2 MOBILITY REDUCTION

Deep in the silicon bulk, the carrier mobility is limited by scattering against phonons and ionized impurity atoms. The vertical electric field, necessary to support the inversion layer, attracts also the carriers towards the surface. At the Si-SiO_2 interface, charge trapped within gate oxide, interface traps and surface roughness provide additional scattering mechanisms that reduce the carriers speed. These effects are more pronounced if the overdrive voltage $V_{GS} - V_{TH}$ is increased, as this results in higher vertical fields. The phenomenon can be modeled by replacing the bulk mobility with an effective mobility given by:

$$\mu_{eff} = \frac{\mu_0}{1 + \theta\,(V_{GS} - V_{TH}) + \theta_B V_{SB}} \tag{2.233}$$

where θ and θ_B are fitting parameters determined by measurements. The value of θ is inversely proportional to the oxide thickness ($\theta = \alpha/t_{ox}$) because thinner oxides results in higher electrical fields for the *same* overdrive voltage. A typical value for α is 1-2 nm/V. Assuming α=1 nm/V and $t_{ox} = 2.2$ nm, (2.233) predicts that with $V_{GS} - V_{TH} = 0.9$ V, the mobility is reduced by 30%. The term $\theta_B V_{SB}$ is introduced for the following reason. Increasing V_{SB} increases the threshold voltage and the term $\theta\,(V_{GS} - V_{TH})$ predicts an increase of the mobility if V_{TH} is higher. However, increasing V_{SB} makes the bulk more negative than the channel, and this contributes in pushing the carriers further towards the surface, exacerbating rather than curing the mobility reduction due to the vertical field.

2.9.3 VELOCITY SATURATION

Velocity saturation is a phenomenon due to high electric field *parallel* to the current flow. In deriving the characteristics of the long channel MOS transistor, it was assumed that the drift speed of the carriers is always proportional to the accelerating field, $v_d = \mu\mathscr{E}$. For higher electric fields, however, it is shown that the carrier velocity reaches a plateau, i.e. saturates at a maximum value v_{dmax}. The description of the phenomenon in term of fundamental physics is very complex, but it can be studied with a simple mathematical model, obtained by changing the expression of v_d as follows:

$$v_d = \frac{\mu\mathscr{E}}{1 + \frac{\mathscr{E}}{\mathscr{E}_C}} \tag{2.234}$$

In the above equation, $\mathscr{E}_C$ defines a critical field, for which the carrier speed is one half of what predicted without taking velocity saturation into account. The effect can be visualized as a reduction in the mobility with the lateral electric field:

$$\mu_{eff} = \frac{\mu}{1+\frac{\mathscr{E}}{\mathscr{E}_C}} \tag{2.235}$$

which again leads to a degradation of the current driving capability of the device. The study of the velocity saturation effect takes a particularly simple form if one assumes that the longitudinal field is so high that the phenomenon occurs already when the transistor operates in the linear region. In this case we can write:

$$I_{DS} = \frac{\mu}{1+\frac{V_{DS}}{L\mathscr{E}_C}} C_{ox} \frac{W}{L} \left[(V_{GS} - V_{TH}) V_{DS} - \frac{V_{DS}{}^2}{2} \right] \tag{2.236}$$

where the lateral field has been calculated as V_{DS}/L. If we now suppose that $V_{DS} \gg L\mathscr{E}_C$ and we neglect the $V_{DS}{}^2/2$ term in (2.236), which is legitimate where we are deep in linear region, we obtain:

$$I_{DS} \approx \mu \mathscr{E}_C C_{ox} W (V_{GS} - V_{TH}) = v_{sat} C_{ox} W (V_{GS} - V_{TH}) \tag{2.237}$$

In the above equation we have put $v_{sat} = \mu \mathscr{E}_C$. Note the following interesting effects which occurs in the velocity saturation regime:

- The current drive capability depends only on W and not anymore on the W/L ratio.
- The distinction between the linear and saturation region vanishes.
- The current increases linearly and not quadratically with the overdrive voltage.
- The transconductance becomes $g_m = v_{sat} C_{ox} W$, so it mainly depends on the channel width.

The effect of vertical and lateral fields on mobility can be combined in a single formula for the drain current [4]:

$$I_{DS} = \frac{1}{2} \mu_0 C_{ox} \frac{W}{L} \frac{(V_{GS} - V_{TH})^2}{1 + \left(\frac{\mu_0}{2 v_{sat} L} + \theta \right) (V_{GS} - V_{TH})} \tag{2.238}$$

Finally, we must observe that the departures of the transistor characteristics from the pure quadratic form can introduce odd-harmonics in the drain current.

2.9.4 DRAIN INDUCED BARRIER LOWERING

The depletion region associated to the drain-bulk junction increases with V_{DS} and in short channel devices it may extend significantly underneath the gate and occupy a large portion of the channel. Since the channel is already partially depleted by the drain action, less effort must be made by the gate to support the depletion layer. Therefore, for the *same* gate-source voltage, more electrons can be attracted in the channel.

This effect manifests itself as a reduction of the effective threshold voltage with the drain bias, hence the name drain-induced barrier lowering (DIBL). The net result is an extra modulation of the current with the drain voltage and a degradation of the output resistance in saturation. On the basis of this description, the threshold reduction due to DIBL should be more pronounced as the gate length is shrunk, while long channel transistors should be almost immune and have a significantly better output resistance. In modern CMOS technologies, however, a *rise* of the threshold voltage at shorter channel length is observed, a phenomenon called Reverse Short Channel Effect (RSCE). Long channel transistors, on the other hand, still displays a significant variation of the current with the drain-source voltage. Known as Drain Induced Threshold Shift (DITS), this effect is caused by a modulation of the threshold by V_{DS} and results in a output resistance worse than the one expected from the traditional channel length modulation and DIBL mechanisms described above. Both RSCE and DITS are attributed to the presence of halo implants, which are introduced to increase the channel doping in proximity of the source and drain junctions, with the purpose of keeping DIBL under control for short channel devices. However, as the gate length is reduced, the halo implants tend to occupy a significant portion of the channel, increasing the transistor threshold, which explains the reverse short channel effect. The impact of the halo doping on long channel devices is less obvious and it has been attributed to the difference existing between the source and drain barriers [16, 17]. A detailed explanation of the phenomenon is beyond our scope and can be found in [17]. The important message to take here is that in a deep submicron technology, increasing the channel length does not necessarily lead to very high output impedance.

2.9.5 HOT CARRIER EFFECTS

Charge carriers accelerated along the channel may acquire enough energy to cause further ionization. The effect is more pronounced for electrons which have higher mobility, hence the synonym "hot electrons effect", which is often used. The primary effect of hot electrons is to create additional electron-hole pairs close to the drain. The electrons are swept to the drain and the holes are collected by the bulk, originating a drain-bulk current, I_{DB}. A few electrons may acquire enough energy to overcome the potential barrier of the oxide. Most electrons that penetrate in the oxide are collected by the gate, giving rise to a gate current. Others might either originate interface state or be trapped in the oxide, modifying in the long term the threshold of the transistor. To minimize the impact of hot electrons, lightly doped source-drain extensions are used. This results in a wider depletion region and hence in a reduced field at the junction. Furthermore, the depletion region protrudes significantly also in the source-drain extension, reducing also the Drain Induced Barrier Lowering. To be effective, these structure must be shallow, which increases their resistance.

2.9.6 GATE LEAKAGE CURRENT

Due to the reduction of the gate oxide, electrons can tunnel through the oxide, originating a gate current. If excessive, this current can compromise the insulation between the gate and the channel, on which the operation of the MOS transistor is based. Gate leakage becomes a concern when the oxide thickness is below 1.5 nm. For this reason, technologies starting from the 45 nm node and below, employ as the gate insulator new materials with higher dielectric constant, Hafnium dioxide being one example. Owing to the higher dielectric constant, the control of the gate over the channel can be maintained with thicker layers, thus keeping tunneling under control. At the time of writing, such ultra-deep submicron processes are not yet widely used in the design of front-end electronics from radiation sensors. In 0.13 μm and 65 nm processes, gate leakage must be watched when many transistors are put in parallel.

REFERENCES

1. S. M. Sze and K. K. Ng. *Physics of Semiconductor Devices*. Wiley, 2007.
2. S. Dimitrijev. *Understanding Semiconductor Devices*. Oxford Series in Electrical and Computer Engineering, 2000.
3. Y. Tsividis. *Operation and Modeling of the MOS Transistor*. McGraw Hill International Editions, 1999.
4. B. Razavi. *Design of CMOS Analog Integrated Circuits*. McGraw Hill International Editions, 2000.
5. Ph. E. Allen and D. R. Holberg. *CMOS Analog Circuit Design*. Oxford University Press, 2002.
6. A. B. Sproul and M. A. Green. Experimental measurement of the intrinsic carrier concentration of silicon. In *Photovoltaic Specialists, IEEE Conference*, 1990.
7. A. B. Sproul and M. A. Green. Intrinsic carrier concentration and minority-carrier mobility of silicon from 77 to 300 K. *Journal of Applied Physics*, 73:1214–1225, 1993.
8. J. A. Ballin et al. Monolithic Active Pixel Sensors (MAPS) in a quadruple well technology for nearly 100% fill factor and full CMOS pixels. *Sensors*, 8:5336–5351, 2008.
9. I. Peric and C. Takacs. Large monolithic particle pixel-detector in high-voltage CMOS technology. *Nucl. Instr. Meth.*, A624:504–508, 2010.
10. Y. Arai et al. Development of SOI pixel process technology. *Nucl. Instr. Meth.*, A636, 2011.
11. W. Snoeys. Monolithic pixel detectors for high energy physics. *Nucl. Instr. Meth.*, A731:125–130, 2013.
12. D. Binkley. *Tradeoffs and Optimization in Analog CMOS Design*. Oxford University Press, 2008.
13. C. C. Enz and E. A. Vittoz. *Charge-Based MOS Transistor Modeling: The EKV Model for Low-Power and RF IC Design*. Wiley, 2006.
14. K. R. Laker and W. M. C. Sansen. *Design of Analog Integrated Circuits and Systems*. McGraw-Hill, 1994.
15. G. E. Moore. Cramming more components onto integrated circuits. *Electronics*, 38:114, 1965.
16. S. Mundanai et al. Analytical modeling of output conductance in long-channel halo-doped MOSFETs. *IEEE Trans. on Electron Devices*, 53:2091–2097, 2006.

17. A. S. Roy, S. P. Mudanai, and M. Stettler. Mechanism of long-channel drain-induced barrier lowering in halo MOSFETs. *IEEE Trans. on Electron Devices*, 58:979–984, 2011.

3 Input Stages

As described in chapter 1, most radiation sensors can be modeled as a current source with a capacitor in parallel. The input stage of the front-end must read the sensor current, which is in general weak, transforming it into a signal strong enough to drive the following circuitry. The first element of the processing chain takes therefore the form of a transimpedance or a current amplifier. For reasons that will become clear in the remainder of the book, the former configuration is often preferred, but the latter in also employed in particular cases. A transimpedance amplifier can be realized by connecting an appropriate network in the feedback path of a high gain voltage amplifier. After a review of basic concepts on the transimpedance topology, in the first part of the chapter we build with a step by step approach the most common architectures which are used as core amplifiers in front-end input stages. We then consider the simpler case in which the feedback network is formed only by a resistor. Finally, in the last section we introduce current-mode amplifiers. In this chapter, we focus on the circuit low-frequency behavior, while the effects of reactive components are discussed in chapter 4.

3.1 TRANSIMPEDANCE AMPLIFIERS

Consider the circuit of Fig. 3.1 a), in which the impedance Z_f is placed in parallel to a voltage amplifier with gain $A \gg 1$ and infinite input impedance. Using the small signal equivalent circuit of Fig. 3.1 b), we can write the following equations:

$$\begin{cases} I_{in} + \frac{V_{in} - V_{out}}{Z_f} = 0 \\ V_{in} = -\frac{V_{out}}{A} \end{cases} \tag{3.1}$$

Solving for V_{out}/I_{in} we find:

$$\frac{V_{out}}{I_{in}} = \frac{A}{1+A} Z_f \tag{3.2}$$

The circuit thus converts an input current signal to a voltage output and, if A is much greater than one, the transimpedance gain is equal to Z_f. It is interesting to calculate the input and output impedance of the circuit. To do so, we can put a test voltage source between the node of interest and ground and we measure the current absorbed by the circuit. The ratio between the imposed voltage and the measured current defines the input impedance. Alternatively, we can force a current into the node and sense the developed voltage. Fig. 3.2 a) shows the circuit that allows us to calculate the input impedance with the first method. Since we are driving the input node of the amplifier with an ideal test voltage signal of amplitude V_t, the output is $-AV_t$ and the current

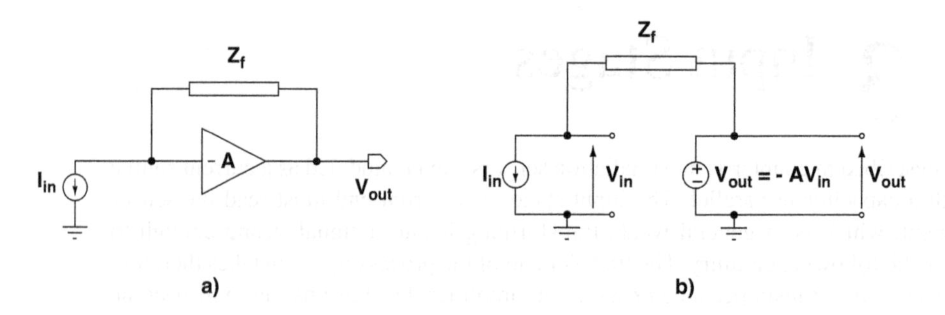

FIGURE 3.1 Ideal transimpedance amplifier and its small signal equivalent circuit.

flowing in Z_f is given by:

$$I_t = \frac{V_t - (-AV_t)}{Z_f} \rightarrow \frac{V_t}{I_t} = Z_{in} = \frac{Z_f}{1+A} \tag{3.3}$$

If A is sufficiently big, the input impedance of the transimpedance amplifier can be made very small and thus the circuit is suitable to read current input signals. We have assumed so far that the core amplifier output impedance is much smaller than Z_f. Fig. 3.2 b) shows a circuit that can be used to calculate the output impedance of the transimpedance amplifier in case the core amplifier has a finite output impedance given by Z_0. The test voltage source is connected to the output, while the input is left floating. Owing to the infinite input impedance of the core amplifier, no current can flow in Z_f , therefore $V_{in} = V_t$ and all the current is sunk by the amplifier output port. In formula, we can write:

$$I_t = \frac{V_t - (-AV_t)}{Z_0} \rightarrow \frac{V_t}{I_t} = \frac{Z_0}{1+A} \tag{3.4}$$

From (3.4) we see that the open-loop output impedance of the core amplifier is dropped by a factor $1+A$ when the feedback element is connected. This usually leads to very low output impedance, which makes the circuit suitable to also drive small loads, confirming that the output port of the circuit behaves as a good voltage source.

3.1.1 THE TRANSIMPEDANCE AMPLIFIER AS A FEEDBACK CIRCUIT

It is interesting to examine the above results with the concepts of feedback theory. To do so, let's first recall a few general notions about negative feedback. From a formal point of view, a feedback amplifier, shown in Fig. 3.3, needs three fundamental building blocks:

- A main amplifier with a gain as high as possible.
- A feedback network that senses the output signal and takes a portion of it back towards the input.

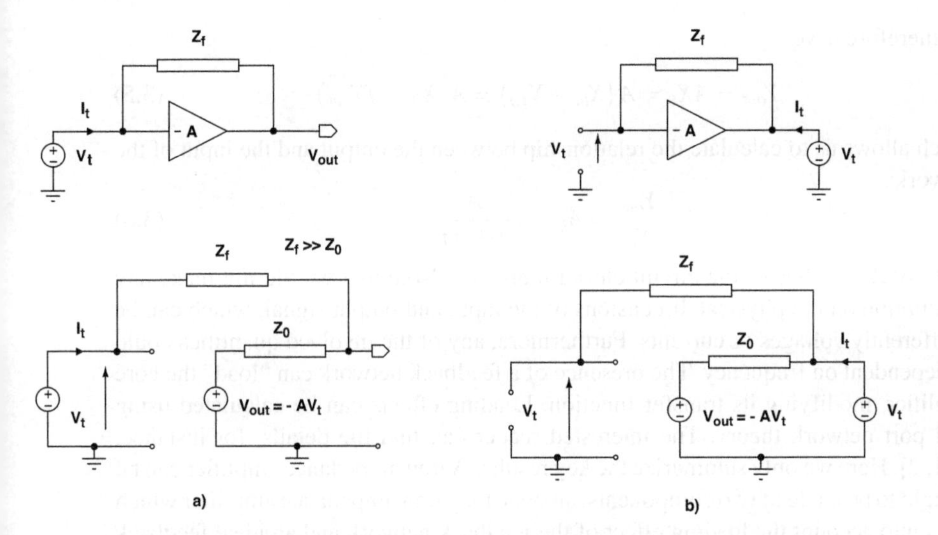

FIGURE 3.2 Circuits to calculate the input and output impedance of a transimpedance amplifier. In the calculation of the input impedance it is supposed that $Z_f \gg Z_0$, so the latter can be neglected.

- A subtracting node, that calculates the difference between the input and the feedback signal, generating an *error signal*. The error signal is then used to drive the main amplifier.

We now put all this in the form of a few equations. Let A be the open loop gain of the main amplifier, i.e. the gain the amplifier has when no feedback is applied. If the input signal X_{in} drives directly the amplifier input, the output becomes $Y_{out} = AX_{in}$. In practical cases, owing to its high gain, the main amplifier saturates if it is used in open-loop. In a feedback system instead, the output of the amplifier is sensed by the feedback network and a portion of it, fY_{out}, is brought back to the subtracting node, which generates the error signal defined as $X_\varepsilon = X_{in} - fY_{out}$. It is this signal which actually drives the main amplifier, so that the condition $Y_{out} = AX_\varepsilon$ must also hold.

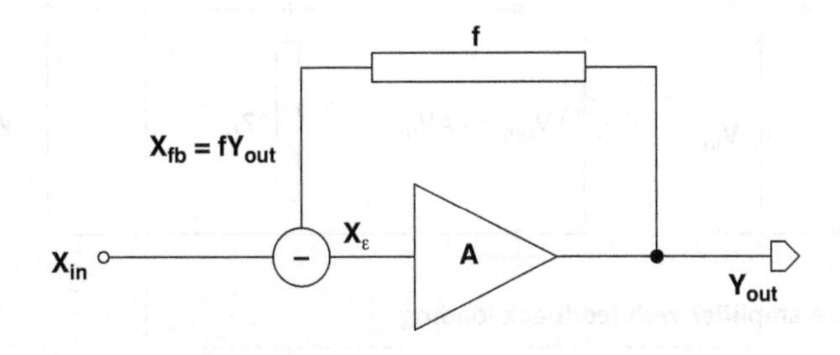

FIGURE 3.3 Generic amplifier with feedback.

We therefore have:

$$Y_{out} = AX_{\varepsilon} = A\left(X_{in} - X_{fb}\right) = A\left(X_{in} - fY_{out}\right) \tag{3.5}$$

which allows us to calculate the relationship between the output and the input of the network:

$$\frac{Y_{out}}{X_{in}} = A_{CL} = \frac{A}{1+Af} \tag{3.6}$$

In (3.6) A_{CL} indicates the circuit closed loop gain. Note that we did not make any assumption on the physical dimensions of the input and output signal, which can be indifferently voltages or currents. Furthermore, any of the involved quantities could be dependent on frequency. The presence of a feedback network can "load" the core amplifier, modifying its transfer function. Loading effects can be calculated using dual port network theory. The interested reader can find the details, for instance, in [1, 2]. Here we only summarize the key results. A transimpedance amplifier can be thought to be made of two components: an open loop transimpedance amplifier which takes into account the loading effect of the feedback network and an ideal feedback network which does not present any load. As shown in Fig. 3.4, the effect of loading is taken into account by connecting an impedance of value Z_f in parallel to both the input and output terminal of the core amplifier. The circuit can also be represented as shown in Fig. 3.5, where the effect of the ideal feedback network is modeled by connecting in parallel with the input port a current source controlled by the output voltage V_{out}, generating a current fV_{out} which is subtracted to the input signal. The open loop transfer function of the new core amplifier can thus be written as:

$$\frac{V_{out}}{I_{in}} = AZ_f = A_T \tag{3.7}$$

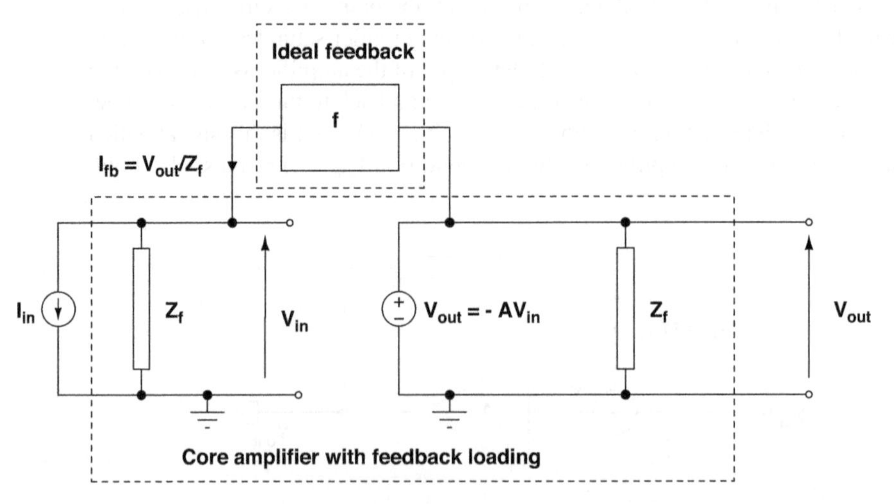

FIGURE 3.4 Circuit including the loading effects of the feedback network on the input and output port of a transimpedance amplifier.

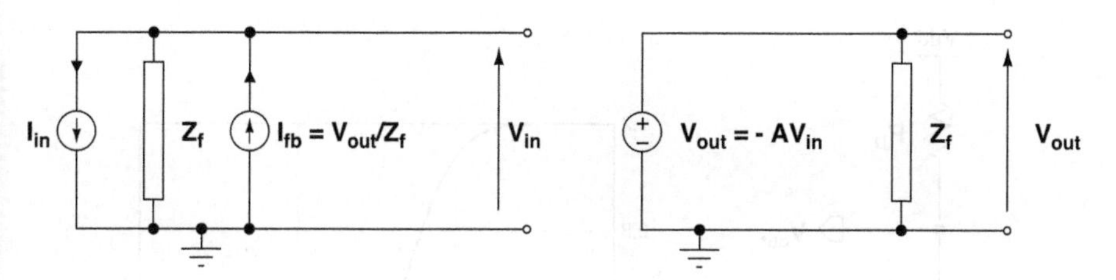

FIGURE 3.5 Alternative model of a transimpedance amplifier including the loading of the feedback network.

Observe that A_T has the units of an impedance. The error signal I_ε is given by:

$$I_\varepsilon = I_{in} - fV_{out} = I_{in} - \frac{V_{out}}{Z_f} \tag{3.8}$$

The output voltage is $V_{out} = A_T I_\varepsilon$:

$$V_{out} = A_T I_\varepsilon = AZ_f \left(I_{in} - \frac{V_{out}}{Z_f} \right) \tag{3.9}$$

Solving the above equation for V_{out}/I_{in} we finally get back the closed loop transfer function already found in (3.2). Two important facts must be remembered about feedback amplifiers. First, if $Af \gg 1$, the closed loop gain can be approximated by $1/f$ and it becomes defined only by the parameters of feedback network. Since $f \leq 1$, the feedback network is a sort of attenuator and can be built with fairly simple circuits. The system closed loop gain therefore depends on a limited number of critical components. Suppose now that the open loop gain of the core amplifier changes by an amount ΔA. The closed loop gain becomes:

$$A_{CL} = \frac{A + \Delta A}{1 + f(A + \Delta A)} \approx \frac{A}{1 + Af} + \frac{\Delta A}{1 + Af} \tag{3.10}$$

from which it is seen that the effect of the gain variation ΔA is approximately reduced by $1 + Af$. It is therefore not necessary that the open loop gain of the core amplifier is precisely controlled, but it is sufficient that it is adequately large. Our purpose in the next sections is to design a voltage amplifier with high gain, suitable to be used as the core amplifier in a transimpedance stage.

3.2 COMMON SOURCE AMPLIFIERS

The MOS transistor can be employed to convert a voltage variation applied between the gate and source terminals into a variation of the drain-source current. The DC input impedance seen looking into the gate is very high. If the device works in saturation, also the impedance measured between the drain and the source is fairly big. A single transistor can then be regarded as an elementary transconductance amplifier. To turn a transconductance amplifier into a voltage amplifier we need, as a first step, to sense

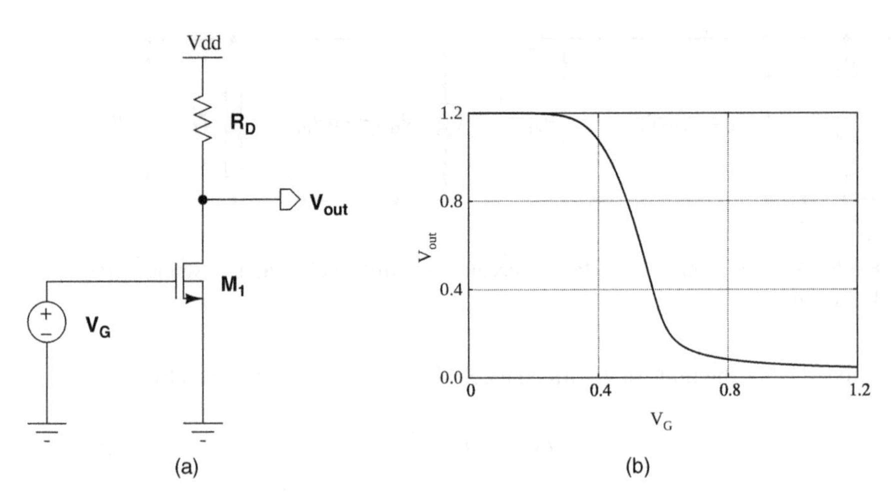

FIGURE 3.6 a) Schematic of an NMOS input common source amplifier. b) DC input-output characteristics.

the current variation at the output with a load that converts it back to a voltage. The simplest type of load is a passive resistor, so we start analyzing the common source amplifier with resistive load, which is depicted in Fig. 3.6 a).

3.2.1 COMMON SOURCE AMPLIFIER WITH RESISTIVE LOAD

We consider for the time being the case of an NMOS input transistor and we suppose that the circuit is powered from a single rail power supply. The source of the transistor is tied to a steady potential that can serve as the small signal ground for both the input and output ports (hence the name common source). The signal is applied to the gate. The load resistor is connected between the positive power supply, V_{DD}, and the device drain, where the output voltage is sensed with respect to ground. To understand the behavior of the circuit, we first make a sweep of the transistor DC gate voltage V_G and we study the response of the output. In the language of circuit simulators, this means performing a DC sweep. Observe that in the circuit of Fig. 3.6 a), V_G coincides with the gate-source voltage V_{GS} of M_1 because the source is tied to ground. When the gate voltage is below the transistor threshold, the device is off and no current flows in R_D. Therefore, there is no voltage drop across it and the output is at V_{DD}. As the gate voltage approaches the threshold, M_1 starts conducting. The current flowing in R_D determines the lowering of the output voltage. Note that when the gate voltage is close to the threshold, the transistor drives only a small current, the voltage drop across R_D is small and the output is still near V_{DD}. The output voltage, measured with respect to ground, coincides with the V_{DS} of M_1. At the turn-on point, the transistor has a small $V_{GS} - V_{TH}$ and a big V_{DS}, so the condition $V_{DS} > V_{GS} - V_{TH}$ is verified and the device is in saturation. As the gate voltage is raised, M_1 drives more and more current and the output voltage decreases. Eventually, the output voltage becomes so

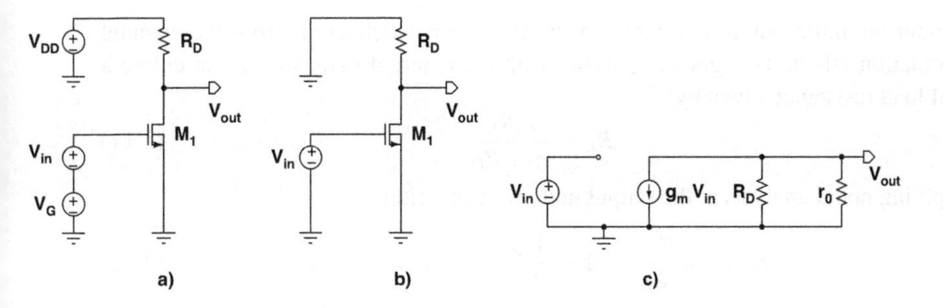

FIGURE 3.7 Derivation of the small signal equivalent circuit of the common source amplifier.

low that the saturation condition is not fulfilled anymore and the transistor is driven in the linear region. At this point, the current is controlled by both V_{GS} and V_{DS}. Fig. 3.6 b) shows the output as a function of the gate-source voltage, which is swept from ground to V_{DD}. We see that there are two regions in which the curve is almost flat, separated by a region of high slope. It is in this region, in which the output voltage is more sensitive to the variation of the input, that the circuit can be used as an amplifier. Note than an increase of the input voltage determines a decrease of the output one, so the amplifier is *inverting*.

If the amplifier loses its sensitivity to the input signal it is said to be saturated. In CMOS architectures, the *saturation* of the amplifier happens when one of its transistors either goes in the *linear* region or is cut-off. Therefore, in computer simulations it is always important to monitor the DC operating points of the individual transistors and make sure that they have enough headroom to handle the expected signal amplitude. The maximum output voltage swing that the circuit can accommodate without saturating is the output dynamic range. This, divided by the gain, yields the *input dynamic range*. Observe that if the slope of the curve in Fig. 3.6 b) becomes steeper, the circuit gain increases while the input dynamic range becomes smaller.

When using an amplifier, we therefore need to properly set DC currents and voltages to guarantee optimal performance with the expected input signal. We will consider later practical techniques for amplifier biasing. For the moment, we control the bias point and feed the signal with ideal voltage sources, as sketched in Fig 3.7 a). Here, the power supply and the source providing the DC bias voltage of the gate are shown explicitly. We can now apply the techniques of small signal analysis to calculate the amplifier gain. First, we have to remember that the small signal equivalent circuit considers only *variations* of quantities inside a circuit. Hence, as a first step, we replace the constant voltage sources with short circuits and the constant current sources with open circuits. This is shown in Fig. 3.7 b), where both the power supply and the voltage source biasing the gate have been replaced with shorts. Note that the circuit shown in Fig. 3.7 b) is only valid for small signal calculation purposes, because if R_D is physically connected to ground no DC current can flow in M_1, turning the device off. Finally, the small signal equivalent circuit of M_1 is introduced, leading to the network shown in Fig. 3.7 c). In the rightmost drawing, we have also taken into account the finite output resistance of the transistor which stems from the channel modulation effect. This goes in parallel with the external load, so we can define a total load resistance given by:

$$R_L = \frac{r_0 R_D}{r_0 + R_D} \tag{3.11}$$

Applying nodal analysis at the output node we can write:

$$g_{m1} V_{in} + \frac{V_{out}}{R_L} = 0 \rightarrow \frac{V_{out}}{V_{in}} = -g_{m1} R_L = -A_v \tag{3.12}$$

The gain of a common source amplifier is thus given by the product between the transconductance of the input transistor and the load resistor, which is the result of the parallel combination of all resistors appearing between the output of the amplifier and the signal ground. As anticipated by the DC transfer curve of Fig. 3.6 b), the "minus" sign means that, the output signal swings in the opposite direction of the input one, hence the amplifier is inverting. We now illustrate these concepts with the following example.

Example 3.1: Analysis of a common source amplifier.

The common source amplifier of Fig. 3.6 is implemented with the following parameters: $(W/L)_1 = 20/1$, $V_{DD} = 2.5$ V, $R_D = 10$ kΩ. The threshold voltage of M_1 is 0.64 V and $\mu_n C_{ox}$ is 165 μA/V^2. Find the input bias voltage that guarantees a DC output voltage of 1.25 V and calculate the gain of the amplifier. Study the accuracy of the small signal model when the input is a negative square pulse of 1 mV and 50 mV amplitude, respectively. Neglect the channel length modulation effect.

* * *

To have the DC output voltage at 1.25 V, we need a voltage drop across R_D of 1.25 V, so the bias current must be 125 μA. The gate source voltage of M_1 necessary to support this current can be calculated as:

$$V_{GS1} = V_{TH1} + \sqrt{\frac{2 L_1 I_{DS1}}{\mu_n C_{ox} W_1}} \tag{3.13}$$

which, with our numbers, yields $V_{GS1} = 0.64 \text{ V} + 0.2752 \text{ V} = 0.9152 \text{ V}$. Note that it is necessary to retain an adequate number of significant digits in the calculations to make a fair comparison between the results provided by the small signal model and the ones obtained through the use of the large signal characteristics of the transistors. The transconductance of M_1 is given by:

$$g_{m1} = \mu_n C_{ox} \frac{W_1}{L_1} (V_{GS1} - V_{TH1}) = \sqrt{2 \mu_n C_{ox} \frac{W_1}{L_1} I_{DS1}} \tag{3.14}$$

which yields $g_{m1} \approx 910$ μS. The voltage gain is given by $9.1 \cdot 10^{-4}$ S $\cdot$ 10000 Ω $= 9.1$. Applying a negative square pulse of 1 mV means that we bring the gate voltage from the quiescent point of 0.9152 V down to 0.9142 V. The small signal model predicts that the output will rise by:

$$\Delta V_{out} = -A_v V_{in} = 9.1\text{mV} \tag{3.15}$$

Note that the "–" sign indicates that the two variations occur in opposite directions. Since we have decreased the input voltage, the output voltage increases. According to the small signal model, the output rises from 1.25 V to 1.2591 V when the signal is applied. Using the full characteristic of the transistor, with a gate voltage of 0.9142 V we have a drain-source current of 124.06 μA. The output voltage, given by $V_{DD} - I_{DS1}R_D$ will then go to 2.5 V – (124.06 μA $\cdot$ 10000 Ω) = 1.2594 V, revealing a reasonable agreement with the results obtained through the small signal equivalent circuit.

With a negative input step of 50 mV the small signal model predicts a positive output variation of 0.455 V. Hence, when the signal is applied the output voltage should swing up from 1.25 V to 1.705 V. The calculation using the full transistor characteristics yields a variation of 0.413 V, which is about 10% smaller. In fact, decreasing the gate voltage we decrease also the device current and its transconductance, thereby reducing the amplifier gain. The reader can verify that a positive input signal will result in a nonlinearity that increases the gain. The small signal equivalent circuit models the amplifier response with linear components, hence it can not predict distortions and eventually the saturation of the amplifier, that need to be studied using the full transistor characteristics. Using the inversion coefficient, it is interesting to evaluate if our initial assumption that the transistor works in strong inversion is justified. The inversion coefficient I_C is given by:

$$I_C = \frac{I_{DS}}{2n\mu_n C_{ox}\phi_T^2 \frac{W}{L}} = \frac{125 \cdot 10^{-6}}{2 \times 1.3 \times 165 \cdot 10^{-6} \times 6.7 \cdot 10^{-4} \times 20} = \frac{125 \cdot 10^{-6}}{5.75 \cdot 10^{-6}} = 21.7 \tag{3.16}$$

Since $I_C > 10$, the original hypothesis that the device is in strong inversion is appropriate. Finally, observe that the transistor parameters given in the example are typical of a 0.35 μm CMOS process with a gate oxide thickness of about 7 nm.

The gain of a common source amplifier with resistive load is modest and inadequate for most practical applications. A first intuitive idea to obtain a bigger gain is to increase the value of the load resistor. However, if we keep the bias current constant, we increase also the voltage drop across R_D, shifting the output DC level downwards and eventually pushing M_1 in the linear region. A good strategy to improve the gain must hence take into account that the DC output voltage should not move from the desired value. If we call $V(R_D)$ the voltage drop across R_D we can write [1]:

$$V(R_D) = I_{DS1}R_D \rightarrow R_D = \frac{V(R_D)}{I_{DS1}} \tag{3.17}$$

We can then express the gain as follows:

$$A_v = -g_{m1}R_D = -\sqrt{2\mu_n C_{ox}\frac{W_1}{L_1}I_{DS1}}R_D = -\sqrt{2\mu_n C_{ox}\frac{W_1}{L_1}}\frac{V(R_D)}{\sqrt{I_{DS1}}} \tag{3.18}$$

The above equation shows that if the output DC voltage is constrained to a fixed value, to increase the gain it is more advantageous to make the load resistor larger, reducing accordingly the bias current. Suppose, for instance, that we need to double the gain. If we multiply the resistor by four, we must also divide the bias current by four in order to keep the DC output voltage at 1.25 V. In doing so, we divide g_{m1} by two, since g_{m1} is proportional to the square root of the current, so the gain is doubled as desired. The reader can easily verify that, using the parameters of the above example, the gain is doubled if R_D is changed to 40 kΩ and I_{DS} is scaled down to 31.25 μA. This method of increasing the gain has however two limits:

- The reduction of the transconductance implies also a reduction of the bandwidth and an increase of the noise, so it might compromise other circuit specifications.
- The method is valid only if M_1 works in strong inversion.

In fact, reducing the bias current we push M_1 towards the moderate and weak inversion regions. When M_1 enters in weak inversion, its transconductance becomes proportional to the bias current, so reducing further the current does not increase anymore the gain. To avoid that M_1 goes out of strong inversion when we reduce the bias current, we should keep the inversion coefficient constant by reducing the transistor aspect ratio. However, using the general expression for g_m, we have:

$$g_m = \frac{I_{DS}}{n\phi_T} \frac{1}{\sqrt{I_C + 0.5\sqrt{I_C} + 1}} \tag{3.19}$$

From the above equation we see that if I_C is kept constant, the transconductance scales linearly with the current, preventing the increase of the gain.

Finally, a comment must be made on the choice of the output DC level. Setting it halfway between V_{DD} and ground, as we have done in our example, is a reasonable choice for a general purpose amplifier, for which the input signal can swing either above or below the input DC level. If the polarity of the input signal is known a priori, one can offset the output DC level to optimize the dynamic range. For instance, if the amplifier has only to process inputs with negative polarity, the output DC levels can be set just above the saturation point of M_1 to leave as much headroom as possible to the output signal, that will swing towards V_{DD}. If, on the other hand, only positive input signals are expected, it is advantageous to set the output DC level as high as possible, because the output signal will swing downwards. The polarity of a radiation sensor signal is known a priori, therefore the DC levels inside the amplifiers can be optimized to yield the best dynamic range.

To increase the gain of our amplifier without reducing its bias current, we would need a component which has no coupling between the current flowing through it and the voltage developed across its terminals by the rest of the circuit, a requirement which is fulfilled by an ideal current source. In this way, we can choose the bias current to optimize the transconductance of the input transistor for bandwidth and noise requirements, with little compromise on the circuit gain. The resulting configuration, the common source amplifier with current source load, is shown in Fig. 3.8 a). In this circuit, we set the bias current in the current source and we adapt the DC gate bias voltage V_{GS} so that M_1 can sink the chosen current. We will discuss later what

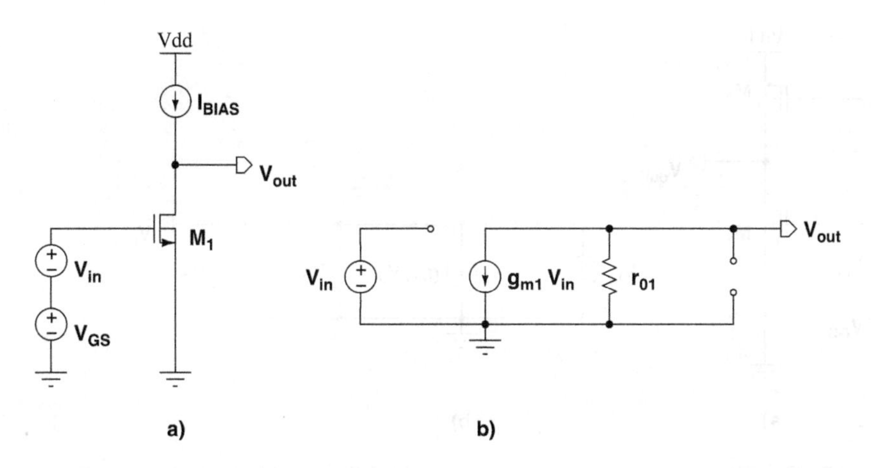

FIGURE 3.8 Common source amplifier with current source load (in a) and its equivalent circuit (in b).

happens when this regulation fails, leading to a mismatch between the current in the driving transistor and the one in the load. Fig. 3.8 b) shows the small signal equivalent circuit. Since the DC current source is ideal, it is replaced by an open circuit and only the output resistance of M_1 is reported in the scheme. The gain is hence given by $A_v = -g_{m1}r_{01}$. Called the *intrinsic transistor gain*, this is the highest voltage gain that can be provided by a single MOS device. The need of realizing with MOS transistors circuits that mimic a current source naturally leads us to the study of current mirrors.

3.3 CURRENT MIRRORS

The natural choice to implement a current source is to use a transistor working in saturation. The circuit is reported in Fig 3.9, which shows also its small signal model. Note that the gate of M_2 is biased by a constant potential and no signal is applied to it. Since its source voltage is also constant, $V_{sg2} = 0$ and the voltage controlled-current source representing the transconductance of M_2 is not active and thus does not enter in the small signal model. Therefore, M_2 contributes to the equivalent circuit only with its finite output resistance r_{02}, which goes in parallel to r_{01} because V_{DD} is a signal ground as well. We therefore expect that the gain of the amplifier is smaller than the intrinsic transistor gain.

Two key questions now arise:

1. How do we bias the gate of M_2 to get the desired current?
2. How can we guarantee that both M_1 and M_2 work in saturation?

To answer the first question, we consider the circuit of Fig. 3.10 in which the gate and drain of the device have been shorted. A time domain simulation to study the transition of the circuit between the off and the on state can be useful to understand its behavior. In circuit simulator jargon this is called a "Transient Analysis" and it employs the full device characteristics, so it takes into account also non-linearities and saturation

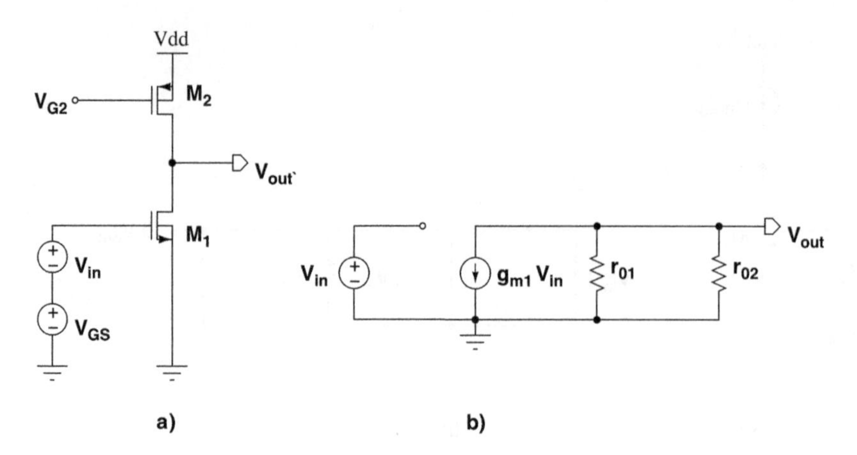

FIGURE 3.9 Common source amplifier with transistor load (in a) and its equivalent circuit (in b).

effects. At $t = 0$, the current source is off. Since the transistor can not drive any current, the condition $V_{SG} < |V_{THP}|$ must be verified and the gate voltage is close to V_{DD}. At a given time, the current source is switched on and it starts driving the nominal current. Due to its parasitic capacitance, the transistor can not react immediately. The current source sinks current from the node, discharging the transistor capacitance and pulling the node down. Because of the short, the gate is also pulled down and V_{SG} increases. When V_{SG} is greater than the threshold, the transistor starts to conduct, providing carriers to the current source. When V_{SG} is such that all the current requested by the current source arrives from V_{DD} through the transistor, an equilibrium point is reached and the voltage at the drain/gate node stays constant. Transistors in which the gate and drain are shorted are called "diode-connected" in analogy to the similar

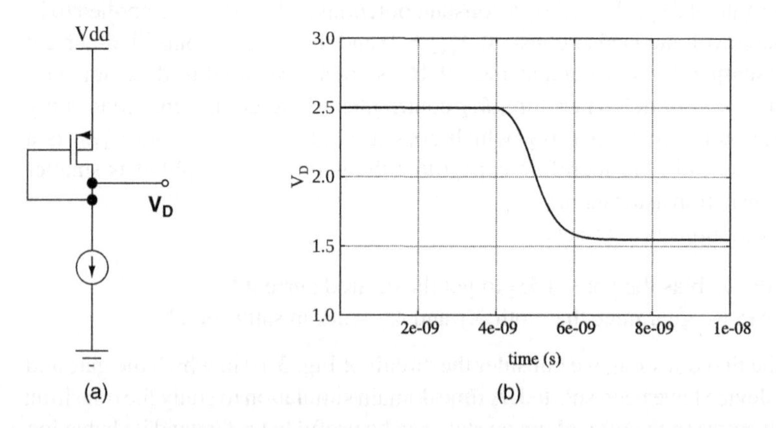

FIGURE 3.10 a) Diode connected PMOS transistor. b) Evolution of the gate voltage when the current source is turned-on.

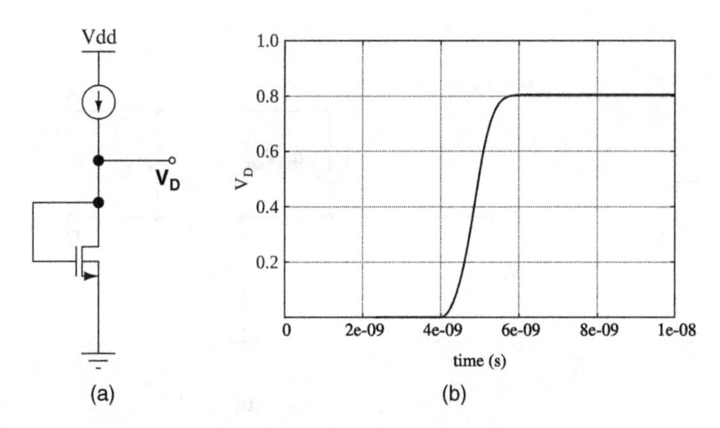

FIGURE 3.11 a) Diode connected NMOS transistor. b) Evolution of the gate voltage when the current source is turned-on.

configuration employed with bipolar devices. However, this is just an analogy since with MOS devices there is no forward biased diode involved in conduction. The same reasoning can be applied to the NMOS case, which is shown in Fig. 3.11, along with the switch-on transient simulation. We see here that V_{GS} settles to the value that allows all the current imposed by the current source to flow to ground. If biased in strong inversion, diode-connected transistors always work in saturation. In fact, for a PMOS transistor we have $V_{SD} = V_{SG} > V_{SG} - |V_{THP}|$ and for an NMOS one $V_{DS} = V_{GS} > V_{GS} - V_{TH}$. In other words, being equal to the gate-source voltage, the gate-drain voltage is always *greater* than the gate-source voltage *minus* the threshold voltage, which is the condition to guarantee the channel pinch-off. For a PMOS, the voltage V_{SG} can be calculated as follows:

$$V_{SG} = |V_{THP}| + \sqrt{\frac{2LI_{SD}}{\mu_p C_{ox} W}} \qquad (3.20)$$

which is the counterpart of (3.13) to be used for the NMOS. It is important to understand that with the configurations of Fig. 3.10 and 3.11 the only gate-source voltage that can be developed is the one which allows the flow of the current imposed by the source.

With the help of Fig. 3.12 a) we can now calculate the *small signal* impedance seen from the drain of diode-connected transistors. The DC current source, I_{BIAS}, sets the proper biasing point. A test signal is then forced by a generator connected between the two points of interest, which, in our case are the drain of the device and a small signal ground. In this example, the test signal is provided by the current source I_t and it determines a small *variation* to the overall transistor current. The variation of the drain voltage V_t is then measured and the ratio V_t/I_t gives the desired quantity. For

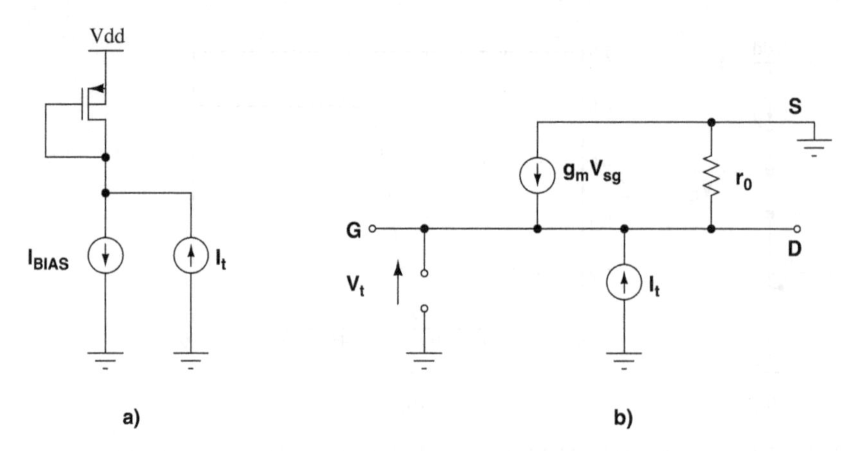

FIGURE 3.12 Circuits for calculating the small signal resistance of a diode-connected PMOS transistor.

the calculation we can use the equivalent circuit of Fig. 3.12 b)[1] . For the PMOS, we can write:

$$-I_t - g_m V_{sg} + \frac{V_t}{r_0} = 0 \tag{3.21}$$

Note that the gate-source voltage is $V_{sg} = V_s - V_g = 0 - V_g = -V_t$ because the source is connected to a constant voltage (so does not experience any variation) and the gate is shorted to the drain, where the voltage is measured. Therefore we have:

$$g_m V_t + \frac{V_t}{r_0} = I_t \rightarrow \frac{V_t}{I_t} = \frac{r_0}{g_m r_0 + 1} \approx \frac{1}{g_m} \tag{3.22}$$

For the NMOS transistor we can use the circuits of Fig. 3.13. In this case, the starting equation is:

$$-I_t + g_m V_{gs} + \frac{V_t}{r_0} = 0 \tag{3.23}$$

Replacing in the above relationship $V_{gs} = V_g - V_s = V_g - 0 = V_t$, one obtains the same value of impedance found for the PMOS. Summarizing, the impedance looking into the drain of a diode-connected transistor, whether it is a PMOS or an NMOS, is given by $1/g_m$, so it is relatively low. Note that in evaluating impedances, one can impose the variation of the current and calculate the variation in voltage that the circuit develops in response to it or vice versa. The context determines which method leads to simpler calculations, as shown in the following example.

---

[1]Remember always that the small signal models consider only *variations* of quantities. Therefore a constant voltage source must be replaced by a short circuit (hence the source of the PMOS transistor appears grounded in the model) and a constant current source is represented as an open circuit.

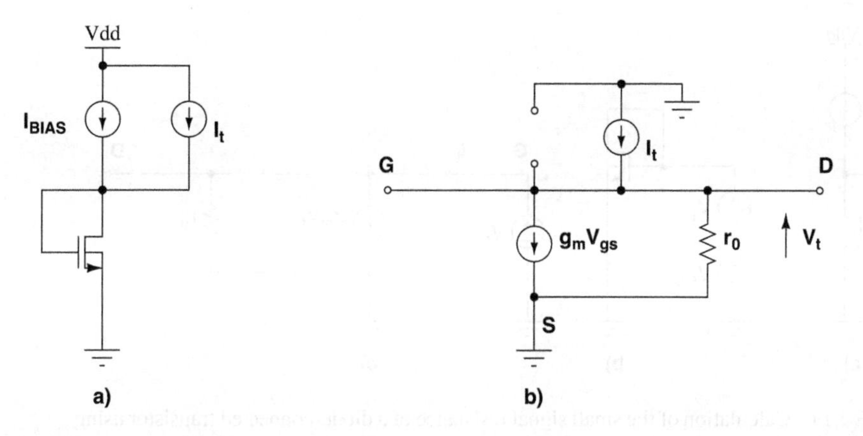

FIGURE 3.13 Circuits for calculating the small signal resistance of a diode-connected NMOS transistor.

Example 3.2: Impedance of a diode-connected transistor.

An NMOS transistor has the following parameters: $W/L = 20/1.5$, $\mu_n C_{ox} = 165\ \mu\text{A/V}^2$, $V_{TH} = 0.64$ V, $\lambda_N = 0.03$. Calculate the impedance seen looking into the drain when the transistor is diode-connected and biased with a current I_{DS} of 30 μA.

* * *

In this example we will use a test voltage source and calculate the current. Equation (3.13) allows us to find the V_{GS} necessary to support the requested bias current. A gate-source voltage 0.805 V is thus obtained. If we connect between gate and ground a voltage source with this value, the transistor will hence drive the same current as if it was connected to the current source. We can now modulate this voltage adding in series the signal source, as shown in Fig. 3.14 b). The small signal equivalent circuit is reported in Fig. 3.14 c). The current developed by the transistor is the sum of two contributions: $g_m V_t$ due to the device transconductance and V_t/r_0 arising from the channel modulation effect. The total current that must be provided by test source therefore is:

$$I_t = g_m V_t + \frac{V_t}{r_0} \tag{3.24}$$

which yields again equation (3.22), that is repeated here for convenience:

$$r_{diode} = \frac{r_0}{1 + g_m r_0} \tag{3.25}$$

The output conductance, g_{ds} is $\lambda_N I_{DS} = 0.9\ \mu$S and the output impedance, $r_0 = 1/g_{ds}$ is 1.1 MΩ. The transconductance is 363 μS. Inserting numbers in (3.25), we obtain that the equivalent impedance is 2748 Ω. If we simplify the formula to $1/g_m$, we get 2755 Ω, which shows that $1/g_m$ is a good approximation to evaluate the impedance seen from the drain of a diode-connected transistor.

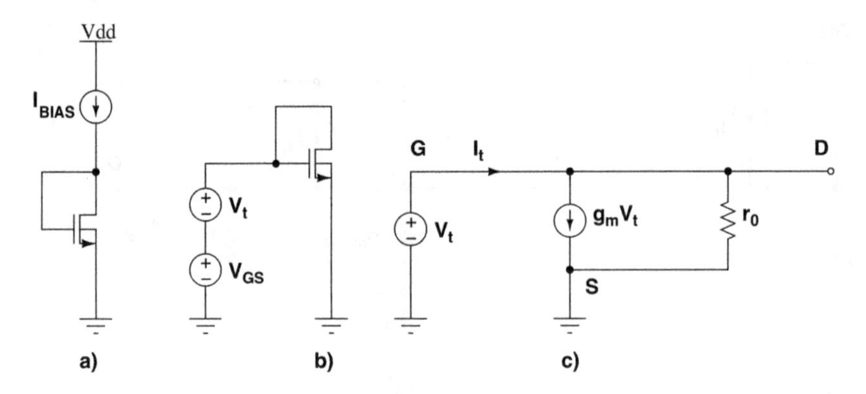

FIGURE 3.14 Calculation of the small signal resistance of a diode-connected transistor using a test voltage source.

In circuit simulations, the impedance can be estimated in two ways. With transient domain analysis, one can employ the circuit of Fig. 3.14 a) or b), apply a test signal, measure the variation in drain current and relate it to the applied test voltage. If this method is used, one must guarantee that the injected signal is small enough so that the bias point is not significantly affected during the test. If this is not the case, non-linearities will be seen because the transient analysis employs the full device characteristics. Furthermore, the signal must be slow enough so that bandwidth limitation is not relevant. The alternative method is to perform an AC analysis, which is done on the small signal equivalent circuit. Remember, however, that in the schematics the transistors must always be properly biased to guarantee the correct results.

The reference current can be copied by connecting one or more transistors in parallel to the diode-connected one, as shown in Fig. 3.15. Consider, for instance, transistors M_1 and M_2 in Fig. 3.15. The ratio between their currents can be written as:

$$\frac{I_{SD1}}{I_{SD2}} = \frac{\frac{1}{2}\left(\frac{W}{L}\right)_1 (V_{SG} - |V_{THP1}|)^2 (1 + \lambda V_{SD1})}{\frac{1}{2}\left(\frac{W}{L}\right)_2 (V_{SG} - |V_{THP2}|)^2 (1 + \lambda V_{SD2})} \tag{3.26}$$

Note that the transistors have the gate and source shorted, so they have the same V_{SG} by construction. If we suppose that the two devices have identical thresholds and the channel modulation effect is negligible, we can write:

$$\frac{I_{SD1}}{I_{SD2}} = \frac{\left(\frac{W}{L}\right)_1}{\left(\frac{W}{L}\right)_2} \tag{3.27}$$

In this ideal case, the ratio between the two currents is equal to the one between the aspect ratios of the devices. In general, the channel length modulation is not negligible and equation (3.27) is rigorously valid only if the two devices have the same drain-source voltage. The drain-source voltage of the reference device M_1 is equal to its V_{SG}

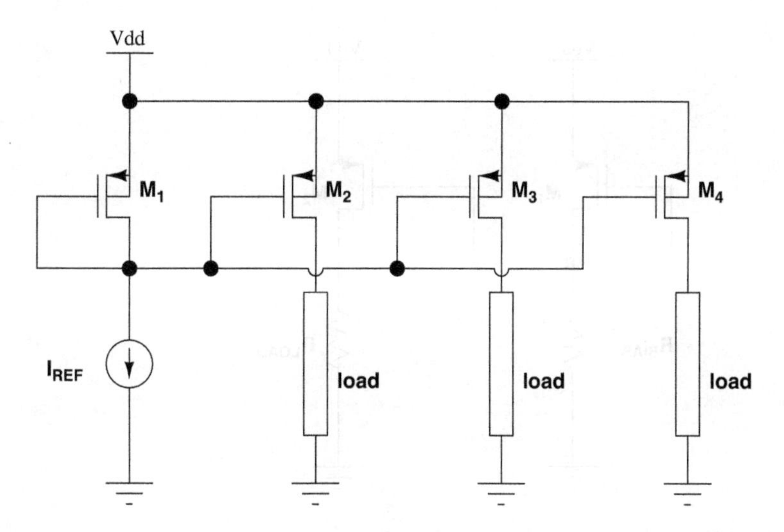

FIGURE 3.15 Example of PMOS current mirrors.

and it can be easily calculated once we know the current flowing in the transistor. As we have discussed above, the short between drain and gate automatically guarantees that M_1 is in saturation. To work as a current source, M_2-M_4 must be in saturation as well. However, their source-drain voltages are determined by the loads, therefore it is not a priori granted that the transistors have enough headroom to stay in saturation. When a transistor is used as a current source, care must be taken to ensure that the load it is connected to does not require a too high voltage drop that pushes the device in the linear region.

In practical implementations, the reference current source must be replaced with physical components. The most simple approach is to use a resistor connected between the drain of the diode-connected transistor and the appropriate power supply rail. These points are further discussed in the next example.

Example 3.3: PMOS current mirror.

Using PMOS transistors, design a current mirror in which the output current is 50 μA and it is five times the reference current. Discuss what happens when the load is a resistor of 20 kΩ and of 50 kΩ, respectively. Use the following device parameters: $\mu_p C_{ox} = 50\ \mu\text{A/V}^2$, $|V_{THP}| = 0.6$ V, $\lambda_P = 0.02\ \text{V}^{-1}$, $V_{DD} = 2.5$ V.

* * *

The circuit is depicted in Fig. 3.16. To guarantee a ratio of five between the output and the reference current we must have: $[(W/L)_2/(W/L)_1]$=5, so that the current in M_1 is 10 μA. We first choose the transistor length. Since the current is small, we can have

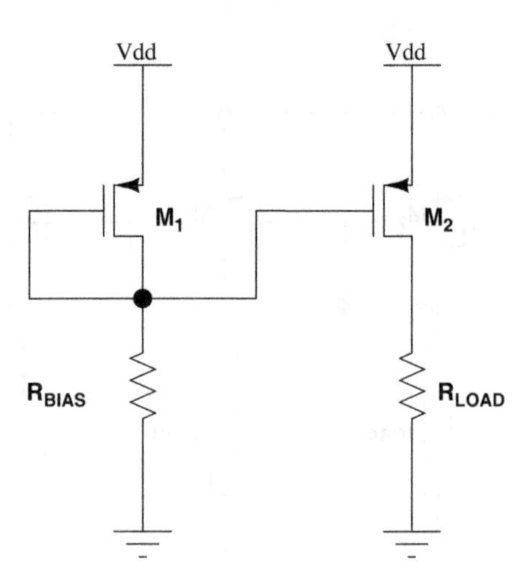

FIGURE 3.16 PMOS current mirror with one output branch.

a fairly long device to minimize the impact of channel length modulation, so we put $L = 2\ \mu$m. If we set $W_1 = 5\ \mu$m, using equation (3.20) we have $V_{SG1} = 1$ V. The source is at V_{DD}, hence the gate must be at 1.5 V. This voltage needs also to be compatible with the value of the bias resistor R_{BIAS}, which is therefore given by 1.5 V/10 μA = 150 kΩ. The aspect ratio of M_2 must be five times that of M_1 and, in principle, can be achieved by scaling the width, the length or both. For instance $(W/L)_2 = 5/0.4$, $(W/L)_2 = 10/0.8$, $(W/L)_2 = 25/2$ all satisfy the requirement. However, to have the two transistors equally sensitive to the channel modulation effect, the length is kept the same and only the channel width is modified. The last option is thus the correct one. When the load resistor is 20 kΩ, the output voltage is 1 V and V_{SD2} is 1.5 V. If we include the channel modulation effect the output current becomes:

$$I_{SD2} = \frac{1}{2}\mu_p C_{ox}\left(\frac{W}{L}\right)_2 (V_{SG2} - |V_{THP}|)^2 (1 + \lambda_P V_{SD2}) \approx 51.5\ \mu A. \tag{3.28}$$

To calculate I_{SD2} the condition that V_{SG1} is equal to V_{SG2} has been used. The fact the two transistors operate with different V_{SD} introduces a systematic error in the accuracy of the current mirror. The drain-source voltage of 1.5 V is greater than the saturation voltage $V_{SG} - |V_{THP}|$ of 0.4 V, so M_2 works in the appropriate region. If the output resistor is now increased to 50 kΩ, the voltage drop across it necessary to support the 50 μA current would be 2.5 V. This is incompatible with the fact that M_2 works in saturation. In practice, if the voltage drop required by the load is too big, the device is pushed in the linear region and its current decreases. When a transistor is not saturated, its current is heavily modulated by the drain-source voltage and its behavior is not compatible with that of a good current source, which should deliver a current independent of the voltage appearing across its terminals.

We have seen in the example that the currents must be scaled by acting only on the transistor width. If we need to multiply a current by a given factor $K > 1$, we are left with two options. The first one is to design a transistor which has the same length of the reference transistor and the gate width K times bigger, drawing its gate as single piece of polysilicon. To achieve the best accuracy, however, the mirroring transistor should be implemented by designing K identical replicas of the reference transistor, with the same gate width *and* length and connecting them in parallel. If we need instead to reduce the current ($K < 1$), the reference device must be laid-out putting in parallel K identical devices, while only one is used in the output branch.

We have focused up to now our study of current mirrors to PMOS-based topologies. It is straightforward to extend the discussion to NMOS-based implementations, that we introduce with the help of a further example.

Example 3.4: NMOS current mirror.

Using NMOS transistors, design a current mirror that delivers an output current of 200 μA. The circuit is loaded by a diode-connected PMOS transistor with a length of 2 μm. Determine the width of M_3 so that the output voltage V_P is 1 V. Size the circuit to obtain an overall power consumption of less than 1 mW. Use the following device parameters: $\mu_n C_{ox} = 165\ \mu\text{A/V}^2$, $V_{THN} = 0.64$ V, $\lambda_N = 0.03\ \text{V}^{-1}$, $\mu_p C_{ox} = 50\ \mu\text{A/V}^2$, $|V_{THP}| = 0.6$ V, $\lambda_P = 0.03\ \text{V}^{-1}$, $V_{DD} = 2.5$ V.

* * *

The circuit is shown in Fig. 3.17. First, we can use the constraint on the power consumption to determine the current in the reference branch. If we set, for instance, the current in M_1 to 50 μA, we have a total power consumption of 250 $\mu\text{A}\times 2.5\ \text{V} = 625\ \mu\text{W}$, which is compatible with our specifications. If we choose as before $(W/L)_1 = 5/2$, we

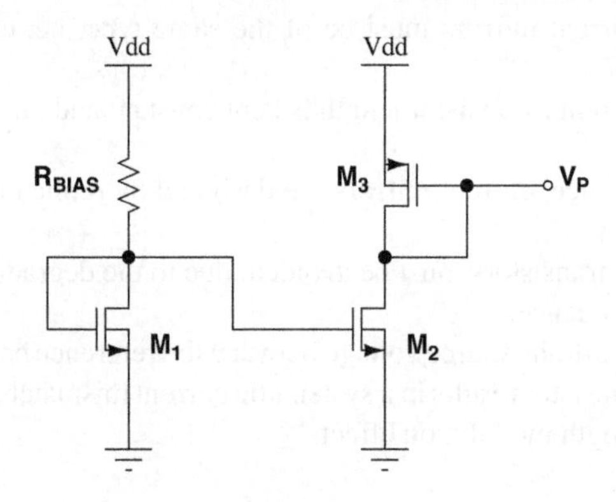

FIGURE 3.17 Example of NMOS current mirror.

can calculate the voltage at the drain node of M_1, which is simply equal to V_{GS1}:

$$V_{GS1} = V_{TH1} + \sqrt{\frac{2L_1 I_{DS1}}{\mu_n C_{ox} W_1}} = 1.13 \text{ V} \tag{3.29}$$

The bias resistor is then given by:

$$R_{BIAS} = \frac{V_{DD} - V_{GS1}}{I_{DS1}} = 27.4 \text{ k}\Omega \tag{3.30}$$

The saturation voltage for M_1 and M_2 is 0.49 V, so M_2 can withstand a drain voltage of 1 V. The V_{SG} of M_3 must be equal to $V_{DD} - V_P = 1.5$ V. The aspect ratio of M_3 can thus be calculated as follows:

$$\left(\frac{W}{L}\right)_3 = \frac{I_{SD3}}{\frac{1}{2}\mu_p C_{ox}\left(V_{SG3} - |V_{THP}|\right)^2} \tag{3.31}$$

Putting in numbers, one obtains $(W/L)_3 \approx 9.87$, so we can choose $W_3 = 20\ \mu$m.

Current mirrors can also be used when it is necessary to reverse the direction of the current flow. Note in fact that the PMOS transistor M_3 in the above example could serve as a reference for a PMOS type current mirror, that will push current towards ground. Let's now summarize what we have discussed so far in this section:

CURRENT MIRROR SUMMARY

- Current mirrors are formed by a reference branch and one or more output branches. An arbitrary number of transistors can be connected to the reference branch.
- Transistors in current mirrors must be of the same type, i.e. either NMOS or PMOS.
- In scaling the currents, transistor length is kept constant and only the width is adapted.
- For best performance, current mirrors are designed by replicating a suitable basic unit.
- Minimum length transistors must be avoided, due to the degradation of the output impedance.
- A difference in the drain-source voltage between the reference branch and the output transistor results in a systematic current mismatch, due to the channel length modulation effect.

The assertion that an "arbitrary" number of transistors can be connected to the reference one stems from the fact that the gate of an MOS transistor does not sink DC current. In very deep submicron CMOS technologies, devices might exhibit a sizable gate leakage current due to the tunneling of the carrier through the very thin gate oxide. If many transistors are put in parallel, the current sunk from the diode-connected load could alter the desired biasing point. Front-end ASICs often incorporates many parallel channels and if a single diode-connected transistor is used to generate the reference bias for all of them, care must be paid also to the effect of the gate leakage current. This issue starts to deserve attention from the 0.13 μm technology node and below, especially in those cases when very small bias currents (in the order of few nA) are necessary. Finally, it must be pointed-out that biasing the reference branch with a resistor to ground or to the power supply is easy, but it makes the circuit sensitive to noise on the power supply and power supply variation. This scheme is used for illustrative purposes and can be practical when first prototypes are implemented. More robust biasing schemes must be adopted for circuits that are intended for the end-user application.

3.4 COMMON SOURCE AMPLIFIERS WITH ACTIVE LOAD

A current mirror can be used as the load of a common source amplifier, as shown in Fig. 3.18. In Fig. 3.18 a) we have an NMOS input stage with a PMOS load and vice versa in Fig. 3.18 b). Both circuits can be studied with the small signal equivalent circuit of Fig. 3.9 b). The gain of the common source amplifier with active load is given by:

$$A_v = -g_{m1}\left(r_{01}//r_{02}\right) = -g_{m1}\frac{r_{01}r_{02}}{r_{01}+r_{02}} \tag{3.32}$$

where g_{m1} is the transconductance of the input transistor and r_{01} and r_{02} are the output resistances of the input and load devices, respectively. In modern deep submicron technologies, the channel length modulation can be quite severe, so the gain for minimum channel devices can be well below 50.

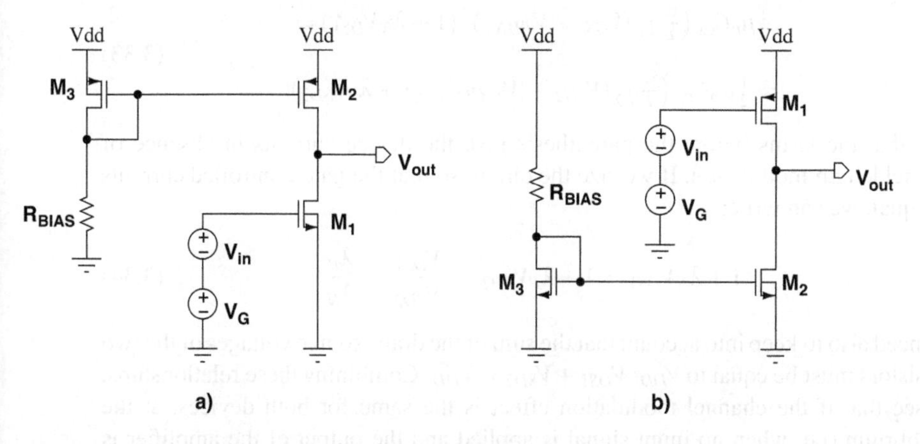

FIGURE 3.18 Common source amplifiers loaded by transistors working as current sources. The NMOS input configuration is shown in a) and the PMOS-input one in b).

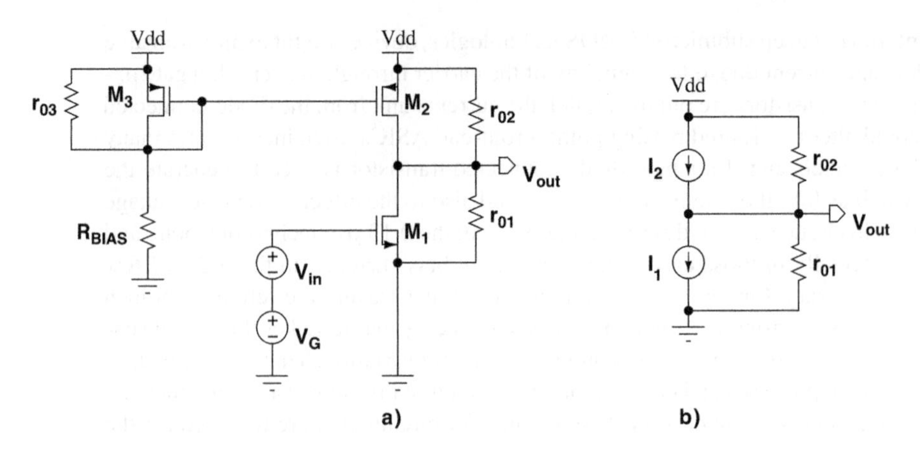

FIGURE 3.19 Circuits to study the effect of mismatch between the bias currents in common source amplifiers. See text for discussion.

With the help of Fig. 3.19, we now examine the problem of biasing a common source amplifier with current mirror load. Conceptually, the bias voltage on the gate of the input transistor should be adapted so that M_1 can accept the current delivered by M_2. For the time being, the input bias network is a simple voltage source connected in series to the gate of M_1. If the output impedance of the two transistors were infinite, we would connect two ideal current sources in series, which is possible if and only if the two currents are equal. In this case, the output voltage is undetermined. However, the transistors are affected by the channel modulation effect. We can split the transistor current in two components, one controlled by the gate-source voltage only and the other one regulated by the channel length modulation. When the amplifier is not connected to any load, the currents in M_1 and M_2 must be equal, so we have:

$$\begin{aligned} &\tfrac{1}{2}\mu_n C_{ox}\left(\tfrac{W}{L}\right)_1 (V_{GS1}-V_{THN1})^2(1+\lambda_N V_{DS1}) = \\ &= \tfrac{1}{2}\mu_p C_{ox}\left(\tfrac{W}{L}\right)_2 (V_{SG2}-|V_{THP2}|)^2(1+\lambda_P V_{SD2}) \end{aligned} \tag{3.33}$$

Note that the terms before the parenthesis give the device currents in absence of channel length modulation. If we size the circuit so that the gate-controlled currents are equal, we can write:

$$1+\lambda_N V_{DS1} = 1+\lambda_P V_{SD2} \rightarrow \frac{V_{DS1}}{V_{SD2}} = \frac{\lambda_P}{\lambda_N} \tag{3.34}$$

We need also to keep into account that the sum of the drain-source voltages of the two transistors must be equal to V_{DD}: $V_{DS1}+V_{SD2}=V_{DD}$. Combining these relationships, we see that if the channel modulation effect is the same for both devices, at the equilibrium (i.e. when no input signal is applied and the output of the amplifier is not connected to any external resistive load) the DC output voltage is $V_{DD}/2$. If the channel length modulation effect is smaller for the PMOS transistor (i.e. $\lambda_P < \lambda_N$) the

DC voltage is pushed downwards and vice versa. Consider now the situation in which the regulation on the gate voltage of M_1 is such that the current M_1 accepts is smaller than the one M_2 would impose. We assume that the mismatch between the currents is small, so we can study the problem with a small signal model, as shown in Fig. 3.19. The only possible path for the extra current is through the channel modulation effect of M_1, which requires that the DC output voltage is increased. However, only little currents can be accommodated through this mechanism, so the output node voltage rises significantly even if the mismatch between the two currents is small. This will eventually push M_2 in the linear region and the current will be regulated by M_1. The opposite is true if the gate voltage of M_1 is bigger than necessary, so that it drives a current larger than the one provided by M_2. In this case, the DC output voltage will lower, driving M_1 in the linear region. The analogy with current sources shown in Fig. 3.19 b) can further help in understanding the process. If $I_2 = I_1$, the output node voltage is determined by the resistors through the voltage divider rule. If $I_2 > I_1$, the extra current $I_2 - I_1$ flows in r_{01} and determine a rise of the output voltage. If $I_2 < I_1$, the current $I_1 - I_2$ is sunk through r_{02}, lowering the output voltage. We can thus conclude that:

When two transistors that should both work in saturation are connected in series and there is a mismatch between their currents, the one that should provide the larger current is driven in linear region.

The following example allows us to appreciate the amount of mismatch between the two currents that can be tolerated.

Example 3.5: Biasing in high gain common source amplifiers.

Consider the common source amplifier shown in Fig. 3.19 a). Calculate the bias voltage that must be applied to the gate of M_1 to fully balance the circuit for a bias current of 50 μA. Explain what happens if an error of 2% is made in setting the required value. Use the following device parameters: $(W/L)_1 = 20/1$, $\mu_n C_{ox} = 165\ \mu A/V^2$, $V_{THN} = 0.64$ V, $\lambda_N = 0.03\ V^{-1}$, $(W/L)_2 = 7/1$, $\mu_p C_{ox} = 50\ \mu A/V^2$, $|V_{THP}| = 0.6$ V, $\lambda_P = 0.03\ V^{-1}$ $V_{DD} = 2.5$ V.

* * *

We first calculate the gate-source voltage needed by M_1 to accept a current of 50 μA:

$$V_{GS1} = V_{THN} + \sqrt{\frac{2 I_{DS} L_1}{\mu_n C_{ox} W_1}} = 0.814 \text{ V} \tag{3.35}$$

A 2% error implies a deviation on V_{GS1} of 16.3 mV. We can estimate its impact by treating it as a small signal, superimposed on the nominal bias voltage. Since we have supposed that $\lambda_N = \lambda_P$, the output resistance of the devices will be equal and given by:

$$r_{01} = r_{02} = \frac{1}{\lambda I_{DS}} = 670\ k\Omega \tag{3.36}$$

The transconductance of the input device is given by:

$$g_{m1} = \mu_n C_{ox} \left(\frac{W}{L}\right)_1 (V_{GS1} - V_{TH1}) = 574\ \mu\text{S} \tag{3.37}$$

The voltage gain is $A_v = -g_{m1}\left(r_{01}//r_{02}\right) = -192$. If our bias voltage is smaller than 2% of the required value it will drive the output node up, but a negative variation of 16.3 mV at the input implies a variation of 3.13 V at the output, that is greater than the positive power supply. In practice, if the error is such that the gate voltage is smaller than required, the circuit will saturate at the positive supply rail, if it is larger, the circuit will saturate with a DC output value close to ground. The sensitivity to the current mismatch is exacerbated if the amplifier gain is increased.

The above example shows that it is impractical to use a high gain amplifier in an open-loop configuration. Any small perturbation of the input (deviation of the bias from the nominal point, temperature drift, noise, etc.) will be in fact multiplied by the gain, thus determining huge variation at the output. A feedback mechanism must hence be used to stabilize the circuit quiescent point.

The output resistance of the common source amplifier, given by the parallel combination of r_{01} and r_{02}, must be as high as possible to provide sufficient voltage gain. If we ask the amplifier to drive an external resistor, this will go in parallel to r_{01} and r_{02}, reducing the amplifier gain. The common source with active load has very high input impedance and fairly high output impedance, so it is a *transconductance* rather than a *voltage* amplifier. Therefore its is desirable to increase the amplifier gain and decrease its output impedance. This will be the goal of the next few sections.

3.5 SOURCE DEGENERATED COMMON SOURCE AMPLIFIERS

Before addressing the increase of the gain and the decrease of the output impedance of common source amplifiers, we go back in this section to a scheme employing resistive loads. The study of this configuration is useful because its small signal equivalent circuit will serve as the basis for analyzing the cascode and source follower topologies, which are instrumental to achieve respectively high gain and reasonably low output impedance. The common source amplifier with source degeneration is depicted in Fig. 3.20 a). The difference with respect to the simple common source topology is that resistor R_S has been inserted between the source of the transistor and ground. The small signal equivalent circuit is reported in Fig. 3.20 b). Note that the bulk effect must now be included because the source and the substrate of the transistor are not at the same potential. In the analysis of the circuit we will proceed as follows: first we compute the transfer function from the input to the output and from the input

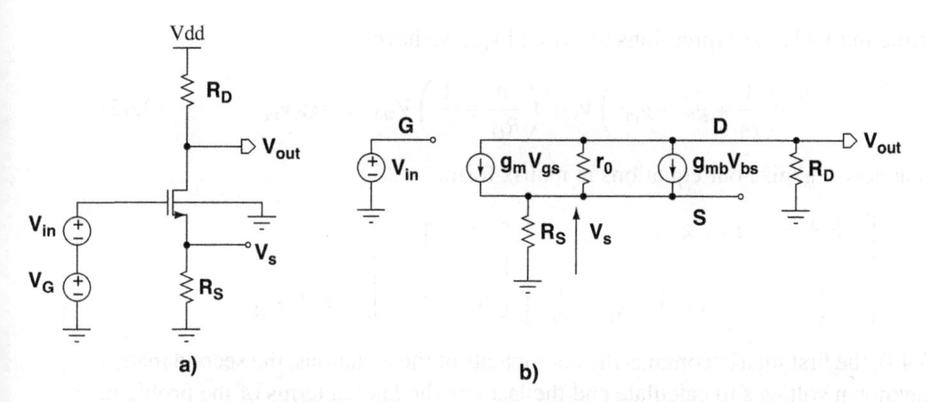

FIGURE 3.20 Common source amplifier with source degeneration: a) Circuit principle b) Small signal equivalent circuit.

to the source. In doing so, we will employ nodal analysis and report calculations in detail. This will also serve to refresh basic concepts about linear network analysis that will be recurrently used in the book. As a second step, we will evaluate the impedance seen from the drain and source nodes. We use here the convention of taking with positive sign the currents leaving the node and with negative sign the currents entering into it. From every node, the algebraic sum of the currents is set to zero. The direction of the current generated by current sources is indicated by the arrow associated to the component symbol. In the circuit of Fig. 3.20 we have two independent nodes, corresponding to source and drain of the transistor. The associated voltages are indicated as V_s and V_{out}, respectively. For the source node we have:

$$-g_m V_{gs} - g_{mb} V_{bs} + \frac{V_s}{R_S} + \frac{V_s - V_{out}}{r_0} = 0 \tag{3.38}$$

We can rearrange the above equation writing the explicit definition of V_{gs} and V_{bs}, i.e. $V_{gs} = V_g - V_s = V_{in} - V_s$ and $V_{bs} = V_b - V_s = 0 - V_s = -V_s$. Note that the bulk of the transistor is kept at ground, so it does not experience any signal variation, but the bulk-source voltage does, because the source is not grounded and its voltage changes during circuit operation. Replacing V_{gs} and V_{bs} in equation (3.38) with the above relationships, we have:

$$-g_m (V_{in} - V_s) + g_{mb} V_s + \frac{V_s}{R_S} + \frac{V_s}{r_0} - \frac{V_{out}}{r_0} = 0 \tag{3.39}$$

which can be reorganized as:

$$\left(\frac{1}{R_S} + \frac{1}{r_0} + g_m + g_{mb} \right) V_s - \frac{V_{out}}{r_0} = g_m V_{in} \tag{3.40}$$

We can now write down the equation for the drain node:

$$\frac{V_{out}}{R_D} + \frac{V_{out} - V_s}{r_0} + g_m V_{gs} + g_{mb} V_{bs} = 0 \tag{3.41}$$

Inserting in (3.41) the expressions of V_{gs} and V_{bs}, we have:

$$-\left(\frac{1}{r_0}+g_m+g_{mb}\right)V_s+\left(\frac{1}{R_D}+\frac{1}{r_0}\right)V_{out}=-g_mV_{in} \tag{3.42}$$

We can now organize our equations in matrix form:

$$\begin{bmatrix} \frac{1}{R_S}+\frac{1}{r_0}+g_m+g_{mb} & -\frac{1}{r_0} \\ -\left(\frac{1}{r_0}+g_m+g_{mb}\right) & \frac{1}{R_D}+\frac{1}{r_0} \end{bmatrix}\begin{bmatrix} V_s \\ V_{out} \end{bmatrix}=\begin{bmatrix} g_mV_{in} \\ -g_mV_{in} \end{bmatrix} \tag{3.43}$$

In (3.43), the first matrix contains the coefficients of the equations, the second matrix the unknown voltages to calculate and the last one the known terms of the problem. From the Cramer's theorem of linear algebra we know that V_s is obtained from the following ratio:

$$V_s=\frac{\Delta_1}{\Delta} \tag{3.44}$$

where Δ_1 is the determinant of the matrix obtained by replacing the first column of the coefficient matrix with the column of the known terms, that is:

$$\Delta_1=\begin{vmatrix} g_mV_{in} & -\frac{1}{r_0} \\ -g_mV_{in} & \frac{1}{R_D}+\frac{1}{r_0} \end{vmatrix}=\frac{g_mV_{in}}{R_D} \tag{3.45}$$

The denominator Δ is the determinant of the coefficient matrix, which is:

$$\Delta=\left(\frac{1}{R_S}+\frac{1}{r_0}+g_m+g_{mb}\right)\left(\frac{1}{R_D}+\frac{1}{r_0}\right)-\frac{1}{r_0}\left(\frac{1}{r_0}+g_m+g_{mb}\right) \tag{3.46}$$

After some algebraic manipulations, we finally obtain:

$$\Delta=\frac{r_0+R_S+R_D+(g_m+g_{mb})\,r_0R_S}{r_0R_SR_D} \tag{3.47}$$

We can now calculate the voltage at the source, V_s:

$$\frac{\Delta_1}{\Delta}=V_s=\frac{g_mV_{in}r_0R_S}{r_0+R_S+R_D+(g_m+g_{mb})\,r_0R_S} \tag{3.48}$$

The output voltage V_{out} is found as:

$$V_{out}=\frac{\Delta_{out}}{\Delta} \tag{3.49}$$

where Δ_{out} is the determinant obtained by replacing in the coefficient matrix the second column with the one of the known terms:

$$\Delta_{out}=\begin{vmatrix} g_m+g_{mb}+\frac{1}{R_S}+\frac{1}{r_0} & g_mV_{in} \\ -\left(g_m+g_{mb}+\frac{1}{r_0}\right) & -g_mV_{in} \end{vmatrix}=-\frac{g_mV_{in}}{R_S} \tag{3.50}$$

Combining the previous results, we finally get:

$$V_{out} = \frac{\Delta_{out}}{\Delta} = -\frac{g_m V_{in} r_O R_D}{r_O + R_s + R_D + (g_m + g_{mb}) r_O R_S} \tag{3.51}$$

The above derivations show that nodal analysis provides a straightforward method to study linear circuits and calculate transfer functions. Unfortunately, hand calculations become quickly impractical and error-prone as the number of nodes in the circuit increases. In this case, one writes down the fundamental equations and uses a symbolic computation program to solve the system. Several of these software also offer specialized tool boxes explicitly designed for circuit analysis. Let's now study the source degenerated amplifier with a practical example.

Example 3.6: Common source amplifier with source degeneration.

Consider the common source amplifier shown in Fig. 3.20. Calculate the bias voltage that must be applied to the gate to have in the input transistor a current of 50 μA. Evaluate the small signal parameters and the gain of the circuit. Use the following device parameters: $(W/L)_1 = 10/1$, $\mu_n C_{ox} = 165\ \mu\text{A/V}^2$, $V_{THN} = 0.64$ V, $\lambda_N = 0.03\ \text{V}^{-1}$, $\gamma = 0.5\ \text{V}^{1/2}$, $\phi_F = 0.3$ V. Assume $V_{DD} = 2.5$ V, $R_D = 20$ kΩ and $R_S = 10$ kΩ.

* * *

We first notice that if the circuit drives 50 μA, the voltage drop across R_S brings the transistor source at 0.5 V. The drain is at $V_{DD} - I_{DS}R_D = 1.5$ V. The transistor has a V_{DS} of 1 V, so it is reasonable to assume that it is in saturation. Since the source is not grounded, the device experience the body effect and its threshold can be calculated as:

$$V_{TH} = V_{TH0} + \gamma\left(\sqrt{2\phi_F + V_{SB}} - \sqrt{2\phi_F}\right) = 0.777\ V \tag{3.52}$$

Using this value of V_{TH}, we can find the V_{GS} needed by M_1 to drive a current of 50 μA:

$$V_{GS} = V_{THN} + \sqrt{\frac{2 I_{DS} L}{\mu_n C_{ox} W}} = 1.02\ V \tag{3.53}$$

The gate voltage must hence be put at the source voltage (0.5 V) plus the needed V_{GS}, so it is 1.02 V+0.5 V=1.52 V. We can now evaluate the small signal parameters. For the transconductance we have:

$$g_m = \sqrt{2\mu_n C_{ox}\frac{W}{L} I_{DS}} = 400\ \mu\text{S} \tag{3.54}$$

The bulk transconductance, g_{mb}, is given by:

$$g_{mb} = \frac{g_m \gamma}{2\sqrt{2\phi_F + V_{SB}}} = 95\ \mu\text{S} \tag{3.55}$$

Finally, the output resistance is:

$$r_0 = \frac{1}{\lambda_N I_{DS}} = 670\ \text{k}\Omega \tag{3.56}$$

We can now use equation (3.51) to calculate the gain. Putting in numbers, we have:

$$A_v = -\frac{4 \cdot 10^{-4} \times 2 \cdot 10^4 \times 0.67 \cdot 10^6}{10^4 + 2 \cdot 10^4 + 0.67 \cdot 10^6 + 4.95 \cdot 10^{-4} \times 10^6 \times 10^4} = -1.33 \tag{3.57}$$

We see that in the denominator, R_S and R_D give a negligible contribution. Using equation (3.48) we can also evaluate the gain from the input to the source node:

$$A_{vs} = \frac{4 \cdot 10^{-4} \times 0.67 \cdot 10^6 \times 10^4}{10^4 + 2 \cdot 10^4 + 0.67 \cdot 10^6 + 4.95 \cdot 10^{-4} \times 0.67 \cdot 10^6 \times 10^4} = 0.67 \tag{3.58}$$

If the transconductance of the device was high enough, the term $(g_m + g_{mb})\, r_0 R_S$ would dominate also over r_0 and the gain at the drain node would simply become:

$$A_v = -\frac{g_m R_D}{(g_m + g_{mb}) R_S} \tag{3.59}$$

Neglecting also the bulk effect, the above expression reduces to:

$$A_v = -\frac{R_D}{R_S} \tag{3.60}$$

Equation (3.60) reveals an interesting point. In case the transconductance is high enough and the body effect is suppressed, the voltage gain at the drain node depends only on the ratio between two resistors. This is very convenient, because in an integrated circuit the ratios between the parameters of components of the same nature are much better controlled than their absolute values. The absolute value of a passive component may change by a few percent on the same chip and by more than 10% between different dies and different production lots. However, the ratio between the values of two carefully laid-out components placed very close by on the same ASIC can be accurate to the per mil level. Whenever it is possible, the key properties of an analog circuit should therefore depend on the ratios between component parameters rather than on their absolute values. As far as amplifiers are concerned, this is only possible with current and voltage configurations. In fact, in these cases the gain is an adimensional number, because we divide two quantities having the same physical units. With transimpedance or transconductance configurations, the ratio between the output and the input signals has the units of either a resistance or a conductance, so it can not be obtained by dividing two quantities measured in the same units.

To suppress the bulk effect, we can use a PMOS transistor and short the bulk to the source. This opportunity is offered in today's technologies by most processes also for NMOS transistors, through the use of triple well devices. Especially in low power applications, the MOS transconductance is not big enough to guarantee that $(g_m + g_{mb}) r_O R_S \gg r_O$ and equation (3.59) may lead to overestimate significantly the gain. Finally, from (3.48) we see that the gain from the input to the source is less than unity. Despite this, taking the output signal at the source can be very useful, as we will discuss later studying source followers.

3.5.1 FEEDBACK IN SOURCE DEGENERATED AMPLIFIERS

A deeper understanding of common source amplifiers can be obtained by applying the concept of negative feedback. Suppose in fact that the output resistance of the transistor, r_O, is infinite and that the bulk effect can be neglected. We can write the gain as:

$$A_v = -\frac{g_m R_D}{1 + g_m R_S} \tag{3.61}$$

Consider now the following term:

$$G_m = \frac{g_m}{1 + g_m R_S} \tag{3.62}$$

The quantity G_m has the unit of a transconductance. If we now make the following substitutions: $g_m = A$, $R_S = f$, we find again our equation of negative feedback amplifier:

$$A_{CL} = \frac{A}{1 + Af} \tag{3.63}$$

This suggests that in the sub-circuit formed by the transistor and the resistor some kind of negative feedback is at work. Conceptually, we can partition its operation in the following steps:

1. The current I_{out} is sensed by the source resistance and converted to a voltage, V_{fb}.
2. This voltage is applied to the source of the transistor, which provides the subtracting node. Note that if V_{in} increases, the current I_{out} increases, but also the voltage drop across R_S does. The source voltage rises and the gate-source voltage of the transistor is reduced, counteracting the effect of V_{in} and providing the negative feedback.
3. The error voltage $V_\varepsilon = V_{in} - V_{fb}$ is then applied to the main amplifier, (the transistor alone) which multiplies it by its open-loop gain (the transconductance g_m).

We can therefore write:

$$I_{out} = g_m V_\varepsilon = g_m (V_{in} - I_{out} R_S) \tag{3.64}$$

and solving for I_{out}/V_{in} we get:

$$\frac{I_{out}}{V_{in}} = G_m = \frac{g_m}{1 + g_m R_S} \tag{3.65}$$

We can consider the common source amplifier as formed by two components: a closed-loop transconductance amplifier, made in turn of the transistor and of the source resistance, plus a load, provided by R_D. Observe that R_D is outside the local feedback loop. It is interesting to note that the simplified analysis provides a deeper insight into the behavior of the circuit with much less calculation with respect to the full nodal analysis. However, insulating the critical elements in a feedback loop may not be trivial and can lead to gross mistakes if it is not done properly. Although more tedious, nodal analysis is safer and takes into account the full complexity of the circuit. The resulting equations might be complex, but a deeper understanding can in general be achieved by studying what happens to the transfer function in limiting cases, where some of its terms become negligible or dominate over the others.

3.5.2 OUTPUT IMPEDANCE OF THE SOURCE DEGENERATED AMPLIFIER

We calculate now the small signal output impedance of the common source amplifier with source degeneration. First, we need to observe that when we evaluate small signal quantities, all constant potentials are set to zero. In this condition, we have two parallel paths leading from the drain of the transistor to ground: one through R_D and one through the structure formed by the transistor and its source degenerating resistor. It is this quantity, that we call here $r_{0,cs}$, which is interesting to evaluate, as the total impedance seen from the drain node is just given by the parallel combination of $r_{0,cs}$ with R_D. Applying feedback theory, one can expect that $r_{0,cs}$ takes the following form:

$$r_{0,cs} = r_0(1+Af) = r_0(1+g_m R_S) \tag{3.66}$$

To calculate explicitly $r_{0,cs}$ we can use the circuits of Fig. 3.21. A test voltage V_t is applied to the transistor drain. The current I_t absorbed by the circuit in response to V_t is measured. The ratio V_t/I_t gives $r_{0,cs}$. During the test, V_t must be the only stimulus applied to the circuit. Note that the circuit of Fig. 3.21 a) is *not* a small signal model, but a circuit that can be used to evaluate the output impedance in a simulation test bench. To make a correct simulation, we need in fact to preserve the biasing points of

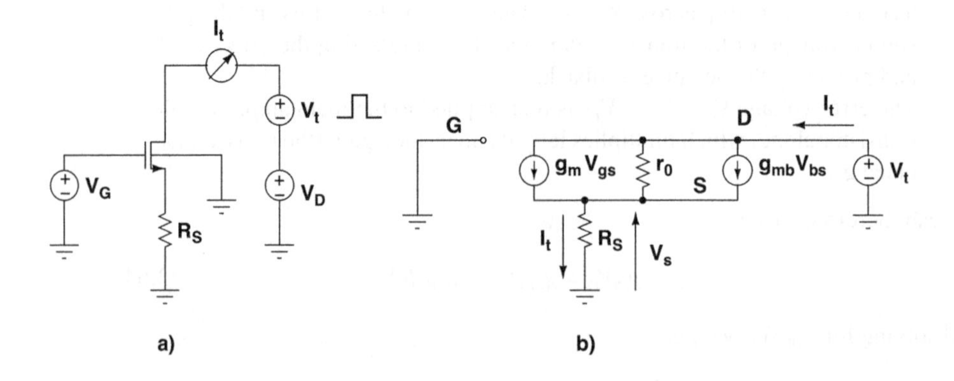

FIGURE 3.21 Circuits for calculating the output impedance of the source-degenerated stage.

the circuit. In example 3.6 we have studied a common source amplifier whose drain voltage was set at 1.5 V and the gate biased at 1.52 V. If we want to run a simulation to evaluate $r_{0,cs}$ for the circuit of this example, these bias voltages must be kept in the circuit. Note that we apply to the gate only the bias generator and not the signal source, since the test signal must be the only stimulus applied to the circuit. At the drain, the test voltage V_t is superimposed to a steady voltage source, V_D that preserves the DC level at which the node seats when the structure is connected to R_D. Fig. 3.21 b) shows the small signal equivalent circuit. This is a model intended to study only how the circuit responds to *variations*, hence constant potentials, having zero variations, must be shorted. For this reason, when we move from Fig. 3.21 a) to Fig. 3.21 b), the gate is connected to ground and only the signal source V_t survives at the drain. Writing down the equation for the drain node we have:

$$\frac{V_t - V_s}{r_0} + g_m V_{gs} + g_{mb} V_{bs} = I_t \tag{3.67}$$

Note that the following conditions hold:

$$\begin{cases} V_{gs} = 0 - V_s \\ V_{bs} = 0 - V_s \\ V_s = I_t R_S \end{cases} \tag{3.68}$$

The first two relationships are obtained by applying the definitions of V_{gs} and V_{bs} and observing that the gate and the bulk do not experience any variation. The last condition in (3.68) stems from the fact that R_S provides the only possible path to ground for the current, since the impedance of the gate is infinite. Therefore, I_t must necessarily flow to ground through R_S, developing across it the voltage drop $I_t R_S$. Inserting the relationships of (3.68) in (3.67) and calculating the ratio V_t/I_t, we get:

$$r_{0,cs} = \frac{V_t}{I_t} = r_0 + R_S + (g_m + g_{mb})\, r_0 R_S \tag{3.69}$$

which is the quantity we were looking for. The first two terms in equation (3.69) could be expected since r_0 and R_S would provide a series path to ground if they were the only components in the circuit. However, the transistor action due to the presence of the current sources represented by g_m and g_{mb} boosts the output impedance introducing the additional term $(g_m + g_{mb})\, r_0 R_S$, which is the dominant one. Equation (3.69) is extremely important and we will use it extensively in the rest of the book. The following example shows how much the source degeneration technique can increase the output impedance.

Example 3.7: Source degeneration and output impedance.

Calculate the output impedance of the circuit discussed in example 3.6.

* * *

In example 3.6 we found the following transistor small signal parameters:

- $r_0 = 670\ \text{k}\Omega$
- $g_m = 400\ \mu\text{S}$
- $g_{mb} = 95\ \mu\text{S}$.

The source resistance R_S was 10 kΩ. Putting numbers in equation (3.69), we get:

$$\begin{aligned} r_{0,cs} &= r_0 + R_S + (g_m + g_{mb})\, r_0 R_S = \\ &= 0.67 \cdot 10^6 + 10^4 + 4.95 \cdot 10^{-4} \times 0.67 \cdot 10^6 \times 10^4 \approx 4\ \text{M}\Omega \end{aligned} \tag{3.70}$$

In this particular case we have a six-fold increase. To further boost the output resistance, we can increase the device transconductance, the degeneration resistance or both.

3.6 CASCODE AMPLIFIERS

With the study of the source-degenerated amplifier, we have laid down the basis to understand one of the most widely used topology in analog integrated circuit design: the cascode. Remember that our main goal is to build amplifiers with gain as high as possible, so that feedback can be applied to implement well controlled transfer functions. Since the gain is the product of the transconductance and the output resistance, increasing the latter leads to configurations with bigger gain without increasing the power consumption. Discussing the common source amplifier, we have seen that increasing the value of the passive load resistor to have higher gain leads to a reduction of the bias current to maintain enough dynamic range at the amplifier output. This issue is addressed by replacing the passive component with a transistor in saturation and using its output resistance as the load. In a similar manner, if we increase the value of R_S in a source degenerated stage, we are forced to scale down the bias current to avoid an excessive voltage drop across it. We can therefore think of replacing the degeneration resistance with a transistor in saturation, as shown in Fig. 3.22. Fig. 3.23 shows the equivalent circuit used to calculate the output impedance. Note that since the gate of M_1 is connected to a DC voltage, the voltage-controlled current source associated to M_1 that represents the device gate transconductance is not active, and M_1 contributes to the small signal equivalent circuit only with its output resistance, r_{01}. The resulting network is identical to the one used for the common source amplifier with passive degeneration resistor. The output impedance of the new configuration is hence given by:

$$r_{out} = r_{01} + r_{02} + (g_{m2} + g_{mb2})\, r_{01} r_{02} \tag{3.71}$$

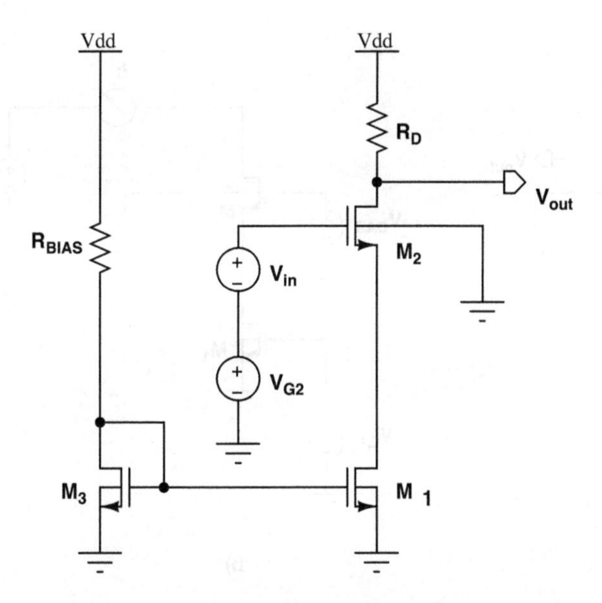

FIGURE 3.22 Circuit replacing the passive source degenerating resistor with an active load.

In the circuit of Fig 3.22, M_1 is biased with the current mirror technique and the signal is applied to the gate of M_2. To make a quick evaluation of the gain V_{out}/V_{in} we can apply the simplified formula of the source degenerated amplifier (equation 3.65):

$$A_v = -G_m R_D = -\frac{g_{m2}}{1+g_{m2}r_{01}} R_D \approx -\frac{R_D}{r_{01}} \leq 1 \qquad (3.72)$$

To obtain the rightmost member of the equation, we made the assumption that $g_{m2}r_{01} \gg 1$, which is in general justified. We see from (3.72) that applying the signal at the gate of M_2 leads to an attenuator, rather than an amplifier. This is because the current source M_1 blocks the possibility of modulating the current in M_2 by changing its gate voltage. We can then modify the circuit as shown in Fig. 3.24 a), obtaining

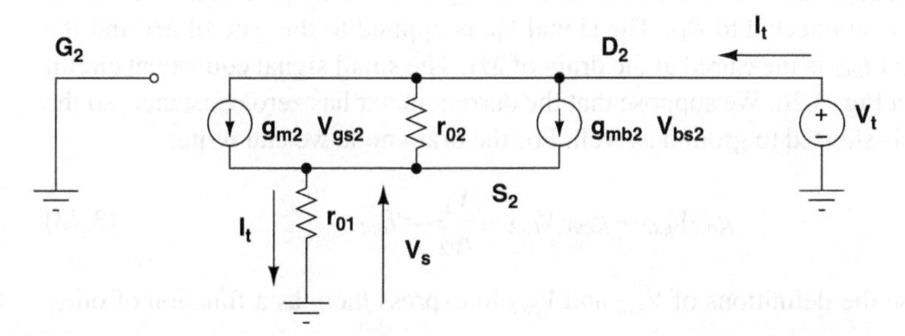

FIGURE 3.23 Small signal equivalent circuit to calculate the output impedance of the configuration of Fig. 3.22.

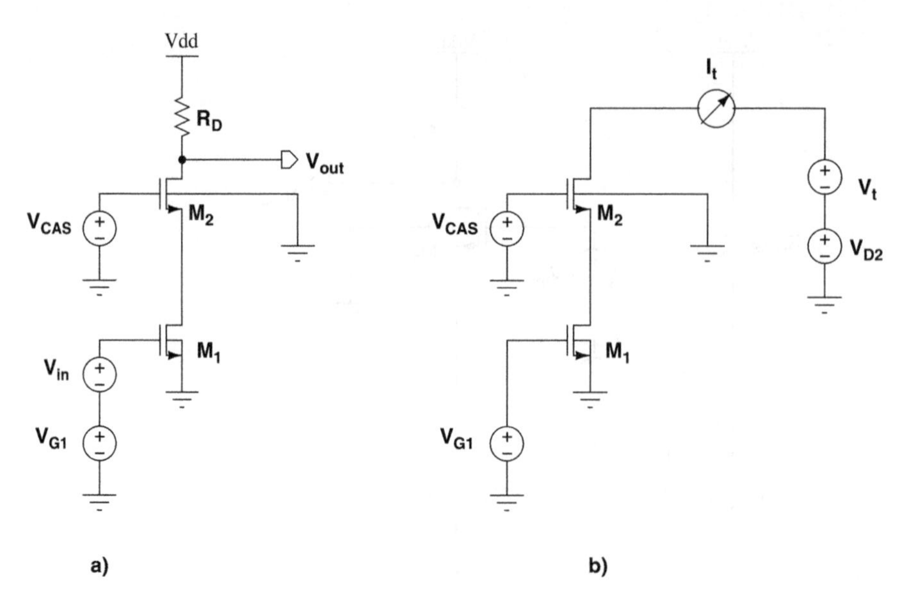

FIGURE 3.24 Common source cascode amplifier (in a) and test bench for evaluating the output impedance (in b).

the well known cascode topology. The input signal is now applied to the gate of M_1, while the gate of M_2 is biased by a constant voltage source. To evaluate the output impedance of the cascode with a circuit simulator we can use the scheme of Fig. 3.24 b). The small signal model of the circuit of Fig. 3.24 b) is identical to the one reported in Fig. 3.23. In fact, when we consider only variations, the voltage source biasing the gate of M_1 must be shorted. The voltage-controlled current source representing the transconductance of M_1 is not active and only r_{01} must be inserted in the model. The output impedance is hence given again by equation (3.71).

Let's now calculate the transconductance G_m of the cascode. Fig. 3.25 shows the schematic that can be used for the simulation. The bias voltages applied the gates of M_1 and M_2 and to the drain of M_2 are chosen so that both transistors are in saturation. The DC voltage source V_{D2} keeps the drain of M_2 to the same potential it has when the cascode is connected to R_D. The signal V_{in} is applied to the gate of M_1 and the output current I_{out} is measured at the drain of M_2. The small signal equivalent circuit is reported in Fig. 3.26. We suppose that the current meter has zero resistance, so the drain of M_2 is shorted to ground as well. For the drain node we can write:

$$g_{m2}V_{gs2} + g_{mb2}V_{bs2} - \frac{V_1}{r_{02}} = I_{out} \tag{3.73}$$

We again use the definitions of V_{gs2} and V_{bs2} to express them as a function of other

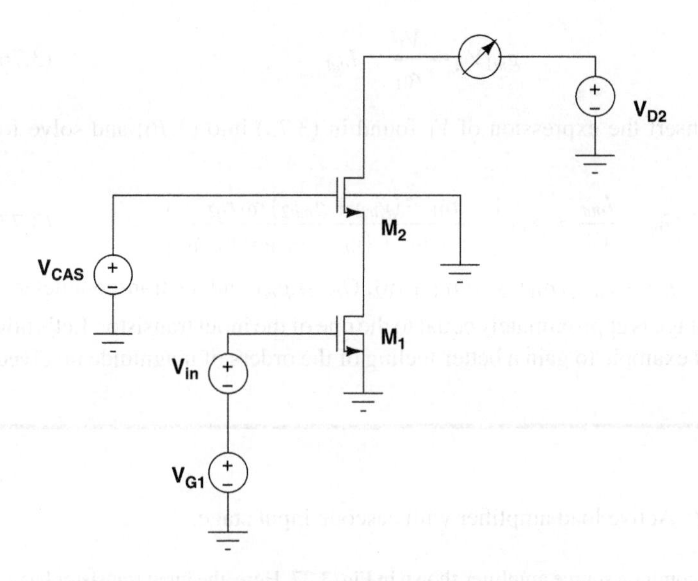

FIGURE 3.25 Simulation test bench for evaluating the G_m of the cascode.

variables in the circuit:

$$\begin{cases} V_{gs2} = V_{g2} - V_{s2} = 0 - V_1 \\ V_{bs2} = V_{b2} - V_{s2} = 0 - V_1 \end{cases} \tag{3.74}$$

In fact, the gate and the bulk are held at constant potential and the source voltage is connected to the node 1, so its voltage is V_1. We insert equation (3.74) in (3.73) and express V_1 as a function of I_{out}:

$$-g_{m2}V_1 - g_{mb2}V_1 - \frac{V_1}{r_{02}} = I_{out} \rightarrow V_1 = -\frac{r_{02}I_{out}}{[1 + (g_{m2} + g_{mb2})r_{02}]} \tag{3.75}$$

All the current must flow to ground through r_{01} and through the current source controlled by g_{m1}. We therefore have:

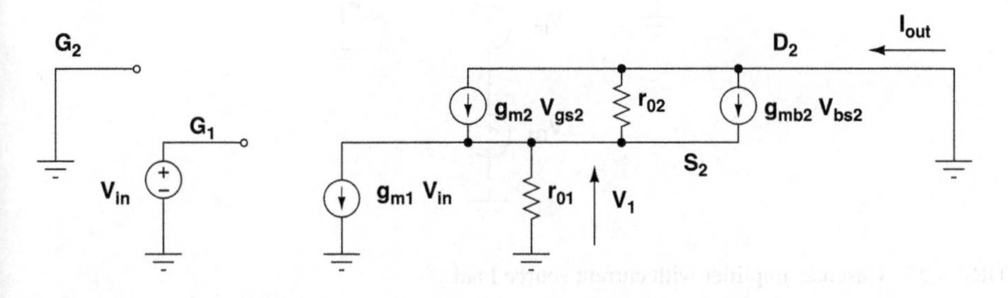

FIGURE 3.26 Small signal equivalent circuit to calculate the G_m of the cascode stage.

$$g_{m1}V_{in} + \frac{V_1}{r_{01}} = I_{out} \tag{3.76}$$

We can now insert the expression of V_1 found in (3.75) into (3.76) and solve for I_{out}/V_{in}:

$$G_m = \frac{I_{out}}{V_{in}} = g_{m1}\left[\frac{r_{01} + (g_{m2} + g_{mb2})\,r_{01}r_{02}}{r_{01} + r_{02} + (g_{m2} + g_{mb2})\,r_{01}r_{02}}\right]. \tag{3.77}$$

Observe that if $(g_{m2} + g_{mb2})\,r_{01}r_{02} \gg r_{01} + r_{02}$, $G_m \rightarrow g_{m1}$ and the transconductance of the cascode stage is approximately equal to the one of the input transistor. Let's now use a numerical example to gain a better feeling of the orders of magnitude involved.

Example 3.8: Active load amplifier with cascode input stage.

Consider the common source amplifier shown in Fig. 3.27. Here, the input transistor has been replaced with a cascode stage. The gate voltages of M_1 and M_2 have been regulated so that all transistors are in saturation and the bias current is 100 μA. Calculate the G_m, the output impedance of the cascode and the gain of the whole amplifier. Use the following device parameters: $(W/L)_1 = 50/1$, $(W/L)_2 = 20/1$, $\mu_n C_{ox} = 165\ \mu\text{A/V}^2$, $V_{THN} = 0.64$ V, $\lambda_N = 0.03\ \text{V}^{-1}$, $(W/L)_3$=$(W/L)_4 = 50/1$, $\mu_p C_{ox} = 50\ \mu\text{A/V}^2$, $|V_{THP}| = 0.6$ V, $\lambda_P = 0.03\ \text{V}^{-1}$, $V_{DD} = 2.5$ V. Assume $g_{mb2} = 0.25 g_{m2}$.

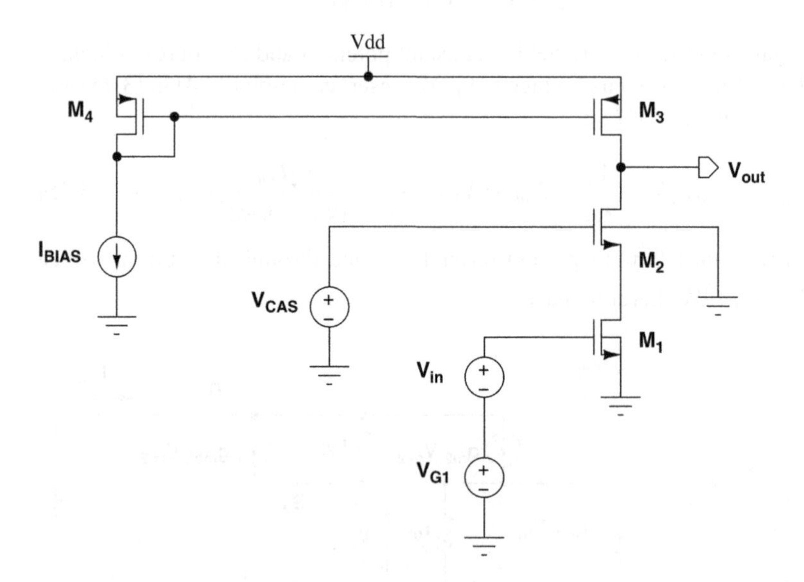

FIGURE 3.27 Cascode amplifier with current source load.

* * *

We first calculate the transconductance of M_1 and M_2. Applying the relationship:

$$g_m = \sqrt{2\mu_n C_{ox}\frac{W}{L}I_{DS}} \tag{3.78}$$

we get: $g_{m1} = 1.3$ mS, $g_{m2} = 800\ \mu$S and $g_{mb2} = 0.25 g_{m2} = 200\ \mu$S. The output resistances are given by:

$$r_0 = \frac{1}{\lambda I_{DS}} \tag{3.79}$$

so we have $r_{01} = r_{02} = r_{03} = 330\,\text{k}\Omega$. The term $(g_{m2} + g_{mb2})\, r_{01} r_{02}$ is equal to 108.9 MΩ, The approximation:

$$r_{01} + r_{02} + (g_{m2} + g_{mb2})\, r_{01} r_{02} \gg r_{01} + r_{02} \tag{3.80}$$

is thus justified, and assuming $G_m \approx g_{m1}$=1.3 mS leads to an error of only 0.3%. The gain of the amplifier is given by:

$$A_v = -G_m R_L \approx -g_{m1} R_L \tag{3.81}$$

where R_L is the load resistance, resulting from the parallel between the output impedance of the cascode and the one of M_3. The output impedance of the cascode is given by (3.71) and is equal to 109.5 MΩ. The output impedance of M_3 hence dominates in defining the value of R_L, which is 329 kΩ. The gain of the amplifier is $-1.3\cdot 10^{-3} \times 329\ \text{k}\Omega = -427.7$. If the cascode transistor M_2 is removed, the gain of the amplifier becomes $A_v = -g_{m1}\,(r_{01} // r_{03}) = 1.3\cdot 10^{-3} \times 165\ \text{k}\Omega = -214.5$.

The above example shows that cascoding a simple common source amplifier with an active load allows us to double the gain with respect to the non-cascoded counterpart, since the impedance of the cascode is so high that it is negligible when compared to the one of the load transistor. To further increase the gain we need to boost also the impedance of the load. This means applying the cascode principle also to M_3.

3.7 CASCODE CURRENT MIRRORS

Let's now consider the circuit of Fig. 3.28. Note that transistors M_3 and M_1 form a current mirror, while M_1 and M_2 implement a cascode. Therefore, the impedance looking into the drain of M_2 is given by equation (3.71). The gate of M_2 is fixed by a bias voltage source. The voltage drop across the load should be small enough so that the potential at the drain of M_2 is still adequate to keep both M_1 and M_2 in saturation. An interesting question to answer is: what is the voltage at the source of M_2? To understand the mechanism that determines this voltage, we can perform again our experiment in which the current is initially zero. If this is the case, M_2 must be off and the source potential must be closer to the gate potential than a threshold voltage: $V_{G2} - V_{S2} < V_{TH2}$, where V_{G2} and V_{S2} are the DC voltages respectively at the gate

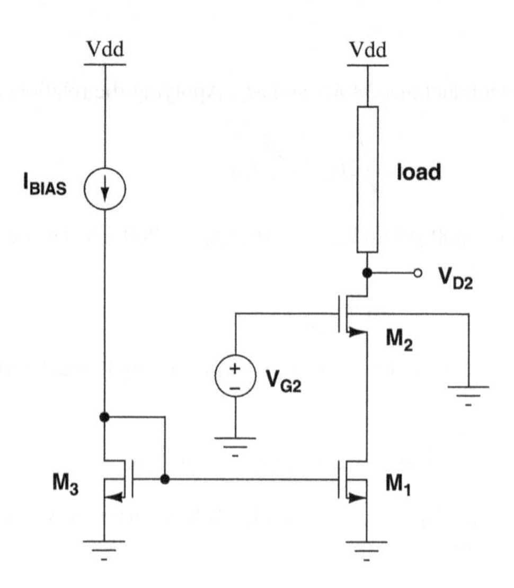

FIGURE 3.28 Principle of the cascode current mirror.

and at the source of M_2. When we switch the current source on, M_1 starts operating is saturation, since the voltage at its drain is close to V_{G2}, so it is relatively high. M_1 sinks current, but M_2 is initially off. Therefore, the carriers needed to support the current of M_1 are initially provided by the parasitic capacitances associated to the source node of M_2, which start being discharged. The source node of M_2 is hence pulled down, increasing its V_{GS} and taking the transistor into conduction. At a given point, the source potential becomes such that the current in M_2 equals the one of M_1. All the charges needed in the unit time by M_1 to support the current I_{DS1} are provided by M_2 and the discharge of the node stops. Suppose now that the gate voltage of M_2 is increased. Due to the presence of the parasitic capacitance at the source, the source potential can not change abruptly, so for a short while the gate-source voltage of M_2 is increased and the transistor drives more current. This extra current can not flow to ground, since the amount of current flowing to ground is limited by M_1, whose gate-source voltage has been left unchanged. Therefore, the potential at the source of M_2 rises till the initial V_{GS2} necessary to support the current imposed by M_1 is restored. Similarly, if V_{G2} is decreased, M_2 initially experiences a reduction of its gate-source potential, decreasing its current. M_1 starts discharging the source node of M_2 till the equilibrium is again established. From these considerations we see that the potential at the source of M_2, "follows" the potential imposed on its gate, shifted by the amount necessary to grant to M_2 the gate-source voltage necessary to supply to M_1 the requested current. Writing this in formulas, we have:

$$\begin{cases} I_{DS1} = I_{DS2} = I_{DS} \\ V_{GS2} = V_{TH2} + \sqrt{\frac{2L_2 I_{DS}}{\mu_N C_{ox} W_2}} \\ V_{S2} = V_{G2} - V_{GS2} \\ V_{S2} = V_{DS1} \end{cases} \tag{3.82}$$

From the last two equations we see that if V_{G2} is set loo low and/or V_{GS2} is too big, the voltage headroom left to M_1 might not be sufficient to allow it to work in saturation. Care must be paid in choosing both the aspect ratio of M_2 and its gate bias voltage to avoid this situation.

Example 3.9: Biasing of cascode mirrors.

Consider the circuit shown in Fig. 3.29. The reference branch sets a current of 40 μA. Determine the gate voltage V_{G2} so that the mirror is as accurate as possible. In case the load is implemented with a passive resistor, determine its maximum acceptable value. Use the following device parameters: $(W/L)_3 = 10/1$, $(W/L)_1 = 20/1$, $(W/L)_2 = 10/0.5$, $\mu_n C_{ox} = 165\ \mu\text{A/V}^2$, $V_{THN} = 0.64$ V, $\lambda_N = 0.03\ \text{V}^{-1}$, $V_{D2} = 1.25$ V. Assume that in the process triple well transistors are available.

* * *

If we have triple well transistors, we can connect the bulk and the source of M_2 and avoid the bulk effect. Since $(W/L)_1 = 2(W/L)_3$, the current mirror is designed to provide a current of 80 μA to the output branch. The V_{GS} of M_2 must be compatible with this current, hence:

$$V_{GS2} = V_{THN} + \sqrt{\frac{2L_2 I_{DS2}}{\mu_n C_{ox} W_2}} = 0.86\ V \tag{3.83}$$

The voltage at the source of M_2, V_{S2}, is given by:

$$V_{S2} = V_{G2} - V_{GS2} = V_{G2} - 0.86\ \text{V} \tag{3.84}$$

To have a precise current mirror, the drain-source voltages of M_3 and M_1 must be equal, so that the channel modulation affects both devices in the same way. Due to the diode connection in M_3, we have $V_{GS3} = V_{DS3}$. Neglecting channel modulation effects in the calculation of V_{GS3}, we have that $V_{GS3} = V_{DS3} = 0.86$ V. To have a precise mirror, also V_{DS1} must be equal to 0.86 V, which implies that V_{S2} must be 0.86 V and $V_{G2} = 0.86 + 0.86 = 1.72$ V. Note that the V_{DSsat1}, i.e. the minimum drain-source voltage needed to keep M_1 in saturation is $V_{DSsat1} = V_{GS1} - V_{THN} = 0.22$ V. Applying the calculated bias, we are therefore oversaturating M_1. The V_{DSsat} of M_2 is also 0.22 V. Observe that all the transistors have identical gate-source voltages because the aspect

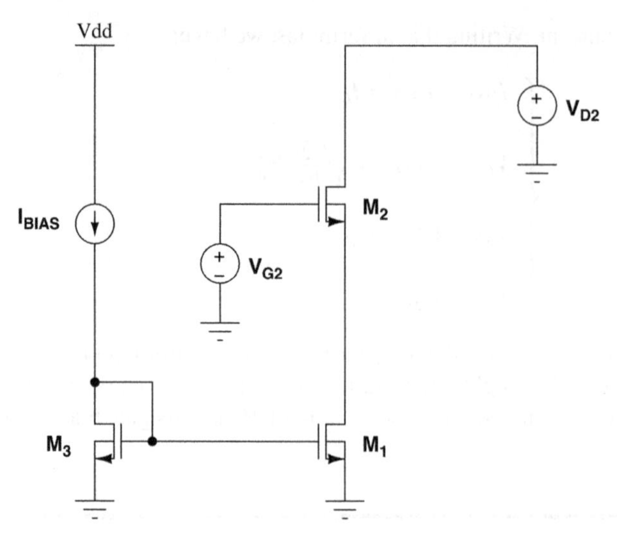

FIGURE 3.29 Cascode current mirror implemented with triple well transistors. Without the triple well option, the connection between the bulk and the source in M_2 is not possible.

ratios have been chosen to maintain the same current densities in all devices. To define the maximum acceptable value for the load resistor we can consider two different cases. In one we privilege the accuracy and in the other the voltage headroom left to the load. If we size the circuit for the best accuracy, we need to maintain V_{DS1} equal to V_{DS3}. Since the transition between the linear and saturated region is not abrupt, we need to leave also some margin to M_2 and we require that V_{DS2} is at least 100 mV above the value that would push M_2 out of saturation. Therefore the minimum acceptable voltage at the drain of M_2 is given by $0.86+0.322=1.182$ V, which implies that at most a voltage drop of $2.5-1.182=1.318$ V can be tolerated across the load. If we accept less accuracy in our mirror, we can decrease V_{GS2} from 1.72 V to 1.182 V. In fact, the source of M_2 follows the gate and by doing so it decreases the drain voltage of M_1 from 0.86 V to 0.322 V, leaving to M_1 only the minimum margin to work in saturation already calculated for M_2. As a result, the minimum acceptable value for V_{D2} is now 0.644 V and the maximum voltage drop available to the load is 1.856 V. Since the drain-source voltage of M_1 is now smaller, we expect that the current is reduced by the channel length modulation effect. We can estimate the amount of this reduction as follows:

$$\Delta I = \frac{1}{2}\mu_n C_{ox}\left(\frac{W}{L}\right)_1 (V_{GS1}-V_{THN})^2 \lambda_N \Delta V_{DS1} \tag{3.85}$$

where $\Delta V_{DS1}=0.86-0.322=0.538$ V. Inserting numbers, we have $\Delta I=1.28\mu$A, so we predict that the current in the output branch will be 78.7 μA. We will study shortly a technique that minimizes the headroom required by a cascode current mirror without compromising its accuracy.

If we do not have triple well transistors, we can not connect the bulk of M_2 to its source and the bulk effect must be considered. Also, neglecting the channel length modulation leads to overestimating slightly V_{GS}. However, the primary purpose of hand calculations is understanding the circuit behavior and getting a first grasp of what to expect from computer simulations. These are the indispensable tools to study in detail and optimize not only the DC levels, but any other circuit parameter in general. To implement V_{G2} is practice the circuit can be modified as shown in Fig. 3.30. Another transistor, M_4 has been inserted on top of M_3. Also M_4 is diode-connected. Furthermore, it has the same length of M_2 and the width is tuned so that its current density is the same of M_2. We can therefore write:

$$V_{DS1} = V_{DS3} + V_{GS4} - V_{GS2} \tag{3.86}$$

Since we have optimized M_4 so that $V_{GS4} = V_{GS2}$, we have that V_{DS3}=V_{DS1} and the mirror attains the best accuracy. In this way, a large headroom is left to M_1. However, the large signal swing will be sustained at the drain of M_2, hence the configuration of Fig. 3.30 is not optimal from the point of view of an efficient usage of the available headroom.

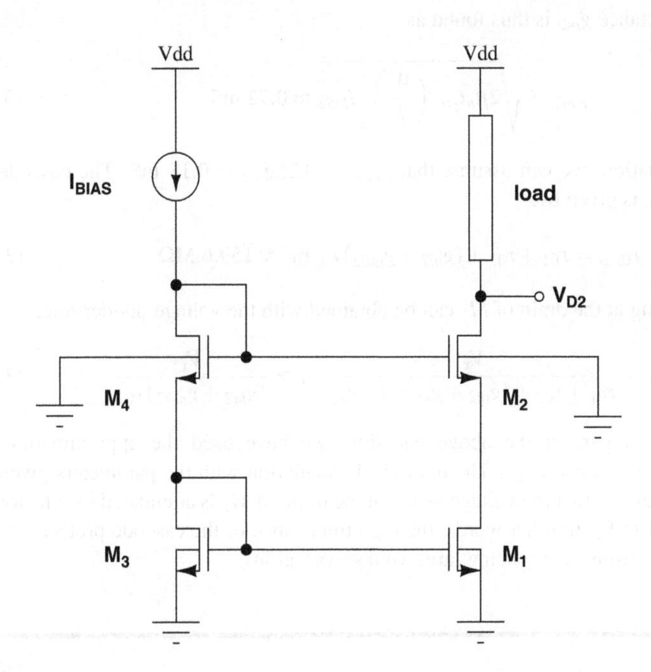

FIGURE 3.30 Cascode current mirror incorporating the biasing of the cascode transistor.

Example 3.10: Shielding effect in cascodes.

Consider the circuit shown in Fig. 3.30, in which the bias current is 40 μA. Suppose that the drain of M_2 experiences a signal of amplitude V_x. Determine the voltage swing at the drain of M_1. Use the following device parameters: $(W/L)_3 = 10/1$, $(W/L)_1 = 20/1$, $(W/L)_2 = 10/0.5$, $(W/L)_4 = 5/0.5$, $\mu_n C_{ox} = 165$ μA/V^2, $V_{THN} = 0.64$ V, $\lambda_N = 0.03$ V^{-1}.

* * *

To address the problem, we suppose that the small signal approximation can be used. A signal V_x at the drain of M_2 originates a current given by V_x/r_{0cas}, where with r_{0cas} we have indicated the output resistance of the cascode, which depends on r_{01}, r_{02}, g_{m2} and g_{mb2}. For the output resistance of the transistors we have:

$$r_{01} = r_{02} = \frac{1}{\lambda_N I_{DS}} = 420 \text{ k}\Omega \tag{3.87}$$

The transconductance g_{m2} is thus found as:

$$g_{m2} = \sqrt{2\mu_n C_{ox}\left(\frac{W}{L}\right)_2 I_{DS2}} \approx 0.72 \text{ mS} \tag{3.88}$$

For a first evaluation, we can assume that $g_{mb2} = 0.25 g_{m2} = 0.18$ mS. The cascode output resistance is given by:

$$r_{0cas} = r_{01} + r_{02} + (g_{m2} + g_{mb2}) r_{01} r_{02} \approx 159.6 \text{ M}\Omega \tag{3.89}$$

The voltage swing at the drain of M_1 can be obtained with the voltage divider rule:

$$V_{ds1} = \frac{V_x}{r_{01} + r_{02} + (g_{m2} + g_{mb2}) r_{01} r_{02}} r_{01} \approx \frac{V_x}{(g_{m2} + g_{mb2}) r_{02}} \tag{3.90}$$

To derive the last part of the above equation, we have used the approximation: $(g_{m2} + g_{mb2}) r_{01} r_{02} \gg r_{01} + r_{02}$. The numerical calculation with the parameters given in the example reveals that the voltage swing at the drain of M_1 is attenuated by a factor 378 with respect to V_x. In other words, the high impedance of the cascode protects the drain node of M_1 from experiencing large voltage variations.

3.7.1 WIDE SWING CURRENT MIRRORS

In the above example we have seen that in a cascode configuration only small variations are experienced at the drain of the driving transistor. It makes therefore sense to bias M_1 almost at the edge of the linear region by lowering the source voltage of M_2.

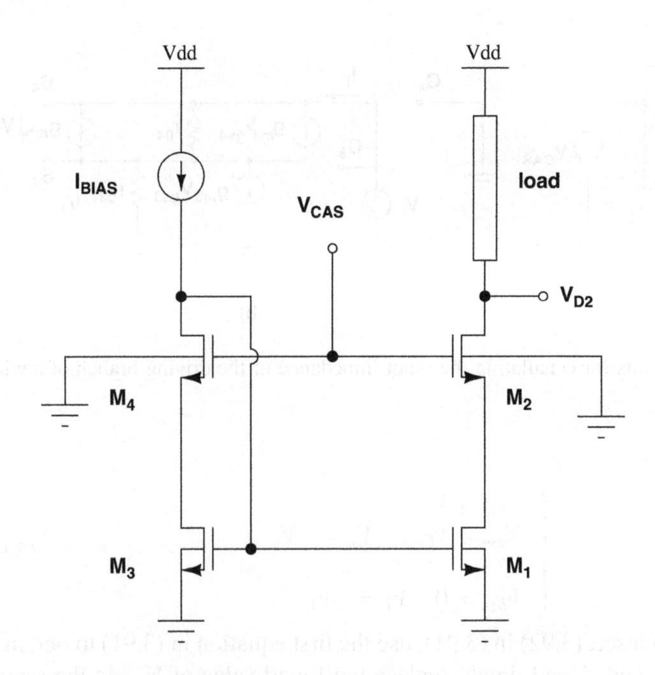

FIGURE 3.31 Wide swing cascode current mirror.

In this way, more headroom is left to M_2, which has to sustain the full signal swing. In example 3.9 we have also seen that it is important to equalize the drain-source voltages of the reference and output transistors (M_3 and M_1 in Fig. 3.30) to obtain the best mirroring accuracy. The circuit that can satisfy both requirements is called a "wide-swing cascode" and it is shown in Fig. 3.31. The key change with respect to the configuration described in Fig. 3.30 is that the gate of M_3 is connected to the drain of M_4, which is still needed to equalize the drain-source voltages of M_3 and M_1. Note that now we have $V_{DS3} + V_{DS4} = V_{GS3}$. In other words, it is not automatically granted that M_3 is in saturation and its gate-source voltage must be large enough to accommodate two saturation voltages. The node V_{CAS} must be connected to a suitable bias voltage. We must now verify that the impedance looking into the drain of M_3 is still adequately small. To to so, we design an appropriate test bench incorporating only the relevant devices, which are M_3 and M_4. As usual, we apply a test voltage V_t between the point of interest and ground and we measure the current I_t absorbed by the circuit. The test bench and its equivalent small signal model are reported in Fig. 3.32 a) and b), respectively. We can write the following set of nodal equations:

$$\begin{cases} g_{m4}V_{gs4} + g_{mb4}V_{bs4} + \frac{V_t - V_1}{r_{04}} = I_t \\ g_{m3}V_{gs3} + \frac{V_1}{r_{03}} - g_{m4}V_{gs4} - g_{mb4}V_{bs4} + \frac{V_1 - V_t}{r_{04}} = 0 \rightarrow g_{m3}V_{gs3} + \frac{V_1}{r_{03}} = I_t \end{cases} \tag{3.91}$$

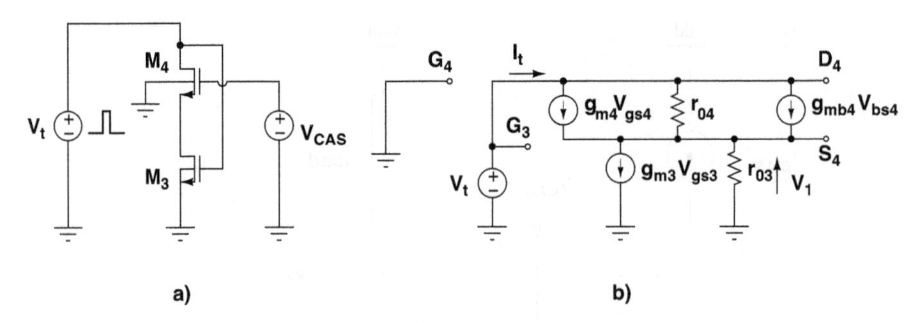

FIGURE 3.32 Circuits for calculating the input impedance of the driving branch of a wide-swing cascode.

We also have that:

$$\begin{cases} V_{gs3} = V_t \\ V_{gs4} = V_{CAS} - V_1 = -V_1 \\ \\ V_{bs4} = 0 - V_1 = -V_1 \end{cases} \tag{3.92}$$

The next step is to insert (3.92) in (3.91), use the first equation in (3.91) to obtain V_1 as a function of I_t and V_t and finally replace the found value of V_{gs4} in the second equation of (3.91) to obtain V_t/I_t. After some algebra, one gets:

$$\frac{V_t}{I_t} = \frac{r_{03} + r_{04} + (g_{m4} + g_{mb4})\, r_{03} r_{04}}{1 + g_{m3}\left[r_{03} + (g_{m4} + g_{mb4})\, r_{03} r_{04}\right]} \tag{3.93}$$

Using now the approximations: $r_{03} r_{04} (g_{m4} + g_{mb4}) \gg r_{03} + r_{04}$, we have that:

$$\frac{V_t}{I_t} \approx \frac{1}{g_{m3}} \tag{3.94}$$

which shows that the impedance looking into the drain of M_4 is defined by the transconductance of M_3. If the gate of M_3 is connected to the drain of M_4, the structure formed by the two devices still provides a low impedance path to the reference current. Interestingly, the above results can be obtained also by applying feedback theory. In fact, when the connection between the gate of M_3 and the drain of M_4 is absent, the two transistors form a standard cascode, as shown in Fig. 3.33 a). The open loop gain of the circuit is given by $G_m r_{cas}$ where:

$$G_m = g_{m3}\left[\frac{r_{03} + (g_{m4} + g_{mb4})\, r_{03} r_{04}}{r_{03} + r_{04} + (g_{m4} + g_{mb4})\, r_{03} r_{04}}\right] \tag{3.95}$$

and r_{0cas} is:

$$r_{0cas} = r_{03} + r_{04} + (g_{m4} + g_{mb4})\, r_{03} r_{04} \tag{3.96}$$

Therefore, the open-loop gain is given by:

$$A_v = -g_{m3}\left[r_{03} + (g_{m4} + g_{mb4})\, r_{03} r_{04}\right] \tag{3.97}$$

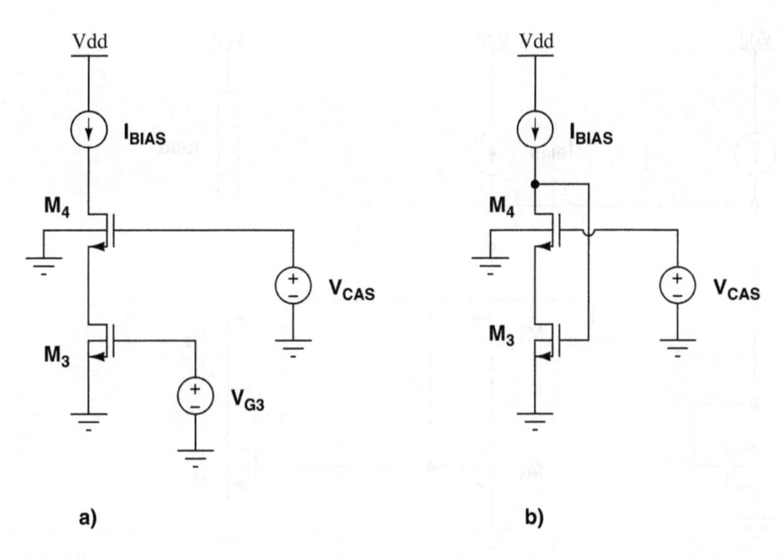

FIGURE 3.33 The reference branch of a wide-swing mirror can be seen as a cascode amplifier with feedback.

In Fig. 3.33 b), the output of the amplifier is short-circuited to the input, performing a unity gain feedback with $f = 1$. The open-loop output resistance is r_{Ocas}, while the closed-loop resistance is given by:

$$r_{0,CL} = \frac{r_{Ocas}}{1 + A_v f} = \frac{r_{03} + r_{04} + (g_{m4} + g_{mb4})\, r_{03} r_{04}}{1 + g_{m3}\left[r_{03} + (g_{m4} + g_{mb4})\, r_{03} r_{04}\right]} \tag{3.98}$$

which coincides with (3.93). Finally, we must have a means to provide the voltage reference V_{CAS}. To do so, we can simply used a diode-connected transistor, which is suitably biased and sized to provide the required voltage level. The resulting circuit is shown in Fig. 3.34.

Example 3.11: Generating the cascode voltage for a wide-swing current mirror.

Using the parameters of example 3.9, size M_5 to generate the appropriate cascode voltage for M_2 and M_4.

* * *

In example 3.9 it was calculated that the minimum gate voltage for M_2 and M_4 was 1.182 V. This is also the gate-source voltage of the biasing transistor. Once the desired V_{GS} is known, the aspect ratio of M_5 can be calculated as:

$$\left(\frac{W}{L}\right)_5 = \frac{I_{DS5}}{\frac{1}{2}\mu_n C_{ox}\left(V_{GS5} - V_{THN}\right)^2} \tag{3.99}$$

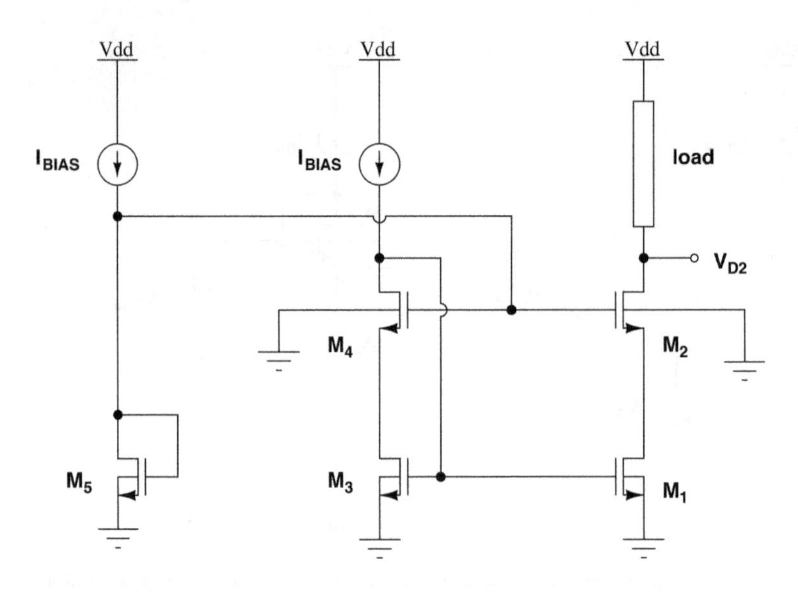

FIGURE 3.34 Generation of the cascode voltage in a wide-swing current mirror.

Assuming we use for M_5 the same current for the reference branch and putting in numbers, we find $(W/L)_5 = 1.65$. As a first step, we can chose $W_5 = 1.6$ and L_5=1. The result can be further refined with the help of computer simulations.

In many applications, it is useful to have current mirrors based on PMOS transistors. The following example clarifies their design.

Example 3.12: Cascode PMOS current mirror.

A cascode current mirror is implemented with PMOS transistors, as shown in Fig. 3.35. The reference current is 20 μA and the load must receive 100 μA. The aspect ratio of the reference transistor M_3 is 5/0.5. Size the transistors in the output branch to deliver the appropriate current to the load. Calculate the maximum voltage the load can develop without compromising the functionality of the mirror. Redesign the circuit employing the wide-swing topology and calculate the output impedance of the mirrors. Use the following device parameters: $\mu_p C_{ox} = 50\ \mu\text{A/V}^2$, $|V_{THP}| = 0.6$ V, $\lambda_P = 0.03\ \text{V}^{-1}$, $V_{DD} = 2.5$ V.

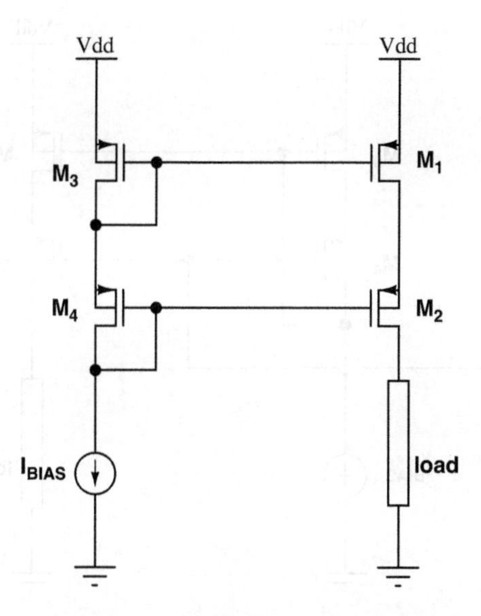

FIGURE 3.35 Cascode current mirror implemented with PMOS transistors.

* * *

The output current $I_{SD1} = I_{SD2}$ is defined by the ratio:

$$\frac{I_{SD1}}{I_{SD3}} = \frac{(W/L)_1}{(W/L)_3} = 5 \rightarrow (W/L)_1 = 25/0.5 \tag{3.100}$$

The voltage at the drain of M_3 is $V_{D3} = V_{DD} - V_{SG3}$ and must be equal to the voltage at the drain of M_1, V_{D1}. Note that V_{D1} and V_{D3} are the voltages measured between the drain of the transistors and ground. To guarantee the equality between V_{D1} and V_{D3}, the gate source voltages of M_4 and M_2 must be equal, in which case we have:

$$V_{D1} = V_{DD} - V_{SG3} - V_{SG4} + V_{SG2} = V_{DD} - V_{SG3} = V_{D3} \tag{3.101}$$

The above equation is in fact true if $V_{SG4} = V_{SG2}$. To have their gate-source voltages equal, M_4 and M_2 must have the same current density, so the relationship between their aspect ratios must be the same as the one between the aspect ratios of M_1 and M_3: $(W/L)_2 = 5\,(W/L)_4$. We observe that this does not necessarily imply that $(W/L)_4$ must be equal to $(W/L)_3$ and that $(W/L)_1$ is equal to $(W/L)_2$. However, this is a convenient choice, since using a limited number of transistors with different sizes can simplify the circuit layout. To calculate V_{D3} we need to know V_{SG3}, which is:

$$V_{SG3} = |V_{THP}| + \sqrt{\frac{2L_3 I_{SD3}}{\mu_p C_{ox} W_3}} \approx 0.883 \text{ V} \tag{3.102}$$

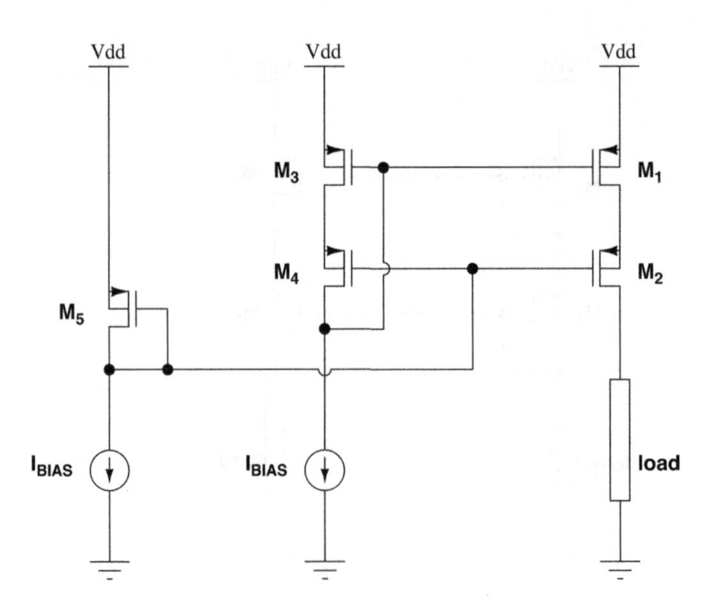

FIGURE 3.36 Wide-swing cascode current mirror implemented with PMOS transistors.

The voltage V_{D3} is hence: $V_{D3} = V_{DD} - V_{SG3} = 2.5\ V - 0.883\ V = 1.617$ V. Applying (3.102), the reader can immediately verify that $V_{SG4} = V_{SG2} = 0.883$ V. Inserting the numbers in (3.101) we have $V_{D2} = 1.617$ V=V_{D3} as requested. The voltage across the load must be low enough so that both M_1 and M_2 work in saturation, therefore we need to have:

$$V_{SD} > V_{SG} - |V_{THP}| \tag{3.103}$$

Since all transistors have been designed to have the same gate-source voltage, their saturation voltages are given by $0.883 - 0.6 = 0.283$ V. The drain-source voltage of M_1 is 0.883 V, so it is largely sufficient to saturate the device. M_2 is pushed in linear region when the voltage at its drain rises above $1.617 - 0.283 = 1.334$ V. In practice, the transition between the saturation and the linear region is smooth, so an extra margin of about 100 mV must be allowed. Hence, the voltage at the drain of M_1 should not rise above 1.234 V. Let's now consider the wide-swing configuration, shown in Fig. 3.36. On the left branch, the voltage at the drain node of M_4 is given by $V_{D4} = V_{DD} - V_{SG3} = 2.5 - 0.883 = 1.617$ V. Since the current in M_4 is fixed, its source voltage follows the gate voltage, i.e.:

$$V_{S4} = V_{D3} = V_{G5} + V_{SG4} \tag{3.104}$$

where with V_{G5} we indicate the gate voltage of M_5 measured with respect to ground. Let's choose V_{D3} so that V_{SD3} is 100 mV above the minimum saturation voltage of 0.283 V, so $V_{D3} = V_{DD} - 0.383 = 2.117$ V. This implies that the gate voltage of M_5 must be at $2.117 - 0.883 = 1.234$ V. The gate voltage of M_5 is given by:

$$V_{G5} = V_{DD} - V_{SG5} \rightarrow V_{SG5} = V_{DD} - V_{G5} = 2.5 - 1.234 = 1.266\ \text{V} \tag{3.105}$$

This allows us to calculate the aspect ratio of M_5:

$$\left(\frac{W}{L}\right)_5 = \frac{I_{DS5}}{\frac{1}{2}\mu_n C_{ox}\left(V_{GS5} - |V_{THP}|\right)^2} \tag{3.106}$$

If we suppose to use for M_5 the same current of the reference branch (20 μA), this yields $(W/L)_5 = 1.8$. For instance, we can choose $L_5 = 0.5$ μm and $W_5 = 0.9$ μm. Observe that to keep both M_1 and M_2 in saturation, we need a minimum voltage drop across them equal to the sum of their saturation voltages, plus eventually the 100 mV margin, hence 766 mV. The voltage at the drain of M_2 can thus rise up to $2.5 - 0.766 = 1.734$ V, so we have gained about 500 mV in swing with respect to the previous scheme.

We can now calculate the output impedance of the mirror, which is identical for both configurations. Neglecting the bulk effect, we need to find the output resistance of M_2 and M_1 and the transconductance of M_1:

$$\begin{cases} r_{01} = r_{02} = \frac{1}{\lambda_P I_{SD}} = \frac{1}{0.03 \times 1 \cdot 10^{-4}} = 330 \text{ k}\Omega \\ g_{m2} = \sqrt{2\mu_p C_{ox}\left(\frac{W}{L}\right)_2 I_{SD2}} = 0.7 \text{ mS} \end{cases} \tag{3.107}$$

The output impedance is given by:

$$r_{Ocas} = r_{01} + r_{02} + g_{m2} r_{01} r_{02} = 69.9 \text{ M}\Omega \tag{3.108}$$

In the reference branch of the wide-swing cascode, the gate-source voltage of M_3 must be greater than two saturation voltages. It is interesting to observe that in some contexts, this condition can be difficult to meet. This is in particular true when a compact mirror controlling a small current has to be implemented in a very deep submicron CMOS technology, as shown in the next example.

Example 3.13: Wide swing cascode at low current.

The wide-swing current mirror of Fig. 3.31 is implemented in a 65 nm CMOS technology, where the NMOS transistors have a $\mu_n C_{ox}$ of 650 μA/V^2 and a threshold voltage of 250 mV. The mirror must deliver a current of 10 nA. Calculate the minimum aspect ratio of M_3 that can keep both M_3 and M_4 in saturation.

* * *

Let's first suppose that we set $(W/L)_3 = 5$ μm/0.5 μm, as done previously. The inversion coefficient I_C allows us to calculate the operating region of the transistor:

$$I_C = \frac{I_{DS}}{2n\mu_n C_{ox}\frac{W}{L}\phi_T^2} = \frac{1 \cdot 10^{-8}}{2 \times 1.3 \times 650 \cdot 10^{-6} \times 10 \times 6.7 \cdot 10^{-4}} = 8.82 \cdot 10^{-4} \tag{3.109}$$

Since $I_C \ll 1$, the device operates deep in weak inversion. We can therefore use the weak inversion characteristics to estimate $V_{GS} - V_{TH}$:

$$I_{DS} = 2n\mu_n C_{ox}\frac{W}{L}\phi_T^2 e^{\frac{V_{GS}-V_{TH}}{n\phi_T}}\left(1 - e^{-\frac{V_{DS}}{\phi_T}}\right) \tag{3.110}$$

If we assume for simplicity that the device works in saturation, we can neglect the dependence on V_{DS} and write $V_{GS} - V_{TH}$ as:

$$V_{GS} - V_{TH} = n\phi_T \ln\left(\frac{I_{DS}}{2n\mu_n C_{ox}\frac{W}{L}\phi_T^2}\right) = n\phi_T \ln I_C \tag{3.111}$$

Putting in numbers, one gets $V_{GS} - V_{TH} = -237$ mV or $V_{GS} = 13$ mV. The voltage and the gate of M_3 and at the drain of M_4 is thus too low to keep the devices in saturation. Let's now change the aspect ratio of M_3 to put it at the center of moderate inversion, with $I_C = 1$. Solving (3.109) for $I_C = 1$, we get $W/L = 8.82 \cdot 10^{-3} \approx 1/113$. Choosing, for instance, $W = 0.2$ μm, the required L is 22.6 μm. We can now calculate the gate source voltage using the general characteristics:

$$I_{DS} = 2n\mu_n C_{ox}\frac{W}{L}\phi_T^2\left[\ln\left(1 + e^{\frac{V_{GS}-V_{TH}}{2n\phi_T}}\right)\right]^2 \tag{3.112}$$

from which we can write $V_{GS} - V_{TH}$ as:

$$V_{GS} - V_{TH} = 2n\phi_T \ln\left(e^{\sqrt{I_C}} - 1\right) \tag{3.113}$$

which yields $V_{GS} - V_{TH} \approx 36$ mV and $V_{GS} = 286$ mV. The example shows that caution must be used when transistors that need to control a small current are drawn in a scaled technology. In this respect, PMOS devices that have a lower current drive capability are preferable. We see also that deep in weak inversion, the assertion that a diode-connected transistor is always is saturation might not be true.

3.8 TELESCOPIC CASCODE AMPLIFIERS

The cascode current mirror can provide the load to a cascode input stage. This obtains amplifiers with much higher gain. In Fig. 3.37 a) the configuration with NMOS input stage and PMOS load is shown. In Fig. 3.37 b) a PMOS input stage with NMOS load is reported. The criteria for choosing one configuration rather than the other will be discussed later in the book. As we have seen, the transconductance of the cascode is basically given by the one of the input transistor. The load is provided by the parallel combination of the output resistance of the two cascode stages. In a first approximation, a cascode can therefore be represented with the same small signal equivalent circuit introduced for the common source amplifier, with g_m given by the

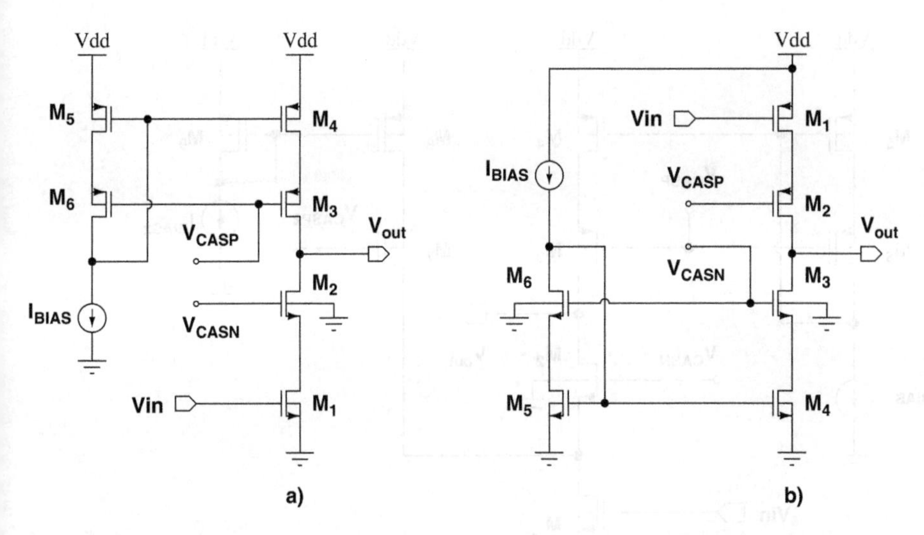

FIGURE 3.37 Telescopic cascode amplifiers. The configuration with NMOS input and PMOS load in shown in a) and the complementary one with PMOS input and NMOS load in b).

transconductance of the input transistor and R_L by the parallel combination of the cascode resistances. The gain can hence be written as:

$$A_v = -g_{m1}R_L = -g_{m1}\left(r_{0casN}//r_{0casP}\right) =$$

$$= -g_{m1}\left[r_{01} + r_{02} + (g_{m2} + g_{mb2})\,r_{01}r_{02}\right] // \left[r_{03} + r_{04} + (g_{m3} + g_{mb3})\,r_{03}r_{04}\right] \quad (3.114)$$

In the circuits of Fig. 3.37 there are four transistors connected between the power supply rails and all of them must work in saturation. If the transistors are in strong inversion, we can assume that a minimum drain-source voltage of 0.2 V is necessary to keep them properly biased. The output can then swing by $V_{DD} - 0.8$ V before pushing one of the devices in the linear region. If the circuit is powered at 2.5 V, we have an available output dynamic range of 1.7 V. The use of thin oxide transistors in very deep submicron technologies limit the power supply voltage to 1 V. The headroom at the cascode output is hence very small, unless the transistors are biased in weak inversion. In this case, a drain-source voltage of 100 mV is sufficient to saturate the devices, and an output dynamic range of 0.5 V÷0.6 V is still possible. The configurations shown in Fig. 3.38 and Fig. 3.39 allow for a better optimization of the cascode amplifier for low power operation. In this approach, the bias current is provided by two independent branches. One branch powers the cascode devices with only a fraction of the total current, while most of the bias current is delivered to the input transistor by the additional branch. Observe that in Fig. 3.38 and Fig. 3.39 the current source formed by $M_7 - M_9$ is designed as completely independent of the one biasing the cascode. The reference branch can of course be shared by the two sources, making the circuit more compact at the expense of some loss in flexibility. Especially in a prototype, it can be useful in fact to tune the two currents independently. Furthermore, in the branch

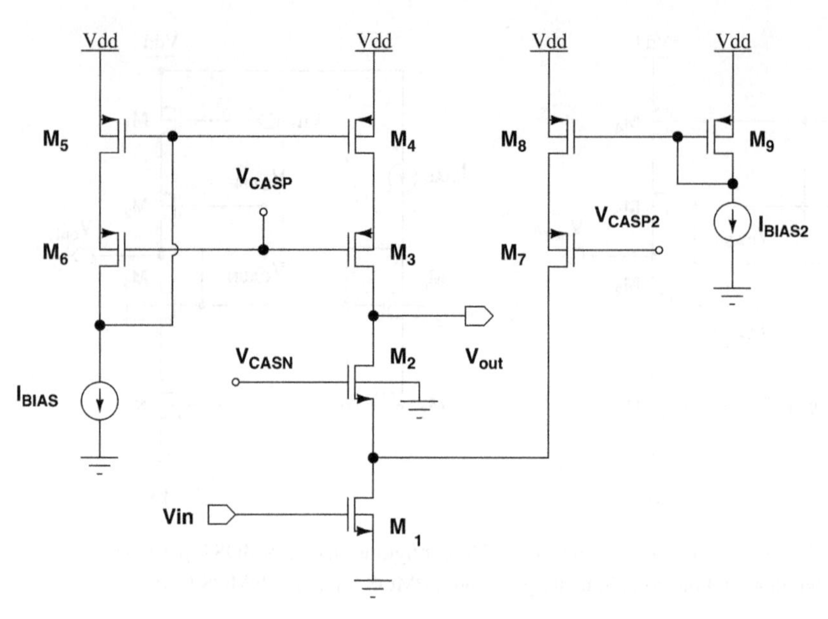

FIGURE 3.38 Telescopic cascode amplifiers with split bias current and NMOS input transistors.

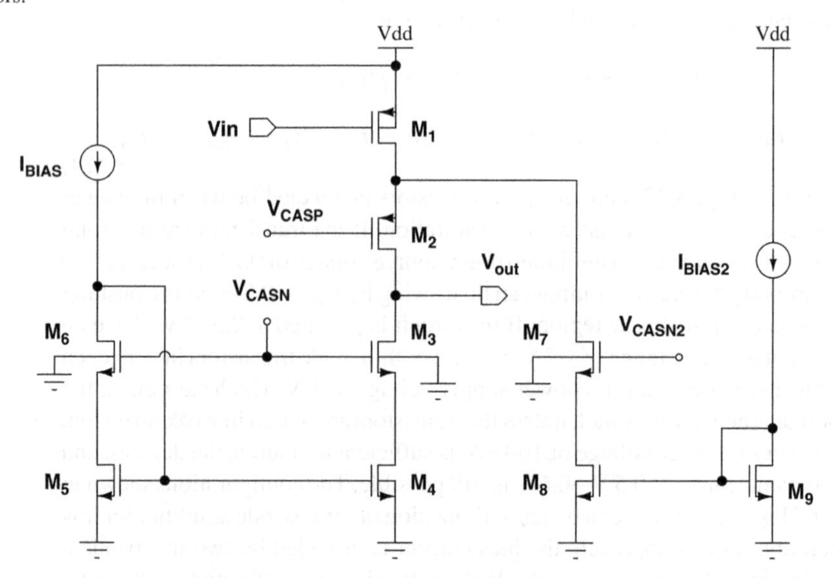

FIGURE 3.39 Telescopic cascode amplifiers with split bias current and PMOS input transistors.

biasing directly the input transistor only three devices are stacked and the voltage at the drain of M_1 exhibits little variation. Therefore, V_{CASP2} can be kept lower than V_{CASP1}, leaving more headroom to M_8, thereby improving its output impedance.

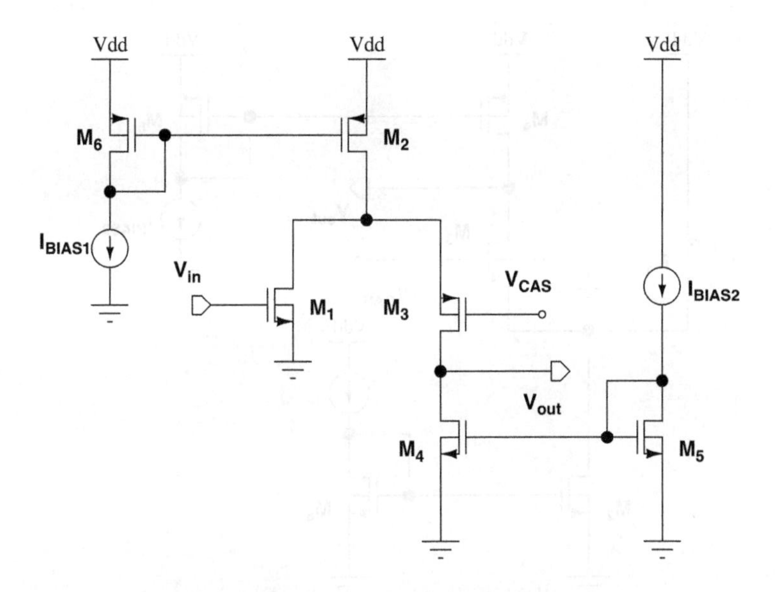

FIGURE 3.40 Folded cascode amplifiers with NMOS input transistor.

3.8.1 FOLDED CASCODE

In the folded cascode the input and the cascode transistor are of complementary types. Fig. 3.40 shows the stage with NMOS input transistor and PMOS cascode, while the complementary configuration is reported in Fig. 3.41. The gain of the circuit can be increased by cascoding both M_2 and M_4. A drawback of this topology is that the current flowing in the input device is the *difference* between the total current provided by M_2 and the bias current of the cascode branch, while in the telescopic cascode with splitting it is the *sum* of the two components. Therefore, for the same total current, in the folded cascode the g_m the input device is smaller, which implies more noise and less bandwidth. If M_2 is not cascoded, the left branch of the circuit only stacks two transistors (M_1 and M_2), so the source of M_1 can be connected to a different potential, reducing the power consumption. This is particularly advantageous in systems employing dual power supply rails (i.e. a positive and a negative supply symmetric with respect to ground), in which case the source of the input device is naturally tied to ground.

3.9 UNBUFFERED TRANSRESISTANCE AMPLIFIERS

The amplifiers studied in the previous section are very commonly employed in the design of front-end for radiation detectors. Cascode amplifiers have however very large gain and they can therefore only be used in feedback configurations. The simplest feedback is obtained by connecting the output to the input with a wire, as done in Fig. 3.42 a). Note however that once feedback is established, the amplifier can not be driven anymore by an ideal voltage source, because the source would override

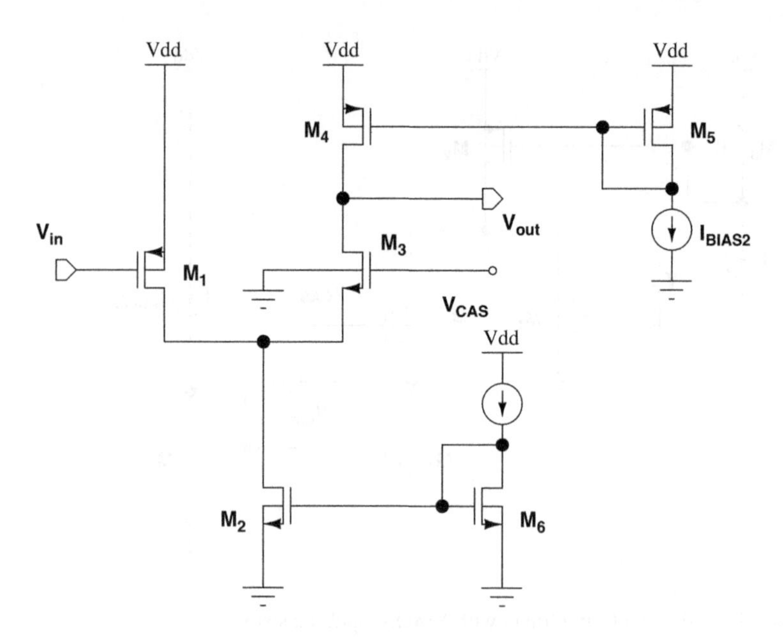

FIGURE 3.41 Folded cascode amplifiers with PMOS input transistor.

any feedback signal, making the loop ineffective. We can then connect to the input a current source. Since the gate of the transistor has a high impedance, the current will flow in the feedback path. However, no voltage drop would be seen across the short-circuit. The connection of Fig. 3.42 a) is therefore only useful to create a low impedance load that can serve as the reference branch of a wide-swing cascode. Replacing the short-circuit with a resistance, we get the circuit of Fig. 3.42 b). If $I_{in} = 0$, the voltage drop across R_f must again be zero because the gate of the input transistor does not sink current. If, for some reason, the current in M_1 is lower than the one imposed by the load, the output voltage rises and so does the input, till the equilibrium is re-established. The resistor connected between the input and the output of the amplifier thus provides a feedback that keeps the circuit properly biased.

To see the response to the input signal, we first consider the circuit shown in Fig. 3.43 a), which shows a transimpedance amplifier with a feedback network implemented with a passive resistor. The small signal equivalent circuit is reported in Fig, 3.43 b). Applying equation (3.2) with $Z_f = R_f$ we have:

$$\frac{V_{out}}{I_{in}} = \frac{A_0}{1+A_0} R_f \tag{3.115}$$

Let's now study what happens if we replace the core amplifier with a cascode. As a first sight, we could surmise that the gain is given by (3.115) just by replacing A_0 with $g_m R_L$. Fig. 3.43 c) shows the circuit with the cascode represented with the simplified

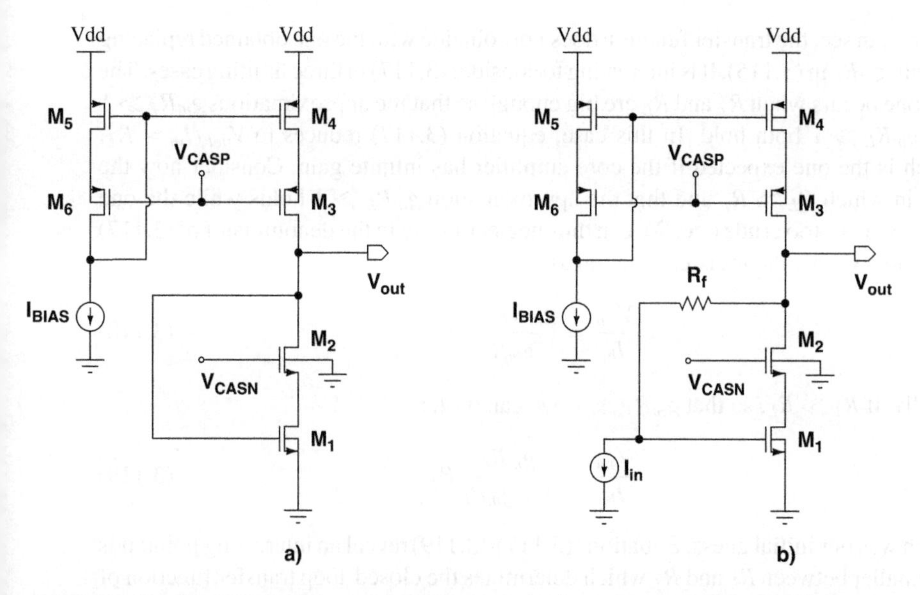

FIGURE 3.42 Cascode amplifiers with feedback to stabilize the circuit DC levels. The circuit in b) forms a transresistance amplifier.

small signal equivalent circuit. The nodal equations now read:

$$\begin{cases} I_{in} + \frac{V_{in}-V_{out}}{R_f} = 0 \\ g_m V_{in} + \frac{V_{out}}{R_L} + \frac{V_{out}-V_{in}}{R_f} = 0 \end{cases} \tag{3.116}$$

Solving for V_{out}/I_{in} we find:

$$\frac{V_{out}}{I_{in}} = \frac{R_L\left(g_m R_f - 1\right) R_f}{R_f\left(1 + g_m R_L\right)} \tag{3.117}$$

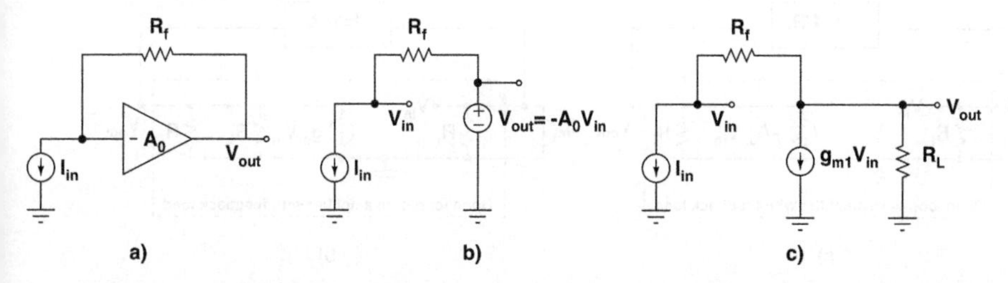

FIGURE 3.43 In a) the concept of a transresistance amplifier built with a high gain voltage amplifier and a feedback resistor is shown. The circuit in b) is the equivalent small signal model of the amplifier in a). The circuit in c) is the simplified equivalent small signal model in case a common-source amplifier is used as the core stage.

As one can see, the transfer function does not coincide with the one obtained replacing A_0 with $g_m R_L$ in (3.115). It is interesting to consider (3.117) in three limiting cases. The first one occurs when R_f and R_L are big enough so that the approximations $g_m R_f \gg 1$ and $g_m R_L \gg 1$ both hold. In this case, equation (3.117) reduces to $V_{out}/I_{in} = R_f$, which is the one expected if the core amplifier has infinite gain. Consider now the case in which $R_L \gg R_f$ and that the approximation $g_m R_L \gg 1$ holds while the one $g_m R_f \gg 1$ is a too crude one. We can thus neglect 1 only in the denominator of (3.117) and approximate the transfer function as:

$$\frac{V_{out}}{I_{in}} = \frac{g_m R_f - 1}{g_m R_f} R_f \tag{3.118}$$

Finally, if $R_f \gg R_L$, so that $g_m R_f \gg 1$, we can write:

$$\frac{V_{out}}{I_{in}} = \frac{g_m R_L}{1 + g_m R_L} R_f \tag{3.119}$$

which was our initial guess. Equations (3.117)-(3.119) reveal an interesting point: it is the smaller between R_L and R_f which determines the closed-loop transfer function of the circuit, as if the two resistors were in parallel. As discussed at the beginning of the chapter, the feedback resistor represents in fact a load for the core amplifier and the transresistance amplifier of Fig. 3.43 a) can be thought as made of two components: an open-loop transresistance amplifier and a feedback network which does not load the amplifier and has a transfer function $1/R_f$, as shown in Fig. 3.44. The open loop transresistance amplifiers incorporates also the loading effect of the feedback network, which is seen from the point of view of the core amplifier as two resistors of value R_f connected one in parallel with the input and another one in parallel to the output. With reference to Fig. 3.44, the open loop transresistance gain is given as:

$$A_{T,OL} = \left.\frac{V_{out}}{I_{in}}\right|_{OL} = R_f g_m \frac{R_L R_f}{R_L + R_f} \tag{3.120}$$

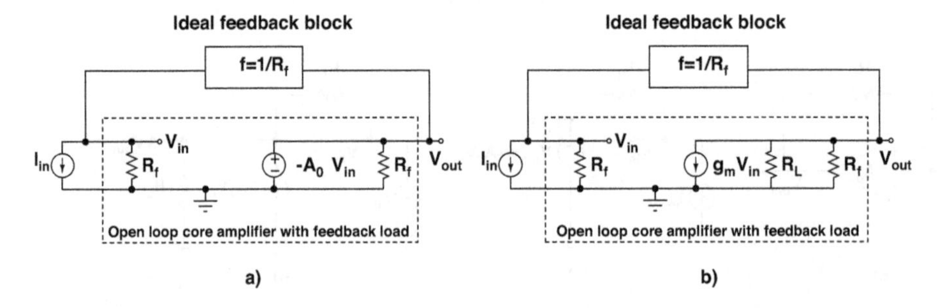

FIGURE 3.44 Model of a transresistance amplifier that considers the load of the feedback network on the core amplifier. The model reported in a) is suitable if the core amplifier is an ideal voltage amplifier. The circuit in b) is appropriate if the core amplifier is a transconductance amplifier.

The closed-loop gain is thus obtained as:

$$A_{T,CL} = \frac{A_{T,OL}}{1 + A_{T,OL}f} = \frac{g_m R_L R_f}{R_L + R_f + g_m R_L R_f} \tag{3.121}$$

If in the above equation we assume that $R_f \gg R_L$, we get back equation (3.119). If, on the other hand, we have $R_L \gg R_f$, we obtain:

$$\frac{V_{out}}{I_{in}} = \frac{g_m R_f}{1 + g_m R_f} R_f \tag{3.122}$$

which is similar, but not identical, to (3.118). The discrepancy arises from the fact that the adopted feedback model implicitly assumes that the feedback network is unilateral, i.e. it generates only a feedback signal proportional to the output voltage, whereas in reality the current through R_f depends also on the input voltage V_{in}. Due to the high gain of the amplifier, V_{in} is usually much smaller than V_{out} and if in the second equation in (3.116) we take $V_{out} - V_{in}R_f \approx V_{out}/R_f$ and solve the system, we get again equation (3.122).

3.10 SOURCE FOLLOWERS

As seen in the previous section, the output impedance of the cascode configuration is high, therefore the amplifier is very sensitive to any load it is connected to. It is thus necessary to "protect" the high impedance node, where the current to voltage conversion is carried out, in order to preserve the circuit performance. To achieve this, we need a circuit which has very high input and reasonably low output impedance. Called *source follower*, the configuration suitable for this purpose is shown in Fig. 3.45. We can study most of the properties of the circuit by reusing results already discussed in previous sections. In fact, in 3.7 we have seen that if we fix both the gate voltage and the drain-source current in an NMOS transistor, the voltage at the source is equal to the one imposed on the gate minus the gate-source voltage needed by M_1 to let the imposed bias current flow, hence the name "source follower". For a PMOS configuration with the same constraints, the source voltage is the gate voltage plus the source-gate voltage needed to support the bias current. The drain of the input transistor is connected to V_{DD} for the NMOS and to ground for the PMOS configuration. In either case, the drain is connected to a signal ground, from which the circuit derives its alternative name of "common drain" amplifier. The equivalent small signal circuit is shown in Fig. 3.46. We can easily recognize that this is a particular case of the common source amplifier with degeneration resistance, where R_D is set to zero and R_S is provided by the output resistance of the current source. The signal is taken at the device source, so the relevant quantity is V_s/V_{in}. We can calculate the gain of our circuit by using equation (3.48) with the substitutions: $V_s = V_{out}$, $r_0 = r_{01}$, $R_D = 0$, $R_S = r_{02}$, which yields:

$$\frac{V_{out}}{V_{in}} = \frac{g_{m1} r_{01} r_{02}}{r_{01} + r_{02} + (g_{m1} + g_{mb1}) r_{01} r_{02}} \tag{3.123}$$

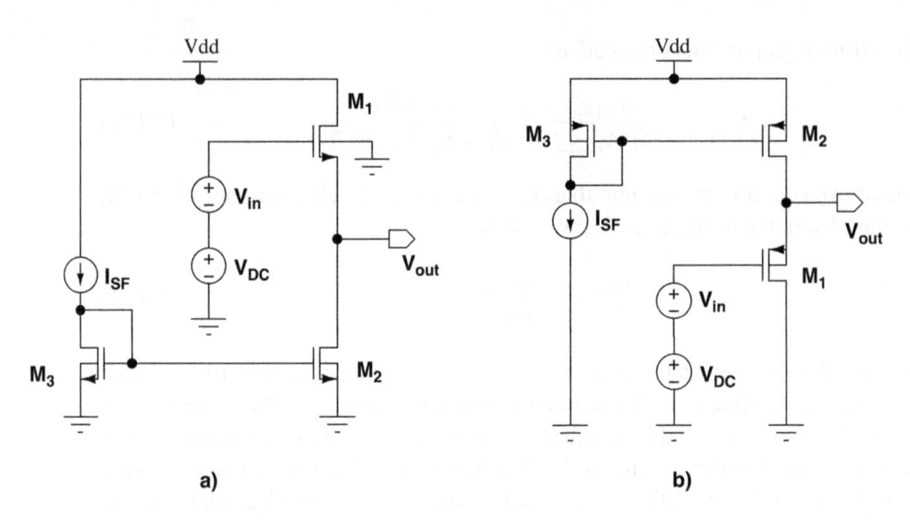

FIGURE 3.45 Source followers. The NMOS-based configuration is shown in a) and the PMOS-based in b).

Under the usual hypothesis that $(g_{m1} + g_{mb1}) r_{01} r_{02} \gg r_{01} + r_{02}$, equation (3.123) reduces to:

$$\frac{V_{out}}{V_{in}} = \frac{g_{m1}}{g_{m1} + g_{mb1}} \tag{3.124}$$

Note that in presence of bulk effect, the gain of the circuit is less than one, while if $g_{mb} = 0$ it attains approximately unity. We need to check now the output impedance, which can be calculated with the help of Fig. 3.47. Observing the figure, we can write:

$$\begin{cases} -g_{m1} V_{gs} - g_{mb1} V_{bs} + \frac{V_t}{r_{01}} = I_t \\ V_{gs} = V_{bs} = 0 - V_t = -V_t \\ (g_{m1} + g_{mb1}) V_t + \frac{V_t}{r_{01}} = I_t \end{cases} \tag{3.125}$$

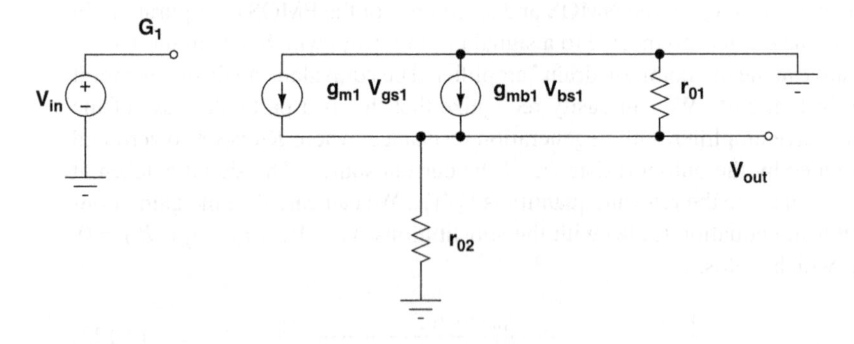

FIGURE 3.46 Source follower small signal equivalent circuit.

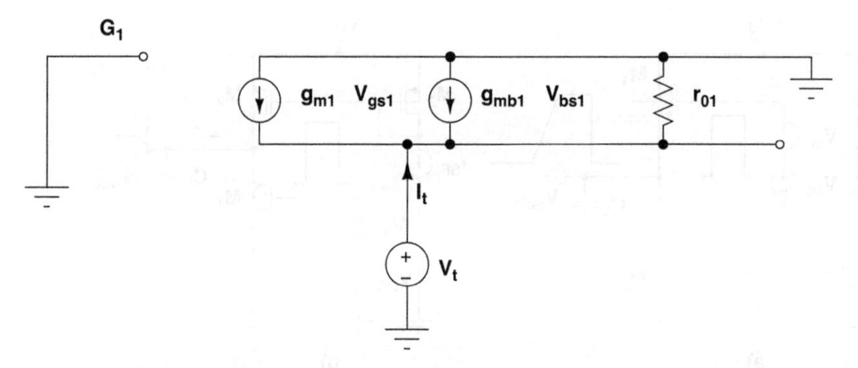

FIGURE 3.47 Small signal equivalent circuit to calculate the source follower output impedance.

from which we can immediately derive the output impedance V_t/I_t as:

$$r_{0sf} = \frac{V_t}{I_t} = \frac{r_{01}}{1+(g_{m1}+g_{mb1})\,r_{01}} \tag{3.126}$$

Using the fact that $(g_{m1}+g_{mb1})\,r_{01} \gg 1$, we have that:

$$r_{0sf} = \frac{1}{g_{m1}+g_{mb1}} \tag{3.127}$$

We notice that suppressing the bulk effect we increase both the gain and the output impedance. In modern technologies, NMOS source followers can in general be implemented with three main options: standard NMOS transistors (with the bulk tied to the chip substrate), triple well NMOS, allowing to short the source and the bulk and "zero threshold" or "natural devices", i.e. NMOS transistors in which the threshold voltage is approximately zero. The last choice avoids the voltage drop of one V_{GS} between the input and the output of the circuit, but usually come with the constraints that the minimum gate length is bigger than the one allowed for standard or triple well devices.

One issue with source follower is their asymmetrical driving capability. Suppose, in fact, that a capacitor is connected to the circuit output, as shown in Fig. 3.48. Consider first the behavior of the NMOS follower when a rectangular pulse is applied and assume the pulse rise time is very fast. Owing to the load capacitor and the intrinsic speed limitation of the circuit, the output voltage can not change instantaneously. For a short while, the gate-source voltage of the transistor is thus increased and the device drives more current, charging the load capacitors. The capability of the transistor to deliver current to the load is primarily limited by its aspect ratio. Consider now the behavior on the trailing edge. Here, when the pulse is applied, the transistor gate-source voltage is reduced together with the current flowing in the device. The capacitor is thus discharged by the current provided by the biasing current source. The maximum rate at which the output capacitor can be discharged is given by I_{SF}/C_L and

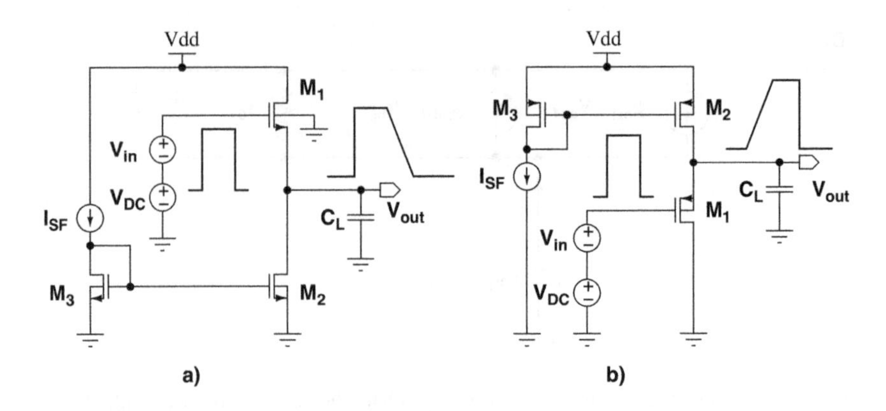

FIGURE 3.48 Illustration of the asymmetric driving capability of source followers.

in case the input signal has a slope greater than this value, the capacitor is discharged with a constant current and the circuit enter into a slew-rate limited operation. NMOS source followers have thus a good driving capability on the rising edge of the signal and a much poorer one on the trailing edge. The same reasoning can be applied to the PMOS source follower and leads to the opposite conclusion, that is a PMOS-based source follower is much more effective if the load has to be discharged, while the current available for the charging phase is limited by the biasing current source.

3.11 BUFFERED CASCODE AMPLIFIERS

The source follower can be used to buffer a high gain cascode stage. The resulting amplifiers are shown in Fig. 3.49 and Fig. 3.50. For simplicity, the driving branches of the biasing current sources have been omitted. The fairly low output impedance of the source follower allows the circuit to drive easily resistive loads and to avoid the loading effects of the feedback resistor when the amplifiers are employed in the transresistance configuration. For this reason, R_f must be connected between the gate of the input transistor and the output of the source follower, as shown in Fig. 3.51. The action of the feedback allows the circuit to self-bias, so that the input transistor accommodates the current imposed by the load. Since in absence of signal no current flows in R_f, at the equilibrium the input and output terminals are equipotential. In the circuit of Fig. 3.51 a), the amplifier DC output is thus equal to V_{GS1}, which, depending on the technology, transistor size and bias current, will be typically between 0.25 and 0.6 V. The circuit has thus more headroom to process signals that determine a positive swing of the output node. For the circuit of Fig. 3.51 b), the output is equal to $V_{DD} - V_{SG1}$ and the amplifier has more headroom to process inputs that cause a negative output waveform. If the range of the input signal is small, both polarities can be accommodated. Furthermore, suitable circuit techniques exist that decouple the input and output DC levels. These methods will be discussed in chapter 6 and in chapter 8. As shown in Fig. 3.52, the cascode amplifier can now be modeled as a voltage amplifier with gain given by $g_{m1}R_L$ and output impedance given by

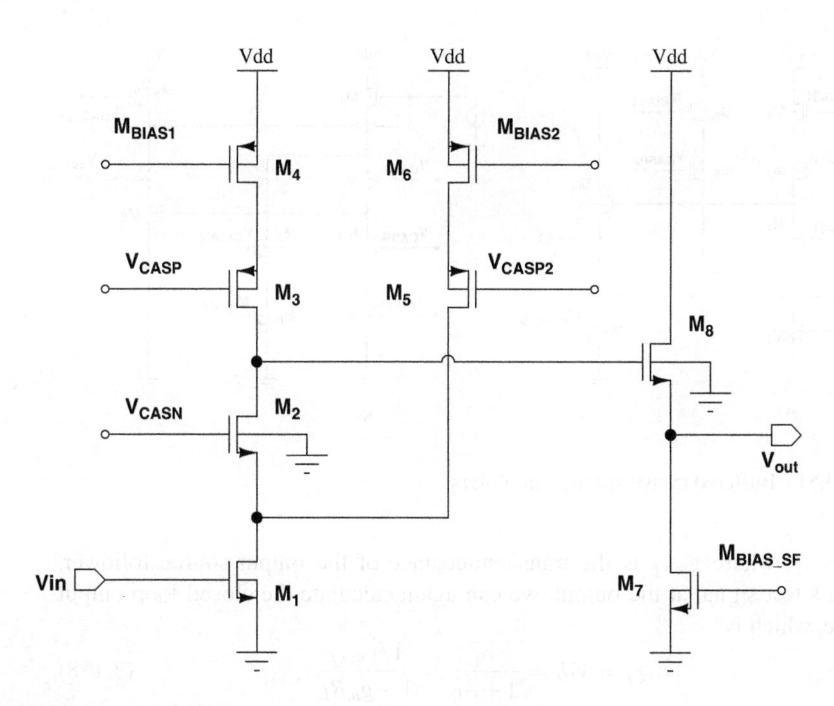

FIGURE 3.49 NMOS-input buffered telescopic cascode amplifier.

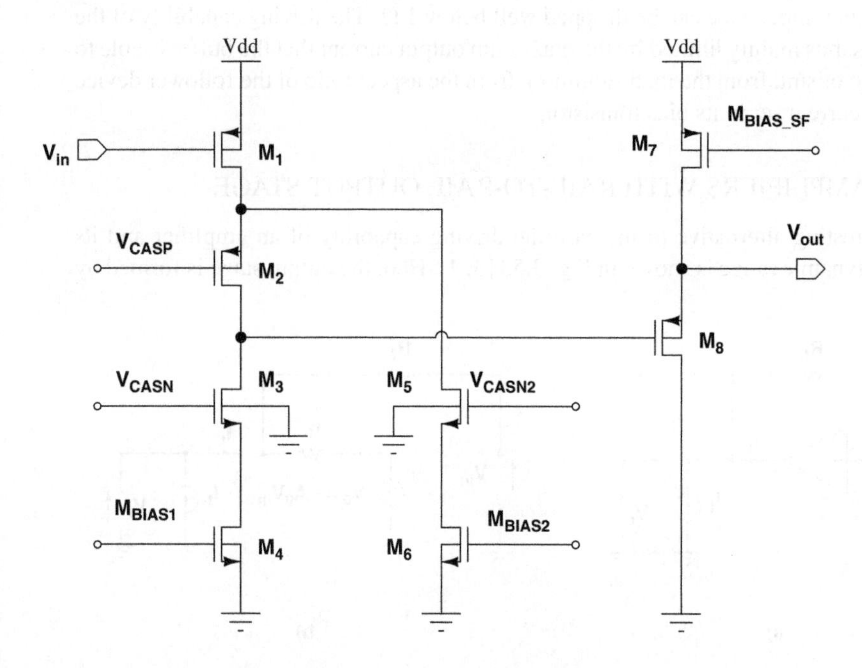

FIGURE 3.50 PMOS-input buffered telescopic cascode amplifier.

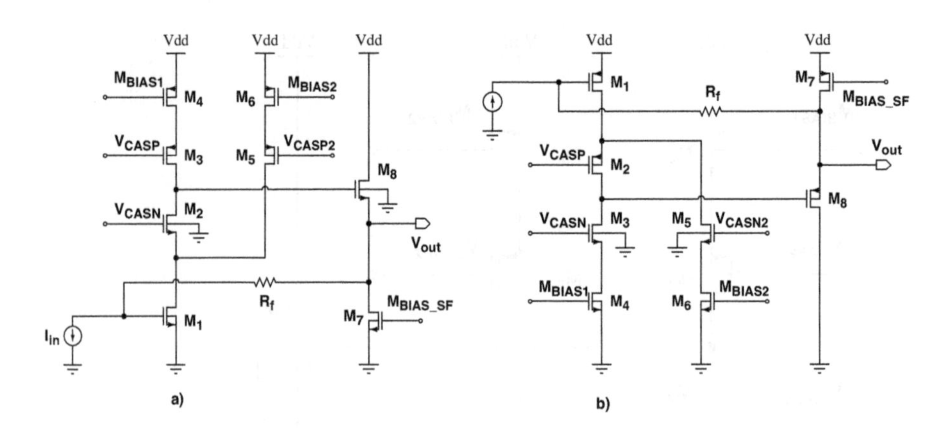

FIGURE 3.51 Buffered transresistance amplifiers.

$r_0 = 1/g_{m,sf}$, where $g_{m,sf}$ is the transconductance of the output source follower. Applying a test signal at the output, we can again calculate the closed-loop output resistance, which is:

$$r_{0,CL} = V_t I_t = \frac{r_0}{1 + A_0} = \frac{1/g_{m,sf}}{1 + g_m R_L} \tag{3.128}$$

If we assume that $g_{m,sf}$ is the kΩ range and the gain is greater than 1000, the closed loop output impedance can be dropped well below 1 Ω. The driving capability of the circuit is thus mainly limited by the maximum output current that the buffer is able to source to or sink from the load stemming from the aspect ratio of the follower device and the current set in its bias transistor.

3.12 AMPLIFIERS WITH RAIL-TO-RAIL OUTPUT STAGE

An interesting alternative to increase the driving capability of an amplifier and its output dynamic range is shown in Fig. 3.53 [3, 4]. Here the output stage is formed by

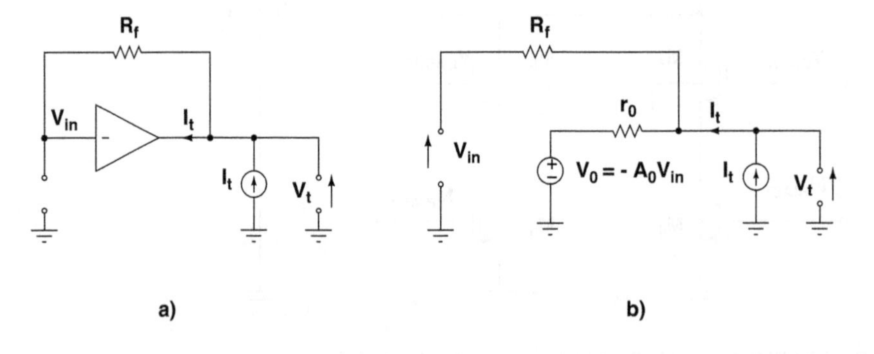

FIGURE 3.52 Circuit to calculate the output impedance of the transresistance amplifier.

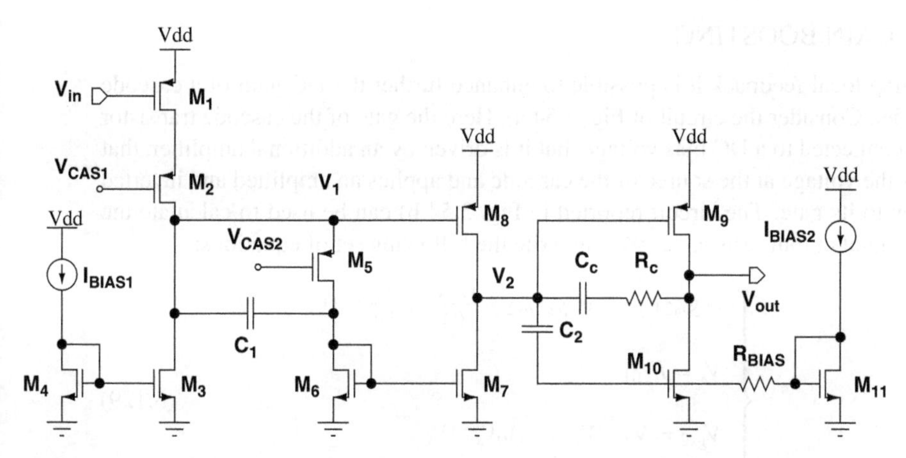

FIGURE 3.53 Single-ended cascode voltage amplifier followed by a rail-to-rail output stage. The circuit is optimized to handle equally well signals of either polarity. © [2013] IEEE. From [3], with permission.

M_9 and M_{10}, which have their source connected respectively to V_{DD} and ground. The output voltage can swing up to $V_{DD} - V_{SD,sat9}$ and down $0 + V_{DS,sat10}$ before pushing M_9 or M_{10} in the linear region, attaining an almost rail-to-rail output dynamic range. The gate of M_9 is DC coupled to the previous stage, while the gate of M_{10} is AC coupled through capacitor C_2. The bias current of M_9 and M_{10} is fixed by M_{11}, whose gate is coupled to the one of M_{10} by a bias resistor. The value of R_{BIAS} must be sufficiently large so that the high pass filter formed by R_{BIAS} and C_2 has a sufficiently low cut-off frequency and the signals of interest are passed from the previous stage without significant attenuation. If V_2 rises, the current in M_9 is reduced and the one in M_{10} is increased, thus the output node is discharged. The opposite is true if V_2 decreases. By an appropriate sizing M_9 and M_{10} can therefore have a symmetric driving capability for signals of either polarity. Observe that since the output stage is inverting, it can not be connected directly to the output of a single stage cascode amplifier, because the overall amplifier would be non-inverting, making the use of negative feedback impossible. A further inversion must thus be introduced in the signal path and this is realized by $M_5 \div M_8$. If V_{in} is decreased, the current in M_1 increases. The extra current can not be absorbed by M_3 and flows in M_5 rising V_1 up. The current is also mirrored by M_6 and M_7, so the current in M_7 increases and the one in M_8 decreases, pulling V_1 down and V_{out} up. If, on the other hand, V_{in} rises, the current in M_5 is decreased. V_1 is pulled down, V_2 is pulled up and V_{out} is lowered, confirming that the overall amplifier is inverting. When the current in M_5 decreases, V_1 can go low enough to switch M_5 off. To avoid the saturation of the amplifier, the bypass capacitor C_1 is added to create a feedforward path towards the gate of M_6. The circuit is thus suitable to handle both negative and positive input signals providing a symmetric response. Resistor R_c and capacitor C_c are used for the frequency compensation of the amplifier, a topic that will be discussed in the next chapter.

3.13 GAIN BOOSTING

Applying local feedback it is possible to enhance further the DC gain of a cascode amplifier. Consider the circuit of Fig. 3.54 a). Here the gate of the cascode transistor is not connected to a DC bias voltage, but it is driven by an additional amplifier, that senses the voltage at the source of the cascode and applies an amplified and inverted version to its gate. The circuit reported in Fig. 3.54 b) can be used to calculate the small signal output resistance. We can write the following set of equations:

$$\begin{cases} -g_{m2}V_{gs2} - g_{mb2}V_{bs2} + \frac{V_s - V_t}{r_{02}} + \frac{V_s}{r_{01}} = 0 \\ V_s = I_t r_{01} \\ V_{gs2} = V_g - V_s = -A_0 V_s - V_s \\ V_{bs2} = 0 - V_s \end{cases} \tag{3.129}$$

Inserting into the first equation the expression of V_s, the definition of V_{gs2} and V_{bs2} and solving for V_t/I_t, we get:

$$\frac{V_t}{I_t} = r_{01} + r_{02} + \left[g_{m2}\left(1 + A_0\right) + g_{mb2}\right] r_{01} r_{02} \tag{3.130}$$

The expression is similar to the one of the simple cascode, with the difference that the transconductance of M_2 is magnified by the factor $(1 + A_0)$. Called "regulated cascode", this method increases significantly the output resistance of the cascode and thus the DC gain of the amplifier. When the technique is employed to this purpose, it is referred to generally as "gain boosting". For the method to be fully effective, the boosting technique must be applied to both the input stage and the load. The gain-boosting amplifier is often implemented as a simple common source stage, as shown in Fig. 3.55. Due to the degraded output conductance of very deep submicron

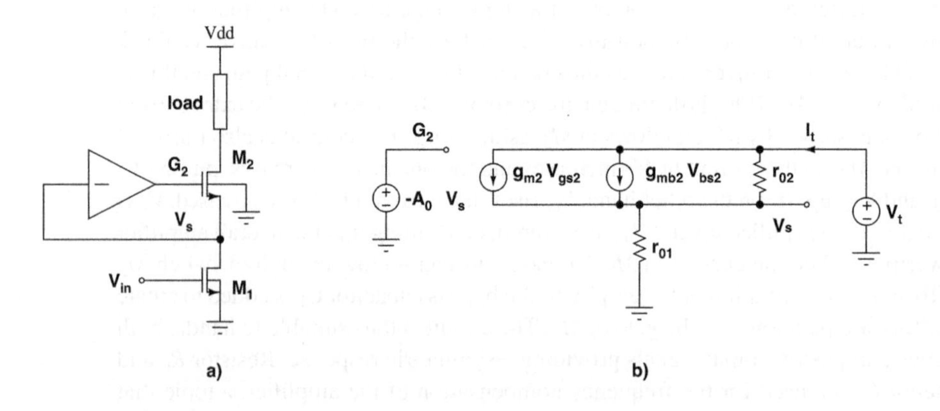

FIGURE 3.54 Gain boosting technique to enhance the cascode output impedance.

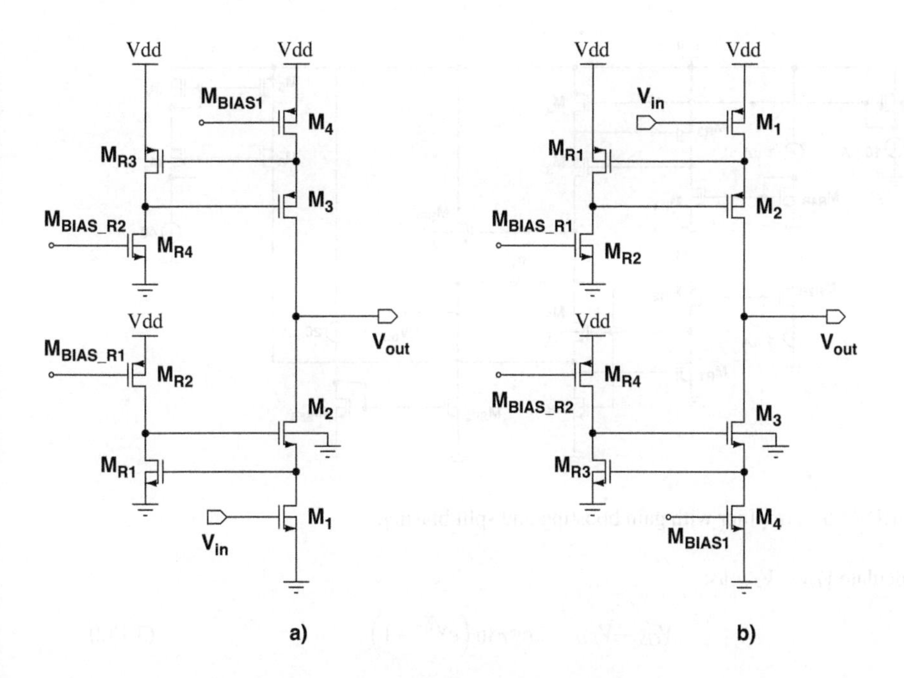

FIGURE 3.55 Cascode telescopic amplifiers with gain boosting.

transistors, the gain boosting technique is expecially useful in circuits implemented in aggressively scaled technologies. The next example discusses an amplifier incorporating the techniques discussed so far.

Example 3.14: Transresistance amplifier with gain boosting.

Fig. 3.56 reports a transimpedance amplifier built with a core amplifier employing both gain boosting and split biasing. The circuit is implemented in a 130 nm technology having the following parameters: $\mu_n C_{ox} = 480\ \mu\text{A/V}^2$, $\mu_p C_{ox} = 120\ \mu\text{A/V}^2$, $V_{THN} = 0.26\ V$, $|V_{THP}| = 0.26$ V. The device sizes are given in table 3.1. Calculate the DC node voltage at the cascode output.

* * *

The total current flowing in the input transistor is the sum of the currents imposed by M_{4B}-M_4 and M_5-M_8 and is 50 μA. The inversion coefficient I_C is given by:

$$I_C = \frac{I_{DS}}{2n\mu_n C_{ox} \frac{W}{L} \phi_T^2} = \frac{5 \cdot 10^{-5}}{2 \times 1.3 \times 4.8 \cdot 10^{-4} \times 66.7 \times 6.7 \cdot 10^{-4}} \approx 0.9 \quad (3.131)$$

With $I_C = 0.9$, the device is close to the center of moderate inversion. We can therefore

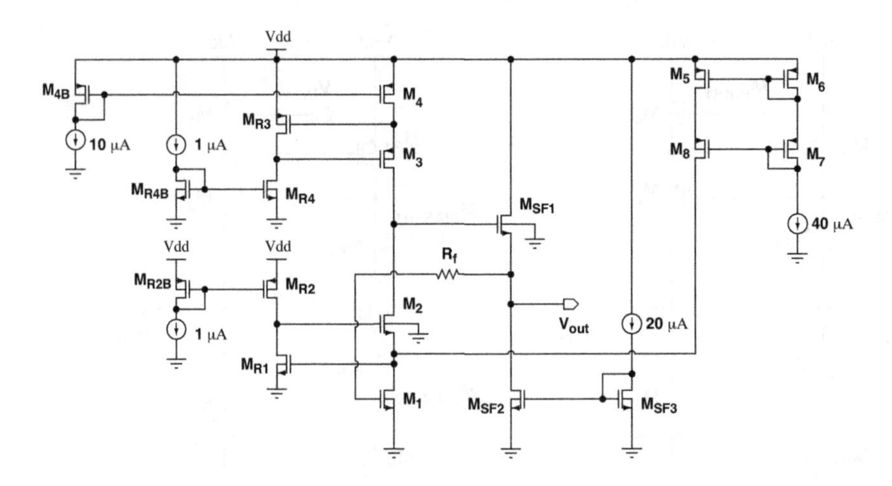

FIGURE 3.56 Amplfier with gain boosting and split biasing.

calculate $V_{GS} - V_{TH}$ as:

$$V_{GS} - V_{TH} = 2n\phi_T \ln\left(e^{\sqrt{I_C}} - 1\right) \tag{3.132}$$

which for $I_C = 0.9$ yields $V_{GS1} - V_{TH1} = 31$ mV and $V_{GS1} = 291$ mV. Since at the equilibrium no current flows in R_f, the same voltage is found also at the source follower output. For M_{SF1} the inversion coefficient is $I_C = 0.36$. Applying again (3.132), one finds $V_{GS} - V_{TH} = -13$ mV, indicating that the device is already approaching weak inversion. The voltage at the output of the cascode amplifier is given by $V_{GS}(M_1) + V_{GS}(M_{SF1}) =$ 538 mV, which is sufficient to keep both M_1 and M_2 in saturation. Observe that the drain-source voltage of M_1 is imposed by M_{R1} and it is equal to the V_{GS} of this transistor. Similarly, the source-drain voltage of M_4 is determined by the source-gate voltage of M_{R3}. For M_{R1}, the inversion coefficient is 0.6, which leads to a V_{GS} of 250 mV. The inversion coefficient of M_{R3} is 2.4 and the source gate voltage is $V_{SG}(M_{R3}) = 349$ mV, hence the drain of M_4 is found at $V_{DD} - V_{SG}(M_{R3}) \approx 1.2 - 0.35 = 0.85$ V. Observe that in our calculation we do not keep into account short channel effects. A discrepancy must thus be expected between these results and the ones obtained with a computer simulation, especially for the shortest devices. For instance, the simulation reveals that the V_{GS1} is 0.35 V. One of the main reasons for the difference is the higher threshold voltage of small gate length transistors due to the reverse short-channel effect.

TABLE 3.1
Device sizes in micron.

$M_1 = 20/0.3$, $M_2 = 5/0.3$, $M_3 = 6/0.3$, $M_4 = M_{4b} = 6/2$
$M_{R1} = M_{R2} = M_{R2B} = M_{R3} = M_{R4} = M_{R4B} = 2/1$
$M_5 = M_6 = M_7 = M_8 = 12/0.5$
$M_{SF1} = 20/0.3$, $M_{SF2} = M_{SF3} = 10/1.2$

TABLE 3.2
Simulated g_m (upper line) and g_{ds} (lower line).

M_1	M_2	M_3	M_4	M_{R1}	M_{R2}	M_{R3}	M_{R4}
$1.1 \cdot 10^{-3}$	$2.3 \cdot 10^{-4}$	$1.5 \cdot 10^{-4}$	$6.4 \cdot 10^{-5}$	$2.6 \cdot 10^{-5}$	$1.6 \cdot 10^{-5}$	$1.6 \cdot 10^{-5}$	$2.5 \cdot 10^{-5}$
$2.1 \cdot 10^{-5}$	$5.3 \cdot 10^{-6}$	$8.7 \cdot 10^{-6}$	$2.7 \cdot 10^{-6}$	$1.7 \cdot 10^{-7}$	$4.1 \cdot 10^{-7}$	$3.9 \cdot 10^{-7}$	$1.8 \cdot 10^{-7}$

It is interesting to evaluate also the gain of the amplifier. Table 3.2 gives the g_m and the g_{ds} of the relevant transistors. Let's first calculate the gain of the gain boosting amplifiers. For the circuit used in the NMOS cascode we have:

$$A_1 = -\frac{g_m(M_{R1})}{g_{ds}(M_{R1}) + g_{ds}(M_{R2})} \approx -45 \tag{3.133}$$

while for the one employed in the PMOS cascode we get:

$$A_2 = -\frac{g_m(M_{R3})}{g_{ds}(M_{R3}) + g_{ds}(M_{R4})} \approx -28 \tag{3.134}$$

The output impedance of the cascodes can be approximated with the following formulas:

$$r_{0,casP} \approx g_{m3}(1 + A_2) r_{03} r_{04} \approx 185 \text{ M}\Omega \tag{3.135}$$

$$r_{0,casN} \approx g_{m2}(1 + A_1) r_{01} r_{02} \approx 95 \text{ M}\Omega \tag{3.136}$$

The gain is given by $A_v = -g_{m1} r_{0,casN} // r_{0,casP} \approx 70.000 = 96.9$ dB. Note that without boosting, the gain would be factor 46 lower, or about 63 dB. Observe also that the gain has been calculated before the source follower, and about 25% of the gain can be lost in the buffer.

3.14 CURRENT-MODE INPUT STAGES

A possible alternative to read a radiation sensor is to use a current amplifier, a circuit that must have a low input impedance and a high output impedance. Consider the scheme shown in Fig. 3.57. The circuit is very similar to a source follower, with two important differences. First, the gate of the input transistor is kept to a steady potential, which is an AC ground, hence the name common gate amplifier. Second, the sensor signal is applied in the form of a current to the source of the input device M_1. The load resistor R_L senses the output current and converts it back to a voltage. We derive now the input and the output impedance of the circuit and its current gain.

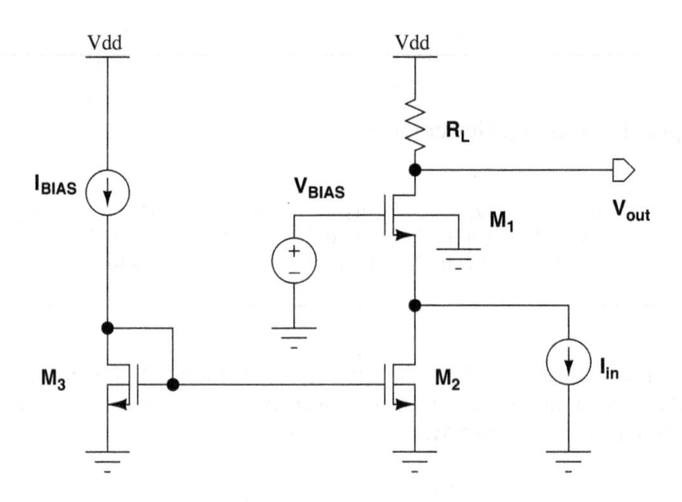

FIGURE 3.57 Common gate amplifier.

The input impedance can be calculated with the circuit of Fig. 3.58, which allows us to write the following set of equations:

$$\begin{cases} -I_t - g_{m1}V_{gs1} - g_{mb1}V_{bs1} + \frac{V_t - V_{out}}{r_{01}} = 0 \\ V_{gs1} = V_{bs1} = -V_t \\ V_{out} = I_t R_L \end{cases} \tag{3.137}$$

Combining the three equations together and solving for V_t/I_t, we get:

$$r_{in} = \frac{V_t}{I_t} = \frac{r_{01} + R_L}{1 + (g_{m1} + g_{mb1})r_{01}} = \frac{r_{01}}{1 + (g_{m1} + g_{mb1})r_{01}} + \frac{R_L}{1 + (g_{m1} + g_{mb1})r_{01}} \tag{3.138}$$

Equation (3.138) shows that the input impedance of the common gate amplifier is the sum of two contributions: the first term is identical to the output resistance of

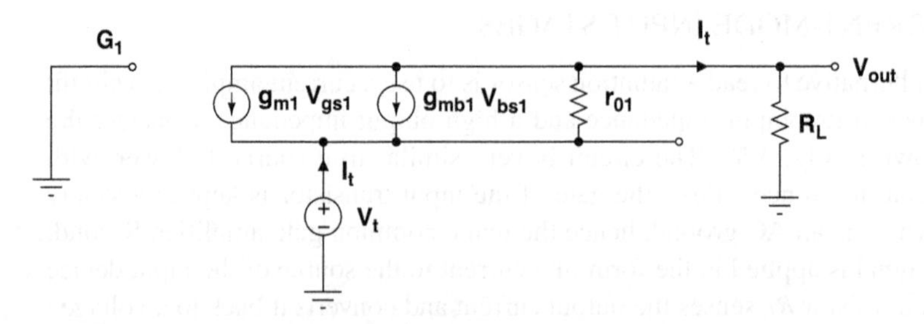

FIGURE 3.58 Small signal equivalent circuit to calculate the input impedance of a common gate amplifier.

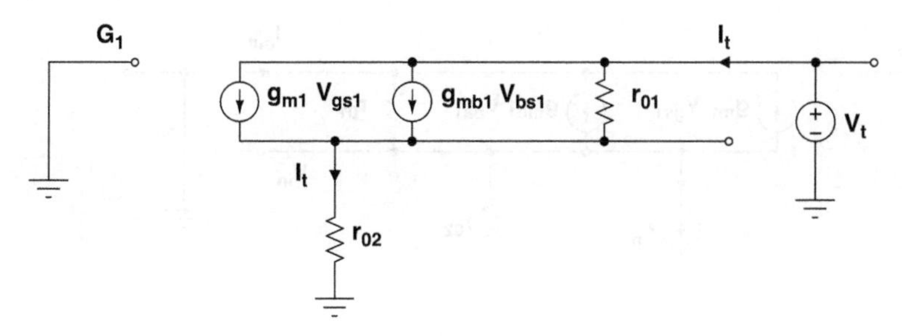

FIGURE 3.59 Small signal equivalent circuit to calculate the output impedance of a common gate amplifier.

the source follower. This comes as no surprise, because in both cases we evaluate the impedance from the source of M_1 and ground and in case $R_L = 0$, the circuit of Fig. 3.58 becomes identical to the one of Fig 3.47. The additional term is equal to the load resistance divided by the factor $[1+(g_{m1}+g_{mb1})r_{01}]$. If the output conductance of the transistor is not too small and R_L is not too large, the input impedance of the common gate amplifier can be approximated by $1/(g_{m1}+g_{mb1})$ and it is thus fairly low, making it a good candidate to read a current from a high impedance source. In the calculation of the input impedance we have not taken into account the output resistance of the biasing transistor M_2, that is seen by the signal source in parallel to the input impedance of the common gate stage. The equivalent circuit model to calculate the output impedance is shown in Fig. 3.59. We notice that the network is the same employed to evaluate the output resistance of the common source amplifier with source degeneration and of the cascode. Therefore, reusing the already known results, we can immediately write the common gate output resistance as:

$$r_{out} = r_{01} + r_{02} + (g_{m1} + g_{mb1})\, r_{01} r_{02} \tag{3.139}$$

The stage has therefore a high output resistance, as it is appropriate to a good current amplifier. Finally, we calculate the current gain of the stage. To do so, we evaluate the short circuit output current, that is the current that the circuit is able to source to a zero-impedance load. The circuit of Fig. 3.60 serves to this purpose and allows us to write the following equations:

$$\begin{cases} I_{in} - g_{m1}V_{gs1} - g_{mb1}V_{bs1} + \frac{V_{in}}{r_{01}} + \frac{V_{in}}{r_{02}} = 0 \\ g_{m1}V_{gs1} + g_{mb1}V_{bs1} - \frac{V_{in}}{r_{01}} - I_{out} = 0 \\ V_{gs1} = V_{bs1} = -V_{in} \end{cases} \tag{3.140}$$

Solving the above system, we can write the ratio I_{out}/I_{in}, which is:

$$\frac{I_{out}}{I_{in}} = \frac{r_{01} + (g_{m1} + g_{mb2})\, r_{01} r_{02}}{r_{01} + r_{02} + (g_{m1} + g_{mb2})\, r_{01} r_{02}} \tag{3.141}$$

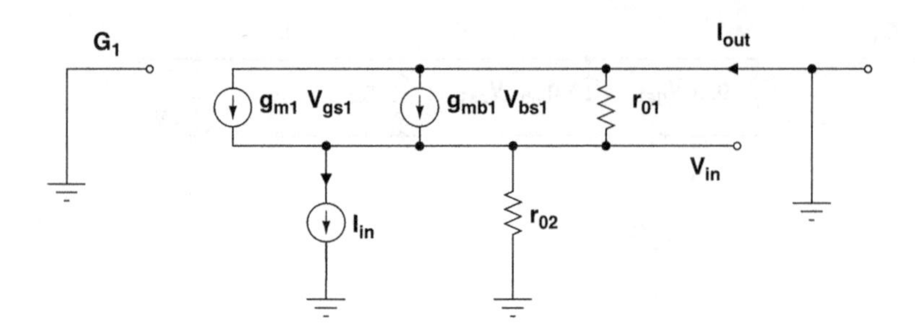

FIGURE 3.60 Calculation of the current gain of the common gate configuration.

With the usual hypothesis that $(g_{m1}+g_{mb1})r_{01}r_{02} \gg r_{01}+r_{02}$, we see that the current gain is about one. The common gate stage is thus a current buffer, that can be used to sense a current from a source and deliver it to a load on a more appropriate impedance level. If we now make the output current flow into a resistor, we get that $V_{out}=I_{out}R_L \approx I_{in}R_L$. To calculate precisely the gain of a common gate amplifier with resistive load, we observe that the load resistance goes in parallel to the output impedance of the stage, resulting in a equivalent resistance given by:

$$r_{eq}=\frac{R_L\left[r_{01}+r_{02}+(g_{m1}+g_{mb1})\,r_{01}r_{02}\right]}{R_L+r_{01}+r_{02}+(g_{m1}+g_{mb1})\,r_{01}r_{02}} \tag{3.142}$$

Multiplying r_{eq} by the expression of I_{out} given in (3.141), we finally get:

$$\frac{I_{out}}{I_{in}}=\frac{\left[1+(g_{m1}+g_{mb1})\,r_{01}\right]r_{02}R_L}{r_{01}+r_{02}+(g_{m1}+g_{mb1})\,r_{01}r_{02}+R_L} \tag{3.143}$$

Although (3.143) is more accurate, the approximation $I_{out}=I_{in}$ is usually adequate. Fig. 3.61 compares an input stage based on a common gate amplifier with the one employing the feedback configuration. We notice that if $R_L=R_f$, the transimpedance gain V_{out}/I_{in} of the two circuits is equivalent. However, in the common gate amplifier R_L is not part of a feedback loop. The two amplifiers have different performances in term of noise and sensitivity to the sensor capacitance, so criteria to choose one configuration over the other will become clear once we will address these two aspects. Finally, it is interesting to note that for the same polarity of the input signal, the polarity of the output voltage is opposite between the two circuits. In fact, in the common gate amplifier if the current in the input transistor increases, so does the current in the load and the output voltage drops. The opposite happens if the current in the input device is decreased. Therefore, the voltages at the input and at the output of the circuit are in phase. If we need to have a common gate amplifier with a high transimpedance gain we could think of using an active load, as shown in Fig. 3.62 a). The first objection to such choice could be that, on the basis of our discussion on the input impedance, the use of a big load resistor causes an increase of the input impedance. Using (3.138), we see that if $R_L=r_{01}$, the input impedance is only doubled, a penalty that in most cases in acceptable. However, there are two more important reasons for which the

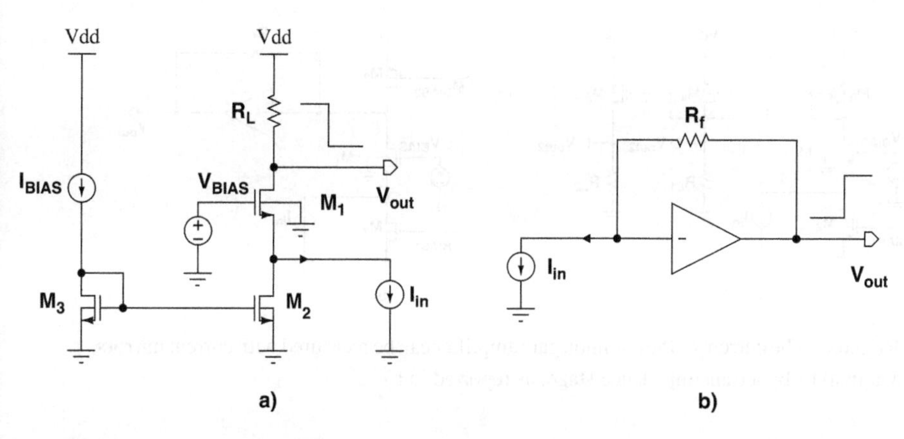

FIGURE 3.61 Comparison between a common gate amplifier, shown in a) and a transimpedance amplifier, shown in b).

circuit of Fig. 3.62 a) is *not* used. First, the load resistor in the common gate stage defines the conversion gain, so in this respect its function is similar to the one of R_f in the transimpedance amplifier. Therefore, this component must be well controlled, whereas in voltage amplifiers with active loads the purpose is only to have a high gain, because feedback is later applied to the circuit. Even more important, from the biasing point of view the combination of M_2 and M_1 can be seen as a cascoded current mirror (formed by M_1 and M_2), loaded by another current mirror, given by M_4. The active load must in fact also work as a current source to provide high gain. We therefore end up with two current sources in series, a situation which is impossible to control in practice without the help of a feedback loop. As for the simple common source amplifier with resistive load, the use of a large passive load resistor to obtain a big conversion gain can be problematic because the DC voltage drop across the resistor can leave too little headroom to M_1. In this case, the circuit of Fig. 3.62 b) can instead

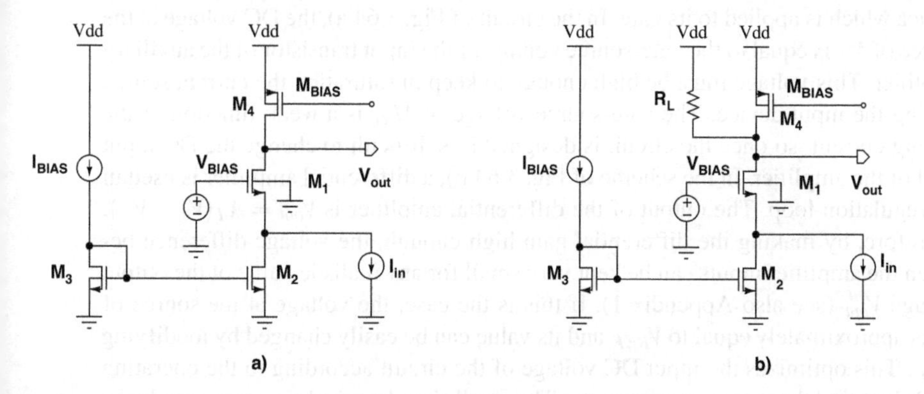

FIGURE 3.62 Common gate amplifier with reduced voltage drop in the load. See text for discussion.

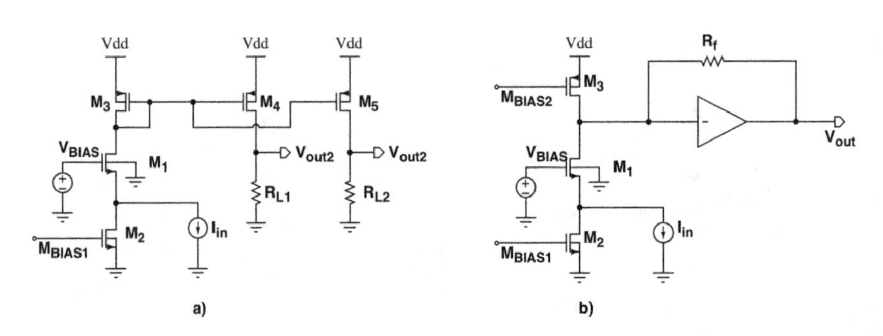

FIGURE 3.63 The current of the common gate amplifier can be measured with current mirrors, as shown in a) or by a transimpedance stage, as reported in b).

be used. Here, M_4 is designed to deliver a current smaller than the one required by M_1, so that only the difference between the drain current of M_1 and the one of M_4 flows in the load. Fig. 3.63 shows two additional methods that can be used to sense the current at the drain of a common gate output stage. In the circuit of Fig. 3.63 a), a diode-connected transistor receives the current, which is then sent to the output branches. This technique is particularly advantageous when the signal must be fanout to multiple destinations. As shown in Fig. 3.63 b), it is also possible to read the current with a standard transimpedance amplifier. The difference between the current in M_1 and M_3 flows in R_f, therefore care must be paid that this does not compromise the functionality of the second stage. In fact, if the DC current flowing in R_f is too big, depending on its direction the output of the transimpedance amplifier can be pushed towards one of the power supply rails, compromising the dynamic range.

An impedance of $1/g_m$ is the lowest that can be obtained from a simple transistor without applying feedback. To drop even further the input impedance, the same strategy adopted to boost the gain of cascode amplifiers can be applied. The concept is shown in Fig. 3.64. In both circuits, an additional amplifier is used to sense the voltage at the source of the input transistor and generate an amplified and inverted replica which is applied to its gate. In the circuit of Fig. 3.64 a), the DC voltage at the source of M_1 is equal to the gate-source voltage of the input transistor of the auxiliary amplifier. This voltage must be high enough to keep in saturation the current source biasing the input device. The gate-source voltage of M_{R1} is a weak function of the biasing current, so once the circuit is designed it is difficult to change the DC input level of the amplifier. In the scheme of Fig. 3.64 b), a differential amplifier is used in the regulation loop. The output of the differential amplifier is $V_{out} = A_d(V_+ - V_-)$. Therefore, by making the differential gain high enough, the voltage difference between the amplifier inputs can be kept very small for any realistic value of the output voltage V_{out} (see also Appendix 1). If this is the case, the voltage at the source of M_1 is approximately equal to V_{REF} and its value can be easily changed by modifying V_{REF}. This optimizes the input DC voltage of the circuit according to the operating conditions and the sensor requirements. The small signal equivalent input impedance

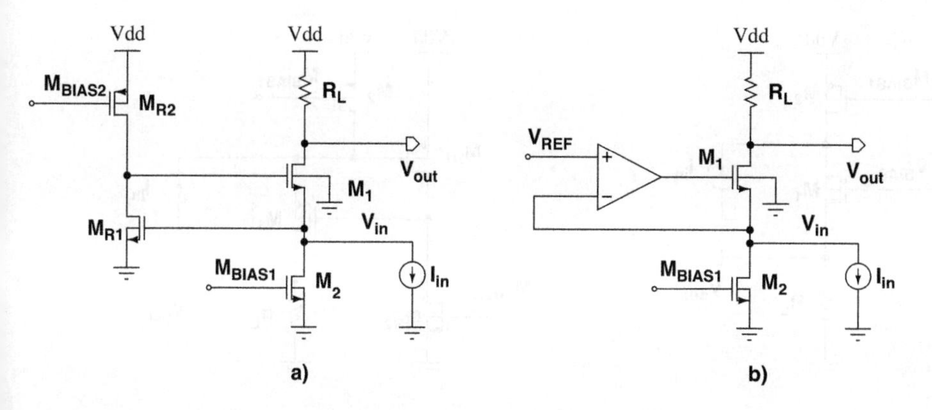

FIGURE 3.64 Regulated common gate amplifiers.

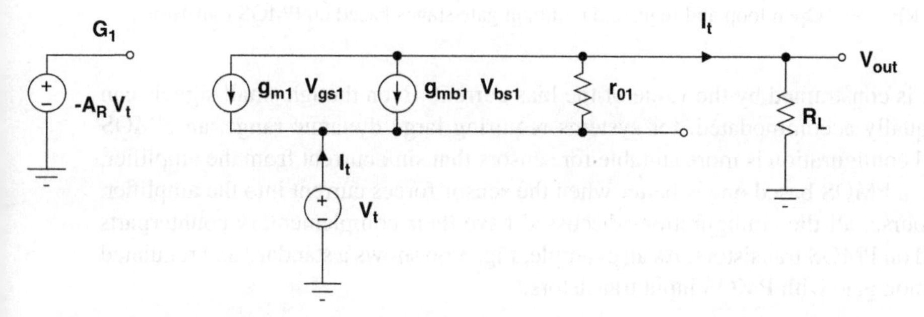

FIGURE 3.65 Equivalent small signal circuit to calculate the small signal input impedance of a regulated common gate amplifier.

can be calculated with the help of Fig. 3.65. We can write:

$$\begin{cases} I_t - g_{m1}V_{gs1} - g_{mb1}V_{bs1} + \frac{V_t}{r_{01}} = 0 \\ V_{gs1} = -AV_t - V_t \\ V_{bs1} = -V_t \end{cases} \tag{3.144}$$

Solving for V_t/I_t, we get:

$$r_{in} = \frac{V_t}{I_t} = \frac{r_{01}}{[1 + (g_{m1}(1+A) + g_{mb1})\, r_{01}]} \approx \frac{1}{g_{m1}(1+A)} \tag{3.145}$$

The input impedance of the circuit is thus approximately divided by the gain of the auxiliary amplifier. Note that the load resistor is still outside the feedback loop.

It must be observed that, like the source follower, also a common gate amplifier is asymmetric. The maximum current that can be sunk from an NMOS input transistor is limited by its aspect ratio, while the maximum current that can be forced into the

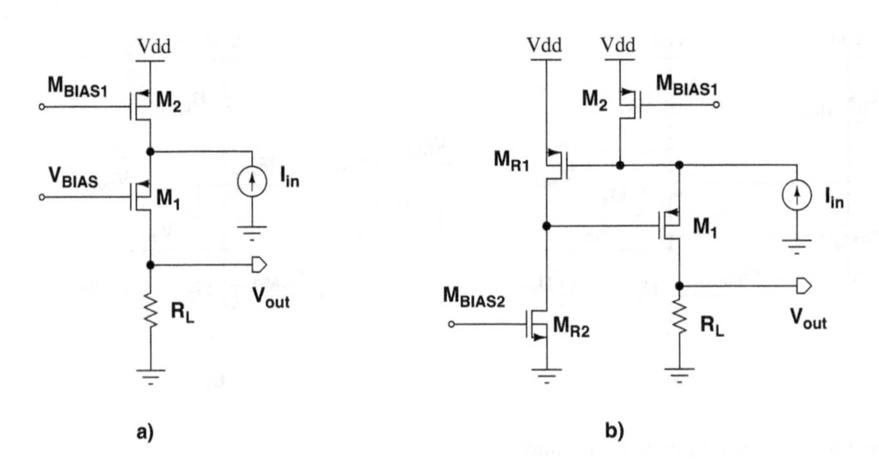

FIGURE 3.66 Open loop and regulated common gate stages based on PMOS transistors.

stage is constrained by the value of the bias current. Even though small signals can be equally accommodated, for systems requiring large dynamic range, an NMOS based configuration is more suitable for sensors that sink current from the amplifier, while a PMOS based one is better when the sensor forces current into the amplifier. Of course, all the configurations discussed have their complementary counterparts based on PMOS transistors. As an example, Fig. 3.66 shows a standard and regulated common gate with PMOS input transistors.

REFERENCES

1. B. Razavi. *Design of Analog CMOS Integrated Circuits.* McGraw-Hill, 2000.
2. P. R. Gray, P. J. Hurst, S. H. Lewis, and R. G. Meyer. *Analysis and Design of Analog Integrated Circuits.* 5th edition, Wiley, 2009.
3. G. De Geronimo et al. VMM1—An ASIC for micropattern detectors. *IEEE Trans. Nucl. Sci.*, 60:2314–2321, 2013.
4. G. De Geronimo et al. Front-end ASIC for a silicon Compton telescope. *IEEE Trans. Nucl. Sci.*, 55:2323–2328, 2008.

4 Input Stages in the Frequency Domain

To complete the study of the circuits introduced in the previous chapter we need to consider also the effect of the capacitances which are unavoidably present in such structures. This implies addressing the frequency behavior of amplifiers, a task which is best achieved by working in the s-domain through the use of the Laplace transforms. This tool should be familiar to the reader from fundamental University courses and adequate treatments of the topic can be found in one of the many textbooks on linear electrical network [1, 2]. The systematic analysis of MOS integrated circuits in the frequency domain is covered in standard microelectronics textbooks [3–7].

4.1 THE COMMON SOURCE AMPLIFIER IN THE FREQUENCY DOMAIN

The analysis of a CMOS circuit in the frequency domain starts by replacing the low-frequency small signal model of all the transistors in the circuit by the ones incorporating the intrinsic device capacitances. Effects due to the signal source and output loads must also be included. With this approach, the resulting network becomes quickly intractable by hand calculations and the complexity of the mathematics overshades the critical insights. Therefore, we consider in the following only the most important contributions. We start our study by examining a more academic example in which the amplifier is driven by an ideal voltage source. This allows us to pin-down some relevant concepts while keeping the algebra simple. We then consider the more realistic case where the input signal is provided by a source with a resistor in series.

4.1.1 VOLTAGE DRIVEN COMMON SOURCE AMPLIFIER

Fig. 4.1 shows the transistor level schematic of the simple common source amplifier with active load together with the most important capacitors. In the output capacitor, C_L, we have lumped all the elements which are connected between the output node and a constant DC potential. These include the drain-bulk capacitances of M_1 and M_2 and the drain-gate capacitance of M_2. In most practical cases, however, C_L is dominated by the input capacitance of the circuit that the amplifier has to drive. We include also in the model the gate-drain and gate-source capacitance of the input transistor M_1, which both play a prominent role in the high frequency performance of the amplifier. In the following, C_{gs} and C_{gd} indicate always the gate-source and gate-drain capacitance of the input device, unless otherwise specified. The load resistance R_L is given by the parallel of the output resistances of M_1 and M_2, i.e $R_L = r_{01}//r_{02}$. The small signal equivalent circuit of the amplifier is reported in Fig. 4.2. To study a circuit in the Laplace domain, the first step is to associate to the reactive components

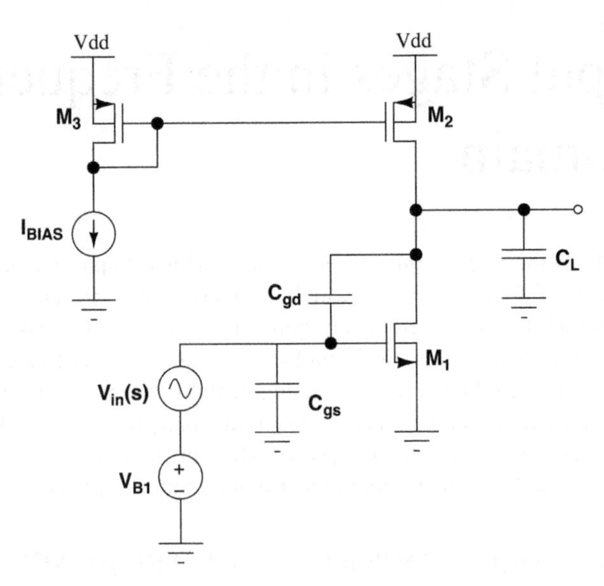

FIGURE 4.1 Common source amplifier with the most relevant capacitors limiting the circuit frequency response.

their complex impedance in the s-domain. For a capacitor, this is given by:

$$Z(s) = \frac{1}{sC} \tag{4.1}$$

Then, we write the nodal equation, in the same way we do for the circuits containing only resistive elements. The analysis of Fig. 4.2 is particularly simple, because we have only one free node. Note that the gate-source capacitance C_{gs} can not play any role in defining the circuit response because the input is driven directly by an ideal voltage source. All the current required by C_{gs} is thus provided by the signal source, that, being ideal, can deliver an unlimited amount of current. The equation for the output node reads:

$$g_{m1}V_{in} + \frac{V_{out}}{R_L} + \frac{V_{out}}{Z_{C_L}(s)} + \frac{V_{out} - V_{in}}{Z_{C_{gd}}(s)} = 0 \tag{4.2}$$

Observe that in writing the nodal equations in the Laplace domain the complex impedance of the capacitor is formally treated as a resistor, so for instance, the current flowing through Z_L, $V_{out}/Z_{C_L}(s)$ is written in the same way as the current flowing

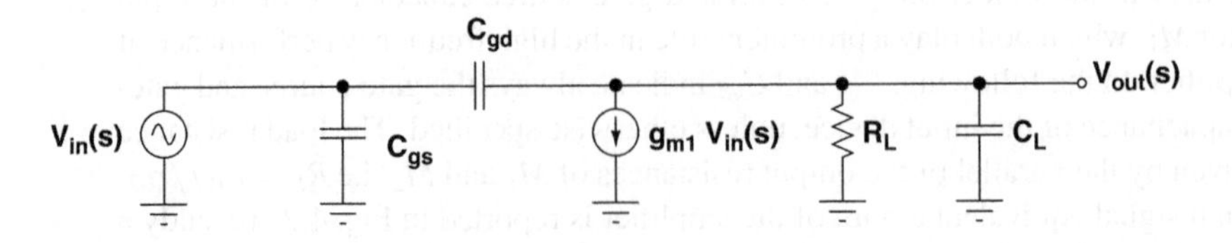

FIGURE 4.2 Small signal equivalent circuit of the common source amplifier of Fig. 4.1.

through R_L, given by V_{out}/R_L. We can now insert into equation (4.2) the explicit definition of the complex impedance, $Z_{C_L} = 1/sC_L$ and $Z_{C_{gd}} = 1/sC_{gd}$, to obtain:

$$g_{m1}V_{in} + \frac{V_{out}}{R_L} + V_{out}sC_L + (V_{out} - V_{in})\,sC_{gd} = 0 \tag{4.3}$$

Solving for V_{out}/V_{in}, yields:

$$\frac{V_{out}}{V_{in}} = -\frac{(g_{m1} - sC_{gd})\,R_L}{1 + sR_L\,(C_L + G_{gd})} = T(s) = \frac{N(s)}{D(s)} \tag{4.4}$$

where with $T(s)$ we have indicated the transfer function of the circuit in the s-domain.

By solving the equation $N(s) = 0$ we find the zeros of the transfer function. In this particular case, we have only one zero given by: $s_z = g_{m1}/C_{gd}$, i.e. the zero is real and is located on the right-half part of the s plane. The value of the zero has an interesting intuitive interpretation, which can be understood with the help of Fig. 4.3. The figure shows the case of a positive variation of V_{in}, but the argument can be easily repeated in case the variation is negative. The signal V_{in} modulates the gate of the input transistor and hence its current through the device transconductance. At the same time, C_{gd} establishes a direct connection between V_{in} and the circuit output, with an impedance which becomes smaller and smaller as the frequency rises. The current through this direct path is $V_{in}sC_{gd}$. When this current equals the one that V_{in} determines between the drain and source of M_1 through the transistor effect, all the current that the transistor sends to ground, $g_{m1}V_{in}$, arrives through C_{gd}, so that all the

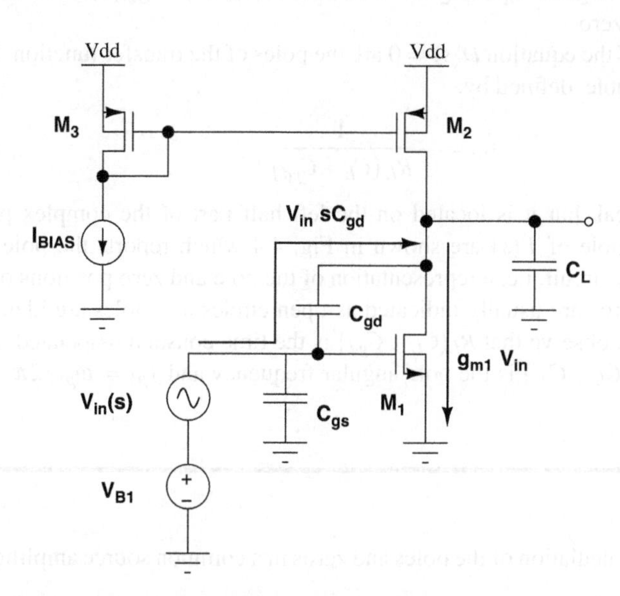

FIGURE 4.3 Origin of the right half-plane zero in a common source amplifier. At the frequency of the zero currents $V_{in}sC_{gd}$ and $g_{m1}V_{in}$ are equal. See text for more details.

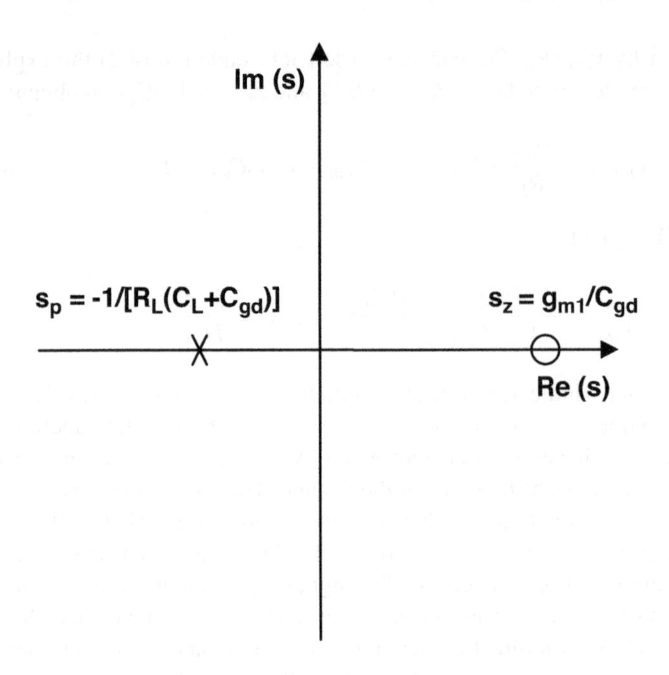

FIGURE 4.4 Pole-zero constellation of the voltage-driven common source amplifier.

effects produced by V_{in} are confined within the transistor itself and do not produce an external measurable signal. Equating the two currents ($V_{in}sC_{gd} = g_{m1}V_{in}$) one obtains the position of the zero.

The solutions of the equation $D(s) = 0$ are the poles of the transfer function. Here we have only one pole, defined by:

$$s_{p1} = -\frac{1}{R_L\left(C_L + C_{gd}\right)} \tag{4.5}$$

Also the pole is real, but it is located on the left half part of the complex plane. The zero and the pole of $T(s)$ are shown in Fig. 4.4 which reports the pole-zero constellation of the circuit, i.e. a representation of the pole and zero positions on the complex plane. Zeros are usually indicated as open circles and poles are identified by crosses. Finally, observe that $R_L(C_L + C_{gd})$ is the time constant associated to the pole, $\omega_{p1} = 1/R_L(C_L + C_{gs})$ is the pole angular frequency and $f_{p1} = \omega_{p1}/2\pi$ is the pole frequency.

Example 4.1: Calculation of the poles and zeros in a common source amplifier.

The circuit of Fig. 4.1 has the following parameters: $I_{BIAS} = 70\ \mu$A, $W_1/L_1 = 20/1$, $W_2/L_2 = W_3/L_3 = 5/1$, $C_L = 100$ fF, $C_{gd} = 10$ fF, $\mu_nC_{ox} = 170\ \mu$A/V^2, $\mu_pC_{ox} =$

60 μA/V^2, $C_{ox} = 5$ fF/μm^2, $\lambda_N = \lambda_P = 0.04$ V^{-1}, $V_{THN} = |V_{THP}| = 0.6$ V, $V_{dd} = 3$ V. Calculate the position of the pole and the zero of the amplifier.

* * *

The first step is to evaluate the load resistor $R_L = r_{01}//r_{02}$. Since M_3 and M_2 have the same aspect ratio, a current of 70 μA flows in M_1 and M_2.Using the given parameters, we have:

$$r_{01} = r_{02} = \frac{1}{\lambda_N I_{BIAS}} = \frac{1}{\lambda_P I_{BIAS}} = 357 \text{ k}\Omega \rightarrow R_L = 178.5 \text{ k}\Omega \tag{4.6}$$

The angular frequency of the pole is:

$$\omega_{p1} = \frac{1}{R_L\left(C_L + C_{gd}\right)} = \frac{1}{1.785 \cdot 10^5 \times 1.1 \cdot 10^{-13}} = 50.93 \text{ MHz} \tag{4.7}$$

To calculate the zero, we need to know the transconductance of M_1. To find the operating region of the transistor, let's first evaluate the inversion coefficient:

$$I_C = \frac{I_{DS}}{2n\mu_n C_{ox}\frac{W_1}{L_1}\phi_T^2} = \frac{7 \cdot 10^{-5}}{2 \times 1.3 \times 1.7 \cdot 10^{-4} \times 20 \times 6.76 \cdot 10^{-4}} = 11.7 \tag{4.8}$$

We can therefore assume that M_1 works in strong inversion and its transconductance is given by:

$$g_{m1} = \sqrt{2\mu C_{ox}\frac{W_1}{L_1}I_{DS}} = 690 \text{ }\mu\text{S} \tag{4.9}$$

The zero is hence located at:

$$\omega_z = \frac{g_{m1}}{C_{gd}} = \frac{6.9 \cdot 10^{-4}}{1 \cdot 10^{-14}} = 69 \text{ GHz} \tag{4.10}$$

From the above example we see that the zero is at much higher frequency than the pole, which is primarily defined by R_L and C_L. To a first approximation, we can hence simplify $T(s)$ as follows:

$$\frac{V_{out}}{V_{in}} = -\frac{g_{m1}R_L}{1 + sR_LC_L} \tag{4.11}$$

Therefore, the voltage-driven common source amplifier has a simple first order low-pass filtering behavior. For $s \rightarrow 0$, the gain reduces to $g_{m1}R_L$, which is the quantity already calculated in the previous chapter without taking into consideration the capacitors. For $s \rightarrow \infty$, we have instead $T(s) = -g_{m1}/sC_L$ and $T(s) = 1$ for $s = g_{m1}/C_L$, which defines the unity gain frequency of the amplifier, i.e. the maximum frequency for which the circuit is able to provide a gain ≥ 1. The primary tool to study the behavior of a circuit in the frequency domain is the Bode plot, which represents the response of the circuit to steady state sinusoidal inputs as a function of frequency. More specifically, the Bode plot is a set of two diagrams, associated to the modulus and phase of the complex transfer function $T(s)$ evaluated for $s = j\omega$. The modulus tells us how much the amplitude of a sinusoid is amplified by the circuit. The phase diagram indicates how much the amplifier changes the phase of the same input sinusoid. The construction of the Bode plots is reviewed in the following example.

Example 4.2: Bode plot of the common source amplifier.

Design the bode plot of the circuit of Fig. 4.1, using the approximation given in (4.11).

* * *

The first step in constructing the plot is to put in (4.11) $s = j\omega$:

$$T(j\omega) = -\frac{g_{m1}R_L}{1 + j\omega R_L C_L} \tag{4.12}$$

Remember in fact that the Laplace transform of $\sin \omega t$ is $\omega/(s^2 + \omega^2)$ and the roots are $s = \pm j\omega$. Limiting s to $j\omega$ implies considering only the cases for which $T(j\omega)$ is a sinusoidal function. Since $T(j\omega)$ is a complex number, it can be studied by examining its modulus and phase. To do so, we can rewrite $T(j\omega)$ as follows:

$$T(j\omega) = -\frac{g_{m1}R_L}{1 + j\omega R_L C_L}\frac{1 - j\omega R_L C_L}{1 - j\omega R_L C_L} = -\frac{g_{m1}R_L}{1 + \omega^2 (R_L C_L)^2} + j\frac{g_{m1}R_L \omega R_L C_L}{1 + \omega^2 (R_L C_L)^2} = a + jb \tag{4.13}$$

where the real part a is given by:

$$a = -\frac{g_{m1}R_L}{1 + \omega^2 (R_L C_L)^2} \tag{4.14}$$

and the imaginary part b is:

$$b = \frac{g_{m1}R_L \omega R_L C_L}{1 + \omega^2 (R_L C_L)^2} \tag{4.15}$$

The modulus then reads:

$$|T(j\omega)| = \sqrt{a^2 + b^2} = \frac{g_{m1}R_L}{\sqrt{1 + \omega^2 \tau_L^2}} \tag{4.16}$$

where, for conciseness, we have put $R_L C_L = \tau_L$. We then plot on a log-log scale the quantity $20\log|T(j\omega)|$, identifying two different regions. In the low frequency one, $\omega^2\tau_L^2 << 1$ and $T(j\omega)| \approx g_{m1}R_L$. With the parameters found in the previous example, we have:

$$20\log|T(j\omega)| = 20\log(6.9 \cdot 10^{-4}\ \text{S} \times 178.5\ \text{k}\Omega) = 41.8\ \text{dB} \tag{4.17}$$

For high frequencies so that $\omega^2\tau_L^2 >> 1$ we can approximate $|T(j\omega)|$ as:

$$|T(j\omega)| = \frac{g_{m1}R_L}{\omega \tau_L} \tag{4.18}$$

Consequently, on a logarithmic scale, we get:

$$20\log|T(j\omega)| = 20\log g_{m1}R_L - 20\log\frac{\omega}{\omega_L} \tag{4.19}$$

where we have replaced τ_L with its reciprocal, the angular frequency $\omega_L = 2\pi f_L$. We see from the above equation that the gain decreases with a slope of 20 dB per decade of frequency, hence it is reduced to 21.8 dB at $\omega = 10\omega_L$ and to 1.8 dB at $\omega = 100\omega_L$. According to our approximation at $\omega = \omega_L$ the gain is still 41.8 dB. However, the exact calculation using equation (4.16) shows that in this point the gain is already reduced to 38.8 dB, i.e. 3 dB below the maximum. The frequency at which the −3 dB point occurs is called the cut-off frequency of the circuit. With the parameters of our example, the angular cut-off frequency is $\omega_L = 1/\tau_L = 56$ MHz (we neglect here the contribution of C_{gd}). In practical circumstances it is usually more relevant the frequency f, describing the number of oscillations per second, which is related to ω by a factor 2π, i.e. $f = \omega/2\pi$, so $f_L = 8.9$ MHz in our case.

Let's now concentrate on the phase of $T(j\omega)$, $\angle T(j\omega)$. From basic algebra we know that the phase of a complex number C with real part a and imaginary part b is given by:

$$\angle C = \tan^{-1}\frac{b}{a} \tag{4.20}$$

Applying this to $T(j\omega)$, we obtain:

$$\angle T(j\omega) = \tan^{-1}\left(-\frac{\omega}{\omega_L}\right) \tag{4.21}$$

Hence, we see that the phase shift is already −5.7° for $\omega = 0.1\omega_L$, it reaches −45° for $\omega = \omega_L$ and tends to −90° for $\omega \to \infty$. To summarize, $T(j\omega)$ describes how sinusoidal inputs are processed by the circuit. In a first order system, i.e. a system whose transfer function contains only one pole, the angular frequency of the pole is given by $\omega_L = 1/\tau_L$, where τ_L is the time constant associated to the pole. Sinusoids with a frequency much smaller than f_L are amplified by the full gain, which for a simple common source amplifier is given by $g_{m1}R_L$, without notable phase shift. One decade before the cut-off frequency, i.e. for $f = 0.1f_L$, the gain is not yet substantially affected, whereas a non-negligible phase shift starts being introduced. Fig. 4.5 and Fig. 4.6 show respectively the magnitude and the phase of $T(j\omega)$ of the considered amplifier. Observe that equation (4.21) does not yield the "absolute" phase shift of the output with respect to the input, but only the additional phase shift with respect to the one that might be present at frequencies much smaller than f_L. In fact, the common source is an inverting amplifier, (as implied by the "minus" sign present in the gain equation). Therefore, the amplitude of a low-frequency sinusoidal signal is multiplied by $g_{m1}R_L$ and, being inverted, it is phase shifted by 180°. The negative phase shift introduced by the pole at high frequencies is subtracted to this low-frequency value, actually bringing the output closer in phase to the input as the frequency of the input signal is increased.

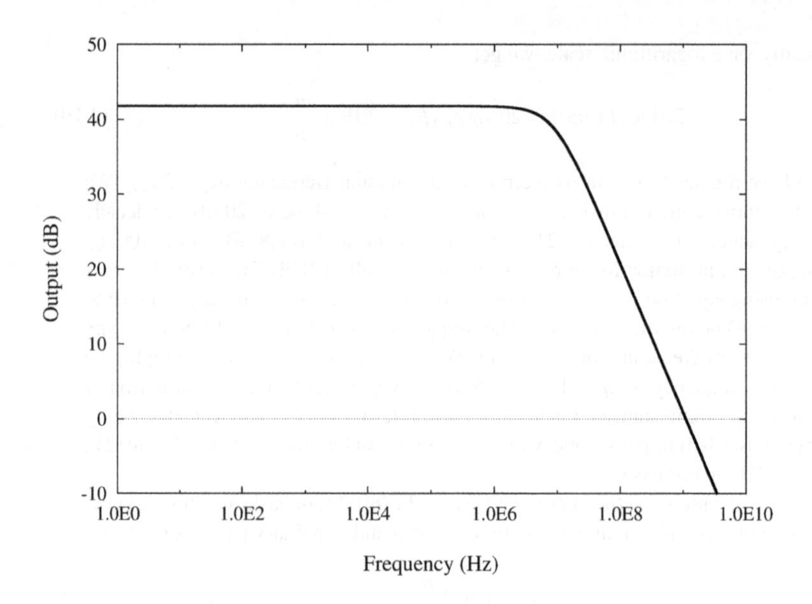

FIGURE 4.5 Bode Plot (magnitude) of the common source amplifier described in the example.

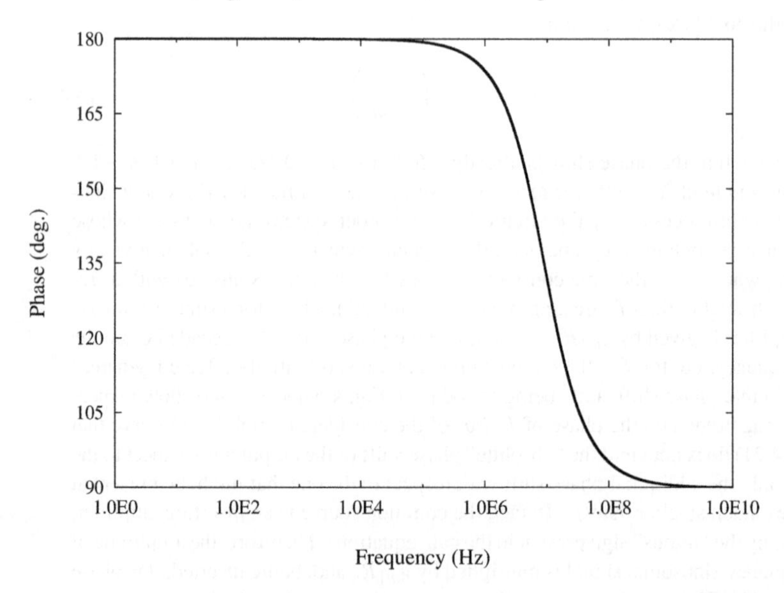

FIGURE 4.6 Bode Plot (phase) of the common source amplifier described in the example.

4.1.2 CURRENT DRIVEN COMMON SOURCE AMPLIFIER

Let's now consider a more realistic case, in which the signal source is not ideal. This is represented by adding a resistor R_S in series to $V_{in}(s)$, as shown in Fig. 4.7.

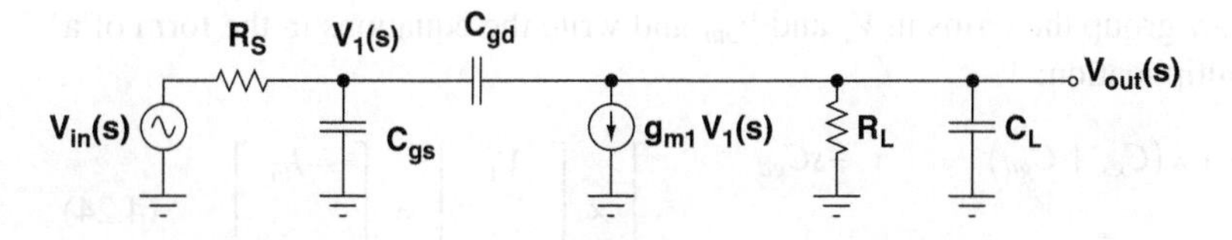

FIGURE 4.7 Common source amplifier driven by a resistive voltage source.

Observe that in this case the controlled voltage source representing the transistor is not anymore driven directly by V_{in}, but by the voltage V_1 appearing across the gate-source capacitance C_{gs}. In addition, R_S limits the maximum current available to charge and discharge C_{gs}, introducing a time constant also at the input and further reducing the speed. The Norton theorem states that a voltage source $V_{in}(s)$ with a resistance R_s in series is equivalent to a current source $I_{in}(s) = V_{in}(s)/R_s$ with the same resistor R_S connected in parallel to I_{in}. The circuit with the Norton transformation of the input source is shown in Fig. 4.8. As discussed in chapter 1, in most cases a radiation sensor can be modeled as a current source. Therefore, the case where R_S is large is of particular interest in our context, since the bigger R_S the closer the signal source behaves as a good current generator. When the input source is represented by its Norton equivalent we speak of a "current driven" amplifier. As already discussed in the previous chapter, a high gain voltage amplifier will never be used to read-out a sensor without an appropriate feedback loop. However, the study of the circuit of Fig. 4.8 is particularly interesting because it allows to pin-down fundamental issues and it will be recurrently used in the following. We now calculate the transfer function $V_{out}(s)/I_{in}(s)$. This time we need to write two nodal equations, one for the input node (which is not anymore constrained by an ideal voltage source) and one for the output one. For the input node we have:

$$I_{in} + \frac{V_1}{R_s} + V_1 sC_{gs} + (V_1 - V_{out})\, sC_{gd} \tag{4.22}$$

For the output node we can write:

$$g_{m1}V_1 + \frac{V_{out}}{R_L} + V_{out}sC_L + (V_{out} - V_1)\, sC_{gd} = 0 \tag{4.23}$$

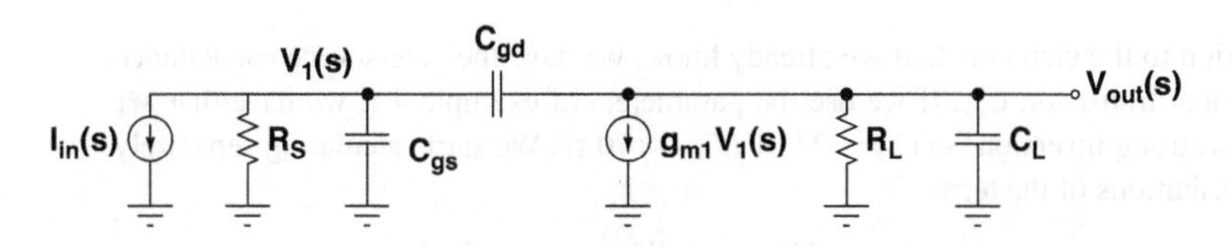

FIGURE 4.8 Common source amplifier with Norton transformation of the input source.

We can now group the terms in V_1 and V_{out} and write the equations in the form of a matrix multiplication:

$$\begin{bmatrix} \frac{1}{R_s}+s\left(C_{gs}+C_{gd}\right) & -sC_{gd} \\ g_{m1}-sC_{gd} & \frac{1}{R_L}+s\left(C_L+C_{gd}\right) \end{bmatrix} \times \begin{bmatrix} V_1 \\ V_{out} \end{bmatrix} = \begin{bmatrix} -I_{in} \\ 0 \end{bmatrix} \tag{4.24}$$

To find the ratio V_{out}/I_{in} we first calculate the determinant of the coefficient matrix:

$$\Delta = \begin{vmatrix} \frac{1}{R_s}+s\left(C_{gs}+C_{gd}\right) & -sC_{gd} \\ g_{m1}-sC_{gd} & \frac{1}{R_L}+s\left(C_L+C_{gd}\right) \end{vmatrix} \tag{4.25}$$

We then evaluate the determinant Δ_{out} of the matrix obtained by replacing the second column with the one of the known terms, i.e.:

$$\Delta_{out} = \begin{vmatrix} \frac{1}{R_s}+s\left(C_{cgs}+C_{gd}\right) & -I_{in} \\ g_{m1}-sC_{gd} & 0 \end{vmatrix} \tag{4.26}$$

We finally obtain $V_{out}(s)/I_{in}(s)$ by taking the ratio Δ_{out}/Δ, which yields:

$$\frac{V_{out}(s)}{I_{in}(s)} = \frac{\left(g_{m1}-sC_{gd}\right)R_S R_L}{s^2 \zeta R_S R_L + s\left[R_L\left(C_L+C_{gd}\right)+R_S\left(C_{gs}+C_{gd}\right)+g_{m1}R_S R_L C_{gd}\right]+1} \tag{4.27}$$

where for conciseness we have put $\zeta = \left(C_{gs}C_L + C_{gs}C_{gd} + C_{gd}C_L\right)$. The above result is rather complicated, so to gain some insight we first evaluate the different components to see if some simplification is possible.

Example 4.3: Capacitances in a current-driven common source amplifier.

Using the parameters of example 4.1, evaluate the coefficient of equation 4.27. Suppose that the source resistance R_S is 1 MΩ.

* * *

In addition to the elements that we already know, we have the gate-source capacitance of the input transistor, C_{gs}. If we use the parameters of example 4.1, we have that M_1 works in strong inversion, so $C_{gs} = 2/3WLC_{ox} \approx 70$ fF. We start calculating separately the contributions of the term s^2:

$$\begin{aligned} C_{gs}C_L &= 7\cdot 10^{-14} \times 1\cdot 10^{-13} = 7\cdot 10^{-27}\ \mathrm{F}^2 \\ C_{gs}C_{gd} &= 7\cdot 10^{-14} \times 1\cdot 10^{-14} = 7\cdot 10^{-28}\ \mathrm{F}^2 \\ C_{gd}C_L &= 1\cdot 10^{-14} \times 1\cdot 10^{-13} = 1\cdot 10^{-27}\ \mathrm{F}^2 \end{aligned} \tag{4.28}$$

We then evaluate the coefficient of s:

$$\begin{aligned}
R_L\left(C_L+C_{gd}\right) &= 1.78\cdot 10^5\times 1.1\cdot 10^{-13} = 1.96\cdot 10^{-8}\ \text{s}\\
R_S\left(C_{gs}+C_{gd}\right) &= 1\cdot 10^6\times 8\cdot 10^{-14} = 8\cdot 10^{-8}\ \text{s}\\
g_{m1}R_LR_SC_{gd} &= 6.9\cdot 10^{-4}\times 1\cdot 10^6\times 1.78\cdot 10^5\times 1\cdot 10^{-14} = 1.22\cdot 10^{-6}\ \text{s}
\end{aligned}\tag{4.29}$$

We see here that despite the small value of C_{gd}, the last time constant is dominant due to the multiplication by the factor $g_{m1}R_LR_S$.

Being a second order polynomial, the denominator of equation (4.27) can also have complex conjugate roots. We will come back later to this point. For the time being, we consider the simplified case in which the zero at the nominator is neglected and the poles are real and distinct, so that the transfer function can be written as:

$$\frac{V_{out}}{I_{in}} = \frac{g_{m1}R_SR_L}{(1+s\tau_1)(1+s\tau_2)}\tag{4.30}$$

where τ_1 and τ_2 are the two time constants associated to the poles that we want to determine. The denominator of the above equation can be written in the following way:

$$D(s) = 1+s(\tau_1+\tau_2)+s^2\tau_1\tau_2 = 1+as+bs^2\tag{4.31}$$

If we assume that $\tau_1 \gg \tau_2$, $D(s)$ can be further simplified as:

$$D(s) \approx 1+s\tau_1+s^2\tau_1\tau_2\tag{4.32}$$

from which we can deduce that τ_1 and τ_2 can be respectively approximated by:

$$\begin{aligned}
\tau_1 &= a \approx g_{m1}R_LR_SC_{gd}\\
\tau_2 &= \frac{b}{a} = \frac{R_SR_L\left(C_{gs}C_{gd}+C_{gs}C_L+C_{gd}C_L\right)}{g_{m1}R_SR_LC_{gd}} = \frac{C_{gs}C_{gd}+C_{gs}C_L+C_{gd}C_L}{g_{m1}C_{gd}}
\end{aligned}\tag{4.33}$$

Consequently, the poles of $T(s)$ are:

$$\begin{aligned}
p_1 &= -\frac{1}{\tau_1} = -\frac{1}{g_{m1}R_LR_SC_{gd}}\\
p_2 &= -\frac{1}{\tau_2} = -\frac{g_{m1}C_{gd}}{C_{gs}C_{gd}+C_{gs}C_L+C_{gd}C_L}
\end{aligned}\tag{4.34}$$

If we use the numbers of our example, we have that p_1 corresponds to an angular frequency of 814 kHz, whereas p_2 is located at 793 MHz, which justifies the assumption that the two poles are well separated. Note that C_{gd} is not the largest capacitor in the circuit, but it dominates the frequency response of the amplifier.

4.1.3 ANALYSIS OF THE COMMON SOURCE AMPLIFIER WITH THE MILLER THEOREM

The Miller theorem calculates the current required to drive an impedance which is connected across a gain stage. Consider the circuit of Fig. 4.9 a), where the impedance Z links the input and output port of an amplifier with gain A_v. The output voltage V_{out} is thus given by $V_{out} = A_v V_{in}$. We provide V_{in} with an ideal voltage source and we ask how much current is required to the signal source to sustain this voltage. Since the output is $A_v V_{in}$, we can write:

$$I_{in} = \frac{V_{in} - A_v V_{in}}{Z} \tag{4.35}$$

from which we have that:

$$\frac{V_{in}}{I_{in}} = Z_{eq1} = \frac{Z}{1 - A_v} \tag{4.36}$$

Observe that if the impedance Z was connected in parallel to the source, it would just drive a current given by V_{in}/Z. However, a variation of V_{in} at the input determines a larger variation of $A_v V_{in}$ on the other terminal of Z which is connected to the output, increasing the effective voltage drop across Z and hence the current necessary to sustain it. We can also calculate the current that must be delivered by the amplifier output port when the signal V_{in} is applied to the input. This is given by:

$$\frac{V_{out}}{I_{out}} = \frac{V_{out} - \frac{V_{out}}{A_v}}{Z} \tag{4.37}$$

which yields:

$$\frac{V_{out}}{I_{out}} = Z_{eq2} = \frac{Z A_v}{A_v - 1} \approx Z \tag{4.38}$$

where the approximation is true if $A_v \gg 1$. The terminal of the impedance connected to the input experience a variation much smaller than V_{out}, therefore the current required to the output port would not be much different if Z was connected directly between the output node and the signal ground. The network formed by the amplifier and the impedance can then be represented as in Fig. 4.9 b). Let's now apply this concept to our common source amplifier. Here the role of Z is played by the gate-drain capacitance

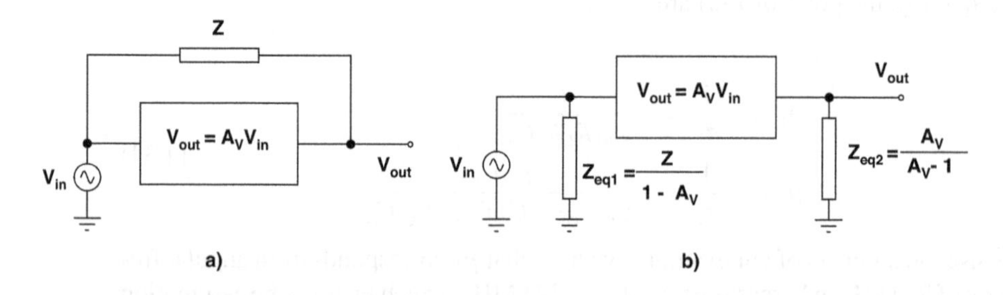

FIGURE 4.9 Illustration of the Miller theorem.

C_{gd}. The gain of the amplifier is given by $A_v = -g_{m1}R_L$, hence, C_{gd} is seen from the input as an impedance of value:

$$\frac{1}{sC_M} = \frac{1}{sC_{gd}(1+g_{m1}R_L)} \tag{4.39}$$

When seen from the input C_{gd} thus appears as a capacitance with a value $1+g_{m1}R_L$ times bigger than its physical value. This effective capacitance, combined with the source resistance R_S, yields a time constant equal to $C_M R_S = C_{gd}(1+g_{m1}R_L)R_S \approx g_{m1}R_L R_S C_{gd}$, where the approximation is valid if $g_{m1}R_L \gg 1$. The denominator of equation (4.27) can also be rewritten as:

$$D(s) = s^2 \zeta R_S R_L + s\left[R_L(C_L+C_{gd}) + R_S C_{gs} + R_S(1+g_{m1}R_L)C_{gd}\right] + 1 \tag{4.40}$$

which puts in evidence the Miller multiplication of C_{gd} by the factor $1+g_{m1}R_L$. With the parameter of our example, the "effective value" of C_{gd} seen from the input is given by: 124×10 fF$=1.24$ pF, which makes it the largest capacitance in the circuit. The use of the Miller representation removes the direct connection between input and output provided by Z, hence it can not predict the zero present in the transfer function.

4.2 FREQUENCY PERFORMANCE OF CASCODE AMPLIFIERS

One of the benefits of the cascode topology is the suppression of the Miller effect acting on the gate-drain capacitance of the input transistor. To understand this point, let's first consider the circuit of Fig. 4.10 a). Now C_{gd} is connected between the gate of M_1 and the source of M_2, so to quantify the Miller effect we need to know the gain between these two points. The detailed analysis of the cascode configuration at low

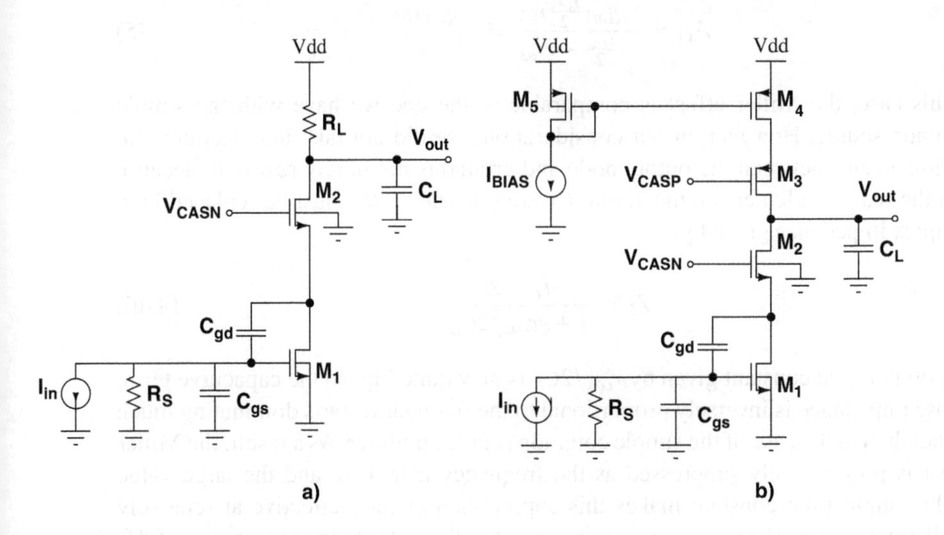

FIGURE 4.10 Current driven cascodes.

frequency was already carried-out in the previous chapter, so here we just report the result. The gain between the gate and drain of M_1 is:

$$A_{v1} = -\frac{g_{m1}\left(r_{02}+R_L\right)r_{01}}{R_L+r_{01}+r_{02}+\left(g_{m2}+g_{mb2}\right)r_{01}r_{02}} \tag{4.41}$$

We can now consider two different situations. The first one occurs when the load resistor R_L is much smaller than the output resistance of M_2, in which case equation (4.41) becomes:

$$A_{v1} = -\frac{g_{m1}r_{01}r_{02}}{r_{01}+r_{02}+\left(g_{m2}+g_{mb2}\right)r_{01}r_{02}} \tag{4.42}$$

The equation can be further simplified if the condition $\left(g_{m2}+g_{mb2}\right)r_{01}r_{02} \gg r_{01}+r_{02}$ is used. This relationship is in general true, so we get:

$$A_{v1} = -\frac{g_{m1}}{g_{m2}+g_{mb2}} \approx -1 \tag{4.43}$$

Applying the Miller theorem, we see that the equivalent value of C_{gd} seen from the amplifier input is $C_{gd}\left(1+A_{v1}\right) \approx 2C_{gd}$, hence the Miller effect is basically suppressed. However, in most cases it is not true that R_L is negligible. In fact, to obtain high DC gain, a cascode load is used, as shown in Fig. 4.10 b). In this case, the value of R_L is given by the parallel of the cascode resistance of the input stage and of the load and R_L is actually much bigger than r_{02}. For simplicity, we can assume that the resistances of the two cascodes are equal, i.e:

$$r_{01}+r_{02}+\left(g_{m2}+g_{mb2}\right)r_{01}r_{02} = r_{03}+r_{04}+\left(g_{m3}+g_{mb3}\right)r_{03}r_{04} = r_{cas} \tag{4.44}$$

which implies that $R_L = r_{cas}/2$. With this hypothesis, we can rewrite the gain as:

$$A_{v1} = -\frac{g_{m1}\frac{r_{cas}}{2}r_{01}}{\frac{r_{cas}}{2}+r_{cas}} = -\frac{g_{m1}r_{01}}{3} \tag{4.45}$$

In this case, the Miller effect is comparable to the one we have with the simple common source. However, in our considerations we did not take into account that the impedance between the output node and ground is not purely resistive, because also the reactive elements lumped into C_L are present, so R_L must be replaced by a complex impedance given by:

$$Z_L = \frac{r_{cas}/2}{1+sr_{cas}/2C_L} \tag{4.46}$$

The output time constant given by $r_{cas}/2C_L$ is now quite big, so the capacitive term, whose impedance is inversely proportional to the frequency, starts dominating much sooner than in the case of the simple common source amplifier. As a result, the Miller effect is progressively suppressed as the frequency increases, and the large value of the output time constant makes this suppression already effective at relatively small frequencies. Finally, we must observe that the gate-drain capacitance of M_2

and M_3 does not suffer from the Miller effect. In fact, the gates of the two transistors are connected to steady bias points, which behave as signal grounds. Hence these capacitances have the same effect as they were directly connected between the circuit output and ground, without any "multiplication". We can thus say that the telescopic cascode amplifier has four main poles in its transfer function. The input pole is given by:

$$p_{in} = -\frac{1}{\left[R_s C_{gs1} + \left(1 + \frac{g_{m1}}{g_{m2}+g_{mb2}}\right) C_{gd1}\right]} \tag{4.47}$$

The pole introduced at the source of the cascode transistor M_2 is given by:

$$p_{cas} = -\frac{g_{m2} + g_{mb2}}{(2C_{gd1} + C_{db1} + C_{sb2} + C_{gs2})} \tag{4.48}$$

Another pole is found at the source of M_3:

$$p_3 = -\frac{g_{m3}}{C_{sg3} + C_{sb3} + C_{db4} + C_{gd4}} \tag{4.49}$$

Finally, the pole at the output can be written as:

$$p_L = -\frac{1}{R_L C_L} \tag{4.50}$$

In a well designed cascode, the frequency response of the circuit is dominated by the input and the output poles. Owing to the suppression of the Miller effect, the cascode configuration thereby greatly improves the frequency response of an amplifier driven by a high impedance source, which is one of the reasons it is the standard topology employed for the input stage of most radiation sensor front-ends.

4.3 FREQUENCY STABILITY OF AMPLIFIERS

In the previous chapter we saw that high gain amplifiers are always used with some type of negative feedback, which is indispensable also to define the DC bias point. A resistor connected around a single-ended cascode stage transforms it into a transimpedance amplifier, which in principle can be used to read a capacitive sensor, as shown in Fig. 4.11. A source follower has been added to buffer the output node and to prevent it from being loaded by the feedback resistor. We now need to make a small signal model which can be easily treated by hand calculations. To do so, we first design the equivalent circuit of the core amplifier, including only the most relevant time constants. These are associated to the nodes with the highest equivalent resistance. In the circuit of Fig. 4.11 we can identify two such nodes, corresponding respectively to the input and output of the circuit. At the input we have in fact the input capacitance of the amplifier plus the sensor capacitance and resistance. All the capacitances connected between the amplifier input and ground have been grouped together into the single term C_T. At the drain of M_2 and M_3 we see the parallel of two cascode resistances, which we want to be as high as possible to have enough

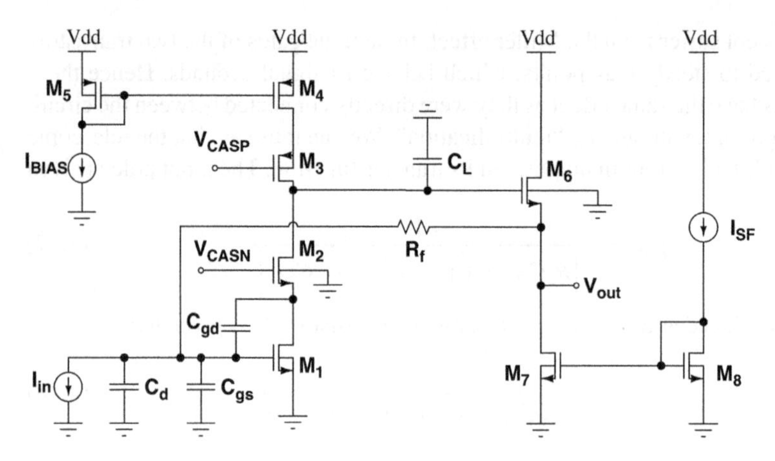

FIGURE 4.11 Transimpedance front-end with the most relevant capacitors affecting the frequency response.

DC gain. This resistance is coupled to the capacitance seen between the output node and the signal ground, that usually are all lumped into the load capacitance "C_L". We then include the feedback network and write the nodal equations. For simplicity, we suppose that the sensor output resistance can be considered infinite. This hypothesis is valid for many sensors and simplifies slightly the calculations, without impairing the general conclusions. We also treat the source follower as an ideal unity gain buffer that copy the output voltage on the high impedance node on a voltage source with zero output impedance. The resulting circuit is shown in Fig. 4.12. The nodal equations are:

$$\begin{cases} V_1\left(sC_T+\frac{1}{R_f}\right)-\frac{V_{out}}{R_f}=-I_{in} \\ \\ g_{m1}R_LV_1+V_{out}\left(1+sC_LR_L\right)=0 \end{cases} \tag{4.51}$$

The transimpedance gain V_{out}/I_{in} can be easily found by solving the system by substitution and is:

$$\frac{V_{out}}{I_{in}}=\frac{g_{m1}R_LR_f}{s^2R_fR_LC_TC_L+s\left(C_TR_f+C_LR_L\right)+1+g_{m1}R_L} \tag{4.52}$$

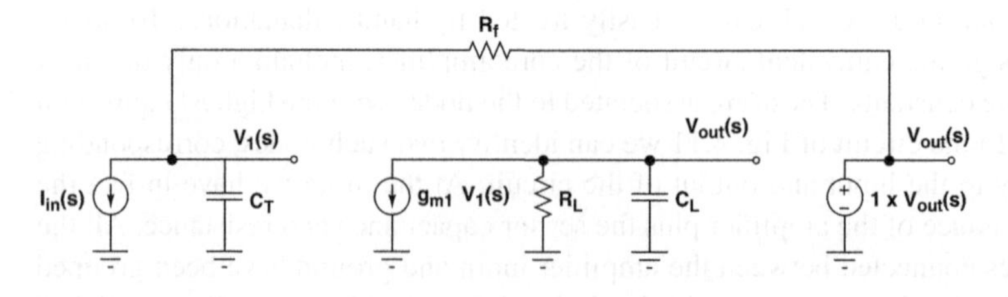

FIGURE 4.12 Small signal equivalent circuit of the transimpedance front-end of Fig. 4.11.

which yields again the already familiar second order transfer function. In our treatment so far we have always assumed that the poles of the encountered transfer functions were real. Now we scrutinize more in detail this hypothesis with the help of the following example.

Example 4.4: Pole locations of a transimpedance amplifier.

The circuit of Fig. 4.11 has the following parameters: $C_T = 10$ pF, $C_L = 500$ fF, $R_L = 1$ MΩ, $R_f = 100$ kΩ. Find the value of g_{m1} for which the poles are real. Assuming $g_{m1} = 1$ mS, calculate the pole position and design the impulse response.

* * *

In order to have real poles, we need that the discriminant of the second order equation at the denominator is positive. Assuming that $g_{m1}R_L \gg 1$ we have:

$$(C_T R_f + C_L R_L)^2 > 4g_{m1} R_L^2 R_f C_T C_L \rightarrow g_{m1} < \frac{(C_T R_f + C_L R_L)^2}{4R_L^2 R_f C_T C_L} \tag{4.53}$$

Putting the numbers in, we get $g_{m1} < 1.12$ μS as the upper limit for g_{m1}. Therefore, with 1 mS we have complex conjugate poles. To design the impulse response in the time domain we must find the inverse Laplace transform of $T(s)$. To do so, let's first consider a second order transfer function in the following form:

$$T(s) = \frac{1}{s^2 + as + b} \tag{4.54}$$

Note that the coefficient of the term s^2 is unitary. If $a^2 - 4b < 0$, the roots are complex conjugate and can be written as:

$$p_{1,2} = Re(p) \pm jIm(p) \tag{4.55}$$

where:

$$\begin{cases} Re(p) = -\frac{a}{2} \\ Im(p) = \frac{1}{2}\sqrt{4b - a^2} \end{cases} \tag{4.56}$$

It can be proven that the inverse Laplace transform of equation (4.56) in case of complex conjugate roots is:

$$T(t) = \frac{e^{Re(p)t}\sin(Im(p)t)}{Im(p)} = \frac{e^{-\frac{a}{2}t}\sin\left(\frac{\sqrt{4b-a^2}}{2}t\right)}{\frac{\sqrt{4b-a^2}}{2}} \tag{4.57}$$

which shows that the impulse response contains a sinusoidal term multiplied by an exponential one. The magnitude of the real part of the pole defines the speed at which the pulse response goes back to zero. The magnitude of the imaginary part determines the oscillation frequency of the sine wave. If the imaginary part is small and the real part is big, the function oscillates at low frequency and converges quickly to zero before the oscillations become visible. If the vice versa is true, the sinusoidal part can make many cycles before the exponential becomes effective in bringing the pulse response back to zero. In the extreme case in which the poles lie on the imaginary axis (a=0), the system oscillates indefinitely. If the discriminant $a^2 - 4b$ is positive, the roots are real and distinct and the pulse response can be written as:

$$T(t) = \frac{e^{-\frac{a}{2}t} \sinh\left(\frac{\sqrt{a^2-4b}}{2}t\right)}{\frac{\sqrt{a^2-4b}}{2}} \tag{4.58}$$

Observe that, formally, equations (4.57) and (4.58) are very similar, with the hyperbolic sine replacing the trigonometric one in the latter. However, since $\sinh(x)$ is defined as:

$$\sinh(x) = \frac{1}{2}\left(e^x - e^{-x}\right) \tag{4.59}$$

equation (4.58) is actually a combination of two exponentials. Let's now consider the three following transfer functions:

$$\begin{gathered} \frac{1}{s^2+5s+4} \\ \frac{1}{s^2+2s+4} \\ \frac{1}{s^2+1s+4} \end{gathered} \tag{4.60}$$

In the first case we have that the roots are given by:

$$p_{1,2} = -\frac{5}{2} \pm \frac{3}{2} \tag{4.61}$$

Applying equation (4.58) and using the explicit definition of $\sinh(x)$ we find that:

$$T(t) = \frac{e^{-\frac{5}{2}t}\left(e^{\frac{3}{2}t} - e^{-\frac{3}{2}t}\right)}{\not{2}} \frac{\not{2}}{3} = \frac{e^{-t} - e^{-4t}}{3} \tag{4.62}$$

For the second equation the roots are complex conjugate:

$$p_{1,2} = -1 \pm j\sqrt{3} \tag{4.63}$$

Therefore, the impulse response can be written as:

$$T(t) = \frac{e^{-t}\sin\sqrt{3}t}{\sqrt{3}} \tag{4.64}$$

and the frequency of oscillations is $f = \omega/2\pi = \sqrt{3}/6.28 = 0.276$ Hz. For the last case, the poles are still complex conjugate with an increased imaginary part and a reduced real one. In fact, the roots are located at:

$$p_{1,2} = -\frac{1}{2} \pm j\frac{\sqrt{15}}{2} \tag{4.65}$$

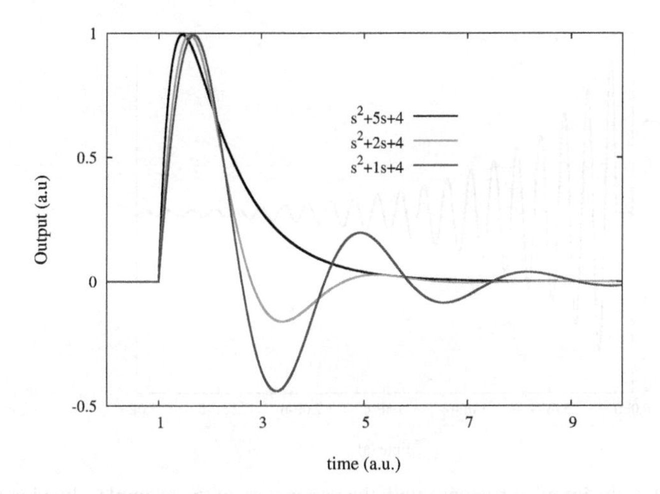

FIGURE 4.13 Impulse responses with real and complex conjugate poles.

Hence the resulting impulse response is:

$$T(t) = \frac{e^{-\frac{t}{2}} sin \frac{\sqrt{15}}{2} t}{\frac{\sqrt{15}}{2}} \tag{4.66}$$

The frequency of oscillations has only moderately increased from 0.275 Hz to 0.308 Hz, but the decay time constant of the exponential has been halved, hence the function converges more slowly to zero and the oscillations become more visible. The three pulse responses are reported in Fig. 4.13. We can now apply the above concepts to the study of our circuit. Using the number of the example, the transfer function reads:

$$T(s) = \frac{1 \cdot 10^8}{5 \cdot 10^{-13} s^2 + 1.5 \cdot 10^{-6} s + 1001} \tag{4.67}$$

We then divide both numerator and denominator by $5 \cdot 10^{-13}$ to have a unitary coefficient in front of s^2, so that equation (4.57) can be directly applied:

$$T(s) = \frac{2 \cdot 10^{20}}{s^2 + 3 \cdot 10^6 s + 2 \cdot 10^{15}} \tag{4.68}$$

To study the impulse response, the input to our transimpedance amplifier must be a current δ-like pulse, $Q_{in}\delta(t)$, whose Laplace transform is simply Q_{in}. If we just take the inverse transform of equation (4.68), we implicitly assume that $Q_{in} = 1$ C, which is a huge charge, leading to an unrealistic large response. With radiation sensors, signal charges range from fraction of fC to several pC, but the gain is often expressed in mV/fC, so taking $Q_{in} = 1$fC is a reasonable choice. Equation (4.68) hence becomes:

$$T(s) = \frac{2 \cdot 10^5}{s^2 + 3 \cdot 10^6 s + 2 \cdot 10^{15}} \tag{4.69}$$

and its inverse Laplace transform is:

$$T(t) = 4.45 \cdot 10^{-3} e^{-1.5 \cdot 10^6 t} \sin(44.72 \cdot 10^6 t) \tag{4.70}$$

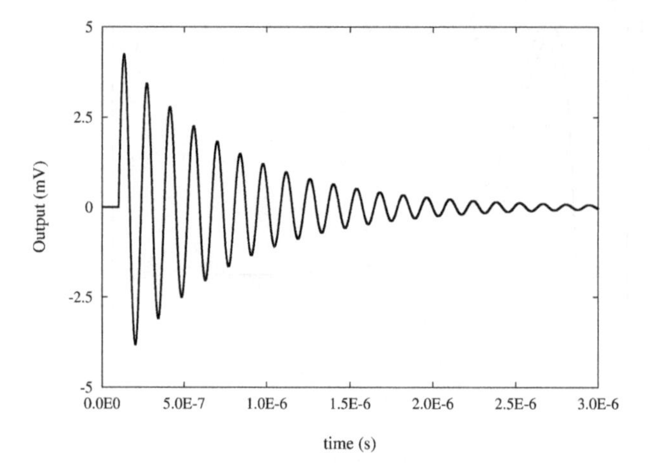

FIGURE 4.14 Plot of the impulse response with the parameters of the example, showing the pronounced ringing.

The sinusoidal component has a frequency of $44.72 \cdot 10^6/6.28 \approx 7.12$ MHz, so we have peaks every 140 ns. The time constant of the exponential is $1/1.5 \cdot 10^6 \approx 670$ ns, which is significantly larger than the oscillation period, therefore we expect that the sinusoid is quite well visible before it gets dumped by the exponential. The plot of $T(t)$ is shown in Fig. 4.14. In the figure, we have applied the signal at $t = 100$ ns and the system takes a few microseconds before it settles back to the baseline. Finally, we must observe that if a is negative (i.e. the real part of the roots is positive), the poles are located in the right-half part of the s plane. In this case, the exponential is divergent. If the poles in the right-half plane are real, the system will diverge exponentially till it reaches one of the power rails. If the poles have an imaginary part, the response will have a divergent oscillatory behavior and oscillations might be sustained indefinitely. In amplifiers, complex conjugate poles can be tolerated if the imaginary part is not too big and in some cases they can be even desirable, while poles with a positive real part must always be avoided.

The *settling time* is defined as the time elapsing between the application of the input stimulus and the instant at which the output reaches its final value within a specified error window, usually symmetric around that value. As seen in the previous example, complex conjugate poles can introduce ringing in the circuit response, thereby increasing its settling time. In a radiation detection system, this can lead to a deterioration of the counting rate capability of the apparatus.

It is customary to write second order transfer functions in the so called normalized form:

$$T_N(s) = \frac{\omega_0^2}{s^2 + s\frac{\omega_0}{Q} + \omega_0^2} \tag{4.71}$$

where ω_0 is the angular frequency associated to the poles and Q is the pole "quality factor". Expressed as a function of ω_0 and Q, the poles of equation (4.71) therefore are:

$$p_{1,2} = -\frac{\omega_0}{2Q} \pm \frac{1}{2}\sqrt{\left(\frac{\omega_0}{Q}\right)^2 - 4\omega_0^2} \rightarrow -\frac{\omega_0}{2Q} \pm \frac{\omega_0}{2Q}\sqrt{1-4Q^2} \tag{4.72}$$

The system is said to be under-damped if the poles are complex conjugate, which happens if the following condition is met:

$$1 - 4Q^2 < 0 \rightarrow Q > \frac{1}{2} \tag{4.73}$$

If $Q = 0.5$ the poles are real coincident, and the circuit is "critically damped", whereas in case the poles are real and distinct the system is "over-damped". A too large value of Q leads to strong and undesirable oscillations in the amplifier pulse response. The roots of the second order equation can also be rewritten as:

$$p_{1,2} = -\frac{\omega_0}{2Q} \pm j\frac{\omega_0}{2Q}\sqrt{4Q^2-1} \tag{4.74}$$

Therefore, knowing the real part R and the imaginary part I of the poles, Q can be calculated as:

$$Q = \frac{1}{2}\sqrt{\left(\frac{I}{R}\right)^2 + 1} \tag{4.75}$$

and its value can be used to infer the system stability.

Up to now we have seen that oscillations are a possible consequence of the presence of complex conjugate poles in the transfer function, but we did not say too much about the causes that are behind the problem and how to cure them. These issues can be understood in the framework of feedback theory. The key concepts can be more effectively explained using simple circuits based on operation amplifiers, in which the roles of the core amplifier and its feedback network can be easily identified. Therefore, in the next section we introduce the frequency behavior of operational amplifiers and the problem of their instability and compensation. This will allow us to gain a deeper insight into the issue of circuit oscillations. After the basic mechanism is discussed, we will switch back to the transimpedance amplifier to finalize its study. Differential and operational amplifiers are ubiquitous circuits and are widely covered in basic electronics courses. We therefore assume that the reader is familiar with the fundamental concepts, which are however summarized in Appendix 1.

4.4 FEEDBACK AND FREQUENCY COMPENSATION OF CMOS OTA

Operational Transconductance Amplifiers (OTAs) are fairly complex circuits and we have to expect that their transfer function contains many poles and zeros. However, as done for the single-ended cascode, we can make simplified models, in which we take into account only the most relevant time constants that are associated to the nodes of highest impedance. In a single stage amplifier only the output node has very high

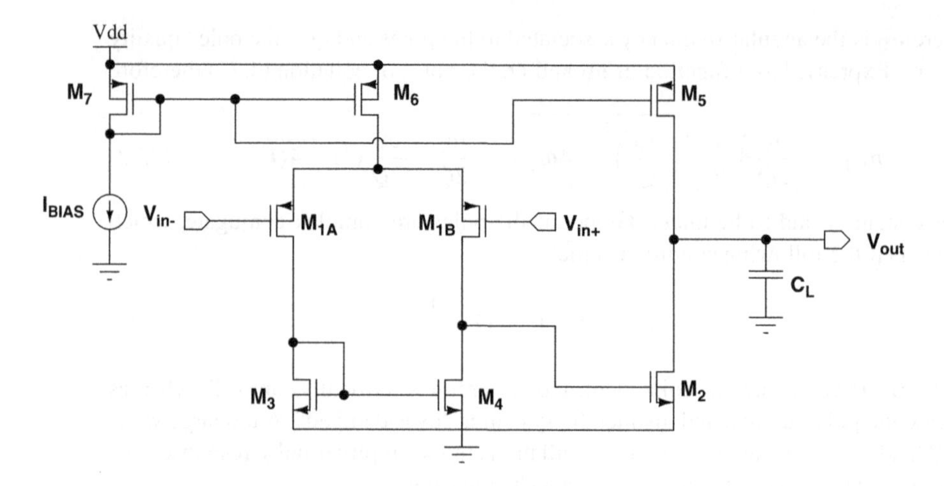

FIGURE 4.15 Two stage CMOS OTA with output load.

impedance, hence the transfer function can be modeled with a single pole associated to this dominant time constant. A two-stage OTA is obtained by cascading two circuits with similar gain. Fig. 4.15 reports the schematic of the classical two-stage CMOS OTA. It consists of a differential amplifier with current mirror load, followed by a common source amplifier. Its equivalent circuit is shown in Fig. 4.16. In this figure, R_1 and C_1 represent the output resistance of the first stage and the capacitance loading the first stage output. R_2 and C_2 represent the same quantities for the second stage. In the small signal model the load capacitance C_L has been incorporated into C_2. C_L represents the capacitance that the amplifier has to drive and usually is the dominant component loading the output stage. Each stage has a single dominant time constant, so the overall transfer function can be written as the product of two single-pole terms:

$$A(s) = \frac{A_1}{(1+s\tau_1)}\frac{A_2}{(1+s\tau_2)} = \frac{A_0}{\left(1+\frac{s}{\omega_1}\right)\left(1+\frac{s}{\omega_2}\right)} \tag{4.76}$$

where A_1 and A_2 are the low-frequency gain of the first and second stage, τ_1 and τ_2 the time constants associated to their high impedance nodes, $A_0 = A_1A_2$ is the overall low frequency gain and $\omega_1 = 1/\tau_1$ and $\omega_2 = 1/\tau_2$ are the pole angular frequencies. Let's now use an amplifier with the transfer function given by (4.76) inside a feedback loop. The general equation relating the closed-loop gain of a feedback system to the parameters of its constituents was derived in the previous chapter under very general conditions, so it is also valid when frequency dependence is considered. We can therefore rewrite it in the following form:

$$G(s) = \frac{A(s)}{1+k(s)A(s)} \tag{4.77}$$

Here $A(s)$ and $k(s)$ are respectively the transfer function of the core amplifier and of the feedback network, while the product $k(s)A(s)$ is the loop gain. For simplicity, we

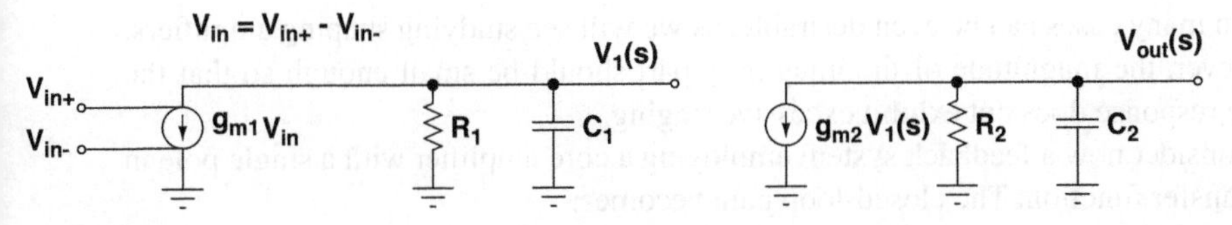

FIGURE 4.16 Small signal model of a two stage OTA incorporating the most relevant capacitances.

suppose that k is independent of frequency and is ≤ 1. The symbol k instead of f has been chosen here to indicate the transfer function of the feedback network. This is done to avoid confusion with the frequency f that will be used in the following. Inserting the expression of $A(s)$ given by (4.76) into (4.77), we get:

$$G(s) = \frac{\frac{A_0}{\left(1+\frac{s}{\omega_1}\right)\left(1+\frac{s}{\omega_2}\right)}}{1+\frac{kA_0}{\left(1+\frac{s}{\omega_1}\right)\left(1+\frac{s}{\omega_2}\right)}} \tag{4.78}$$

which can be simplified as:

$$G(s) = \frac{A_0}{\left(1+\frac{s}{\omega_1}\right)\left(1+\frac{s}{\omega_2}\right)+kA_0} = \frac{A_0}{D(s)} \tag{4.79}$$

Let's now find the poles of the circuit by solving the equation $D(s) = 0$. This yields:

$$s^2 + s\,(\omega_1+\omega_2) + (1+kA_0)\,\omega_1\omega_2 = 0 \tag{4.80}$$

which has the following roots:

$$\frac{-(\omega_1+\omega_2) \pm \sqrt{(\omega_1+\omega_2)^2 - 4\,(1+kA_0)\,\omega_1\omega_2}}{2} \tag{4.81}$$

To avoid a priori sinusoidal terms in the pulse response, we require that the poles are real, that means:

$$(\omega_1+\omega_2)^2 \geq 4\,(1+kA_0)\,\omega_1\omega_2 \tag{4.82}$$

Looking at equations (4.81) and (4.82) we see that the stronger the feedback, the higher is the risk of having complex conjugate poles with a significant imaginary part. In fact, $k = 0$ means that no feedback is applied and solving equations (4.81) we find that the poles angular frequencies are ω_1 and ω_2. This is obvious because in absence of feedback we are just left with the transfer function $A(s)$ of the core amplifier. On the opposite, if $k = 1$ the feedback is very strong because all the output signal is brought back towards the input and fed to the subtracting node, thus maximizing the risk of instability. It is hence the feedback which can turn real poles into complex conjugate ones and an amplifier that is stable with $k = 1$ will be also stable with $k \leq 1$. As noted above, complex conjugate poles in closed-loop amplifiers are not a priori forbidden and in many cases can be even desirable, as we will see studying shaping amplifiers. However, the magnitude of the imaginary part should be small enough so that the pulse response does not exhibit excessive ringing.

Consider now a feedback system employing a core amplifier with a single pole in its transfer function. The closed-loop gain becomes:

$$G(s) = \frac{\frac{A_0}{1+\frac{s}{\omega_1}}}{1+\frac{kA_0}{1+\frac{s}{\omega_1}}} = \frac{A_0}{1+\frac{s}{\omega_1}+kA_0} = \frac{A_0}{1+kA_0}\frac{1}{1+\frac{s}{\omega_1(1+kA_0)}} \tag{4.83}$$

The above relationship can also be written as:

$$G(s) = \frac{G_0}{1+\frac{s}{\omega_{CL}}} \tag{4.84}$$

where G_0 and ω_{CL} indicates, respectively, the low-frequency gain and the angular frequency of the system once feedback has been applied. We see that no matter how k is big, the closed-loop circuit has only one real pole as well, just shifted forward in frequency. If we multiply G_0 by ω_{CL} we get:

$$\frac{A_0}{1+kA_0}\omega_1\left(1+kA_0\right) = A_0\omega_1 \tag{4.85}$$

which is the well-known rule stating that the gain-bandwidth product of a first order system is constant. Both the core amplifier and the closed-loop system based on it have identical unity gain frequencies, so the Bode plot of both circuits must cross the zero-dB axis in the same point. Since they are both first order systems, the frequency roll-off after the pole is found displays always a slope of -20 dB/decade. However, the closed-loop system has a lower gain, thus to reach the same end-point the roll-off must start at a higher frequency and the gain in kept constant over a wider frequency range. The concept is illustrated in Fig. 4.17, which compares the Bode plots of the open and closed-loop circuits.

4.4.1 PHASE MARGIN

A more quantitative metric for stability can be derived by studying the modulus of $G(s)$ for sinusoidal components, i.e. for $s = j\omega$:

$$|G(j\omega)| = \left|\frac{A(j\omega)}{1+kA(j\omega)}\right| = \frac{|A(j\omega)|}{|1+kA(j\omega)|} \tag{4.86}$$

If there is a value of ω for which $|kA(j\omega)| = -1$, then $|G(j\omega)|$ diverges. Since $kA(j\omega)$ is a complex number, requiring that it is equal to -1 implies satisfying simultaneously two conditions: its absolute value is equal to 1 and its phase is $-180°$. To quantify how a system is close to this critical condition, we measure the phase of $G(j\omega)$ at the frequency for which $|G(j\omega) = 1$. The closer the phase is to $-180°$, the more the system gets prone to instability. The amount of phase rotation which is necessary to

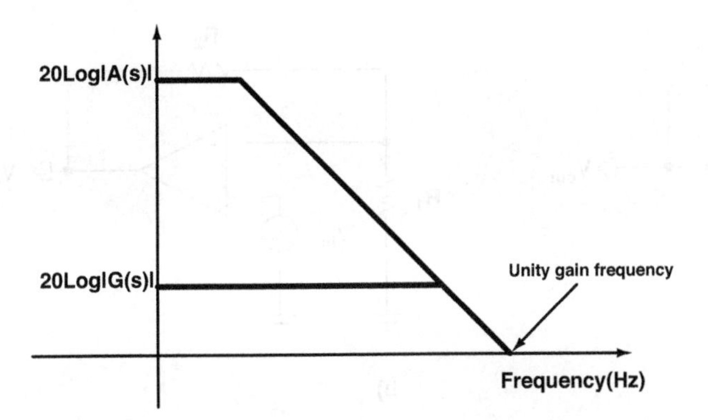

FIGURE 4.17 Gain-bandwidth trade-off in a first order system.

reach the critical value is called the "phase margin". For example, if we have in our loop gain a frequency ω_u for which $|kA(j\omega_u)| = 1$ and $\angle|kA(j\omega_u)| = -120°$ the phase margin is 60°. The amount of phase margin considered acceptable depends on the application. With front-end amplifiers, a phase margin above 70° is usually desirable. The concept of phase margin is further illustrated in the following example.

Example 4.5: Phase margin of a two-stage amplifier.

A two stage amplifier has two poles located at $f = 1$ kHz and $f = 10$ kHz. The DC gain is 10.000. Determine the phase margin when the amplifier is used to implement a unity gain buffer and when it is used to make a non-inverting voltage amplifier with gain 100.

* * *

The voltage buffer, obtained by applying unitary feedback to an op-amp, is shown in Fig. 4.18 a). The signal is connected to the non inverting input. The output is shorted to the inverting terminal, which implements the subtracting function. To realize the negative feedback, the input and output signals must be in phase, so that they get actually subtracted. Since the full output signal is fed-back, $k = 1$ and the closed-loop gain is given by:

$$G(s) = \frac{A(s)}{1 + A(s)} \tag{4.87}$$

Writing explicitly the relationships between the op-amp terminals and taking into consideration the frequency dependence, we have:

$$V_{out}(s) = A(s)\,(V_{in} - V_{out}) \tag{4.88}$$

Solving for V_{out}/V_{in}, we get equation (4.87). Suppose we apply to our circuit a sinusoidal signal. This, propagating through the op-amp, encounters the circuit poles and gets both attenuated and delayed, i.e. its phase is changed. If the transfer function of the

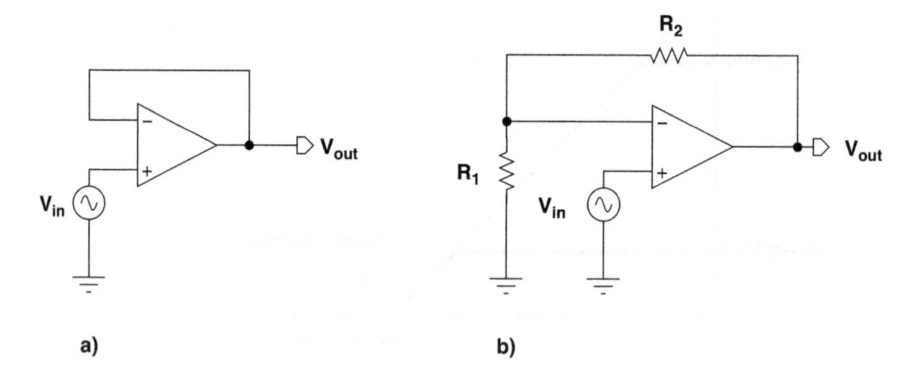

FIGURE 4.18 Voltage buffer (in a) and non-inverting amplifier (in b).

operational amplifier contains only one pole, the phase shift will be at most 90°. If $A(s)$ has at least two poles, at some frequency the phase shift reaches $-180°$, so a sinusoid of this critical frequency applied to the "+" terminal is inverted at the output. Therefore, the difference in $V_{in} - V_{out}$ becomes actually a sum. If the gain $A(s)$ is still greater than one when this happens, the quantity is amplified and a regeneration process is started, bringing the circuit into oscillation. Therefore, the phase shift experienced by the signal while propagating in the feedback loop can turn a negative feedback at low frequency into a positive feedback at higher ones. We need to make here two important observations: in general an amplifier has to process signals which are more complicated than a simple sinusoid. However, thanks to the Fourier analysis, complex waveforms can be represented as a superposition of sinusoidal components of different frequencies. It is sufficient that some of the components in the signal fall in the critical region to trigger oscillations. Furthermore, even though the signal does not contain critical frequencies, the noise will, because noise is in general very broadband, and the circuit may start oscillating under the effects of its own noise. Therefore, the condition that the gain is still greater than one while the phase shift is already approaching $-180°$ must be avoided a priori.

Our amplifier has two poles, so there is the risk of having an unstable circuit once the feedback loop is closed. Since $k = 1$, the study of the loop-gain reduces to that of the transfer function of the core amplifier itself, which can be written as:

$$A(s) = \frac{A_1}{\left(1+\frac{s}{\omega_1}\right)}\frac{A_2}{\left(1+\frac{s}{\omega_2}\right)} = \frac{A_0}{\left(1+j\frac{f}{f_1}\right)\left(1+j\frac{f}{f_2}\right)} \tag{4.89}$$

In the above equation we have made the substitutions $A_1A_2 = A_0$, $s = j\omega = 2\pi f$, $\omega_1 = 2\pi f_1$ and $\omega_2 = 2\pi f_2$. In polar form, a complex number C with modulus $|C|$ and phase $\angle C$ can be written as:

$$C = |C|e^{j\angle C} \tag{4.90}$$

and the product between two complex numbers C_1 and C_2 becomes:

$$C_1C_2 = |C_1||C_2|e^{j(\angle C_1 + \angle C_2)} \tag{4.91}$$

The transfer function $A(s)$ is the product of two single pole terms. The first one, $A_1(f) = A_1/(1+jf/f_1)$ has the following modulus and phase:

$$|A_1(f)| = \frac{A_1}{\sqrt{1+\left(\frac{f}{f_1}\right)^2}}$$
$$\angle A_1(f) = \tan^{-1}\left(-\frac{f}{f_1}\right) \tag{4.92}$$

Similarly, for $A_2(f) = A_2/(1+jf/f_2)$ we have:

$$|A_2(f)| = \frac{A_2}{\sqrt{1+\left(\frac{f}{f_2}\right)^2}}$$
$$\angle A_2(f) = \tan^{-1}\left(-\frac{f}{f_2}\right) \tag{4.93}$$

Consequently, their product can be written as:

$$A(f) = A_1(f)A_2(f) = \frac{A_1A_2}{\sqrt{1+\left(\frac{f}{f_1}\right)^2}\sqrt{1+\left(\frac{f}{f_2}\right)^2}}e^{j\left[\tan^{-1}\left(-\frac{f}{f_1}\right)+\tan^{-1}\left(-\frac{f}{f_2}\right)\right]} \tag{4.94}$$

Let's now evaluate $|A(f)|$ and $\angle A(f)$ in a few significant points. We know that the gain at low frequency is 10000, i.e. 80 dB. Evaluating (4.94) on $f = f_1$ (i.e. $f/f_1 = 1$ and $f/f_2 = 0.1$) we have that $|A(f)| = 77$ dB and $\angle A(f) = -50.7°$. In $f = f_2$, $|A(f)| = 57$ dB and $\angle A(f) = -129.3°$. Observe that from $f = f_1$ to $f = f_2$ the gain has decreased with a slope of -20 dB/decade, which means that the second pole does not affect substantially the gain before $f = f_2$. However, at $f = f_2$ the phase rotation induced by the second pole is already $-45°$, which, summed to the $-84.3°$ of the first pole yields the total of $-129.3°$. At $f = 100$ kHz $= 10f_2 = 100f_1$, the gain is $|A(f)| = 20$ dB. In one decade of frequency we have now lost 40 dB, because each pole contributes with a gain reduction of -20 dB/decade. To reach the unity gain frequency (which means zero dB) we need to lose an additional 20 dB. Since the slope is now 40 dB/decade, this means half a decade. So the zero dB axis is crossed at a frequency:

$$100\text{ kHz} \times 10^{0.5} = 316.2\text{ kHz} \tag{4.95}$$

The phase at this frequency is given by:

$$\tan^{-1}(-316.2) + \tan^{-1}(-31.62) = -89.8° - 88.19° = -178° \tag{4.96}$$

The phase margin is only 2°, which is very far even from the minimum value usually considered acceptable for an amplifier, 45°. We can then conclude that the amplifier is not suitable to be used in a unity gain configuration.

Consider now the case of the same amplifier employed in the non-inverting configuration with closed-loop gain of 100, shown in Fig. 4.18 b). Here the feedback transfer function f is given by $R_1/(R_1+R_2) = 0.01$. We need therefore to study the loop gain:

$$kA(f) = \frac{0.01A_0}{\sqrt{1+\left(\frac{f}{f_1}\right)^2}\sqrt{1+\left(\frac{f}{f_2}\right)^2}}e^{j\left[\tan^{-1}\left(-\frac{f}{f_1}\right)+\tan^{-1}\left(\frac{f}{f_2}\right)\right]} \tag{4.97}$$

Reasoning as before, we see that the loop gain is 37 dB for $f = f_1$ and 17 dB for $f = f_2$. After f_2 the slope becomes -40 dB/decade. To reach the zero-dB axis with a slope of -40 dB decade we need $17/40 = 0.425$ decades. Hence, the unitary loop gain is reached at a frequency $10^{0.424} f_2 = 26.7$ kHz. At this frequency, the total phase shift is $-89.5°$ due to the first pole and $-69.5°$ due to the second pole, for a total shift of $-159°$, which leaves a phase margin of $21°$. This is better than before, but it is still far from an acceptable condition. It is interesting to evaluate for both cases the quantity $|G(f_{UG})|$, i.e. the magnitude of the closed-loop gain at the frequency for which $|kA(jf)|$ becomes unity and compare it to the low frequency value. The modulus of $G(jf)$ is given by:

$$|G(jf_{UG})| = \frac{|A(jf_{UG})|}{|1 + kA(jf_{UG})|} \tag{4.98}$$

For the voltage buffer, $k = 1$ and $|A(jf_{UG})| = 1$. We must then evaluate the following quantity:

$$\left|\frac{1}{1 + A(f_{UG})}\right| = \left|\frac{1}{1 + 1 \cdot e^{-j178°}}\right| \tag{4.99}$$

Using the Euler formula, the denominator can be written as $1 + e^{-j178} = \cos(-178) + j\sin(-178) = 6.1 \cdot 10^{-4} - j0.035$. For a complex number we have that:

$$\left|\frac{1}{a + jb}\right| = \frac{1}{\sqrt{a^b + b^2}} \tag{4.100}$$

Therefore, the value of $|G(jf)|$ at $f = f_{UG}$ becomes

$$|G(jf_{UG})| = \left|\frac{1}{6.1 \cdot 10^{-4} - j0.035}\right| = \frac{1}{\sqrt{(6.1 \cdot 10^{-4})^2 + (0.035)^2}} = 28.57 \tag{4.101}$$

From a "well-behaved" system we would expect that $G(jf)$ stays constant to the nominal value till f_{UG} is reached, and then decreases. Equation (4.101) shows instead that the gain rises while approaching f_{UG}, reaching the peak at $f = f_{UG}$. In the Laplace domain, a step voltage signal of amplitude V_0 is represented as V_0/s, therefore the plot is the inverse Laplace transform of:

$$V_{out}(s) = \frac{1}{s} G(s) = \frac{1}{s} \frac{A(s)}{1 + A(s)} \tag{4.102}$$

The gain-peaking at $f = f_{UG}$ is the frequency domain implies an emphasis on the sinusoidal components present is the step signal with a frequency close to f_{UG}, which results in significant ringing. For the circuit with $k = 0.01$ we have that $|kA(jf_{UG})| = 1 \rightarrow |A(jf_{UG})| = 1/k = 100$. The modulus of $G(jf)$ at $f = f_{UG}$ is hence given by:

$$|G(jf_{UG})| = \frac{100}{|1 + 1 \cdot e^{-j159°}|} \tag{4.103}$$

Proceeding as before, we get

$$|G(jf_{UG})| = \frac{1}{k} 275 = 2.75 \tag{4.104}$$

which shows that the gain at $f = f_{UG}$ has a peak of 2.75 times the desired value of $1/k$, which is achieved at low frequency. We see that increasing the phase margin of the loop reduces the gain peaking at $f = f_{UG}$.

4.4.2 FREQUENCY COMPENSATION

The above example shows that the use of an operational amplifier with two poles in its transfer function separated only by one decade of frequency leads to closed-loop circuits with unsatisfactory stability performance. In an op-amp built by cascading two building blocks with similar gain and bandwidth, it is natural that the most important time constants are relatively close by. Before feedback can be applied, the open-loop transfer function must be modified and it is important to understand how the poles should be placed to achieve a given phase margin. Suppose that we want to use an op-amp in unity gain feedback. In this case the loop gain $kA(s)$ coincides with the open-loop gain of the core amplifier because $k = 1$. To have a phase margin, for instance, of 45° we can allow that $A(s)$ has a total phase rotation of $-135°$ when its modulus reaches the zero-dB axis. Remember now that each pole in the transfer function contributes with a maximum phase shift of $-90°$ and that the shift introduced by a given pole is already $-45°$ at the frequency of the pole. This means that to have $\angle A(jf) = -135°$ at $f = f_{UG}$ we can at most have two poles before $f = f_{UG}$: one located exactly at $f = f_{UG}$ which contributed with $-45°$ and another before f_{UG} that gives the $-90°$ shift. This situation is shown in Fig. 4.19. The DC gain of the amplifier determines how much the first pole must be located before the second one. The low-frequency gain A_0 in fact stays constant till the first pole is found at $f = f_1$. At this frequency, the phase rotation is already $-45°$ and the gain starts dropping with

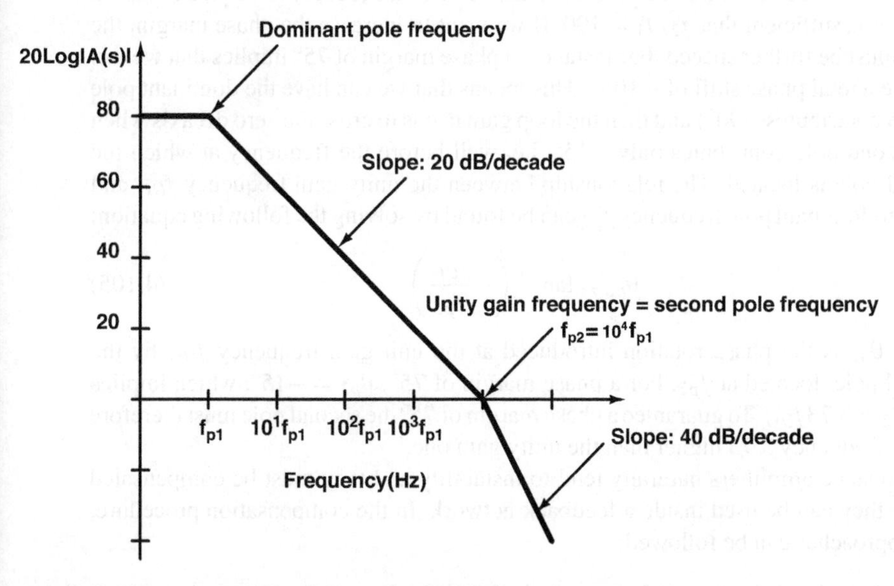

FIGURE 4.19 An amplifier with one dominant pole and a second pole located at the unity gain frequency allows for a closed-loop gain of 1 with 45° phase margin.

a slope of -20 dB/decade. To reach the zero dB axis, we need $n = A_0/20$ decades of frequency. Since we want to have a total phase rotation of $-135°$, the frequency of the second pole, f_2 must be equal to f_{UG}, so the second pole must be located n decades after the first one. For example, if A_0 is 80 dB we need four decades to reach the zero dB axis with a slope of -20 dB/decade. If the first pole p_1 is located f_1 =1 kHz, to guarantee a phase margin of $-45°$ the second one must be placed four decades beyond, i.e at 1×10^4 kHz or 10 MHz. If, on the other hand, A_0 is just 40 dB, two decades are sufficient, so $f_2 = f_{UG} = 100$ kHz. The low-frequency closed-loop performance are related to the magnitude of A_0, which in general is chosen as high as possible and in practical cases always exceeds 60 dB. From this discussion we see that the two dominant poles must be spaced several decades in frequency to guarantee a reasonable phase margin. The problem is less severe for smaller loop gain, which implies a bigger closed-loop gain. Fig. 4.20 shows the Bode plot of an amplifier with 80 dB open-loop gain and two poles separated by only two decades in frequency. For a closed-loop gain of 1 ($k = 1$), such an amplifier has a phase margin of only 5.7°. The situation improves if the amplifier is used for a closed-loop gain of 100, ($k = 0.01$), which implies a loop gain of $kA(s) = 0.01A(s)$, which is also depicted in the figure. In this case, when the loop gain crosses the zero-dB axis, the second pole is found and the phase margin is hence 45°. Note that if $k \neq 1$, what must be studied to predict the phase margin is the Bode plot of the loop gain $kA(s)$ and not the transfer function of the core amplifier $A(s)$ alone. For the case just discussed, $k = 0.01$ and if k does not depend of frequency, the pole positions are the ones of $A(s)$. The DC loop gain is $0.01 \times 10000 = 40$ dB, hence, after the first pole is found we need only two decades to reach the zero dB axis. As a consequence, for a phase margin of 45° it is sufficient that $f_2/f_1 = 100$. If we want to increase the phase margin, the poles must be further spaced. For instance, a phase margin of 75° implies that we can tolerate a total phase shift of $-105°$. This means that we can have the dominant pole (which contributes $-90°$) and then the loop gain needs to cross the zero dB axis when the second pole contributes only $-15°$, i.e. well before the frequency at which the second pole is located. The relationship between the unity gain frequency f_{UG} and the non-dominant pole frequency f_{p2} can be found by solving the following equation:

$$\theta_{p2} = \tan^{-1}\left(-\frac{f_{UG}}{f_{p2}}\right) \tag{4.105}$$

where θ_{p2} is the phase rotation introduced at the unit gain frequency f_{UG} by the second pole, located at f_{p2}. For a phase margin of 75°, $\theta_{p2} = -15°$, which implies that $f_{p2} = 3.73 f_{UG}$. To guarantee a phase margin of 75° the second pole must therefore be at a frequency 3.73 higher than the unity gain one.

Two stage amplifiers naturally tend to instability and they must be compensated before they can be used inside a feedback network. In the compensation procedure, two approaches can be followed:

- The gain of the loop is known a priori. In this case, the compensation can be optimized to provide the necessary phase margin at the required gain. An amplifier compensated in this way might not be stable when it is used with

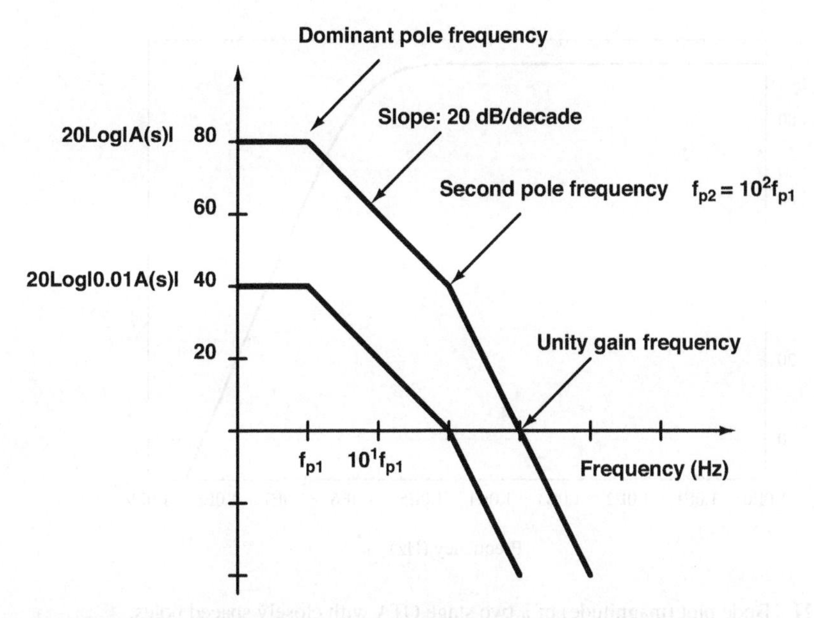

FIGURE 4.20 Pole location allowing a closed-loop gain of 100 with a phase margin of 45°. The loop gain $kA(s) = 0.01A(s)$ is also shown.

a lower closed-loop gain (remember that a *greater* loop gain $kA(s)$ implies a *lower* closed-loop gain $G(s)$).

- The amplifier is compensated for the most severe condition, i.e. $k = 1$. In this case it will be automatically stable also for smaller values of k, although this "over-compensation" results in sub-optimal speed when the amplifier is used with smaller values of k. In the following we limit our discussion to this second case.

We consider now a two stage amplifier whose small signal equivalent circuit is reported in Fig. 4.16. R_1 and C_1 represents the output resistance and the load capacitance of the first stage. R_2 and C_2 are the similar quantities for the second stage. We suppose that the circuit has the following parameters $g_{m1} = 1$ mS, $R_1 = 100$ kΩ, $C_1 = 1$ pF, $g_{m2} = 10$ mS, $R_2 = 50$ kΩ, $C_2 = 10$ pF. First, we design the Bode plot of the circuit. The DC gain A_0 is given by:

$$A_0 = g_{m1} R_1 g_{m2} R_2 = 50000 = 94 \text{ dB} \quad (4.106)$$

The pole of the first stage is located at:

$$f_{p1} = \frac{1}{2\pi R_1 C_1} = 1.59 \text{ MHz} \quad (4.107)$$

The pole associated to the second stage is found at:

$$f_{p2} = \frac{1}{2\pi R_2 C_2} = 318.5 \text{ kHz} \quad (4.108)$$

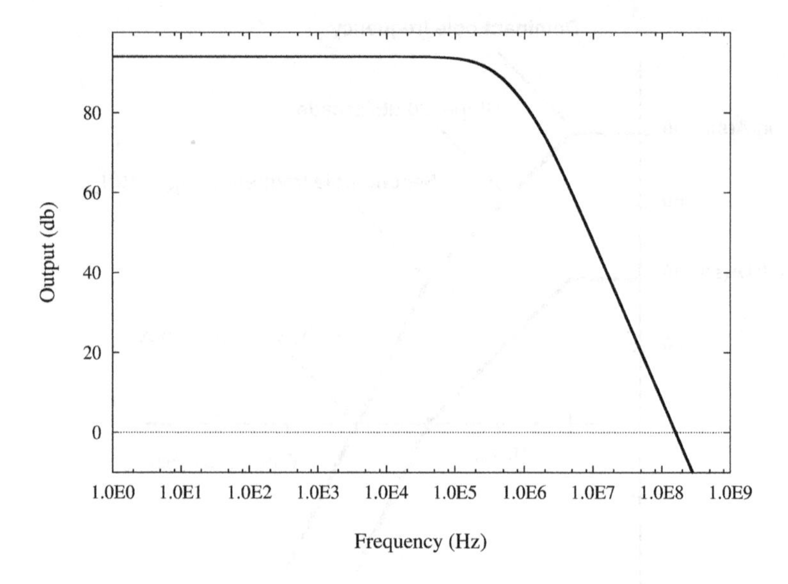

FIGURE 4.21 Bode plot (magnitude) of a two stage OTA with closely spaced poles.

Due to the large capacitance and comparable resistance, f_{p2} is smaller than f_{p1}, so f_{p2} defines the frequency of the dominant pole. Regardless of the stage they belong to, we call f_d and f_{nd} the poles located respectively at the lower and at higher frequency (the suffix *d* here means "dominant" and *nd* "non-dominant"). We want to make the amplifier stable for a closed-loop gain of 1, so that it can be used also as a voltage buffer. Therefore, we study directly the transfer function $A(s)$. Fig. 4.22 reports the magnitude of $A(jf)$. The gain is constant till the frequency $f_d = 318.5$ kHz, after which it starts decreasing with the slope of -20 dB/decade. The second pole is found already at 1.59 MHz, after only 0.5 decade, so the gain is reduced only by 10 dB, reaching the value of 84 dB. After this point, the gain decreases with a slope of -40 dB/decade, therefore 2.1 decades are necessary to cross the 0 dB axis, i.e. to reach the unitary gain. This happens at a frequency given by $10^{2.1} \times 1.59 \cdot 10^6 = 200$ MHz. We can already anticipate that since the two poles are very close by, the amplifier will not be stable in unity gain configuration. In fact, the phase of $A(f)$ at the unity gain frequency is given by:

$$\angle A(2 \cdot 10^8) = \tan^{-1}\left(-\frac{2 \cdot 10^8}{3.18 \cdot 10^5}\right) + \tan^{-1}\left(-\frac{2 \cdot 10^8}{1.59 \cdot 10^6}\right) = -179.46^\circ \quad (4.109)$$

so the phase margin is zero. The phase of the circuit is shown in Fig. 4.21. Observe that the poles are so close that in practice it is difficult to disentangle their effects from the Bode plot.

To compensate the amplifier we need to split the two poles apart. A first obvious solution may seem to shift the already dominant pole to lower frequencies, leaving the second pole unchanged, so let's try to apply this strategy first. If we do not move

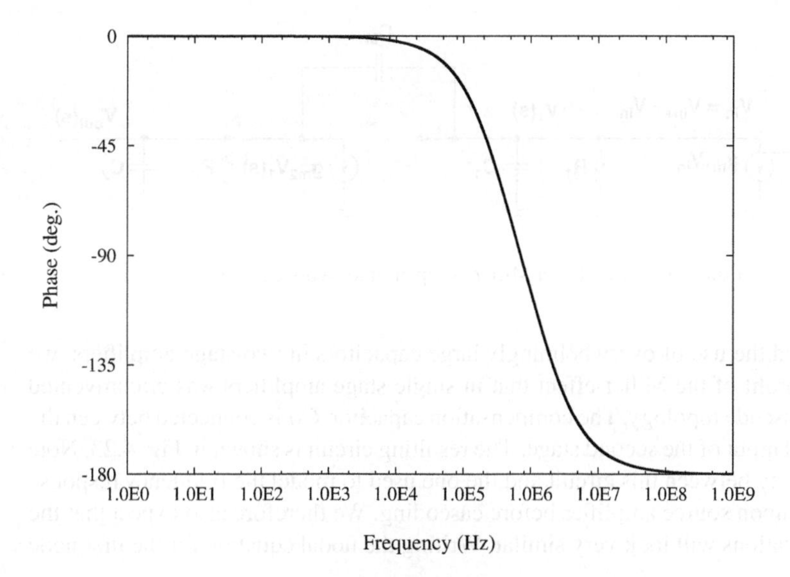

FIGURE 4.22 Bode plot (phase) of a two stage OTA with closely spaced poles.

the second pole, we have to anticipate the first one so that the $|A(f)|$ crosses the zero dB axis no later than the frequency of the second pole, which is located at 1.59 MHz. From our previous discussion, this would ensure a phase margin of 45°. Since the DC gain in 94 dB and after the first pole the roll-off is -20 dB/decade, the first pole must be placed 4.7 decades before the second one, or at $1.59 \cdot 10^6/10^{4.7} = 31.72$ Hz. To have such a low frequency, we need a time constant $R_2C_2 = 5$ ms. Given that R_2 is 50 kΩ, this would require a capacitor of 100 nF in parallel to C_2. Such a large capacitor can only be an external component, and the resulting amplifier has in addition a very poor bandwidth, with a unity gain frequency which decreases from 200 MHz before compensation to a mere 1.59 MHz. This compensation strategy is therefore very inefficient for multi-stage amplifiers.[1] There is however a case in which this approach is used and it is in the case of single stage amplifiers. In configurations such as telescopic or folded cascode, there is in fact only a very high impedance node inside the circuit, to which the dominant pole is associated. We have so far modeled these circuits as single pole amplifiers, that would be a priori stable. However, this is an abstraction, because also the low impedance nodes (for instance those associated with the diode connected current mirrors) contribute with their time constants, which are however located at much higher frequency than the dominant one. Adding then a small capacitor to the output shifts the dominant pole to lower frequencies, increasing the phase margin if necessary.

[1] One exception is the case of Low Drop-Out Regulators (LDO), which are compensated with a large external capacitor at the output that serves also as a battery to supply current to the load during transients.

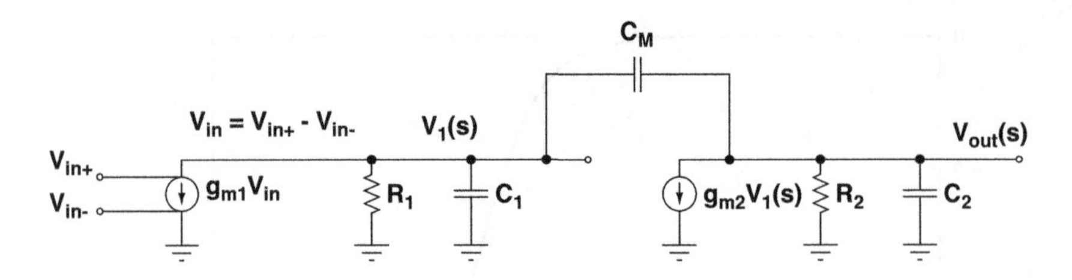

FIGURE 4.23 Equivalent circuit of a Miller compensated two-stage OTA.

To avoid the use of overwhelmingly large capacitors in two-stage amplifiers, we can take profit of the Miller effect that in single stage amplifiers was circumvented with the cascode topology. The compensation capacitor, C_M is connected between the output and input of the second stage. The resulting circuit is shown in Fig. 4.23. Note the similarity between this circuit and the one used to model the frequency response of the common source amplifier before cascoding. We therefore also expect that the circuit equations will look very similar. Writing the nodal equation for the first node we have:

$$g_{m1} V_{in} + \frac{V_1}{R_1} + V_1 s C_1 + (V_1 - V_{out}) s C_M = 0 \tag{4.110}$$

For the output node we get:

$$g_{m2} V_1 + \frac{V_{out}}{R_2} + V_{out} s C_2 + (V_{out} - V_1) s C_M = 0 \tag{4.111}$$

We can now organize the two equations in the usual matrix form:

$$\begin{bmatrix} \frac{1}{R_1} + s(C_1 + C_M) & -sC_M \\ g_{m2} - sC_{CM} & \frac{1}{R_2} + s(C_2 + C_M) \end{bmatrix} \times \begin{bmatrix} V_1 \\ V_{out} \end{bmatrix} = \begin{bmatrix} -g_{m1} V_{in} \\ 0 \end{bmatrix} \tag{4.112}$$

which makes fully apparent the equivalent of the two circuits, with the only difference that the independent current source is replaced by a dependent one. We can therefore write the ratio V_{out}/V_{in} as:

$$\frac{g_{m1} g_{m2} R_1 R_2 (1 - sC_M/g_{m2})}{s^2 (C_1 C_2 + C_1 C_M + C_2 C_M) + s\left[R_2 C_2 + R_2 C_M + R_1 C_1 + R_1 (1 + g_{m2} R_2) C_M\right] + 1} \tag{4.113}$$

As expected we find again the term $(1 + g_{m2} R_2) C_M$ which indicates that the Miller effect is at play and "magnifies" C_M, which has an effective value much greater than its physical one. Observe that the Miller connection is a form of local feedback put around the second stage of the amplifier, so in principle it could result in complex conjugate poles. However, since we will place a global feedback around our amplifier, we would like to have two reals and adequately split poles. We must therefore estimate the minimum value of C_M to have only real poles in the transfer function. To do so, we assume that the term $g_{m2} R_1 R_2 C_M$ dominates the coefficient of s and that

C_2C_M dominates the one of s^2. The latter assumption is reasonable, since C_2 is the greatest capacitance we have in our circuit. Equation (4.113) can hence be simplified as follows:

$$\frac{V_{out}}{V_{in}} = \frac{g_{m1}g_{m2}R_1R_2(1 - sC_M/g_{m2})}{s^2R_1R_2C_2C_M + sg_{m2}R_1R_2C_M + 1} \tag{4.114}$$

To have real poles we need to verify the following condition:

$$g_{m2}^2R_1^2R_2^2C_M^2 > 4R_1R_2C_2C_M \rightarrow C_M > \frac{4C_2}{g_{m2}^2R_1R_2} \tag{4.115}$$

Putting in the numbers of our circuit, we get $C_M > 8 \cdot 10^{-17}$ F. Therefore, any practical value of C_M will yield real poles and the value of the Miller capacitance will be dictated by the need of having the two poles far away enough from each other to guarantee the requested phase margin. Observe also that a minimum of capacitance is always provided by the gate-drain capacitance of the input transistor of the second stage (M_2 in Fig. 4.15). Knowing that the poles are real, we can use the approximation already employed before to estimate the pole frequencies, which, comparing the denominator to a generic second order polynomial $as^2 + bs + 1$ allowed us to estimate the pole angular frequencies as:

$$\begin{aligned}\omega_1 &\approx -\frac{1}{g_{m2}R_1R_2C_M} \\ \omega_2 &\approx -\frac{g_{m2}}{C_2}\end{aligned} \tag{4.116}$$

where to estimate the pole positions we have used the simplified transfer function given in (4.114). Let's assume for the time being to choose a tentative value for C_M of 0.5 pF, which is a pretty small one in an integrated circuit and evaluate the magnitude of the new poles. With the parameters used up to now we get $f_1 = \omega_1/2\pi = 6.34$ kHz and $f_2 = \omega_2/2\pi = 159$ MHz. We can now make a few interesting considerations:

- The Miller compensation moves *both poles* in the transfer function. The pole associated to the first stage is pushed to lower frequencies and becomes the dominant pole. The pole associated to the second stage is shifted to higher frequencies and becomes the non-dominant pole.This effect is known as *pole splitting*.
- For any realistic values of the Miller capacitance, C_M, the two poles are real.
- In first approximation, the non-dominant pole does not depend on the Miller capacitance, but only on the ratio between the transconductance of the second stage and its load capacitance. For the same load capacitance, the second pole can be pushed to higher frequency by increasing g_{m2}, which means increasing the power consumption of the amplifier.

For the case we are considering, we can now choose the Miller capacitance to have the desired phase margin. The position of the second pole has been calculated to be 159 MHz. The DC gain of the amplifier is $g_{m1}R_1g_{m2}R_2 = 94$ dB. We have already seen that for a phase margin of 45° the dominant pole must be located 4.7 decades before the non-dominant one, to allow the crossing of the zero-dB axis exactly at f_2. Therefore, f_1 must be equal to $159 \cdot 10^6/10^{4.7} = 3.17$ kHz. This corresponds to an angular frequency $\omega_1 = 2\pi f_1 = 19.9$ kHz. We can then use equation (4.116) to calculate C_M, that, with our parameters, is 1 pF. Observe that for a reasonable phase margin and DC gain of the amplifier, only one pole is allowed before the zero dB axis is crossed. Therefore, in the region between the first pole and the zero dB axis, the gain slope is −20 dB/decade. Once the unity gain frequency has been calculated from the desired phase margin, the position of the first pole is thus determined by $\omega_1 = \omega_{UG}/A_0$.

Example 4.6: Miller capacitance calculation.

A two stage CMOS OTA has the following parameters: $g_{m1} = 0.5$ mS, $R_1 = 200$ kΩ, $C_1 = 1$ pF, $g_{m2} = 4$ mS, $R_2 = 30$ kΩ, $C_2 = 20$ pF. Determine the value of the Miller compensation capacitance C_M for the following phase margins: 45°, 60° and 75°, assuming that the amplifier can be used also with unitary closed-loop gain.

* * *

Let's first calculate the DC gain of the amplifier:

$$A_0 = g_{m1}g_{m2}R_1R_2 = 0.5 \cdot 10^{-3} \times 2 \cdot 10^5 \times 4 \cdot 10^{-3} \times 3 \cdot 10^4 = 12000 = 81.6 \text{ dB}. \quad (4.117)$$

After compensation, the angular frequency of the second pole is $\omega_{p2} = g_{m2}/C_2 =$ 200 MHz. For a phase margin of 45°, this is also the unity gain frequency of the amplifier. The frequency of the dominant pole is $\omega_{p1} = \omega_{UG}/A_0 = 200 \cdot 10^6/12000 = 16.7$ kHz. The Miller capacitor C_M can then be calculated as:

$$C_M = \frac{1}{g_{m2}R_2R_1\omega_{p1}} \quad (4.118)$$

which leads to $C_M = 2.5$ pF.

Alternatively, one can observe that, for a phase margin of 45°, $\omega_{UG}/A_0 = \omega_{p1}$ can be written as:

$$\omega_{p1} = \frac{g_{m2}}{C_2}\frac{1}{g_{m1}g_{m2}R_1R_2} \quad (4.119)$$

On the other hand, the position of ω_{p1} is given by the effective value of C_M multiplied by the output resistance of the first stage:

$$\omega_{p1} = \frac{1}{g_{m2}R_2C_MR_1} \quad (4.120)$$

Combining the two above equations, the following expression for C_M is obtained:

$$C_M = C_2 \frac{g_{m1}}{g_{m2}} \tag{4.121}$$

which allows an immediate calculation of C_M for a phase margin of 45° and puts in evidence that in a stable system the transconductance of the second stage must be greater than the one of the first stage.

For a phase margin of 60°, the second pole can only contribute 30° of phase shift at the unity gain frequency. Therefore, we have:

$$\tan^{-1}\left(-\frac{\omega_{UG}}{\omega_{p2}}\right) = -30° \tag{4.122}$$

from which we obtain $\omega_{UG} = 0.58\omega_{p2} = 116$ MHz. The first pole must thus be located at 9.7 kHz, which, according to equation (4.118), requires $C_M = 4.3$ pF. Finally, for a phase margin of 75°, the phase shift at the unity gain can only be 15°, hence

$$\tan^{-1}\left(-\frac{\omega_{UG}}{\omega_{p2}}\right) = -15° \rightarrow \omega_U G = 0.27\omega_{p2} = 53.6 \text{ MHz} \tag{4.123}$$

The dominant pole is therefore placed at $\omega_{p1} = 5.36 \cdot 10^7/12000 = 4400$ kHz, which is achieved with $C_M = 9.5$ pF.

4.4.3 EFFECT OF THE RIGHT-HALF PLANE ZERO

We have so far neglected the effect of the zero located at the numerator $N(s)$ of the transfer function. Unfortunately, owing to its location in the s-plane, such a zero can have an adverse affect on circuit stability. To understand why, let's study the magnitude and the phase of the numerator as a function of frequency. We have:

$$N(s) = g_{m1}g_{m2}R_1R_2(1 - sC_M/g_{m2}) = A_0(1 - sC_M/g_{m2}) \rightarrow A_0(1 - j\omega C_M/g_{m2}) \tag{4.124}$$

The modulus of $N(j\omega)$ is:

$$|N(j\omega)| = A_0\sqrt{1 + \left(\frac{\omega C_M}{g_{m2}}\right)^2} \tag{4.125}$$

The phase is instead:

$$\angle N(j\omega) = \tan^{-1}\left(-\frac{\omega}{\omega_z}\right) \tag{4.126}$$

where $\omega_z = g_{m2}/C_M$ is the zero angular frequency. We see from the above equations that the zero rises the value of the numerator, thus increasing the magnitude of the full transfer function, while introducing a negative phase shift. Combined together, the two effects reduce the phase margin and can compromise the system stability. In fact, on the one hand, the magnitude of $|A(j\omega)|$ is increased, on the other the phase rotation is incremented, enhancing the chance that the $|A(j\omega)|$ crosses the zero dB axis with insufficient phase margin.

Example 4.7: Effect of the RHP zero on the phase margin

Re-evaluate the phase margins of example 4.6 keeping into account the effect of the RHP zero.

* * *

For $C_M = 2.3$ pF, (phase margin 45°), the zero is located at $\omega_z = g_{m2}/C_M = 4\cdot 10^{-3}/2.3\cdot 10^{-12} = 1.6$ GHz. This frequency is much beyond the unity gain bandwidth, so the effect on the magnitude before $\omega_{UG} = 200$ MHz is negligible. The additional phase rotation introduced by the zero at ω_{UG} is given by:

$$\angle A(j\omega)|_{\omega_z} = \tan^{-1}\left(-\frac{\omega_{UG}}{\omega_z}\right) = -7.1^\circ \tag{4.127}$$

This reduces the phase margin to 37.9°, below the bare acceptable minimum. For $C_M = 4.3$ pF, ω_z is reduced to 930 MHz, while ω_{UG} is 116 MHz. Applying again equation (4.127) we obtain the same additional phase shift of -7.1° bringing the total phase margin to 52.9°. Finally, for $C_M = 9.5$ pF, $\omega_{UG} = 52.8$ MHz, $\omega_z = 421$ MHz and the contribution of the zero to the phase is again -7.1° which reduces the phase margin to 67.9°.

In the above example the zero degrades the phase margin always by the same amount, but this should not come as a surprise. In fact, the unity gain frequency after compensation is given by $\omega_{UG} = g_{m1}/C_M$, while the zero is placed at $\omega_z = g_{m2}/C_M$. The ratio between the unity gain and the zero frequency is then given by:

$$\frac{\omega_{UG}}{\omega_z} = \frac{g_{m1}}{g_{m2}} \tag{4.128}$$

From the above considerations, we see that to have a large bandwidth we need a high g_{m1}/C_M ratio, while to have a good phase margin an adequate g_{m2}/g_{m1} is necessary. Since the transconductance depends on the device current, fast amplifiers demand high power consumption.

4.4.4 ADVANCED COMPENSATION TECHNIQUES

We hereafter briefly mention more advanced compensation techniques which enable the design of amplifiers with greater bandwidth and/or phase margin. For each configuration, we just report without proof the positions of the poles and zeros. The reader

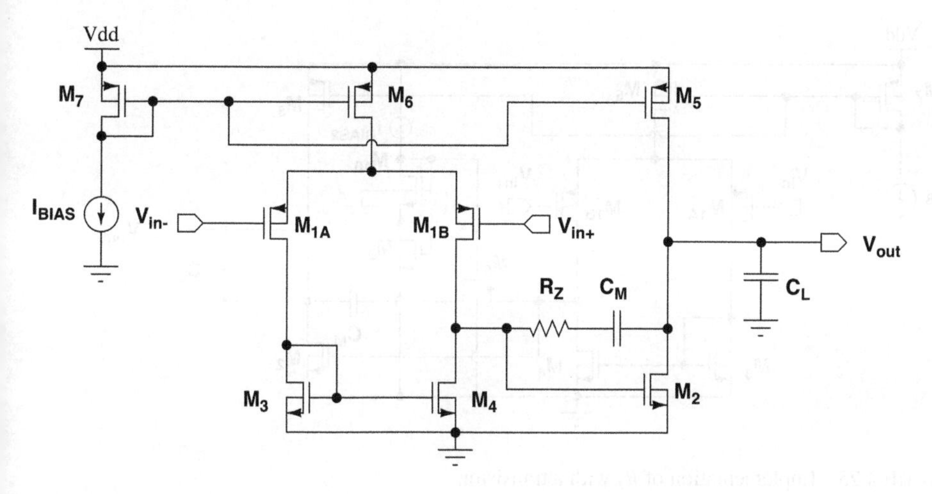

FIGURE 4.24 OTA Miller with resistor to shift the RHP zero.

can find a more detailed discussion of these topics in [3] and [8]. All the methods have in common the goal of removing the RHP zero, improving the amplifier stability. The simplest approach is to insert a resistor R_z in series with C_M as shown in Fig. 4.24. The presence of R_z shifts the zero to a new position, given by:

$$\omega_z \approx \frac{1}{C_M\left(\frac{1}{g_{m2}} - R_z\right)} \tag{4.129}$$

By choosing $R_z = 1/g_{m2}$ the zero is displaced to infinity, while with $R_z > 1/g_{m2}$ is moved in the left-half plane. In principle, the frequency of the zero could be chosen equal to the one of the non-dominant pole, which requires the following value of R_z:

$$\frac{1}{C_M\left(\frac{1}{g_{m2}} - R_z\right)} = \frac{g_{m2}}{C_L} \rightarrow R_z = \frac{1}{C_M} \frac{C_M + C_L}{g_{m2}} \tag{4.130}$$

In this case the non-dominant pole and the zero cancel each other (pole-zero cancellation). While this might seem attractive, achieving a perfect pole-zero cancellation which is stable against process fluctuations and environmental changes (such as temperature variation) is not trivial. Furthermore, the optimal value of R_z depends also on the load. This poses an additional issue in case the OTA is used to implement an output stage that has to drive off-chip loads which might not be precisely known a priori and can change between different experimental set-ups. The consequence of an imperfect pole-zero cancellation, arising when the pole and the zero are close by in frequency but do not overlap exactly, is to create a pole-zero doublet that may lead to undesirably long settling times. This problem is discussed, for instance, in [9] and references therein. Resistor R_z is often implemented as a transistor working in the

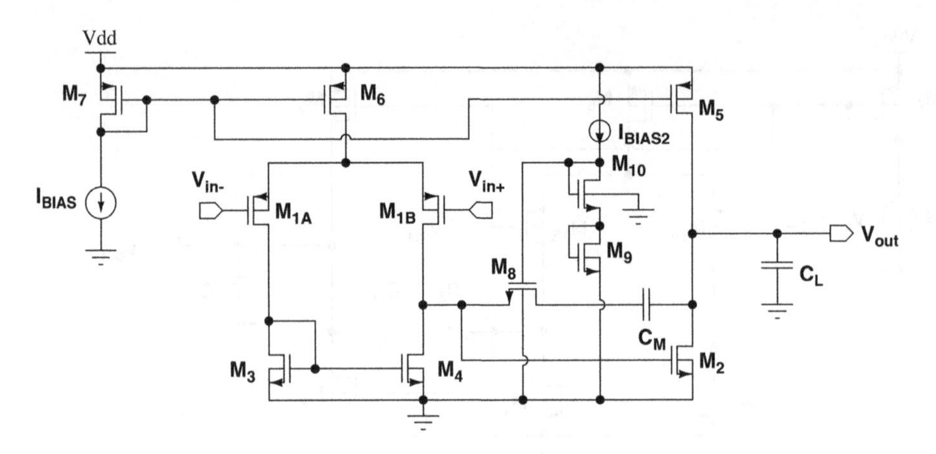

FIGURE 4.25 Implementation of R_z with a transistor.

linear region, as shown in Fig. 4.25. The resistor value is given by:

$$R_z = \frac{1}{\mu_n C_{ox} \left(\frac{W}{L}\right)_8 (V_{GS} - V_{TH})_8} \tag{4.131}$$

The current in M_2 is fixed, hence the source potential of M_8 is defined by the V_{GS} of M_2. If the threshold of M_2 changes because of process fluctuations or temperature variations, its V_{GS} changes as well. It is important that the gate voltage of M_8 track these variations to keep $(V_{GS} - V_{TH})_8$ and hence R_z constant. This is the purpose of the biasing circuit shown in the figure. In fact, it is expected that the thresholds of M_9 and M_{10} will approximately change like the one of M_2. The current in M_9 and M_{10} is fixed as well with a current source, so if the threshold of the NMOS transistors rises, both the source and gate voltage of M_8 will increase, reducing the variation of the equivalent resistance. The same is true if the threshold of the devices decreases. Better compensation schemes are shown in Fig. 4.26 and Fig. 4.27, respectively. The circuit of Fig. 4.26 shifts the zero to very high frequencies by cutting the direct forward path established by the compensation capacitor C_M. Note that C_M is driven from the output through the source follower M_8, biased by the current source M_9. In this manner, C_M is put in series to the gate-source capacitance of M_8 and if the latter is smaller, the equivalent capacitance directly seen between the gate of M_2 and the output is significantly reduced. The poles and zero of the circuit are respectively given by:

$$\omega_z = \frac{g_{m8}}{C_M} \tag{4.132}$$

$$\omega_d = \frac{1}{g_{m2} R_2 R_1 C_M} \tag{4.133}$$

$$\omega_{nd} = \frac{g_{m2}}{C_L} \tag{4.134}$$

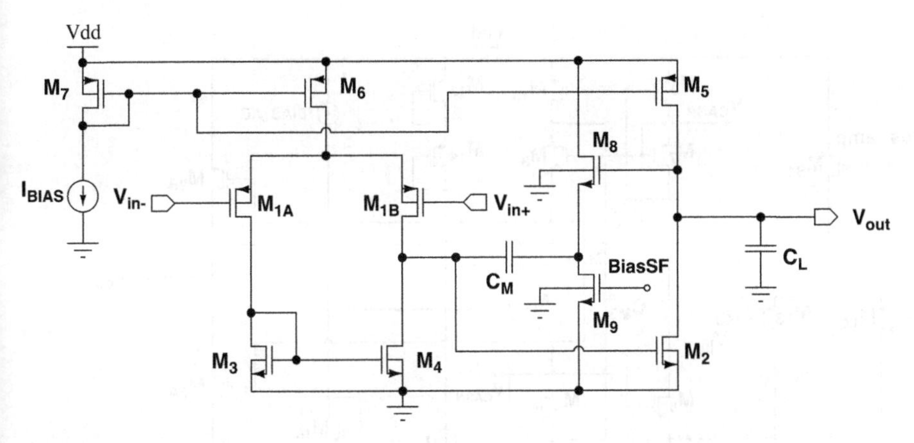

FIGURE 4.26 Use of a source follower to suppress the feed-forward part in a Miller compensated OTA.

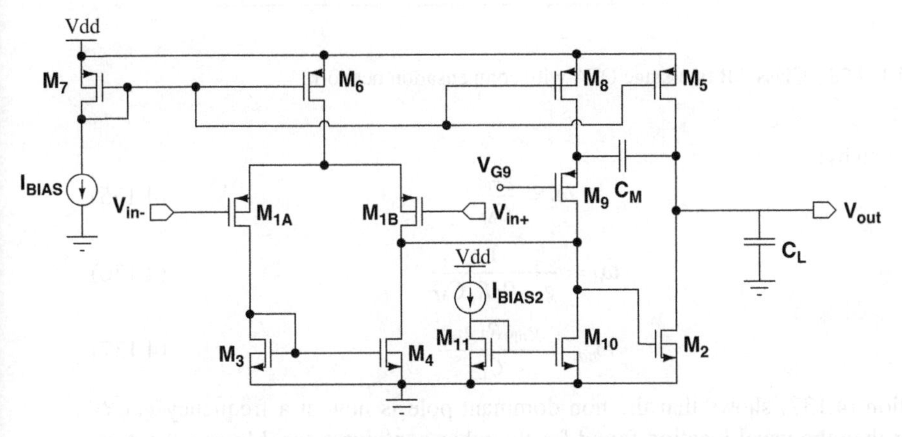

FIGURE 4.27 Use of a common gate stage in the compensation loop to eliminate the RHP zero.

where R_1 and R_2 are as usual the equivalent small signal resistance of the first and second stage. We observe that the zero has been moved to the left-half plane, where it is not so harmful to the stability, whilst the dominant and non-dominant poles are not affected. If a standard NMOS transistor is used to implement M_8, the source voltage of M_8 is equal to $V_{out} - V_{GS8}$ and must be large enough to allow M_9 to work in saturation. This introduces a limitation on the minimum value of V_{out} which makes this configuration not particularly appealing for modern CMOS technologies working with power supplies down to 1 V. A possible alternative is to implement M_8 as a native threshold transistor, i.e. a device with a threshold close to zero. The circuit of Fig. 4.27 offers however a better approach. Here the output signal is AC coupled to the source of M_9, hence avoiding the aforementioned problem. Furthermore, this configuration pushes the non-dominant pole to higher frequencies. The zero and the poles are in

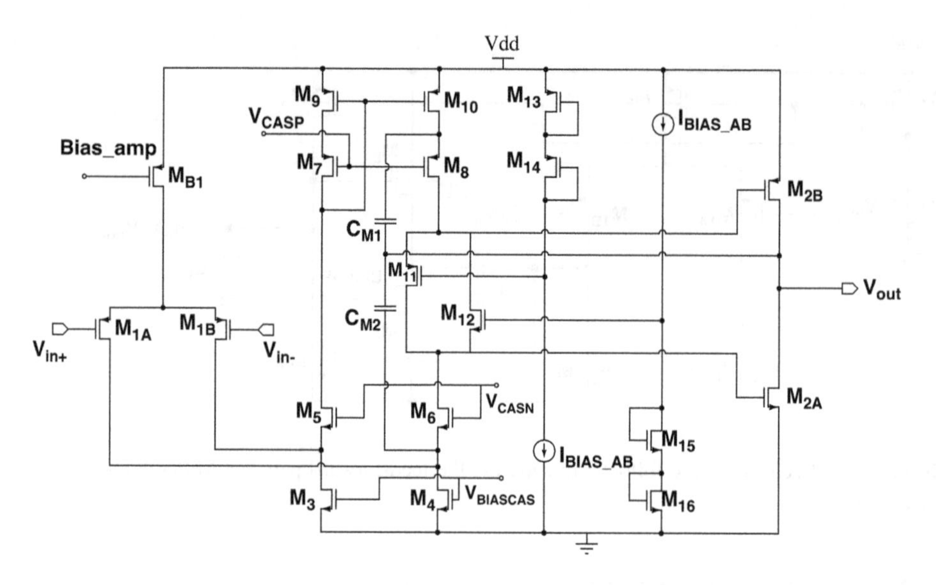

FIGURE 4.28 Class AB two-stage OTA with compensation network.

fact given by:

$$\omega_z = \frac{g_{m8}}{C_M} \tag{4.135}$$

$$\omega_d = \frac{1}{g_{m2}R_2R_1C_M} \tag{4.136}$$

$$\omega_{nd} = \frac{g_{m9}R_1g_{m2}}{C_L} \tag{4.137}$$

Equation (4.137) shows that the non-dominant pole is now at a frequency $g_{m9}R_1$ higher than the usual location found for the other configurations. Moving the non-dominant pole to higher frequencies allows in-turn to reduce C_M, increasing the unity gain frequency of the amplifier. In amplifiers with a cascode input stage, C_M can be connected to the source of the cascode devices of the first stage. The concept is demonstrated in Fig. 4.28, which shows a folded-cascode input stage followed by a class AB output. Before concluding this session on CMOS OTA it is important to mention that for all the circuits described the compensation capacitance limits also the slew rate of the amplifier. In fact, the slew rate is given by $SR = I_{SS}/C_M$, where I_{SS} is the bias current of the input differential pair.

4.5 FREQUENCY STABILITY AND COMPENSATION OF FRONT-END AMPLIFIERS

We can now revert to the study of the transimpedance amplifier of Fig. 4.11, analyzing the stability issues with the feedback concepts just discussed. We first consider a more idealized case in which the output buffer is modeled as an ideal voltage-controlled

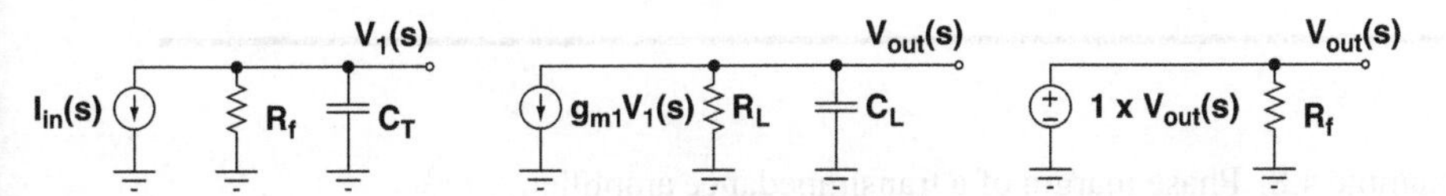

FIGURE 4.29 Model for calculating the high-frequency open-loop gain of a transimpedance amplifier.

voltage source. The open-loop amplifier has thus only two poles, one located at the input and one associated with the internal high impedance node. We then relax the hypothesis considering more realistic output stages.

4.5.1 TRANSIMPEDANCE AMPLIFIER WITH IDEAL OUTPUT BUFFER

To study the circuit stability, we need to write an expression for the loop gain and calculate the phase margin. This requires two steps. The first one is to separate the core amplifier from its feedback network. The transfer function of the core amplifier is then calculated and multiplied by the one of the feedback to obtain the loop gain. In the previous chapter, we have seen that a transimpedance amplifier can be thought as made of two components: a core open-loop transimpedance stage and a feedback network with gain $k = 1/R_f$. The small signal equivalent circuit of the core amplifier is reported in Fig. 4.29. Observe that in the figure, R_f is removed from the feedback path because we want to study the open-loop transfer function. A resistor of value R_f is however connected in parallel to the input and output ports of the circuit to take properly into account the loading effect of the feedback device. We neglect in the following the effect of R_f on the amplifier output because we suppose that the buffer has a negligible output impedance and can provide all the necessary current to the load. The open-loop transimpedance gain can be easily calculated as:

$$A_T = \frac{V_{out}(s)}{I_{in}(s)} = \frac{g_{m1}R_LR_f}{(1+sC_TR_f)(1+sC_LR_L)} \tag{4.138}$$

The loop gain $kA(s)$ is thus given by:

$$kA(s) = \frac{1}{R_f}\frac{g_{m1}R_LR_f}{(1+sC_TR_f)(1+sC_LR_L)} = \frac{g_{m1}R_L}{(1+sC_TR_f)(1+sC_LR_L)} \tag{4.139}$$

This is the quantity whose phase margin must be studied to assess the stability of the full amplifier. The closed-loop transimpedance gain is then found as:

$$T(s) = \frac{A_T(s)}{1+kA_T(s)} = \frac{g_{m1}R_LR_f}{s^2R_LR_fC_TC_f + s\left(C_TR_f + C_LR_L\right) + 1 + g_{m1}R_L} \tag{4.140}$$

which is the expression we already know. The following example discusses typical values of phase margin achieved by an uncompensated amplifier in a few representative situations.

Example 4.8: Phase margin of a transimpedance amplifier.

For the circuit of Fig. 4.29 assume the following parameters: $g_{m1} = 1$ mS, $R_L = 2$ MΩ, $C_T = 10$ pF, $C_L = 0.5$ pF. Evaluate the phase margin of the circuit when it is used in a transimpedance configuration with the following values of R_f: 5 kΩ, 100 kΩ and 200 MΩ.

* * *

The loop gain has two poles. The first one is associated to the internal high-gain node of the amplifier and has a time constant $R_L C_L$. The second pole is found at the input, with time constant $R_f C_T$, where C_T is the total capacitance shunting the amplifier input. The DC gain is $g_{m1} R_L = 1 \cdot 10^{-3} \times 2 \cdot 10^6 = 2000 = 66$ dB. The time constant $R_L C_L$ is equal to 1 μs, which yields a pole angular frequency of 1 MHz. The magnitude of the input pole, all the other conditions being equal, is determined by the value of R_f. Consider first the case $R_f = 5$ kΩ. Such a small value of feedback resistance can be found in applications where fairly large signals need to be processed with the greatest possible speed. Common examples are photodiode amplifiers used in optical communication links or front-end circuits reading-out devices such as photomultipliers or micro-channel plates in high rate applications. For $R_f = 5$ kΩ, the input pole time constant is 50 ns and the angular frequency 20 MHz. Therefore, in this case the internal amplifier pole is the dominant one and the two singularities are separated by 1.3 decades. At $\omega = 1$ MHz ($f = 159$ kHz) the gain start decreasing with an initial slope of −20 dB/decade. When the second pole is found, the gain has decreased by 26 dB, reaching 40 dB. At this point, we are more than one decade above the frequency of the first pole, which has already introduced a full −90° phase rotation. After the second pole is met, the slope increases to −40 dB/decade and at 200 MHz the loop gain crosses the zero-dB axis. However, this happens already one decade of frequency beyond the second pole that contributes to the phase rotation with an additional −90°, thus the phase margin is basically zero. Consider now the second case with $R_f = 100$ kΩ. Values of R_f in the range of hundreds of kΩ can be found in fast front-ends employed with semiconductor detectors providing relatively small signals, such as silicon microstrip and pixel sensors. Now the time constants of the two poles are the same, so we have two real coincident poles at $\omega = 1$ MHz. With a slope of −40 dB/decade, the unity gain axis is crossed 1.65 decades beyond the pole frequency, so the phase rotation is again −180° leaving no phase margin left. Finally, for $R_f = 200$ MΩ, the dominant pole is the one at the input, which has now an angular frequency of 500 Hz. The zero-dB axis is crossed at 1 MHz, exactly in correspondence of the second pole. Therefore, the system has a phase margin of 45°. Feedback resistors in the MΩ and above are typical of "charge sensitive amplifiers" which are the most popular input stage for front-end systems used in nuclear and high-energy physics applications.

We have already noted the similarity between the equivalent circuit of the transimpedance amplifier and the one of two-stage CMOS OTA. We can then surmise that also for the transimpedance amplifier the frequency compensation can be ob-

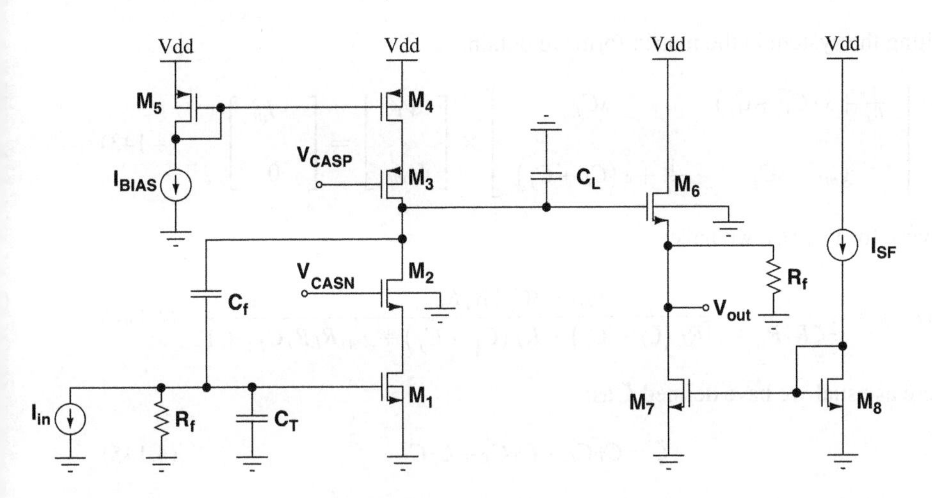

FIGURE 4.30 Single stage cascode amplifier with frequency compensation.

tained exploiting the Miller effect and connecting a capacitor between the input and the high-gain node of the amplifier. The resulting transistor level circuit is shown in Fig. 4.30 and its small signal model is reported in Fig. 4.31. Observe that also in this case we have connected R_f in parallel to the input and output port of the amplifier to keep properly into account in the calculations the loading effects of the feedback network. For reasons that will become clear later the Miller compensating capacitance has been called C_f rather than C_M as we did for the CMOS OTA. We can now find the transfer function of the amplifier after compensation, writing the nodal equation for the circuit of Fig. 4.31. For the input node we have:

$$I_{in} + V_1 sC_T + \frac{V_1}{R_f} + (V_1 - V_{out})\, sC_f = 0 \tag{4.141}$$

For the output node we get instead:

$$g_{m1} V_1 + \frac{V_{out}}{R_L} + \frac{V_{out}}{sC_L} + (V_{out} - V_1)\, sC_f = 0 \tag{4.142}$$

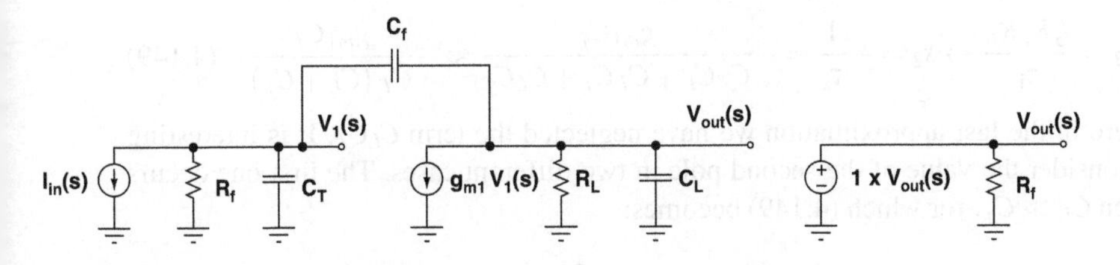

FIGURE 4.31 Small signal equivalent model of the circuit of Fig. 4.30.

Writing the system in the matrix form we obtain:

$$\begin{bmatrix} \frac{1}{R_f}+s\left(C_T+C_f\right) & -sC_f \\ g_{m1}-sC_f & \frac{1}{R_L}+s\left(C_L+C_f\right) \end{bmatrix} \times \begin{bmatrix} V_1 \\ V_{out} \end{bmatrix} = \begin{bmatrix} -I_{in} \\ 0 \end{bmatrix} \tag{4.143}$$

Solving for V_{out}/I_{in}, we have:

$$A_T(s) = \frac{\left(g_{m1}-sC_f\right)R_fR_L}{s^2\zeta R_fR_L + s\left[R_L\left(C_L+C_f\right)+R_f\left(C_T+C_f\right)+g_{m1}R_LR_fC_f\right]+1} \tag{4.144}$$

where as usual we have defined ζ as:

$$\zeta = C_TC_L + C_TC_f + C_LC_f \tag{4.145}$$

In general, the coefficient of s is dominated by the term $g_{m1}R_LR_fC_f$, thus (4.144) can be rewritten as:

$$A_T(s) = \frac{\left(g_{m1}-sC_f\right)R_fR_L}{s^2\zeta R_fR_L + sg_{m1}R_LR_fC_f+1} \tag{4.146}$$

Equation (4.144) and its simplified version (4.146) represent the open-loop trans-impedance gain after compensation. The loop gain is $kA_T(s) = A_T(s)/R_f$ and it has the same poles of $A_T(s)$, hence studying the position of the poles of $A_T(s)$ we can infer the stability of the circuit when also R_f will be connected between input and output. We can again assume that the poles are real and that one pole is at much lower frequency than the other. In this case, we can write the denominator of $A_T(s)$, $D(s)$, as:

$$D(s) = (1+s\tau_1)(1+s\tau_2) = s^2\tau_1\tau_2 + s(\tau_1+\tau_2)+1 \approx s^2\tau_1\tau_2 + s\tau_1 + 1 \tag{4.147}$$

where we have assumed that $\tau_1 \gg \tau_2$. By comparing the coefficients of (4.147) and of the denominator of (4.144) we can infer the values of the time constants and hence the pole positions of $A_T(s)$:

$$\tau_1 = g_{m1}R_LR_fC_f \rightarrow s_1 = -\frac{1}{\tau_1} = -\frac{1}{g_{m1}R_LR_fC_f} \tag{4.148}$$

$$\tau_2 = \frac{\zeta R_fR_L}{\tau_1} \rightarrow s_2 = -\frac{1}{\tau_2} = -\frac{g_{m1}C_f}{C_TC_L+C_TC_f+C_LC_f} \approx -\frac{g_{m1}C_f}{C_T\left(C_L+C_f\right)} \tag{4.149}$$

where in the last approximation we have neglected the term C_LC_f. It is interesting to consider the value of the second pole in two different cases. The first one occurs when $C_L \gg C_f$, for which (4.149) becomes:

$$s_2 = -\frac{g_{m1}}{C_L}\frac{C_f}{C_T} \tag{4.150}$$

Here the non-dominant pole is located at g_{m1}/C_L, which is the unity gain frequency of a voltage-driven common-source amplifier, scaled by the ratio C_f/C_T. The opposite alternative is when $C_f \gg C_L$, for which we get:

$$s_2 = -\frac{g_{m1}}{C_T} \tag{4.151}$$

In both cases, we see that increasing the sensor capacitance and thus C_T, we reduce the frequency of the non-dominant pole, bringing the system towards a more unstable condition.

Example 4.9: Phase margin of a compensated transimpedance amplifier.

Using the parameters of example 4.8, find the phase margin of the transimpedance amplifier after a compensation capacitance C_f of 0.5 pF has been inserted in the circuit.

* * *

Let's start with $R_f = 5$ kΩ. Applying equations (4.148) and (4.149) with the parameters given in example 4.8, we find $\omega_1 = 200$ kHz and $\omega_2 = 50$ MHz. After ω_1 the gain drops with −20 dB/decade and its value is 18 dB when ω_2 is found. The slope then becomes −40 dB/decade and the unity gain axis is crossed 0.5 decade above ω_2, i.e. at about 158 MHz. At this frequency, the total phase rotation is given by:

$$\tan^{-1}\left(-\frac{1.58\cdot 10^8}{2\cdot 10^5}\right)+\tan^{-1}\left(-\frac{1.58\cdot 10^8}{5\cdot 10^7}\right) = -162.4° \tag{4.152}$$

hence the phase margin is 180° − 162.4° = 17.6°. Repeating the calculations for $R_f =$ 100 kΩ, we find $\omega_1 = 10$ kHz, while ω_2 is always at 50 MHz. Now the first pole is anticipated and the gain drops start at 10 kHz, crossing the zero dB axis at 20 MHz. Calculating the phase margin with the new values we find a phase margin of 68.2°. Finally, for $R_f = 200$ MΩ, the dominant pole is at 5 Hz and the phase margin is 90°.

We can now consider the closed-loop gain of the full amplifier, depicted in Fig. 4.32. Applying the feedback formula, we have:

$$T(s) = \frac{A_T(s)}{1+kA_T(s)} \tag{4.153}$$

Replacing the expression of $A(s)$ we get:

$$T(s) = \frac{(g_{m1}-sC_f)\,R_fR_L}{s^2\zeta R_fR_L + s\left[R_LC_L + R_fC_T + (1+g_{m1}R_L)C_fR_f\right] + 1 + g_{m1}R_L} \tag{4.154}$$

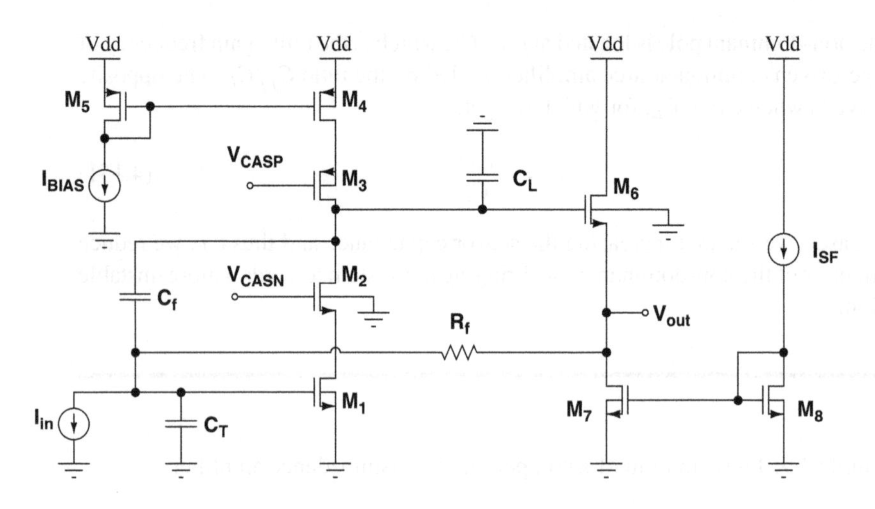

FIGURE 4.32 Full transimpedance amplifier with the compensating capacitor C_f.

As usual, we can use the approximate form, assuming that $g_{m1}R_L \gg 1$:

$$T(s) = \frac{\left(g_{m1} - sC_f\right) R_f R_L}{s^2 \zeta R_f R_L + s g_{m1} R_L C_f R_f + g_{m1} R_L} \tag{4.155}$$

In the transfer function we have also a zero in the right half plane, located at g_{m1}/C_f. It is interesting to note that, in this case, both the position of the non-dominant pole and of the zero are determined by the transconductance of the input stage. Therefore, the ratio between the frequency of the zero and the one of the non-dominant pole is given by:

$$\frac{\omega_z}{\omega_2} = \frac{C_T\left(C_L + C_f\right)}{C_f^2} \tag{4.156}$$

If $C_f \gg C_L$, the above ratio reduces to C_T/C_f. For $C_T/C_f > 10$, the zero is located one decade beyond the non-dominant pole, and contributes to the phase shift by a negligible amount. For reasons that will become clear in the next chapters, this situation is easily verified by radiation sensor front-ends, hence techniques to shift the position of the zero are usually not necessary and the zero at the numerator can be neglected in most cases. Dividing now both the numerator and denominator of (4.155) by $g_{m1}R_L$ and omitting the zero, $T(s)$ can be rewritten as:

$$T(s) = \frac{R_f}{s^2 \frac{\zeta R_f}{g_{m1}} + sR_f C_f + 1} \tag{4.157}$$

To have real poles in the closed-loop transfer function, the following inequality must be satisfied:

$$R_f C_f > \frac{4\left(C_T C_L + C_T C_f + C_f C_L\right)}{g_{m1} C_f} \tag{4.158}$$

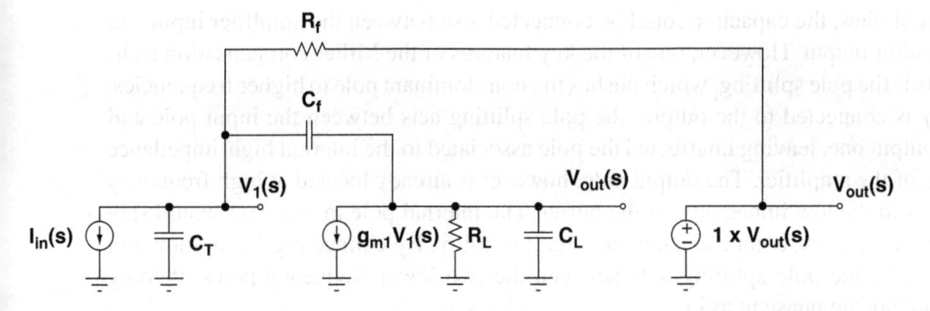

FIGURE 4.33 Small signal model of a complete transimpedance amplifier.

The condition takes a particular simple form in case $C_T C_f$ is the dominant term in the parenthesis, reducing to:

$$R_f C_f > \frac{4C_T}{g_{m1}} \tag{4.159}$$

The above relationship shows that to have real poles in the closed-loop transfer function, the time constant $R_f C_f$ associated to the feedback loop must be at least four times the ratio between the total input capacitance and the transconductance of the input transistor.

It is interesting to study the circuit of Fig. 4.32 by drawing its small signal equivalent circuit, reported in Fig. 4.33, and performing the nodal analysis. Here an important point must be stressed. The circuits of Fig. 4.29-4.31 were "partial" circuits designed to calculate the open-loop gain of the amplifier while taking into account the loading effects of the feedback network. To do so, R_f was removed from the feedback and connected between the input and output of the circuit and ground. However, when we make a model of the full circuit, all the components must be represented in their true positions. For this reason, in Fig. 4.33 R_f is only connected between input and output, as it is the case in the physical circuit. The transfer function that we calculate with the nodal analysis is already the closed-loop one. For the circuit of Fig. 4.33 the nodal equations read:

$$I_{in} + V_1 sC_T + \frac{(V_1 - V_{out})}{R_f} + (V_1 - V_{out})\, sC_f = 0 \tag{4.160}$$

$$g_{m1} V_1 + \frac{V_{out}}{R_L} + \frac{V_{out}}{sC_L} + (V_{out} - V_1)\, sC_f = 0 \tag{4.161}$$

Writing the system in the matrix form we obtain:

$$\begin{bmatrix} \frac{1}{R_f} + s\left(C_T + C_f\right) & -\left(\frac{1}{R_f} + sC_f\right) \\ g_{m1} - sC_f & \frac{1}{R_L} + s\left(C_L + C_f\right) \end{bmatrix} \times \begin{bmatrix} V_1 \\ V_{out} \end{bmatrix} = \begin{bmatrix} -I_{in} \\ 0 \end{bmatrix} \tag{4.162}$$

The reader can easily verify that solving for V_{out}/I_{in} one gets back (4.154). An additional comment concerns the way C_f is connected. In fact, from the signal processing point of view, the capacitor could be connected also between the amplifier input and the buffer output. However, one of the key features of the Miller compensation technique is the pole splitting, which pushes the non-dominant pole to higher frequencies. If C_f is connected to the output, the pole splitting acts between the input pole and the output one, leaving unaffected the pole associated to the internal high-impedance node of the amplifier. The output pole, however is already located at high frequency thanks to the low impedance of the buffer. The internal pole in this case would stay at low frequency, compromising the circuit stability. By connecting C_f as shown in Fig. 4.32, the pole splitting acts between the two lower frequency poles, thereby enhancing the phase margin.

4.5.2 INPUT IMPEDANCE

It is interesting to study the input impedance Z_{in} of the circuit of Fig. 4.32. To find Z_{in}, we just need to divide the input voltage V_1 by the input current I_{in}. The voltage V_1 can be calculated from the matrix equations given in (4.162). The input impedance can be approximated as:

$$Z_{in} = \frac{V_1}{I_{in}} = \frac{\left[1+sR_L\left(C_L+C_f\right)\right]R_f}{s^2\zeta R_f R_L + s g_{m1} R_L R_f C_f + 1 + g_{m1} R_L} \tag{4.163}$$

We can identify in Z_{in} three different frequency regions. The first one is for $s \to 0$, which yields:

$$Z_{in} = \frac{R_f}{1+g_{m1}R_L} \tag{4.164}$$

This is the well known expression of the input resistance when frequency dependence is not considered. For $s > 1/R_f C_f$ the term in s becomes greater than $g_{m1}R_L$. The term in s and the one in s^2 are equal if:

$$s^2\zeta R_L R_f = s g_{m1} R_L R_f C_f \to s_H = \frac{g_{m1}C_f}{C_T C_L + C_T C_f + C_L C_f} \approx \frac{g_{m1}C_f}{C_T\left(C_L+C_f\right)} > \frac{1}{R_f C_f} \tag{4.165}$$

When $1/(R_f C_f) < s < s_H$ the coefficient of s dominates and the input impedance becomes:

$$Z_{in} = \frac{C_L+C_f}{g_{m1}C_f} \tag{4.166}$$

In case $C_L \gg C_f$ the above equation reduces to:

$$Z_{in} = \frac{1}{g_{m1}}\frac{C_L}{C_f} \tag{4.167}$$

If, on the other hand, $C_f \gg C_L$, we get:

$$Z_{in} \approx \frac{1}{g_{m1}} \tag{4.168}$$

The above equations show that in the intermediate frequency range, the input impedance is inversely proportional to the transconductance of the input transistor, eventually scaled up by the ratio C_L/C_f. This has an intuitive explanation because, as the frequency increases, C_f short-circuits the gate of the input device to the output, transforming the circuit in a diode-connected stage. Finally, above $s = s_M$ the term s^2 starts dominating and for $s \gg s_M$ we have:

$$Z_{in} = \frac{C_L + C_f}{s\left(C_T C_L + C_T C_f + C_f C_L\right)} \approx \frac{1}{sC_T} \tag{4.169}$$

At very high frequencies, C_T shunts the circuit input directly to ground and the signals are not anymore processed by the amplifier. It is also interesting to consider the effect of C_f on Z_{in}. If $C_f = 0$, equation (4.163) becomes:

$$Z_{in} = \frac{(1 + sR_L C_L) R_f}{s^2 R_f R_L C_T C_L + 1 + g_{m1} R_L} \approx \frac{(1 + sR_L C_L) R_f}{s^2 R_f R_L C_T C_L + g_{m1} R_L} \tag{4.170}$$

In this case, the term s^2 starts dominating when:

$$s^2 R_f R_L C_L C_T = g_{m1} R_L \rightarrow s_H = \sqrt{\frac{g_{m1}}{R_f C_T C_L}} \tag{4.171}$$

At intermediate frequencies, the input impedance can be approximated as:

$$Z_{in} = \frac{sR_L C_L R_f}{1 + g_{m1} R_L} \approx \frac{sR_f C_L}{g_{m1}} \tag{4.172}$$

From the above equation we see that, in absence of C_f, the input impedance at intermediate frequency rises with frequency, displaying an inductive behavior. An additional benefit of C_f thus consists in keeping the input impedance low in the frequency range typically of interest for a radiation detector front-end.

4.5.3 TRANSIMPEDANCE AMPLIFIER WITH REAL OUTPUT BUFFER

The hypothesis adopted so far of an ideal output buffer is of course a crude approximation, as it can be easily understood by considering the small signal model of a source follower that incorporates the most important parasitic capacitors, shown in Fig. 4.34. The signal source is represented with an input resistance in series, R_s. In case of a transimpedance amplifier, R_s is the resistance seen looking into the high impedance node of the amplifier, i.e. coincides with R_L of Fig. 4.33 and it is typically very high. The nodal equations are:

$$\begin{cases} \frac{V_1 - V_{in}}{R_s} + V_1 sC_{gd} + (V_1 - V_{out}) sC_{gs} = 0 \\ V_{out} sC_{out} + (V_{out} - V_1) sC_{gs} - g_{m1} V_{gs1} = 0 \\ V_{gs1} = V_1 - V_{out} \end{cases} \tag{4.173}$$

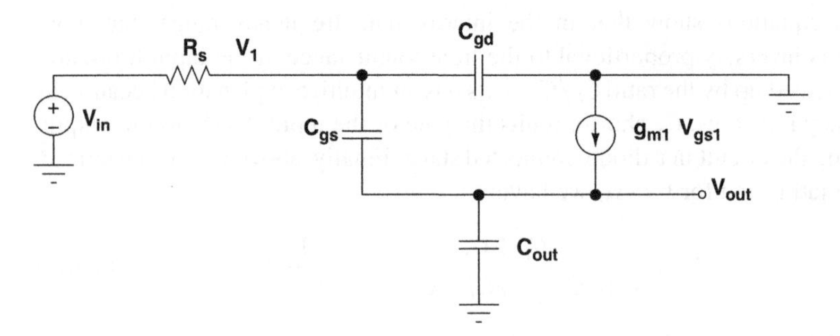

FIGURE 4.34 Small frequency small signal model of a source follower. For simplicity, the bulk transconductance and the output resistance of the transistor are omitted.

where V_1 is the voltage on the transistor gate. Solving the above system for V_{out}/V_{in} yields:

$$\frac{V_{out}}{V_{in}} = \frac{g_{m1} + sC_{gs}}{s^2 R_s \left(C_{gs}C_{out} + C_{gd}C_{out} + C_{gd}C_{gs}\right) + s\left(g_{m1}R_sC_{gd} + C_{gs} + C_{out}\right) + g_{m1}} \tag{4.174}$$

Although in the model the bulk transconductance and the MOS output resistance were not considered, the resulting transfer function is already quite complex. The second order denominator suggests that the poles could also become complex conjugate, resulting in a ringing in the buffer response. The design of source followers as output stages of front-end amplifiers raises an interesting trade-off. In fact, increasing the device gate width, reduces its low frequency output impedance, which is proportional to $1/g_{m1}$, but increases at the same time the capacitive load that the follower presents to the preceding stage, thus affecting its speed. The transconductance improves till the follower transistor enters into the weak inversion region, thus increasing further the gate width beyond this point does not decrease anymore the buffer output impedance, but still worsens the capacitive load on the high impedance node.

As we have seen in the previous chapter, the source follower has an asymmetric driving capability. For example, in an NMOS follower, the current available to discharge the output node is limited by the current fixed in the biasing mirror, but the current that can be sourced to charge up the node is defined by the transistor aspect ratio. The opposite is true for the PMOS configuration. As we will study more in detail in the next section, the typical response of a front-end input stage has a very fast rise time, limited by the amplifier bandwidth, followed by a slow return to the baseline defined by the time constant R_fC_f. In this case, the slew rate required to the buffer is much greater in one of the two fronts. Therefore, apart from using the appropriate follower type, one should also pay attention that the device aspect ratio is sufficiently large to guarantee an appropriate driving capability on the larger signals. Once the power and thus the current in the buffer is fixed, the minimum gate length is usually chosen. An optimal gate width then exists that yields the best speed as a compromise between the buffer strength and the bandwidth limitation imposed on

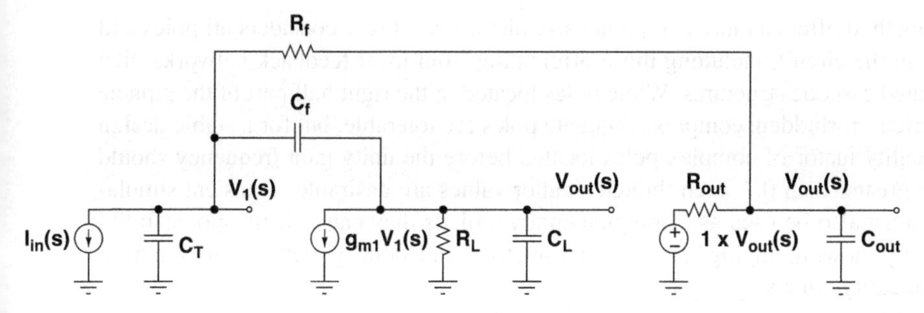

FIGURE 4.35 Small signal model of a transimpedance amplifier including the non-zero resistance of the output buffer.

the preceding stage. This optimum can be easily found with the help of parametric computer simulations.

Incorporating the detail source follower model in the equivalent circuit of a transimpedance stage is cumbersome, but one should at least consider the buffer output resistance, as shown in Fig. 4.35. Note that now an additional pole is found at the output node. Fig. 4.36 shows a modified version of the circuit. Here, two capacitors are used, one connected between the input and the high impedance point and another between the input and the output node. This nested configuration improves the circuit stability, because the C_{f2} realizes a splitting between the input and the output poles.

The extra node added to the circuit by the finite output resistance makes tedious to solve the nodal equations by hand. While simplified models provide useful insights, computer simulations become thus necessary to study the behavior of more realistic circuits. The phase margin can be estimated by simulating the loop gain. However, breaking a feedback loop is always a delicate and error prone operation. With the circuit simulator, a pole-zero analysis of the close-loop transfer function can be performed, checking the presence of complex conjugate roots and their quality factors.

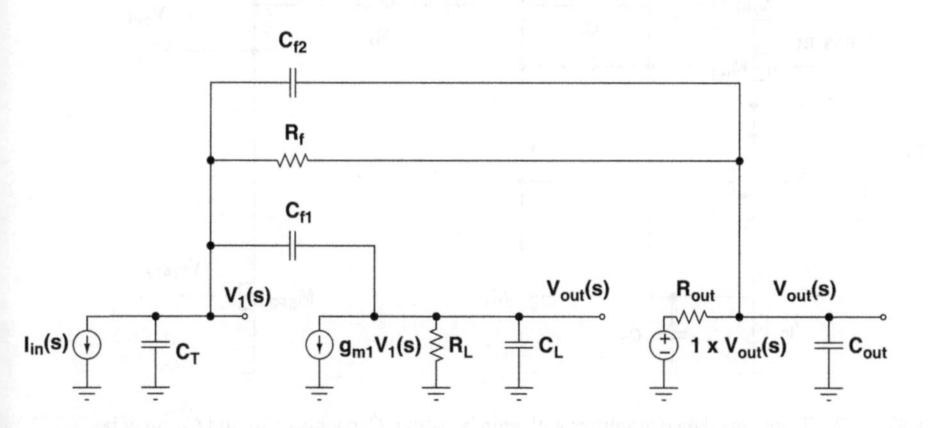

FIGURE 4.36 Transimpedance amplifier with split feedback capacitor.

This method offers a more comprehensive picture because it considers all poles and zeros in the circuit, including those originating from local feedback networks, like regulated cascode structures. While poles located in the right half part of the s-plane are strictly forbidden, complex conjugate poles are tolerable, but for a stable design the quality factor of complex poles located before the unity gain frequency should be no greater than 0.7, even though smaller values are desirable. Transient simulations must also be used as a complementary tool, as they can quickly spot stability issues by showing ringing in the output waveform and identify problems arising from non-linearity effects.

4.5.4 INPUT STAGES WITH GAIN BOOSTING

The gain boosting technique can be used to increase the equivalent resistance of both the input cascode and the load, thereby improving the DC gain of the core amplifier. Fig. 4.37 reports an input stage employing this technique. The gain boosting amplifiers form local feedback loops and their frequency stability must also be ensured. The phase margin of the boosting loops can be improved by loading their outputs with compensating capacitors, as shown in the figure.

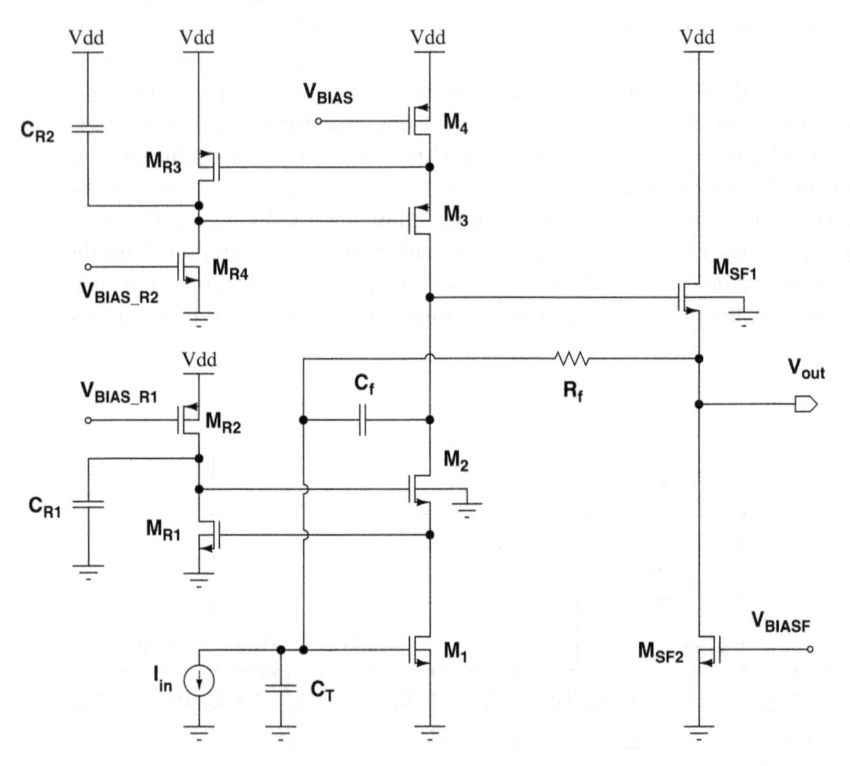

FIGURE 4.37 Transimpedance amplifier with gain boosting. Capacitors C_{R1} and C_{R2} may be necessary to improve the phase margin of the cascode boosting loops.

4.6 THE CHARGE SENSITIVE AMPLIFIER

We now study the transfer function of the transimpedance amplifier in a few interesting limiting cases. The first one is when R_f is very big, so that the "1" in the denominator of (4.157) becomes negligible already at very low frequency. Equation (4.157) can thus be approximated as:

$$T(s) \approx \frac{R_f}{s^2 \frac{\zeta R_f}{g_{m1}} + sR_f C_f} = \frac{1}{sC_f \left(1 + s\frac{\zeta}{g_{m1} C_f}\right)} = \frac{1}{sC_f} \frac{1}{(1 + s\tau_r)} \tag{4.175}$$

where we have introduced the rise time constant τ_r defined as:

$$\tau_r = \frac{\zeta}{g_{m1} C_f} = \frac{C_T C_L + C_T C_f + C_L C_f}{g_{m1} C_f} \tag{4.176}$$

The circuit transfer function is thus the product between the transfer function of an ideal integrator ($1/sC_f$) and the one of a single-pole low pass filter. If the input signal is a Dirac-delta conveying the total charge Q_{in}, the output voltage is simply given by $V_{out}(s) = Q_{in} T(s)$. Taking the Inverse Laplace Transform, we obtain in the time domain:

$$V_{out}(t) = \frac{Q_{in}}{C_f} \left(1 - e^{-\frac{t}{\tau_r}}\right) \tag{4.177}$$

The above equation shows that for $R_f \to \infty$ the circuit responds to a Dirac-delta input with a voltage step at the output of amplitude Q_{in}/C_f, which is reached with a time constant τ_r, determined by the sensor and the core amplifier parameters. In case a generic input stimulus $I_{in}(s)$ is applied, we can write:

$$V_{out}(s) = \frac{1}{sC_f} \frac{I_{in}(s)}{(1 + s\tau_r)} = \frac{1}{C_f} \frac{1}{s} F(s) \tag{4.178}$$

The division by s in the Laplace domain corresponds to the integration operator in the time domain, thus the circuit integrates anyway the input signal. If this is a pulse of limited duration, the internal bandwidth limitation of the amplifier and the shape of the input signal determine the rise time of the output waveform and its shape while it swings towards the final value, given by Q_{in}/C_f. Once this is attained, the output voltage is however proportional to the total input charge, which is one of the primary variable of interest in a radiation detection system. This is one of the reasons for which the integrator, usually referred to in the context of nuclear pulse processing as the "Charge Sensitive Amplifier (CSA)", is the most common topology for implementing front-end input stages. Observe that, in the ideal case, the output voltage only depends on the detector charge and on value of C_f, which is a well controlled circuit element, and it is not influenced by the value of the detector capacitance due to the "virtual ground" action of the high gain voltage amplifier.

In a continuous time system, the presence of a DC feedback path between input and output is nevertheless necessary to allow the amplifier to self-bias, therefore, although very high, R_f can not be infinite and its effect on the signal must be considered. We can thus revert to equation (4.157) and rewrite the denominator supposing that it has two real and widely spaced poles:

$$D(s) = (1+s\tau_f)(1+s\tau_r) = 1+s(\tau_f+\tau_r)+s^2\tau_f\tau_r \approx 1+s\tau_f+s^2\tau_f\tau_r \quad (4.179)$$

where we have made the hypothesis that time constant τ_f is much greater than the amplifier internal time constant τ_r. Comparing equation (4.179) with (4.157), we have:

$$\tau_f = R_f C_f \quad (4.180)$$

$$\tau_r\tau_f = \frac{\zeta R_f}{g_{m1}} \rightarrow \tau_r = \frac{C_T C_L + C_T C_f + C_L C_f}{g_{m1} C_f} \quad (4.181)$$

The transfer function can thus be rewritten as:

$$T(s) = \frac{R_f}{(1+s\tau_r)(1+s\tau_f)} \quad (4.182)$$

where $\tau_f = R_f C_f$ is the time constant of the feedback loop. If we suppose that the core amplifier is very fast, τ_r can be neglected and $T(s)$ further simplifies to:

$$T(s) = \frac{R_f}{1+sC_f R_f} \quad (4.183)$$

Here we can distinguish two frequency regimes. For $sR_fC_f \ll 1$, the transimpedance gain is simply equal to R_f. The amplifier hence multiplies the input current by R_f without introducing any modification to the signal shape. For $sR_fC_f \gg 1$ equation (4.183) becomes:

$$T(s) = \frac{1}{sC_f} \quad (4.184)$$

which is again the transfer function of the ideal integrator. The boundary between the two regimes is for $sR_fC_f = 1$, which means $R_f = 1/sC_f$. For smaller values of s, the impedance of C_f is greater than R_f and since the two are in parallel, the signal current flows preferentially through R_f. At high frequency, the impedance of the capacitance becomes smaller, the signal flows through C_f and gets integrated. If we take the Inverse Laplace Transform of (4.184) we get for a Dirac-delta input:

$$V_{out}(t) = \frac{Q_{in}}{C_f} e^{-\frac{t}{\tau_f}} \quad (4.185)$$

In this case, the output reaches immediately the peak value given by Q_{in}/C_f, which then fades away with a time regulated by τ_f. Resistor R_f thus provides a path that discharges the feedback capacitor, allowing the restoration of the baseline. If R_f is big enough, the discharge of C_f is slow. Fig. 4.38 shows the output of a CSA with $\tau_r = 0$ and different values of τ_f. Observe that very large values of the feedback time constant are necessary to make the circuit response close to the one of an ideal integrator, that would respond to a Dirac-delta with a step of infinite duration at the output.

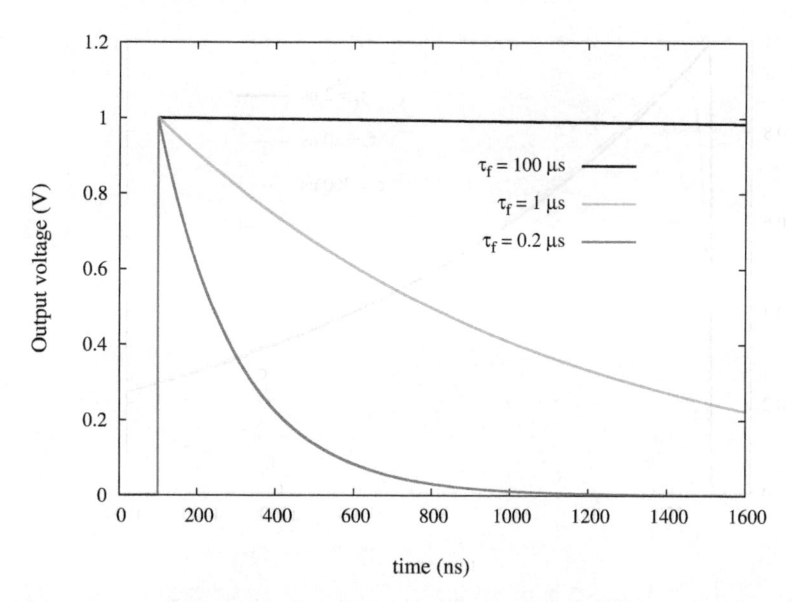

FIGURE 4.38 Impulse response of a charge sensitive amplifier with no internal bandwidth limitation ($\tau_r = 0$) and different values of the feedback time constant. Q_{in}/C_f is taken equal to 1 for simplicity.

In most cases, τ_r can not be neglected, because to keep it very small we need small capacitance and a high value of the input device transconductance, which implies large power consumption. The impulse response of the circuit in thus given by the Inverse Laplace Transform of (4.182), which reads:

$$V_{out}(t) = Q_{in}\frac{R_f}{\tau_r - \tau_f}\left(e^{-\frac{t}{\tau_r}} - e^{-\frac{t}{\tau_f}}\right) = \frac{Q_{in}}{C_f}\frac{\tau_f}{\tau_r - \tau_f}\left(e^{-\frac{t}{\tau_r}} - e^{-\frac{t}{\tau_f}}\right) \tag{4.186}$$

where the last member has been obtained by multiplying and dividing the term before the parenthesis by C_f. Differentiating the above expression, we find the time T_p at which the output voltage reaches its peak:

$$\frac{dV_{out}}{dt} = 0 \rightarrow \frac{1}{\tau_f}e^{-\frac{T_p}{\tau_f}} = \frac{1}{\tau_r}e^{-\frac{T_p}{\tau_r}} \rightarrow T_p = \frac{\tau_f\tau_r}{\tau_r - \tau_f}\ln\frac{\tau_r}{\tau_f} \tag{4.187}$$

We can now insert the expression of T_p given by (4.182) in (4.181) to find the value of the peak voltage. To do so, observe that the following relationship holds:

$$e^{-\frac{T_p}{\tau_r}} = e^{-\frac{1}{\tau_r}\frac{\tau_f\tau_r}{\tau_r-\tau_f}\ln\frac{\tau_r}{\tau_f}} = \left(e^{ln\frac{\tau_f}{\tau_r}}\right)^{\frac{\tau_f}{\tau_r-\tau_f}} = \left(\frac{\tau_f}{\tau_r}\right)^{\frac{\tau_f}{\tau_r-\tau_f}} \tag{4.188}$$

In a similar way, the second exponential can be manipulated so that:

$$e^{-\frac{T_p}{\tau_f}} = \left(\frac{\tau_f}{\tau_r}\right)^{\frac{\tau_r}{\tau_r-\tau_f}} \tag{4.189}$$

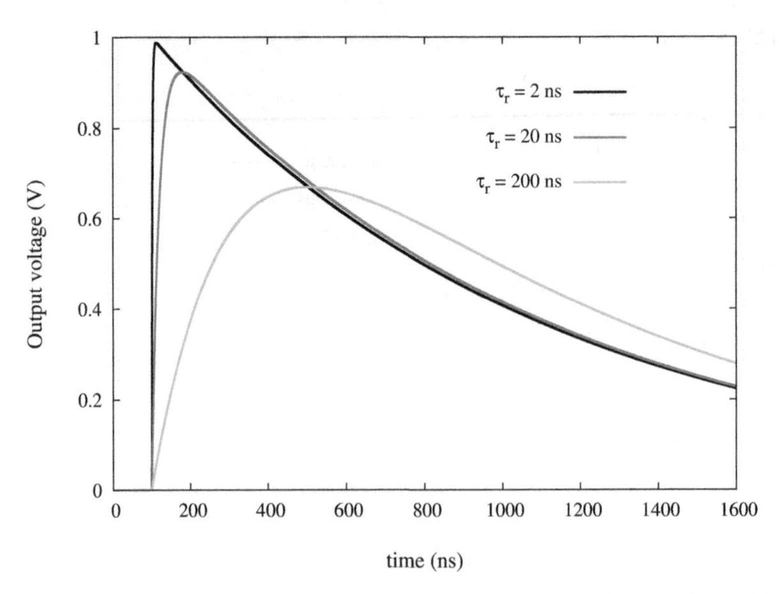

FIGURE 4.39 Impulse response of a charge sensitive amplifier for different values of the rise time constant τ_r. The feedback time constant was fixed to 1 μs. Q_{in}/C_f is assumed equal to 1.

The difference between the two exponential functions in (4.181) can hence be written as:

$$\left(\frac{\tau_f}{\tau_r}\right)^{\frac{\tau_f}{\tau_r-\tau_f}} - \left(\frac{\tau_f}{\tau_r}\right)^{\frac{\tau_r}{\tau_r-\tau_f}} = \left(\frac{\tau_f}{\tau_r}\right)^{\frac{\tau_r}{\tau_r-\tau_f}} \left[\left(\frac{\tau_f}{\tau_r}\right)^{\frac{\tau_f-\tau_r}{\tau_r-\tau_f}} - 1\right] = \\ = \left(\frac{\tau_f}{\tau_r}\right)^{\frac{\tau_r}{\tau_r-\tau_f}} \left(\frac{\tau_r}{\tau_f} - 1\right) = \frac{\tau_r-\tau_f}{\tau_f} \left(\frac{\tau_f}{\tau_r}\right)^{\frac{\tau_r}{\tau_r-\tau_f}} \tag{4.190}$$

Finally, multiplying the above expression by the term in front of the parenthesis in (4.181), we get:

$$V_{out,peak} = \frac{Q_{in}}{C_f}\left(\frac{\tau_f}{\tau_r}\right)^{\frac{\tau_r}{\tau_r-\tau_f}} \tag{4.191}$$

The step amplitude is thus modulated by a term depending on the ratio between the rise and discharge time constants. For instance, the peak output voltage attains 95.4% of its theoretical value for $\tau_f/\tau_r = 100$ and 99.3% for $\tau_f/\tau_r = 1000$. Fig. 4.39 shows the effect of τ_r on the impulse response of the CSA. In the plot, τ_f has been fixed to 1 μs, and three different waveforms have been generated corresponding respectively to τ_r =2 ns, 20 ns and 200 ns. Note that for $\tau_r = 20$ ns (which corresponds to a ratio τ_r/τ_f of 50), the delay and the attenuation on the peak already become clearly visible. The effect of bandwidth limitation in CSA can be better understood when the pulse shaper is considered, so this aspect will be further discussed in chapter 6.

4.7 FREQUENCY PERFORMANCE OF CURRENT-MODE INPUT STAGES

Fig. 4.40 reports the scheme of a common gate amplifier. The small signal equivalent circuit including the most relevant capacitors is drawn in Fig. 4.41, which allows us to write the following set of nodal equations:

$$\begin{cases} g_{m1}V_{gs1} + \frac{V_{out}}{R_L} + V_{out}sC_L = 0 \\ \frac{V_1}{r_{02}} + V_1 sC_T + V_1 sC_{gs} - g_{m1}V_{gs1} = I_{in} \\ V_{gs1} = 0 - V_1 \end{cases} \tag{4.192}$$

For reasons that will become clear when we will discuss the regulated cascode performance, we have not included the gate-source capacitance of the input transistor in C_T, which as usual incorporates the sensor capacitance C_d plus any other contribution appearing in parallel with it. Solving the above system, we can write the circuit transfer function as:

$$\frac{V_{out}}{I_{in}} = \frac{g_{m1}R_L r_{02}}{(1+sC_L R_L)\left[1+s(C_T+C_{gs})\,r_{02}+g_{m1}r_{02}\right]} \tag{4.193}$$

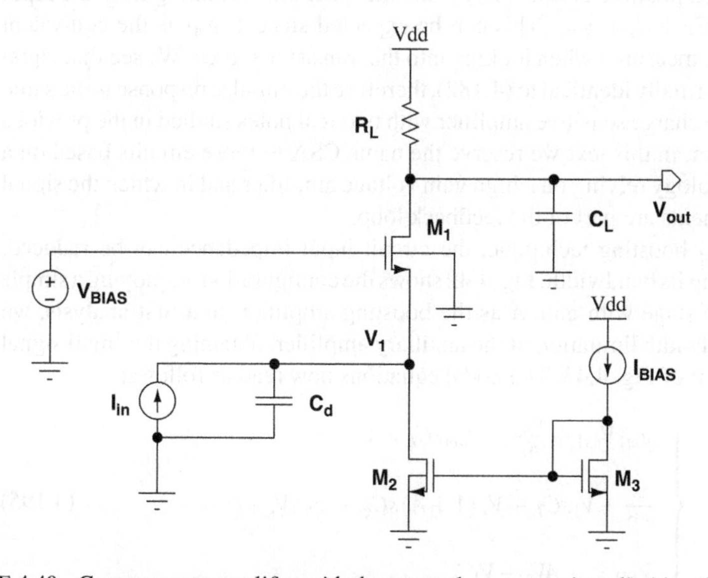

FIGURE 4.40 Common gate amplifier with the most relevant capacitors limiting the circuit bandwidth. The capacitor connected to the output, C_L, can represent the parasitics connected to the node or it can be a component inserted on purpose to shape the signal. If this is the case, C_L is usually connected in parallel to R_L.

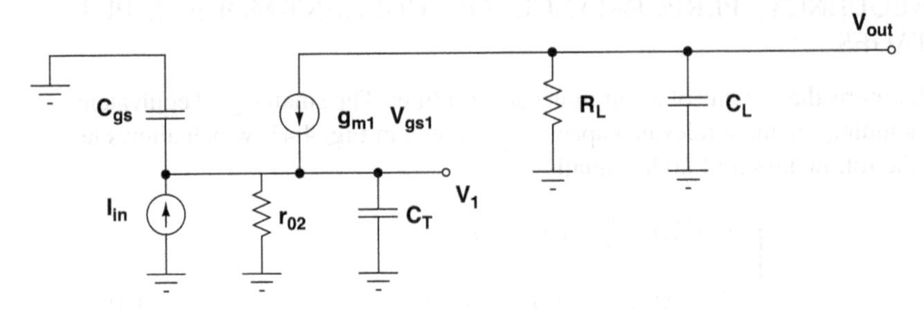

FIGURE 4.41 Equivalent small signal model of the common gate amplifier of Fig. 4.40. The capacitor C_T is the sum of the detector capacitance C_d plus any other parasitic capacitance connected between the input and ground.

Assuming $g_{m1}r_{02} \gg 1$ and factorizing the term $g_{m1}r_{02}$ in the denominator, we get:

$$\frac{V_{out}}{I_{in}} = \frac{R_L}{(1+sR_LC_L)\left(1+s\frac{C_T+C_{gs}}{g_{m1}}\right)} \tag{4.194}$$

Equation (4.194) shows that the common gate amplifier has two real poles, one associated to the output time constant R_LC_L and the other one stemming from the input time constant $(C_T+C_{gs})/g_{m1}$. This can be expected since $1/g_{m1}$ is the equivalent resistance of M_1 measured when looking into the transistor source. We see that equation (4.194) is formally identical to (4.182), therefore the impulse response is the same as the one of the charge sensitive amplifier with two real poles studied in the previous section. However, in this text we reserve the name CSA to those circuits based on a closed-loop topology relying on a high gain voltage amplifier and in which the signal processing elements are part of the feedback loop.

Using the g_m boosting technique, the circuit input impedance can be reduced, further enhancing its bandwidth. Fig. 4.42 shows the configuration employing a simple common source stage with gain A as the boosting amplifier. In a first analysis, we neglect the bandwidth limitation of the auxiliary amplifier, obtaining the small signal equivalent circuit of Fig. 4.43. The nodal equations now read as follows:

$$\begin{cases} g_{m1}V_{gs1} + \frac{V_{out}}{R_L} + V_{out}sC_L = 0 \\ \frac{V_1}{r_{02}} + V_1sC_T + V_1(1+A)sC_{gs} - g_{m1}V_{gs1} = I_{in} \\ V_{gs1} = -AV_1 - V_1 \end{cases} \tag{4.195}$$

Solving again for V_{out}/I_{in}, we get:

$$\frac{V_{out}}{I_{in}} = \frac{g_{m1}(1+A)r_{02}R_L}{[1+g_{m1}r_{02}(1+A)+sr_{02}[C_T+(1+A)C_{gs}]](1+sR_LC_L)} \tag{4.196}$$

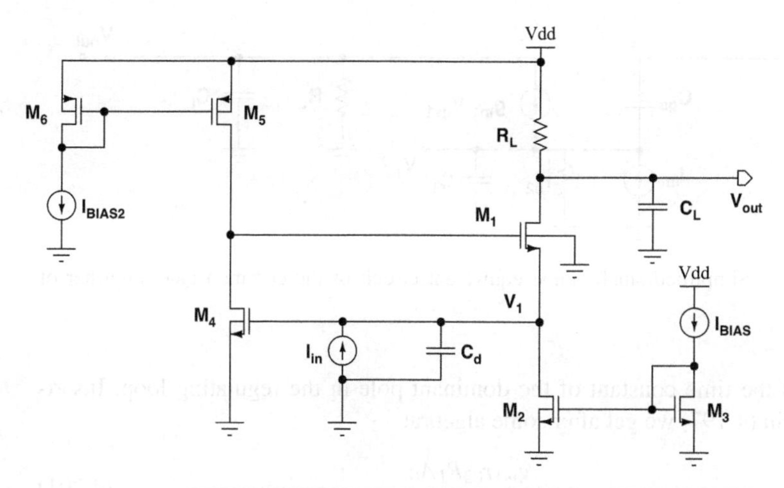

FIGURE 4.42 Regulated common gate amplifier with capacitors limiting the frequency performance.

Assuming now that $A \gg 1$ and that $g_{m1} r_{02} A \gg 1$, we can approximate (4.196) with:

$$\frac{V_{out}}{I_{in}} = \frac{g_{m1} R_L r_{02} A}{[g_{m1} r_{02} A + s r_{02} (C_T + A C_{gs})] (1 + s C_L R_L)} \tag{4.197}$$

Dividing numerator and denominator by $g_{m1} r_{02} A$, we finally obtain:

$$\frac{V_{out}}{I_{in}} = \frac{R_L}{\left[1 + s\frac{(C_T + A C_{gs})}{g_{m1} A}\right] (1 + s R_L C_L)} \tag{4.198}$$

It is interesting to study the above transfer function under two limiting conditions. The first one occurs when $C_T \gg A C_{gs}$, i.e. when the input capacitance is very large. In this case, (4.198) becomes:

$$\frac{V_{out}}{I_{in}} = \frac{R_L}{\left(1 + s\frac{C_T}{g_{m1} A}\right) (1 + s C_L R_L)} \tag{4.199}$$

which can be derived from (4.194) by neglecting C_{gs} and replacing g_{m1} with $A g_{m1}$. As expected, the gain boosting multiplies the transistor transconductance by the gain of the servo amplifier. This suggests that the input time constant can become very small, offering a large bandwidth even in case the input capacitance is large. The complementary situation is found when $A C_{gs} \gg C_T$, i.e. in case of small input capacitance. The input time constant is then given by C_{gs1}/g_{m1}, which corresponds to the cut-off frequency of the input device.

The auxiliary amplifier however has also its own intrinsic speed limitation. In a simplistic approximation, the boosting amplifier can be represented with a first order transfer function given by:

$$A(s) = \frac{A_0}{1 + s\tau_R} \tag{4.200}$$

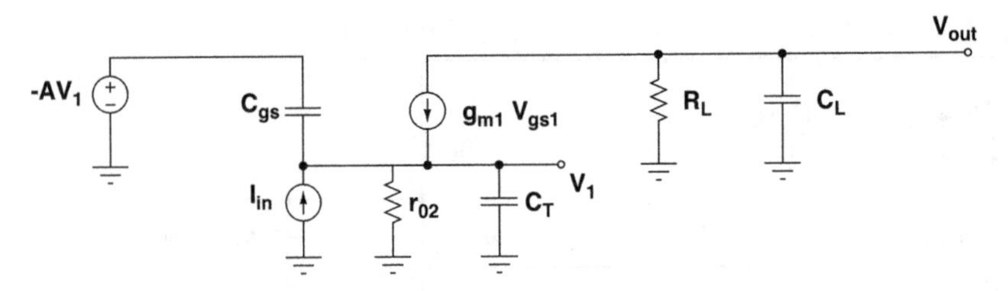

FIGURE 4.43 Simplified small signal equivalent circuit of the common gate amplifier of Fig. 4.42.

where τ_R is the time constant of the dominant pole in the regulating loop. Inserting (4.200) in (4.197) we get after some algebra:

$$\frac{V_{out}}{I_{in}} = \frac{g_{m1} r_{02} R_L A_0}{\left[s^2 r_{02} C_T \tau_R + s r_{02}\left(C_T + A_0 C_{gs}\right) + g_{m1} r_{02} A_0\right]\left(1 + s R_L C_L\right)} \tag{4.201}$$

The denominator thus consists of a real pole multiplied by a second order term, that can have complex conjugate roots. To have only real poles the following condition must be fulfilled:

$$\left(C_T + A_0 C_{gs}\right)^2 > 4 g_{m1} A_0 C_T \tau_R \tag{4.202}$$

We can again distinguish two limiting situations for which the study of (4.202) is particularly simple. If $C_T \gg A_0 C_{gs}$, (4.202) becomes:

$$C_T > 4 g_{m1} A_0 \tau_R \tag{4.203}$$

If, on the opposite, $C_T \ll A_0 C_{gs}$, we have:

$$A_0 C_{gs}^2 > 4 g_{m1} C_T \tau_R \rightarrow C_T < \frac{A_0 C_{gs}^2}{4 g_{m1} \tau_R} \tag{4.204}$$

The last two equations show that complex conjugate roots can be avoided if the sensor capacitance is either very small or sufficiently big, while they can appear for intermediate values of C_T. In any case, the stability of the system is favored by small values of τ_R and of g_{m1}. An important point to note is that in the transimpedance amplifier discussed in the previous section the stability could be compromised for too large values of C_T. Current-mode stages are thus particularly suitable to build fast interfaces for sensors having a very large capacitance. It must also be observed that, due to its bandwidth limitation, the gain boosting amplifier becomes progressively less effective as the frequency rises, which implies that the input impedance of the circuit rises at high frequency, approaching the open-loop value $1/g_{m1}$.

REFERENCES

1. L. S. Bobrow. *Elementary Linear Circuit Analysis.* 2nd ed., Oxford University Press, 1995.

2. N. H. Sabah. *Electric Circuits and Signals*. CRC Press, Taylor & Francis Group, 2008.
3. B. Razavi. *Design of Analog CMOS Integrated Circuits*. McGraw-Hill, 2000.
4. P. E. Allen and D. R. Holberg. *CMOS Analog Circuits Design*. 2nd ed., Oxford University Press, 2002.
5. D. Johns and K. Martin. *Analog Integrated Circuit Design*. Wiley, 1996.
6. P. R. Gray, P. J. Hurst, S. H. Lewis, and R. G. Meyer. *Analysis and Design of Analog Integrated Circuits*. Wiley, 5th edition, 2009.
7. R. J. Baker. *CMOS Circuit Design, Layout and Simulations*. Wiley-IEEE Press, 3rd edition, 2010.
8. J. Huijsing. *Operational Amplifiers: Theory and Design*. Springer, 2nd edition, 2011.
9. G. Palumbo and S. Pennisi. *Feedback Amplifiers: Theory and Design*. Springer, 2002.

5 Noise

Signals delivered by radiation sensors are often weak, and in some cases they may contain only a few hundred electrons. Such small pulses can be easily over-shaded by the intrinsic current and voltage fluctuations occurring in the read-out circuitry. It is therefore not surprising that many studies have been done to understand the causes and minimize the effects of front-end electronic noise. In this book, we will be mainly concerned with noise arising from unavoidable physical phenomena, whose effects can not be entirely suppressed, but only minimized by appropriate signal processing techniques. This type of noise must not be confused with the interference noise, caused by the electro-magnetic coupling between the external environment and the measuring system. The latter can in principle be completely eliminated by proper shielding and grounding techniques [1–4], even though achieving a complete isolation is not trivial.

A rigorous mathematical treatment of noise is lengthy and complex and in this chapter we provide only a concise overview of the topic. In the first section, the key concepts are introduced, while in the second one the expressions of thermal and shot noise spectral densities are derived. The main noise sources found in MOS devices are reviewed in section 3 and examples of noise calculations in practical circuits are discussed in section 4. The concept of optimum noise filtering is introduced in section 5. A more detailed coverage of the matter can be found in [5–8].

5.1 FUNDAMENTAL CONCEPTS

Consider the signal shown in Fig. 5.1. The plot represents the typical baseline of a front-end amplifier, i.e. the circuit output observed when no external signal is applied. To make the fluctuations more visible, the DC offset has been subtracted. We see from the figure that the voltage changes erratically with time, making it impossible to infer from the past history of the waveform its exact value at any particular future instant. Another useful representation of noise is reported in Fig. 5.2. Here the signal has been observed for a sufficiently long time and the captured samples have been used to build a histogram that shows how many times the output voltage falls in a particular bin. The larger the noise, the wider the distribution becomes, therefore the noise can be quantified by measuring the variance and the rms of the amplitude distribution of the signal under study. Even though its exact value at a particular time can not be anticipated, noise can thus be characterized with statistical methods and its average properties can be in general calculated and predicted. In the following, we will be mainly concerned with stationary processes, for which statistical parameters such as the mean and the variance are constant in time. Let's now give a few mathematical definitions which are fundamental in noise analysis. The average value of a signal $x(t)$ is defined as:

$$< x(t) >= \lim_{T \to \infty} \frac{1}{T} \int_{-T/2}^{T/2} x(t)dt \tag{5.1}$$

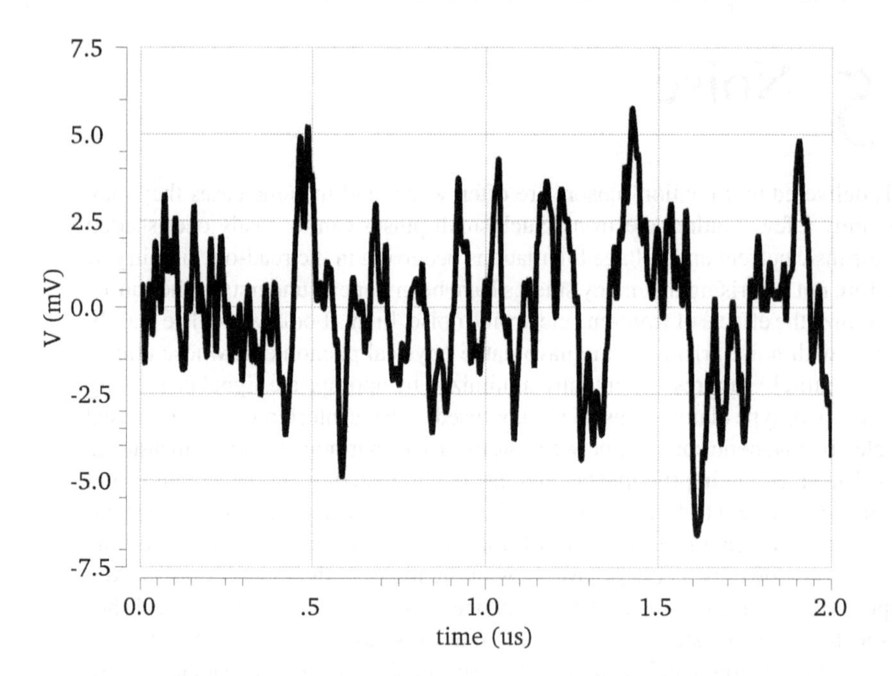

FIGURE 5.1 Example of noise affecting the baseline of a front-end amplifier.

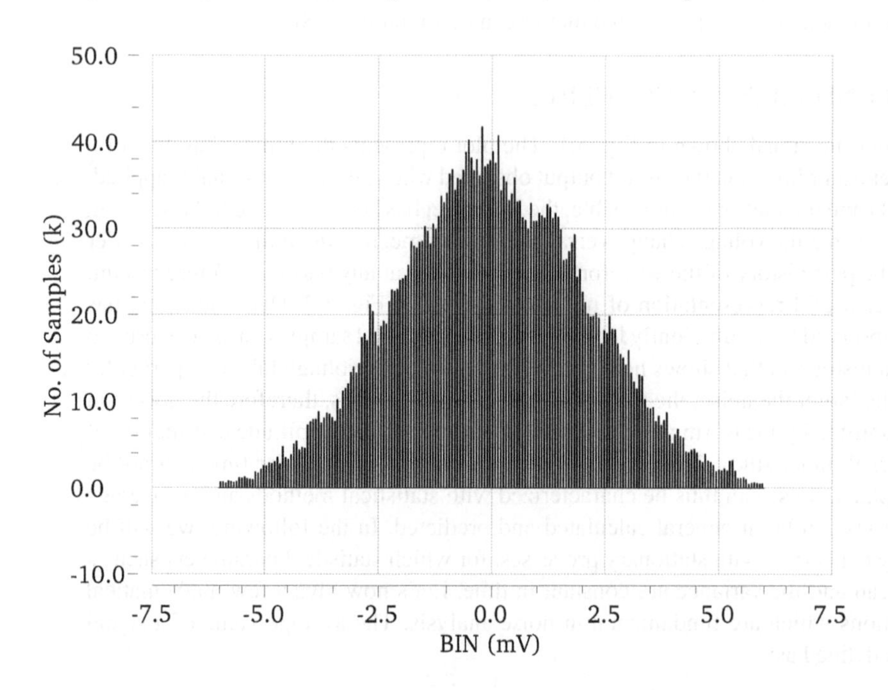

FIGURE 5.2 Amplitude distribution of the signal shown in Fig. 5.1

We assume that if no input stimulus is applied to the circuit and the DC offset is subtracted, $< x(t) >$ is zero. For our purposes, a more interesting quantity is the average value of $x(t)^2$, which is:

$$< x(t)^2 >= \lim_{T\to\infty} \frac{1}{T}\int_{-T/2}^{T/2} x(t)x^*(t)dt = \lim_{T\to\infty} \frac{1}{T}\int_{-T/2}^{T/2} |x(t)|^2 dt \quad (5.2)$$

where $x^*(t)$ is the complex conjugate of $x(t)$. The variance σ^2 is defined as:

$$\sigma^2 =< x(t)^2 > - < x(t) >^2 \quad (5.3)$$

and coincides with $< x(t)^2 >$ when the signal average is zero. If $x(t)$ is a current or a voltage, $< x(t)^2 >$ can be interpreted as the average power dissipated by the signal across a resistance of 1 Ω, hence $< x(t)^2 >$ is also called the average power and indicated as P_{av}.

Another useful mathematical entity is the autocorrelation function, which is a tool to measure the similarity between a given signal and its time-shifted replica. In this text, we will employ the following definition:

$$R_{xx}(\tau) = \lim_{T\to\infty} \frac{1}{T}\int_{-T/2}^{T/2} x(t)x^*(t-\tau)d\tau \quad (5.4)$$

Observe that $R_{xx}(0)$ coincides with the average power defined in (5.2). Let's now consider the truncated Fourier transform of $x(t)$, that is:

$$X_T(f) = \int_{-T/2}^{T/2} x(t)e^{-j2\pi ft}dt \quad (5.5)$$

and define the quantity $S_{xx}(f)$ as:

$$S_{xx}(f) = \lim_{T\to\infty} \frac{1}{T}|X_T(f)|^2 \quad (5.6)$$

The Wiener-Kintchine theorem states that the above limit exists and the following relationships holds:

$$S_{xx}(f) = \int_{-\infty}^{\infty} R_{xx}(\tau)e^{-j2\pi f\tau}dt \quad (5.7)$$

$$R_{xx}(\tau) = \int_{-\infty}^{\infty} S_{xx}(f)e^{j2\pi f\tau}df \quad (5.8)$$

In other words, $R_{xx}(\tau)$ and $S_{xx}(f)$ forms a pair of Fourier transforms. A rigorous proof of the theorem is not trivial and it is omitted here for brevity. We will instead use the above results to understand how noise is processed by linear electrical networks. Before taking this step, observe that if we put $\tau = 0$ in (5.8) we get:

$$R_{xx}(0) = \int_{-\infty}^{\infty} S_{xx}(f)df = P_{av} \quad (5.9)$$

Using Fourier analysis, the signal $x(t)$ can be thought as the superposition of sinusoids of different frequencies. Equation (5.9) shows that $S_{xx}(f)$ can be interpreted as the power spectral density of the signal $x(t)$ and it describes how the total power contained in $x(t)$ is distributed among the different harmonics. If $x(t)$ is a voltage, the associated power spectral density is measured in units of V^2/Hz and its square-root in units of V/$\sqrt{\text{Hz}}$. If $x(t)$ is a current, the unit of the power spectral density and of its square-root are respectively A^2/Hz and A/$\sqrt{\text{Hz}}$.

Suppose now that a signal $x(t)$, to which the power spectral density $S_{xx}(f)$ is associated, is presented to the input of a linear network, characterized by a transfer function $H(s)$ in the Laplace domain and an impulse response $h(t)$ in the time domain. Applying the convolution theorem, the output $y(t)$ can be written as:

$$y(t) = \int_{-\infty}^{\infty} x(t-\theta)h(\theta)d\theta \tag{5.10}$$

Consider now the autocorrelation function of the output, $R_{yy}(\tau)$, which reads:

$$R_{yy}(\tau) = \lim_{T\to\infty} \frac{1}{T} \int_{-T/2}^{T/2} y(t)y^*(t-\tau)dt \tag{5.11}$$

Using the convolution theorem to express explicitly $y(t)$ and $y^*(t)$ as a function of $x(t)$, we get:

$$R_{yy}(\tau) = \lim_{T\to\infty} \frac{1}{T} \int_{-T/2}^{T/2} \int_{-\infty}^{\infty} \int_{-\infty}^{\infty} x(t-\theta_1)h(\theta_1)x^*(t-\tau-\theta_2)h^*(\theta_2)d\theta_1 d\theta_2 dt \tag{5.12}$$

Exchanging the integration and the limit, we can rearrange the above equation as:

$$R_{yy}(\tau) = \int_{-\infty}^{\infty} \int_{-\infty}^{\infty} h(\theta_1)h^*(\theta_2)d\theta_1 d\theta_2 \lim_{T\to\infty} \frac{1}{T} \int_{-T/2}^{T/2} x(t-\theta_1)x^*(t-\tau-\theta_2)dt \tag{5.13}$$

We focus first on the limit, which can be calculated by putting $t-\theta_1 = t'$, which implies $t = t' + \theta_1$ and $dt = dt'$. Performing the change of variables, we have:

$$\lim_{T\to\infty} \frac{1}{T} \int_{-T/2}^{T/2} x(t-\theta_1)x^*(t-\tau-\theta_2)dt = \lim_{T\to\infty} \frac{1}{T} \int_{-T/2}^{T/2} x(t')x^*(t'-\tau+\theta_1-\theta_2)dt' \tag{5.14}$$

Applying the definition of the autocorrelation function, we can recognize in the right-most part of the above equation the autocorrelation of $x(t)$, with the shifting variable τ replaced by $\tau-\theta_1+\theta_2$, that means:

$$\lim_{T\to\infty} \frac{1}{T} \int_{-T/2}^{T/2} x(t')x^*(t'-\tau+\theta_1-\theta_2)dt' = R_{xx}(\tau-\theta_1+\theta_2) \tag{5.15}$$

Inserting the above expression in (5.13) we can write:

$$R_{yy}(\tau) = \int_{-\infty}^{\infty} \int_{-\infty}^{\infty} h(\theta_1)h^*(\theta_2)R_{xx}(\tau-\theta_1+\theta_2)d\theta_1 d\theta_2 \tag{5.16}$$

We can now calculate the spectral density of the output, which is the Fourier transform of $R_{yy}(\tau)$:

$$S_{yy}(f) = \int_{-\infty}^{\infty} R_{yy}(\tau)e^{-j2\pi f\tau}d\tau \tag{5.17}$$

Introducing in the above equation the definition of $R_{yy}(\tau)$ given in (5.16), we get:

$$S_{yy}(f) = \int_{-\infty}^{\infty}\int_{-\infty}^{\infty}\int_{-\infty}^{\infty} h(\theta_1)h^*(\theta_2)R_{xx}(\tau-\theta_1+\theta_2)e^{-j2\pi f\tau}d\theta_1 d\theta_2 d\tau \tag{5.18}$$

To solve the integral, we can now make the following substitution: $\tau-\theta_1+\theta_2 = p$, $d\tau = dp$:

$$S_{yy}(f) = \int_{-\infty}^{\infty}\int_{-\infty}^{\infty}\int_{-\infty}^{\infty} h(\theta_1)h^*(\theta_2)R_{xx}(p)e^{-j2\pi f(p+\theta_1-\theta_2)}d\theta_1 d\theta_2 dp \tag{5.19}$$

Splitting the exponential, we can divide the above integral in the product of three independent ones:

$$S_{yy}(f) = \int_{-\infty}^{\infty} h(\theta_1)e^{-j2\pi f\theta_1}d\theta_1 \int_{-\infty}^{\infty} h^*(\theta_2)e^{j2\pi f\theta_2}d\theta_2 \int_{-\infty}^{\infty} R_{xx}(p)e^{-j2\pi fp}dp \tag{5.20}$$

The first integral is the Fourier transform of the impulse response, that means the circuit transfer function $H(f)$ in the frequency domain. The second integral is the complex conjugate of the transfer function, $H^*(f)$. Therefore, the product between the first two integrals yields the modulus square of the transfer function, $|H(f)|^2$. The last integral is the Fourier transform of the auto-correlation of the input, which is the power spectral density of the input signal $S_{xx}(f)$. We can thus finally write:

$$S_{yy}(f) = S_{xx}(f)|H(f)|^2 \tag{5.21}$$

The above equation shows an extremely important result: if a signal with power spectral density $S_{xx}(f)$ is fed to the input of a linear network with transfer function $H(f)$, the power spectral density of the output $S_{yy}(f)$ is simply obtained multiplying $S_{xx}(f)$ by the modulus square of $H(f)$. So far, we have considered spectral densities defined between $-\infty$ and $+\infty$. However, in case $x(t)$ is real, the spectrum for $f < 0$ is a mirror image of the one for $f > 0$. The two sided spectral density $S_{xx}(f)$ can then be replaced by a single sided one with value $2S_{xx}(f)$ and the integration is carried out between 0 and $+\infty$. Equation (5.21) gives the recipe to calculate noise in a circuit by operating in the frequency domain. First, the transfer function of the network is evaluated by mean of circuit analysis. Second, the input-referred noise spectral density is calculated by considering the contributions of all devices in the network and properly referring them to the circuit input. Finally, (5.21) is applied. Integrating (5.21) one gets the average power of the output noise:

$$P_{av,out} = \langle y(t)^2 \rangle = \int_{-\infty}^{\infty} S_{xx}(f)|H(f)|^2 df \tag{5.22}$$

By taking the square-root of the above relationship the rms of the output noise is obtained. Suppose that we have two noise sources each producing separately an output

$y_1(t)$ and $y_2(t)$. Applying the superposition principle, we can write the total output noise power as:

$$< y(t)^2 >=< (y_1(t)+y_2(t))^2 >= \lim_{T\to\infty}\frac{1}{T}\int_{-T/2}^{T/2}(y_1(t)+y_2(t))^2 dt \tag{5.23}$$

The limit can be further expanded as:

$$\lim_{T\to\infty}\frac{1}{T}\int_{-T/2}^{T/2} y_1(t)^2 dt + \lim_{T\to\infty}\frac{1}{T}\int_{-T/2}^{T/2} y_2(t)^2 dt + 2\lim_{T\to\infty}\frac{1}{T}\int_{-T/2}^{T/2} y_1(t)y_2(t)dt \tag{5.24}$$

If the two noise sources are uncorrelated, the last integral is zero, thus we have:

$$< y(t)^2 >=< y_1(t)^2 > + < y_2(t)^2 > \tag{5.25}$$

and:

$$y_{rms} = \sqrt{< y_1(t)^2 > + < y_2(t)^2 >} \tag{5.26}$$

The result can be extended to arbitrary number n of uncorrelated sources: the effect of each one can be calculated with (5.22) and the total noise is then obtained by summing in quadrature all the contributions, as shown in (5.25) and (5.26).

5.2 THERMAL AND SHOT NOISE SPECTRAL DENSITIES

Thermal and shot noise are the two most prominent noise sources found in electronic circuits. In this section, we give simplified derivations of their power spectral densities.

5.2.1 THERMAL NOISE

One of the primary noise sources in electronic circuits is due to the thermal agitation of charge carriers in conductors. These erratic movements generate currents and voltages that have zero average value but result in instantaneous fluctuations of signals within a circuit. The problem was originally studied by Johnson [9] and Nyquist [10] and explained quantitatively through the principles of thermodynamics and statistical mechanics. The spectral density of thermal noise can in fact be derived by applying the theory of black-body radiation to the system shown in Fig. 5.3, formed by two resistors connected by an ideal transmission line with characteristic impedance $Z = R$ [1]. The system is kept at thermal equilibrium, so the temperature is the same in every point and, for simplicity, the problem is discussed in one dimension only. In each resistor, the thermal energy forces the carriers to move in a disordered way and the resulting currents are translated into voltages by the resistor itself. Each resistor thus generates voltage signals at different frequencies. These signals propagate through the line and reach the other resistor, transferring power to it. Let's study, for instance, the noise

[1] The reader not familiar with the concept of transmission lines can find a clear treatment of the topic in [11].

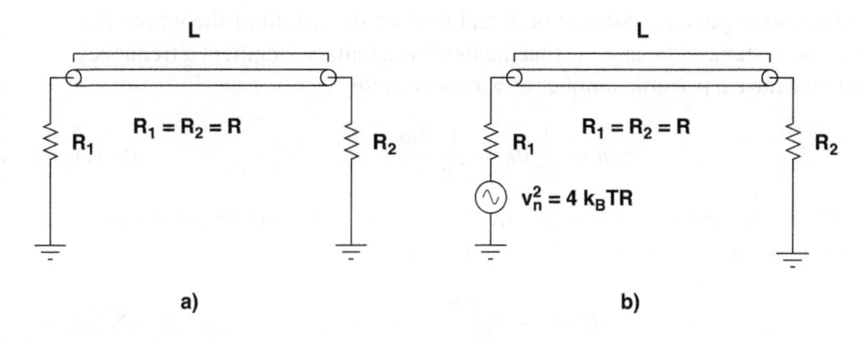

FIGURE 5.3 Two identical resistors of value R are connected by an ideal transmission line of arbitrary length L and with characteristic impedance $Z = R$. The system is kept at thermal equilibrium. The noise generated by each resistor can be calculated applying the Planck formula of black body radiation.

power generated by resistor R_1 and transferred to resistor R_2. Of course, we could do also the opposite. The purpose of our exercise is to calculate the power spectral density of the individual resistor, and due to the full symmetry of the system, it is indifferent which component is considered the "source" and which the "destination". The noise can be viewed as a superposition of sinusoidal signals of different frequencies. For a particular angular frequency ω, the corresponding wave can be written as:

$$v(t) = v_0 e^{j(kx-\omega t)} \tag{5.27}$$

where $k = 2\pi/\lambda$ and λ is the wave length. Suppose that the signal has a value v_1 at a point $x = x_1$ and at a time $t = t_1$. Since the wave oscillates, it will take again the same value in the future. If we call x_2 and t_2 the value of x and t when the value is repeated, we can write:

$$kx_2 - \omega t_2 = kx_1 - \omega t_1 \rightarrow k\Delta x = \omega \Delta t \tag{5.28}$$

Taking the limit for $\Delta t \rightarrow 0$, we get $v = dx/dt = \omega/k$. The signal thus propagates in the line with a speed $v = \omega/k$ and it reaches the other resistor. We choose the origin of the x axis so that $x = 0$ at the beginning of the transmission line. In case of a perfectly terminated line, no reflection is observed, therefore the signals in the line must respect the condition $v(0) = v(L)$, where L is the line length. Consider first a wave in which the period is equal to the time T necessary to travel the distance L. The boundary condition $v(0) = v(L)$ implies that the following relationship holds:

$$kL - \omega T = 0 \rightarrow kL = 2\pi \tag{5.29}$$

where we have used the fact that $\omega = 2\pi/T$. This boundary condition is however also fulfilled by all waves with a period which is an integer multiple of T, hence we can write:

$$kL = 2\pi n \tag{5.30}$$

The above equation puts a constraint on k and thus on the length of the waves that can exist in the system. If we assume that the line has a unitary length, in a frequency interval $d\omega$ we have a possible number of waves given by:

$$dn = \frac{1}{2\pi}dk = \frac{1}{2\pi}\frac{d\omega}{v} \tag{5.31}$$

Each possible wave represents a degree of freedom of the system and the mean energy per oscillation mode is given by the Planck formula:

$$E(\omega) = \frac{\hbar\omega}{e^{\frac{\hbar\omega}{k_B T}} - 1} \tag{5.32}$$

where $\hbar = h/2\pi$ is the reduced Planck constant and k_B is the Boltzmann constant. For a length dL and a time dt, the power can be written as:

$$P = \frac{1}{2\pi}\frac{d\omega}{v}\frac{dL}{dt}E(\omega) = \frac{1}{2\pi}E(\omega)d\omega = E(2\pi f)df \tag{5.33}$$

where we have used the fact that $dL/dt = v$ and we have replaced in the rightmost member ω with $2\pi f$. Considering the resistor as a black-body, this is the power that the component is able to absorb, which is also the power that the object emits when it is brought at a given temperature. Each resistor can then be treated as a voltage source, that generates waves at different frequencies that travel through the line till they reach the other resistor, where they are absorbed. Consider for instance the situation shown in Fig. 5.3 b), where we have represented the noise generated by the first resistor as a voltage source connected in series to it. The presence of a voltage source v in the loop produces a current given by $v/2R$, to which a power dissipation of $R < i^2 >$ is associated. The power can also be written as:

$$P = R < i^2 > = \frac{1}{4R} < v^2 > = \frac{1}{4R}\int_0^{\infty} S_{V,th}(f)df \tag{5.34}$$

where the relationship between the amplitude and the power spectral density has been used. Combining (5.34) and (5.32) we can write for a power generated in the infinitesimal frequency interval df:

$$\frac{1}{4R}S_{V,th}(f)df = E(2\pi f)df \rightarrow S_{V,th}(f) = 4RE(2\pi f) \tag{5.35}$$

Introducing now $\omega = 2\pi f$ in the expression of $E(\omega)$ given in (5.32), we finally get:

$$S_{V,th}(f) = \frac{4Rhf}{e^{\frac{hf}{k_B T}} - 1} \tag{5.36}$$

It is interesting to evaluate the frequency for which the argument of the exponential is unitary, which is:

$$f_u = \frac{k_B T}{h} \tag{5.37}$$

Substituting in the above expression the values of $k_B = 1.38 \times 10^{-23}$ J/K and $h = 6.626 \times 10^{-34}$ J·s, we get $f = 6.248 \times 10^{12} \approx 6.25$ THz. This frequency is well beyond the maximum operating frequency of present-day front-end circuits, therefore we can assume that the approximation $hf \ll k_B T$ always holds. The exponential in the denominator of (5.36) can thus be replaced with a first order approximation, hence we can write:

$$S_{V,th}(f) = \frac{4Rhf}{1 + \frac{hf}{k_B T} - 1} = 4k_B TR \tag{5.38}$$

The above equation expresses the well-known power spectral density of the thermal voltage noise generated by a resistor of value R and it shows that in the cases of practical interest the noise is uniformly spread at all frequencies, that is, its spectrum is "white". An alternative derivation of the thermal noise spectral density in (5.38), valid for the classical regime, can be found in [12]. Observe however that when the frequency approaches f_u, the use of the complete expression in (5.36) becomes mandatory to predict correctly the radiated power, which diverges if the quantum corrections are not taken into account. To derive the units of $S_{V,th}$, note that the Boltzmann constant is expressed in [J][K]$^{-1}$, therefore the product $k_B T$ has the unit of energy. Remember that the energy stored in a capacitor is CV^2, hence energy can also be expressed in unit of [C][V]^2. Therefore, the units of $S_{V,th}$ are [C][R][V]^2 and since the product of a resistor and a capacitor gives a time constant, $S_{V,th}$ has the unit of [V]^2[T] ([T] denoting here the unit of time), or, equivalently [V]^2/[f]. Therefore, when electrical signals are involved, spectral densities are measured in units of [V]^2[Hz]$^{-1}$ or [A]^2[Hz]$^{-1}$ and not in units of power over frequency and they are *numerically* equal to the power that the source would dissipate into a resistor of 1 Ω. As a consequence, the result of the integration in (5.22) has the unit of [V]^2 or [A]^2.

In circuit analysis, a noise voltage spectral density is usually indicated for brevity with e_n^2 or v_n^2, while a current spectral density is represented as i_n^2. In the following, to avoid ambiguities, we will indicate with v_n^2 and i_n^2 the spectral densities (measured respectively in V^2/Hz and in A^2/Hz) and with $< v^2 >$ and $< i^2 >$ the quantities after frequency integration. An additional suffix will help in identifying the type of noise when necessary. As shown in Fig. 5.4, for calculation purposes a noisy resistor can be modeled as an ideal, noise-free resistor within series a voltage source with a spectral density $v_n^2 = 4k_B TR$. Alternatively, the Norton equivalent can be used and the voltage source can be transformed into a current source whose spectral density is $i_n^2 = 4k_B T/R$. Note that to obtain the appropriate spectral density of the current source, v_n^2 must be divided by R^2 and not by R.

5.2.2 SHOT NOISE

Shot noise stems from the fact that the electrical current is quantized in fundamental packets, the electrons, each having a charge q and it is present, for instance, when a photo-current is generated by an incident beam of light or when a potential barrier needs to be crossed, like in diodes. Even though the current is stationary, at the microscopic level there are fluctuations in the number of generated electrons or in the

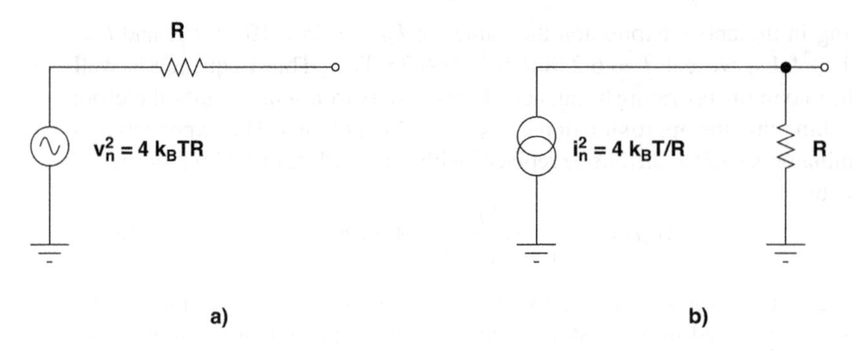

FIGURE 5.4 For calculation purposes, a noisy resistor can be modeled as an ideal, noise-free resistor with a voltage source connected in series having a spectral density of $4k_BTR$. The circuit in a) can also be transformed into its Norton equivalent, reported in b), where the noise generator is represented as current source in parallel with the resistor. For brevity, voltage and current spectral densities are indicated next to the component respectively as v_n^2 and i_n^2.

number of carriers that cross the junction. Each carrier has thus a given probability of being stripped from an atom (in case of a photo-current) or of crossing the junction and the individual events are independent of each other and described by Poisson statistics. With these assumptions, if we integrate the current for a time T, we get the average number of electrons given by:

$$N = \frac{I_0 T}{q} \tag{5.39}$$

where I_0 is the DC current. For a Poisson process, the variance is given by N and the rms by $\sqrt{N}$, therefore, the observed number of electrons in an arbitrary time T fluctuates with an rms given by $\sqrt{N}$. This determines in turn a fluctuation in the value of the current, whose rms can be written as:

$$I_{rms} = \frac{\sqrt{N}q}{T} \tag{5.40}$$

Squaring the above expression and introducing in it the value of N given in (5.39) we get the variance of the current:

$$< i(t)^2 >= \frac{q}{T} I_0 \tag{5.41}$$

Calculating the average current implies integrating it for the time T and then divide the result by T. This operation can be modeled as a filtering algorithm, in which the current is processed by a circuit having the following impulse response:

$$\begin{aligned} h(t) &= \tfrac{1}{T} \text{ for } |t| < \tfrac{T}{2} \\ h(t) &= 0 \text{ everywhere else} \end{aligned} \tag{5.42}$$

Applying the results obtained in the previous section, we can work in the frequency domain and write the noise at the filter output as:

$$S_{yy}(f) = S_{xx}(f)|H(f)|^2 \tag{5.43}$$

where $S_{xx}(f)$ is the unknown spectral density of the shot noise that we want to determine. Since the current is modeled as a sum of individual current pulses generated without correlation between each other, the spectrum is expected to be white. The Fourier transform $H(f)$ of $h(t)$ is given by:

$$H(f) = \frac{\sin(\pi T f)}{\pi T f} \tag{5.44}$$

The output power density can thus be expressed as:

$$S_{yy}(f) = S_{xx}(f) \int_{-\infty}^{\infty} \frac{\sin^2(\pi T f)}{(\pi T f)^2} df \tag{5.45}$$

The integral can be solved by making the substitution $Tf = r$, which yields:

$$S_{yy}(f) = S_{xx}(f) \frac{1}{T} \int_{-\infty}^{\infty} \frac{\sin^2(\pi r)}{(\pi r)^2} dr \tag{5.46}$$

Since:

$$\int_{-\infty}^{\infty} \frac{\sin^2(\pi r)}{(\pi r)^2} dr = 1 \tag{5.47}$$

the output spectral density is simply given by:

$$S_{yy}(f) = S_{xx}(f) \frac{1}{T} \tag{5.48}$$

Equating (5.48) to (5.41) we finally find the value of $S_{xx}(f)$:

$$S_{xx}(f) = qI_0 \tag{5.49}$$

Observe that $S_{xx}(f)$ is measured in units of [C][A], which is equivalent to saying $[\mathrm{A}]^2[\mathrm{T}] = [\mathrm{A}]^2[\mathrm{Hz}]^{-1}$. We can in fact write:

$$[\mathrm{C}][\mathrm{A}] = \frac{[\mathrm{C}]}{[\mathrm{T}]}[\mathrm{A}][\mathrm{T}] = [\mathrm{A}^2][\mathrm{T}] \tag{5.50}$$

The spectral density found in (5.49) is bilateral. The unilateral spectral density, to be used when the integration is carried-out between 0 and $+\infty$, is $2qI_0$, which is the common expression used for shot noise in circuit analysis, where the current noise power spectral density is usually indicated as i_n^2.

5.3 NOISE IN MOS TRANSISTORS

In MOS transistors, several noise sources are at play. The two most prominent ones are the thermal noise due to the resistance of the conductive channel and the flicker noise. Additional contributions come from the resistance of the gate material and of the device bulk. A classical discussion of noise in MOS transistors is given by Van der Ziel in [13, 14], while a modern introduction can be found in [15].

5.3.1 CHANNEL THERMAL NOISE IN MOS TRANSISTORS

A simplified derivation of the noise spectral density in an MOS device can be provided through a resistor analogy. As seen in the previous section, the current noise spectral density generated by a resistor is given by:

$$i_n^2 = \frac{4k_BT}{R} \tag{5.51}$$

Let's start with a slab of material with length L, section A and containing n carriers per unit volume. Using the second Ohm law, the resistance of the slab can be written as:

$$R = \rho\frac{L}{A} = \frac{1}{nq\mu}\frac{L}{A} \tag{5.52}$$

Multiplying and diving the rightmost part of the equation by L, we can introduce in the denominator the slab volume AL and thus the total charge contained in the bar, which is $nqAL$. The resistance can therefore be expressed in the following way:

$$R = \frac{L^2}{\mu Q_M} \tag{5.53}$$

where μ is the carrier mobility and Q_M is the amount of mobile charge. The noise spectral density can hence be rewritten as:

$$i_n^2 = 4k_BT\frac{\mu Q_M}{L^2} \tag{5.54}$$

In a transistor, we can indentify L with the channel length. The problem thus reduces to the calculation of the mobile charge which is present in the transistor channel. Since the latter is in general non-homogeneous, the computation of Q_M is carried-out by dividing the channel into small uniform segments and integrating over all possible contributions. Although rigorous mathematical derivations are quite complex, a simple expression can be obtained for a device working in strong inversion and in saturation. Consider in fact a transistor biased with a stationary drain current I_{DS}, so that in each particular time the total charge in the channel is constant and given by Q_M. For an MOS in saturation and in strong inversion, Q_M was calculated in chapter 2 and it is:

$$Q_M = \frac{2}{3}C_{ox}WL(V_{GS} - V_{TH}) \tag{5.55}$$

Inserting the value of Q_M in (5.54), we obtain:

$$i_n^2 = 4k_BT\frac{2}{3}\mu C_{ox}\frac{W}{L}(V_{GS} - V_{TH}) \tag{5.56}$$

The device transconductance in strong inversion is:

$$g_m = \mu C_{ox}\frac{W}{L}(V_{GS} - V_{TH}) \tag{5.57}$$

which, introduced in (5.56), yields:

$$\overline{i_n^2} = 4k_B T \frac{2}{3} g_m \tag{5.58}$$

The charge Q_M present in the channel changes with the level of inversion, therefore, a more general noise expression is:

$$\overline{i_n^2} = 4k_B T \gamma g_m \tag{5.59}$$

As discussed in chapter 2, γ measures the level of inversion of the channel and is related to the inversion coefficient I_C by:

$$\gamma = \frac{1}{2} + \frac{1}{6} \frac{I_C}{I_C + 1} \tag{5.60}$$

In weak inversion $I_C \to 0$ and γ can be approximated by $1/2$, whereas in strong inversion, where $I_C \gg 1$, γ can be taken equal to $2/3$, as anticipated in (5.58). In short channel devices, an increase in noise beyond what predicted by pure thermal noise consideration is often observed. Called excess noise, the effect is modeled by introducing a correction coefficient in the expression of the spectral density:

$$\overline{i_n^2} = 4k_B T \alpha_w \gamma g_m \tag{5.61}$$

where α_w is called the excess noise factor. Several mechanisms have been proposed in the past to explain this additional noise, including peculiar effects of short channel devices like hot electrons. However, recent works invoke more fundamental physical reasons. An interesting idea is that the phenomenon is caused by the fact that, during their transit in a very short channel, the carriers may not experience enough collisions to reach the thermal equilibrium, which is the assumption at the root of the standard derivation of the thermal noise spectral density. The carriers are injected in the channel through the source-channel potential barrier. The current thus has a shot noise component (stemming from the random crossing of the barrier), which is only partially suppressed by the channel resistance [12, 16]. Other explanations can however also be found [17]. Additional interesting material on channel thermal noise is provided in [18, 19].

It is interesting to observe that, at high frequency, the current noise in the channel couples directly to the gate through the gate capacitance, generating an input noise current. This *induced* gate noise is frequency dependent and can be approximated as [14]:

$$\overline{i_{ng}^2} = 4k_B T \frac{g_m}{5} \delta \left(\frac{f}{f_T} \right)^2 \tag{5.62}$$

where f_T is the transistor cut-off frequency given by:

$$f_T = \frac{g_m}{2\pi C_{gs}} \tag{5.63}$$

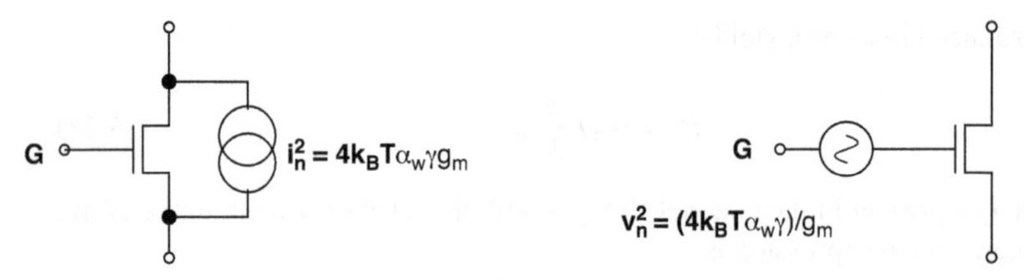

FIGURE 5.5 In the MOS transistor the thermal noise originates from the thermal agitation of the carriers in the channel, which induce a fluctuation in the transistor current. The effect can be modeled as a current source connected between drain and source or by an equivalent voltage source connected in series to the device gate.

In long channel devices, the parameter δ can be taken equal to $4/3$. From (5.62), one can see that i_{ng}^2 becomes relevant only when the signal frequency is a significant fraction of the transistor cut-off frequency. At the typical frequencies employed in front-end design the condition $f \ll f_T$ is in general respected, making induced-gate noise a secondary phenomenon.

The transistor can be seen as an elementary transconductance amplifier that converts the gate-source voltage into a drain-source current. Using (5.21) we can thus transform the output current spectral noise density into an input voltage one dividing by the modulus square of the gain. In this particular case, the gain is simply provided by the device transconductance, which is a real number, so we have:

$$v_n^2 = 4k_B T \alpha_w \gamma \frac{1}{g_m} \tag{5.64}$$

This representation can be convenient in circuit noise analysis. Observe that what we are making here is just a model. The noise is physically produced in the channel, but, to simplify calculations, we treat it as if it was caused by an equivalent source connected in series to the gate. Therefore, the situation is completely different from the case of the induced gate noise, where there is a physical influence through the gate capacitance between the channel and the gate. Fig. 5.5 shows the two equivalent representations of the channel thermal noise, that can of course be applied to either PMOS or NMOS devices, hence a generic symbol is used in the figure. Note that the polarity of the source is arbitrary because noise is a random signal. Observe also that the device is accessed from the outside world after the noise sources, which are considered internal to the transistor.

5.3.2 FLICKER NOISE

Flicker noise gives an important contribution to the overall noise of the MOS transistor and may pose a fundamental limit to the performance of high resolution systems. Flicker noise is usually modeled as a voltage source connected in series to the transistor gate. To avoid ambiguities, from now on we will indicate with v_{nf}^2 the power spectral density of flicker noise and with v_{nw}^2 the one of thermal noise, which is white in the frequency range of our interest. Traditionally, two different mechanisms have been proposed to explain the presence of a noise with power spectrum decreasing with frequency. In the mobility fluctuation model, or Hoodge model [20–22], it is assumed that the carriers mobility changes randomly with time due to phonon scattering and interaction with the crystal lattice. This, in turn, affects the device current. Using the Hoodge model, the following spectral density can be derived:

$$v_{nf}^2 = \frac{K_{fa}(V_{GS} - V_{TH})}{C_{ox}WL}\frac{1}{f} \tag{5.65}$$

where K_{fa} is a constant typical of a given process and device type. The McWorther model [23–25] explains instead the $1/f$ noise with a change in the number of carriers in the current flow. This change is due to traps located at Si-SiO_2 interface, that can temporarily capture and then release the charge carriers. This "number fluctuation" model leads to the following spectral noise density:

$$v_{nf}^2 = \frac{K_{fb}}{C_{ox}^2 WL}\frac{1}{f} \tag{5.66}$$

The fact that increasing the bias current flicker noise often increases for PMOS transistors, while it stays stable for NMOS ones have led several authors to propose the mobility fluctuation as the primary cause of $1/f$ noise in PMOS and the carrier number fluctuation as the root of $1/f$ noise in NMOS devices. Unified descriptions incorporating both models have also been proposed [26]. The spectral density can be written for instance as:

$$v_{nf}^2 = \frac{K_{fc}}{C_{ox}WL}\frac{1}{f}\left(1 + \frac{V_{GS} - V_{TH}}{V_f}\right)^2 \tag{5.67}$$

where V_f is a parameter that models the increase of the noise with the overdrive voltage, which is predicted by the Hoodge model. Experimental measurements also show that K_f increases significantly when the gate length is shrunk, possibly because the trap density increases when the source and drain implants are fabricated very close by. Furthermore, the exponent of the $1/f$ term has also deviations from unity. The flicker spectral density thus takes the form:

$$v_{nf}^2 = \frac{K_f(I_C, L)}{C_{ox}WL}\frac{1}{f^{\alpha_f}} \tag{5.68}$$

If the previous parametrization is used with $\alpha_f = 1$, K_f is measured in Joules. The value of K_f is very technology dependent, but a typical value for an NMOS transistor is $3 \cdot 10^{-24}$ J. The value for PMOS transistor is often at least one order of magnitude smaller. An important point is that $1/f$ noise is inversely proportional to the gate area, thus its impact is more severe for small devices.

In chapter 2 we have studied that the mobile charge stored in the transistor channel when the device is in saturation is given by:

$$Q_M = \frac{2}{3}C_{ox}WL(V_{GS} - V_{TH}) \tag{5.69}$$

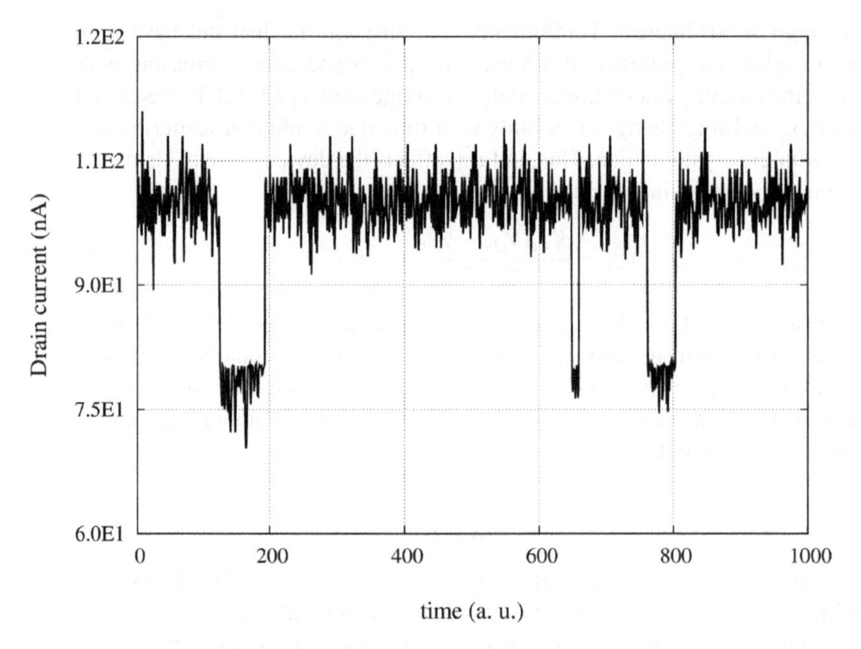

FIGURE 5.6 Random telegraph noise. In very small area transistors, the removal of a single carrier may already create a measurable fluctuation of the device current.

Using the above equation, it is easy to calculate that in a deep-submicron device with $t_{ox} = 2$ nm, $W = 160$ nm and $L = 80$ nm biased with an overdrive voltage $V_{GS} - V_{TH}$ of 0.2 V, about 180 electrons populate the channel at any given instant. In such small devices, only one trapping center may be found in the gate oxide. However, the removal of a single carrier from the current flow can already determine a sizable variation of the measured current, which appears disturbed by a series of random pulses superimposed on its average value, as shown in Fig. 5.6. Called Random Telegraph Noise (RTN) [27, 28], the phenomenon is of particular concern for very deep-submicron technologies and has thus attracted considerable interest [29–34]. This further shows that very small area devices should be avoided in noise-sensitive applications.

5.4 NOISE CALCULATIONS IN CIRCUITS

Knowing the expressions of the most important noise power spectral densities and the rule given in (5.21) that connects the input to the output noise, we can now evaluate the noise performance of practical circuits. To illustrate the method, we start with a very simple example, so that the calculations can be easily done by hand. We then study the most interesting case of the input stage discussed in the previous chapters. To facilitate the prompt identification of the type of noise, in the following we will indicate with v_{nw}^2 a white voltage spectral density (usually of thermal origin), and v_{nf}^2

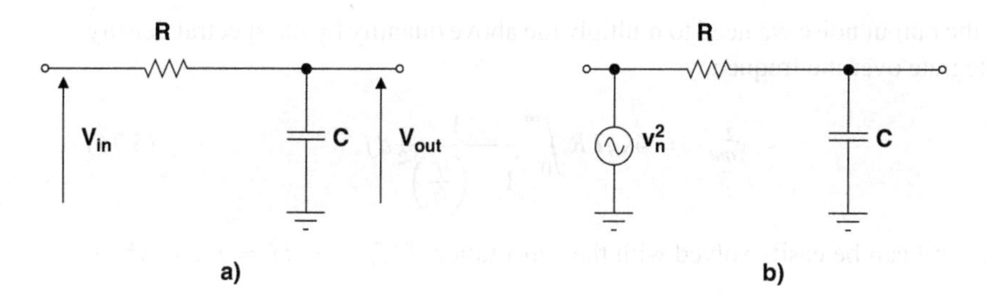

FIGURE 5.7 a) Simple *RC* low pass filter. b) Circuit model to calculate the output noise.

the flicker noise spectral density. The current noise will be represented by i_n^2 and we will always assume that it has a white spectrum. The context will clarify if i_n^2 is of shot or thermal origin.

5.4.1 NOISE IN AN RC LOW PASS FILTER

Consider the simple *RC* low pass filter shown in Fig. 5.7 a). To calculate the noise, we must go through three steps. First, we need to identify the relevant noise sources, calculate their spectral densities and transform them into an equivalent noise source connected to the input. In complex circuits, the last operation can be non-trivial. Second, we must find with the method of circuit analysis the transfer function and its modulus square. Third, applying (5.21) and integrating over all possible frequencies, we get the output noise. In case of a simple *RC* circuit the only noise source is provided by the resistor that generates the thermal noise $4k_BTR$, because a capacitor does not generate noise. The system can then be modeled as shown in Fig. 5.7 b), with the noise source connected at its input. Written in the Laplace domain, the transfer function of a single pole low pass filter with unity gain reads:

$$\frac{V_{out}(s)}{V_{in}(s)} = \frac{1}{1+sRC} \tag{5.70}$$

The cut-off frequency is:

$$f_T = \frac{1}{2\pi RC} \tag{5.71}$$

Putting now in (5.70) $s = j\omega = j2\pi f$ and introducing the expression of f_T, we get:

$$H(f) = \frac{1}{1+j\frac{f}{f_T}} \tag{5.72}$$

The modulus square can be calculated by multiplying both numerator and denominator by $1 - jf/f_T$ and taking the modulus of the resulting function and it is:

$$|H(f)|^2 = \frac{1}{1+\left(\frac{f}{f_T}\right)^2} \tag{5.73}$$

To get the output noise we need to multiply the above quantity by the spectral density and integrate over the frequency:

$$< v_{out}^2 >= 4k_BTR \int_0^\infty \frac{1}{1+\left(\frac{f}{f_T}\right)^2} df \tag{5.74}$$

The integral can be easily solved with the substitution $f/f_T = x$, $df = f_T dx$, which yields:

$$< v_{out}^2 >= 4k_BTRf_T \int_0^\infty \frac{1}{1+x^2} dx \tag{5.75}$$

Since we have that:

$$\int_0^\infty \frac{1}{1+x^2} dx = \tan^{-1}(x)\Big|_0^\infty = \frac{\pi}{2} \tag{5.76}$$

we can write the output noise as:

$$< v_{out}^2 >= 4k_BTRf_T \frac{\pi}{2} = \frac{k_BT}{C} \tag{5.77}$$

where the rightmost member is obtained by inserting the explicit definition of f_T. Despite its simplicity, the above equation shows two interesting points. Consider first the expression before simplification: the noise is written as the product of the input spectral density times the cut-off frequency of the circuit multiplied by a correction factor, $\pi/2$ is this particular case. The product $f_T \cdot (\pi/2)$, which depends only on the filter parameters, is called the noise bandwidth of the system and it is in general larger than the signal bandwidth, defined by f_T. Physically, this has the following meaning. It is conventionally assumed that beyond the cut-off frequency the signals are suppressed. The circuit gain however does not drop immediately to zero when f_T is reached, but it decreases with a roll-off which depends on the number of poles and zeros which are present in the transfer function. Since noise is a broad-band signal, it contains also many frequencies beyond f_T and these are partially transmitted to the output, giving a non-negligible contribution to the output noise.[2] The amount of noise which is embarked beyond the cut-off point depends on the shape of the transfer function, but to have a noise bandwidth as close as possible to the signal bandwidth, the frequency role-off should be as steep as possible. Therefore higher order filters have greater frequency slope and thus smaller noise bandwidth. After simplification, (5.77) shows that the output noise does not actually depend on the resistor but only on the capacitor value. While the resistor is the physical noise source, it is the capacitor that determines how much noise is seen at the output. This has a simple explanation: if we increase or decrease the value of the resistor, we increase or decrease its noise, but we change correspondingly also f_T, thus keeping constant the noise seen at the output. The kT/C noise is particularly important for sampling cells obtained with switches and capacitors, where the noise originates from the channel resistance of the switch.

[2] Remember that for the signal calculation the transfer function $H(f)$ is used, while for the evaluation of noise power $|H(f)|^2$ must be employed.

5.4.2 NOISE IN A SINGLE STAGE FRONT-END

We now study the noise performance of the transimpedance input stage described in the previous chapters, using as a support the transistor-level schematic reported in Fig. 5.8. Following the same procedure adopted for the simple low pass filter, we must identify the relevant noise spectral densities, refer them to the system input and calculate their effects on the circuit output. The analysis carried-out in this section is particularly important because it provides the basis to understand the noise behavior of multi-stage front-end amplifiers that are discussed in chapter 6.

In a radiation detection system, noise sources can be schematically divided between intrinsic, i.e. belonging to the amplifier itself, and extrinsic, originating in the sensor and its biasing network. In principle, all devices forming the core amplifier and its feedback network introduce noise, but usually only a few of them have a sizable impact on the system resolution. In chapter 8 it will be shown that a design effort is always made to limit as much as possible the total noise to the one produced by the input transistor. We therefore start by analyzing the contribution of this device, indicated as M_1 in Fig. 5.8. We have seen in the previous section that the resistive channel of the MOS produces a current noise given in (5.61), that can be transformed into a voltage source connected in series with the transistor gate:

$$v_{nw1}^2 = 4k_BT\alpha_{w1}\gamma_1\frac{1}{g_{m1}} \tag{5.78}$$

The flicker noise was already given in the form of an input voltage. For the time being, we will use the simpler representation of the spectral density:

$$v_{nf1}^2 = \frac{K_f}{C_{ox}WL}\frac{1}{f} \tag{5.79}$$

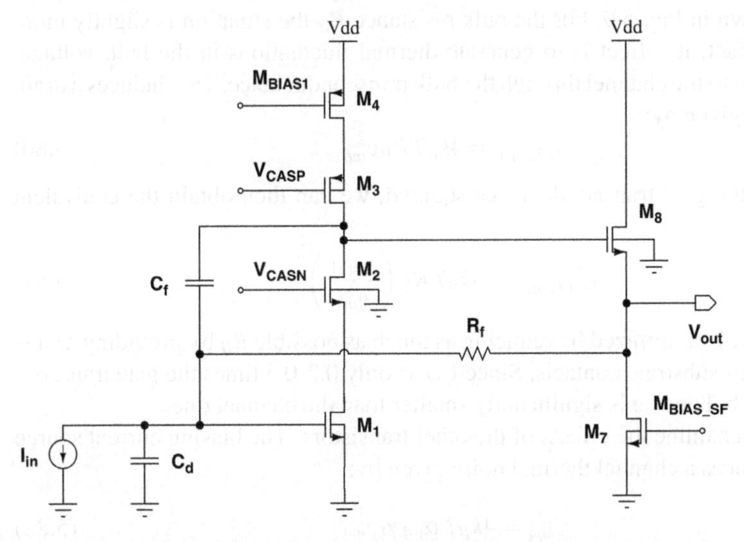

FIGURE 5.8 Single stage front-end amplifier. The most important intrinsic noise sources limiting the amplifier resolution are due to the input and the load transistors and to the feedback resistor R_f. See text for discussion.

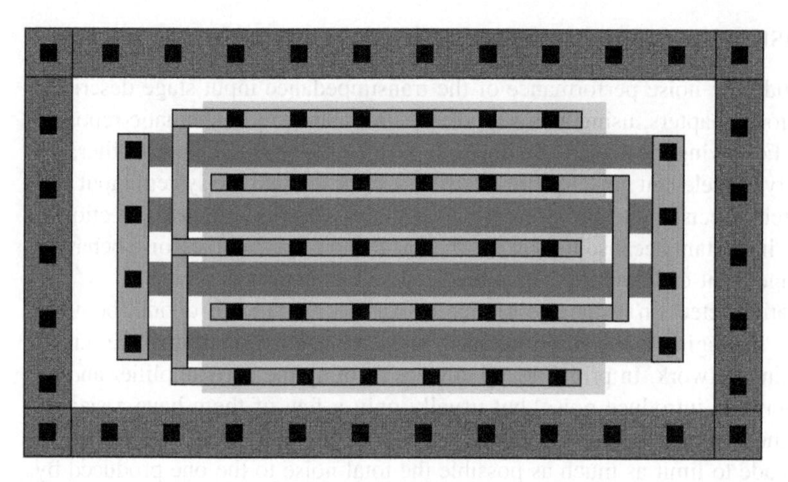

FIGURE 5.9 Typical layout of a large transistor. The gate (represented in dark gray) is divided into more fingers connected in parallel. To reduce the resistivity, the fingers are also connected at both ends with metal strips. A ring of substrate contacts provide a low-impedance connection to the bulk.

We have now to inspect more in detail the two additional contributions stemming from the gate and the bulk resistance. If the gate material has a total resistance R_G, it produces also thermal noise that can be modeled by a voltage source of value $4k_BTR_G$ connected in series to the input. To minimize the gate resistance, wide transistors are usually split in more devices which are then connected in parallel. The two end-points of each finger are also short-circuited with a metal strip to further reduce R_G. A typical layout is shown in Fig. 5.9. For the bulk resistance R_B the situation is slightly more complex. In fact, its effect is to generate thermal fluctuations in the bulk voltage, which couples to the channel through the bulk transconductance. This induces a drain current noise given by:

$$i_{n1,bulk}^2 = 4k_BTR_Bg_{mb1}^2 \tag{5.80}$$

Dividing by the gate transconductance squared, we can then obtain the equivalent input voltage:

$$v_{nw1,bulk}^2 = 4k_BTR_B\left(\frac{g_{mb1}}{g_{m1}}\right)^2 \tag{5.81}$$

The bulk noise is minimized by reducing as much as possible R_B by providing a suitable number of substrate contacts. Since g_{mb} is only 0.2-0.3 times the gate transconductance, the bulk noise is significantly smaller than the channel one.

Let's now examine the effects of the other transistors. The biasing current source M_4 also produces a channel thermal noise given by:

$$i_{n4}^2 = 4k_BT\alpha_{w4}\gamma_4g_{m4} \tag{5.82}$$

This contribution can be referred to the input dividing again by g_{m1}^2, thus we get:

$$v_{nw4}^2 = 4k_BT\alpha_{w4}\gamma_4\frac{g_{m4}}{g_{m1}^2} \quad (5.83)$$

The above equation shows that to minimize the input referred noise of the biasing current source, its transconductance must be much smaller than the one of the input transistor. To keep into account also the flicker noise of M_4, we must remember that the $1/f$ spectral density is usually given as a voltage. We thus need first to transform it into a current, multiplying it by g_{m4}^2, and then we refer it to the amplifier input dividing the result by g_{m1}^2. We therefore get:

$$v_{nf4,in}^2 = \frac{K_f}{C_{ox}W_4L_4}\frac{1}{f}\left(\frac{g_{m4}}{g_{m1}}\right)^2 \quad (5.84)$$

which shows that flicker noise of the current source is suppressed by the factor $(g_{m4}/g_{m1})^2$ when referred to the amplifier input. It is finally interesting to evaluate the noise of the cascode transistors g_{m2} and g_{m3}. If we consider, for instance, M_2, its thermal noise spectral density can be expressed as:

$$i_{n,2}^2 = 4k_BT\alpha_{w2}\gamma_2 g_{m2,eq} \quad (5.85)$$

where $g_{m2,eq}$ is the *equivalent* transconductance of M_2. Remember now that a cascode transistor can be seen as a source degenerated amplifier in which the role of the degeneration resistance is played by the output resistance of M_1, r_{01}. The equivalent transconductance of the cascode is thus:

$$g_{m2,eq} = \frac{g_{m2}}{1+g_{m2}r_{01}} \approx \frac{1}{r_{01}} \ll g_{m1} \quad (5.86)$$

The above equation shows that the source degeneration effect strongly suppresses the equivalent transconductance of the cascode transistors and thus their noise contribution, which can in general be neglected.

The feedback resistor R_f generates a voltage noise $4k_BTR_f$ which can be represented in series with the component. To refer it to the input we have to divide by the resistor transfer function squared, i.e. by R_f^2. We get in this way a noise current, which is represented by a noise generator connected in parallel with the input.

We can now inspect the contributions external to the core amplifiers. In semiconductor sensors, the leakage current I_L produces shot noise, that can be modeled with a current source with a spectral density of $2qI_L$ connected in parallel with the input. To optimize the noise performance, the feedback resistor should be chosen large enough so that its contribution is smaller than the one of the sensor leakage current. The value $R_{f,eq}$ for which the noise generated by R_f equals the sensor-induced one is given by:

$$\frac{4k_BT}{R_{f,eq}} = 2qI_L \rightarrow R_{f,eq} = \frac{2k_BT}{qI_L} \approx \frac{50\text{ mV}}{I_L} \quad (5.87)$$

Known also as the 50 *mV rule*, the above relationship allows us to evaluate quickly the minimum value of R_f to be used. For instance, if I_L is 1 nA, R_f should be larger than 50 MΩ. Such large value resistors can only be implemented using transistors, because a passive component would be prohibitively big. Equation (5.87) shows that in the typical input stage one uses large values of R_f, thus demanding most of the signal processing to the feedback capacitor and operating the first stage in the "charge sensitive" mode.

In chapter 1, we have seen that the sensor is often biased by connecting a resistor R_{HV} in series with the high voltage power supply. Such a resistor introduces also noise that, when referred to the system input, has a spectral density given by $4k_BT/R_{HV}$. Additional noise can be generated by the resistance R_I of the interconnection, which is found in series with the input terminal. This contribution can hence be modeled by a voltage source with spectral density $4k_BTR_I$, connected in series to the amplifier input.

We see that, when referred to the input, the noise can be represented either by voltage sources connected in series with the input or by current sources connected in parallel. This is what justifies the terminology of "series noise" and "parallel noise" reserved to the two categories of noise sources. Parallel noise has thermal or shot origin and thus its spectrum is white. All parallel noise sources can then be lumped into a single generator, i_n^2. The series noise is either of thermal or flicker origin and the two sources must be kept distinct as they are processed differently by the front-end transfer function.

Once we have identified the noise sources, we can make a simplified system model, as shown in Fig. 5.10. Note that the series noise sources originating from the core amplifier are represented inside the feedback loop as they are intrinsic to the core amplifier itself. All the input referred white noises are lumped into a single generator v_{nw}^2 which is the sum of all the white noise spectral densities that are modeled as voltage source in series with the input. The same is done for the flicker and parallel noise generators. In noise calculations, it is relevant to know all the capacitance which is connected between the amplifier input and ground, which will be indicated as C_T and given by the following expression [35]:

$$C_T = C_d + C_{gs1} + C_{ds1} \tag{5.88}$$

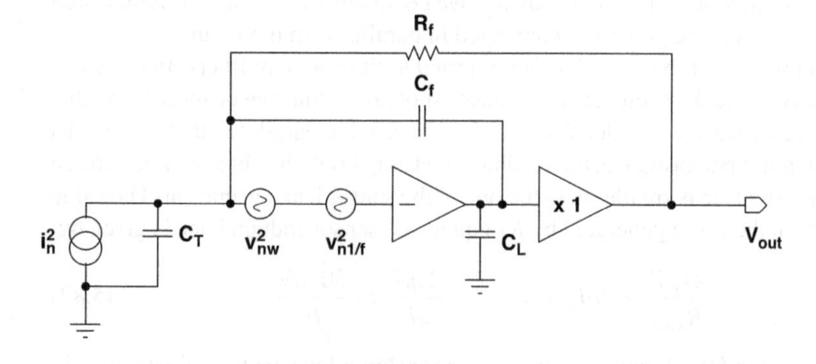

FIGURE 5.10 Simplified model to calculate the noise of the amplifier of Fig. 5.8.

where C_{gs1} and C_{ds1} are respectively the gate-source and drain-source capacitance of the input transistor. To proceed with the calculations, we need first to know the amplifier transfer function. As discussed in chapter 4, a circuit like the one in Fig. 5.8 has usually a second order transfer function given by:

$$T(s) = \frac{(g_{m1} - sC_f)\, R_f R_L}{s^2 \zeta R_f R_L + s g_{m1} R_L C_f R_f + g_{m1} R_L} \tag{5.89}$$

where ζ is a sum of products between C_T, C_f and C_L:

$$\zeta = C_T C_L + C_T C_f + C_L C_f \tag{5.90}$$

To keep the formalism simple, we neglect the zero at the numerator and divide both numerator and denominator by $g_{m1} R_L$ to get:

$$T(s) = \frac{R_f}{s^2 \zeta \frac{R_f}{g_{m1}} + s R_f C_f + 1} \tag{5.91}$$

We assume also that only the first term in (5.90) is relevant so ζ can be approximated by $C_T C_L$ and the poles in equation (5.91) are real and widely spaced, so that the denominator can be written as:

$$D(s) = (1 + s\tau_f)(1 + s\tau_r) = 1 + s(\tau_f + \tau_r) + s^2 \tau_f \tau_r \approx 1 + s\tau_f + s^2 \tau_f \tau_r \tag{5.92}$$

where we have assumed that τ_f is much larger than τ_r. By comparing (5.92) with (5.91) we see that $\tau_f = R_f C_f$ while τ_r is obtained from:

$$\tau_f \tau_r = \zeta \frac{R_f}{g_{m1}} \rightarrow \tau_r = \frac{C_T C_L}{g_{m1} C_f} \tag{5.93}$$

Note that, in a typical front-end amplifier, R_f is usually chosen to be large to minimize its noise contribution, therefore the approximation $\tau_f \gg \tau_r$ is usually justified. We can thus write the amplifier transfer function as:

$$T(s) = \frac{R_f}{(1 + s\tau_f)(1 + s\tau_r)} \tag{5.94}$$

and we will use the explicit definitions of τ_f and τ_r when appropriate. The system impulse response, obtained from the Inverse Laplace Transform of (5.94), is:

$$V_{out}(t) = \frac{Q_{in} R_f}{\tau_f - \tau_r} \left(e^{-\frac{t}{\tau_f}} - e^{-\frac{t}{\tau_r}} \right) \tag{5.95}$$

where Q_{in} is the input signal charge. If $\tau_f \gg \tau_r$, the pulse response is simply given as:

$$V_{out} = \frac{Q_{in}}{C_f} e^{-\frac{t}{\tau_f}} \tag{5.96}$$

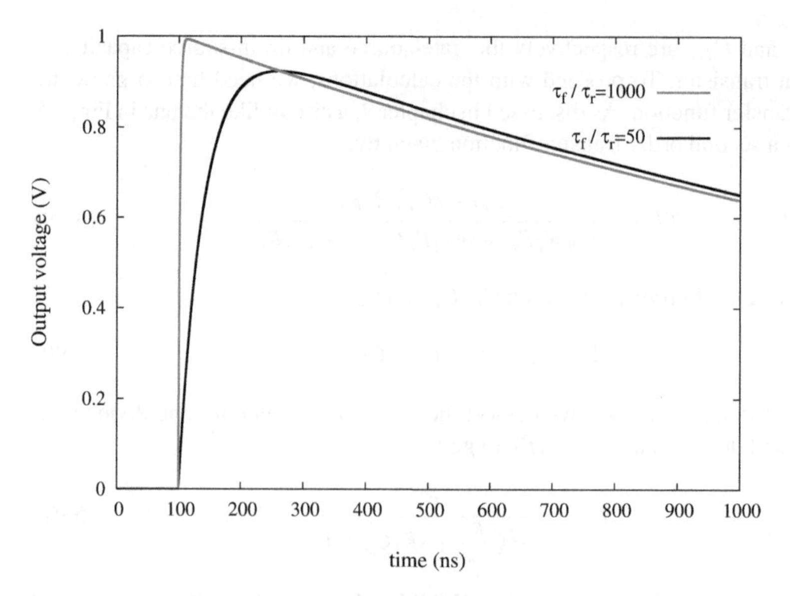

FIGURE 5.11 Impulse response of a transimpedance amplifier with two real and distinct poles. If the core amplifier is very fast, the rise time constant τ_r becomes negligible with respect to the feedback time constant $R_f C_f$ and the impulse response can be approximated with that of a single pole transfer function given in equation (5.96).

Fig. 5.11 shows the impulse response for $Q_{in}/C_f = 1$, $\tau_f = 2\ \mu$s and τ_r equal respectively to 2 ns and 40 ns.

We consider now separately the effect of the three noise sources. Let's start from parallel noise. Observe that the noise generator is connected in parallel with the input exactly as the signal, therefore the noise is processed by the same transfer function as the signal. The power of the output noise voltage can be written as:

$$< v_{out}^2 >_{ni} = i_n^2 R_f^2 \int_0^{\infty} \left| \frac{1}{(1+s\tau_f)(1+s\tau_r)} \right|^2 df \tag{5.97}$$

The calculation of the integral yields:

$$\int_0^{\infty} \left| \frac{1}{(1+s\tau_f)(1+s\tau_r)} \right|^2 df = \frac{1}{4(\tau_f+\tau_r)} \approx \frac{1}{4\tau_f} \tag{5.98}$$

The output noise can thus be written as:

$$< v_{out}^2 >_{ni} = \frac{i_n^2}{4} \frac{R_f}{C_f} \tag{5.99}$$

It is interesting to consider two separate cases. The first one is when there is no leakage current and only the amplifier feedback resistor contributes to the noise. In this case

i_n^2 is given by:

$$i_n^2 = \frac{4k_BT}{R_f} \tag{5.100}$$

Inserting (5.100) into (5.99) we get:

$$< v_{out}^2 >_{ni} = \frac{k_BT}{C_f} \tag{5.101}$$

The noise of the feedback resistor is thus transferred to the output as k_BT/C_f, so its effect on the output depends only on the value of the feedback capacitor. The other interesting case is when only the detector shot noise is present, thus i_n^2 is given by:

$$i_n^2 = 2qI_L \tag{5.102}$$

where I_L is the sensor leakage current. Introducing (5.102) in (5.99) we obtain:

$$< v_{out}^2 >_{ni} = \frac{qI_L}{2}\frac{R_f}{C_f} \tag{5.103}$$

As discussed in chapter 1, it is customary and convenient to refer the front-end noise to the input and express it in units of the electron charge. To do so, we need to divide the above equation by the square of the signal given by one electron and then take the square-root of the result. Since a single electron yields a peak output voltage of value q/C_f, we get for the general case:

$$ENC_i^2 = \frac{i_n^2}{4}\frac{R_fC_f}{q^2} \tag{5.104}$$

If the noise is only due to the feedback resistor, dividing (5.101) by q^2/C_f^2 and taking the square root, we get the following value for the ENC:

$$ENC_i = \frac{1}{q}\sqrt{k_BTC_f} \tag{5.105}$$

whereas when the noise is dominated by the sensor leakage current we can write:

$$ENC_i = \sqrt{\frac{I_L\tau_f}{2q}} \tag{5.106}$$

In the circuit that we are considering, the contribution of the detector shot noise can be reduced by making faster the discharge of the feedback capacitor. However, this has also additional drawbacks. It was proven in the previous chapter that, for a transfer function with two real poles, a more exact expression of the peak output voltage is given by:

$$V_{out,peak} = \frac{Q_{in}}{C_f}\left(\frac{\tau_f}{\tau_r}\right)^{\frac{\tau_r}{\tau_r-\tau_f}} \tag{5.107}$$

The above equation shows that the approximation that the peak voltage is Q_{in}/C_f only holds if $\tau_r \ll \tau_f$ (see also Fig. 5.11), thus decreasing τ_f to improve the parallel noise performance also reduces the peak voltage. To avoid that, also τ_r must be decreased, which can be obtained by increasing g_{m1} and thus burning more power in the amplifier.

We can now study the effect of the white voltage noise, represented by the source v_{nw}^2. This noise is converted into a current $v_{nw}^2 s^2 C_T^2$ by the impedance that shunts the amplifier input, represented by the capacitor C_T. The corresponding voltage current noise is then processed by the amplifier transfer function. Putting as usual $s = j2\pi f$, we can write:

$$< v_{out}^2 >_{nw} = v_{nw}^2 R_f^2 C_T^2 \int_0^\infty \left| \frac{j2\pi f}{(1 + j2\pi f \tau_f)(1 + j2\pi f \tau_r)} \right|^2 df \tag{5.108}$$

Solving the integral we get:

$$\int_0^\infty \left| \frac{j2\pi f}{(1 + j2\pi f \tau_f)(1 + j2\pi f \tau_r)} \right|^2 df = \frac{1}{4\tau_f \tau_r (\tau_f + \tau_r)} \approx \frac{1}{4\tau_f^2 \tau_r} \tag{5.109}$$

where the approximation $\tau_f \gg \tau_r$ has again been used. Inserting the value of the integral into (5.108) and using the explicit definition of τ_r, we obtain:

$$< v_{out}^2 >_{nw} = v_{nw}^2 C_T^2 R_f^2 \frac{1}{4R_f^2 C_f^2 \frac{C_T C_L}{g_{m1} C_f}} \tag{5.110}$$

The expression takes a simpler form if we suppose that the noise is dominated by the contribution of the input transistor. Assuming that the excess noise coefficient is one, we get:

$$< v_{out}^2 >_{nw} = 4k_B T \gamma \frac{1}{g_{m1}} C_T^2 R_f^2 \frac{1}{4R_f^2 C_f^2 \frac{C_T C_L}{g_{m1} C_f}} = \gamma k_B T \frac{C_T}{C_f C_L} \tag{5.111}$$

The above equation reveals an interesting result: the noise depends only on a ratio of capacitances and not on the transconductance of the input transistor. This is because g_{m1} determines both the input noise and the amplifier bandwidth, hence increasing g_{m1} decreases the noise but increases also the bandwidth over which the noise gets integrated. Dividing (5.111) by $(q/C_f)^2$ and taking the square-root we get the input-referred rms noise in units of the electron charge:

$$ENC_w = \frac{1}{q} \sqrt{\gamma k_B T \frac{C_T C_f}{C_L}} \tag{5.112}$$

Finally, we inspect the effect of $1/f$ noise, assuming that it is dominated by the contribution of the input transistor, for which we use the simplified spectral density:

$$v_{nf}^2 = \frac{K_f}{C_{ox} W_1 L_1} \frac{1}{f} \tag{5.113}$$

Since this is a voltage, we must first convert it into a current and then process it with the front-end transfer function. We therefore have:

$$i^2_{nf,in} = \frac{K_f}{C_{ox}W_1L_1}\frac{1}{f}s^2C_T^2 \tag{5.114}$$

Note that the term $1/f$ is not squared in the conversion because it is already part of the power spectral density. The output noise voltage due to flicker noise thus reads:

$$< v^2_{out} >_{nf}= \frac{K_f}{C_{ox}W_1L_1}C_T^2R_f^2\int_0^{\infty}\frac{1}{f}\left|\frac{j2\pi f}{(1+j2\pi f\tau_f)(1+j2\pi f\tau_r)}\right|^2 df \tag{5.115}$$

The value of the integral is:

$$\int_0^{\infty}\frac{1}{f}\left|\frac{j2\pi f}{(1+j2\pi f\tau_f)(1+j2\pi f\tau_r)}\right|^2 df = \frac{\ln\left(\frac{\tau_r}{\tau_f}\right)}{\tau_r^2-\tau_f^2} \approx \frac{1}{\tau_f^2}\ln\left(\frac{\tau_f}{\tau_r}\right) \tag{5.116}$$

Inserting (5.116) in (5.115) and substituting τ_r and τ_f with their definitions, we get:

$$< v^2_{out} >_{nf}= \frac{K_f}{C_{ox}W_1L_1}\frac{C_T^2}{C_f^2}\ln\left(R_fC_f\frac{g_{m1}C_f}{C_TC_L}\right) \tag{5.117}$$

Dividing again by $(q/C_f)^2$ and taking the square-root, we arrive at the flicker component of the ENC:

$$ENC_f = \frac{C_T}{q}\sqrt{\frac{K_f}{C_{ox}W_1L_1}}\sqrt{\ln\left(R_fC_f\frac{g_{m1}C_f}{C_TC_L}\right)} \tag{5.118}$$

The total noise is obtained by summing in quadrature the three contributions, that is:

$$ENC = \sqrt{ENC_w^2+ENC_i^2+ENC_f^2} \tag{5.119}$$

Observe that only for parallel noise the function whose modulus is integrated over frequency to yield the output noise coincides with the signal transfer function. In other words, in noise calculations one must be careful in using the appropriate transfer function for the particular noise source under consideration. Finally, we must note that the direct cascode employed in Fig. 5.8 is just one of the many possible configurations that can be adopted for the core amplifier. The noise analysis can however be easily extended to any other topology once the key noise contributors (mainly the input and the load device) have been properly identified.

5.5 NOISE FILTERING AND OPTIMIZATION OF THE SIGNAL-TO-NOISE RATIO

In the previous section we have studied the noise performance of a representative circuit, which is very often employed as the first element of the front-end chain in a radiation detection system. We can now ask ourselves the following question: can we further manipulate our signals to improve the Signal-to-Noise Ratio (SNR)? Answering this question will lead us to the concept of the match filter and, more in general, to the need of adding additional blocks after the input stage.

5.5.1 THE MATCHED FILTER

We start considering a linear time-invariant network with impulse response $h(t)$, which, in the Laplace domain, becomes $H(s)$. If an input signal $x_{in}(s)$ is fed to the network, its output becomes:

$$y_{out}(s) = x_{in}(s)H(s) \tag{5.120}$$

In the following, we will put $s = j\omega = j2\pi f$, so that we can use the Fourier transforms. Our output signal thus becomes:

$$y_{out}(j2\pi f) = x_{in}(j2\pi f)H(j2\pi f) \tag{5.121}$$

For simplicity, we assume also that $x_{in}(t)$ and $y_{out}(t)$ are real functions. Suppose now that our input signal is corrupted by a noise term, represented as $s_n(t)$ in the time domain, so that the overall input signal can be written as:

$$x'_{in}(t) = x_{in}(t) + s_n(t) \tag{5.122}$$

If the noise source has a power spectral density given by $S_{xx}(f)$, we can quantify its effect on the output measuring the variance of the output signal, which is:

$$< y_{out}^2 >= \int_{-\infty}^{\infty} S_{xx}(f)\,|H(j2\pi f)|^2\,df \tag{5.123}$$

Using the inverse Fourier transform, we can write y_{out} in the following form:

$$y_{out}(t) = \int_{-\infty}^{\infty} x_{in}(j2\pi f)H(j2\pi f)e^{j2\pi f t}\,df \tag{5.124}$$

The value of the output signal at a generic instant t_m is:

$$y_{out}(t_m) = \int_{-\infty}^{\infty} x_{in}(j2\pi f)H(j2\pi f)e^{j2\pi f t_m}\,df \tag{5.125}$$

We can therefore define the signal-to-noise ratio for the signal observed at t_m as:

$$SNR^2 = \frac{\left[\int_{-\infty}^{\infty} x_{in}(j2\pi f)H(j2\pi f)e^{j2\pi f t_m}\,df\right]^2}{\int_{-\infty}^{\infty} S_{xx}(f)\,|H(j2\pi f)|^2\,df} \tag{5.126}$$

Let's now consider the following identity:

$$x_{in}(j2\pi f)H(j2\pi f)e^{j2\pi f t_m} = [S_{xx}(f)]^{\frac{1}{2}}\,H(j2\pi f)\frac{x_{in}(j2\pi f)e^{j2\pi f t_m}}{[S_{xx}(f)]^{\frac{1}{2}}} \tag{5.127}$$

We recall the Schwartz inequality which, for two generic complex functions $f_1(j2\pi f)$ and $f_2(j2\pi f)$ can be written as:

$$\left|\int_{-\infty}^{\infty} f_1(j2\pi f) f_2(j2\pi f) df\right|^2 \leq \int_{-\infty}^{\infty} |f_1(j2\pi f)|^2 df \int_{-\infty}^{\infty} |f_2(j2\pi f)|^2 df \quad (5.128)$$

Applying (5.128) to (5.127), we get:

$$\left|\int_{-\infty}^{\infty} x_{in}(j2\pi f) H(j2\pi f) e^{j2\pi f t_m} df\right|^2 = \left[\int_{-\infty}^{\infty} x_{in}(j2\pi f) H(j2\pi f) e^{j2\pi f t_m} df\right]^2 \leq$$

$$\leq \int_{-\infty}^{\infty} S_{xx}(f) |H(j2\pi f)|^2 df \int_{-\infty}^{\infty} \frac{|x_{in}(j2\pi f)|^2 \left|e^{j2\pi f t_m}\right|^2}{S_{xx}(f)} df \quad (5.129)$$

Note that the first line of (5.129) holds because we suppose that $y_{out}(t)$ is a real function of t. Observe also that we can write:

$$e^{j2\pi f t_m} = \cos(j2\pi f t_m) + j\sin(2\pi f t_m) \rightarrow |e^{j2\pi f t_m}| = \cos^2(j2\pi f t_m) + \sin^2(2\pi f t_m) = 1 \quad (5.130)$$

which implies also that:

$$\left|e^{j2\pi f t_m}\right|^2 = 1 \quad (5.131)$$

Combining (5.131) and (5.129) and inserting the result in (5.126) we finally get:

$$SNR^2 \leq \int_{-\infty}^{\infty} \frac{|x_{in}(j2\pi f)|^2}{S_{xx}(f)} df \quad (5.132)$$

The above equation shows that the signal to noise ratio after the linear network has an upper limit which is equal to the integral of the ratio between the modulus square of the input signal and the power spectral density of the noise. The equality in (5.132) is verified if $H(j2\pi f)$ takes the following form:

$$H(j2\pi f) = a \frac{x_{in}^*(j2\pi f)}{S_{xx}(f)} e^{-j2\pi f t_m} \quad (5.133)$$

where a is an arbitrary constant. It can be verified in fact that inserting (5.133) in (5.126) one obtains again (5.132). Equation (5.133) thus yields the transfer function of the linear network that maximizes the signal-to-noise ratio and its determination implies the knowledge of both the input signal and of the noise spectral density. Finding $H(j2\pi f)$ becomes particularly simple if the noise is white, which means that $S_{xx}(f)$ is constant. If we put $S_{xx}(f) = K^2$ in (5.133) we get:

$$H(j2\pi f) = a \frac{x_{in}^*(j2\pi f)}{K^2} e^{-j2\pi f t_m} \quad (5.134)$$

Note that we write here K^2 and not K to emphasize that this is a *power* spectral density. To find the time-domain representation of the impulse network, $h(t)$, we need to compute the inverse Fourier transform of (5.134), which is:

$$h(t) = \frac{a}{K^2}\int_{-\infty}^{\infty} x_{in}^*(j2\pi f)e^{-j2\pi f t_m}e^{j2\pi f t}df = \frac{a}{K^2}\int_{-\infty}^{\infty} x_{in}^*(j2\pi f)e^{j2\pi f(t-t_m)}df \tag{5.135}$$

We can now make the substitution $t - t_m = \tau$ and recall that the inverse Fourier transform of $x_{in}^*(j2\pi f)$ is $x_{in}(-t)$ to write:

$$h(t) = \frac{a}{K^2}\int_{-\infty}^{\infty} x_{in}^*(j2\pi f)e^{j2\pi f \tau}df = \frac{a}{K^2}x_{in}(-\tau) = b x_{in}(t_m - t) \tag{5.136}$$

Observe that since a is an arbitrary constant, also $b = a/K^2$ becomes an arbitrary constant. The above equation shows that, in order to maximize the signal-to-noise ratio in case of white noise, the signal should be filtered by a system which has an impulse response equal to the mirror image of the signal itself, calculated with respect to the measuring instant t_m. Such a network is called a "matched filter". In the following, we will apply the matched filter concept to the case of a typical detector front-end.

5.5.2 OPTIMUM FILTER FOR ENERGY MEASUREMENTS

Fig. 5.12 shows the input stage studied up to now followed by an additional linear network. Our objective in this paragraph is to find the transfer function $H(s)$ of the full chain so that the energy released in the sensor by a quantum of radiation is measured with the best possible accuracy. In the figure, two noise generators are also shown, representing respectively the series and the parallel noise contributions. In the following, we neglect the flicker noise and we introduce a few simplifying hypotheses with the aim of reducing the computational complexity. First, we assume that the detector signal is fast enough to be modeled by a Dirac-delta, i.e. $I_{in}(t) = Q_{in}\delta(t)$, where Q_{in} is the total charge released in the sensor by the impinging particle. Second, we consider the feedback resistance R_f to be large enough, so that its effect on the signal processing can be neglected and the input stage can be treated as an ideal charge

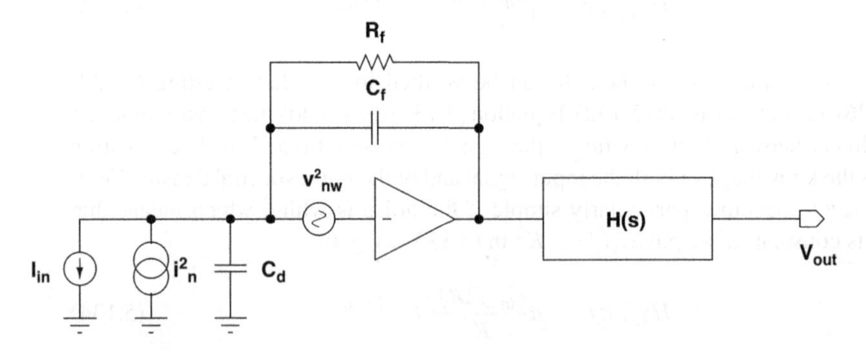

FIGURE 5.12 A transimpedance stage is followed by a network with transfer function $H(s)$. The purpose of the additional filter is to optimize the signal-to-noise ratio.

sensitive amplifier. It is also convenient to introduce an equivalent noise resistance for series and parallel noise defined respectively as following:

$$R_s = \frac{v_{nw}^2}{4k_B T} \tag{5.137}$$

and

$$R_p = \frac{4k_B T}{i_n^2} \tag{5.138}$$

Without loss of generality, we assume that the system output is a voltage. In the Laplace domain, the output signal can be written as:

$$V_{out}(s) = Q_{in} H(s) \tag{5.139}$$

In the time domain, we have instead:

$$V_{out}(t) = Q_{in} h(t) \tag{5.140}$$

From the two equations above, we see that the output voltage is proportional to the input charge and that the latter can be inferred by taking a single sample of the output waveform at a generic instant t_m. Of course, the best measurement is done if t_m is the time at which $V_{out}(t)$ reaches its maximum. Using the matched filter concepts, we want now to determine the impulse response $h(t)$ that yields the best possible SNR. As a first step, we calculate the noise spectral densities at the output of the first stage. For the series noise we have:

$$v_{out1,s}^2 = v_{nw}^2 \left| \frac{C_T}{C_f} \right|^2 \tag{5.141}$$

where C_T is, as usual, the sum of the detector capacitance C_d and of any other capacitance shunting the amplifier input to ground. For the parallel noise we get:

$$v_{out1,p}^2 = i_n^2 \left| \frac{1}{sC_f} \right|^2 \tag{5.142}$$

Substituting in the above equation $s = j\omega$, we obtain:

$$v_{out1,p}^2 = \frac{i_n^2}{\omega^2 C_f^2} \tag{5.143}$$

In this and in the next few equations we use the variable ω instead of f to simplify the notation. The total noise spectral density at the charge integrator output is:

$$S(\omega) = v_{out1,s}^2 + v_{out1,p}^2 = \left(\frac{C_T}{C_f}\right)^2 \left(v_{nw}^2 + \frac{i_n^2}{\omega^2 C_T^2}\right) = A^2 \left(v_{nw}^2 + \frac{i_n^2}{\omega^2 C_T^2}\right) \tag{5.144}$$

where A is equal to the ratio C_T/C_f. We note that at the output of the first stage the total noise spectral density is not white anymore. Therefore, the results derived in the previous paragraph, that were obtained assuming a white noise spectrum, are not immediately applicable. We can however insert an additional filter with a transfer function $H_w(j\omega)$ so that at its output the noise spectral density becomes independent of frequency again, that is:

$$N^2 = S(\omega)\,|H(j\omega)|^2 \tag{5.145}$$

where N is a constant. Inserting in (5.145) the expression of $S(\omega)$ given in (5.144) we have:

$$|H_w(j\omega)|^2 = \frac{N^2}{S(\omega)} = \frac{N^2\omega^2 C_T^2}{A^2\left(i_n^2 + v_{nw}^2\omega^2 C_T^2\right)} \tag{5.146}$$

This condition is verified by a network whose transfer function has the following expression:

$$H_w(j\omega) = \frac{j\omega C_T N}{A\left(i_n + j\omega C_T v_{nw}\right)} \tag{5.147}$$

Calculating the modulus square of (5.147) and multiplying it by $S(\omega)$ we obtain in fact the constant term N^2. Equation (5.147) can be written in a more explicative form. First, we divide both the numerator and the denominator by i_n to get:

$$H_w(j\omega) = \frac{1}{A}\frac{j\omega C_T \frac{N}{i_n}}{\left(1 + j\omega C_T \frac{v_{nw}}{i_n}\right)} \tag{5.148}$$

We now multiply and divide the numerator by v_{nw}:

$$H_w(j\omega) = \frac{1}{A}\frac{j\omega C_T \frac{N}{i_n}\frac{v_{nw}}{v_{nw}}}{\left(1 + j\omega C_T \frac{v_{nw}}{i_n}\right)} \tag{5.149}$$

Factorizing the term N/v_{nw} in the numerator, we finally have:

$$H_w(j\omega) = \frac{1}{A}\frac{N}{v_{nw}}\frac{j\omega C_T \frac{v_{nw}}{i_n}}{\left(1 + j\omega C_T \frac{v_{nw}}{i_n}\right)} \tag{5.150}$$

Using now the expressions of v_{nw} and i_n in terms of R_s and R_p we can write:

$$\frac{v_{nw}}{i_n} = \sqrt{R_s R_p} \tag{5.151}$$

Therefore the product:

$$\tau_c = C_T\frac{v_{nw}}{i_n} = C_T\sqrt{R_s R_p} \tag{5.152}$$

has the unit of a time constant. Consider now the value of the parallel noise at the first stage output at the angular frequency $\omega_c = 1/\tau_c$. This is:

$$v_{out1,p}^2\Big|_{\omega=\omega_c} = \frac{i_n^2}{\omega_c^2 C_f^2} = \frac{i_n^2}{\frac{i_n^2}{v_{nw}^2}\frac{1}{C_T^2}C_f^2} = \left(\frac{C_T}{C_f}\right)^2 v_{nw}^2 = v_{out1,s}^2 \tag{5.153}$$

Therefore, τ_c identifies the frequency at which the series and the parallel noise give identical contributions at the integrator output. This particular frequency is called the *noise corner frequency* and should not be confused with the frequency, called with the same name, for which the thermal and the flicker noise of an MOS transistor are equal. Using τ_c, $H_w(j\omega)$ can be written as:

$$H_w(j\omega) = H_0 \frac{j\omega\tau_c}{1 + j\omega\tau_c} \tag{5.154}$$

where the constant H_0 is given by:

$$H_0 = \frac{1}{A}\frac{N}{v_{nw}} \tag{5.155}$$

The above equation tells us that the noise at the integrator output can be made white again by filtering it with a high pass filter having a time constant equal to the noise corner time constant. The impulse response of the charge integrator plus the high pass filter is given by:

$$H_1(s) = \frac{Q_{in}}{sC_f} H_0 \frac{s\tau_c}{1 + s\tau_c} \tag{5.156}$$

In the time domain, this becomes:

$$h_1(t) = \frac{Q_{in}}{C_f} H_0 e^{-\frac{t}{\tau_c}} u(t) \tag{5.157}$$

where $u(t)$ is the unit step function. Note that $h_1(t)$ reaches its maximum at $t = 0$ and then decays exponentially. If we consider a generic measuring time t_m we need to shift our signal forward by t_m, which yields:

$$h_1(t - t_m) = \frac{Q_{in}}{C_f} H_0 e^{-\frac{t-t_m}{\tau_c}} u(t - t_m) \tag{5.158}$$

We can also replace in (5.158) H_0 with its explicit expression to get:

$$h_1(t - t_m) = \frac{Q_{in}}{C_T}\frac{N}{v_{nw}} e^{-\frac{t-t_m}{\tau_c}} u(t - t_m) \tag{5.159}$$

Since at the output of the high pass filter the noise is white, we can apply the result on the matched filter and conclude that the SNR is optimized by further processing the signal with a network whose impulse response $h_2(t)$ is the mirror image in time of $h_1(t)$:

$$h_2(t - t_m) = b e^{\frac{t-t_m}{\tau_c}} u(t_m - t) \tag{5.160}$$

where b is an arbitrary constant. To obtain the overall circuit response we need to calculate the convolution integral of $h_1(t)$ and $h_2(t)$, that is:

$$V_{out}(t) = \int_{-\infty}^{\infty} h_1(\alpha) h_2(t - \alpha) d\alpha \tag{5.161}$$

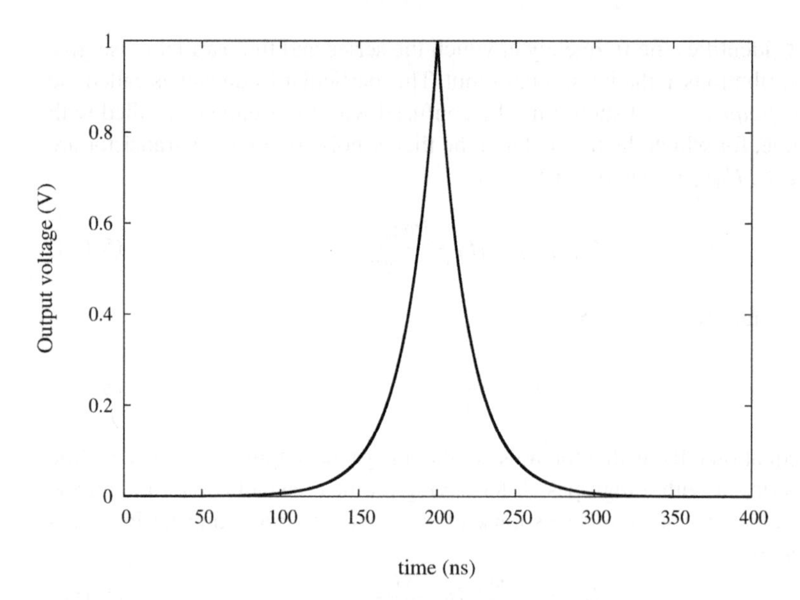

FIGURE 5.13 In energy measurements, the effect of white series and parallel noise is minimized if the impulse response of the processing chain is an indefinite cusp. The decay time constant of the exponential must be set equal to the time constant τ_c, which identifies the frequency at which the series and the parallel noise gives equal contributions at the first stage output.

Using the explicit definitions of $h_1(t)$ and $h_2(t)$ we get:

$$V_{out}(t) = b\frac{Q_{in}}{C_T}\frac{N}{v_{nw}}\int_{-\infty}^{\infty} e^{-\frac{\alpha}{\tau_c}}u(\alpha)e^{-\frac{\alpha-t-t_m}{\tau_c}}u(\alpha-t+t_m)d\alpha \quad (5.162)$$

The integration of (5.162) yields the following expression:

$$V_{out}(t) = b\frac{\tau_c}{2}\frac{Q_{in}}{C_T}\frac{N}{v_{nw}}\left[e^{-\frac{t-t_m}{\tau_c}}u(t-t_m)+e^{\frac{t-t_m}{\tau_c}}u(t_m-t)\right] \quad (5.163)$$

The resulting impulse response function of the full chain is the indefinite cusp and it is shown in Fig. 5.13. In this particular example, the peak voltage has been normalized to one, the constant τ_c was fixed to 20 ns and the measuring time was chosen at $t = 200$ ns from the origin of the time axis. Note that, in principle, the input must be applied at $t = -\infty$ and the output signal grows exponentially till the maximum is reached at $t = t_m$. Immediately after $t = t_m$, the system starts relaxing back to the baseline following an exponential decay law. A single measurement thus requires an infinite time, making this approach not feasible in practice. However, the indefinite cusp is interesting because it sets the lower possible noise that can be achieved as far as white noise is concerned and it provides therefore a benchmark against which the performance of realistic filters can be compared.

The SNR is defined as the ratio between the peak amplitude and the rms output voltage measured when no input signal is applied. The former can be deduced from equation (5.163) evaluating it for $t = t_m$ and remembering that $u(0) = 1/2$, which yields:

$$V_{out,max} = b\frac{\tau_c}{2}\frac{Q_{in}}{C_T}\frac{N}{v_{nw}} \tag{5.164}$$

The total output noise can be calculated observing that at the output of the first filter the noise spectral density is white and equal to N^2. Therefore, we can just multiply N^2 by $|H_2(j2\pi f)|^2$ and integrate over the frequency:

$$< V_{out}^2 >= N^2 \int_0^\infty |H_2(j2\pi f)|^2 \, df \tag{5.165}$$

The integral can also be solved in the time domain applying the Parseval theorem which states that:

$$\int_{-\infty}^{\infty} |F(j2\pi f)|^2 \, df = \int_{-\infty}^{\infty} [f(t)]^2 \, dt \tag{5.166}$$

where $F(j2\pi f)$ is the Fourier transform of $f(t)$. Since $|H_2(j2\pi f)|^2$ is an even function of f, we have:

$$\int_{-\infty}^{\infty} |H_2(j2\pi f)|^2 \, df = 2\int_0^\infty |H_2(j2\pi f)|^2 \, df = \int_{-\infty}^{\infty} [h_2(t)]^2 \, dt \tag{5.167}$$

Applying the above result we can thus write:

$$N^2 \int_0^\infty |H_2(j2\pi f)|^2 \, df = \frac{1}{2}N^2 \int_{-\infty}^{\infty} [h_2(t)]^2 \, dt = \frac{1}{2}N^2 \int_{-\infty}^{t_m} b^2 e^{\frac{2(t-tm)}{\tau_c}} \, dt \tag{5.168}$$

The integration in the time domain is stopped at $t = t_m$ because $h_2(t)$ is zero for $t > t_m$. The evaluation of the integral is straightforward and yields:

$$< V_{out}^2 >= \frac{1}{2}N^2 b^2 \frac{\tau_c}{2} \rightarrow \sqrt{< V_{out}^2 >} = V_{out,rms} = \frac{Nb}{2}\sqrt{\tau_c} \tag{5.169}$$

Dividing (5.164) by (5.169) we get the signal-to-noise ratio on the peak:

$$\text{SNR} = \frac{Q_{in}}{C_T}\frac{\sqrt{\tau_c}}{v_{nw}} \tag{5.170}$$

The above equation shows an interesting result: the SNR of a system is proportional to the ratio between the charge created in the sensor by the ionizing event and the total capacitance connected by the amplifier input and ground. In general, the dominant contribution to C_T is given by the sensor capacitance C_d. For a given type of radiation, also the released charge Q_{in} depends on the sensor properties. Therefore, the ratio Q_{in}/C_d can be taken as a metric to evaluate the ultimate energy resolution that a given radiation sensor can offer. Introducing now in (5.170) the expression of τ_c given in (5.152) we finally get:

$$\text{SNR} = \frac{Q_{in}}{\sqrt{C_T v_{nw} i_n}} \tag{5.171}$$

The ENC is defined at the input charge that yields a signal-to-noise ratio of 1, therefore:

$$\text{ENC} = \sqrt{C_T v_{nw} i_n} \tag{5.172}$$

We can also write:

$$ENC^2 = C_T^2 \frac{v_{nw}^2}{\tau_c} \tag{5.173}$$

Equations (5.172)-(5.173) give the minimum ENC that can be obtained as far as white parallel and series noise are concerned and provide a reference to compare the performance of filters that can be realized in practice. Observe that in (5.172) the ENC is expressed in unit of charge and that v_{nw} and i_n are the *square roots* of the power spectral densities.

5.5.3 OPTIMUM FILTER FOR TIMING MEASUREMENTS

In many applications, the time of occurrence of an event must be captured with good precision. High accuracy timing systems routinely achieve better than 100 ps resolution and research is ongoing to further improve their performance. As discussed in chapter 1, the relevant quantity to be optimized in order to obtain good timing is the noise-to-slope ratio of the signal. Consider now the derivative of the normalized indefinite cusp. This can be written as:

$$\frac{dh(t)}{dt} = \begin{cases} e^{t-t_m} \text{ for } t < t_m \\ -e^{-(t-t_m)} \text{ for } t > t_m \end{cases} \tag{5.174}$$

The resulting curve, shown in Fig. 5.14 has infinite slope at $t = t_m$. Therefore, the time at which the derivative of the cusp changes sign would provide in principle a jitter-free timing measurement, implying that an additional derivation of the signal must be performed if the best timing accuracy is searched for. Note that if $F(s)$ is the Laplace transform of $f(t)$, the Laplace transform of df/ft is $sF(s)$. In analog circuits, the derivative of a signal is obtained by processing the signal itself with a high pass filter, which has a transfer function given by $s\tau/(1+s\tau)$ and thus performs the derivation only on those frequencies for which $s\tau \ll 1$. The slope of bipolar signals obtained in this way is thus smaller than the one resulting from the mathematical derivation of the function describing the unipolar signal and the jitter performance is correspondingly worse.

The noise minimization methods described in this section are not realizable in practice, but they further elucidate two important concepts that were anticipated in chapter 1:

- To optimize the SNR of a radiation detection system, the input stages discussed in the previous chapters must be followed by additional circuits that filter the signal and minimize the noise.
- Different types of filters must be used for energy and timing measurements.

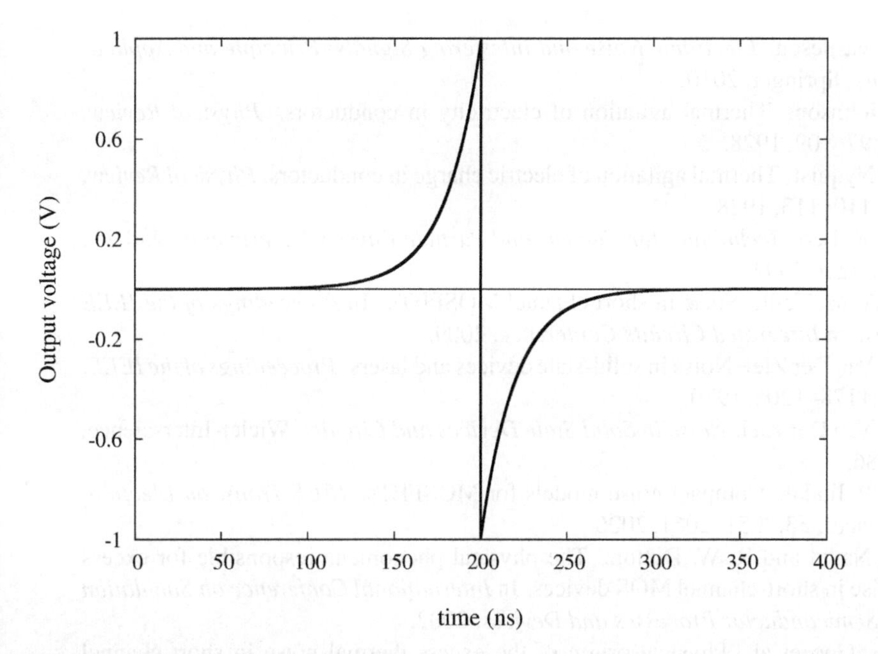

FIGURE 5.14 The derivative of the indefinite cusp has infinite slope at $t = t_m$. This would result in a time pick off with zero jitter.

Before closing this chapter, we should also observe that we have derived the optimum filter in the most simple situation. To get more realistic results, one should take into account additional constraints, such as the fact that in a practical case the duration of the impulse response should be limited and the noise spectrum may not be white. Further insights on these aspects and more general treatments of the problem can be found in [36–40].

REFERENCES

1. H. W. Ott. *Noise Reduction Techniques in Electronic Systems.* 2nd ed., Wiley-Interscience, 1988.
2. H. W. Ott. *Electromagnetic Compatibility Engineering*. John Wiley & Sons, 2009.
3. R. Morrison. *Grounding and Shielding: Circuits and Interference*. John Wiley & Sons, 2007.
4. H. Spieler. *Semiconductor Detector Systems*. Oxford Science Publications, 2005.
5. W. B. Davenport and W. L. Root. *An Introduction to the Theory of Random Signals and Noise*. McGraw-Hill, 1958.
6. W. C. Van Etten. *Introduction to Random Signals and Noise*. Wiley, 2005.
7. S. Engelberg. *Random Signals and Noise: A Mathematical Introduction*. CRC Press, 2007.

8. G. Vasilescu. *Electronic Noise and Interfering Signals: Principle and Applications*. Springer, 2010.
9. J. Johnson. Thermal agitation of electricity in conductors. *Physical Review*, 32:97–109, 1928.
10. H. Nyquist. Thermal agitation of electric charge in conductors. *Physical Review*, 32:110–113, 1928.
11. W. R. Leo. *Techniques for Nuclear and Particle Physics Experiments*. 2nd ed., Springer, 1994.
12. J. A. McNeill. Noise in short channel MOSFETs. In *Proceedings of the IEEE Custom Integrated Circuits Conference*, 2009.
13. A. Van Der Ziel. Noise in solid-state devices and lasers. *Proceedings of the IEEE*, 58:1178–1206, 1970.
14. A. Van Der Ziel. *Noise in Solid State Devices and Circuits.* Wieler-Interscience, 1986.
15. R. P. Jindal. Compact noise models for MOSFETs. *IEEE Trans. on Electron. Devices*, 53:2051–2061, 2006.
16. R. Navid and R. W. Dutton. The physical phenomena responsible for excess noise in short-channel MOS devices. In *International Conference on Simulation of Semiconductor Processes and Devices*, 2002.
17. J-S. Goo et al. Physical origin of the excess thermal noise in short channel MOSFETs. *IEEE Electron. Device Letters*, 22:101–103, 2001.
18. G. Tan, C. H. Chen, B. Hung, P. Lei, and C. S. Yeh. Channel thermal noise and its scaling impact on deep sub-100 nm MOSFETs. In *International Conference on Noise and Fluctuations*, 2011.
19. C. H. Chen, D. Chen, R. Lee, L. Peiming, and D. Wan. Thermal noise modeling of nano-scale MOSFETs for mixed-signal and RF applications. In *Custom Integrated Circuit Conference*, 2013.
20. F. N. Hoodge. Discussions of recent experiments on 1/f noise. *Physica*, 60:130–144, 1976.
21. F. N. Hoodge and L. K. J. Vandamme. Lattice scattering causes 1/f noise. *Physics Letters*, 66A:315–316, 1978.
22. F. N. Hoodge. 1/f noise sources. *IEEE Trans. on Electron. Devices*, 37:1926–1935, 1994.
23. S. Christensson, I. Lundstrom, and C. Svensson. Low frequency noise in MOS transistors - I Theory. *Solid-state Electronics*, 11:797–812, 1968.
24. S. Christensson, I. Lundstrom, and C. Svensson. Low frequency noise in MOS transistors - II Experiments. *Solid-state Electronics*, 11:813–820, 1968.
25. G. Ghibaudo. A simple derivation of Reimbold's drain current spectrum formula for flicker noise in MOSFETs. *Solid-state Electronics*, 30:1037–1038, 1987.
26. K. K. Hung, P. K. Ko, C. Hu, and Y. C. Cheng. A unified model for the flicker noise in metal-oxide-semiconductor field-effect transistors. *IEEE Trans. on Electron. Devices*, 37:654–665, 1990.
27. K. K. Hung, P. K. Ko, C. Hu, and Y. C. Cheng. Random telegraph noise of deep-submicrometer MOSFETs. *IEEE Electron. Device Letters*, 11:90–92, 1990.

28. S. Kogan. *Electronic Noise and Fluctuations in Solids* Cambridge University Press, 2008.
29. C. Leyris, F. Martinez, M. Valenza, A. Hoffmann, J. C. Vildeuil, and F. Roy. Impact of random telegraph signal in CMOS image sensors for low-light levels. In *European Solid-State Circuits Conference*, pages 376–379, 2006.
30. S. Realov and K. L. Shepard. Random telegraph noise in 45-nm CMOS: Analysis using an on-chip test and measurement system. In *International Electron Devices Meeting*, 2010.
31. N. Zanolla, D. Siprak, M. Tiebout, P. Baumgartner, E. Sangiorgi, and C. Fiegna. Reduction of RTS noise in small-area MOSFETs under switched bias conditions and forward substrate bias. *IEEE Trans. on Electron Devices*, 57:1119–1128, 2010.
32. P. Martin-Gonthier and P. Magnan. Novel readout circuit architecture for CMOS image sensors minimizing RTS noise. *IEEE Electron Device Letters*, 32:776–778, 2011.
33. M. Jamal Deen, S. Majumder, O. Marinov, and M. M. El-Desouki. Random telegraph signal noise in CMOS active pixel sensors. In *IEEE International Conference on Noise and Fluctuations*, 2011.
34. M. Nour et al. Variability of random telegraph noise in analog MOS transistors. In *IEEE International Conference on Noise and Fluctuations*, 2013.
35. W. M. C. Sansen and Z. Y. Chang. Limits of low noise performance of detector readout front-ends in CMOS technology. *IEEE Trans. Circ. Syst.*, 37:1375–1382, 1990.
36. V. Radeka. Low-noise techniques in detectors. *Annual Review of Nuclear and Particle Science*, 38:217–277, 1988.
37. E. Gatti and P. F. Manfredi. Processing the signals from solid-state detectors in elementary-particle physics. *Rivista Del Nuovo Cimento*, 9:1–146, 1986.
38. E. Gatti, A. Geraci, and G. Ripamonti. Timing of pulses of any shape with arbitrary constraints and noises: optimum filters synthesis method. *Nucl. Instr. Meth.*, A457:347–355, 2001.
39. E. Gatti, M. Sampietro, and P. F. Manfredi. Optimum filters for detector charge measurements in presence of 1/f noise. *Nucl. Instr. Meth.*, A287:513–520, 1990.
40. A. Pullia. How to derive the optimum filter in presence of arbitrary noises, time-domain constraints, and shaped input signals: A new method. *Nucl. Instr. Meth.*, A397:414–425, 1997.

6 Time Invariant Shapers

The use of multi-stage architectures increases the flexibility and performance of front-end amplifiers. In these topologies, the signal processing is distributed between the "preamplifier" that provides the interface to the sensor and the "pulse shaper" that, as its name implies, generates an output signal with a well defined shape and determines the signal-to-noise ratio. In this text, we will reserve the name "front-end amplifier" to the preamplifier-shaper combination.

"Shaping" implies manipulating and altering the frequency content of the original waveform, therefore a pulse shaper is primarily an analog filter, a type of circuit widely covered in the literature [1–7]. The peculiarities found in the field of radiation sensors lead however to implementation details that deserve a dedicated treatment. The subject of pulse processing for radiation detectors has been developed over a period of more than seventy years, with first specialized books appearing as early as in 1953 [8], when electronics was still based on vacuum tubes. Modern treatments can be found in [9] and in [10]. The topic is also addressed, from the user point of view, in most textbooks on radiation detectors (see the references of chapter 1 for a list). The original research material is published in journal papers and conference proceedings, a small sample of which is included in the bibliography of this chapter.

The preamplifier and the shaper might not be the most complex building blocks to be designed on a modern front-end ASIC. However, they set the ultimate limits to the measurement quality and the factors affecting their performance must be properly understood. In this chapter, we concentrate on systems employing only linear, time invariant networks that can be easily studied with the help of the Laplace transforms. In the first part of the chapter we focus on the methods to achieve a given signal shape, while in the second part we discuss noise calculations. The effects of non-idealities occurring in the building blocks of a front-end amplifier are examined in the last section.

6.1 IDEAL CHARGE SENSITIVE AMPLIFIERS

A multi-stage front-end often employs as the input stage the Charge Sensitive Amplifier (CSA) introduced in chapter 4 and shown schematically in Fig. 6.1. Remember that a CSA is a form of transimpedance amplifier in which the feedback resistor, necessary to define the amplifier DC operating point, is big enough to give a minor and ideally negligible contribution to the signal processing. For the time being, we study the transfer function of our circuits under the following assumptions:

- The input signal can be approximated with a δ-like pulse: $I_{in}(t) = Q_{in}\delta(t)$, where $\delta(t)$ is the Dirac δ function and Q_{in} is the total charge conveyed by the pulse.
- The core amplifiers have infinite gain and bandwidth.
- The feedback resistor R_f of the input stage is infinite.

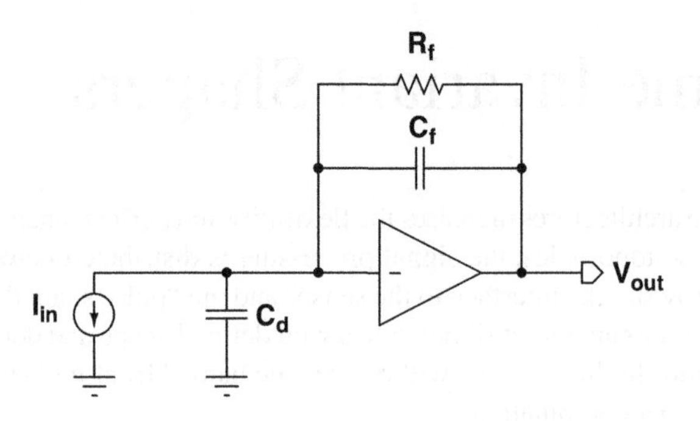

FIGURE 6.1 Charge Sensitive Amplifier (CSA).

The CSA thus behaves as an ideal integrator and its output voltage is given by:

$$V_{out} = \frac{1}{C_f} \int i(t)dt \tag{6.1}$$

In case the input is a δ-like stimulus, the output becomes:

$$V_{out} = \frac{Q_{in}}{C_f} u(t) \tag{6.2}$$

where $u(t)$ is the unit step function, which is the integral of the Dirac-δ and is defined as:

$$\begin{cases} u(t) = 1 \text{ for } t \geq 0 \\ u(t) = 0 \text{ for } t < 0 \end{cases} \tag{6.3}$$

For an easier visualization, the signals can be shifted by a proper amount of time from the origin by making the following substitutions[1]:

$$\begin{cases} Q_{in}\delta(t) \rightarrow Q_{in}\delta(t - t_0) \\ u(t) \rightarrow u(t - t_0) \end{cases} \tag{6.4}$$

Fig. 6.2 shows the response of an ideal CSA when the input signal and the feedback capacitor have been tuned to provide an output signal of 1 V. In case $R_f = \infty$, the CSA

---

[1] For simplicity, in the text we will write in general the equations in the time domain assuming $t_0 = 0$. However, for visualization purposes, the associated plots are shifted forward in time by an appropriate amount that can be easily deduced by the graph itself. The reader must remember to apply this shift when reproducing the curves.

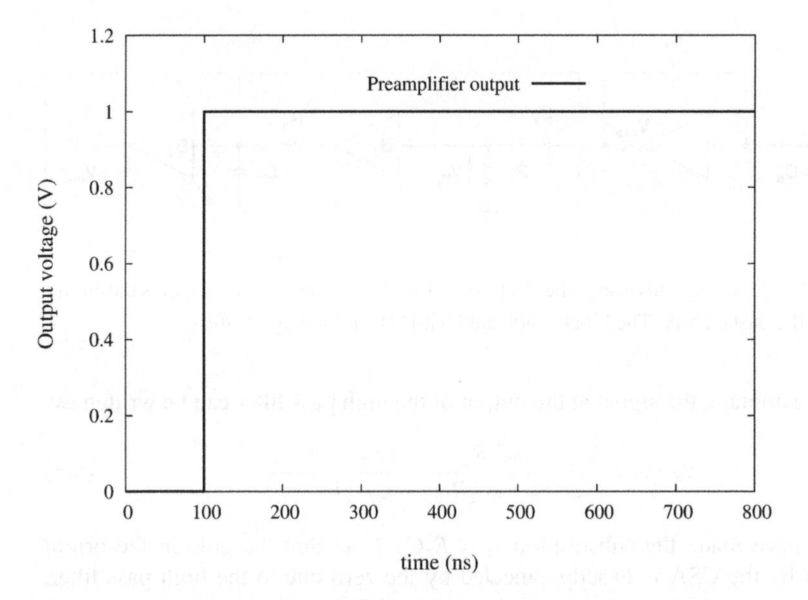

FIGURE 6.2 Response of an ideal CSA to a Dirac-δ input.

output is therefore a voltage step, which is then used to drive the pulse shaper. It is important to remember that the idealized input signal for the full front-end chain is a Dirac current pulse, whereas the idealized input stimulus for the shaper alone is a voltage step with zero rise time.

Note that in Fig. 6.2 the baseline is assumed to be 0 V. In a practical circuit, this is in general possible if the device is powered by a symmetrical power supply scheme, so that the ground is found half-way between the positive and negative rail. In many applications, single-rail solutions are preferred for simplicity, so the reference level (the baseline) will be above zero. The polarity of the output signal can be either positive (as in the figure) or negative with respect to the steady level. Since the actual baseline value and the signal polarity are irrelevant for our discussions in this chapter (unless otherwise specified), for simplicity we assume that the considered waveforms have a positive swing from a zero quiescent point.

6.2 THE CR-RC SHAPER

We now examine the simplest type of pulse shaper. Shown in the dashed box of Fig. 6.3, it consists of a high pass filter followed by a low pass one. The filters are isolated by a voltage buffer that decouples the two time constants, while the output buffer drives the load presented to the front-end amplifier by the following circuits. In this drawing, the feedback resistor of the CSA has been omitted because we neglect for the time being its contribution to the signal processing. We call V_{csa} the amplitude of the step delivered by the CSA in response to a Dirac-delta presented to its input. In

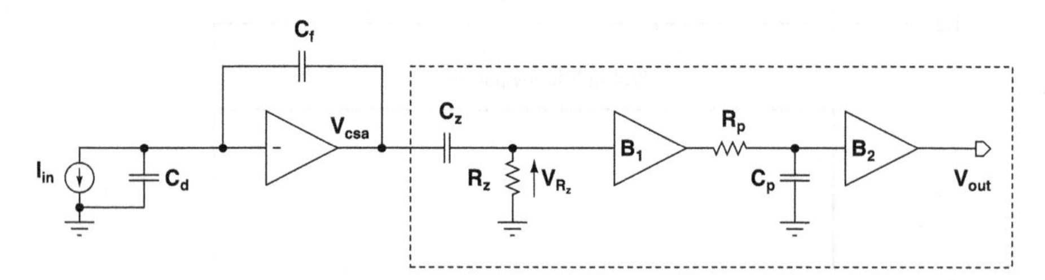

FIGURE 6.3 Charge sensitive amplifier followed by a CR-RC shaper. The shaper is the circuit included in the dashed box. The blocks indicated with "B" are voltage buffers.

the Laplace domain, the signal at the output of the high pass filter can be written as:

$$V_{R_z}(s) = \frac{V_{csa}}{s}\frac{sC_zR_z}{(1+sC_zR_z)} = \frac{Q_{in}}{C_f}\frac{\tau_z}{(1+s\tau_z)} \tag{6.5}$$

where we have made the substitution $\tau_z = R_zC_z$. Note that the pole in the origin introduced by the CSA is exactly canceled by the zero due to the high pass filter. Taking the inverse Laplace transform of equation (6.5) we obtain:

$$V_{R_z}(t) = \frac{Q_{in}}{C_f}e^{-t/\tau_z} \tag{6.6}$$

Fig. 6.4 shows the CSA and the low pass filter outputs. In the plot it has been assumed that $Q_{in}/C_f = 1$ and the signal has been shifted by 100 ns from the origin ($t \rightarrow t - 100$). We observe here that the output of the low pass filter reaches immediately the maximum value given by Q_{in}/C_f and then decays following an exponential law regulated by the time constant τ_z. The primary role of the high pass filter is hence to cut the constant or slowly varying components of the CSA output, creating a signal that goes back to the baseline before the next pulse arrives. In many applications, the charge released in the sensor is one of the primary information of interest and it can be extracted by capturing the peak height. However, as we see from Fig. 6.4, the signal at the high pass filter output starts fading away immediately after the peak value has been reached. This implies that the circuit capturing the peak must be very fast and accurate as a jitter in the sampling time can introduce a significant error in the measurement. It is therefore desirable that the signal has slower variations around its maximum. To smooth a waveform we need to filter-out its high frequency components, and this is one of the purposes of the low pass filter. The second, and even more important role, is to optimize the signal-to-noise ratio and it will become clear later when we will treat noise calculations. The low pass filter has a time constant $\tau_p = R_pC_p$. Since a high pass filter behaves as a derivator for low frequency signals and a low pass filter as an integrator for high frequency ones, τ_z and τ_p are also known as the derivation and the integration time constants. The transfer function of the full chain can thus be written as:

$$V_{out}(s) = \frac{Q_{in}}{Cf}\frac{\tau_z}{(1+s\tau_z)(1+s\tau_p)} \tag{6.7}$$

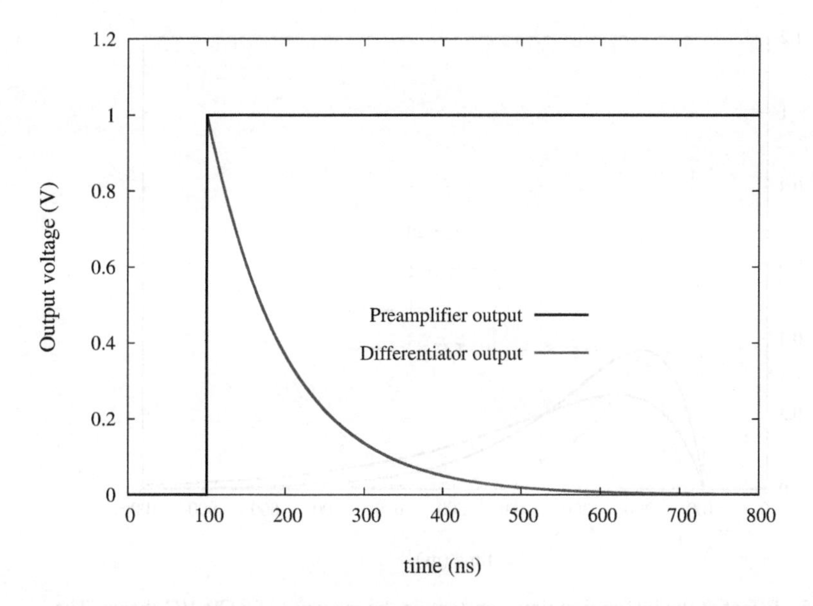

FIGURE 6.4 CSA and high pass filter outputs.

Note that the above equation can be derived simply by multiplying the transfer function of the two stages because the buffers isolate the two filters, avoiding that the two time constants interfere with each other. Calculating the inverse Laplace transform, we obtain the signal representation in the time domain:

$$V_{out}(t) = \frac{Q_{in}}{C_f} \frac{\tau_z}{\tau_z - \tau_p} \left(e^{-\frac{t}{\tau_z}} - e^{-\frac{t}{\tau_p}} \right) \tag{6.8}$$

which is valid for $\tau_z \neq \tau_p$. For the particular case in which the two time constants are equal we have, respectively:

$$V_{out}(s) = \frac{Q_{in}}{C_f} \frac{\tau}{(1 + s\tau)^2} \tag{6.9}$$

in the Laplace domain and:

$$V_{out}(t) = \frac{Q_{in}}{C_f} \left(\frac{t}{\tau} \right) e^{-\frac{t}{\tau}} \tag{6.10}$$

in the time domain. In the last two equations we have put $\tau_z = \tau_p = \tau$. These expressions are valid if the buffers have unity gain. If a constant gain $G = B_1 B_2$ (i.e. independent of frequency) is introduced along the chain, equations (6.7)-(6.10) must be multiplied by G as well.

To understand the optimal choice of the two time constants we can perform the following exercises. First, we keep the derivation time constant fixed and change the integration time constant. The result is shown in Fig. 6.5. We observe from the plot

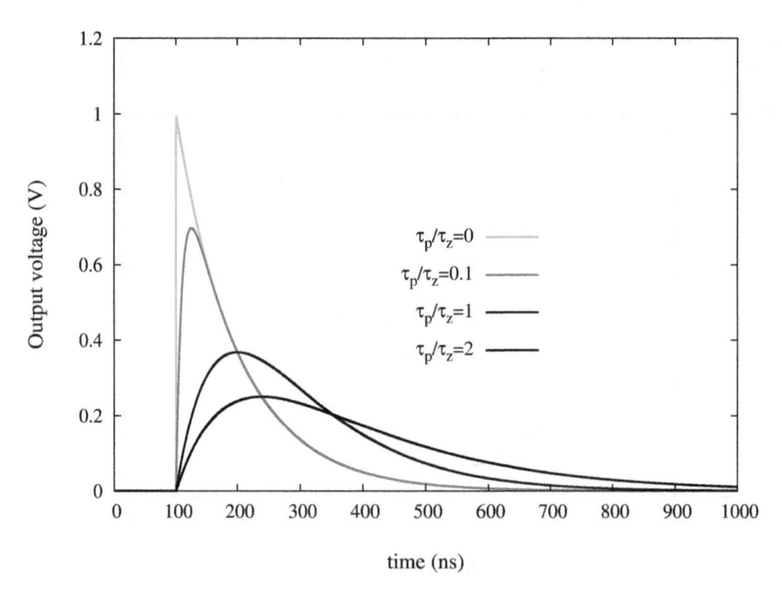

FIGURE 6.5 Effect of the integration time constant on the response of a CR-RC shaper. The derivation time constant is fixed to 100 ns.

that increasing τ_p makes the signal smoother and decreases its amplitude. However, when τ_p becomes greater than τ_z it starts also dominating the signal duration. In fact, for $\tau_p \gg \tau_z$, equation (6.8) can be approximated as:

$$V_{out}(t) \approx \frac{Q_{in}}{C_f}\frac{\tau_z}{\tau_p}e^{-\frac{t}{\tau_p}} \tag{6.11}$$

and τ_p dominates the signal decay. The complementary exercise consists in keeping the integration time constant fixed and sweeping τ_z. The result is shown in Fig. 6.6. We see here that when $\tau_z = \infty$ there is no derivation of the signal. In this case, in fact, equation (6.8) can be approximated as:

$$V_{out}(t) \approx \frac{Q_{in}}{C_f}\left(1 - e^{-\frac{t}{\tau_p}}\right) \tag{6.12}$$

The step at the preamplifier output is just low pass filtered by the integrator, and results in a smoother output signal with a 10% to 90% rise time of $2.2\tau_p$ (220 ns in our example). Decreasing the derivation time constant, the signal starts to be cut both in duration and in amplitude. When τ_z becomes smaller than τ_p, only the amplitude is further reduced, since it is the slower time constant that dominates the return to the baseline. These considerations suggest that equal time constants provide the best compromise as they maximize the signal amplitude for a *given pulse duration*.

Let's now consider the circuit with $\tau_z = \tau_p$ in more detail. Its response in the time domain is given in equation (6.10). By taking the derivative and equating it to zero

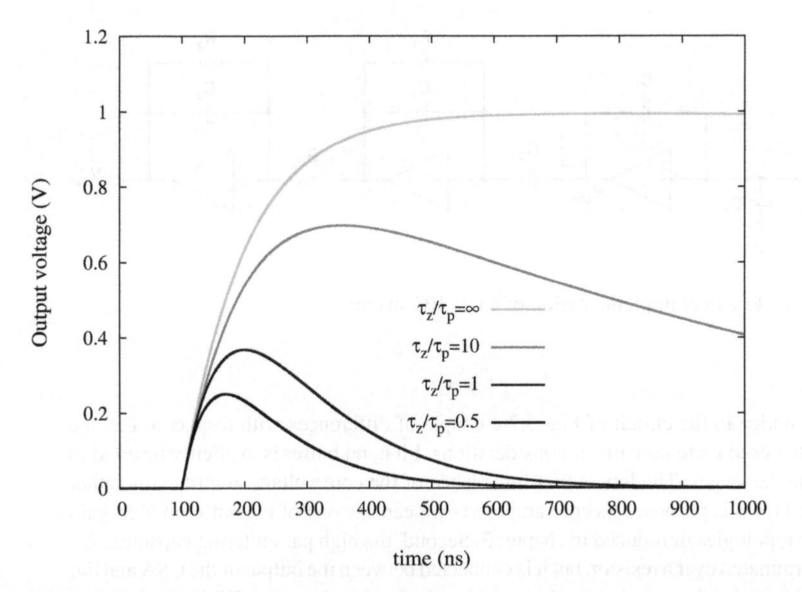

FIGURE 6.6 Effect of the derivation time constant on the response of a CR-RC shaper. The integration time constant is fixed to 100 ns.

we find the time at which the signal reaches its peak:

$$\frac{dV_{out}}{dt} = \frac{1}{\tau}e^{-\frac{t}{\tau}} - \frac{t}{\tau^2}e^{-\frac{t}{\tau}} = 0 \rightarrow t_{max} \equiv T_p = \tau \tag{6.13}$$

The time T_p is called the "peaking time" of the pulse shaper and corresponds to the time taken by the output signal to swing from the baseline to the maximum amplitude. For a CR-RC shaper, this is equal to the value of the integration and derivation time constants. The magnitude of the peak is obtained by putting in equation (6.10) $t = \tau$, which yields:

$$V_{out,max} = \frac{Q_{in}}{C_f}\frac{1}{e} \tag{6.14}$$

Note that the output voltage is reduced by a factor $1/e$ with respect to the maximum reached at the CSA output. If necessary, this gain loss can be recovered by inserting a suitable amplification along the chain.

Example 6.1: Practical design of a CR-RC shaper.

Consider the circuit shown in Fig. 6.7. Prove that its transfer function implements a CR-RC shaper. Assuming that C_f is 100 fF, size the other components for a peaking time of 50 ns. Suppose that the circuit has to process signals up to 10 fC and that the core amplifiers provide a maximum output swing of 1 V.

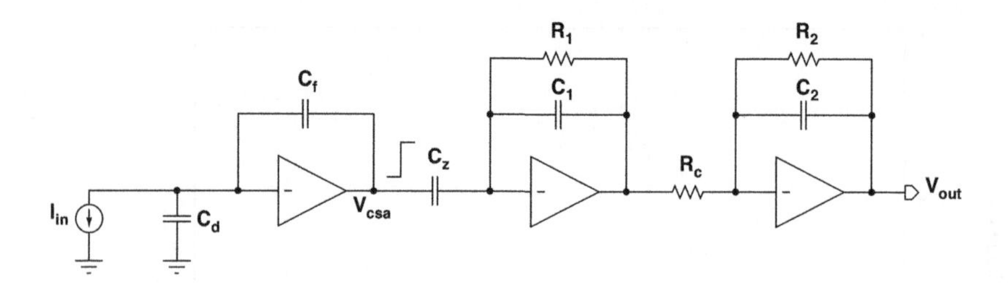

FIGURE 6.7 Practical implementation of a CR-RC shaper.

* * *

We can notice in the circuit of Fig. 6.7 a couple of differences with respect to the one of Fig. 6.3 used up to now in our considerations. First, no buffer is explicitly inserted to decouple the stages. The low output impedance of the core voltage amplifiers is rather exploited for this purpose. As core amplifiers we can use one of the buffered high-gain cascode topologies introduced in chapter 3. Second, the high pass filtering capacitor, C_z, is not terminated over a resistor, but it is connected between the output of the CSA and the input of the second stage, that can be considered as a virtual ground. While propagating from the input to the output, the signal undergoes several conversions between the current and the voltage domains, so it is instructive to build the transfer function step by step. First, the current signal coming from the sensor is converted to a voltage at the CSA output. The CSA output voltage is hence transformed back to a current flowing in C_z:

$$I_{C_z}(s) = \frac{I_{in}(s)}{\cancel{s}C_f}\cancel{s}C_z = I_{in}(s)\frac{C_z}{C_f} \tag{6.15}$$

Note that before being presented to the second stage, the detector signal is scaled by the ratio C_z/C_f without altering its shape. To have a current gain in the first stage we thus need C_z bigger than C_f. The second stage is a transimpedance amplifier, with a complex impedance Z_1 placed in the feedback path of the core amplifier:

$$Z_1 = R_1 // \frac{1}{sC_1} = \frac{R_1}{(1+sC_1R_1)} \tag{6.16}$$

At the output of the second stage the signal is converted to a voltage:

$$V_{out2}(s) = I_{in}\frac{C_z}{C_f}\frac{R_1}{(1+sC_1R_1)} \tag{6.17}$$

Finally, V_{out2} is again converted to a current by resistor R_c and transformed to a voltage by the feedback impedance of the last stage:

$$V_{out}(s) = \frac{V_{out2}}{R_c}\frac{R_2}{(1+sC_2R_2)} \tag{6.18}$$

Inserting into the above equation the expression of V_{out2} we obtain the full transfer function:

$$V_{out}(s) = I_{in}\frac{C_z}{C_f}\frac{R_1}{(1+s\tau_1)}\frac{1}{R_c}\frac{R_2}{(1+s\tau_2)} \tag{6.19}$$

To implement a CR-RC shaper we must require that the two time constants are equal, i.e. $\tau_1 = \tau_2 = \tau$. If we further suppose that the input signal has a δ-like shape and contains a charge Q_{in}, we can write:

$$V_{out}(s) = Q_{in} \frac{C_z}{C_f} R_1 \frac{R_2}{R_c} \frac{1}{(1+s\tau)^2} \tag{6.20}$$

Taking the inverse Laplace transform, we have:

$$V_{out}(t) = Q_{in} \frac{C_z}{C_f} R_1 \frac{R_2}{R_c} \frac{1}{\tau} \left(\frac{t}{\tau}\right) e^{-\frac{t}{\tau}} \tag{6.21}$$

Since $\tau = R_1 C_1 = R_2 C_2$ we can insert either expression of τ into the above equation to further simplify it. Using $\tau = R_1 C_1$ we get:

$$V_{out}(t) = \frac{Q_{in}}{C_f} \frac{C_z}{C_1} \frac{R_2}{R_c} \left(\frac{t}{\tau}\right) e^{-\frac{t}{\tau}} \tag{6.22}$$

In a CR-RC shaper the value of the time constant defines also the peaking time T_p, which is requested to be 50 ns. The choice of R_1 and C_1 depends on the type of passive components available in the process. Assuming that high value polysilicon resistors are available, we can select $R_1 = 100$ kΩ and $C_1 = 0.5$ pF. In principle, R_2 and C_2 might have different values as long as their product is 50 ns. However, practical reasons suggest to minimize the number of different components inside a circuit, so we may put $R_2 = R_1$ and $C_2 = C_1$. The feedback capacitor C_f is assigned and results in a gain of 10 mV/fC at the CSA output. We have to accommodate a maximum input signal 10 fC within the swing of 1 V allowed by the core amplifiers. The total gain must then be 100 mV/fC. The peak amplitude is given by:

$$V_{out,peak} = \frac{Q_{in}}{C_f} \frac{C_z}{C_1} \frac{R_2}{R_c} \frac{1}{e} \tag{6.23}$$

Therefore, we need to have:

$$\frac{C_z}{C_1} \frac{R_2}{R_c} = 10e \approx 27.18 \tag{6.24}$$

Now we have to choose how to distribute the gain between the two stages. In general, at the output of a given stage, the signal should be as large as permitted by the dynamic range of the core amplifier. In this way, the signal is less sensitive to the noise and other imperfections introduced by the following stage. In the time domain, the output of the second stage is:

$$V_{out2}(t) = \frac{Q_{in}}{C_f} \frac{C_z}{C_1} e^{-\frac{t}{\tau}} \tag{6.25}$$

To exploit as much as possible the output dynamic range of the core amplifiers, we can choose $C_z/C_1 = 10$, so that we have already a gain of 100 mV/fC from the CSA input to second stage output. If the last stage has unity gain, the peak amplitude would be divided by e. Therefore, by putting the gain of the final stage equal to 2.718, we recover this factor and we fully exploit also the output dynamic range of the last amplifier. Since we have already selected $C_1 = 0.5$ pF and $R_2 = 100$ kΩ, we have that $C_z = 5$ pF and $R_c = 36.8$ kΩ. The shaper output for a signal of 10 fC is shown in Fig. 6.8. The input signal is injected at t=100 ns. As expected, the output attains its maximum after 50 ns, reaching the desired peak value of 1 V.

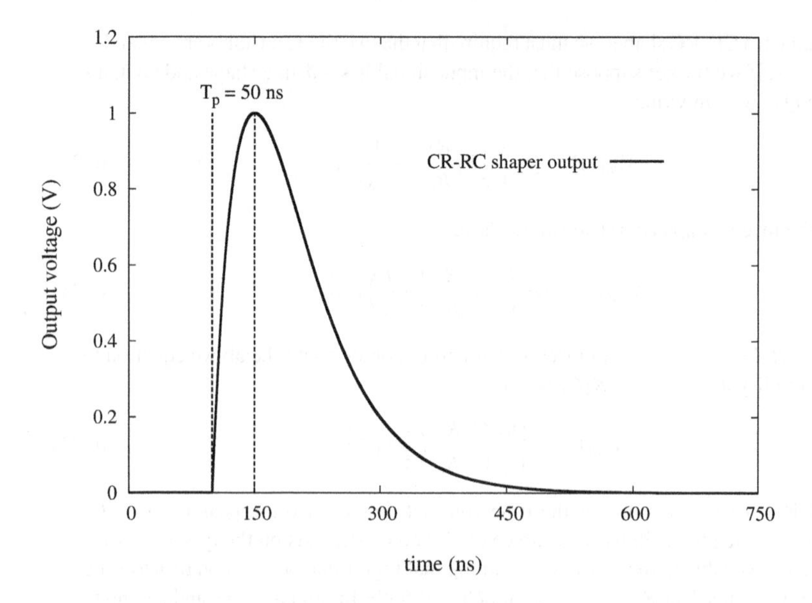

FIGURE 6.8 Output of the shaper discussed in example 6.1 for an input charge of 10 fC.

Remember now that the inverting voltage amplifier is actually obtained by converting the voltage to a current which is then fed to a transimpedance stage. Another way of looking at the processing chain of Fig. 6.7 is hence to interpret it as charge sensitive amplifier followed by two inverting voltage amplifiers, one with gain $-Z_1/C_1$ and the other with gain $-Z_2/R_c$. This allows us to grasp immediately the signal polarity at the outputs of the different stages. In fact, due to the double inversion, the signal at the shaper output has the same polarity of the one generated by the CSA. If the detector sinks its current from the input node, as shown in Fig. 6.7, the CSA output moves upwards starting from its baseline, while the output of the second stage has a negative-going swing. If we make a purely linear model, the second stage output will go below zero. In a practical circuit this can be achieved by using a dual power supply. Suppose, for instance, that we implement our front-end in a 0.25 μm CMOS technology, that can withstand a maximum power supply of 2.5 V. Using $V_{DD} = +1.25$ V and $V_{SS} = -1.25$ V, we can keep the output quiescent points of all building blocks at zero. In this case, the second stage output swings from 0 to -1 V and third stage output from 0 to $+1$ V. Note that here we do not make full use of the maximum possible output dynamic range of each stage, that in a well designed amplifier can almost be rail-to-rail. Consider now the case in which the same circuit must be implemented in a 0.13 μm process, which tolerates a maximum power supply of 1.2 V. Now, to achieve the desired voltage swing of 1 V we need to exploit much more efficiently the maximum dynamic range offered by the technology. This means, in practice, that the baseline of the second stage must be positioned close to the positive supply rail.

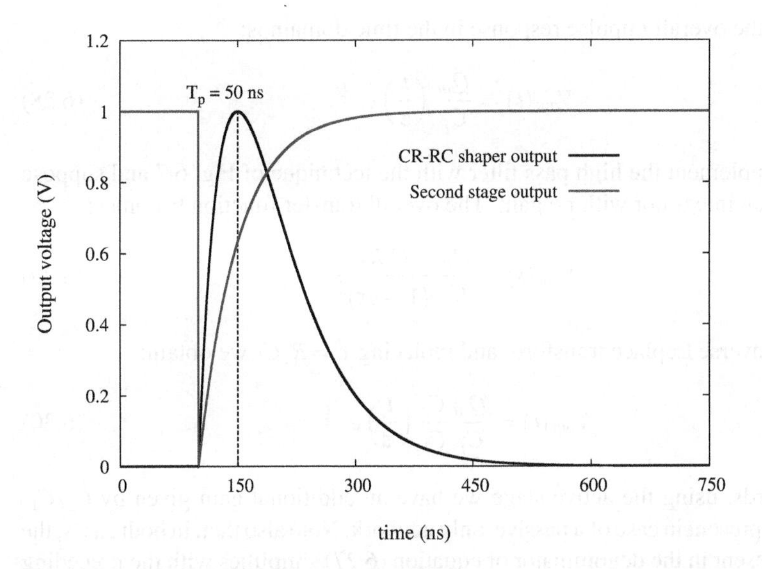

FIGURE 6.9 Output of the second and of the last stage of the circuit of Fig. 6.7.

The third stage output moves again in the positive direction, and its quiescent point must be set close to the negative rail. In this case, the connection between the two stages is not trivial, as a big differences between the output DC point of one stage and the input DC level of the next one can not be tolerated by the circuit of Fig. 6.7. In fact, if no other actions are taken, each stage in the chain will have equal output and input DC levels, hence the output of the second stage would be close to 1 V and the input of the third one close to zero. This determines a DC current given by 1 V/36 k$\Omega = 27.8\ \mu$A, that, entering in R_2 would determine in it a voltage drop of 2.78 V. In practice, the output of the last stage would be clipped at zero, but the core amplifier would be deeply saturated, losing its high gain performance. In designing a practical working circuit, we therefore need to master appropriately the DC levels and the coupling between the different building blocks, which requires additional circuitry. We will come back to this point in section 6.6. Fig. 6.9 shows on the same plot the output of the second and third stages of the chain of Fig. 6.7. Observe that the second stage output starts from 1 V and swings downward and that both signals have the same peak amplitude, as discussed in the example. Finally, we can notice another difference in the implementation of the shapers in Fig. 6.3 and in Fig. 6.7. The transfer function of the circuit of Fig. 6.3 in fact is:

$$V_{out}(s) = \frac{Q_{in}}{C_f} \frac{\tau}{(1+s\tau)^2} \tag{6.26}$$

The inverse Laplace transform of the function $1/(1+s\tau)^2$ is:

$$\frac{1}{\tau}\left(\frac{t}{\tau}\right)e^{-\frac{t}{\tau}} \tag{6.27}$$

As a result, the overall impulse response in the time domain is:

$$V_{out}(t) = \frac{Q_{in}}{C_f}\left(\frac{t}{\tau}\right)e^{-\frac{t}{\tau}} \tag{6.28}$$

Let's now implement the high pass filter with the technique of Fig. 6.7 and suppose that we use an integrator with no gain. The overall transfer function becomes:

$$V_{out}(s) = \frac{Q_{in}}{C_f}\frac{C_z R_1}{(1+s\tau)^2} \tag{6.29}$$

Taking the inverse Laplace transform and replacing $\tau = R_1 C_1$ we obtain:

$$V_{out}(t) = \frac{Q_{in}}{C_f}\frac{C_z}{C_1}\left(\frac{t}{\tau}\right)e^{-\frac{t}{\tau}} \tag{6.30}$$

In other words, using the active stage we have an additional gain given by C_z/C_1, which is not present in case of a passive-only network. Note also that, in both cases, the "extra" τ present in the denominator of equation (6.27) simplifies with the preceding terms to yield either 1 or C_z/C_1.

6.3 CR-RC^N SHAPERS

We study now how the transfer function of the shaper is modified by adding further integrators. The number of integrators define the order of the shaper. For the time being, we consider the case in which the integrators are implemented with buffered low pass filters with unity gain. The considerations made above about the matching of the derivation and integration time constants are still valid, so we suppose that all the filters used in the circuit have the same time constant τ. The transfer function of a generic CR-RC^N shaper contains $n+1$ poles, n introduced by the integrators and one by the differentiator and in the Laplace domain it can be written as:

$$V_{out}(s) = \frac{Q_{in}}{C_f}\frac{\tau}{(1+s\tau)^{n+1}} \tag{6.31}$$

in case we have a high pass filter implemented as in Fig. 6.3 or:

$$V_{out}(s) = \frac{Q_{in}}{C_f}\frac{C_z R_1}{(1+s\tau)^{n+1}} \tag{6.32}$$

if we use the approach of Fig. 6.7. The response in the time domain is respectively given by:

$$V_{out}(t) = \frac{Q_{in}}{C_f}\frac{1}{n!}\left(\frac{t}{\tau}\right)^n e^{-\frac{t}{\tau}} \tag{6.33}$$

and

$$V_{out}(t) = \frac{Q_{in}}{C_f}\frac{C_z}{C_1}\frac{1}{n!}\left(\frac{t}{\tau}\right)^n e^{-\frac{t}{\tau}} \tag{6.34}$$

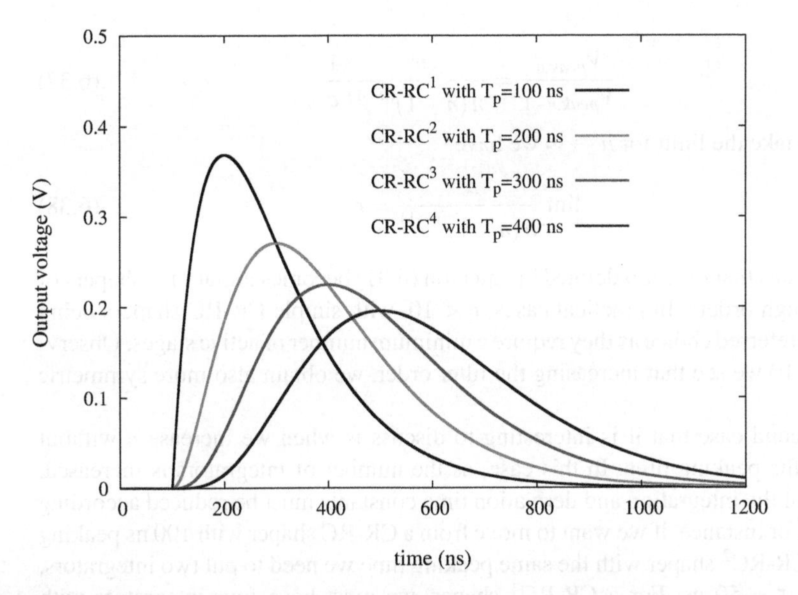

FIGURE 6.10 Outputs of CR-RC^n shapers of different orders with the same derivation and integration time constants τ of 100 ns. The peaking time for each waveform is indicated on the graph.

We can now take the derivative of $V_{out}(t)$ and equate it to zero to find the peaking time. Using, for instance, $V_{out}(t)$ as defined by equation (6.33) we have:

$$\frac{dV_{out}}{dt} = \frac{Q_{in}}{C_f}\frac{1}{n!}\left[\frac{n}{\tau}\left(\frac{t}{\tau}\right)^{n-1} e^{-\frac{t}{\tau}} - \left(\frac{t}{\tau}\right)^{n}\frac{1}{\tau}e^{-\frac{t}{\tau}}\right] = 0 \rightarrow T_p = n\tau \tag{6.35}$$

which shows that in a CR-RC^n shaper the peaking time is given by the integration time constant multiplied by the number of integrators. The peak amplitude is:

$$V_{out,max} = \frac{Q_{in}}{C_f} G \frac{n^n}{n!} e^{-n} \tag{6.36}$$

where G indicates the overall additional gain that might be inserted in the chain. We now consider two cases. The first one is when we increase the number of integration stages without modifying the integration and derivation time constants, and it is shown in Fig. 6.10. To obtain these plots, we have used the expression of V_{out} given in (6.33), assuming that the voltage step provided by the CSA is unitary, i.e $Q_{in}/C_f = 1$. We see that increasing the order of the shaper, the peaking time increases according to equation (6.35), while the peak amplitude follows the rule given in (6.36). Observe that if no additional gain is provided, the signal is more and more attenuated as the order of the shaper is increased. However, for bigger values of n, the amount of attenuation is less pronounced when we move from a given order to the next. In fact, the relationships between the peak amplitudes of two shapers of generic order n and

$n-1$ is:

$$\frac{V_{peak,n}}{V_{peak,n-1}} = \frac{n^n}{n\,(n-1)^{(n-1)}}\frac{1}{e} \tag{6.37}$$

If we now take the limit for $n \to \infty$ we have:

$$\lim_{n\to\infty} \frac{n^n}{n\,(n-1)^{(n-1)}} = e \tag{6.38}$$

which implies that the ratio defined in equation (6.37) becomes unitary for shapers of arbitrary high orders. In practical cases, $n < 10$, with simple CR-RC shapers being often the preferred choice as they require a minimum number of active stages. Observing Fig. 6.10 we see that increasing the filter order, we obtain also more symmetric signals.

The second case that it is interesting to discuss is when we increase n without changing the peaking time. In this case, as the number of integrators is increased, the value of the integration and derivation time constants must be reduced according to (6.35). For instance, if we want to move from a CR-RC shaper with 100 ns peaking time to a CR-RC^2 shaper with the same peaking time we need to put two integrators, each with $\tau = 50$ ns. For a CR-RC^4 shaper, we must have four integrators with $\tau = 25$ ns. This is shown in Fig. 6.11 which reports the impulse response of the three shapers. To allow a better comparison between the different pulse shapes, the amplitudes have been normalized to 1 V. Observe that increasing the shaper order for the same peaking time we obtain a faster return to zero. This is because all time

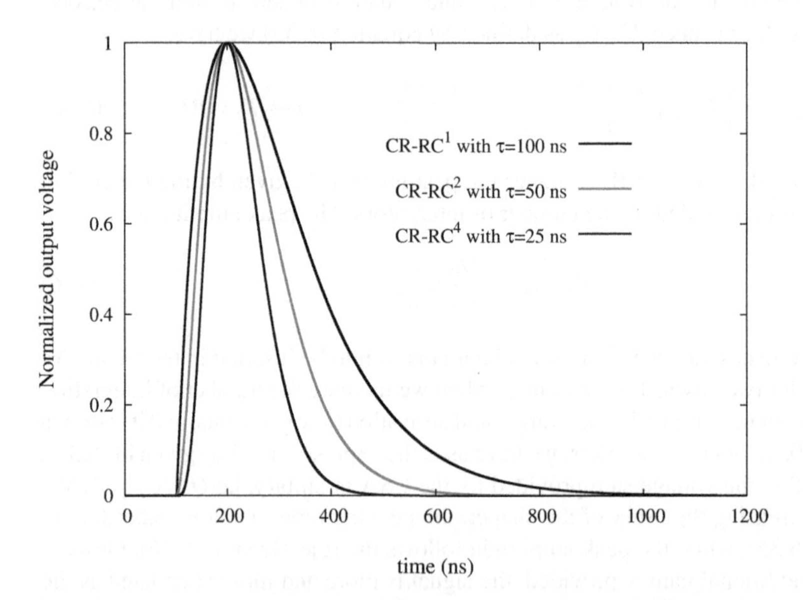

FIGURE 6.11 Outputs of CR-RC^n shapers of different orders with the same peaking time. The time constants of each integrator are indicated in the plot. Note that τ is equal to the peaking time T_p only for the CR-RC filter.

constants, including the derivation one, have been shortened. We thus conclude that, for a given peaking time, higher order shapers allow for a faster return to the baseline and are a better choice for high rate applications.

Example 6.2: Design of a CR-RC^4 shaper.

Consider the shaper with one differentiator and four integrators reported in Fig. 6.12. Write the transfer function and size the passive components to achieve a gain of 100 mV/fC and a peaking time of 50 ns. Suppose that the core amplifiers allow for a maximum output swing of 1 V and that C_f is 100 fF.

* * *

We have used on purpose here the same parameters of example 6.1 to allow a direct comparison between the two topologies. First, we write the output voltage in the Laplace domain, which reads:

$$V_{out}(s) = \frac{I_{in}(s)}{C_f} C_z \frac{R_1}{R_{c1}} \frac{R_2}{R_{c2}} \frac{R_3}{R_{c3}} \frac{R_4}{R_{c4}} \frac{R_5}{(1+s\tau)^5} \tag{6.39}$$

We can now take the inverse Laplace transform to find the response in the time domain. We assume that the input signal is a δ-like pulse, conveying a total charge Q_{in}.

$$V_{out}(t) = \frac{Q_{in}}{C_f} C_z \frac{R_1}{R_{c1}} \frac{R_2}{R_{c2}} \frac{R_3}{R_{c3}} \frac{R_4}{R_{c4}} R_5 \frac{1}{24} \frac{1}{\tau} \left(\frac{t}{\tau}\right)^4 e^{-\frac{t}{\tau}} \tag{6.40}$$

Inserting $\tau = R_1 C_1$ in the above equation we arrive at:

$$V_{out}(t) = \frac{Q_{in}}{C_f} \frac{C_z}{C_1} \frac{R_2}{R_{c1}} \frac{R_3}{R_{c2}} \frac{R_4}{R_{c3}} \frac{R_5}{R_{c4}} \frac{1}{24} \left(\frac{t}{\tau}\right)^4 e^{-\frac{t}{\tau}} \tag{6.41}$$

To get a peaking time of 50 ns, the time constant τ of each integrator must be equal to 12.5 ns. If we fix as before the value of the resistors to 100 kΩ the value of capacitors C_1-C_5 are 125 fF. The CSA gives a gain of 10 mV/fC. If we have no gain in the shaper, its output peak amplitude, reached at $t = 4\tau$, would be 0.195 times the one delivered by the CSA or 1.95 mV/fC. To reach 100 mV/fC we therefore need a total gain of 51.28. If we choose $C_z/C_1 = 10$, we need a total gain of 5.128 from the four integrators. As before, we can optimize the gain of each stage in such a way that the output voltage

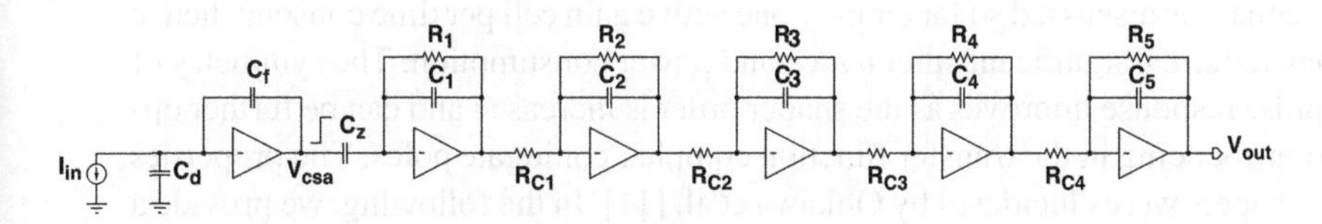

FIGURE 6.12 Front-end with CR-RC^4 shaper.

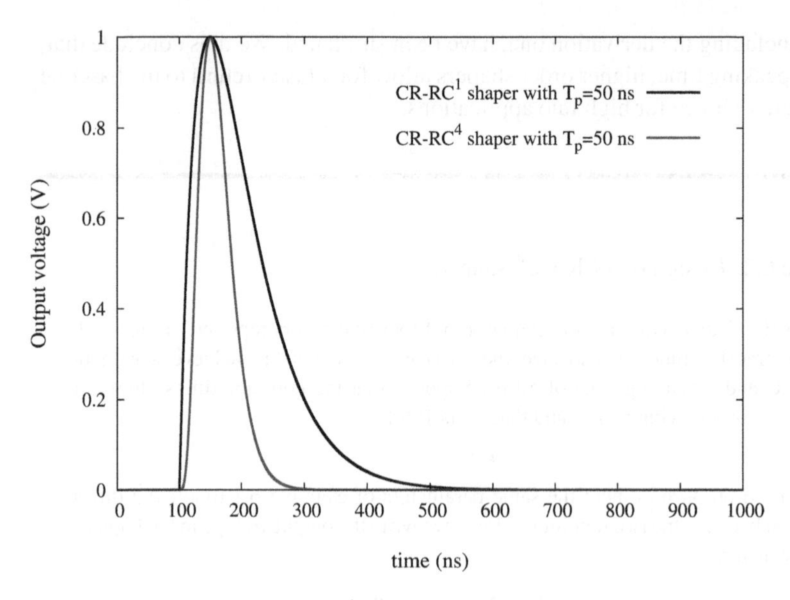

FIGURE 6.13 Response of the CR-RC^4 shaper compared to the one of a CR-RC network with the same peaking time.

spans the full dynamic range. Alternatively, we can equally distribute the gain across the stages. In this case, the gain per stage is given by $\sqrt[4]{5.128} \approx 1.505$. This solution has the disadvantage that we may not make an optimal use of the dynamic range of all stages, but it minimizes the different types of integrator design. With this choice, the peak voltage attained for the expected maximum input signal of 10 fC would respectively be 0.1 V after the CSA and then, moving towards the output, 1 V, 0.553 V, 0.613 V, 0.763 V and, finally, 1 V. Fig. 6.13 shows the shaper response for the maximum input signal of 10 fC. For comparison, the output of the circuit discussed in example 6.1 is also drawn. Observe the much faster return to the baseline of the higher order circuit.

6.4 SHAPERS WITH COMPLEX CONJUGATE POLES

Due to the similarity between their impulse response and a Gaussian waveform, CR-CR^n filters are commonly known as "Semi-Gaussian shapers". The straightforward implementations discussed so far employ one active gain cell per time constant, hence they may require a significant silicon area and power consumption. The symmetry of the impulse response improves as the shaper order is increased and can be further optimized introducing in the transfer function complex conjugate poles. The properties of such shapers were elucidated by Ohkawa et al. [11]. In the following, we provide a summary of the original Ohkawa's arguments. The basic idea is to find a network that can be implemented with physical components and whose impulse response yields

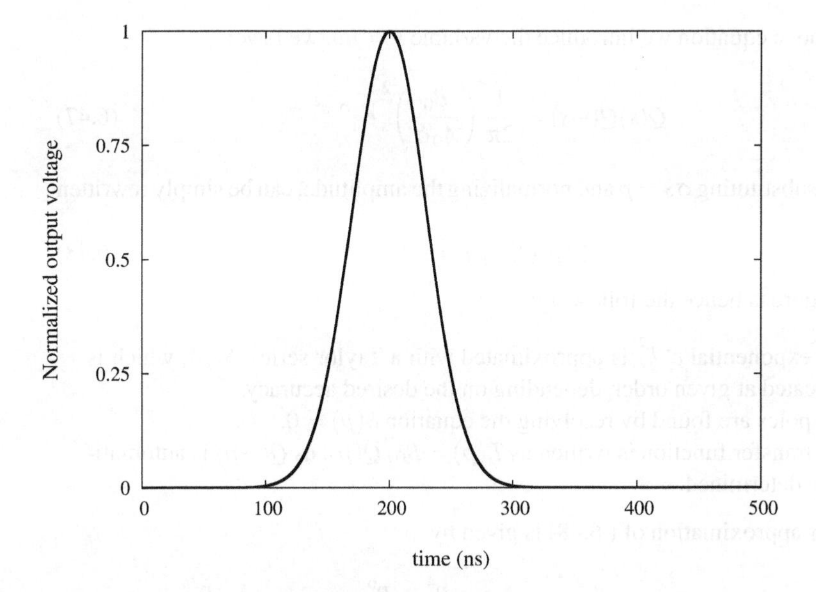

FIGURE 6.14 Theoretical Gaussian pulse shape with $\sigma = 30$ ns.

the best approximation to a Gaussian shape:

$$f(t) = A_0 e^{-\frac{1}{2}\frac{t^2}{\sigma^2}} \tag{6.42}$$

where σ is the standard deviation of the Gaussian. Fig. 6.14 shows this idealized impulse response, obtained assuming a σ of 30 ns. Note the perfect symmetry between the rise and fall times of the signal. As a first step, we switch to the frequency domain by considering the Fourier transform of equation (6.42), which reads:

$$T(j\omega) = A_0\sqrt{2\pi}\sigma e^{-\frac{1}{2}\sigma^2\omega^2} \tag{6.43}$$

We now search for a network whose transfer function contains only poles and exhibits an impulse response approximating at best our ideal Gaussian pulse. Such a transfer function can be written as:

$$T(j\omega) = \frac{B_0}{Q(j\omega)} \tag{6.44}$$

Therefore, the problem reduces to finding the best expression for $Q(j\omega)$. We can then write:

$$\frac{1}{T(j\omega)} = \frac{Q(j\omega)}{B_0} = \frac{1}{A_0\sqrt{2\pi}\sigma} e^{\frac{1}{2}\sigma^2\omega^2} \rightarrow Q(s) = \frac{B_0}{A_0\sqrt{2\pi}\sigma} e^{\frac{1}{2}\sigma^2\omega^2} \tag{6.45}$$

We can now take the absolute values of both quantities:

$$|Q(j\omega)|^2 = Q(j\omega)Q(-j\omega) = \frac{1}{2\pi}\left(\frac{B_0}{A_0\sigma}\right)^2 e^{\sigma^2\omega^2} \tag{6.46}$$

If in the above equation we introduce the variable $s = j\omega$, we have:

$$Q(s)Q(-s) = \frac{1}{2\pi}\left(\frac{B_0}{A_0\sigma}\right)^2 e^{-\sigma^2 s^2} \tag{6.47}$$

which, by substituting $\sigma s = p$ and normalizing the amplitude, can be simply rewritten as:

$$Q(p)Q(-p) = e^{-p^2} \tag{6.48}$$

The procedure is hence the following:

1. The exponential e^{-p^2} is approximated with a Taylor series, $S(p)$, which is truncated at given order, depending on the desired accuracy.
2. The poles are found by resolving the equation $S(p) = 0$.
3. The transfer function is written as $T(p) = B_0/Q(p)$, as $Q(-p)$ is automatically determined.

The Taylor approximation of (6.48) is given by:

$$e^{-p^2} = S(p) = Q(p)Q(-p) = 1 - p^2 + \frac{p^4}{2!} - \frac{p^6}{3!} + \cdots + (-1)^n \frac{p^{2n}}{n!} \tag{6.49}$$

The interesting cases are for higher orders ($n \geq 3$), for which the roots of the above polynomial are found numerically. It is however instructive to consider the two lowest orders to illustrate the method without cumbersome mathematical computations. For $n = 1$ we have:

$$S(p) = Q(p)Q(-p) = 1 - p^2 = (1+p)(1-p) \tag{6.50}$$

Therefore, $Q(p) = (1+p)$ and our transfer function is given by:

$$T(p) = \frac{B_0}{(1+p)} \tag{6.51}$$

which is a first order low pass function with only one real pole. If we now stop to $n = 2$, we have:

$$Q(p)Q(-p) = 1 - p^2 + \frac{p^4}{2!} \tag{6.52}$$

As the reader can verify by direct inspection, this polynomial can be factorized in the following way:

$$S(p) = \frac{1}{\sqrt{2}}\left(\sqrt{2} + \sqrt{2+2\sqrt{2}}p + p^2\right)\frac{1}{\sqrt{2}}\left(\sqrt{2} - \sqrt{2+2\sqrt{2}}p + p^2\right) \tag{6.53}$$

The function $Q(p)$ is hence:

$$Q(p) = \frac{1}{\sqrt{2}}\left(\sqrt{2} + \sqrt{2+2\sqrt{2}}p + p^2\right) \tag{6.54}$$

Calculating the roots of the above polynomial, we get:

$$p_{1,2} = -\frac{\sqrt{2+2\sqrt{2}}}{2} \pm \frac{1}{2}\sqrt{2-2\sqrt{2}} \tag{6.55}$$

Since the discriminant of the equation is negative, the solutions are complex conjugate and can be rewritten as:

$$p_{1,2} = -\frac{\sqrt{2+2\sqrt{2}}}{2} \pm j\frac{\sqrt{2}}{2}\sqrt{\sqrt{2}-1} \tag{6.56}$$

Note that what we have found are the zeroes of the polynomial $Q(p)$ that will become the poles of the overall transfer function $B_0/Q(p)$. From the above discussion, we see that the quest of the best approximation to a Gaussian response leads to the introduction of complex conjugate poles in the filter. Observe also that complex conjugate poles come in pairs, so when we have a system with an even number of poles (i.e. shapers with order 2, 4, etc.) they can all be complex conjugate, whereas if the number of poles is odd (n=3, 5, etc.) the extra pole must be real. The calculations of the roots of $Q(p)$ for $n > 2$ is carried out with numerical methods. Table 6.1 gives the pole positions from $n = 2$ to $n = 7$. The real part of the pole is indicated with R_i and the imaginary one with I_i. R_0 indicates the position of the real pole for odd-order shapers. The meaning and use of the table will be clarified later, when we will discuss practical implementations. Looking at table 6.1 we see that if we increase the filter order we need to recalculate the positions of *all* the poles.

The term e^{-p^2} can also be examined from a different perspective. In fact, it can be proven that the following relationship holds:

$$e^{-p^2} = \lim_{n\to\infty}\left(1 - \frac{p^2}{n}\right)^n \tag{6.57}$$

TABLE 6.1

Pole positions for approximating the Gaussian filter. The number n represents the total number of poles.

	n=2	n=3	n=4	n=5	n=6	n=7
R_0		1.2633		1.4767		1.6610
R_1	1.0985	1.1491	1.3554	1.4167	1.5601	1.6230
I_1	0.4551	0.7864	0.3278	0.5979	0.2687	0.5008
R_2			1.1810	1.2037	1.4614	1.4950
I_2			1.0604	1.2995	0.8330	1.0455
R_3					1.2207	1.2344
I_3					1.5145	1.7113

Therefore, we can write for $Q(p)$ the following expression:

$$Q(p)Q(-p) = \left(1 - \frac{p^2}{n}\right)^n \rightarrow Q(p) = \left(1 + \frac{p}{\sqrt{n}}\right)^n \tag{6.58}$$

Let's now take the case $n = 2$ and remember that the variable p is related to s by the relationship: $p = \sigma s$. If we make the choice $\sigma = \sqrt{2}RC$, we get:

$$Q(s) = (1 + sRC)^2 \rightarrow T(s) = \frac{B_0}{(1 + sRC)^2} \tag{6.59}$$

which is the transfer function of a CR-RC shaper. The above considerations suggest that a Gaussian filter can also be approximated by cascading RC filters with real poles. However, equation (6.57) also shows that an infinite number of stages is necessary to implement a true Gaussian shaper. For the same order, transfer functions with complex conjugate poles thus yield a better approximation. Consider now a second order low pass filter with unity gain at low frequencies. Its transfer function can be written as:

$$T(s) = \frac{c}{as^2 + bs + c} \tag{6.60}$$

We see from the above relationship that for $s \rightarrow 0$ the gain becomes one. The roots of the denominator are given by:

$$p_{1,2} = -\frac{b}{2a} \pm \frac{1}{2a}\sqrt{b^2 - 4ac} \tag{6.61}$$

If we want complex conjugate poles, we need to require that $b^2 < 4ac$ and we can rewrite the roots as:

$$p_{1,2} = -\frac{b}{2a} \pm j\frac{1}{2a}\sqrt{4ac - b^2} \tag{6.62}$$

It is then straightforward to verify that equation 6.60 can be rewritten as:

$$T(s) = \frac{R^2 + I^2}{\left[(s + R)^2 + I^2\right]} \tag{6.63}$$

where R is the real part of the pole (taken without the "-" sign) and I is the imaginary part. A filter with two complex conjugate poles can be realized by applying negative feedback to a network containing two real and distinct poles. Higher order filters can then be obtained by cascading an adequate number of these fundamental units. If we do so, we can simply write the total transfer function of the filter as the product of the ones of the individual cells, given in equation 6.63. Therefore, the transfer function of a Gaussian shaper with an even number of poles is written as:

$$T(s) = \frac{\prod_{k=1}^{n/2}\left(R_k^2 + I_k^2\right)}{\prod_{k=1}^{n/2}\left[(s + R_k)^2 + I_k^2\right]} \tag{6.64}$$

For odd transfer functions containing also a real pole we have:

$$T(s) = \frac{p_0}{(s+p_0)} \frac{\prod_{k=1}^{(n+1)/2} (R_k^2 + I_k^2)}{\prod_{k=1}^{(n+1)/2} \left[(s+R_k)^2 + I_k^2 \right]} \tag{6.65}$$

with $p_0 = 1/\tau_0$, where τ_0 is the time constant associated to the real pole. Note that in equation (6.64), the sum stops to $n/2$, if n is the filter order. In fact, each stage contributes *two* complex conjugate poles, so, for example, two stages are needed to get a filter with four complex conjugate poles. A similar consideration applies to equation (6.65), where the presence of the extra real pole is taken into account in the summation indexes.

Let's now recall here a few concepts that were anticipated in chapter 4. Discussing circuit stability, we have seen in fact that it can be convenient to write a second order transfer function in the normalized form:

$$T_N(s) = \frac{\omega_0^2}{s^2 + \frac{\omega_0}{Q} + \omega_0^2} \tag{6.66}$$

where ω_0 is the angular frequency associated to the poles and Q is the "quality factor". The poles of equation (6.66) therefore are:

$$p_{1,2} = -\frac{\omega_0}{2Q} \pm \frac{1}{2}\sqrt{\left(\frac{\omega_0}{Q}\right)^2 - 4\omega_0^2} \rightarrow -\frac{\omega_0}{2Q} \pm \frac{\omega_0}{2Q}\sqrt{1-4Q^2} \tag{6.67}$$

The condition that the poles are complex conjugate is thus satisfied if:

$$1 - 4Q^2 < 0 \rightarrow Q > \frac{1}{2} \tag{6.68}$$

The poles can be also be written as:

$$p_{1,2} = -\frac{\omega_0}{2Q} \pm j\frac{\omega_0}{2Q}\sqrt{4Q^2 - 1} \tag{6.69}$$

The above equation makes apparent that the ratio between the imaginary and the real part depends only on Q, which can be in turn calculated once I and R are known

$$\frac{I}{R} = \sqrt{4Q^2 - 1} \rightarrow Q = \frac{1}{2}\sqrt{\left(\frac{I}{R}\right)^2 + 1} \tag{6.70}$$

From (6.69) we see that the frequency ω_0 affects both the real and the imaginary part in the same way and can be regarded as a "scale factor" determining the actual position of the poles on the frequency axis and, with that, the effective speed of the system. The quality factor Q, on the other hand, gives the relative magnitude of the real and the imaginary part, if any, of the poles and hence the particular shape of the impulse response. Observe also that the following relationship exists between ω_0, R, and I:

$$\omega_0 = \sqrt{\left(\frac{\omega_0}{2Q}\right)^2 + \left(\frac{\omega_0}{2Q}\sqrt{4Q^2-1}\right)^2} = \sqrt{R^2+I^2} \tag{6.71}$$

Table 6.1 can hence be interpreted as follows: the table gives the positions of the poles that, for a given order, results in the best approximation to a Gaussian in $p = 0$ (Taylor's series). The absolute value of the poles actually depends on the frequency of interest, that, in general, is quite far from the origin. Therefore, the value of R *and* I will change from case to case and will be in general very different from the ones reported in table 6.1, but their *ratio*, in case we seek for the best approximation to a Gaussian response, must always match the ones of the poles given in the table. The next example illustrates how to use the data of table 6.1 to calculate Q.

Example 6.3: Quality factor for complex conjugate poles in a Gaussian shaper.

Calculate the quality factors for the complex conjugate poles needed to implement a Gaussian shaper with $n = 2$ and $n = 3$.

* * *

Looking at table 6.1, for the case $n = 2$ we have that the real part R is 1.0985 and the imaginary part I is 0.455. We therefore have:

$$\frac{I}{R} = \sqrt{4Q^2-1} = \frac{0.455}{1.0985} = 0.4142 \rightarrow Q = 0.541 \tag{6.72}$$

For the case $n = 3$, from the same table we see that $R = 1.149$ and $I = 0.7864$ which leads to $Q = 0.606$.

Finally, we need to locate the poles on the s-plane, which requires the knowledge also of ω_0. Since for pulse shapers the relevant quantity that characterizes the speed is the peaking time, it is useful to have a relation between the latter and ω_0. This can be found taking the inverse Laplace transform and studying the response in the time domain. For the reader convenience, the product between the angular frequencies and the peaking time T_p is given in table 6.2 for the most common cases.

Several topologies are available to implement in practice pairs of complex conjugate poles. In the following we consider three representative options. Fig. 6.15 shows the schematic of shaper implementing one real and two complex conjugate poles. The transfer function of the circuit can be expressed as:

$$T(s) = T_0(s)T_1(s) \tag{6.73}$$

TABLE 6.2
Relationship between angular frequency and peaking time for most common Gaussian shapers. The number n represents the total number of poles.

	n=2	n=3	n=4	n=5	n=6	n=7
$\omega_0 T_p$		1.793		2.945		3.758
$\omega_1 T_p$	1.031	1.976	2.471	3.066	3.400	3.842
$\omega_2 T_p$			2.812	3.532	3.612	4.128
$\omega_3 T_p$					4.178	4.775

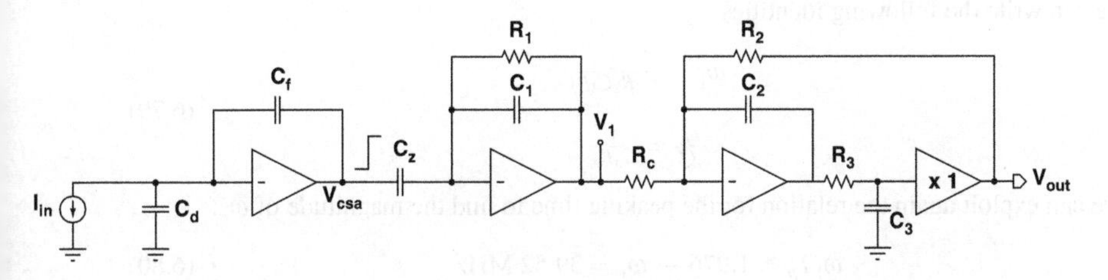

FIGURE 6.15 Pulse shaper implementing one real and two complex conjugate poles.

$T_0(s)$ is the transfer function from the input node to V_1 and it is given by:

$$T_0(s) = \frac{I_{in}}{C_f} \frac{C_z R_1}{(1 + sC_1 R_1)} \tag{6.74}$$

$T_1(s)$ is the transfer function from V_1 to the output. Observe that the last stage is a unity gain buffer. Writing the nodal equations, after some algebra one gets [11]:

$$T_1(s) = -\frac{V_{out}}{V_1} = \left(-\frac{R_2}{R_c}\right) \frac{1}{1 + C_2 R_2 s + C_2 R_2 C_3 R_3 s^2} \tag{6.75}$$

Therefore, the overall response is given by:

$$T(s) = \frac{I_{in}}{C_f} \frac{C_z R_1}{(1 + sC_z R_1)} \left(-\frac{R_2}{R_c}\right) \frac{1}{1 + C_2 R_2 s + C_2 R_2 C_3 R_3 s^2} \tag{6.76}$$

Example 6.4: Sizing the components of a pulse shaper with one real and two complex conjugate poles.

Consider the circuit reported in Fig. 6.15. Find the component values so that the peaking time is 50 ns.

* * *

As we can see from equation (6.76), the transfer function implements one real and, potentially, two complex conjugate poles. We need then to use table 6.1 and table 6.2 with $n = 3$. We start calculating the position of the real pole. From table 6.2 we see that for a peaking time of 50 ns, we need to have:

$$\omega_0 T_p = 1.793 \rightarrow \omega_0 = \frac{1.793}{50 \cdot 10^{-9}} = 35.86 \text{ MHz}. \tag{6.77}$$

Hence, the time constant $R_1C_1 = 1/\omega_0$ must be 27.86 ns. If we compare T_1 to the normalized second order response:

$$T_{1N} = \frac{\omega_1^2}{s^2 + \frac{\omega_1}{Q}s + \omega_1^2} \tag{6.78}$$

we can write the following identities:

$$\begin{aligned} \omega_1^2 &= \frac{1}{C_2R_2C_3R_3} \\ \frac{\omega_1}{Q} &= \frac{1}{C_3R_3} \end{aligned} \tag{6.79}$$

We can exploit again the relation for the peaking time to find the magnitude of ω_1:

$$\omega_1 T_p = 1.976 \rightarrow \omega_1 = 39.52 \text{ MHz}. \tag{6.80}$$

We can now use the value of Q calculated in the previous example for $n = 3$ to find R_3C_3

$$C_3R_3 = \frac{Q}{\omega_1} = \frac{0.606}{39.52 \cdot 10^6} = 15.33 \text{ ns} \tag{6.81}$$

Finally, the last missing time constant, R_2C_2, is found as:

$$C_2R_2 = \frac{1}{\omega_1^2 \times 15.36 \cdot 10^{-9}} = 41.77 \text{ ns} \tag{6.82}$$

Once we have determined all the time constants, we need to choose the values of resistors and capacitors. If we select $C_2 = C_3 = 1$ pF, we find, respectively, $R_2 = 41760\ \Omega$ and $R_3 = 15330\ \Omega$, all sizes that can be easily implemented in an integrated circuit. Observe also that R_c and the ratio R_2/R_c do not interfere with the pole locations, hence R_c can be optimized to give the desired gain without the need of recalculating the value of the other components to keep the pulse shape unaltered. This can be particularly convenient in systems requiring a programmable gain to adapt the front-end to different working conditions.

Another possibility of implementing a shaper with one real and two complex conjugate poles is shown in Fig. 6.16. Here, the transfer function from V_1 to the output can be written as:

$$T_1(s) = -\frac{\frac{1}{R_2R_4C_2C_3}}{s^2 + s\left(\frac{1}{R_2C_2} + \frac{1}{R_3C_2} + \frac{1}{R_4C_2}\right) + \frac{1}{R_3R_4C_2C_3}} \tag{6.83}$$

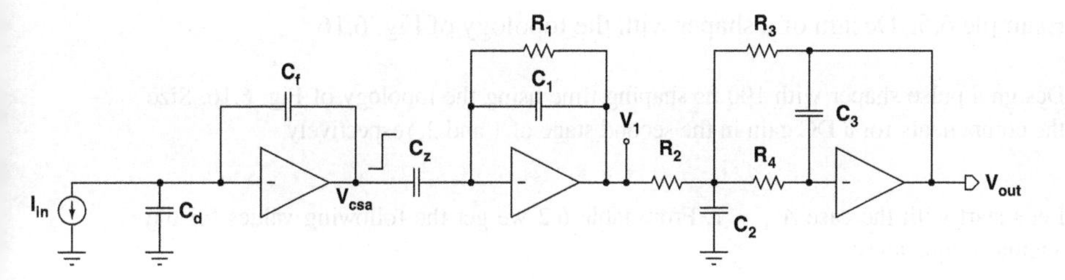

FIGURE 6.16 Alternative implementation of shaper with complex conjugate poles.

Multiplying and dividing the numerator of the above equation by R_3 we can rewrite it as:

$$T_1(s) = -T_0 \frac{\omega_1^2}{s^2 + \left(\frac{\omega_1}{Q}\right) + \omega_1^2} \tag{6.84}$$

where:

$$A_{v1} = \frac{R_3}{R_2} \tag{6.85}$$

$$\omega_1^2 = \frac{1}{R_3 R_4 C_2 C_3} \tag{6.86}$$

$$\frac{1}{Q} = \sqrt{\frac{C_3}{C_2}} \left(\frac{\sqrt{R_3 R_4}}{R_1} + \sqrt{\frac{R_3}{R_4}} + \sqrt{\frac{R_4}{R_3}} \right) \tag{6.87}$$

Observe that R_3 and R_2 define the DC gain, but they intervene also in determining ω_1 and the ratio (through Q) between the real and the imaginary part. So a modification of the DC gain requires a readjustment of the other components to keep the pulse shape unaffected. Since we have more variables than equations, we need to define additional constraints. We can leave C_3 and A_{v1} as free parameters and write:

$$R_2 = \frac{1}{2 A_{v1} \omega_1 Q C_3} \tag{6.88}$$

$$R_3 = \frac{1}{2 \omega_1 Q C_3} \tag{6.89}$$

$$R_4 = \frac{1}{2 \omega_1 Q C_3 (1 + A_{v1})} \tag{6.90}$$

$$C_2 = 4 Q^2 (1 + A_{v1}) C_3 \tag{6.91}$$

Note that only R_3 does not depend on the choice of the DC gain A_{v1}.

Example 6.5: Design of a shaper with the topology of Fig. 6.16.

Design a pulse shaper with 100 ns shaping time using the topology of Fig. 6.16. Size the components for a DC gain in the second stage of 1 and 2, respectively.

* * *

Let's start with the case $A_{v1} = 1$. From table 6.2 we get the following values for the angular frequencies:

$$\omega_0 T_p = 1.793 \rightarrow \omega_0 = \frac{1.793}{100 \cdot 10^{-9}} = 17.93 \text{ MHz} \tag{6.92}$$

$$\omega_1 T_p = 1.976 \rightarrow \omega_1 = \frac{1.976}{100 \cdot 10^{-9}} = 19.76 \text{ MHz} \tag{6.93}$$

If we choose $C_3 = 2$ pF, we have:

$$C_2 = 4Q^2 \cdot 2 \cdot 2 \text{ pF} = 5.87 \text{ pF} \tag{6.94}$$

Using table 6.1 we can in fact calculate that $Q = 0.606$. We can now find the value of the resistors:

$$R_2 = R_3 = \frac{1}{2\omega_1 Q C_3} = 20880\ \Omega \tag{6.95}$$

$$R_4 = \frac{1}{2\omega_1 Q C_3 \cdot 3} = 10440\ \Omega \tag{6.96}$$

If we now re-size the circuit for A_{v1}=2, using equations (6.88) to (6.91) we get: $C_3 =$ 8.8 pF, $R_2 = 10440\ \Omega$, $R_3 = 20880\ \Omega$ and $R_4 = 6960\ \Omega$.

We consider now another topology which is often adopted in front-end electronics design to realize a pair of complex conjugate poles. The circuit employs a T-bridge in the feedback network and a shaper using this scheme is shown in Fig. 6.17. The transfer function from node 1 to V_{out} can be written as:

$$\frac{V_{out}}{V_1} = \frac{R_3 + R_4}{R_2} \frac{1 + s(R_3//R_4)C_3}{s^2 C_2 C_3 R_3 R_4 + sC_2(R_3 + R_4) + 1} \tag{6.97}$$

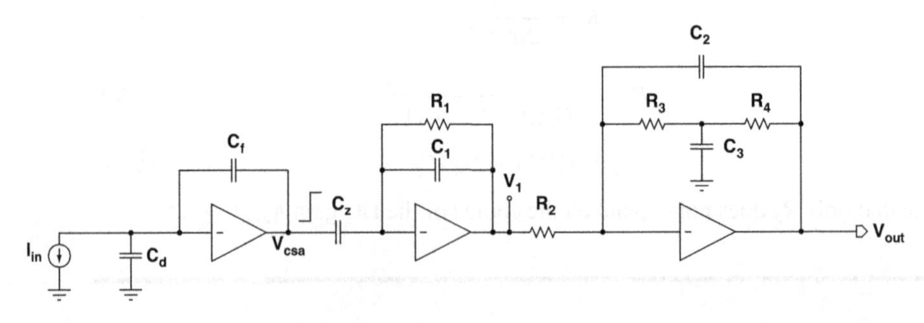

FIGURE 6.17 T-bridge feedback to implement one zero and a pair of complex conjugate poles.

We can divide both the numerator and the denominator by $C_2C_3R_2R_3$ to put the transfer function in its normalized form. If we do so, we can write ω_1 and Q as follows:

$$\omega_1^2 = \frac{1}{C_2C_3R_3R_4} \tag{6.98}$$

$$\frac{\omega_1}{Q} = \frac{C_2(R_3+R_4)}{C_2C_3R_3R_4} \tag{6.99}$$

We have again more variables than equations. A first reasonable assumption is to put $R_3 = R_4 = R$, which allows us to simplify the equation of ω_1 and Q:

$$\omega_1^2 = \frac{1}{C_2C_3R^2} \tag{6.100}$$

and

$$\frac{\omega_1}{Q} = \frac{2}{C_3R} \tag{6.101}$$

We can now observe that $C_3R = 2Q/\omega_1$ and introduce this relationship in (6.100) to get:

$$C_2R = \frac{1}{2Q\omega_1} \tag{6.102}$$

and:

$$C_3R = \frac{2Q}{\omega_1} \tag{6.103}$$

Finally, if we divide (6.103) by (6.102) we get:

$$\frac{C_3}{C_2} = 4Q^2 \rightarrow Q = \frac{1}{2}\sqrt{\frac{C_3}{C_2}} \tag{6.104}$$

From the above equation, we see that the value of Q is only determined by the ratio of C_3 and C_2. Their absolute values, together with the value of R are then chosen to locate the poles on the s plain and achieve the desired peaking time. Note that in this transfer function, the gain depends on the ratio $(R_3+R_4)/R_2$ and R_2 does not intervene in the determination of the poles and the zero of the stage. Therefore the DC gain can be set and changed without altering the shaping. Observe also that the transfer function also includes a zero, located at $s = -(R_3//R_4)C_3$. The frequency of the zero is close to the one of the poles and can not be neglected, therefore table 6.2 is not directly applicable to find the relationship between ω_1 and the peaking time. The extra zero can be canceled by introducing a further pole in the same position, as shown in Fig. 6.18 and discussed in the following example.

Example 6.6: Determination of the pole position with a T-bridge network.

For the circuit of Fig. 6.17 determine the position of the poles so that the peaking time is 100 ns and the network has only one real and two complex-conjugate poles.

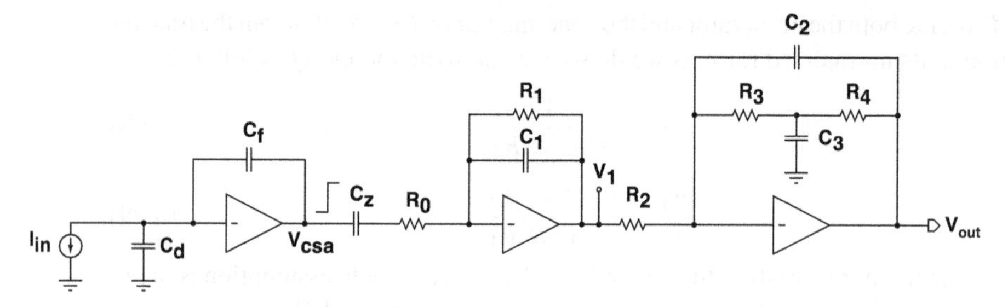

FIGURE 6.18 Shaper employing a T-bridge-feedback and pole-zero cancellation to achieve one real and two complex conjugate poles.

* * *

In Fig. 6.18 a resistor R_0 has been introduced in series to C_z. The current fed to the second stage now becomes:

$$I_z(s) = I_{in}(s)\frac{1}{\not{s}C_f}\frac{\not{s}C_zR_0}{(1+sC_zR_0)} \tag{6.105}$$

In this way, a further pole has been introduced in the transfer function, which now reads:

$$\frac{V_{out}(s)}{I_{in}(s)} = \frac{1}{C_f}\frac{C_zR_0}{(1+sC_zR_0)}\frac{R_1}{(1+sC_1R_1)}\frac{R_3+R_4}{R_2}\frac{1+s(R_3//R_4)C_3}{s^2C_2C_3R_3R_4+sC_2(R_3+R_4)+1} \tag{6.106}$$

Using the condition on the peaking time, we have:

$$\omega_1 T_p = 1.976 \rightarrow \omega_1 = 19.76 \text{ MHz} \tag{6.107}$$

We already know that if we have one real and two complex conjugate poles, the value of Q is 0.606. Applying equation (6.104) we find that the ratio C_3/C_2 must be 1.469. So we can choose $C_2 = 1$ pF and $C_3 = 1.47$ pF. Assuming that $R_3 = R_4 = R$, we can use the expression of one of the two time constants to find R. For example:

$$C_2R = \frac{1}{2Q\omega_1} = \frac{1}{2\times 0.606\times 19.76\cdot 10^6} = 41.75 \text{ ns} \rightarrow R = 41.75 \text{ k}\Omega. \tag{6.108}$$

The zero is located at $(R_3//R_4)C_3 = R/2C_3 = 30.68$ ns. We can then set $C_zR_0 = 30.68$ ns to cancel out the zero. If we use our previous parameters for the other components and suppose that $C_z = 5$ pF, we get $R_0 = 6136\,\Omega$. Note that the zero could be used to cancel the effect of other poles already present in the network and sometimes zeros in the transfer function are exploited to cancel unwanted components in the input signal. Introducing R_0 as done in Fig. 6.18 is practical as it requires only a small extra resistor. However, the impact of this additional component on the overall noise should not be overlooked.

Further interesting readings on filters with elaborated transfer function can be found in [12, 13].

6.5 NOISE CALCULATIONS IN TIME INVARIANT SHAPERS

The purpose of this section is to show how to determine the Equivalent Noise Charge (ENC) of a front-end amplifier once the noise sources are known and properly referred to the system input. In doing so, we will also derive some broad criteria for the design of pulse shapers. The problem of sizing circuit components for optimal noise performance will be dealt with in chapter 8. Noise calculations for CR-RC^n shapers are discussed in [9], [14], [15] and [16]. The method is the same already studied in the previous chapter for the single stage amplifier. The first step is to identify the relevant noise generators and represent them as current or voltage sources properly connected to the system input, as shown in Fig. 6.19. Remember that, in the jargon of front-end designers, a noise term modeled by a voltage source placed in series to the amplifier input is known as *series noise*. A contribution represented by a current source put in parallel to the amplifier input is called *parallel noise*. The capacitance C_T encompasses the sensor capacitance plus any other capacitance seen between the amplifier input and ground, including the gate capacitance of the input device. In the frequency domain a power spectral density, S_n, is associated to each noise source. The effect of S_n on the amplifier output is then calculated by multiplying the power spectral density by the modulus square of the appropriate transfer function and integrating the result over the whole frequency spectrum. The quantity thus obtained is the power of the output waveform generated by the considered noise source. In the cases of our interest, this coincides with the variance of the output signal corrupted by noise. In formula:

$$< v_{out}{}^2 >= \int_0^{\infty} S_n \left|T_n\left(j2\pi f\right)\right|^2 df \tag{6.109}$$

Note that $T_n\left(j2\pi f\right)$ is the transfer function from the particular noise source to the output and it might or might not coincide with the signal transfer function. If more uncorrelated noise sources are at play, all the square contributions are summed at the output:

$$< v_{out,tot}{}^2 >=< v_{out1}{}^2 > + < v_{out2}{}^2 > + \cdots + < v_{outn}{}^2 > \tag{6.110}$$

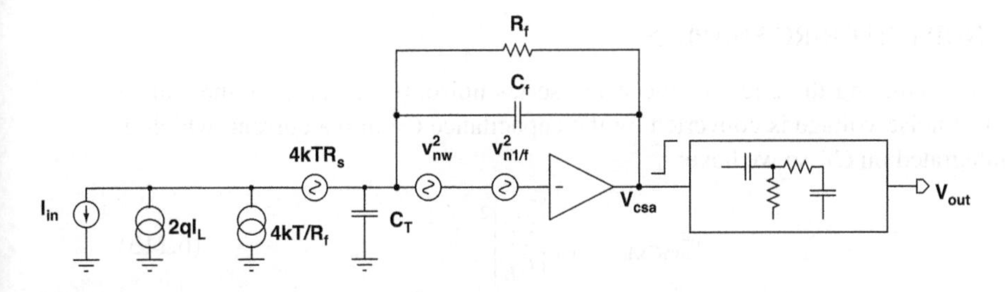

FIGURE 6.19 Noise sources in a front-end amplifier.

The square root of the above expression gives the output rms noise, which is the quantity usually measured at the test bench:

$$v_{out,rms} = \sqrt{< v_{out1}{}^2 > + < v_{out2}{}^2 > + \cdots + < v_{outn}{}^2 >} \tag{6.111}$$

The peak output signal generated by a Dirac-delta pulse containing a single electron is then calculated. The Equivalent Noise Charge (ENC) is finally obtained dividing the rms output noise by the peak amplitude of this one-electron signal. In the above notation, we implicitly assumed that the quantity at the output is a voltage. Of course, the method does not change for systems whose output is a current.

In a multi-stage system, the signal is amplified as it travels along the chain, becoming progressively stronger in comparison to the intrinsic electronics noise. In a well designed amplifier, the noise should therefore be dominated by the one of the input stage, which operates on weaker signals. To appreciate this point better, consider as an example the noise generated by the feedback resistance of the second stage of the circuit of Fig. 6.7. When referred to the input of the stage, the noise has a power spectral density given by:

$$i_{n1}^2 = \frac{4k_BT}{R_1} \tag{6.112}$$

However, to refer i_{n1}^2 to the global input, we must still divide (6.112) by the current gain observed between the input of the second and the input of the first stage, which is given by $|C_z/C_f|^2$. For instance, if $R_1 = R_f$ and $C_z = 10C_f$, when referred to the channel input i_{n1}^2 becomes one hundred times smaller than the noise power spectral density generated by R_f. For this reason, starting from the second stage much smaller resistors can be used, which can be implemented with passive components. Similar arguments apply also to the noise generated inside the core amplifiers of the stages following the first one. This does not mean that the noise introduced by the shaper itself can be neglected a priori. For a well designed system, however, the key noise contributions are found in the sensor, in the first stage input transistor, in its load device and in the CSA feedback resistor and they were already analyzed in detail in chapter 5.

6.5.1 NOISE IN CR-RC SHAPERS

We start calculating the effect of the white series noise, represented by the sourse $v_{nw}{}^2$. The noise voltage is converted by the capacitance C_T into a current, which is then integrated on C_f, so we have:

$$v_{nw,CSA}^2 = v_{nw}^2 \left| \frac{C_T}{C_f} \right|^2 \tag{6.113}$$

The output noise of the CSA is then processed by the transfer function of the pulse shaper. For a CR-RC filter, this is

$$T(s) = \frac{s\tau}{(1+s\tau)^2} \tag{6.114}$$

We suppose here that the shaper does not introduce gain. Our objective is to calculate the ENC and any gain would equally affect both the signal and the noise, therefore it is not relevant for our purpose. We need to make an integration in the frequency domain, so we put $s = j\omega = j2\pi f$. The time constant τ is the reciprocal of the pole angular frequency, i.e. $\tau = 1/\omega_0 = 1/2\pi f_0$. We can therefore rewrite equation (6.114) as:

$$T(j2\pi f) = \frac{j\left(\frac{f}{f_0}\right)}{\left(1+j\frac{f}{f_0}\right)^2} \tag{6.115}$$

The noise output voltage is given by:

$$< v_{out}^2 >_{nw} = v_{nw}^2 \int_0^\infty \left|\frac{C_T}{C_f}\right|^2 |T(j2\pi f)|^2 \, df \tag{6.116}$$

The modulus squared of $T(j2\pi f)$ is:

$$|T(j2\pi f)|^2 = \frac{f^2}{\left(f_0 - \frac{f^2}{f_0}\right)^2 + 4f^2} \tag{6.117}$$

Introducing this result in equation (6.116), we get:

$$< v_{out}^2 >_{nw} = v_{nw}^2 \left|\frac{C_T}{C_f}\right|^2 \int_0^\infty \frac{f^2}{\left(f_0 - \frac{f^2}{f_0}\right)^2 + 4f^2} df = v_{nw}^2 \left(\frac{C_T}{C_f}\right)^2 \frac{\pi}{4} f_0 \tag{6.118}$$

The above quantity can also be rewritten in the following way:

$$< v_{out}^2 >_{nw} = v_{nw}^2 \left(\frac{C_T}{C_f}\right)^2 \frac{\pi}{4} \frac{1}{2\pi RC} = v_{nw}^2 \left(\frac{C_T}{C_f}\right)^2 \frac{1}{8} \frac{1}{T_p} \tag{6.119}$$

where we have used the fact that in a CR-RC shaper the peaking time T_p is equal to the integration/derivation time constants. The shaper impulse response in the time domain is given by:

$$V_{out}(t) = \frac{Q_{in}}{C_f}\left(\frac{t}{\tau}\right) e^{-\frac{t}{\tau}} \tag{6.120}$$

Therefore, the peak output voltage for an input signal of one electron is:

$$V_{out,peak} = \frac{q}{C_f}\frac{1}{e} \tag{6.121}$$

where q is the electron charge, ($q = 1.6022 \cdot 10^{-19}$ C). Dividing (6.119) by the square of (6.121) we get:

$$ENC_w^2 = v_{nw}^2 C_T^2 \frac{e^2}{8}\frac{1}{T_p} \rightarrow ENC_w = v_{nw} C_T \sqrt{\frac{e^2}{8}\frac{1}{T_p}} \tag{6.122}$$

We see from the above equation that the ENC due to series white noise increases linearly with the total capacitance shunting the amplifier input and decreases with the square root of the peaking time.

Let's now consider the $1/f$ noise. By making the substitution:

$$A_f = \frac{K_f}{C_{ox}WL} \tag{6.123}$$

we can write the flicker noise spectral density as:

$$v_{nf}^2 = \frac{A_f}{f} \tag{6.124}$$

The voltage produced at the CSA output thus becomes:

$$v_{n1/f,CSA}^2 = \frac{A_f}{f}\left|\frac{C_T}{C_f}\right|^2 \tag{6.125}$$

Hence, the total noise power at the shaper output is given by:

$$< v_{out}^2 >_{n1/f} = A_f\left(\frac{C_T}{C_f}\right)^2 \int_0^{\infty} \frac{1}{f} \frac{f^2}{\left(f_0 - \frac{f^2}{f_0}\right)^2 + 4f^2} df = A_f\left(\frac{C_T}{C_f}\right)^2 \frac{1}{2} \tag{6.126}$$

We can now divide (6.126) by the peak voltage due to one electron signal given by (6.121) to get the ENC contribution of the flicker noise:

$$ENC_f^2 = A_f C_T^2 \frac{e^2}{2} \rightarrow ENC_f = C_T\sqrt{A_f \frac{e^2}{2}} \tag{6.127}$$

Equation (6.127) tells us that the equivalent noise charge due to flicker noise sources is proportional to the total input capacitance, like the one due to the series white noise, but it is independent of the peaking time.

Finally, we can evaluate the effect of parallel noise. The noise at the CSA output due to ${i_n}^2$ is:

$$v_{ni,CSA}^2 = i_n^2\left|\frac{1}{j2\pi f C_f}\right|^2 = \left(\frac{i_n}{2\pi f C_f}\right)^2 \tag{6.128}$$

Processing this noise with the shaper transfer function, we get:

$$< v_{out}^2 >_{ni} = \frac{i_n^2}{4\pi^2 C_f^2} \int_0^{\infty} \frac{1}{f^2} \frac{f^2}{\left(f_0 - \frac{f}{f_0}\right)^2 + 4f^2} df = \frac{i_n^2}{16\pi f_0 C_f^2} \tag{6.129}$$

We introduce now in the above result the expression of the peaking time:

$$f_0 = \frac{1}{2\pi RC} = \frac{1}{2\pi T_p} \rightarrow < v_{out}^2 >_{ni} = \frac{i_n^2 T_p}{8C_f} \tag{6.130}$$

Finally, dividing by the square of the peak voltage due to one electron signal, we have:

$$ENC_i^2 = \frac{i_n^2}{q^2}\frac{e^2}{8}T_p \rightarrow ENC_i = \frac{i_n}{q}\sqrt{\frac{e^2}{8}T_p} \qquad (6.131)$$

The above equation reveals that the effect of the parallel noise is independent of the input capacitance and is proportional to the square-root of the peaking time. The total output noise is obtained by summing the squares of all contributions:

$$ENC^2 = ENC_w^2 + ENC_f^2 + ENC_i^2 = \frac{v_{nw}^2}{q^2}\frac{e^2}{8}C_T^2\frac{1}{T_p} + \frac{A_f}{q^2}\frac{e^2}{2}C_T^2 + \frac{i_n^2}{q^2}\frac{e^2}{8}T_p \qquad (6.132)$$

Example 6.7: Noise evaluation of a CR-RC shaper.

A CR-RC pulse shaper has a peaking time of 1 μs. Assuming an NMOS input transistor, estimate the total ENC with the following parameters: $g_{m1} = 1$ mS, $I_L = 2$ nA, $K_f = 1 \cdot 10^{-24}$ V^2·F, $R_f = 10$ MΩ, $C_{ox} =$ 5fF/μm^2, $C_T = 2$ pF, $W_1 = 500$ μm, $L_1 = 0.5$ μm. (W_1, L_1 and g_{m1} refer respectively to the gate width, length and transconductance of the CSA input device). Calculate the rms output noise voltage if the front-end has a gain of 5 mV/fC.

* * *

First, we need to make a guess about the operating region of the input transistor. The capacitance density given is typical of a 0.25 μm CMOS process, for which a $\mu_n C_{ox}$ of 220 – 250 μA/V^2 can be expected. Assuming the lower value, we can calculate the weak-to-strong inversion boundary for the input transistor:

$$I_{WI,SI} = 2\mu_n C_{ox}\phi_T^2\frac{W}{L} = 2\times 220\cdot 10^{-6}\times 1.4\times 6.76\cdot 10^{-4}\times 100 = 416\ \mu\text{A} \qquad (6.133)$$

If we calculate g_{m1} with the strong inversion formula we expect for the above current a transconductance given by:

$$g_{m1} = \sqrt{2\mu_n C_{ox}\frac{W}{L}I_{DS}} = 13.5\ \text{mS} \qquad (6.134)$$

This is significantly bigger than 1 *mS*, which means that to achieve a current of 1 *mS* the input transistor is biased with a current smaller than the one given by (6.133), so it is legitimate to assume that the device works in weak inversion. Since the transition between the weak and strong inversion region is quite broad the device will most probably be in moderate inversion. Therefore, using simple weak inversion formulas may lead to underestimating the noise. However, this can be a reasonable starting point, to be then cross-checked and refined with computer simulations. In weak inversion, the noise spectral density due to the device thermal noise is given by:

$$v_{nw}^2 = 2k_B T\frac{1}{g_m} = 8.3\cdot 10^{-18}\frac{\text{V}^2}{\text{Hz}} \qquad (6.135)$$

The parallel noise due to the leakage current is $2qI_L = 2\times 1.6022\cdot 10^{-19}\times 2\cdot 10^{-9} = 6.4\cdot 10^{-28}$ A^2/Hz, while the one originating from the CSA feedback resistor is $4k_BT/R_f = 1.6\cdot 10^{-27}$ A^2/Hz. The total parallel noise is hence $i_n^2 = 2.24\cdot 10^{-27}$ A^2/Hz. Finally, we need to calculate the flicker noise coefficient A_f:

$$A_f = \frac{1\cdot 10^{-24}}{5\cdot 10^{-3}\times 500\cdot 10^{-6}\times 0.5\cdot 10^{-6}} = 8\cdot 10^{-13}\text{V}^2 \tag{6.136}$$

We can now apply (6.132) to get the total ENC:

$$ENC^2 = 1197+462+80817 = 82476 \rightarrow ENC = \sqrt{82476} = 287 \text{ electrons rms} \tag{6.137}$$

If we suppose that there is no additional gain in the filter, the peak output voltage is given by Q_{in}/eC_f. To have a gain of 5 mV/fC, we need $C_f = 73.5$ fF. Using equations(6.119),(6.126) and (6.128) we can calculate the rms output noise voltage as:

$$V_{n,out} = \sqrt{v_{nw}{}^2\left(\frac{C_T}{C_f}\right)^2\frac{1}{8}\frac{1}{T_p} + A_f\frac{1}{2}\left(\frac{C_T}{C_f}\right)^2 + \frac{i_n^2}{C_f^2}\frac{1}{8}T_p} \approx 230\ \mu\text{V} \tag{6.138}$$

Note that the method employed to calculate the ENC is impractical in a measurement set-up, as the signal due to a single electron is undetectable because it is overshaded by noise. Therefore, at the test bench one injects a known stimulus well above the noise level. For instance, if we use a signal of 4 fC we get an output of 20 mV. The measured ENC is then obtained as:

$$\frac{V_{out,rms}}{V_{peak}}\times(\text{Number of electrons in the test signal}) = \frac{0.23\text{ mV}}{20\text{ mV}}\times 25000 = 287\ e^{-}\text{rms} \tag{6.139}$$

The above example leads to a few important considerations. First, we see that with the chosen peaking time, the parallel noise source dominates over the other contributions. In fact, if we neglect the thermal and flicker noise, the ENC becomes 284 electrons, which shows that the two non-dominant sources contribute to the total noise only at the 1% level. However, series and parallel noise are weighted in opposite ways by the peaking time. Therefore, if we decrease T_p, we reduce the effect of parallel noise and enhance the one of series noise. We expect to reach an optimum when the two sources give identical contributions. To find the optimal peaking time, we can write:

$$\frac{\partial ENC^2}{\partial T_p} = 0 \rightarrow -\frac{v_{nw}^2}{q^2}\frac{e^2}{8}C_T^2\frac{1}{T_p^2} + \frac{i_n^2}{q^2}\frac{e^2}{8} = 0 \rightarrow T_{p,opt} = \frac{v_{nw}}{i_n}C_T \tag{6.140}$$

If we use the parameters of the previous example, we find that the optimal peaking time is 121.74 ns. Recalculating the noise with the new value of T_p, we have:

$$ENC = \sqrt{9843+462+9843} = 142 \text{ electrons rms} \tag{6.141}$$

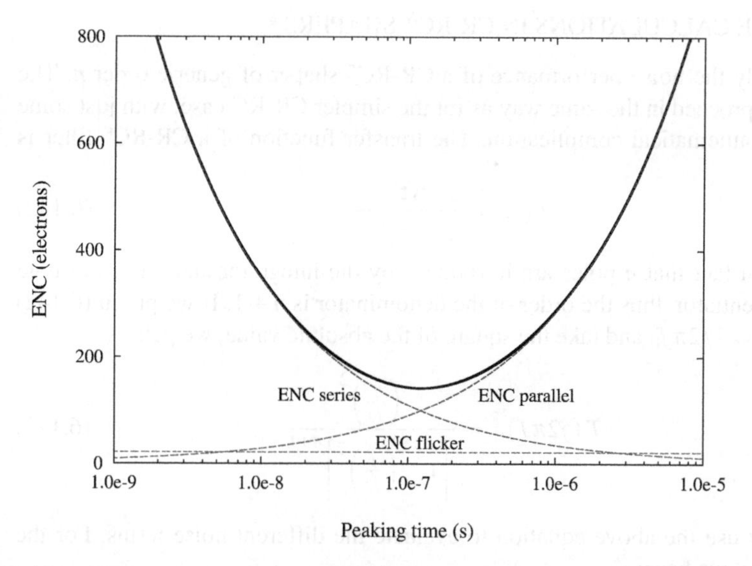

FIGURE 6.20 ENC versus peaking time. The circuit parameters used to obtain the plot are given in example 6.7.

which is substantially smaller than the value found with $T_p = 1\ \mu$s. As expected, the parallel and thermal noise have now equal weight. In the calculation, we have used the theoretical optimal value of the peaking time to demonstrate the procedure. Of course, in a practical implementation, one would not target such a precise number. Note, however, that using a "round" value of 100 ns will result in a negligible noise increase (143 vs 142 electrons), because T_p enters under square root in the ENC definition. In fact, an improvement of the ENC by a factor of two has requested a ten-fold peaking time reduction. The other important comment concerns flicker noise, which, in our example, gives the smallest and almost negligible contribution. This is a representative situation, because the $1/f$ noise becomes usually the limiting factor only when small input capacitance, combined with low leakage current and modest event rates, allow to push the ENC due to thermal and parallel noise well below the 100 electrons range. However, especially in very deep-submicron technologies, inappropriate device sizing may result in significantly increased $1/f$ component so also this noise source must always be very carefully considered in the design phase. Finally, observe that in our example we have taken into account only the noise of the input transistor as it should happen in a well optimized amplifier. However, having the input device dominating the noise performance is the *target* of the design process but it is not guaranteed a priori. A careless sizing of other components in the chain may lead them to add significant noise, sometimes even greater than the one of the input device. Fig. 6.20 reports the ENC plot versus peaking time for the system described in the example and shows that the minimum noise is actually achieved at the intersection between the series and parallel noise curve.

6.5.2 NOISE CALCULATIONS IN CR-RC^N SHAPERS

We now study the noise performance of a CR-RC^n shaper of generic order n. The calculations proceed in the same way as for the simpler CR-RC case, with just some additional mathematical complication. The transfer function of a CR-RC^n filter is given by:

$$T(s) = \frac{s\tau}{(1+s\tau)^{n+1}} \tag{6.142}$$

Remember in fact that n poles are introduced by the integrator and one extra pole by the differentiator, thus the order of the denominator is $n+1$. If we put in (6.142) $s = j2\pi f$, $\tau = 1/2\pi f_0$ and take the square of the absolute value, we get:

$$|T(j2\pi f)|^2 = \frac{\left(\frac{f}{f_0}\right)^2}{\left[1+\left(\frac{f}{f_0}\right)^2\right]^{n+1}} \tag{6.143}$$

We can now use the above equation to evaluate the different noise terms. For the thermal noise, we have:

$$< v_{out}^2 >_{nw} = v_{nw}^2 \left(\frac{C_T}{C_f}\right)^2 \int_0^\infty \frac{\left(\frac{f}{f_0}\right)^2}{\left[1+\left(\frac{f}{f_0}\right)^2\right]^{n+1}} df \tag{6.144}$$

To solve the integral, we can make the following substitution:

$$\left(\frac{f}{f_0}\right)^2 = u \rightarrow df = \frac{1}{2} u^{-\frac{1}{2}} f_0 du \tag{6.145}$$

With the above change of variable, equation (6.144) now reads:

$$< v_{out}^2 >_{nw} = v_{nw}^2 \left(\frac{C_T}{C_f}\right)^2 f_0 \int_0^\infty \frac{u^{\frac{1}{2}}}{(1+u)^{n+1}} du \tag{6.146}$$

The integral can be solved using the beta function $B(x,y)$, recalling that:

$$B(m+1,\alpha+1) = \int_0^\infty \frac{u^m}{(1+u)^{m+\alpha+2}} du \tag{6.147}$$

By comparing (6.147) with (6.146), we see that:

$$m = \tfrac{1}{2}$$
$$m+\alpha+2 = n+1 \rightarrow \alpha = n - \tfrac{3}{2} \tag{6.148}$$

Therefore, we have:

$$\int_0^\infty \frac{u^{\frac{1}{2}}}{(1+u)^{n+1}} du = B\left(\frac{3}{2}, n-\frac{1}{2}\right) \tag{6.149}$$

Finally, we need to remember the relationships between the peaking time, T_p, the time constant RC and the cut-off frequency f_0:

$$T_p = nRC$$

$$f_0 = \frac{1}{2\pi RC} = \frac{n}{2\pi T_p}$$

Combining all the above results, we obtain for the output noise voltage:

$$< v_{out}^2 >_{nw} = v_{nw}^2 \left(\frac{C_T}{C_f}\right)^2 \frac{B\left(\frac{3}{2}, n-\frac{1}{2}\right)}{4\pi T_p} n \tag{6.150}$$

For an n order shaper the impulse response in the time domain is:

$$V_{out}(t) = \frac{Q_{in}}{C_f} \frac{1}{n!} \left(\frac{t}{\tau}\right)^n e^{-\frac{t}{\tau}} \tag{6.151}$$

The peak is reached at $T_p = n\tau = nRC$, hence the output for one electron charge is:

$$V_{out,q} = \frac{q}{C_f} \frac{n^n}{n!} e^{-n} \tag{6.152}$$

Dividing equation (6.150) by the square of (6.152), we get:

$$ENC_w^2 = v_{nw}^2 \frac{C_T^2}{q^2} \frac{B\left(\frac{3}{2}, n-\frac{1}{2}\right)}{4\pi T_p} n \frac{n!^2 e^{2n}}{n^{2n}} \tag{6.153}$$

We can now calculate the effect of $1/f$ noise. In this case we have:

$$< v_{out}^2 >_{n1/f} = A_f \left(\frac{C_T}{C_f}\right)^2 \int_0^\infty \frac{1}{f} \frac{\left(\frac{f}{f_0}\right)^2}{\left[1+\left(\frac{f}{f_0}\right)^2\right]^{n+1}} df = A_f \left(\frac{C_T}{C_f}\right)^2 \frac{1}{2n} \tag{6.154}$$

Dividing again by the square of the gain for one electron, we obtain:

$$ENC_{n1/f}^2 = A_f \frac{C_T^2}{q^2} \frac{1}{2n} \frac{n!^2 e^{2n}}{n^{2n}} \tag{6.155}$$

Finally, we need to evaluate the amount of parallel noise:

$$< v_{out}^2 >_{ni} = \frac{i_n^2}{C_f^2} \int_0^\infty \frac{1}{f^2} \frac{\left(\frac{f}{f_0}\right)^2}{\left[1+\left(\frac{f}{f_0}\right)^2\right]^{n+1}} df = \frac{i_n^2}{C_f^2 f_0^2} \int_0^\infty \frac{1}{f_0^2} \frac{1}{\left[1+\left(\frac{f}{f_0}\right)^{n+1}\right]} df \tag{6.156}$$

We can reuse the substitutions of (6.145) to rewrite the above equation in the following form:

$$< v_{out}^2 >_{ni} = \frac{i_n^2}{C_f^2} \frac{1}{f_0} \int_0^\infty \frac{u^{-\frac{1}{2}}}{(1+u)^{n+1}} du \tag{6.157}$$

This can be expressed using again the definition of the beta function, noting that:

$$m = -\tfrac{1}{2}$$
$$\alpha = n - \tfrac{1}{2} \tag{6.158}$$

The noise output voltage is then:

$$< v_{out}^2 >_{ni} = \frac{i_n^2}{C_f^2} \frac{B\left(\frac{1}{2}, n+\frac{1}{2}\right)}{4\pi n} T_p \tag{6.159}$$

where the relationship between the cut-off frequency and the peaking time has again been introduced. Finally, dividing equation (6.159) by the gain square for one electron, we obtain:

$$ENC_i^2 = i_n^2 \frac{B\left(\frac{1}{2}, n+\frac{1}{2}\right)}{4\pi n} T_p \frac{n!^2 e^{2n}}{n^{2n}} \tag{6.160}$$

This more generalized analysis confirms the results already obtained with the simpler CR-RC shaper. Using the noise indexes, the ENC equation can be put in a more compact form:

$$ENC^2 = \frac{1}{q^2}\left(v_{nw}^2 N_w C_T^2 \frac{1}{T_p} + N_f A_f C_T^2 + i_n^2 N_i T_p\right) \tag{6.161}$$

We can now write a more general expression also for the optimal peaking time. In fact, differentiating equation (6.161) with respect to T_p we get:

$$\frac{\partial ENC^2}{\partial T_p} = 0 \rightarrow T_{p,opt} = \frac{v_{nw}}{i_n} C_T \sqrt{\frac{N_w}{N_i}} \tag{6.162}$$

This formula shows that the optimal peaking time depends also on the shaper noise indexes. Observe that for the CR-RC shaper, the two indexes are equal ($N_w = N_i = e^2/8$), which "masked" the dependency when we evaluated $T_{p,opt}$ for this filter. The noise calculations can be carried-out with the same technique for any time invariant shaper, including the ones containing complex conjugate poles, at the price of more cumbersome mathematics as the number of poles increases. However, the use of equation (6.161) allows for a fast estimation of the noise performance once the noise indexes are known. For this reason, the indexes of the most recurrent shapers have been collected in table 6.3 and table 6.4 for filters with real and complex conjugate poles, respectively. In table 6.3, n indicates the number of integrators, hence $n = 1$ is the CR-RC shaper, $n = 2$ is the CR-RC^2 and so on. In table 6.4, n is the total number of poles in the transfer function, therefore $n = 2$ corresponds to a shaper with no real and two complex conjugate poles, $n = 3$ indicates a filter with one real and two complex conjugate poles and so forth. The two tables lead to a few important observations. First, there is no substantial difference in the noise coefficients between filters with only real poles and filters with complex conjugate ones. Second, the flicker noise coefficient is a weak function of the shaper order, therefore $1/f$ noise is not the primary concern

TABLE 6.3
Noise indexes of CR-RC^n shapers. n indicates the number of integrators in the chain.

	n=1	n=2	n=3	n=4	n=5	n=6	n=7
N_w	0.92	0.85	0.93	1.02	1.10	1.19	1.27
N_f	3.69	3.42	3.32	3.27	3.25	3.23	3.22
N_i	0.92	0.63	0.52	0.45	0.40	0.36	0.34

in choosing the type of filter. For shapers with real poles, the thermal noise index, N_w has minimum for $n = 3$ and then slowly increases. Hence, all other parameters being equal, a seventh order shaper has about 20% more noise that a second order filter. However, the parallel noise coefficient decreases significantly going to higher orders, even though the advantage becomes progressively smaller as n increases. We can see from table 6.3 that a shaper with $n = 7$ offers a reduction in parallel noise of 1.5 times with respect to the simple CR-RC case. The evolution of N_w and N_i with the shaper order has an intuitive explanation. For the same peaking time, in fact, higher order shapers have larger bandwidths (which in principle implies that more series noise is embarked) but steeper frequency roll-off, which makes the filter more selective and partially compensates for the greater bandwidth. On the other hand, the system behaves as an integrator after the cut-off point is reached, so displacing the cut-off to higher frequencies makes the circuit integrate less current noise, thereby reducing its impact on the overall ENC. As far as the ENC in concerned, the optimal peaking time is the one that minimizes both series and parallel noise. However, an additional constraint on T_p may come from the event rate, that limits the pulse duration. In the previous section we have seen that higher order filters offer a faster return to the baseline for the same peaking time (see Fig. 6.11). This implies that, for a fixed pulse duration, a longer peaking time can be afforded, while the potential increase of the parallel noise due to the longer peaking time is mitigated by the reduction in N_i. As a result, higher order shapers tend to offer overall better performance. However, the extra area and power consumption required to implement the additional poles must always be weighted against the potential benefits. Since noise trades with power, targeting an adequate ENC rather than the lowest possible one may be a better choice from the system point of view. The most important results obtained with the noise analysis are summarized hereafter.

TABLE 6.4
Noise indexes of shapers with complex conjugate poles. n is the total number of poles in the transfer function.

	n=2	n=3	n=4	n=5	n=6	n=7
N_w	0.93	0.85	0.91	0.96	1.10	1.04
N_f	3.70	3.39	3.33	3.27	3.26	3.21
N_i	0.88	0.61	0.51	0.45	0.42	0.40

SUMMARY OF NOISE PROPERTIES OF PULSE SHAPERS

- The ENC due to white series noise is proportional to the total input capacitance and inversely proportional to the square-root of the peaking time.
- The ENC due to parallel noise is independent of the input capacitance and proportional to the square-root of the peaking time.
- An optimum peaking time can thus be found that minimizes the effect of series white and parallel noise.
- The ENC due to series flicker noise is proportional to the input capacitance and independent of the peaking time.
- All noise contributions depend also on a coefficient which is specific of the particular type of pulse shaper. Called "the shaper noise index", this coefficient depends only on the functional form of the transfer function, i.e. on the number of poles and zeros and their relative positions, but not on their absolute location on the s plane.

6.5.3 NOISE SLOPE

The same front-end amplifier may be used with different types of sensors and the sensor capacitance is one of the key ingredients in determining the achievable SNR. The amplifier characterization must therefore include, both in simulation and in measurements, a curve showing the ENC trend versus the input capacitance. Taking the square-root of (6.161) and calculating the limit for $C_T \to \infty$ we get:

$$\lim_{C_T \to \infty} ENC = \frac{1}{q}\sqrt{\frac{v_{nw}^2}{T_p}N_w + N_f A_f C_T} \tag{6.163}$$

The above equation shows that, for sufficiently high values of C_T, thermal and flicker noise dominate and the ENC grows linearly with C_T. From equation (6.161), we see

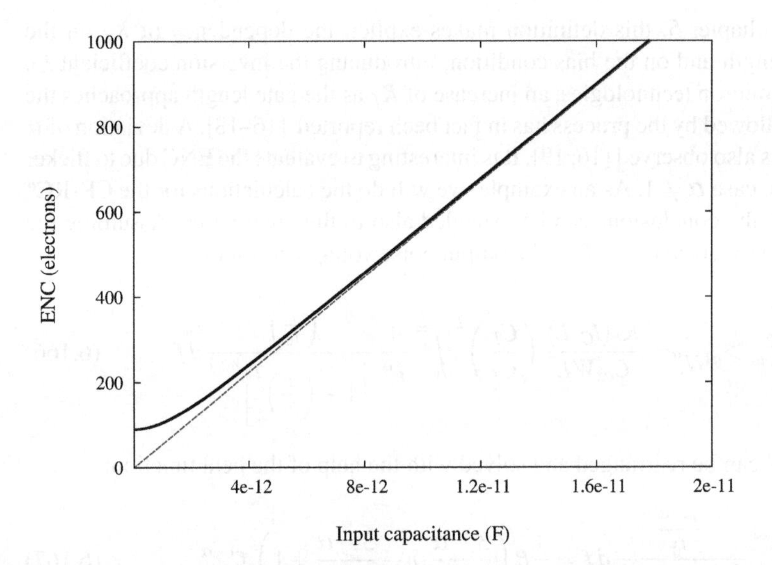

FIGURE 6.21 ENC trend versus input capacitance. The plot was obtained with the data of example 6.7 for a peaking time of 100 ns.

instead that parallel noise sets the minimum achievable ENC when C_T approaches zero. The quantity:

$$N_S = \frac{1}{q}\sqrt{\frac{v_{nw}^2}{T_p}N_w + N_f A_f} \tag{6.164}$$

is called the noise slope and, in principle, it is measured in units of electrons/F. If we put (6.164) in the parameters of example 6.7 we get a slope $2.043 \cdot 10^{13}$ electrons/F. Measuring N_S in these units is inconvenient because 1 F is an unrealistic large capacitance. As a consequence, the noise slope is usually given in electrons/pF, obtained multiplying (6.164) by $1 \cdot 10^{-12}$. The system of our example has thus a slope of 20 electrons/pF. Fig. 6.21 shows the "noise versus capacitance" curve built with the data of example 6.7. The plot makes apparent the linear increase of the ENC for sufficiently large values of input capacitance. In the frequent case in which thermal noise is dominating, the noise slope can be reduced by increasing the transconductance of the input transistor and/or extending the peaking time.

6.5.4 EFFECT OF $1/F^{\alpha}$ NOISE

A more accurate modeling of the flicker noise requires a refinement of the spectral density used so far, which can be modified as follows:

$$v_{n1/f^{\alpha}}^2 = \frac{K_f(I_C, L)}{C_{ox}WL}\frac{1}{f^{\alpha}} \tag{6.165}$$

As seen in chapter 5, this definition makes explicit the dependency of K_f on the transistor length and on the bias condition, introducing the inversion coefficient I_C. In deep submicron technologies, an increase of K_f as the gate length approaches the minimum allowed by the process has in fact been reported [16–18]. A deviation of α from unity is also observed [16, 19]. It is interesting to evaluate the ENC due to flicker noise for the case $\alpha \neq 1$. As an example, we will do the calculations for the CR-RC^n shapers, but the conclusions can be extended also to the other cases. Assuming the spectral density given in (6.165), the output noise voltage becomes:

$$< v_{out}^2 >_{n1/f^\alpha} = \frac{K_f(I_C,L)}{C_{ox}WL}\left(\frac{C_T}{C_f}\right)^2 \int_0^\infty \frac{1}{f^\alpha}\frac{\left(\frac{f}{f_0}\right)^2}{\left[1+\left(\frac{f}{f_0}\right)^2\right]^{n+1}}df \tag{6.166}$$

The integral can be rearranged and solved with the help of the beta function:

$$\int_0^\infty \frac{\frac{f^{2-\alpha}}{f_0^2}}{\left[1+\left(\frac{f}{f_0}\right)^2\right]}df = \frac{1}{2}B\left(\frac{3-\alpha}{2}, n-\frac{3-\alpha}{2}+1\right)f_0^{1-\alpha} \tag{6.167}$$

Replacing now $f_0 = 1/2\pi RC$, $RC = T_p/n$ and dividing by the shaper gain, we get:

$$ENC_{1/f^\alpha}^2 = \frac{1}{q^2}\frac{K_f(I_C,L)}{C_{ox}WL}C_T^2 N_{1/f^\alpha}(\alpha,n)\frac{1}{T_p^{1-\alpha}} \tag{6.168}$$

where the noise index $N_{1/f^\alpha}(\alpha,n)$ incorporates all the quantities depending only on the pulse shaper. The key point here is that the deviation of α from unity makes the flicker noise component of the ENC dependent on the peaking time. This refined noise model becomes necessary to make accurate predictions of $1/f$ noise in contexts where it is the factor limiting the system resolution. For most applications, however, a careful choice of gate length and transistor area is sufficient to make the flicker contribution negligible with respect to the ones of thermal and parallel noise.

6.5.5 ALTERNATIVE FORMALISM FOR NOISE CALCULATIONS

It is worth mentioning that alternative notations for noise calculations are found in the literature. In fact, instead of using the frequency variable f one could use $\omega = 2\pi f$. Since $df = d\omega/2\pi$, an integral in $d\omega$ and one in df are related by the following relationship:

$$\int_0^\infty |T(j2\pi f)|^2 df = \frac{1}{2\pi}\int_0^\infty |T(j\omega)|^2 d\omega \tag{6.169}$$

Using ω, the flicker noise spectral density becomes:

$$v_{nf}^2 = \frac{K_f}{C_{ox}WL}\frac{1}{f} = \frac{2\pi K_f}{C_{ox}WL}\frac{1}{\omega} \tag{6.170}$$

The ENC due to $1/f$ noise is then written as:

$$ENC_f^2 = \frac{N_f' 2\pi K_f}{C_{ox} WL} \tag{6.171}$$

Hence, the ficker noise index is redefined as $N_f' = N_f/2\pi$. In our calculations, we have first evaluated the noise at the CSA output and we have then considered the effect of the pulse shaper. We can also observe that the following terms:

$$\begin{aligned} & v_{nw}^2 \omega^2 C_T^2 \\ & v_{nf}^2 \omega^2 C_T^2 = \frac{2\pi K_f}{C_{ox} WL} \omega C_T^2 = A_f \omega C_T^2 \end{aligned} \tag{6.172}$$

obtained by dividing the voltage noise by the impedance of C_T, give the currents produced in C_T by the noise voltage sources. The noise at the output can thus be obtained by multiplying these currents by $|A(j\omega)|^2$ and integrating in $d\omega$. Observe that $A(j\omega)$ represents the transfer function of the entire front-end amplifier, i. e. the product of the transfer functions of the CSA and the pulse shaper. The output noise can then be rewritten as:

$$< v_{out}^2 >_{nw} = v_{nw}^2 C_T^2 \frac{1}{2\pi} \int_0^\infty \omega^2 |A(j\omega)|^2 d\omega \tag{6.173}$$

$$< v_{out}^2 >_{n1/f} = A_f C_T^2 \frac{1}{2\pi} \int_0^\infty \omega |A(j\omega)|^2 d\omega \tag{6.174}$$

$$< v_{out}^2 >_{ni} = i_n^2 \frac{1}{2\pi} \int_0^\infty |A(j\omega)|^2 d\omega \tag{6.175}$$

Depending on the context, one formalism can be more advantageous than the other, but of course both lead to the same final results.

6.5.6 NOISE SIMULATIONS

Analytical noise calculations provide a first grasp of the noise circuit performance and indicate basic trends. A more quantitative assessment requires computer simulations, which take into account all active and passive devices in the amplification chain and use noise models more sophisticated than those that can be mastered in a pure analytical approach. Two types of noise simulations are available in advanced CAD programs. The standard approach is the AC noise simulation, which employs the small signal equivalent circuit model (hence it is conceptually similar to what we have done in this section) and returns the rms noise at a given point. In general it is also possible to obtain a ranking of the noise contributors and of the dominating type of noise (white or $1/f$) for each device. This is particularly useful to avoid wrong sizing of devices (e.g. using too small area transistors that might contribute excessive flicker noise, even if they are further down in the amplification chain). The complementary method is the transient noise simulation, in which the noise is added to a time domain simulation, producing as output oscilloscope-grade waveforms in

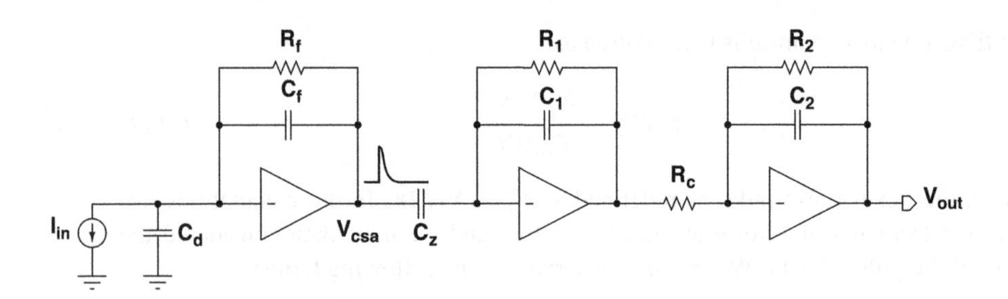

FIGURE 6.22 CR-RC shaper with finite feedback resistor in the CSA.

which the nominal signal is corrupted by noise. This approach is more computation-intensive, but it becomes necessary when noise has to be evaluated for non-linear circuits, since distortion is not considered in the small-signal analysis.

6.6 POLE-ZERO CANCELLATION AND BASELINE CONTROL

We have assumed so far that the feedback resistor of the charge sensitive amplifier does not give any contribution to the signal shape. This is an over-simplification, because even large values of R_f result in visible effects on the impulse response of the front-end. For simplicity, we study this issue for the case of the CR-RC shaper shown in Fig. 6.22, but the conclusions we will derive are qualitatively valid for unipolar shapers of any order. The following parameter values, resulting in a gain of 100 mV/fC with a δ-like input, are used in the calculations:

- $R_f = 20\ \text{M}\Omega$
- $C_f = 100\ \text{fF}$
- $C_z = 5\ \text{pF}$
- $R_1 = R_2 = 100\ \text{k}\Omega$
- $C_1 = C_2 = 500\ \text{fF}$
- $R_c = 36.79\ \text{k}\Omega$.

Before proceeding further with our considerations, it is necessary to observe that a resistor in the MΩ range and above is not usually implemented as a passive component because it would take too much silicon area and exhibit large parasitic capacitance. This can interfere with the circuit stability by introducing additional and undesired poles in the circuit. Therefore, the DC feedback path of the CSA must be provided with transistors and R_f must then be interpreted as the small signal equivalent resistance of this active network. In the shaping stages, on the other hand, the signal swing is substantial and linearity must be maintained over a wider range. Here resistors are very often implemented as true passive devices, but the values required are normally well below 500 kΩ. When a finite value of R_f is considered, the feedback impedance of the CSA is given by the parallel between R_f and $1/sC_f$:

$$Z_f = \frac{R_f}{1 + sC_f R_f} \tag{6.176}$$

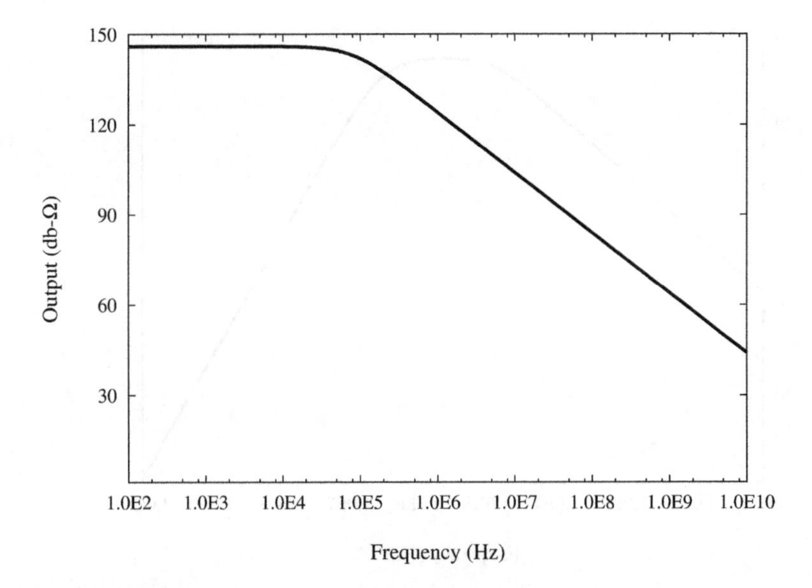

FIGURE 6.23 Bode plot of a CSA with finite feedback resistor.

Consequently, the full transfer function of the front-end amplifier in the Laplace domain becomes:

$$V_{out}(s) = I_{in}(s)\frac{R_f}{(1+s\tau_f)}\frac{sC_zR_1}{(1+s\tau_{sh})^2}\frac{R_2}{R_c} \tag{6.177}$$

where $\tau_f = R_fC_f$ is the time constant associated to the feedback network of the CSA, while $\tau_{sh} = R_1C_1 = R_2C_2$ is the shaping time constant. Observe that the finite value of R_f has moved the pole in the CSA transfer function from the origin to a new value of s, given by:

$$s_{CSA} = -\frac{1}{R_fC_f} \tag{6.178}$$

Therefore, the pole of the CSA and the one of the differentiator do not cancel each other any longer. Both effects can be easily seen considering the Bode plots of the CSA and of the full amplifier, shown in Fig. 6.23 and Fig. 6.24, respectively. As we see from Fig. 6.23, the CSA has a first-order low pass filter behavior. With a feedback capacitor of 100 fF and a feedback resistor of 20 MΩ, the cut-off frequency is $f_{CSA} = 1/2\pi R_fC_f \approx 79.6$ kHz, after which the gain drops with a slope of 20 dB/decade. If $s\tau_f \gg 1$, the "1" in the denominator of equation (6.176) can be neglected, so the feedback impedance of the CSA reduces to the ideal value $1/sC_f$ and the pole-zero cancellation is again effective. Since the zero has been left in the origin, the gain of the full amplifier rises with a slope of 20 dB/decade. When the CSA pole is found, the effect of the zero in canceled and the gain remains constant until the roll-off of 40 dB/decade introduced by the two real coincident shaping poles manifests at a frequency of $f_{sh} = 1/2\pi\tau_{sh} \approx 3.2$ MHz. In other words, the front-end amplifier becomes a band-pass filter. The cut-off of the low portion of the frequency spectrum

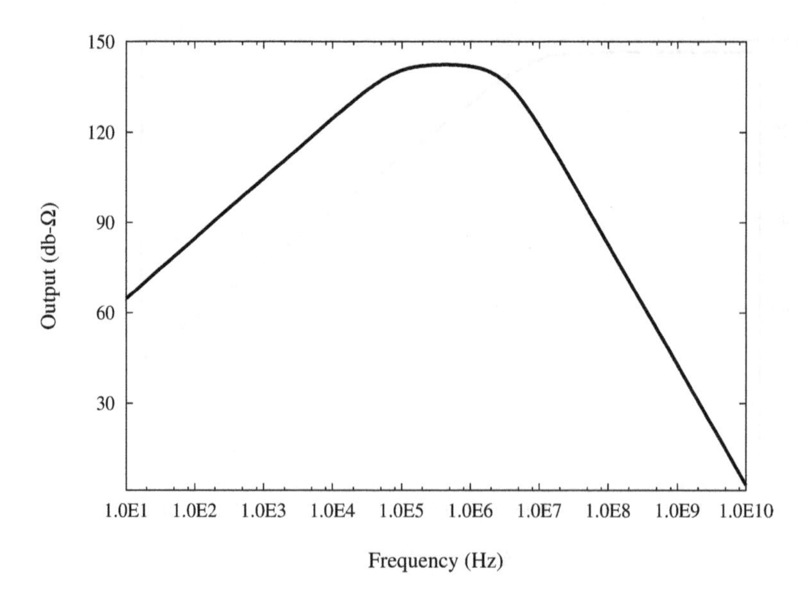

FIGURE 6.24 Bode plot of a CR-RC shaper with finite feedback resistor in the CSA.

has several advantages as the impact of DC and slow variations occurring in the CSA is suppressed. However, it has also potentially undesirable consequences on the signal shape. To understand this point, we take the inverse Laplace transform of equation (6.177) which, under the hypothesis that the input is a δ-like signal, reads:

$$V_{out}(t) = Q_{in}\left[\frac{C_z R_f R_1 R_2}{R_c \tau_{sh}\left(\tau_f - \tau_{sh}\right)}\left(\frac{t}{\tau_{sh}}\right)e^{-\frac{t}{\tau_{sh}}} + \frac{C_z R_f R_1 R_2}{R_c\left(\tau_{sh} - \tau_f\right)^2}\left(e^{-\frac{t}{\tau_{sh}}} - e^{-\frac{t}{\tau_f}}\right)\right] \tag{6.179}$$

The above expression shows that we have introduced an extra contribution in the already familiar CR-RC pulse shape. Since $\tau_f > \tau_{sh}$, in the rightmost part of the equation the second exponential always prevails on the first one and the new term is subtracted to the main signal. The resulting impulse response is represented in Fig. 6.25. Here we see that the waveform goes well below the baseline, before returning to the quiescent point in a time scale defined by the product $R_f C_f$. Called undershoot, this negative tail extends significantly the signal duration, reducing the rate capability of the system. If we reverse the sign of the input charge, we change the polarity of both the peak and its tail, turning the undershoot into an overshoot. The presence of the undershoot (overshoot) stems from the fact that we are feeding a unipolar pulse (the output of the CSA) to a network (the shaper) that contains a capacitor in series. This capacitor blocks any DC component coming from the first stage, so the CSA signal can not modify the DC level at the shaper output. As a consequence, the shaper response can not be purely unipolar, but must have a bipolar nature, so that the net signal contribution to the output DC value is zero. The most important effect of the undershoot is the baseline drift at high rates. Fig. 6.26 shows this effect for pulses with a rate of 1 MHz

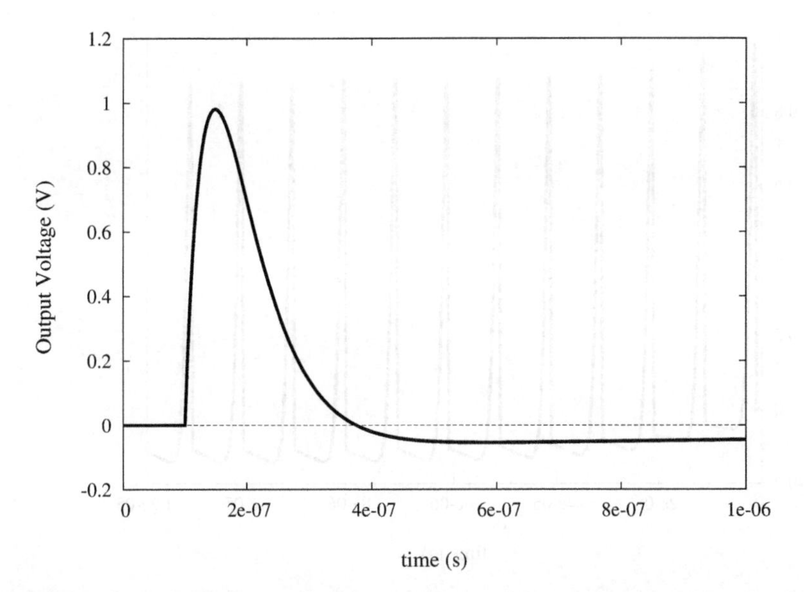

FIGURE 6.25 CR-RC impulse response with finite CSA feedback resistance. Note the undershoot after the signal peak.

applied to our circuit. As we can see from the plot, the baseline drift downwards till the new level that yields a zero average value of the output is reached. If we had to process events regularly spaced in time, we could just wait that the baseline settles to its appropriate value before start taking valid measurements. Unfortunately, this is not the case with radiation sensors, because the events are randomly distributed in time, usually following Poisson statistics. When the interval between pulses is not uniform, the baseline fluctuates up and down rather than reaching a stable value. In case this effect is not properly mastered, such fluctuations can be much larger than those induced by the intrinsic system noise, leading to a significant performance degradation.

Before studying the solutions to this problem, it is interesting to observe that there is a trade-off between the undershoot (overshoot) duration and its amplitude. In fact, for values of t larger than the shaping time constant τ_{sh} we can approximate the undershoot as:

$$V_u(t) \approx -Q_{in}\frac{C_z R_f R_1 R_2}{R_c \tau_f^2} e^{-\frac{t}{\tau_f}} = -\frac{Q_{in}}{C_f}\frac{C_z}{C_f}\frac{R_1 R_2}{R_c R_f} e^{-\frac{t}{\tau_f}} \tag{6.180}$$

Therefore, increasing the time constant in the CSA feedback we reduce the undershoot amplitude, but we extend its duration. In CMOS technologies equivalent values of feedback resistance as high as 1 GΩ can be reached using transistors biased in weak inversion. Equation (6.180) allows us to calculate that in this case the undershoot measured at $t = 1\mu$s drops from 45 mV to 0.9 mV, becoming undetectable in most practical cases. With our settings, 1 mV corresponds in fact to a charge of 65 electrons, a value well below the ENC of most systems. The reduction of the undershoot

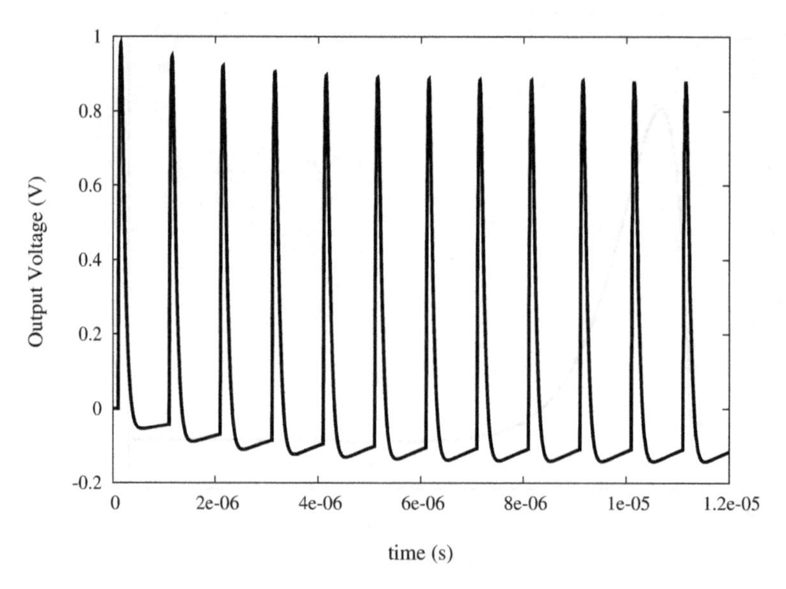

FIGURE 6.26 Example of baseline drift induced by a train of pulses occurring at regular intervals.

amplitude, however, does not solve the issue of baseline drift, which is intrinsically due to the AC coupling between the preamplifier and the shaper. Fig. 6.27 shows the baseline response to a train of pulses when R_f is changed to 1 GΩ. To show the trend without drawing too many pulses, the baseline has been sampled every microsecond just before a new signal is applied. From the plot one may surmise that the baseline drift is smaller, but this is not true. In fact, the amplitude of the output pulses was reduced from 1 V to 100 mV, otherwise the CSA would saturate. Therefore, the final baseline drift measured in percentage of the signal amplitude is the same, but it is reached with a longer time constant, determined by the new value of R_fC_f. Intuitively, for a given pulse amplitude and rate, the baseline must shift by the same amount to keep constant the average output value. However, since at each pulse the undershoot is smaller, more steps are needed to reach the appropriate level. Despite reducing the undershoot on the single pulse, increasing R_f worsens the rate capability, because it makes the CSA more prone to pile-up by discharging its feedback capacitor at a slower pace. This discussion also shows that the study of a front-end amplifier must not be limited to the characterization of the single impulse response. With the help of computer programs, the circuit behavior should be inspected for times longer than the bigger time constant present in the network, feeding input signals with a realistic rate. We emphasize once more here that using regularly spaced pulses is easier from the simulation point of view, but it does not represent accurately a physical situation, where the events occur randomly in time and with varying amplitudes. For an accurate assessment of the rate-dependent effects, both amplitude and time of arrival distribution must be incorporated in the simulation.

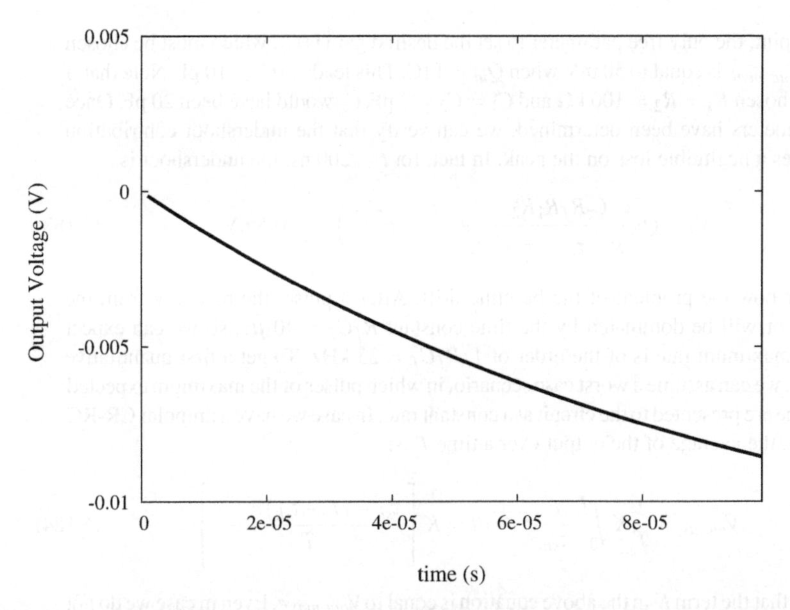

FIGURE 6.27 Baseline drift with R_f set to 1 GΩ. A ten-fold reduction of the input signal with respect to the case $R_f = 1\text{M}\Omega$ was necessary to avoid saturation of the CSA before the baseline trend could be observed.

Example 6.8: Maximum event rate in presence of signal undershoot.

The front-end amplifier of Fig. 6.22 is designed with a CSA feedback resistance of 200 MΩ and a feedback capacitance of 200 fF. Determine the other circuit components so that the peaking time is 200 ns and the gain is 50 mV/fC. Estimate the maximum event rate that the circuit can tolerate, knowing that the ENC is 620 electrons rms.

* * *

To simplify the problem, we assume that the second term in equation (6.179) is negligible when the peaking time is reached. The peak voltage then becomes:

$$V_{out,peak} \approx Q_{in} \frac{C_z R_f R_1 R_2}{R_c \tau_{sh} \left(\tau_f - \tau_{sh}\right)} \frac{1}{e} \tag{6.181}$$

To achieve the desired peaking time we need to have $R_1C_1 = R_2C_2 = 200$ ns. In many technologies, resistors much above 200 kΩ are impractical, because they are large and affected by substantial parasitic capacitance. If we choose $R_1 = R_2 = 200$ kΩ, we need to have $C_1 = C_2 = 1$ pF. If we want to exploit as much as possible the dynamic range of all stages, the ratio R_2/R_c must be maintained equal to e, which leads to $R_c \approx 73.5$ kΩ. We can also use the fact that $\tau_{sh} = R_1C_1 = R_2C_2$ to simplify equation (6.181):

$$V_{out,peak} = Q_{in} \frac{C_z}{C_1} \frac{R_f R_2}{R_c} \frac{1}{\left(\tau_f - \tau_{sh}\right)} \frac{1}{e} \tag{6.182}$$

At this point, the only free parameter to set the desired gain is C_z, which must be chosen so that $V_{out,peak}$ is equal to 50 mV when $Q_{in} = 1$ fC. This leads to $C_z = 10$ pF. Note that if we had chosen $R_1 = R_2 = 100$ kΩ and $C_1 = C_2 = 2$ pF, C_z would have been 20 pF. Once the parameters have been determined, we can verify that the undershoot contribution introduces a negligible loss on the peak. In fact, for $t = 200$ ns, the undershoot is:

$$V_u = Q_{in}\frac{C_z R_f R_1 R_2}{R_c\left(\tau_{sh}-\tau_f\right)^2}\left(e^{-\frac{t}{\tau_{sh}}} - e^{-\frac{t}{\tau_f}}\right) = -0.5\text{mV} \tag{6.183}$$

Consider now the problem of the baseline drift. After a pulse, the recovery from the undershoot will be dominated by the time constant $R_f C_f = 40\ \mu$s, so we can expect that the maximum rate is of the order of $1/R_f C_f = 25$ kHz. To get a first quantitative estimate, we can assume a worst case scenario, in which pulses of the maximum expected amplitude are presented to the circuit at a constant rate. In case we have a unipolar CR-RC response, the average of the output over a time T is:

$$V_{out,av} = \frac{1}{T}K\int_0^T \frac{t}{\tau_{sh}}e^{-\frac{t}{\tau_{sh}}}dt = K\left[\frac{\tau_{sh}-\left(T+\tau_{sh}\right)e^{-\frac{T}{\tau_{sh}}}}{T}\right] \tag{6.184}$$

Observe that the term K in the above equation is equal to $V_{out,max}e$. Even in case we do not expect any baseline drift, the maximum interval between pulses must be significantly greater than the peaking time to avoid pile-up at the output. If we choose $T = 10\tau_{sh}$ the term modulated by the exponential becomes negligible, so we can approximate the average as:

$$V_{out,av} = \frac{K}{T}\tau_{sh} \tag{6.185}$$

To keep the DC value constant, the baseline would shift in direction opposite to the peak by a quantity given by equation (6.185). We can further require that the maximum baseline drift does not exceed the rms noise at the output. With a gain of 50 mV/fC, a noise of 620 electrons rms results in an output rms voltage of 5 mV. The last input that we miss is the maximum output voltage. Assuming that the maximum expected signal has a peak of 1 V(which with our case corresponds to an input of 20 fC), putting all numbers in (6.185) and solving for T, gives $T \approx 100\ \mu$s, which corresponds to a maximum rate of 10 kHz. This is only a first order estimation. On the one hand, in fact, a baseline random fluctuation of the order of the rms noise can be considered too high, since it worsens the noise performance by 40%. On the other hand, not all the pulses will have the maximum amplitude, so in this respect the estimate is too conservative. While the example shows how to make a quick assessment of the problem, a realistic evaluation should take into account both the amplitude and time of arrival statistics of the input pulses.

6.6.1 POLE-ZERO CANCELLATION

A first and very common approach to address the baseline drift issue is to shift the position of the zero introduced by C_z so that its frequency matches again exactly the one of the CSA pole. This is done by adding an extra resistor, R_z, in parallel to C_z.

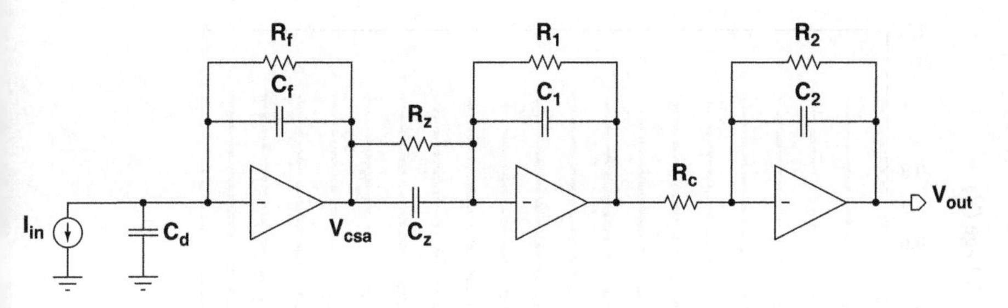

FIGURE 6.28 Front-end with pole-zero cancellation.

The resulting circuit is shown in Fig. 6.28. The transfer function of the circuit can be written as:

$$V_{out}(s) = I_{in}(s)\frac{R_f}{(1+sC_fR_f)}\frac{(1+sC_zR_z)}{R_z}\frac{R_1R_2}{R_c\,(1+s\tau_{sh})^2} \tag{6.186}$$

where, as usual, it has been assumed that $R_1C_1 = R_2C_2 = \tau_{sh}$. If R_z is chosen so that $R_zC_z = R_fC_f$, the terms $1+sR_fC_f$ and $1+sR_zC_z$ cancel out and the transfer function reduces to:

$$V_{out}(s) = I_{in}(s)\frac{R_f}{R_z}\frac{R_1R_2}{R_c\,(1+s\tau_{sh})^2} \tag{6.187}$$

The Bode plot obtained from a system with $R_f = 20$ MΩ, $C_f = 100$ fF, $C_z = 5$ pF, $R_z = 400$ kΩ, $R_1 = R_2 = 100$ kΩ and $C_1 = C_2 = 500$ fF is shown in Fig. 6.29. We observe here that the circuit has a strict low pass filter behavior, so it amplifies any

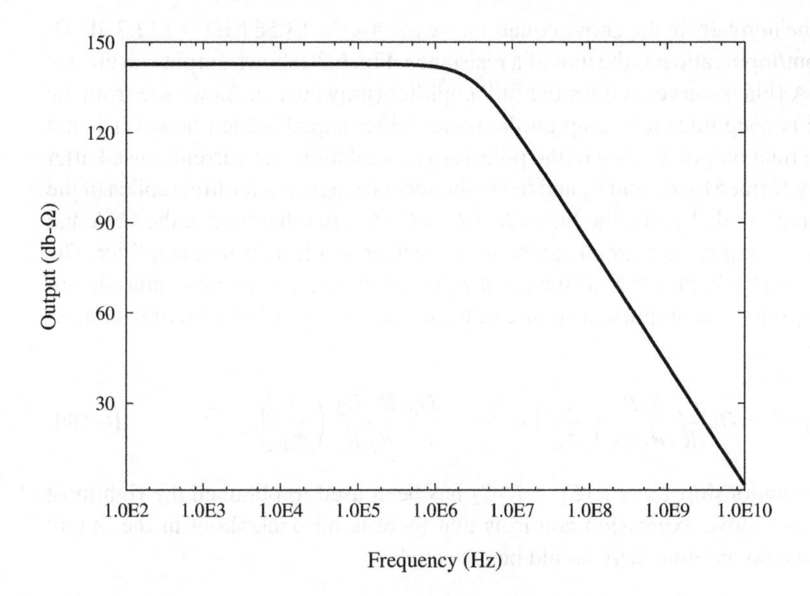

FIGURE 6.29 Bode plot of a front-end with pole-zero cancellation.

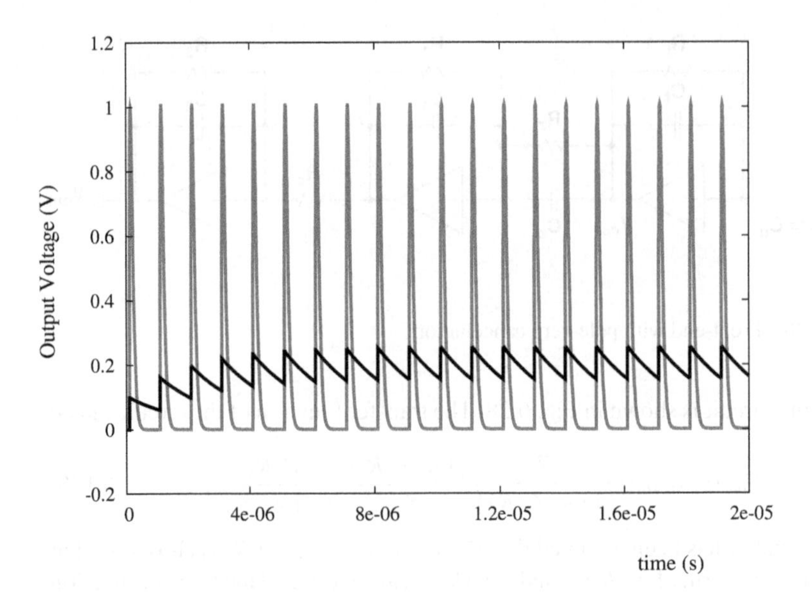

FIGURE 6.30 CSA and front-end outputs in system with pole-zero cancellation.

signal starting from DC. The low-frequency gain is obtained from (6.187) by putting $s = 0$, which yields:

$$A_{DC} = \frac{V_{out}}{I_{in}} = \frac{R_f}{R_z}\frac{R_1}{R_c}R_2 \tag{6.188}$$

Inserting the numbers in the above equation, we get $A_{DC} = 13.56\,\text{M}\Omega = 142.7$ dB-Ω, as the output/input ratio has the unit of a resistance. Fig. 6.30 shows output waveforms for the CSA (black curve) and for the full amplifier (gray curve). As we see from the plot, there is a significant pile-up on the preamplifier signal, which however is not seen in the final output. Owing to the pole-zero cancellation, the current sensed after the network formed by R_f and C_z and fed to the second stage, is a faithful replica of the sensor current, scaled-up by the factor $R_f/R_z = C_z/C_f$. In other words, the CSA and the pole-zero suppression circuit can be seen together as a fast current amplifier. The signal can overlap in the CSA, as long as the pile-up does not lead to the saturation of its core amplifier. The impulse response of the circuit of Fig. 6.28 in the time domain is given by:

$$V_{out}(t) = Q_{in}\frac{R_f}{R_z}\frac{R_1 R_2}{R_c \tau_{sh}}\left(\frac{t}{\tau_{sh}}\right)e^{-\frac{t}{\tau_{sh}}} = \frac{Q_{in}}{C_1}\frac{R_f}{R_z}\frac{R_2}{R_c}\left(\frac{t}{\tau_{sh}}\right)e^{-\frac{t}{\tau_{sh}}} \tag{6.189}$$

where the relationship $\tau_{sh} = R_1C_1 = R_2C_2$ has been used to obtained the rightmost member. The above expression confirms that there is no undershoot in the output signal, hence no baseline drift should be expected.

6.6.2 BASELINE HOLDER AND BASELINE RESTORERS

The pole-zero cancellation makes the circuit sensitive to DC or low frequency variations occurring at its input. In many applications, like the ones involving semiconductor radiation sensors, the detector has a leakage current that must be absorbed by the front-end amplifier without degrading the system performance. As an example, the circuit employed to illustrate the pole-zero technique has a low frequency transimpedance gain of 13.5 MΩ. In this case, an input DC current of 1 nA produces an output voltage shift of 13.5 mV, which is well above the noise floor of most systems. The amount of current to be tolerated depends on the size of the collection electrode and the applied bias voltage, but values in the range 0.1 − 2 nA per channel are commonly encountered. If the sensor is exposed to a radiation field that damages the device bulk, the leakage current increases significantly and may exceed 1 μA/channel in some cases.

In practical circuits, the deviation of the DC levels of the stages from their nominal values is also a concern. In our discussion so far, we have considered the core amplifiers as ideal building blocks, assuming that the input and output voltages in absence of signals were zero. However, if the amplifiers are implemented with single ended cascode topologies biased from a single rail power supply, the DC input and output voltages would be one V_{GS} above ground or one V_{SG} below V_{DD}, depending whether an NMOS or a PMOS is used as the input transistor. In the front-end chain of Fig. 6.30, the difference between the output DC voltage of one stage and the input DC voltage of the next is amplified as much as the signal. For example, if there is a difference ΔV of 10 mV between the DC output voltage of the second stage and the DC input voltage of the third one, a current given by $\Delta V/R_c$ flows in R_c and R_2, determining a shift of the output voltage of $\Delta V_{out} = -\Delta V \times (R_2/R_c) = -27.2$ mV from the nominal baseline. This example shows that in practical circuits, AC coupling between the different building blocks is desirable or in many cases even mandatory. Our previous discussion about baseline drift suggests however that techniques more elaborate than a simple blocking capacitor are required. One possible approach is the baseline holder and it is sketched in Fig. 6.31. The circuit in Fig. 6.31 a) is a scheme suitable to compensate a DC current entering into the stage, while the one of Fig. 6.31 b) is useful to compensate a current sunk from the input. To understand the principle, consider the circuit of Fig. 6.31 a). In absence of the additional feedback network formed by the differential amplifier A_2 and transistor M_1, a current entering into the input node would flow in R_1, lowering the output voltage. The quantity $V_{ref} - V_{out}$ is however sensed and amplified by A_2, which rises the voltage at the gate of M_1. The transistor then sinks the input current, minimizing the portion of it that flows through R_1. In this way, the DC output voltage of the stage can be locked to the reference voltage fed to the differential amplifier, provided that the loop gain is high enough. The differential amplifier and the transistor can be considered as a single transconductance amplifier, with transconductance G_m given by $G_m = g_{m1}A_{d2}$ where g_{m1} is the transconductance of M_1 and A_{d2} is the differential voltage gain of A_2. To make the analysis more quantitative, we suppose that A_2 has a single pole transfer

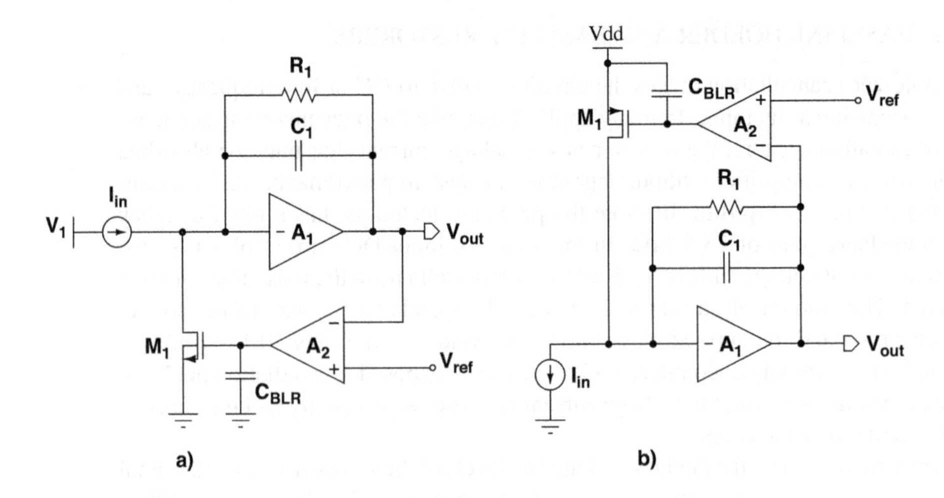

FIGURE 6.31 Baseline holder principle.

function given by:

$$A_{d2}(s) = \frac{A_{d20}}{(1+s\tau_L)} \tag{6.190}$$

The capacitor C_{BLR} is inserted to keep the bandwidth of the differential stage small enough, as we want that this additional feedback loop processes only low-frequency signals. The transconductance G_m can be written as:

$$G_m = \frac{A_{d20}g_{m1}}{(1+s\tau_L)} = \frac{G_{m0}}{(1+s\tau_L)} \tag{6.191}$$

where G_{m0} is the low-frequency overall transconductance. We start analyzing the circuit with the assumption that the time constant τ_L is much bigger that $\tau_1 = R_1C_1$, so that we can consider only the resistive part of the feedback impedance of the main amplifier A_1. Writing the equation for the input node we have:

$$-I_{in} + \frac{V_{in}-V_{out}}{R_1} + I(M_1) = 0 \rightarrow I_{in} = I(M_1) - \frac{V_{out}}{R_1} \tag{6.192}$$

If the input node of A_1 is treated as a virtual ground (i.e. we assume that $V_{in} = 0$), the small signal current flowing in M_1 can be written as:

$$I(M_1) = -G_mV_{out} \tag{6.193}$$

Inserting this expression of $I(M_1)$ in equation (6.192) and solving for V_{out}/I_{in} we get:

$$\left|\frac{V_{out}}{I_{in}}\right| = \frac{R_1}{1+G_mR_1} = \frac{R_1}{1+\frac{G_{m0}R_1}{1+s\tau_L}} \rightarrow \frac{R_1}{1+G_{m0}R_1}\frac{1+s\tau_L}{1+s\frac{\tau_L}{1+G_{m0}R_1}} \tag{6.194}$$

Let's now examine equation (6.194) in more detail. At low frequency ($s = 0$), the transimpedance gain becomes:

$$\frac{V_{out}}{I_{in}} = \frac{R_1}{1 + G_{m0}R_1} \tag{6.195}$$

To appreciate the orders of magnitude involved, assume that $R_1 = 200$ kΩ, $g_{m1} = 100\ \mu$S and $A_{d0} = 1000$. With these numbers, the low frequency gain becomes 10 Ω. This implies that a variation in the input DC current of 1 μA produces a change in the output voltage of only 10 μV. In absence of the additional feedback, the current would entirely flow in R_1, resulting in a change at the output of 0.2 V. As the frequency rises, the zero is found at $f_l = 1/2\pi\tau_L$. From this point, the gain rises with a slope of 20 dB/decade till the pole is met at a frequency $1 + G_{m0}R_1$ greater than the one of the zero. Above the pole frequency, the gain can be approximated as:

$$\frac{V_{out}}{I_{in}} \approx \frac{R_1}{G_{m0}R_1} \frac{s\tau_L}{\frac{s\tau_L}{G_{m0}R_1}} = R_1 \tag{6.196}$$

where the condition $G_{m0}R_1 \gg 1$ has also been used. High frequency signals are then again amplified by R_1. This separation in frequency is important, because the G_m feedback must only compensate undesired DC or close to DC components.

Example 6.9: Analysis of a stage with baseline holder.

Consider the circuit of Fig. 6.31 a). Suppose that the amplifier A_1 is implemented as a single-ended cascode stage and requires an input DC voltage of 0.6 V. If V_{ref} is set to 1.5 V, explain what happens for the following DC values of I_{in}: $I_{in} = 0$, $I_{in} = 3\ \mu$A entering into the input node, $I_{in} = 3\ \mu$A exiting from the input node and $I_{in} = 6\ \mu$A exiting from the input node. The value of R_1 is 200 kΩ.

* * *

Let's start considering the situation for $I_{in} = 0$. Note first that without the baseline holder, in this case the output DC voltage would be equal to the input one because the input of the main core amplifier does not sink or source any current. The low-frequency feedback acts so that the difference between the input voltages of the differential amplifier is instead minimized, making the output DC voltage track the reference voltage V_{ref}. To bring the output voltage to 1.5 V, a current given by:

$$I(R_f) = \frac{1.5\text{ V} - 0.6\text{ V}}{200\text{ k}\Omega} = 4.5\mu\text{A} \tag{6.197}$$

must flow in R_1. This current must be sunk from the input of A_1 towards ground, so it can be provided by M_1. If now we have an additional DC current of 3μA entering into the node, the feedback acts so that the gate voltage of M_1 is increased and the transistor sinks a total current of 7.5 μA. The limit for currents entering into the input node is basically defined by the current drive capability of M_1. The higher the current, the bigger the voltage at the gate of M_1 will be and for big currents the value required could be so high that the transistors in the differential amplifiers do not have enough headroom to work in saturation any longer. Hence, M_1 must be appropriately sized so that it can sink the maximum expected DC current. Consider now the case of a current of 3 μA sunk from the input. We still need a total current of 4.5 μA flowing in R_1 to keep the output DC voltage at 1.5 V. Since 3 μA are already provided from the current source, M_1 must only deliver the difference and the feedback reduces its current to 1.5 μA. The current of M_1 can be reduced, but its polarity can not be reversed. If the sunk current now becomes 6 μA, M_1 is switched off. The 6 μA current flows in R_1 rising the output voltage to 1.8 V, provided that this value can be tolerated by A_1 without saturating. The DC output voltage is not locked to V_{ref} any longer, but it is only determined by the input current. This analysis shows that a unipolar baseline holder may compensate currents of the "wrong" polarity, but in general to a much less extent than the ones of the polarity it has been designed for.

It is interesting to study the response of circuit described by equation (6.194) to a sudden change in the input DC current. To do so, we model the current variation as a step, represented in the Laplace domain by I_{in0}/s, where I_{in0} is the step amplitude. Neglecting the time constant R_1C_1, the response of the circuit can be approximated as:

$$V_{out}(s) = \frac{I_{in0}}{s} \frac{R_1}{1+G_{m0}R_1} \frac{1+s\tau_L}{1+s\frac{\tau_L}{1+G_{m0}R_1}} \tag{6.198}$$

Taking the inverse Laplace transform, we have in the time domain:

$$V_{out}(t) = \frac{R_1 I_{in0}}{1+G_{m0}R_1} \left[\frac{(\tau_L - \tau_b)}{\tau_b} e^{-\frac{t}{\tau_b}} + 1 \right] \tag{6.199}$$

In the above equation, we have indicated as τ_b the time constant of the pole introduced by the baseline holder, given by:

$$\tau_b = \frac{\tau_L}{1+G_{m0}R_1} \tag{6.200}$$

Considering that $\tau_L \gg \tau_b$, and inserting the expression of τ_b in equation (6.199), we can rewrite it as:

$$V_{out}(t) = I_{in0} \left(R_1 e^{-\frac{t}{\tau_b}} + \frac{R_1}{1+G_{m0}R_1} \right) \tag{6.201}$$

Therefore, at $t = 0$ the current step is initially fully amplified by R_1. Then the first term decays exponentially to zero with time constant τ_b, leaving only the second term, which is the suppressed DC gain already calculated. The response to a step input current of 1 μA for a baseline holder with the same parameters used in the example is shown in Fig. 6.32. We see from equation (6.201) and from the plot, that τ_b defines the time-scale necessary to the circuit to recover the baseline after a sudden change has occurred at the input. The response to an input step is similar to the one

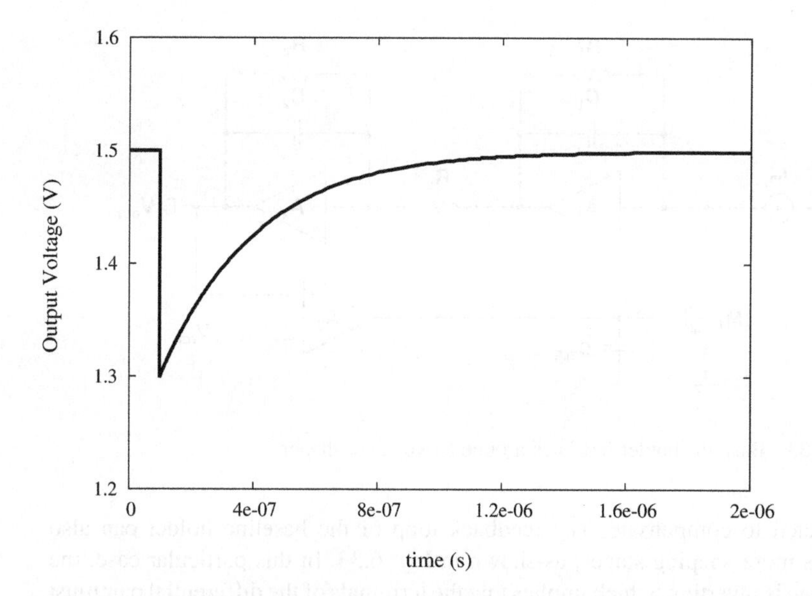

FIGURE 6.32 Response of a linear baseline holder to a current input step.

of the shaper high pass filtering stage. The DC loop gain $G_{m0}R_1$ defines both the attenuation at low frequency and the position of the pole. If the latter is located at too high frequencies, the high pass filter affects the fast input signals, introducing a tail. There is hence a trade-off between the need of suppressing low-frequency components and the one of avoiding undesirable differentiation of high-frequency signals.

Let's now consider a more realistic case, in which we keep into account also the feedback capacitance C_1. We can rewrite the baseline holder transfer function by replacing R_1 with the complex impedance $Z_1 = R_1/(1+sC_1R_1)$:

$$\frac{V_{out}}{I_{in}} = \frac{R_1(1+s\tau_L)}{s^2\tau_1\tau_L + s(\tau_1+\tau_L) + 1 + G_{m0}R_1} \tag{6.202}$$

Thus, when C_1 is considered, we have a second order transfer function which could also have complex conjugate poles and can be potentially unstable. To avoid complex conjugate roots, we need to require that:

$$(\tau_L+\tau_1)^2 > 4\tau_1\tau_L(1+G_{m0}R_1) \tag{6.203}$$

If we suppose that $\tau_L \gg \tau_1$ and that $G_{m0}R_1 \gg 1$, the above conditions can be approximated by:

$$\tau_L > 4\tau_1 G_{m0}R_1 \tag{6.204}$$

which links the low frequency time constant, the shaping time constant and the loop gain. We see that an excessive loop gain may also take the circuit to instability. If the current in M_1 increases, so does its transconductance and thus the loop gain, therefore the stability of a baseline holder must be checked for the maximum current

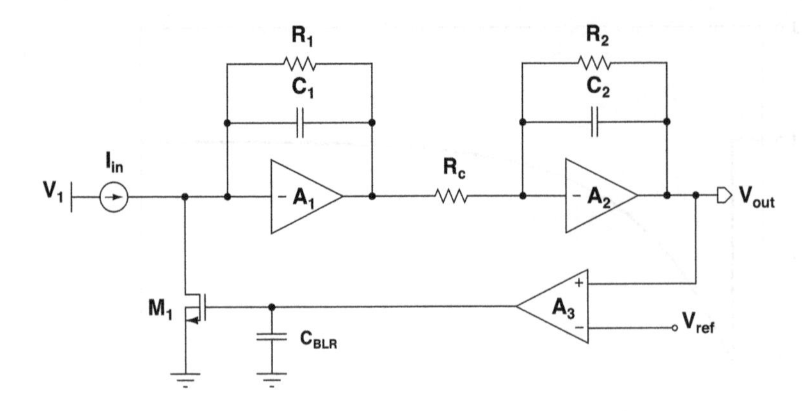

FIGURE 6.33 Baseline holder feedback around a two-stage shaper.

it is expected to compensate. The feedback loop of the baseline holder can also encompass more shaping stages, as shown in Fig. 6.33. In this particular case, the second stage is inverting, which implies that the terminals of the differential pair must be exchanged with respect to Fig. 6.31 in order to maintain the negative feedback. For the circuit of Fig. 6.33, the DC loop gain is given by:

$$G_0 = A_{d30} g_{m1} R_1 \frac{R_2}{R_c} \tag{6.205}$$

If τ_{sh} is the shaping time constant, the BLH should have a dominant time constant, τ_L which satisfies two conditions:

$$\begin{cases} \tau_L \gg G_0 \tau_{sh} \\ \tau_L \gg \tau_{lnd} \end{cases} \tag{6.206}$$

where τ_{lnd} is the first non-dominant internal pole of the BLH. The circuit described so far simply mimics an AC coupling, even though it offers two clear advantages with respect to a simple series capacitor. First, the cut-off frequency can be set not only by sizing the capacitor C_{BLH}, but also the current in the differential stage. Biasing this at low currents, makes it possible to have low cut-off frequencies even at moderate values of C_{BLH}. Second, the output DC voltage can be locked to a desired level to optimize the dynamic range of the stage and the coupling to the following ones. However, due to the high pass filtering transfer function, the circuit still introduces an undershoot when driven by unipolar signals, which implies that we run again into the problem of baseline drift. The issue is addressed by supplementing the network with a non-linear element which is able to distinguish between fast pulses (which must be left unaffected by the baseline holder) and slow variations, which must be compensated for. In CMOS technologies, this effect can be achieved by introducing in the chain a stage with a severe slew-rate limitation [20]. The concept is represented in Fig. 6.34 where a unity gain buffer dumped by a capacitor at the output has been introduced in front of the differential pair. The main purpose of this capacitor is not

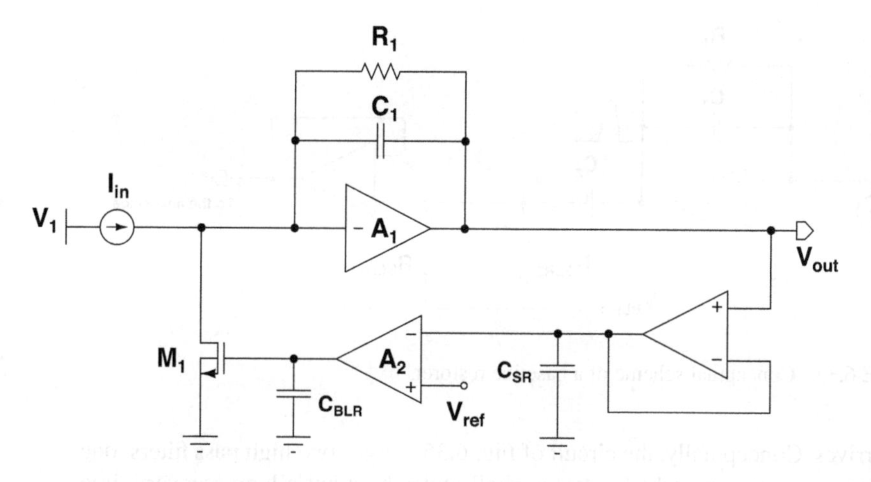

FIGURE 6.34 Baseline holder with slew rate limitation to suppress baseline drift at high rates.

to cut the bandwidth, but to limit the ability of the baseline holder to follow fast transients. In fact, the slew rate of the buffer is given by $SR = I_{max}/C_{SR}$, where I_{max} is the maximum current available to the buffer to charge and discharge C_{SR}. Suppose, for instance that SR is limited to 20000 V/sec and that the output of the front-end delivers pulses with a maximum amplitude of 1 V and a peaking time of 200 ns. Due to the slew rate limitation, in 200 ns the output of the buffer can only move by $\Delta V = SR\dot{T}_p = 4$ mV. Therefore, a fast pulse is suppressed before being presented to the section of the baseline holder implementing the AC coupling, thereby reducing the baseline drift at high rates. Slow variations are not affected by the slew rate limitation and they are fully compensated for by the feedback loop. Note that, in practical cases, the differential pair will also have its own slew rate limitation, which implies that even without the extra buffer the baseline drift at high frequency is less severe than the one predicted by the pure linear model. A exhaustive discussion of the baseline holder technique can be found in [20]. Transistor level implementations will be presented in chapter 8.

6.6.3 BASELINE RESTORERS

An alternative method to the baseline holder is the baseline restorer, which is a time-variant circuit. The concept is shown in Fig. 6.35. Here, the first stage represents a generic chain delivering unipolar pulses. Resistor R_{BLR1} establishes a DC path from the reference voltage V_{BLR} to the buffer input. The switch S must be open when the signal arrives. If the time constant $C_z R_{BLR1}$ is much bigger than the pulse peaking time, a single pulse is passed over the capacitor with a minimal undershoot. However, without S and R_{BLR2}, the circuit is a simple high pass filter and baseline drift would occur at high rates. To avoid this, after the pulse, S is closed. R_{BLR2} is much smaller than R_{BLR1}, so that the time constant $C_z R_{BLR2}$ allows for a quick recovery of the baseline to the reference level V_{BLR}. The switch is then open again before the next

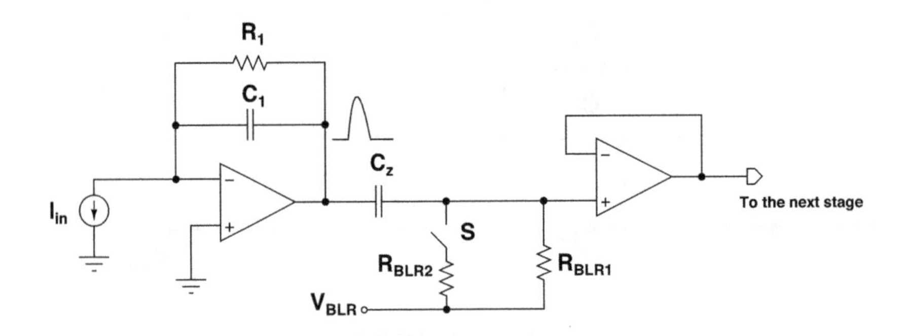

FIGURE 6.35 Conceptual scheme of a baseline restorer.

pulse arrives. Conceptually, the circuit of Fig. 6.35 merges two high pass filters, one with long time constant, which gives a small undershoot but a long recovery time with another with short time constant, which offers short recovery time but bigger undershoot and visible differentiation of the signal. To promptly recover the DC level without affecting the unipolar pulse shape, the system swaps between the former when a pulse is present, and the latter after the pulse peak is over. In this case, the baseline restoration circuit is made active only at selected times by switches controlled by an appropriate circuit (not shown in Fig 6.35), therefore we speak of a "gated baseline restorer". Alternatively, the switch can be replaced by a non-linear element which allows only unidirectional current flow, so that it becomes active only when the buffer input goes below the reference voltage. Baseline restorers were introduced quite early in the history of nuclear electronics [21, 22]. Fundamental design issues are described, for instance, in [23] and the effects of baseline restoration on the noise performance have been studied in [24]. A review of different techniques can be found in [25], while [26] and [27] provide examples of modern implementations.

6.6.4 BIPOLAR SHAPERS

A complementary approach to the use of baseline restorers or baseline holders consists of introducing one or more additional high pass filters in the signal path. This leads to a network with an impulse response with a lobe going significantly below or above the baseline, but returning quickly to the steady state. The simplest implementation of the concept is shown in Fig. 6.36. A filter with this topology is known as a CR^2-RC shaper, but the technique can be applied also to systems with higher order integrators (hence the name CR^2-RC^n) or in filters having complex conjugate poles. If we choose the time constants so that $R_1C_1 = R_2C_2 = R_3C_3 = \tau$ we can write the transfer function as:

$$V_{out}(s) = I_{in}(s)\frac{R_f}{R_z}\frac{R_1}{R_c}R_2\frac{s\tau}{(1+s\tau)^3} \tag{6.207}$$

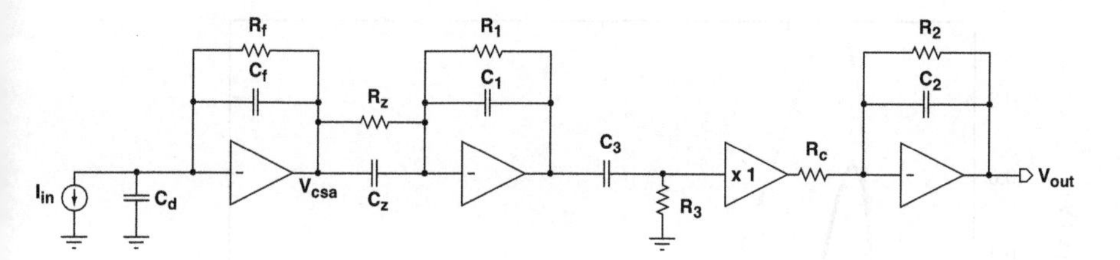

FIGURE 6.36 Schematic of a CR^2-RC shaper.

The impulse response of the system in the time domain therefore is:

$$V_{out}(t) = \frac{Q_{in}}{C_1}\frac{R_f}{R_z}\frac{R_2}{R_c}\left[\frac{t}{\tau} - \frac{1}{2}\left(\frac{t}{\tau}\right)^2\right]e^{-\frac{t}{\tau}} \quad (6.208)$$

Noting that in a system with pole-zero cancellation $R_f/R_z = C_z/C_f$, we can also write:

$$V_{out}(t) = \frac{Q_{in}}{C_f}\frac{C_z}{C_1}\frac{R_2}{R_c}\left[\frac{t}{\tau} - \frac{1}{2}\left(\frac{t}{\tau}\right)^2\right]e^{-\frac{t}{\tau}} \quad (6.209)$$

Differentiating the above expression and equating the result to zero we find two solutions, both proportional to the time constant τ:

$$\tau_1 = \tau\left(2-\sqrt{2}\right) \quad (6.210a)$$

$$\tau_2 = \tau\left(2+\sqrt{2}\right) \quad (6.210b)$$

The first solution, τ_1, corresponds to the maximum of the bipolar pulse, while the second one to its minimum, as we can see in Fig. 6.37. The plot shows the impulse response delivered by a filter employing the same component values already used in our previous examples, i.e. $R_f = 20$ MΩ, $C_f = 100$ fF, $R_z = 400$ kΩ, $C_z = 5$ pF, $R_1 = R_2 = R_3 = 100$ kΩ, $C_1 = C_2 = C_3 = 500$ fF. In this simulation, the input charge was 10 fC. Note that the amplitudes of the negative and positive parts of the pulse are now comparable, therefore CR^2-RC circuits are also known as bipolar shapers. Since the time constant τ is 50 ns, using equation (6.210) we can easily calculate that the maximum and the minimum must respectively occur 29.3 ns and 170.7 ns after the signal start. For comparison, the figure reports also the impulse response of the CR-RC counterpart with $\tau = 50$ ns. We can conclude that the signal of the CR^2-RC filter has the same duration of the one provided by a CR-RC shaper with identical time constants. However, the bipolar waveform has zero area, which means that no baseline drift will occur at high rates. This is apparent in Fig. 6.38, which reports the circuit response to input pulses with 1 MHz frequency.

The peak amplitude can be calculated from equation (6.208) by putting $\tau_1 = \tau\left(2-\sqrt{2}\right)$ and is 62% of the one of the corresponding CR-RC case. The noise indexes of the CR^2-RC can be evaluated with the same methodology employed for

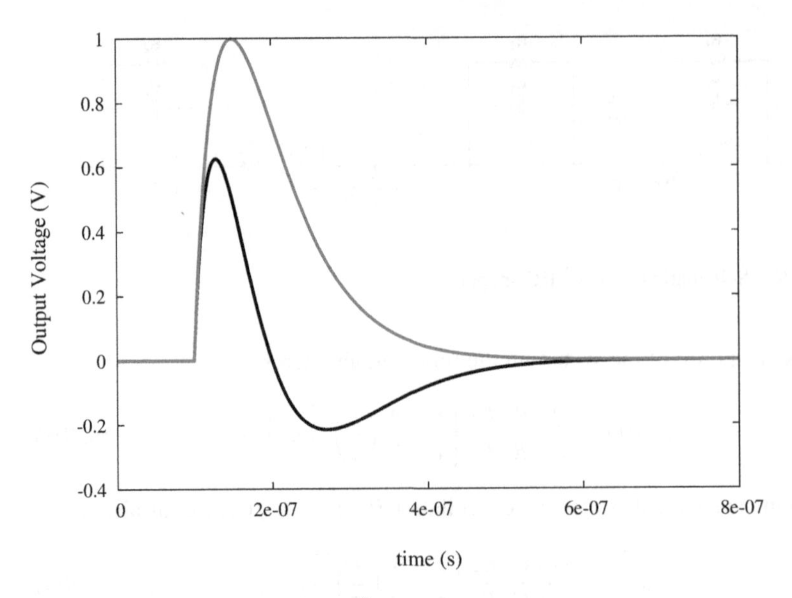

FIGURE 6.37 Impulse response of a CR^2-RC shaper (curve in black). For comparison, the response of the CR-RC pulse before differentiation is also shown.

the CR-RC^n filters and are: $N_w = 0.93$, $N_f = 3.7$ and $N_1 = 0.88$. Indexes for higher order bipolar shapers can be found, for instance, in [16]. The noise indexes for the CR^2-RC and CR-RC shapers are very similar. However, the almost 40% reduction in the peak height of the CR^2-RC filter leads to a worse ENC figure. Another potential disadvantage of the bipolar pulse shape is that the negative lobe occupies a significant portion of the output dynamic range of the amplifier. Despite these shortcomings, CR^2-RC^n filters are robust and simpler to implement than circuits with baseline holders or baseline restorers. Therefore, they are an attractive solution for high rate applications in case achieving the minimum theoretical noise is not indispensable. Finally, Fig. 6.39 reports the Bode plot of the circuit, which further illustrates the band-pass nature of this type of shaper. The T-bridge network introduced earlier can also be conveniently used in implementing bipolar shapers. Consider first the circuit shown in Fig. 6.40. The transfer function can be written as:

$$\frac{V_{out}(s)}{I_{in}(s)} = (R_3 + R_4)\frac{1 + s(R_3//R_4)C_3}{s^2 C_2 C_3 R_3 R_4 + sC_2(R_3 + R_4) + 1} \tag{6.211}$$

Dividing both numerator and denominator by $C_2C_3R_3R_4$ and factorizing the coefficient of s in the denominator, we can rewrite the transfer function in the following way:

$$\frac{V_{out}(s)}{I_{in}(s)} = \frac{1}{C_2}\frac{(s+b)}{s^2 + bs + c} \tag{6.212}$$

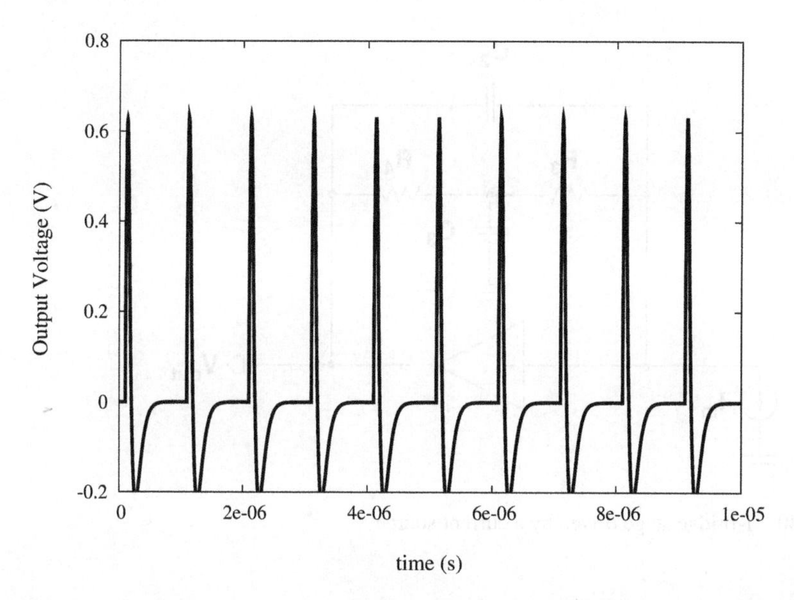

FIGURE 6.38 Impulse response of a CR^2-RC with 50 ns time constant (29 ns peaking time) to a train of 1 MHz pulses.

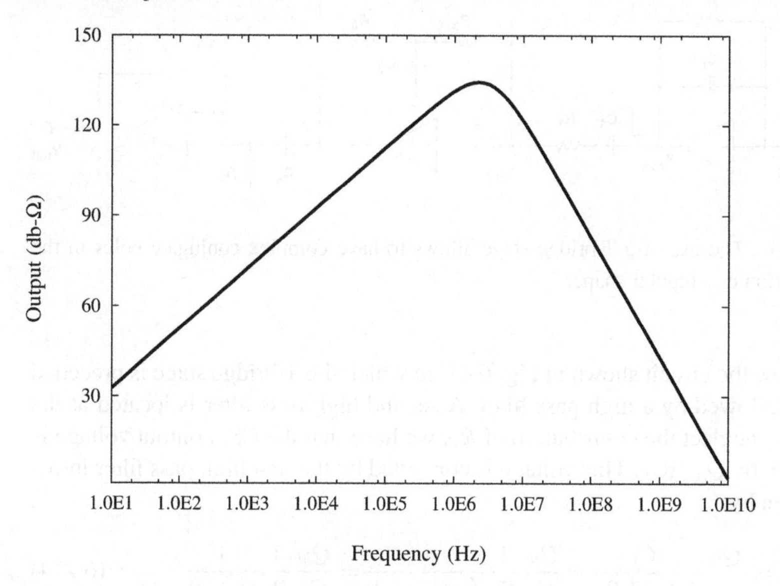

FIGURE 6.39 Bode plot of a CR^2-RC shaper. (Circuit parameters are given in the text).

where the coefficients b and c are respectively:

$$\begin{cases} b = \frac{1}{(R_3//R_4)C_3} \\ c = \frac{1}{C_2C_3R_3R_4} \end{cases} \tag{6.213}$$

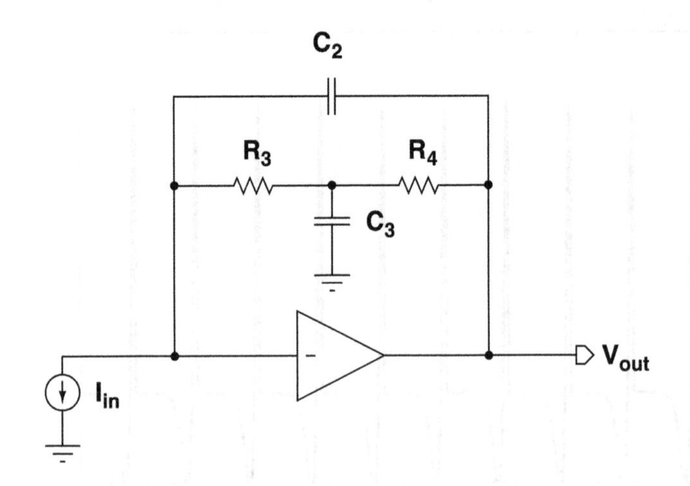

FIGURE 6.40 T-bridge stage driven by a current source.

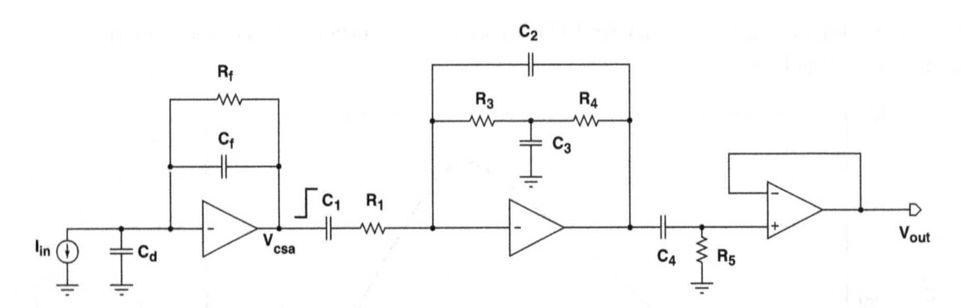

FIGURE 6.41 The use of a T-bridge stage allows to have complex conjugate poles in the transfer function of a bipolar shaper.

Consider now the circuit shown in Fig. 6.41, in which the T-bridge stage is preceded by a CSA followed by a high pass filter. A second high pass filter is located at the output. If we neglect the contribution of R_f, we have that the CSA output voltage is simply given by Q_{in}/sC_f. This voltage is converted by the first high pass filter into a current given by:

$$I(s) = \frac{Q_{in}}{sC_f}\frac{sC_1}{1+sC_1R_1} = \frac{Q_{in}}{C_f}\frac{1}{R_1}\frac{1}{\left(\frac{1}{C_1R_1}+s\right)} = \frac{Q_{in}}{C_f}\frac{1}{R_1}\frac{1}{(s+a)} \tag{6.214}$$

If we call V_T the voltage at the output of the T-bridge stage, after the second high pass filter we get:

$$V_{out}(s) = V_T(s)\frac{sR_5C_4}{(1+sR_5C_4)} = V_T(s)\frac{s}{\left(\frac{1}{R_5C_4}+s\right)} = V_T(s)\frac{s}{(s+d)} \tag{6.215}$$

Combining all the terms, we can write the global system response as:

$$V_{out}(s) = \frac{Q_{in}}{C_f}\frac{1}{R_1C_2}\frac{s(s+b)}{(s+a)(s+d)(s^2+bs+c)} \tag{6.216}$$

From the above equation we see that by choosing $b = d$ we can cancel the zero introduced by the T-bridge network and we are left with a zero in the numerator that yields the bipolar response. Furthermore, by properly selecting the coefficient b and d, we can also introduce complex conjugate poles that makes the return to the baseline faster. Other band-pass bi-quadratic cells can also be used to introduce complex conjugate poles in the transfer function. Bipolar shapers are treated in detail in [28]. An example of implementation in CMOS technology can be found in [29].

6.7 FRONT-END WITH TRANSIMPEDANCE INPUT STAGE

The transfer function of a CSA with finite feedback resistor is given by:

$$T_{CSA}(s) = \frac{R_f}{1+sR_fC_f} = \frac{R_f}{1+s\tau_f} \tag{6.217}$$

Taking the inverse Laplace transform, we get for a Dirac-delta input:

$$V_{out}(t) = \frac{Q_{in}R_f}{\tau_f}e^{-\frac{t}{\tau_f}} = \frac{Q_{in}}{C_f}e^{-\frac{t}{\tau_f}} \tag{6.218}$$

Observe that the pulse shape is identical to the one given by a CSA with infinite feedback resistance followed by a high pass filter (see equation (6.6)). Therefore, if R_f is kept small enough, the CSA has a fast return to the baseline. Adding for instance an additional stage identical to the first one, as shown in Fig. 6.42, results in a circuit with a CR-RC-like transfer function:

$$T(s) = \frac{R_fR_{sh}}{R_c}\frac{1}{(1+s\tau_{sh})^2} \tag{6.219}$$

where we have assumed that $\tau_f = \tau_{sh}$. Note that the input stage provides also the first shaping pole. More elaborated shapers can of course be employed. Although a CSA is a particular case of transimpedance amplifier, in the language of radiation detectors it is said the input stage works in "transimpedance mode" when the feedback capacitor of the input stage is discharged fast enough to make superfluous the use of a high pass filter to shorten the pulse. The drawback of the approach is that a small feedback resistance increases the parallel noise. Therefore, the method is particularly suited when high event rates and/or high leakage current in the sensor make necessary the

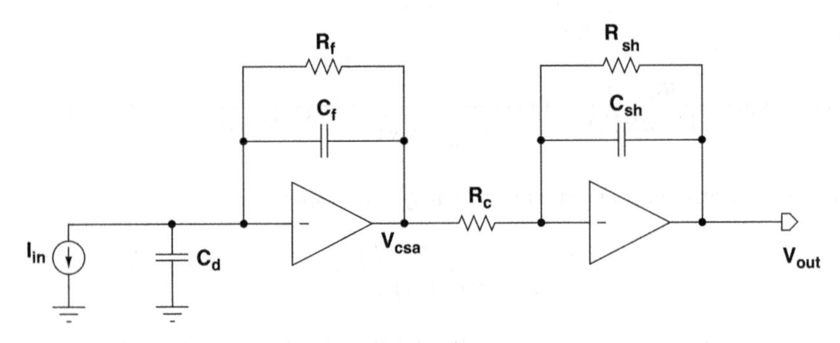

FIGURE 6.42 Front-end amplifier with input stage working in "transimpedance mode". If the feedback resistance R_f is sufficiently small, the return to the baseline of the first stage is fast enough and the high pass filter can be avoided.

use of short shaping times. Observe the input stage may also have complex conjugate poles. In this case, the first stage may already provide a satisfactory pulse shape and the signal, if necessary, is further amplified with large bandwidth voltage amplifiers. Examples of circuits with the input stage working in transimpedance mode can be found, for instance, in [30–34].

6.8 GAIN AND BANDWIDTH LIMITATIONS IN CHARGE SENSITIVE AMPLIFIERS (CSA)

We have assumed so far that the core stages used in the implementation of our front-end chains were ideal. However, the gain and bandwidth limitations of these circuits must be considered and they are particularly important for the charge sensitive amplifier as this stage provides the direct interface to the sensor.

6.8.1 EFFECTS OF FINITE GAIN IN THE CSA

Using the circuit of Fig. 6.43, we start studying the response of a CSA in which the core amplifier has yet infinite bandwidth, but a finite gain A_0. The input node can not be considered as a virtual ground any longer, so the nodal equation for the input now reads:

$$I_{in}(s) + V_{in}sC_T + (V_{in} - V_{out})\, sC_f = 0 \tag{6.220a}$$

$$V_{in} = -\frac{V_{out}}{A_0} \tag{6.220b}$$

Combining the two relationships and solving for V_{out}/I_{in} one gets:

$$\frac{V_{out}}{I_{in}} = -\frac{A_0}{s\left[C_T + (1+A_0)\,C_f\right]} \tag{6.221}$$

To find again the ideal result $V_{out}/I_{in} = 1/sC_f$ we need to satisfy the two following conditions:

$$A_0 \gg 1 \tag{6.222a}$$

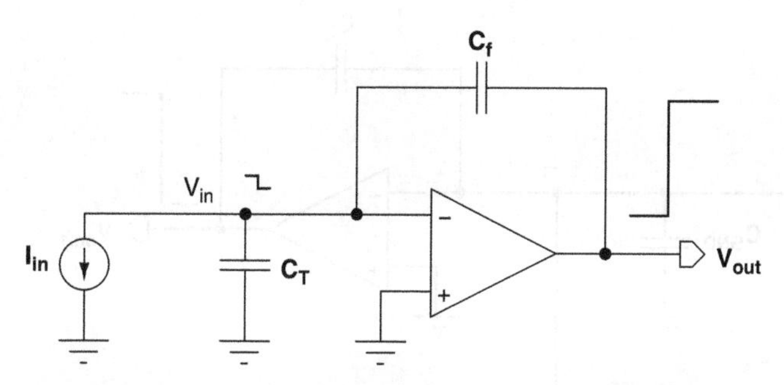

FIGURE 6.43 Charge sensitive amplifier with finite DC gain in the core stage. In presence of a signal, the input node swings as well, although to a much less extent than the output one.

$$(1+A_0)\,C_f \gg C_T \tag{6.222b}$$

The gain of the amplifier should not only be much above unity, but the feedback capacitance multiplied by the open-loop gain must be much greater than the total capacitance seen between the amplifier input and ground. This can be understood using the Miller theorem, which represents the term $(1+A_0)C_f$ as a capacitance in parallel to C_T. Since both capacitances are in parallel, they will accumulate a charge given respectively by $Q_T = V_{in}C_T$ and $Q_f = V_{in}(1+A_0)C_f$. The sum of these two charges must be equal to the one contained in the input signal, i.e. $Q_T + Q_f = Q_{in}$. However, only Q_f, being physically stored in C_f, contributes to the output signal, while the rest is lost to further processing. The problem is even more serious because in multichannel sensors, at least a fraction of the sensor capacitance stems from the coupling between one channel and its neighbors. The lower the amplifier gain, the higher the swing of the input voltage will be, coupling part of the signal to the neighbors and originating cross-talk. This situation is sketched in Fig. 6.44. In the figure, C_{GND} represents the portion of the input capacitance that can be referred to the signal ground while C_M is the mutual capacitance coupling the two stages. A simulation of two CR-RC shapers with the inputs stages connected as described in Fig. 6.44 is reported in Fig. 6.45. The plot shows the output of the channel that does not receive the pulse for two different gains of the core amplifier. In the simulation G_{GND} was set to 2 pF and C_M to 5 pF. The output of the channel receiving the signal was 1 V, so the cross-talk in the worst case in about 5%. Note that it is important to prevent the CSA from saturating, because in this case the gain drops and the charge in directly integrated on the input node, rising significantly the input voltage and exacerbating the cross-talk.

6.8.2 EFFECT OF CSA BANDWIDTH LIMITATION

We now reconsider the effects of a finite bandwidth in the CSA, summarizing the key results that were obtained in chapter 4. A typical implementation would use as core amplifier one of the cascode stages introduced in chapter 3 and whose frequency

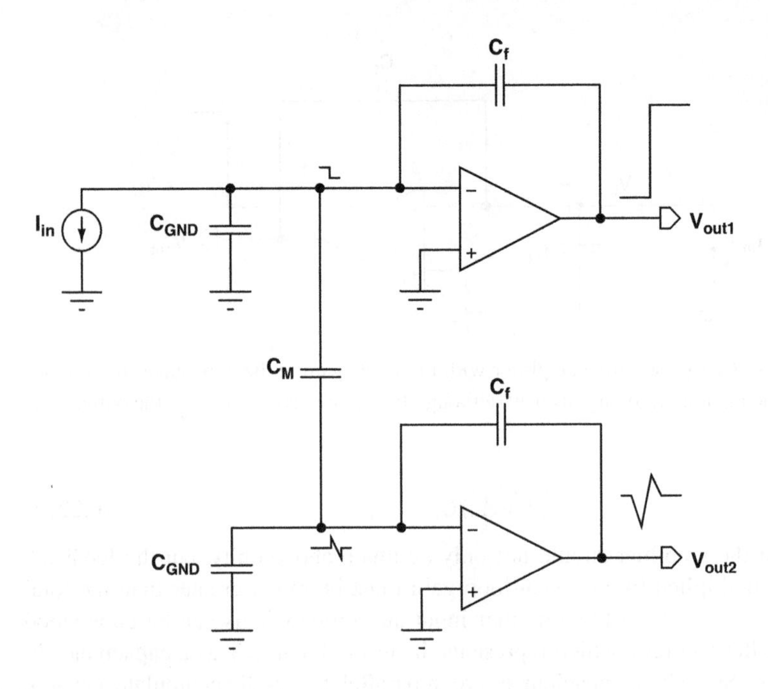

FIGURE 6.44 Circuit for simulating the cross-talk between two channels.

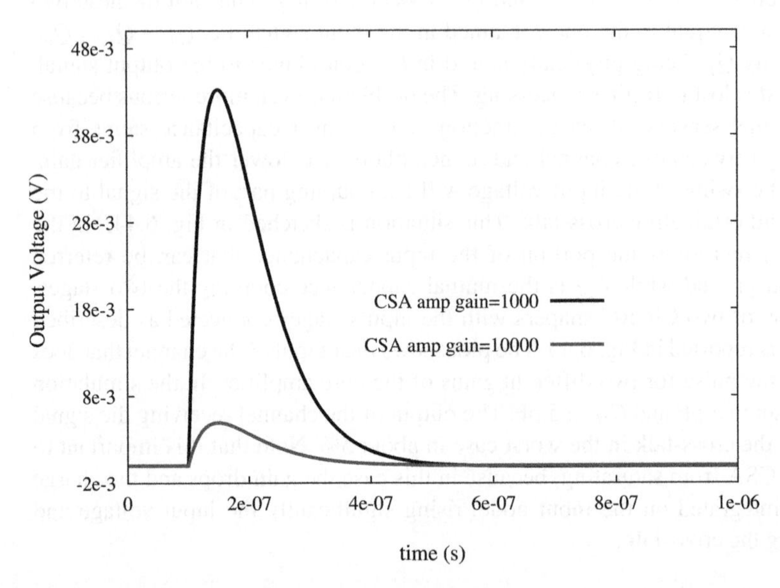

FIGURE 6.45 Simulation of cross-talk between two channels with two different gains of the CSA core amplifier.

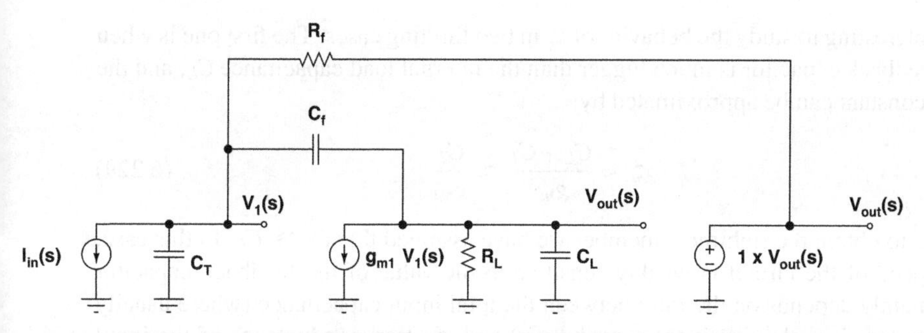

FIGURE 6.46 Small signal equivalent circuit of a CSA to study the effects of bandwidth limitations.

behavior was examined in chapter 4. The simplified small signal equivalent circuit is represented in Fig. 6.46, where g_{m1} is the transconductance of the input transistor, while R_L and C_L are, respectively, the equivalent load resistance and capacitance. The transfer function of the circuit of Fig. 6.46 can be written as:

$$\frac{V_{out}}{I_{in}} = \frac{\left(g_{m1} - sC_f\right) R_f R_L}{s^2 \zeta R_f R_L + s\left[R_L C_L + R_f C_T + (1 + g_{m1} R_L) C_f R_f\right] + 1 + g_{m1} R_L} \tag{6.223}$$

where ζ is given by:

$$\zeta = C_T C_L + C_T C_f + C_L C_f \tag{6.224}$$

Neglecting the zero in the numerator and dividing both numerator and denominator by $g_{m1} R_L$ we get:

$$\frac{V_{out}}{I_{in}} = \frac{R_f}{s\frac{\zeta R_f}{g_{m1}} + sR_f C_f + 1} \tag{6.225}$$

which has real poles if the following condition is met:

$$R_f C_f > \frac{4\left(C_T C_L + C_T C_f + C_L C_f\right)}{g_{m1} C_f} \tag{6.226}$$

From the above equation, we see that having a large feedback resistance in the CSA is not only useful for noise and signal processing purposes, but it also helps in keeping the circuit stable. Complex conjugate poles can however be tolerated, provided that they do not lead to excessive ringing in the CSA response.

When the poles are widely separated, (6.225) can be approximated as:

$$\frac{V_{out}}{I_{in}} = \frac{R_f}{(1 + s\tau_r)(1 + s\tau_f)} \tag{6.227}$$

where $\tau_f = R_f C_f$ and τ_r is:

$$\tau_r = \frac{C_L C_T + (C_T + C_L) C_f}{g_{m1} C_f} \tag{6.228}$$

It is interesting to study the behavior of τ_r in two limiting cases. The first one is when the feedback capacitor is much bigger than the internal load capacitance C_L, and the time constant can be approximated by:

$$\tau_r \approx \frac{C_L + C_T}{g_{m1}} \approx \frac{C_T}{g_{m1}} \tag{6.229}$$

where to obtain the rightmost member we have assumed that $C_T \gg C_L$. In this case, the speed of the circuit is weakly sensitive to the value of the feedback capacitor and mainly depends on the ratio between the total input capacitance (where usually the sensor gives the dominant contribution) and the transconductance of the input transistor. On the opposite, for very small value of C_f ($C_f \ll C_L$) we can neglect the second term in the numerator of equation (6.228) and write:

$$\tau_r = \frac{C_L}{g_{m1}} \frac{C_T}{C_f} \tag{6.230}$$

In this case, the speed of the circuit is limited by the ratio between the sensor and the feedback capacitance, rather than by the sensor capacitance alone. The above equations show that if the sensor capacitance is increased, also the transconductance of the input MOS must be augmented to preserve the CSA speed. Serving sensors with large capacitances thus requires significant power to maintain adequate speed and noise performance in the front-end.

With two real poles, the circuit response in the time domain is:

$$V_{out}(t) = Q_{in} \frac{R_f}{\tau_r - \tau_f} \left(e^{-\frac{t}{\tau_r}} - e^{-\frac{t}{\tau_f}} \right) = \frac{Q_{in}}{C_f} \frac{\tau_f}{\tau_r - \tau_f} \left(e^{-\frac{t}{\tau_r}} - e^{-\frac{t}{\tau_f}} \right) \tag{6.231}$$

whereas the peak voltage is given by:

$$V_{out,peak} = \frac{Q_{in}}{C_f} \left(\frac{\tau_f}{\tau_r} \right)^{\frac{\tau_r}{\tau_r - \tau_f}} \tag{6.232}$$

Small τ_f / τ_r ratios thus lead to a significant attenuation of the voltage which is presented to the pulse shaper, reducing the overall circuit gain below its theoretical value. In principle, in a continuous reset system, the feedback resistor R_f starts immediately discharging the capacitor. However, if the pace at which C_f is reset is much smaller than the one at which the charge accumulates in it, the ideal peak given by Q_{in}/C_f is reached before the discharge process gives a measurable contribution and the exponential decay tail becomes visible. If, on the other hand, the two competing processes act on comparable time scales, only a fraction of the maximum theoretical output voltage can be obtained. This amplitude loss is called ballistic deficit and is found whenever the mechanisms determining the signal formation and those intended to reset the system act simultaneously and with comparable speed. In this specific case, it is the finite bandwidth of the CSA that limits the signal formation time. These considerations show that the intrinsic preamplifier rise-time must be short (1% or less) with respect to the time constant of the CSA feedback loop to avoid reducing too much the

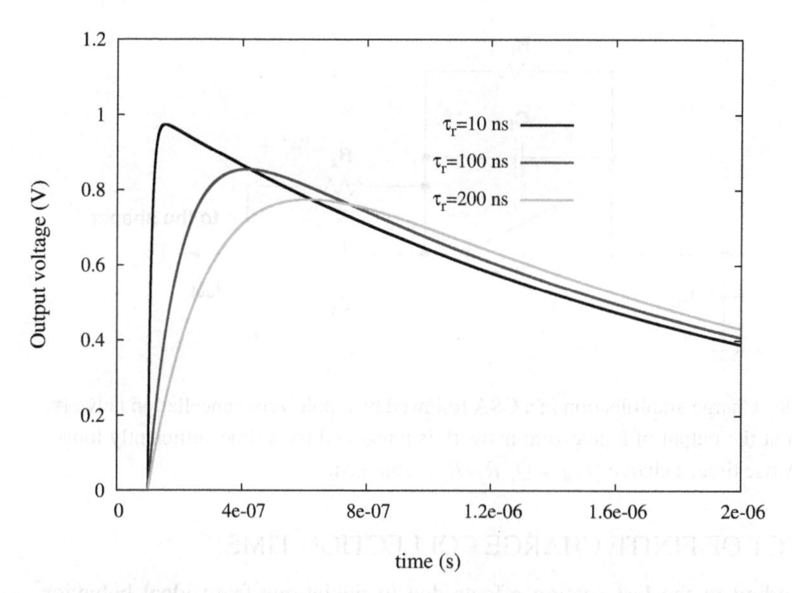

FIGURE 6.47 Response of a CSA with finite rise time constants. The loop feedback constant $R_f C_f$ was fixed to 2 μs.

output signal amplitude. The reduction of the peak amplitude with increasing value of τ_r is shown again in Fig. 6.47.

Consider now a CSA with finite rise time followed by a pole-zero cancellation network, as shown in Fig. 6.48. For a Dirac-delta input, the output current $I_{out}(s)$ is given by:

$$I_{out}(s) = Q_{in} \frac{R_f}{(1+s\tau_r)(1+s\tau_f)} \frac{(1+s\tau_z)}{R_z} = Q_{in} \frac{R_f}{R_z} \frac{1}{1+s\tau_r} \tag{6.233}$$

where the pole-zero matching condition $R_f C_f = R_z C_z$ has been applied. The current in the time domain is:

$$I_{out}(t) = \frac{Q_{in}}{\tau_r} \frac{R_f}{R_z} e^{-\frac{t}{\tau_r}} \tag{6.234}$$

Integrating the above expression between 0 and ∞ one gets the total output charge Q_{out}:

$$Q_{out} = Q_{in} \frac{R_f}{R_z} = Q_{in} \frac{C_z}{C_f} \tag{6.235}$$

Therefore, if the peaking time of the shaper is longer than τ_r, the circuit of Fig. 6.48 can be seen as a charge multiplier that magnifies the input charge by a factor R_f/R_z. If the peaking time is comparable to τ_r or shorter, the charge gain is smaller than what is predicted by (6.235). As will be shown in the next section, the result can be easily generalized to the case where the input signal is not a Dirac-delta.

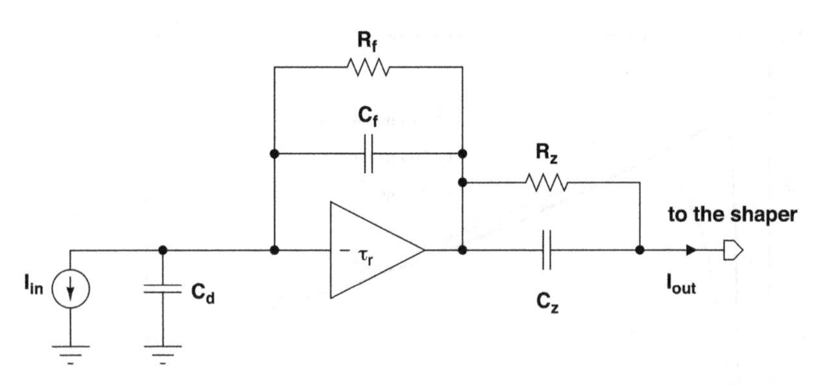

FIGURE 6.48 Charge amplification in a CSA followed by a pole-zero cancellation network. If the current at the output of a pole-zero network is integrated for a time sufficiently longer than the CSA rise time, a charge $Q_{out} = Q_{in}R_f/R_z$ is obtained.

6.9 EFFECT OF FINITE CHARGE COLLECTION TIME

We have studied in the last section effects due to deviations from ideal behavior occurring in the front-end amplifier building blocks. However, also in the sensor itself several phenomena lead to discrepancies between the actual device characteristics and the idealized model used so far in our calculations. In particular, in a real sensor, the charge collection time is finite. Furthermore, depending on the detector size and geometry, the current pulse may have a rather complex shape. To understand the effect of the non-zero charge collection time on the front-end electronics response with a simple mathematical model, we can replace the δ-like input pulse with an exponential waveform:

$$I_{in}(t) = I_0 e^{-\frac{t}{\tau_s}} \tag{6.236}$$

where I_0 is the peak current and τ_s is a constant defining the sensor charge collection time. The total charge delivered by the pulse is given by:

$$Q_{in} = \int_0^{\infty} I_0 e^{-\frac{t}{\tau_s}} dt = I_0 \tau_s \tag{6.237}$$

The Laplace transform of our input signal is:

$$I_{in}(s) = \frac{I_0 \tau_s}{(1 + s\tau_s)} \tag{6.238}$$

Let's first consider the response of the CSA alone, assuming that the circuit has infinite bandwidth, hence $\tau_r = 0$ and its transfer function is given by:

$$T_{CSA} = \frac{R_f}{1 + sC_f R_f} = \frac{R_f}{1 + s\tau_f} \tag{6.239}$$

Multiplying (6.238) and (6.239) we get at the CSA output:

$$V_{out,CSA}(s) = I_0 \tau_s \frac{R_f}{(1 + s\tau_f)(1 + s\tau_s)} \tag{6.240}$$

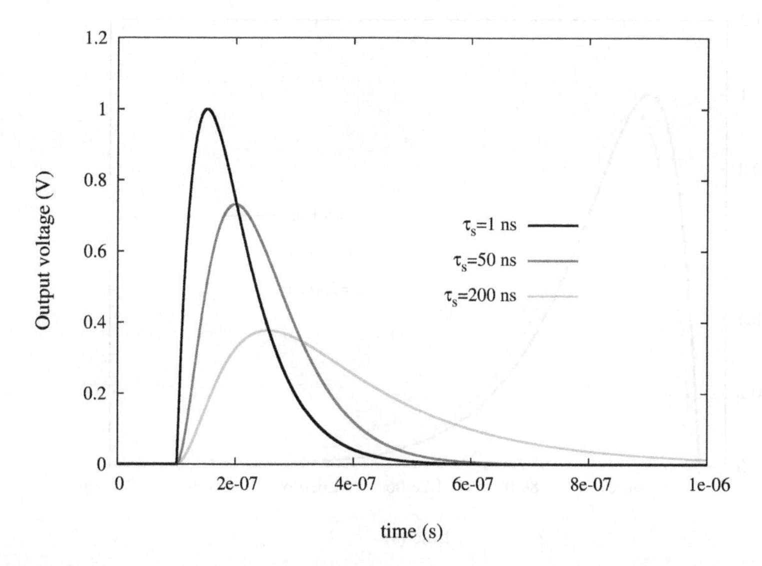

FIGURE 6.49 Response of a CR-RC shaper with 50 ns peaking time to exponential current pulses with different decay times.

which shows that the overall response has again two real poles. Therefore, the same considerations done in the previous paragraph on the output peak loss when τ_f is not much greater than τ_r, apply also when the CSA feedback time constant is not much longer that the sensor charge collection time. The CSA output will thus reach a maximum value smaller than Q_{in}/C_f and this will reflect in a gain loss also at the shaper output. To see the effect on the full chain we multiply $I_{in}(s)$ by the impulse response of the CR-RC network:

$$V_{out}(s) = \frac{I_0 \tau_s}{(1+s\tau_s)} \frac{K}{C_f} \frac{\tau}{(1+s\tau)^2} \tag{6.241}$$

where K indicates the gain, if any, introduced in the filter. The inverse Laplace transform of $V_{out}(s)$ yields:

$$V_{out}(t) = \frac{Q_{in}}{C_f} K \left[\frac{t}{(\tau - \tau_s)} e^{-\frac{t}{\tau}} + \frac{\tau_s}{(\tau - \tau_s)^2} \left(e^{-\frac{t}{\tau_s}} - e^{-\frac{t}{\tau}} \right) \right] \tag{6.242}$$

It is interesting to study how $V_{out}(t)$ changes as a function of the signal speed. To do so, we sweep τ_s and adjust I_0 to maintain constant the total charge fed into the system. The results for a peaking time of 50 ns are shown in Fig. 6.49. We see that when the decay time is very short compared to the peaking time we get the familiar response already observed with a δ pulse. As the decay time increases, the peaking time gets longer and the peak smaller. We are again in presence of a form of ballistic deficit, but now it is the sensor itself that limits the signal formation time. Therefore, the effect is present also with a fully ideal front-end. Note also that we have assumed that the CSA is ideal

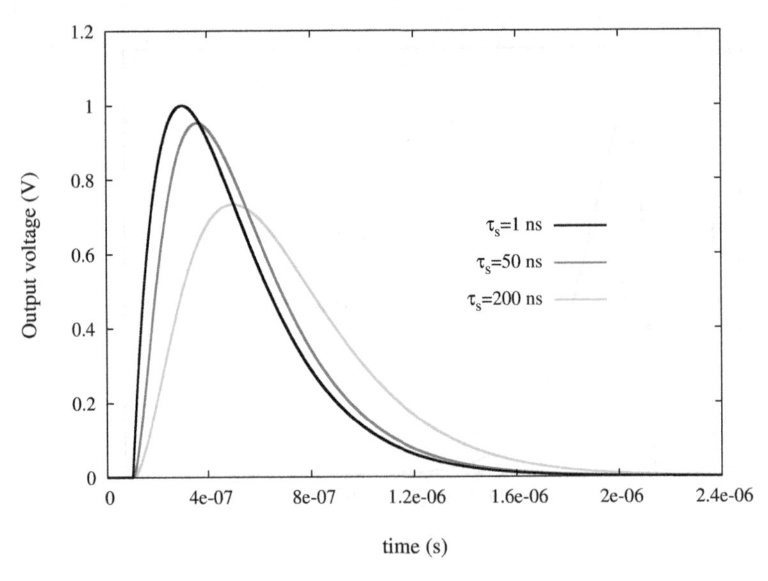

FIGURE 6.50 Response of a CR-RC shaper with 200 ns peaking time to exponential current pulses with different decay times.

(i.e. $\tau_r = 0$ and $\tau_f = \infty$). However, the derivation time constant of the shaper cuts the signal, introducing an effect similar to the one found when the feedback resistance of the CSA is not big enough. The most prominent result of the ballistic deficit is to reduce the signal-to-noise ratio on the peak. Remember, in fact, the noise of front-end amplifiers is given in terms of ENC, which is calculated dividing the output rms noise by the amplitude of a reference pulse, obtained as the response to a δ-like input. If the peak produced by the sensor actual signal is smaller, the SNR for the same total charge contained in the pulse is degraded. Equation (6.242) shows that if $\tau \gg \tau_s$, $V_{out}(t)$ goes back to the standard CR-RC response. Fig. 6.50 reports the response to the same input signals of a shaper with $\tau = 200$ ns. The longer time constant in the shaper now mitigates significantly the effect of the ballistic deficit, at the expense of a longer signal duration and a poorer performance at high rates. As we have discussed at the beginning of this chapter, higher order shapers offer the advantage of a bigger peaking time, without extending in proportion the overall signal duration. This is confirmed by Fig. 6.51. The plot reports the ratio between the pulse width and the peaking time as a function of the shaper order. Two curves have been drawn, one for CR-RC^n shapers and one for unipolar filters with complex conjugate poles. We see from the figure that lower order circuits offer a significantly worse performance in terms of ratio between pulse width and peaking time. Going to higher orders, the advantage becomes progressively smaller. We note also that shapers with complex conjugate poles have systematically better performance than filters containing only real poles. To see the advantage of higher order systems on a practical example, Fig. 6.52 compares the impulse response of a CR-RC and a CR-RC^3 shaper. Both

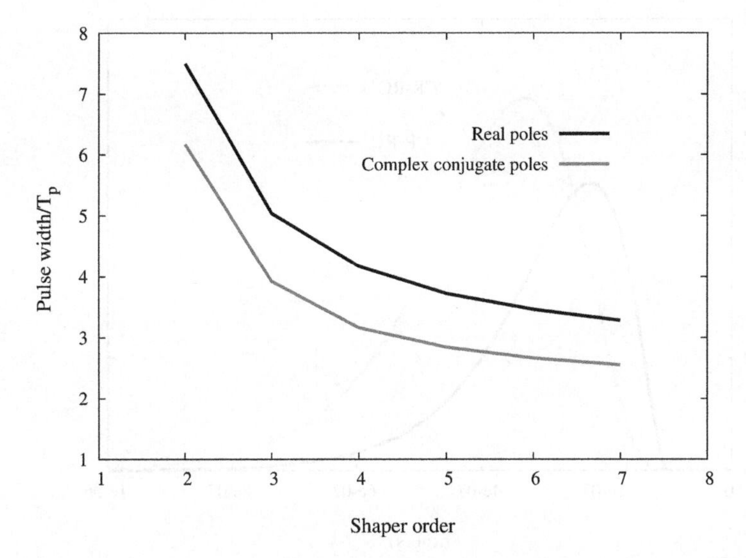

FIGURE 6.51 Ratio between the pulse width and the peaking time for shapers of different orders.

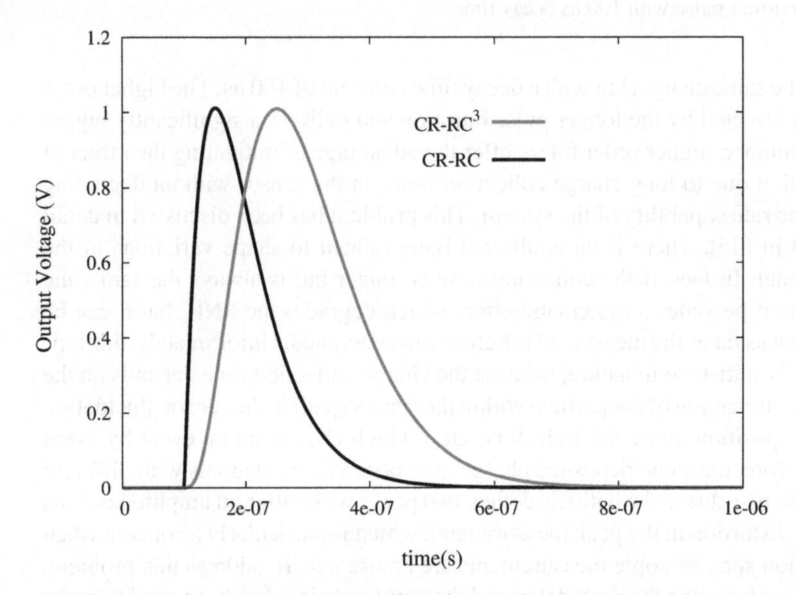

FIGURE 6.52 Comparison between the impulse response of a CR-RC shaper with 50 ns peaking time and the one of a CR-RC^3 shaper with 150 ns peaking time.

circuits employ in the differentiator and in the integrator stages a time constant of 50 ns, which implies a peaking time of 50 ns for the CR-RC filter and of 150 ns for the CR-RC^3 one. Fig. 6.53 shows the response of the two circuits to a current pulse

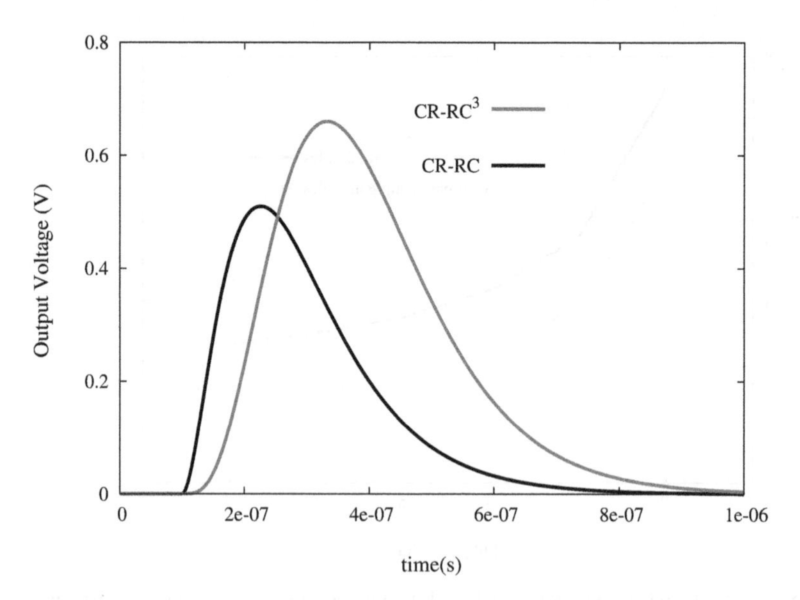

FIGURE 6.53 Comparison between the response of a CR-RC and a CR-RC^3 shaper to an exponential current pulse with 100 ns decay time.

delivering the same charge, but with a decay time constant of 100 ns. The higher order filter is less affected by the longer pulse duration and delivers a significantly bigger peak. In summary, higher order filters offer the advantage of mitigating the effect of ballistic deficit due to long charge collection times in the sensor without degrading too much the rate capability of the system. This problem has been discussed in detail in [16] and in [35]. There is an additional issue related to shape variations in the detector signal. In fact, if the collection time is longer but is always the same, the ballistic deficit becomes a systematic effect which degrades the SNR, but it can be calibrated out as far as the measure of the charge in concerned. Unfortunately, this type of variation is statistical in nature, because the charge collection time depends on the detail of the interaction of the particle within the sensor (path in the sensor, fluctuation in energy deposition along the trajectory, etc.) which change on an event by event basis. Therefore, the same deposited charge may originate two signals with different shapes and hence, due to the ballistic deficit, two peaks with different amplitudes. This generates a distortion in the peak measurement, which is particularly a concern when high precision spectroscopic measurements are envisaged. To address this problem, the correlation between the peak delay and the amplitude loss has been used to make corrections on the peak amplitude [36], [37]. Another possibility is to reconstruct the measurement of the charge by sampling and integrating the full waveform. In fact, the output voltage $V_{out}(t)$ can also be obtained applying the convolution theorem:

$$V_{out}(t) = I_{in}(t) * h(t) = \int_{-\infty}^{\infty} I(u)h(t-u)du \tag{6.243}$$

where $h(t)$ is the system response to a δ-like input. If we now calculate the integral of the convolution we get:

$$\begin{aligned} \textstyle\int_{-\infty}^{\infty} I_{in}(t) * h(t)dt &= \textstyle\int_{-\infty}^{\infty} \left[\int_{-\infty}^{\infty} I(u)h(t-u)du\right] dt = \\ &= \textstyle\int_{-\infty}^{\infty} I(u) \left[\int_{-\infty}^{\infty} h(t-u)dt\right] du \end{aligned} \tag{6.244}$$

The above integral can be further processed to obtain:

$$\int_{-\infty}^{\infty} I_{in}(t) * h(t)dt = \left[\int_{-\infty}^{\infty} I(u)du\right]\left[\int_{-\infty}^{\infty} h(t)dt\right] = Q_{in}\int_{-\infty}^{\infty} h(t)dt \tag{6.245}$$

which proves that the integral of the output pulse is still proportional to the total charge contained in the input signal.

A few sensors deliver signals which are a combination of fast and slow components. Such is the case, for instance, of some gaseous detectors, in which the fast moving electrons give a prompt signal while the ions, being much slower, generate a long tail. If the tail can be described by an exponential function, when transposed in the Laplace domain it generates a pole (see equation (6.241)) that can be suppressed by introducing a zero at the same frequency in the shaper transfer function. This "tail cancellation" is a form of pole-zero compensation. In gaseous detector, mixtures of different gases are usually employed, therefore more "tails" have to be compensated for. Examples of circuits employing such strategy can be found in [38–40]. The precision of the tail correction depends on the accurate matching between the zeros of the filter and the decay time constants of the sensor, which is not trivial to achieve. For this reason, whenever possible, it might be more effective to just sample the front-end signal and correct for the tails in the digital domain with numeric filters. An example of this approach is described in [41].

REFERENCES

1. M. E. Van Valkenburg. *Analog Filter Design.* Oxford University Press, 1995.
2. S. Franco. *Design with Operational Amplifiers and Analog Integrated Circuits.* 3rd ed., McGraw-Hill, 2001.
3. A. Williams and F. Taylor. *Electronic Filter Design Handbook.* 4th ed., McGraw-Hill, 2006.
4. R. Schaumann, H. Xiao, and M. Van Valkenburg. *Design of Analog Filters.* 2nd ed., Oxford University Press, 2009.
5. R. Rau and M. N. S. Swamy. *Modern Analog Filters Analysis and Design: A Practical Approach.* Wiley-CH, 2010.
6. E. Tlelo-Cuautle (editor). *Integrated Circuits for Analog Signal Processing.* Springer, 2012.
7. P. V. Ananda Mohan. *VLSI Analog Filters: Active RC, OTA-C, and SC.* Birkhauser, 2012.
8. A. B. Gillespie. *Signal, Noise and Resolution in Nuclear Counter Amplifiers.* Pergamon Press, 1953.

9. H. Spieler. *Semiconductor Detector Systems*. Oxford Science Publications, 2005.
10. K. Iniewski (editor). *Medical Imaging: Principles, Detectors, and Electronics*. Wiley, 2009.
11. S. Ohkawa, M. Yoshizawa, and K. Husimi. Direct synthesis of the Gaussian filter for nuclear pulse amplifiers. *Nucl. Instr. Meth.*, 138:85–92, 1976.
12. C. H. Nowlin. Pulse shaping for nuclear pulse amplifiers. *IEEE Trans. Nucl. Sci.*, 17:226–241, 1970.
13. E. Fairstein. Linear unipolar pulse-shaping networks: current technology. *IEEE Trans. Nucl. Sci.*, 37:382–397, 1990.
14. W. M. C. Sansen and Z. Y. Chang. Limits of low noise performance of detector readout front-ends in CMOS technology. *IEEE Trans. Circ. Syst.*, 37:1375–1382, 1990.
15. P. Seller. Summary of thermal, shot and flicker noise in detectors and readout circuits. *Nucl. Instr. Meth.*, A426:538–543, 1999.
16. G. De Geronimo. Low-Noise Electronics for Radiation Sensors. *Medical Imaging: Principles, Detectors, and Electronics, cit.*
17. M. Manghisoni, L. Ratti, V. Re, V. Speziali, and G. Traversi. Resolution limits in 130 nm and 90 nm CMOS technologies for analog front-end applications. *IEEE Trans. Nucl. Sci.*, 54:531–537, June 2007.
18. G. De Geronimo and P. O'Connor. MOSFET optimization in deep submicron technologies for charge amplifiers. *IEEE Trans. Nucl. Sci.*, 52:3223–3232, Dec. 2005.
19. M. Manghisoni, L. Ratti, V. Re, V. Speziali, and G. Traversi. Survey of noise performances and scaling effects in deep submicrometer CMOS devices from different foundries. *IEEE Trans. Nucl. Sci.*, 54:531–537, June 2007.
20. G. De Geronimo, P. O'Connor, and C. Grosholtz. A CMOS baseline holder (BLH) for readout ASICs. *IEEE Trans. Nucl. Sci.*, 47:818–822, June 2000.
21. L. B. Robinson. Reduction of baseline shift in pulse amplitude measurements. *Rev. Sci. Instr.*, 32:1057, July 1961.
22. R. L. Chase and L. R. Poulo. A high-precision DC restorer. *IEEE Trans. Nucl. Sci.*, NS-14:83–88, 1967.
23. E. Fairstein. Gated baseline restorer with adjustable asymmetry. *IEEE Trans. Nucl. Sci.*, NS-22:463–466, February 1975.
24. V. Radeka. The effect of baseline restoration on the signal-to-noise ratio in pulse amplitude measurements. *Rev. Sci. Instr.*, 38:1397–1403, 1967.
25. A. F. Arbel. *Analog Signal Processing and Instrumentation*. Cambridge University Press, 1984.
26. C. Arnaboldi and G. Pessina. A very simple baseline restorer for nuclear applications. *Nucl. Instr. Methods*, A512:129–135, 2003.
27. A. Pullia, D. Maiocchi, G. Bertuccio, and S. Caccia. A compact VLSI DC restorer for multichannel X-γ ray detectors. *IEEE Trans. Nucl. Sci.*, 52:1643–1646, October 2005.
28. E. Fairstein. Bipolar pulse shaping revisited. *IEEE Trans. Nucl. Sci.*, 44:424–428, 1997.

29. J-F. Pratte et al. Front-end electronics for the RatCAP mobile animal PET scanner. *IEEE Trans. Nucl. Sci.*, 51:1318–1323, 2004.
30. J. Kaplon and W. Dabrowski. Fast CMOS binary front end for silicon strip detectors at LHC experiments. *IEEE Trans. Nucl. Sci.*, 52:2713–2720, 2005.
31. E. Martin et al. The 5ns peaking time transimpedance front end amplifier for the silicon pixel detector in the NA62 Gigatracker. In *IEEE NSS-MIC Conference Records*, pages 381–388, 2009.
32. P. Jarron, F. Anghinolfi, E. Delagne, W. Dabrowski, and L. Scharfetter. A transimpedance amplifier using a novel current-mode feedback loop. *Nucl. Instr. Meth.*, A377:435–439, 1996.
33. G. Anelli et al. A high-speed low-noise transimpedance amplifier in a 0.25 μm CMOS technology. *Nucl. Instr. Meth.*, A512:117–128, 2003.
34. M. Chiosso, O. Cobanoglu, G. Mazza, D. Panzieri, and A. Rivetti. A fast binary front-end ASIC for the RICH detector of the COMPASS experiment at CERN. In *IEEE NSS-MIC Conference Records*, pages 1495–1500, 2008.
35. B. W. Loo, F. S. Goulding, and D. Gao. Ballistic deficits in pulse shaping amplifiers. *IEEE Trans. Nucl. Sci.*, 35:114–118, Feb. 1988.
36. G. F. Knoll. *Radiation Detection and Measurement.* Wiley, 2000.
37. F. S. Goulding and D. A. Landis. Ballistic deficit correction in semiconductor detector spectrometers. *IEEE Trans. Nucl. Sci.*, 35:119–124, Feb. 1988.
38. A. Kandasamy, E. O'Brien, P. O'Connor, and W. Von Achen. A monolithic preamplifier-shaper for measurement of energy loss and transition radiation. *IEEE Trans. Nucl. Sci.*, 46:150–155, June 1999.
39. N. Lam, F. M. Newcomer, R. Van Berg, and H. H. Williams. Implementation of the ASDBLR straw tube readout ASIC in DMILL technology. *IEEE Trans. Nucl. Sci.*, 48:1239–1243, August 2001.
40. C. Posch, E. Hazen, and J. Oliver. MDT-ASD, CMOS front-end for ATLAS MDT. *ATLAS Muon Note, ATL-MUON-2002-003*, 2002.
41. R.R. Bosch, A. J. De Parga, B. Mota, and L. Musa. The ALTRO chip: a 16-channel A/D converter and digital processor for gas detectors. *IEEE Trans. Nucl. Sci.*, 50:2460–2469, Dec. 2003.

7 Time Variant Shapers

In some applications, the readout of radiation sensors requires signal processing schemes different from those allowed by linear, time invariant filters. The need of overcoming the limitations of CR-RC-like networks lead to the development of systems in which the transfer function is changed during the circuit operation. The purpose of this chapter is to provide an overview of such "time variant" filters. First, the basic concept is presented by discussing a classical example: the reduction of ballistic deficit in large Germanium (Ge) detectors. Then, the methods to evaluate in the time domain the noise performance of time variant shapers are reviewed and Correlated Double Sampling (CDS) is discussed as a case study. The CDS technique finds in fact widespread use in integrated front-end electronics, in particular in the readout of monolithic CMOS sensors.

Understanding the basics of noise calculations in the time domain is crucial for several reasons. First, it is the most direct tool to quantify the ENC of time variant networks, at least as far as white series and parallel noise sources are concerned. Second, it is the technique employed in a number of key papers dealing with the subject of noise analysis and minimization in electronics for radiation sensors. Third, the time domain view provides an alternative and in some cases more intuitive perspective on the central issue of noise. In the last part of the chapter a few significant examples of time variant filters implemented in CMOS technologies are discussed.

7.1 BALLISTIC DEFICIT AND THE GATED INTEGRATOR

In the previous chapter, the concept of ballistic deficit was introduced. It was also shown that when ballistic deficit occurs, the integral of the output of a linear, time-invariant filter is still proportional to the signal charge. In the sixties of the last century, scientists were challenged by the need of reading large Germanium detectors at the highest possible rates. Fig. 7.1 shows the concept of such a sensor. A cylindrical Germanium crystal is grown with a purity as high as possible. In order to form the rectifying junction necessary to deplete the semiconductor, the outer surface is doped with polarity opposite to the one of the bulk. The inner part around the cylinder axis is removed and the remaining surface is properly processed to provide the second contact, where the front-end amplifier is connected. The semiconductor bulk can be n^- type, as shown in Fig. 7.1 a). In this case, the rectifying contact is p^+ and the inner collecting electrode is n^+. In case the bulk is p^-, the doping of the rectifying contact and of the collecting electrodes are respectively n^+ and p^+, as shown in Fig. 7.1 b). To reduce the leakage current, Ge detectors are operated at liquid Nitrogen temperature (77 K). In the early days of semiconductor detectors, spectroscopy grade sensors were obtained by drifting Lithium (Li) atoms in the Germanium bulk. In fact, the residual impurities in the Ge crystal tend to be p-type. Inserted in the semiconductor bulk, the Li atoms form interstitial sites and behaves as donors, compensating the native p type

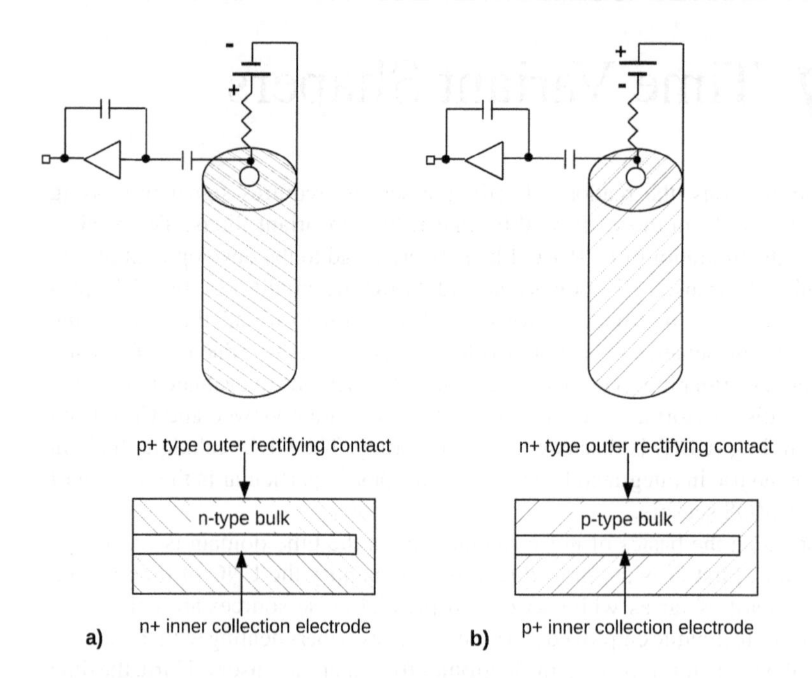

FIGURE 7.1 Examples of close-ended coaxial Germanium detectors.

impurities present in the crystal and yielding a material with properties close to those of an intrinsic semiconductor. Li-drifted sensors needed to be kept always cooled even when not used, because the high mobility of Li in Germanium at room temperature would cause a redistribution of the Li atoms and disrupt the sensor performance. From the early eighties, Hyper-Pure Germanium (HPGe) became available thanks to the floating zone purification technique [1], which makes it possible to achieve Ge crystals of excellent purity. These sensors need to be cooled down only during operation to keep the leakage current low, but they can be stored at room temperature, which greatly simplifies their use. For this reason, HPGe detectors have superseded the Lithium doped ones. An in-depth description of these detectors can be found in [2].

Due to their high density and large signal yield, Ge sensors are still today the preferred tool for high precision γ-ray spectroscopy. However, the large detector size necessary to have high efficiency implies a charge collection time well above the microsecond. Furthermore, large spread in signal shape can be observed for the same deposited energy, depending on where the interaction takes place within the crystal. In the sixties, where fast digitizer where not available, appropriate analog techniques had to be developed to minimize the impact of the ballistic deficit problem without incurring in severe event loss penalties. The straightforward approach of using semi-Gaussian shapers with long integration times compared to the signal collection time is not a good option in this case, because peaking times of tens of microseconds would be necessary to make the peak output voltage insensitive to the variations of

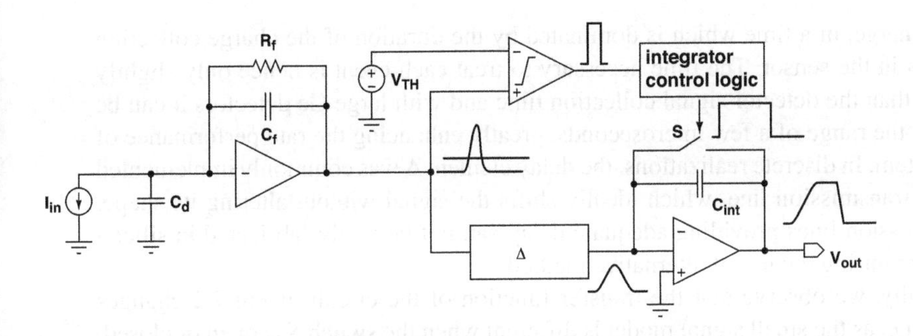

FIGURE 7.2 Principle of the gated integrator.

the input signal shape. The overall output pulse may then last close to one hundred microseconds limiting the rate capability of the system. This fostered the development of time variant filters, in which the circuit parameters are changed during operation, in an effort to combine good energy resolution with adequate speed. One possible method, the gated integrator, is illustrated in Fig. 7.2. In this scheme, the signal is split after the first stage in a fast and a slow path. The fast signal is fed to a comparator, which allows to detect the presence of an event and drives a digital network that generates a control pulse of the desired width. A delayed copy of the signal is also presented to an integrator. Observe that in the circuit of Fig. 7.2 the signal at the output of the delay element Δ must be a current. The timing of the circuit must be such that right before the delayed signal appears at the integrator input, the switch S is opened and the signal current is integrated over C_{int}. If the peaking time of the time-invariant network preceding the integrator is kept short in comparison to the charge collection time, the time duration of the signal is dominated by the sensor response. Consider now the case in which photons of the same energy interact in different parts of the crystal. The released charge will always be the same, but the signal shape might change depending on the interaction point. If the integration window is long enough, at the output of the integrator we will observe signals with different rise time, but all of them will reach at the end the same plateau given by Q_{in}/C_{int}. For this to happen, the integration phase must last longer that the duration of the slower expected input pulse, i.e. longer than the maximum expected charge collection time in the sensor. After the full signal has been collected, the integrator output voltage is sampled and digitized and the baseline is restored by closing the switch S and discharging C_{int}. Observe that after all the signal has been collected, the integrator output remains constant till the circuit is reset by shorting C_{int}. This is in contrast to what happens in CR-RC^n networks, where the signal starts fading away immediately after the peak has been reached. The circuit impulse response is also sketched in Fig. 7.2. We see that the output signal has a constant portion after the maximum. For this reason, it is said that the shaper response has a “flat-top”. For reasons that will become clear in the next sections, “trapezoidal filters” is another common name for these circuits. A key reference that discusses in more detail this type of signal processing is [3]. Note that with the gated integrator we obtain an output voltage directly proportional to the input charge, in a time which is dominated by the duration of the charge collection process in the sensor. The time necessary to treat each event is hence only slightly longer than the detector signal collection time and with large Ge detectors it can be kept in the range of a few microseconds, greatly enhancing the rate performance of the system. In discrete realizations, the delay element Δ was commonly implemented with a transmission line, which ideally shifts the signal without altering its shape. Transmission lines providing adequate delay can not be easily fabricated in silicon chips, requiring the use of alternative methods.

Finally, we observe that the transfer function of the circuit in Fig 7.2 changes over time, as the small signal model is different when the switch S is open or closed. However, when the signal is present, S must be open in order to allow the integration, i.e. any signal should find the circuit in the same state (or set of states) and it is processed always in the same manner. Noise pulses, on the other hand, come in at any time and their effect is different if the filter is in one configuration or in the other. Therefore, time variant filters are such for the noise rather than for the signal [3].

7.2 NOISE ANALYSIS IN THE TIME DOMAIN

We now focus on the problem of evaluating the noise of a gated integrator, using a time-domain approach. This will lead us to an alternative method for noise calculations that can be used with any type of time variant and time-invariant filters, providing additional insights on how noise affects the performance of a front-end amplifier. Time-domain noise calculations are elucidated in a number of fundamental papers [4–8], which are a "must read" for any engineer involved in the design of front-end electronics for radiation sensors. This type of analysis is particularly well suited to the study of white series and parallel noise sources and for simplicity we will limit our treatment to those cases. However, more general approaches incorporating also flicker and colored noise have been described in the literature [9, 10]. Recently, methods to calculate the noise of time variant filters in the frequency domain have also been discussed [11].

We attack the problem by studying first the noise of a time-invariant filter, so that we can compare the time and the frequency domain methodologies. Since the purpose is to pin-down the fundamental concepts while minimizing the computational effort, we do the exercise on a simple CR-RC shaper. Once the basic concepts are clarified, we extend the analysis to time variant filters, using the gated integrator as a practical example.

7.2.1 TIME-DOMAIN NOISE ANALYSIS OF THE CR-RC FILTER

We start from the simplest case, which is the calculation of parallel noise. We can think of parallel noise as a series of elementary current pulses arriving at the amplifier input. These current pulses have the same nature of the sensor signal (which is also a current pulse), therefore each noise event is processed by the front-end transfer function in the same way as a normal signal. The pulses are randomly distributed in time and the average distance between two consecutive pulses is much shorter than the system peaking time. Therefore, the responses to the individual pulses will pile-up at the output, causing a measurable perturbation on the baseline. We now ask ourselves the following question: what is the noise that we have to expect when we measure the output of our front-end at a generic time T_m? To find the answer, we consider first the effect of a single noise pulse. The impulse response of a CR-RC shaper is given by:

$$V_{out}(t) = \frac{Q_{in}}{C_f}\frac{t}{\tau}e^{-\frac{t}{\tau}} \tag{7.1}$$

where Q_{in} is the charge contained in the input signal. Observe that the above equation gives the response of the system to a δ-like signal which is presented at its input at $t = 0$. Suppose now that an elementary noise pulse of charge q arrives at the input exactly at the measurement time T_m (remember always that T_m is an arbitrary instant in time). The response of the circuit to this pulse is given by:

$$V_{outn}(t) = \frac{q}{C_f}\frac{t - T_m}{\tau}e^{-\frac{t-T_m}{\tau}} \tag{7.2}$$

Hence, $V_{outn}(T_m) = 0$. In other words, a noise pulse occurring at the system input *exactly* when we measure its output can not corrupt our measurement. This intuitively makes sense, because the amplifier can not react instantaneously to an input, but it will only develop *later* a full response. Therefore, the noise pulse does not affect the particular measurement we are considering, but it might affect other measurements done at a later time. We thus need to worry only about *all* the noise events that have stimulated the input *before* our measurement time. Consider now a noise pulse which was present at the input β seconds before the measure. The response of the filter to such a pulse can be written as:

$$V_{outn} = \frac{q}{C_f}\left[\frac{t-(T_m-\beta)}{\tau}\right]e^{-\frac{[t-(T_m-\beta)]}{\tau}} \tag{7.3}$$

Therefore, the value of the output at the measurement time T_m is obtained by putting $t = T_m$ in equation (7.3) and yields:

$$V_{outn}(\beta) = \frac{q}{C_f}\frac{\beta}{\tau}e^{-\frac{\beta}{\tau}} \tag{7.4}$$

We have hence constructed a relationship that gives us the output of the front-end as a function of the time β elapsed between the measuring time and the instant in the past at which a particular noise pulse has appeared at the input. For a noise event occurring exactly at the measurement time, $\beta = 0$ and no effect is seen at the output. Consider now the case $\beta = T_p = \tau$, where T_p is the peaking time that for a CR-RC shaper is equal to the integration/derivation time constant τ. This implies that the noise pulse was present at the input T_p seconds before we made our observation. In this case, its contribution to the measurement is given by:

$$V_{outn}(T_p) = \frac{q}{C_f}\frac{1}{e} \tag{7.5}$$

Hence, a noise impulse occurring at the input exactly $T_p = \tau$ seconds before we make the measurement gives the maximum contribution, because the system has developed a full response to it by the time we perform the observation. If $\beta > T_p$, the response starts already fading away. A noise event which has occurred at a time greater than T_p in the past becomes progressively less harmful as β increases, since $V_{outn} \to 0$ if $\beta \to \infty$. We can now rewrite equation (7.4) by inserting the explicit definition of β:

$$\beta = T_m - t \tag{7.6}$$

Note that T_m is the measuring time and t is the time at which a generic noise pulse has occurred *in the past.* i.e. before T_m. Equation (7.4) now reads:

$$V_{outn}(t) = \frac{q}{C_f}\frac{T_m - t}{\tau}e^{-\frac{T_m - t}{\tau}}u(T_m - t) \tag{7.7}$$

Observe that for $t = T_m$, $V_{outn}(T_m) = 0$. For an event occurred T_p seconds before the measuring time, we have to put $t = T_m - T_p = T_m - \tau$ and we get again our maximum response. In equation (7.7) we have introduced explicitly the unit step function $u(T_m - t)$ to emphasize that in a physical system, the response to a given pulse is zero before the signal is applied. Considering a function different from zero for times greater than T_m would imply that events occurring in the future of the measurement can influence the measurement itself. For simplicity, we will in general omit the term $u(T_m - t)$ in the following. We can rewrite equation (7.7) as $V_{outn}(t) = qW(t)$ where q is the signal charge and $W(t)$ depends only on the characteristics of the pulse processing network. $W(t)$ has a maximum given by $1/(C_f \cdot e)$. Observe that the product $qW(t)$ gives a voltage. If we divide $W(t)$ by its peak value, which is the charge to voltage conversion gain, we obtain:

$$W_N(t) = e\frac{T_m - t}{\tau}e^{-\frac{T_m - t}{\tau}}u(T_m - t) \tag{7.8}$$

$W_N(t)$ is now a number, and when we multiply it by q we have still a charge. Both $W(t)$ and $W_N(t)$ "weight" the contribution of an input noise pulse as a function of its distance in time from the measurement, hence the name "weighting functions". $W(t)$ allows us to calculate the noise at the output, while its normalized form $W_N(t)$ can be used if we want to evaluate directly the input referred noise in units of charge. Observe that the weighting function is obtained from the impulse response of the system given in equation (7.1) by changing t in $-t$ (i.e. mirroring it with respect to the y axis) and by shifting it forward by the measurement time T_m. The concept of weighting function is central to the noise calculation in the time domain, so let's practice it on a concrete case.

Example 7.1: Weighting function of a CR-RC shaper

Design the weighting function of a CR-RC shaper with $T_p = 1\ \mu$s.

* * *

First, let's design the impulse response of the system. To do so, we inject a δ-like signal at the input of the front-end amplifier and we choose the signal injection time as the origin of our time axis. Hence negative time values simply mean events which occurred before the signal injection. Since our shaper has a peaking time of 1 μs, the output will reach its maximum excursion at $t = 1$ μs. If we sample our output waveform only once, this is the value that we are interested to capture, as it has the greatest signal-to-noise ratio. Therefore, our measuring time is $t = 1$ μs. We can write the response of our shaper to a pulse occurring at a generic point in time as:

$$V_{out}(t) = \frac{Q_{in}}{C_f}\frac{(t-\alpha)}{\tau}e^{\frac{-(t-\alpha)}{\tau}} \tag{7.9}$$

where α is the time at which the signal appears at the input, measured with respect to the conventional origin of the time axis. For the signal, $\alpha = 0$ and since we want to evaluate it at the peak, we do the measurement at $t = \tau = 1$ $\mu\text{s} = T_m$. Consider a noise pulse occurring at $t = 1$ μs. The system response to this noise pulse is described by equation (7.9) by putting $\alpha = 1$ μs and $t = T_m$. Hence, for $t = 1$ μs no effect is observed on the output and $W(t) = 0$. Consider now a noise pulse which is present at the input 200 ns before T_m. It's contribution to the output is given by equation (7.9) with $\alpha = 800$ ns and $t = T_m = 1$ μs:

$$V_{outn,200\ ns} = \frac{q}{C_f} \times 0.2 \times 0.82 = \frac{q}{C_f} \times 0.164 \tag{7.10}$$

Thus, the weighting function value is $(1/C_f) \times 0.164$. Remember that if we want to work with normalized quantity, we have to divide this number by the system peak gain which is $1/(C_f \cdot e)$. Therefore, the value of the normalized weighting function is 0.445. This means that a noise pulse which has occurred 200 ns before the measuring time affects the measurement with only one half of its "maximum potential". If the noise pulse appears at the input simultaneously with the signal, $\alpha = 0$, $W(t) = 1/(C_f \cdot e)$, $W_N(t) = 1$, i.e. the noise pulse makes the maximum damage to the measurement. Finally, consider a pulse which was present at the input 4 μs before the measurement. This means that $\alpha = -3$ μs, $W(t) = (1/C_f) \times 0.073$ and $W_N(t) = 0.199$. In other words, the earlier the noise pulse has occurred before the signal, the less harmful it will be to the measurement. Fig. 7.3 shows the system impulse response and its noise weighting function. Both functions have been normalized to give a unitary output peak. We see here that the weighting function is a mirrored image of the impulse response, shifted forward by the measuring time T_m. Of course, the choice of the time origin is arbitrary, albeit some choices can be more convenient than others. However, this just implies a translation in the same direction of both the impulse response and the weighting function. As an example, Fig. 7.4 shows the impulse response and the weighting function for an event occurring in the same system 20 μs after the instant conventionally chosen as reference for the time measurement. The interesting event occurs at $t = 20$ μs and the front-end reaches its peak at 21 μs. The weighting function starts from zero at $t = 21$ μs and grows "backwards" to indicate the impact of the noise pulses as a function of their time of occurrence with respect to the instant of the measurement.

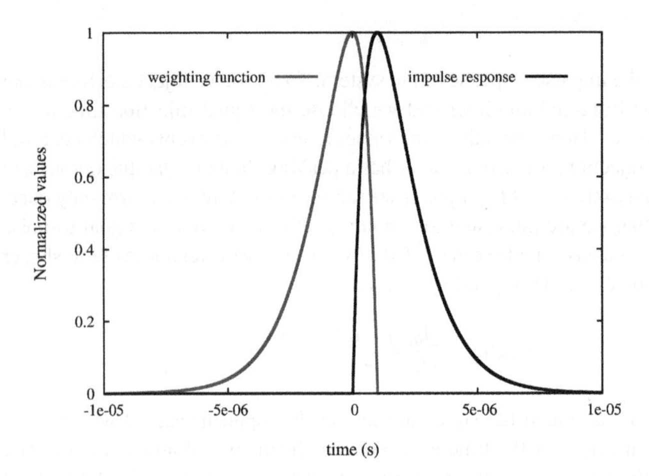

FIGURE 7.3 Impulse response and noise weighting function for a CR-RC shaper with peaking time $T_p = 1\ \mu$s, designed for a signal occuring at the circuit input at $t = 0$

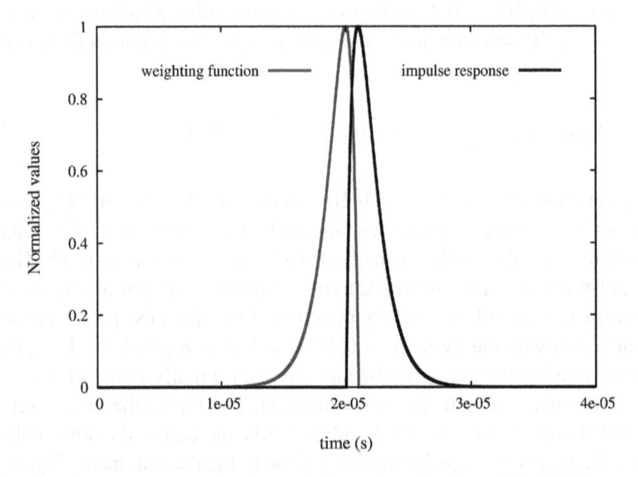

FIGURE 7.4 Impulse response and noise weighting function for a CR-RC shaper with peaking time $T_p = 1\ \mu$s, designed for a signal arriving at the input at $t = 20\ \mu$s.

Up to now we have discussed what happens in case of single noise pulses. This was useful to derive in an intuitive way the concept of the weighting function. Now we use this concept to calculate the overall noise. Remember that all what we have discussed so far applies when the noise can be modeled as current pulses directly appearing at the front-end input, i.e. for the case of parallel noise. The overall noise can be represented by a sequence of pulses of equal magnitude and randomly distributed in time, assuming that they follow a Poisson distribution. The rate of the pulses defines the magnitude of the noise: the higher the rate, the greater the measured noise will be. Suppose that n_s is the rate of the noise process, i.e. the average number of noise pulses per second. In a time dt we have on average $n_s dt$ pulses. Since we are considering a Poisson process, this will be also the variance of the number of pulses. Each pulse gives a contribution to the output voltage given by $qW(t)$. The variance on the output voltage determined by the $n_s dt$ pulses occurring in an infinitesimal interval dt is then given by $n_s q^2 [W(t)]^2 dt$. The total variance at the output can be obtained by integrating over all possible time intervals:

$$< v_{out}^2 >_{ni} = n_s q^2 \int_{-\infty}^{+\infty} [W(t)]^2 dt \tag{7.11}$$

If we divide equation (7.11) by the peak gain squared, we obtain the variance of the input charge measured in [Coulomb]^2. This is equivalent to replacing in equation (7.11) $W(t)$ with $W_N(t)$:

$$\sigma_q^2 = n_s q^2 \int_{-\infty}^{+\infty} [W_N(t)]^2 dt \tag{7.12}$$

where with σ_q^2 we have expressed the noise as a variance of the input charge. Observe now that we can manipulate $n_s q^2$ as follows:

$$n_s q^2 = n_s \cdot q \cdot q = qI = \frac{1}{2} 2qI \tag{7.13}$$

where we have put $n_s q = I$ because this quantity can be interpreted as the average current associated with the process. Hence, (7.12) can be rewritten as:

$$\sigma_q^2 = 2qI \frac{1}{2} \int_{-\infty}^{+\infty} [W_N(t)]^2 dt \tag{7.14}$$

In the above equation we have reintroduced the spectral density $2qI$ associated with the shot noise. We can therefore say that the effect of parallel noise in the time domain can be calculated by multiplying the noise spectral density by the integral of the weighting function squared. For a CR-RC shaper, we have:

$$W_N(t) = e \frac{(T_m - t)}{\tau} e^{-\frac{(T_m - t)}{\tau}} \rightarrow [W_N(t)]^2 = e^2 \frac{(T_m - t)^2}{\tau^2} e^{-2\frac{(T_m - t)}{\tau}} \tag{7.15}$$

To evaluate the integral $[W_N(t)]^2$, we make the following substitution: $T_m - t = t'$, which implies $dt = -dt'$. The integration limits become $t' = 0$ for $t = T_m$ and $t' = \infty$ for $t = -\infty$. We observe also that $W_N(t)$is zero after T_m as noise pulses from the future can not influence the present measurement. We hence limit the integration from $-\infty$ to T_m, writing:

$$\int_{-\infty}^{T_m} [W_N(t)]^2 dt = \int_{\infty}^{0} [W_N(t')]^2 (-dt') = \int_{0}^{\infty} [W_N(t')]^2 dt' \tag{7.16}$$

We therefore need to calculate the following integral:

$$\int_{0}^{\infty} [W_N(t')]^2 dt' = e^2 \int_{0}^{\infty} \left(\frac{t'}{\tau}\right)^2 e^{-2\frac{t'}{\tau}} dt' \tag{7.17}$$

which can be solved by applying iteratively the method of integration by parts. In fact we have:

$$\int_0^\infty \left(\frac{t'}{\tau}\right)^2 e^{-2\frac{t'}{\tau}} dt' = \left[-\left(\frac{t'}{\tau}\right)^2 \frac{\tau}{2} e^{-\frac{2t'}{\tau}}\right]_0^\infty + \int_0^\infty \frac{t'}{\tau} e^{-\frac{2t'}{\tau}} dt' \tag{7.18}$$

The term in square brackets in (7.18) is zero, and the integral can be again evaluated by parts:

$$\int_0^\infty \frac{t'}{\tau} e^{-\frac{2t'}{\tau}} dt' = \left[-\left(\frac{t'}{\tau}\right) \frac{\tau}{2} e^{-\frac{2t'}{\tau}}\right]_0^\infty + \frac{1}{2}\int_0^\infty e^{-\frac{2t'}{\tau}} dt' \tag{7.19}$$

The contribution in square brackets is again zero, so we are left with:

$$\frac{1}{2}\int_0^\infty e^{-\frac{2t'}{\tau}} dt' = \left[\frac{1}{2}\left(-\frac{\tau}{2}\right) e^{-\frac{2t'}{\tau}}\right]_0^\infty = \frac{\tau}{4} \tag{7.20}$$

Summarizing, we can hence write:

$$\int_{-\infty}^{T_m} [W_N(t)]^2 dt = e^2 \int_0^{+\infty} \left(\frac{t'}{\tau}\right)^2 e^{-2\frac{t'}{\tau}} dt' = e^2 \frac{\tau}{4} = e^2 \frac{T_p}{4} \tag{7.21}$$

where we have used the fact that for a CR-RC shaper $\tau = T_p$. Inserting equation (7.21) in (7.14) we have:

$$\sigma_q^2 = 2qI\frac{e^2}{8}T_p = 2qIN_iT_p \tag{7.22}$$

This is the expression that was already found in the previous chapter with the frequency domain calculations, in which $e^2/8$ represents the shaper noise index of the CR-RC filter for parallel noise. Note that equation (7.22) gives the noise variance expressed in $[\text{Coulomb}]^2$. If we want to get the ENC measured in electrons, we need to divide (7.22) by the charge of one electron squared, and take the square-root of the result. The parallel noise index in the time domain can hence be defined as:

$$N_i = \frac{1}{2T_p}\int_{-\infty}^{+\infty} [W_N(t)]^2 dt \tag{7.23}$$

The above equation can be generalized to higher order shapers and provides a way to calculate the noise index using a time-domain representation of the circuit response. Note that for generality the integral is extended from $-\infty$ to $+\infty$, but the context will suggest the appropriate integration limits.

We now take the further step, which is the time domain calculation of the series white noise. Also this noise source can be modeled as a sequence of δ-like pulses. A δ-like voltage signal can be defined in a similar way as done in the previous chapters for δ-like current pulses. We start with a rectangular voltage pulse of amplitude V_0 and duration Δt. Such a pulse can be represented as a difference of two unit step functions:

$$V(t) = V_0\left[u(t) - u(t-\Delta t)\right] \tag{7.24}$$

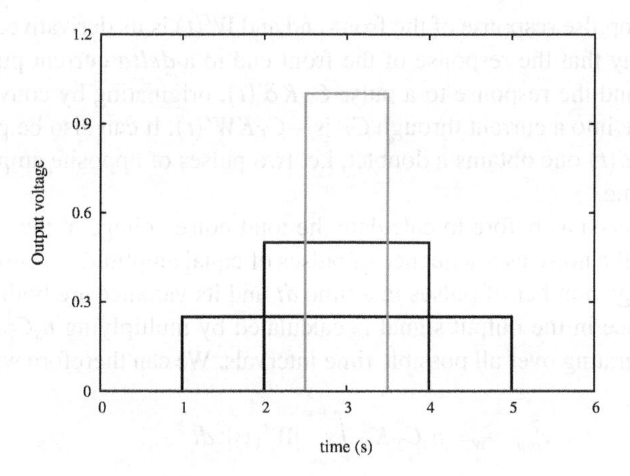

FIGURE 7.5 Succession of voltage rectangular pulses of unitary area.

We multiply and divide the above equation by Δt to get:

$$V(t) = V_0 \Delta t \frac{u(t) - u(t - \Delta t)}{\Delta t} \tag{7.25}$$

The product $K = V_0 \Delta t$ has units of $[V] \cdot [T]$ and represents the area of the pulse. Consider now a succession of pulses of constant area, but of progressively smaller duration and higher amplitude, as shown in Fig. 7.5, which, for simplicity, reports the case of three pulses with unitary area. Taking the limit for $\Delta t \to 0$ and $V_0 \to \infty$, so that $V_0 \Delta t = K$ stays constant we have:

$$\lim_{\Delta t \to 0} K \frac{u(t) - u(t - \Delta t)}{\Delta t} = K\delta(t) \tag{7.26}$$

We must now convert our voltage pulse into a current, which is then processed by the transfer function of our front-end. The total capacitance shunting the input, C_T, is the circuit element that provides the voltage to current conversion, i.e.:

$$i(t) = C_T \frac{dV}{dt} = C_T K \delta'(t) \tag{7.27}$$

Consider now the response of our circuit to such pulses. In the time domain, we need to use the convolution theorem, so we have:

$$V_{out}(t) = C_T K \int_{-\infty}^{+\infty} \delta'(t) W(t - \tau) d\tau \tag{7.28}$$

The above integral can be solved by parts, yielding:

$$\begin{aligned} &C_T K \textstyle\int_{-\infty}^{+\infty} \delta'(t) W(t - \tau) d\tau = \\ &= C_T K \left[\delta(t) W(t - \tau)\big|_{-\infty}^{+\infty} - \textstyle\int_{-\infty}^{\infty} \delta(t) W'(t - \tau) d\tau \right] = -C_T K W'(t) \end{aligned} \tag{7.29}$$

where $W(t)$ is the impulse response of the front-end and $W'(t)$ is its derivative. Summarizing, we can say that the response of the front-end to a *delta* current pulse of charge q is $qW(t)$ and the response to a pulse $C_T K\delta'(t)$, originating by converting a delta voltage pulse into a current through C_T is $-C_T KW'(t)$. It can also be proven that differentiating $\delta(t)$ one obtains a doublet, i.e. two pulses of opposite amplitude closely spaced in time.

We can now proceed as before to calculate the total noise voltage at the output. We represent again the noise as a sequence of pulses of equal amplitude occurring at a rate n_s. The average number of pulses in a time dt and its variance are both given by $n_s dt$. The variance in the output signal is calculated by multiplying $n_s C_T^2 K^2$ by $[W'(t)]^2 dt$ and integrating over all possible time intervals. We can therefore write:

$$< v_{out}^2 >_w = n_s C_T^2 K^2 \int_{-\infty}^{+\infty} [W'(t)]^2 dt \tag{7.30}$$

Observe now that the quantity $n_s K^2$ is measured in units of $[\mathrm{V}]^2[\mathrm{T}]=[\mathrm{V}]^2[\mathrm{Hz}]^{-1}$ so it can be interpreted as a noise spectral density. Actually, it can be proven that $n_s K^2 = \frac{1}{2} v_{nw}^2$ [4], where v_{nw}^2 is the spectral density for white noise. If we want to determine the input referred noise charge, we can just use the derivative of $W_N(t)$ and write:

$$\sigma_{qs}^2 = \frac{1}{2} v_{nw}^2 C_T^2 \int_{-\infty}^{+\infty} [W_N'(t)]^2 dt \tag{7.31}$$

Let' now apply this formulation to the particular case of the CR-RC shaper, for which $W_N(t)$ reads:

$$e\frac{T_m - t}{\tau} e^{-\frac{T_m - t}{\tau}} \tag{7.32}$$

Remember also that $W_N(t) = 0$ for $t \geq T_m$. We can again make the substitution $T_m - t = t'$, observing that, for a generic function of t, we have:

$$\frac{df}{dt} = \frac{df}{dt'}\frac{dt'}{dt} = -\frac{df}{dt'} \tag{7.33}$$

where we have used the fact that in our particular case $t' = T_m - t$, so $dt'/dt = -1$. We thus obtain:

$$W_N'(t) = -\frac{1}{\tau} e^{-\frac{t'}{\tau}} + \frac{t'}{\tau^2} e^{-\frac{t'}{\tau}} \tag{7.34}$$

Using the variable t' we again change the integration limits from $(-\infty, T_m)$ to $(0, +\infty)$, thereby obtaining:

$$\int_0^{\infty} [W_N'(t')]^2 dt' = \frac{1}{\tau^2} \int_0^{\infty} e^{-\frac{2t'}{\tau}} dt' + \frac{1}{\tau^4} \int_0^{\infty} t'^2 e^{-\frac{2t'}{\tau}} dt' - \frac{2}{\tau^3} \int_0^{\infty} t' e^{-\frac{2t'}{\tau}} dt' \tag{7.35}$$

The first integral yields:

$$\frac{1}{\tau^2} \int_0^{\infty} e^{-\frac{2t'}{\tau}} dt' = \left[-\frac{1}{2\tau} e^{-\frac{2t'}{\tau}} \right]_0^{\infty} = \frac{1}{2\tau} \tag{7.36}$$

Let's now evaluate the last term with an integration by parts:

$$-\frac{2}{\tau^3}\int_0^{\infty} t' e^{-\frac{2t'}{\tau}} dt' = -\frac{2}{\tau^3}\left[t'\left(-\frac{\tau}{2}\right)e^{-\frac{2t'}{\tau}}\right]_0^{\infty} - \frac{1}{\tau^2}\int_0^{\infty} e^{-\frac{2t'}{\tau}} dt' = -\frac{1}{2\tau} \tag{7.37}$$

We thus see that the first and the last integrals cancel each other, so only the middle one gives a net contribution. The surviving term was already calculated in equation (7.20), so we have:

$$\frac{1}{\tau^2}\int_0^{\infty}\left(\frac{t'}{\tau}\right)^2 e^{-\frac{2t'}{\tau}} dt' = \frac{1}{\tau^{\cancel{2}}}\frac{\cancel{\tau}}{4} = \frac{1}{4\tau} \tag{7.38}$$

Therefore, we can write:

$$\int_{-\infty}^{+\infty} [W_N'(t)]^2 dt = \frac{e^2}{4}\frac{1}{T_p} \tag{7.39}$$

Combining now equations (7.31) and (7.39), we obtain:

$$\sigma_{qs}^2 = v_{nw}^2 C_T^2 \frac{e^2}{8}\frac{1}{T_p} = v_{nw}^2 C_T^2 N_w \frac{1}{T_p} \tag{7.40}$$

which gives the input-referred noise charge due to white series noise and is again the familiar result already found in the previous chapter. If we compare equation (7.39) to equation (7.40) we can define the noise index for series white noise, N_w, as:

$$N_w = \frac{1}{2}T_p \int_{-\infty}^{+\infty} W_N'(t)^2 dt \tag{7.41}$$

In an attempt to make the treatment more understandable, we have used the CR-RC shaper as a support, since it has a simple transfer function which reduces the computational effort. However, the results given in equations (7.23) and (7.41) are very general, because in deriving the noise formulas we did not make any special assumption on the nature of $W(t)$ and its first derivative. In a time-invariant filter, the weighting function is obtained from the impulse response by a "mirror and shift" operation. Furthermore, in the noise calculation, we need to find the areas of $[W(t)]^2$ and of $[W'(t)]^2$ which are numerically equal to the ones of the impulse response squared and its derivative squared. In practice, to evaluate the noise indexes of a time-invariant shaper, one can proceed in the following way. A δ-like test signal is injected into the front-end, using for example a voltage step across a capacitor in series to the input. The impulse response is acquired with a digital scope. Knowing the input charge, the peak gain can also be calculated, so that the impulse response can be normalized. The derivative of the impulse response is then easily obtained and equations (7.23) and (7.41) can be employed to evaluate the noise indexes. At this point, the concept of weighting function and its distinction from the impulse response may seem just a pedagogical aid to understanding the effect of noise, without any real implications on practical calculations. However, the distinction between the weighting function and the impulse response is essential, because in time variant filters, the two are in general *different*.

7.2.2 TIME-DOMAIN NOISE ANALYSIS OF THE GATED INTEGRATOR

Now that we have clarified the concept of weighting function on an already familiar circuit, we can extend the results to the gated integrator. First, we build the weighting function of the circuit of Fig. 7.2 step by step and then we evaluate its noise performance. In doing so, we make the following simplifying assumptions:

- we consider that both the signal and the noise can be modeled as δ-like pulses;
- we assume that the prefilter Δ is a time-invariant network delivering a rectangular current pulse as a response to a Dirac-δ;
- we suppose that the circuit is reset immediately after the measurement time, and that the reset time is negligible.

Fig. 7.6 shows the relevant waveforms of the circuit in response to a signal. The timing of the circuit is adjusted so that the switch S opens (see Fig. 7.2) and the integrator becomes active when the current pulse is presented to its input. The integrator output is a voltage ramp because we are integrating a constant current. When the input current is over, the integrator output stays flat. At some point, the output voltage is measured (e.g. by sampling and digitizing it with an ADC) and S can be closed to reset the integrator and prepare the circuit for a new measurement. Since we assume to feed a δ-pulse to the input of the overall circuit, the integrator output waveform in Fig. 7.6 is the impulse response of the complete system. Observe that there is a shift between the δ-like pulse and the prefilter response, because we are considering here the output of the branch that feeds the delayed copy of the signal to the gated integrator. Let's now take the first step in building the circuit weighting function. We consider a noise pulse which arrives at the measurement time T_m. The time-invariant part of the circuit develops a full response to it, which starts at $t = T_m$. However, the

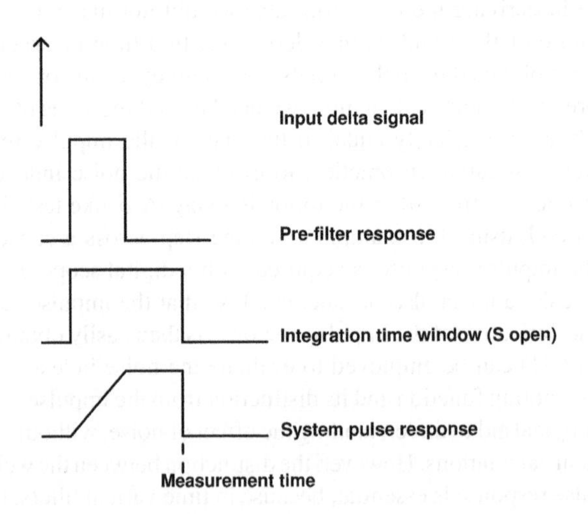

FIGURE 7.6 Relevant waveforms in a gated-integrator circuit.

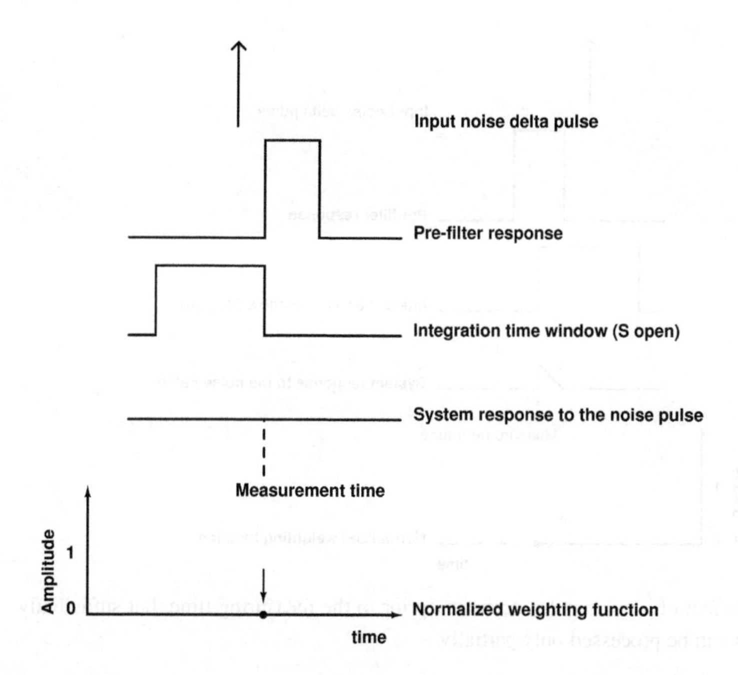

FIGURE 7.7 Effect of a noise pulse occurring at the measuring time in a gated integrator.

signal is zero for $t = T_m$ and the integrator is reset immediately afterwards. Therefore, this pulse can not corrupt the measurement and the weighting function for $t = T_m$ is zero, as represented by the point in the graph in the bottom of Fig. 7.7. It is important to observe that here the activation of the integration window is not synchronized with the noise pulse, because the timing of the integrator is regulated by the signal deposited by the impinging particle, while the noise pulses can arrive at any time during the filter operation. Consider now the second case, shown in Fig. 7.8. Here the noise pulse arrives before the measurement time, but it is still close enough to T_m that the response of the time-invariant pre-filter can only be partially integrated. The integrator output hence develops only a partial response to it. In this and in the next figures, we report the current value of the weighting function, indicated by an arrow, together with the values previously calculated. This allows us to see how the weighting function develops looking backwards in time from T_m. The next step is to consider a pulse which occurs at a time such that the pre-filter response entirely falls inside the integration window. In this case, the response to the noise pulse is fully integrated and determines at the circuit output the greatest possible amplitude. For pulses falling in this window, the weighing function assumes its maximum value, that we have normalized to one. Fig. 7.9 and Fig. 7.10 show, respectively, the latest and the earliest pulse (always with respect to the measuring time) that can be fully integrated, generating the maximum response and hence the maximum damage to the measurement. If a pulse arrives before the integration is started, but not too early, the response of the pre-filter may still partially fall inside the integration window and

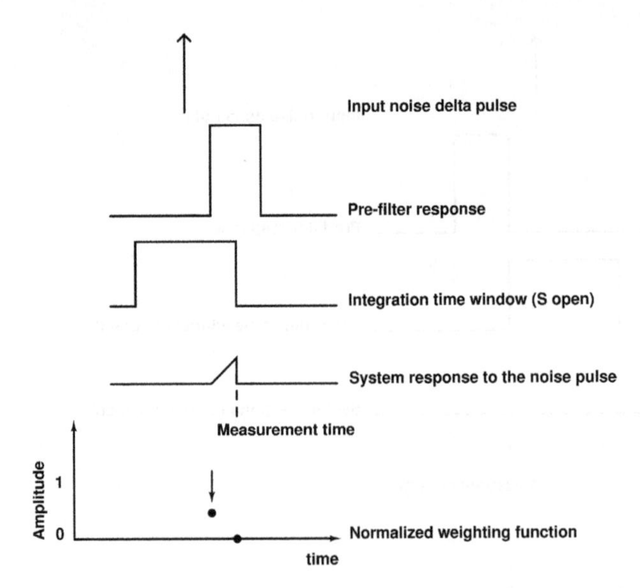

FIGURE 7.8 Effect of a noise pulse occurring prior to the measuring time, but sufficiently close to it that it can be processed only partially.

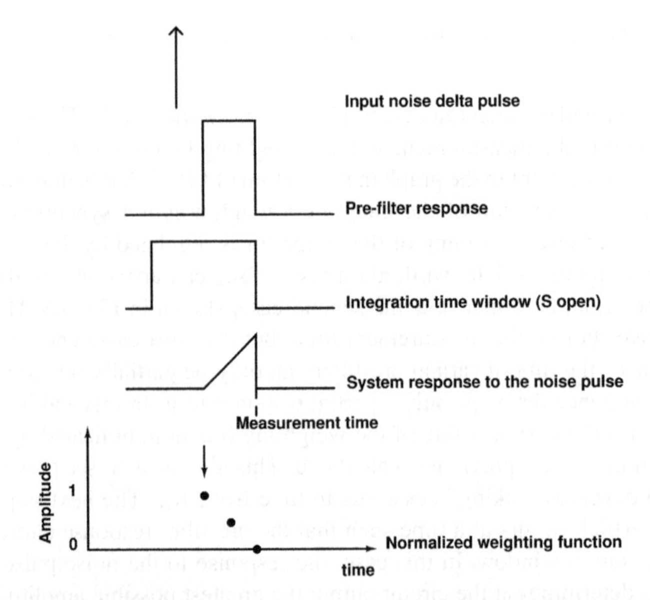

FIGURE 7.9 Latest noise pulse whose response falls completely inside the integration window, generating the maximum response at the output.

is partially integrated. This situation is depicted in Fig. 7.11, which shows that the weighting function is now decreasing. Finally, a noise event might arrive so early that the prefilter response to it does not have any overlap with the integration window. The

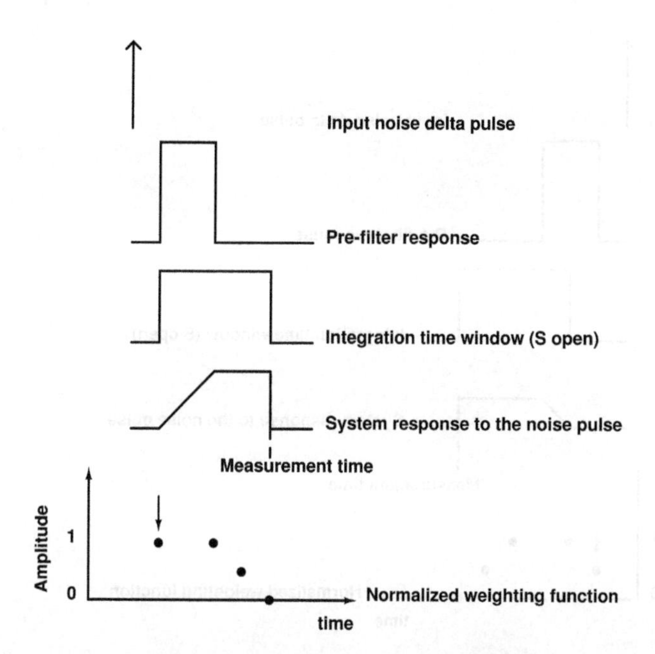

FIGURE 7.10 Earliest noise pulse whose response falls completely inside the integration window, generating the maximum response at the output.

overall system response to such an event is thus zero and the weighting function is again back to zero, as shown in Fig. 7.12. With our step-by-step approach, we have drawn the weighting function only at selected points, to show its trend. Clearly, the function is continuous, as reported in Fig. 7.13, which also shows its first derivative. Owing to the shape of its weighting function, the gated integrator is also called a "trapezoidal filter". Observe that the weighting function can not be any more derived from the system impulse response by a "mirror and shift" operation, and using the latter to calculate the noise would lead to incorrect results. Furthermore, different circuit configuration may result in the same weighting function. Note also that all the reasoning that we have done so far is for noise pulses which are treated by the system in the same way as the signal, i.e. for current noise pulses or parallel noise. However, from our previous discussion, we know that the effect of series noise is taken into account by $W'(t)$, thus when $W(t)$ is known, we have all the ingredients to calculate the noise. We can now evaluate the parallel and series noise indexes of the gated integrator. If we call T_1 the time width of the integration window and T the length of the rectangular input signal fed to the integrator, the duration of the flat top is given by $T_1 - T$. For the weighting function, we can use the following piece-wise

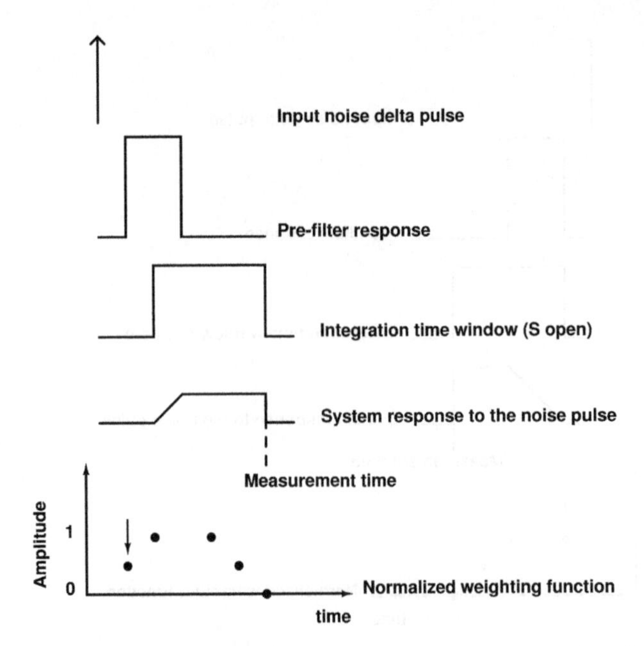

FIGURE 7.11 Pulses arriving before the integration starts produce only a fraction of the maximum possible voltage.

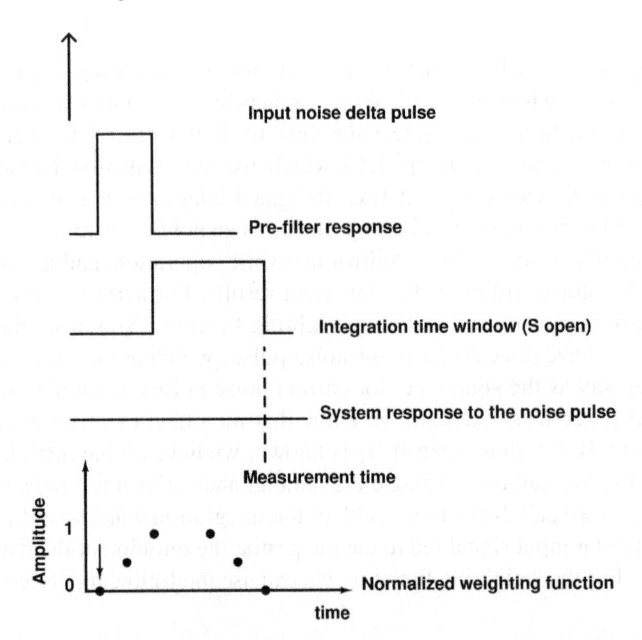

FIGURE 7.12 If a pulse arrives early enough, its effect is already over when the integration starts.

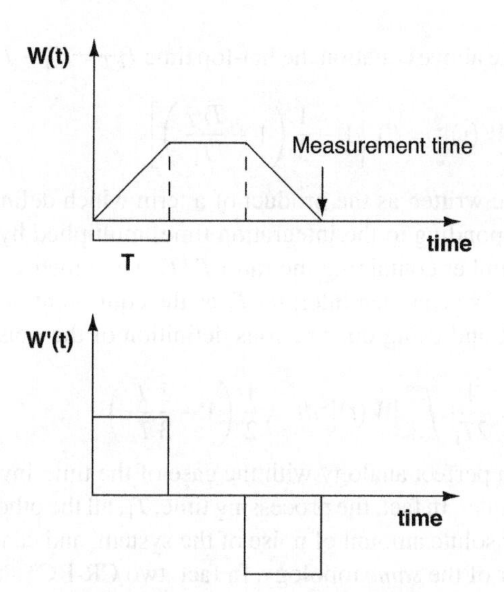

FIGURE 7.13 Trapezoidal weighting function and its derivative.

definition:

$$\begin{cases} W_1(t) = \frac{t}{T} \text{ for } 0 < t \leq T \\ W_2(t) = 1 \text{ for } T < t \leq T_1 \\ W_3(t) = 1 - \frac{(t-T_1)}{T} \text{ for } T_1 < t \leq T_1 + T \end{cases} \tag{7.42}$$

We can now square and integrate separately the three parts. For the first one we have:

$$\int_0^T [W_1(t)]^2 dt = \int_0^T \left(\frac{t}{T}\right)^2 dt = \left[\frac{T}{3}\left(\frac{t}{T}\right)^3\right]_0^T = \frac{T}{3} \tag{7.43}$$

The contribution of the second piece is simply given by:

$$[W_2(t)]^2 = \int_T^{T_1} dt = T_1 - T \tag{7.44}$$

The last term requires a little bit more algebra:

$$[W_3(t)]^2 = \int_{T_1}^{T_1+T} dt + \int_{T_1}^{T_1+T} \frac{(t-T_1)^2}{T^2} dt - \frac{2}{T} \int_{T_1}^{T_1+T} (t-T_1)\, dt \tag{7.45}$$

The three integrals are however straightforward to calculate and lead to $[W_3(t)]^2 = T/3$. Summing up all the contributions, we can write:

$$[W(t)]^2 = T_1 - \frac{T}{3} = T_1\left(1 - \frac{1}{3}\frac{T}{T_1}\right) \tag{7.46}$$

We can also introduce in the above equation the flat-top time $T_{FT} = T_1 - T$ and rewrite it as:

$$[W(t)]^2 = T_1\left[1 - \frac{1}{3}\left(1 + \frac{T_{FT}}{T_1}\right)\right] \tag{7.47}$$

We see that $[W(t)]^2$ can be written as the product of a term which defines the absolute time scale (T_1, corresponding to the integration time) multiplied by the term in parenthesis, which is a number containing the ratio T/T_1, i.e. a metric of the shape of the weighting function. We can then interpret T_1 as the equivalent of the peaking time of a CR-RC^n shaper, and using our previous definition of the noise index, we can write:

$$N_i = \frac{1}{2T_1}\int_{-\infty}^{\infty}[W(t)]^2 dt = \frac{1}{2}\left(1 - \frac{1}{3}\frac{T}{T_1}\right) \tag{7.48}$$

We see that there is here a perfect analogy with the case of the time-invariant filters studied in the previous chapter. In fact, the processing time, T_1, all the other conditions being equal, defines the absolute amount of noise of the system, and can be different from one filter to the other of the *same* topology. In fact, two CR-RC^n shapers of the same order and different peaking times have different noise performance, although they have the same noise index. In the same way, two trapezoidal shapers with identical ratio T/T_1 have the same noise index, but different total noise depending on the integration time T_1. The following numerical example should clarify this point better.

Example 7.2: Parallel noise performance of a gated integrator

A gated integrator is used to read a sensor with a leakage current of 1 nA. Calculate the noise performance with the following filter parameters: integration time T_1 of 1 μs and flat top of 100 ns; T_1=10μs and $T = 9\mu$s; $T_1 = 2\mu$s and $T = 500$ ns.

* * *

In the first case, we know the flat top, which implies that $T = T_1 - T_{FT} = 900$ ns. The shaper noise index can be found applying (7.48) and is $N_i = 0.35$. The charge variance is then calculated as:

$$\sigma_q^2 = 2qIN_iT_1 = 2 \times 1.6 \cdot 10^{-19} \times 1 \cdot 10^{-9} \times 1 \cdot 10^{-6} \times 0.35 = 1.12 \cdot 10^{-34}\ \mathrm{C}^2 \tag{7.49}$$

This gives the variance of the input charge due to the noise. Taking the square-root and dividing by the electron charge, we obtain the ENC expressed in electrons, which is 66 electrons rms. The second shaper is just a ten-fold "scaled-up" version of the first one. Therefore, since the ratio T/T_1 does not change, it has the same noise index, equal to 0.35. Changing the value of the integration time in equation (7.49) and recalculating the noise, we get an ENC of 209 electrons rms. Finally, the third option has different T/T_1 ratio and thus a different noise index. Using equation (7.48), we get $N_i = 0.417$. Introducing the new noise index and the integration time in (7.49) and calculating the ENC as described above, we find 102 electrons rms. As we have already seen with time-invariant filters, longer integration times give more emphasis to parallel noise sources.

We can now calculate the contribution of the series noise. All that we have to do is to differentiate $W(t)$, square the result and evaluate its integral. Taking the derivative of $W(t)$ we get:

$$\begin{cases} W_1'(t) = \frac{1}{T} \\ W_2'(t) = 0 \\ W_3'(t) = -\frac{1}{T} \end{cases} \tag{7.50}$$

Calculating now the integrals, we find:

$$[W'(t)]^2 = \int_0^T \frac{1}{T^2}dt + \int_{T_1}^{T_1+T} \frac{1}{T^2}dt = \frac{2}{T} \tag{7.51}$$

If we now apply the definition of noise index given in equation (7.41), replacing as done before T_p with T_1, we obtain:

$$N_w = \frac{T_1}{T} \tag{7.52}$$

The above equation tells us that the longer the "ramp-up" to the flat top, the lesser the impact of the series white noise. This could have been expected since slow circuits can work with a smaller bandwidth, which limits the contribution of the white series noise component. In high rate applications, where the overall pulse duration must be limited, series noise is usually the primary concern. Therefore, trapezoidal weighting functions with a short flat top are often used in these contexts, because they make the best use of the short available time with a limited circuit bandwidth. Before taking our next step and studying the correlating double sampling method, we can derive a general mathematical rule to build up weighting functions. We have obtained the weighting function of the gated integrator by sliding the prefilter impulse response into the integration window and calculating the results. This is nothing more than performing the operation of convolution between the two functions, the one describing the prefilter response and the one describing the integrator. In mathematical terms, we have:

$$W(t) = \int_{-\infty}^{+\infty} W_1(\tau)W_2(t-\tau)d\tau \tag{7.53}$$

where $W(t)$ is the total weighting function and $W_1(t)$ and $W_2(t)$ are the ones of the two subsystems (in our example the preamplifier with rectangular impulse response and the gated-integrator stage). The paper by Goulding [5] provides a systematic comparison between the noise performance of different weighting functions.

7.3 CORRELATED DOUBLE SAMPLING

Correlated double sampling consists in measuring the same output twice at different times, subtracting the first sample from the second one. The technique is very effective

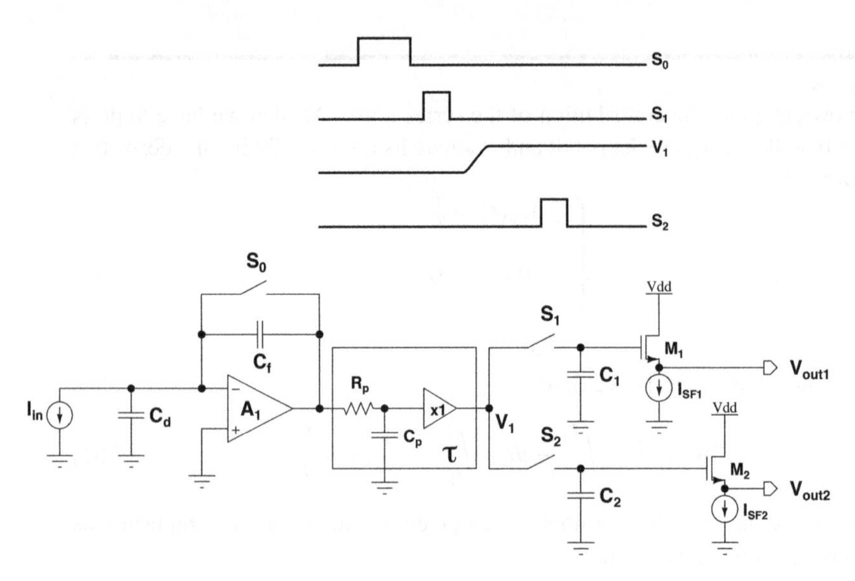

FIGURE 7.14 Principle of correlating double sampling. The output of the front-end amplifier is sampled at two different instants. The difference between the two samples is then calculated. The method removes DC or slowly variable components.

when small signals superimposed on large DC or slowly-changing components have to be detected. This approach was adopted in the first custom VLSI chips designed to readout microstrip sensors in high energy physics applications [12–14] and it is today widely employed in monolithic CMOS radiation sensors. Fig. 7.14 shows the concept applied to a simple front-end, consisting of an integrator followed by a buffered low pass filter. The switch S_0 is first closed to reset the circuit and establish a suitable DC bias point in the integrator. The switch is then opened and the filter output V_1 is sampled on C_1 by connecting the switch S_1. After a time T, the output is sampled again on C_2 by opening S_2. The difference between the voltage stored in C_2 and the one sampled by C_1 contains the useful information. By measuring this difference, any component which is common to both measurements, such as a DC offset, is hence suppressed. The impact of low-frequency noise is greatly mitigated as well, because it can induce only small variations on V_1 in the time elapsing between the two snap-shots. Note that when S_0 is closed for reset, the channel resistance of the transistor generates thermal noise, which is seen at the amplifier output as a random flunctuation with an rms given by k_BT/C_f. When S_0 is open, the instantaneous value of this fluctuation is "frozen" at the amplifier output and since the latter is sampled twice, the *reset noise* is also removed by the CDS action. It is instructive to calculate the noise performance of a CDS circuit applying the weighting function concept. To do so, we need first to know the impulse response of the front-end. We suppose that the core amplifier A_1 is fast enough, so that its intrinsic bandwith limitation can be neglected. The low pass filter provides a well defined time constant $\tau = R_pC_p$ that limits the circuit bandwith and thus the series noise contribution at high frequency.

In the Laplace domain, the front-end impulse response once S_0 is open is given by:

$$V_{out}(s) = \frac{Q_{in}}{sC_f} \frac{1}{(1+s\tau)} \tag{7.54}$$

whereas in the time domain it reads:

$$V_{out}(t) = \frac{Q_{in}}{C_f}\left(1 - e^{-\frac{t}{\tau}}\right) \tag{7.55}$$

To simplify the algebra, we choose the origin of the time axis so that the first sample is taken at $t = 0$ and the second one at $t = T$, where T is the time interval between two consecutive samples. In addition, we neglect the effect of S_0 and we consider the operation of the integrator only after S_0 has been opened, modeling this stage as a time-invariant filter. (Alternatively, the first stage can be thought of as a true time-invariant filter, in which S_0 is replaced by a feedback resistor that slowly discharges C_f and is so big that it can be neglected in the calculations.) We suppose also that the time during which S_1 and S_2 are closed for sampling is much shorter than T, so that the sampling procedure can be assimilated to an instantaneous measurement of V_1, done first at $t = 0$ and then at $t = T$. As usual, we first consider the case of parallel noise, deriving the weighting function $W(t)$ and we then take its derivative to assess the impact of series noise. For a time invariant filter, the weighting function is obtained from the impulse response, by mirroring and shifting it forward by the measuring time. Therefore, the weighting function of the first stage is:

$$W_0(t) = 1 - e^{\frac{t}{\tau}} \tag{7.56}$$

for a measurement done at $t = 0$, whereas for a measurement performed at $t = T$ we can write:

$$W_T(t) = 1 - e^{\frac{t-T}{\tau}} \tag{7.57}$$

The noise events occurring before $t = 0$ will affect both samples. Consider for simplicity the case in which the sampling time interval is chosen equal to τ, i.e. $T = \tau$ (we will see later that this can be an appropriate choice). Suppose now that a current noise pulse appears at the amplifier input at a time 1.5τ before the first sample is taken. The weight of this noise pulse on the first measurement is given by equation (7.56) by putting $t = -1.5\tau$ and is 0.78. The same noise pulse is sampled again on C_2 $T = \tau$ seconds later. Its weight on the second measurement can be obtained from equation (7.57) with $t = -1.5\tau$ and $T = \tau$, which yields 0.92. The value of the plateau is q/C_f, where q is the charge contained in the noise pulse and it is reached asymptotically. When we take the first sample on C_1, a time 1.5τ has passed since the instant in which the noise pulse has appeared at the input, and V_1 has reached a value of $0.78q/C_f$. Due to the choice of our sampling interval, we store again the value of V_1 on C_2 τ seconds later, i.e. 2.5τ seconds after the pulse has been presented to the integrator, when the value reached by V_1 is $0.92q/C_f$. Since at the end we subtract the two samples, this particular noise pulse gives a net contribution of $0.92 - 0.78 = 0.14$. The noise is not completely canceled, because the system has evolved between the two sampling times. We thus conclude that the total weighting function for noise events occurring before the first sample is not given simply by W_0, but it is found using the composite weighting function given by the difference between (7.57) and (7.56):

$$W_1(t) = W_T(t) - W_0(t) = 1 - e^{\frac{t-T}{\tau}} - \left(1 - e^{\frac{t}{\tau}}\right) = e^{\frac{t}{\tau}}\left(1 - e^{-\frac{T}{\tau}}\right) \tag{7.58}$$

This portion of the weighting function must be squared and integrated from $-\infty$ to zero. We must then include also the contribution of the noise pulses occurring after the first sample has already been taken, i.e. after $t = 0$. These noise events influence only the second measurement, so W_T as given by equation (7.57) must be used, integrating its square from 0 to T. Finally, once both samples have been acquired they can not be influenced any more by the noise appearing at the integrator input, hence $W(t)$ is zero for $t > T$. Observe that every current pulse at the integrator input originates a step voltage at the output. However, the noise events are of either polarity, thus making the output baseline fluctuate around its average value. If the noise events were always of the same polarity, they would eventually drive the integrator out of its intended operating region. The sensor leakage current has instead a given polarity, hence, being integrated on C_f, it causes a drift in the first stage output voltage. Care must therefore be paid to handling the leakage current appropriately, so that it does not disrupt the measurement.

Let's now study the noise performance of the complete CDS circuit. The first step is evaluating the parallel noise contribution, which implies calculating the following integral:

$$Q_n^2 = \frac{1}{2} i_n^2 \int_{-\infty}^{+\infty} [W(t)]^2 dt \tag{7.59}$$

Using equations (7.57) and (7.58) we get:

$$\frac{1}{2}\int_{-\infty}^{+\infty} [W(t)]^2 dt = \frac{1}{2}\int_0^T \left(1 - e^{\frac{t-T}{\tau}}\right)^2 dt + \frac{1}{2}\int_{-\infty}^{0} \left[e^{\frac{t}{\tau}}\left(1 - e^{-\frac{T}{\tau}}\right)\right]^2 dt \tag{7.60}$$

For the integral of the first part of the weighting function we have:

$$\int_0^T \left(1 - e^{\frac{t-T}{\tau}}\right)^2 dt = \int_0^T dt + \int_0^T e^{2\frac{t-T}{\tau}} dt - 2\int_0^T e^{\frac{t-T}{\tau}} dt \tag{7.61}$$

The computation of all the above integrals is elementary and yields:

$$\int_0^T \left(1 - e^{\frac{-t}{\tau}}\right)^2 dt = T - \frac{\tau}{2} e^{-\frac{2T}{\tau}} + \frac{\tau}{2} + 2\tau e^{-\frac{T}{\tau}} - 2\tau \tag{7.62}$$

Observe that the integral of equation (7.61) can be transformed by making the following substitutions: $t = T - t'$, $dt = -dt'$, $t = 0 \rightarrow t' = T$, $t = T \rightarrow t' = 0$, after which it becomes:

$$\int_0^T \left(1 - e^{\frac{t-T}{\tau}}\right)^2 dt = \int_T^0 \left(1 - e^{\frac{-t'}{\tau}}\right)^2 (-dt') = \int_0^T \left(1 - e^{\frac{-t'}{\tau}}\right)^2 dt' \tag{7.63}$$

The integration of the second piece of (7.60) is also straightforward and leads to:

$$\left(1-e^{-\frac{T}{\tau}}\right)^2\int_{-\infty}^{0} e^{\frac{2t}{\tau}} = \frac{\tau}{2}+\frac{\tau}{2}e^{-\frac{2T}{\tau}} - \tau e^{-\frac{T}{\tau}} \tag{7.64}$$

The substitution $t-T=-t'$ can also be used for this second integral, with the effect that the integration must be carried out between T and $+\infty$. Summing up the contributions of (7.64) and (7.63), we get:

$$\int_{-\infty}^{\infty}[W(t)]^2 dt = T+\tau\left(e^{-\frac{T}{\tau}}-1\right) \tag{7.65}$$

To calculate the noise charge, consider that the signal peak is not necessarily given by Q_{in}/C_f, because this would be true if the signal was sampled at $T \gg \tau$. The maximum possible difference between the voltage stored in C_2 and the one captured by C_1 is obtained when the signal is applied at 0 and sampled at T, for which it is given by $\frac{Q_{in}}{C_f}\left(1-e^{-\frac{T}{\tau}}\right)$. In defining $W(t)$, we have already normalized only for $1/C_f$, so we need still to divide equation (7.65) by the factor $\left(1-e^{-\frac{T}{\tau}}\right)^2$ (remember in fact that we are considering here squared quantities). Combining all the terms together, we can write the input noise charge due to parallel noise as:

$$\sigma_{qp}^2 = i_n^2\tau\frac{1}{2\left(1-e^{-\frac{T}{\tau}}\right)}\left(\frac{T/\tau}{1-e^{-\frac{T}{\tau}}}-1\right) \tag{7.66}$$

We have now to calculate the series noise. First, we differentiate $W(t)$, obtaining:

$$W_1'(t) = \frac{1}{\tau}e^{\frac{t}{\tau}}\left(1-e^{-\frac{T}{\tau}}\right) \text{ for } \infty < t < 0 \tag{7.67}$$

$$W_T'(t) = -\frac{1}{\tau}e^{\frac{t-T}{\tau}} \text{ for } 0<t<T \tag{7.68}$$

We then evaluate the integrals of the squares:

$$\int_{-\infty}^{0}[W_1'(t)]^2 dt = \frac{1}{2\tau}\left(1+e^{-\frac{2T}{\tau}}-2e^{-\frac{T}{\tau}}\right) \tag{7.69}$$

$$\int_{0}^{T}[W_T'(t)]^2 dt = \frac{1}{2\tau}-\frac{1}{2\tau}e^{-\frac{2T}{\tau}} \tag{7.70}$$

Summing (7.69) and (7.70), we get:

$$\int_{-\infty}^{0}[W_1'(t)]^2 dt + \int_{0}^{T}[W_T'(t)]^2 dt = \frac{1}{\tau}\left(1-e^{-\frac{T}{\tau}}\right) \tag{7.71}$$

The last step is to divide by $\left(1-e^{-\frac{T}{\tau}}\right)^2$ to get the noise in charge units:

$$\sigma_{qws}^2 = \frac{1}{2}v_{nw}^2 C_T^2\int_{-\infty}^{+\infty}[W'(t)]^2 dt = \frac{1}{2}v_{nw}^2 C_T^2\frac{1}{\tau}\frac{1}{1-e^{-\frac{T}{\tau}}} \tag{7.72}$$

It is interesting to analyze this result in more detail. If we have a sampling time T much shorter than the rise time constant of the integrator, we actually reduce the signal-to-noise ratio. In fact, in case the second sample is taken too early, the output signal is still close to the baseline. Therefore, the difference between the two acquired values is small, while they are nevertheless affected by noise. If the sampling time T is much larger than τ, the input referred white noise becomes instead:

$$\sigma_{qws}^2 = \frac{1}{2}\frac{1}{\tau}C_T^2 v_{nw}^2 \tag{7.73}$$

We can now compare this result with the case of a single measurement done at a time T, for which the weighting function is:

$$W(t) = 1 - e^{\frac{t-T}{\tau}} \tag{7.74}$$

The series noise is given by:

$$\sigma_{qws,T}^2 = \frac{1}{2}C_T^2 v_{nw}^2 \int_{-\infty}^{T} [W'(t)]^2 dt = \frac{1}{2}C_T^2 v_{nw}^2 \frac{1}{2\tau} \tag{7.75}$$

Dividing equation (7.75) by (7.73) and taking the square-root we get:

$$\frac{\sigma_{qws,T}}{\sigma_{qws}} = \frac{1}{\sqrt{2}} \tag{7.76}$$

In fact, if $T \gg \tau$, it means that the noise is sampled in a fully uncorrelated way, in which case by taking the difference between two samples, we get the noise of a single measurement multiplied by $\sqrt{2}$, as expected for cases in which two measurements affected by uncorrelated noise sources are summed or subtracted. We can finally write the total white series and parallel noise of the CDS circuit as:

$$\sigma_{qtot}^2 = \sigma_{qp}^2 + \sigma_{qws}^2 = i_n^2 \tau N_i + v_{nw}^2 C_T^2 \frac{1}{\tau} N_w \tag{7.77}$$

where the noise indexes have been defined as:

$$N_i = \frac{1}{2\left(1 - e^{-\frac{T}{\tau}}\right)} \left(\frac{T/\tau}{1 - e^{-\frac{T}{\tau}}} - 1\right) \tag{7.78}$$

for parallel noise and

$$N_w = \frac{1}{2\left(1 - e^{-\frac{T}{\tau}}\right)} \tag{7.79}$$

for series noise. We see from equation (7.77) that the series noise is inversely proportional to the time constant of the integrator and the parallel noise is directly proportional to it. The integrator time constant τ plays hence a similar role to the peaking time of a CR-RC^n shaper and there is an optimal value of τ which minimizes the noise. This can be found by imposing the condition that series and parallel noise give equal contributions:

$$\frac{C_T^2 v_{nw}^2}{I_n^2} \frac{1}{2\left(1-e^{-\frac{T}{\tau}}\right)} = i_n^2 \tau \frac{1}{2\left(1-e^{-\frac{T}{\tau}}\right)} \left(\frac{T/\tau}{1-e^{-\frac{T}{\tau}}} - 1 \right) \tag{7.80}$$

which leads to:

$$\tau^2 = \frac{C_T^2 v_{nw}^2}{i_n^2} \frac{1-e^{-\frac{T}{\tau}}}{T/\tau - 1 + e^{-\frac{T}{\tau}}} \tag{7.81}$$

By inserting equation (7.81) into (7.77) and minimizing the resulting noise charge, one can find the following set of optimal conditions [15]:

$$\begin{cases} \tau^2 = 1.65 \frac{v_{nw}^2 C_T^2}{i_n^2} \\ \frac{T}{\tau} = 1.036 \end{cases} \tag{7.82}$$

The noise optimization of the system can thus proceed as follows: first, the optimal integrator time constant is defined by knowing the input capacitance and the current and voltage noise spectral densities. The time between samples T is then chosen basically equal to the resulting value of τ.

To conclude this part on noise on CDS systems, we briefly mention their performance with flicker noise. It can be shown [11] that the modulus of the flicker noise weighting function for a CDS system with sampling time T is:

$$|W_{1/f}(\omega, T)| = |(2\sin(\pi f T))| \left| \left(\frac{1/\tau}{1/\tau^2 + \omega^2} \right) \right| \tag{7.83}$$

The above quantity stems from the product of two contributions. The sinusoidal term represents the direct effect of the CDS process. Observe that $2sin(\pi f T)$ is zero for $f = 0$ and grows as f increases till the maximum value is reached at $f = 1/2T$. This shows that the sampler behaves as a high pass filter for low-frequency signals, suppressing the lower portion of the $1/f$ noise spectrum. The function then repeats in frequency with a period of $2/T$. The second term in equation (7.84) arises from the low pass filtering effect of the bandwidth-limited integrator, which cuts the high frequency part of the spectrum.

7.4 TIME VARIANT FILTERS IN CMOS TECHNOLOGY

Time variant filters were originally adopted in the first integrated front-ends fabricated in technologies with minimum feature size well above 1 μm [12]. The high operating voltage of these early technologies, combined with the large area of the switches made parasitic charge injection a major issue and in most applications time variant architectures were abandoned in favor of solutions based on continuous time filters. The advent of deep-submicron processes, capable of fabricating fast and very small switches driven by control signals as low as 1 V makes time variant networks again interesting in a number of applications, in particular those in which the radiation quanta are so closed in time that the individual pulses can not be disentangled and an integrating readout approach becomes necessary.

One of the issues in the use of time variant filters is to synchronize the circuit operation with the incoming radiation quanta, so that the system is found in the appropriate state when the signal arrives. In many applications involving particle accelerators, the time at which the events have to be expected is known and the front-end control signals can be tuned accordingly. However, in several other circumstances the arrival time of the impinging particles is not predictable, a common example being provided by the observation of the decay of a radioactive source. In these cases, other strategies have to be devised. One method, already illustrated in Fig. 7.2, consists in generating two pulses, a fast one providing the trigger and a slow one, that conveys the information on the charge/energy of the event. This approach has the disadvantage that the hit is validated by the fast output, that is more noisy, compromising the detection efficiency of small signals [16]. A possible alternative is to exploit the high integration density offered by modern VLSI technologies to realize more identical blocks that work in a time-interleaved configuration. Each block cycles between different states, but the timing of the operations is arranged so that there is always one element able to process correctly the signal. An example of this implementation is described in [16]. The circuit is reproduced in Fig. 7.15. The input current pulse is first integrated by a charge sensitive amplifier, which produces an output voltage step. The voltage at the amplifier output is then translated into a current by a transconductor. The circuit cycles between four states. In the first one, the switch T_1 is in position A and the current is integrated for a period ΔT_A. In the second phase, T_1 is moved to ground and the circuit is in the idle state for a time ΔT_B. T_1 is then connected to point C and a second integration takes place for a duration ΔT_C. Finally, the switch is steered back

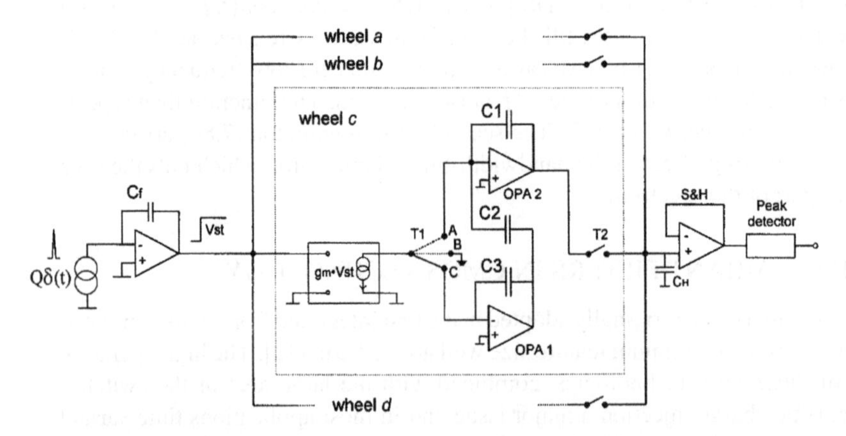

FIGURE 7.15 Principle of a time variant filter implementing a trapezoidal weighting function and a time-interleaved operation. From [16], reprinted with permission.

to ground and the output of A_2 is held constant while it is stored in the sampled and hold circuit. The integrators are then reset and a new cycle can start. Observe that when the current source is connected to point A the voltage measured at the output of the operational amplifier A_2 is given by:

$$V_{outA} = \frac{I_{gm}}{sC_1} \tag{7.84}$$

where I_{gm} indicates the current delivered by the transconductor. When the current source is connected to point C the output becomes:

$$V_{outC} = -\frac{I_{gm}}{sC_3}\frac{C_2}{C_1} \tag{7.85}$$

Therefore, if the choice $C_2 = C_3$ is made, the gain of the circuit in the two phases is equal in magnitude and opposite in phase. Suppose that the baseline current in the transconductor is I_{BL} and a full cycle is performed without any signal. After a time ΔT_A during which T_1 is in position A, the output voltage is given by:

$$V_{outA} = \frac{I_{BL}}{C_1}\Delta T_A \tag{7.86}$$

When the wait interval has elapsed, the current is again integrated by steering T_1 in position C. The final voltage value is hence given by:

$$V_{out} = V_{outA} + V_{outC} = \frac{I_{BL}}{C_1}\Delta T_A - \frac{I_{BL}}{C_3}\frac{C_2}{C_1}\Delta T_B \tag{7.87}$$

If the conditions $C_2 = C_3$ and $\Delta T_A = \Delta T_B$ are met, no net signal is measured at the output as the voltages produced by the two integrations cancel each other. Suppose now that the pulse arrives at a time ΔT_p after the first integration has already started. This implies an increase in the transconductor current by the amount $g_m V_{st}$ when the pulse is detected. The total output voltage is then given by:

$$V_{out} = \frac{I_{BL}}{C_1}\Delta T_p + \frac{I_{BL} + g_m V_{st}}{C_1}(\Delta T - \Delta T_p) - \frac{I_{BL} + g_m V_{st}}{C_1}\Delta T = -\frac{g_m V_{st}}{C_1}\Delta T_p \tag{7.88}$$

Observe that ΔT_p measures the distance between the pulse arrival time and the instant at which T_1 is moved in position A. Therefore, the later the pulse arrives in the first integration window, the greater the resulting output voltage will be. By repeating the above reasoning it is easy to see that pulses which arrive during the wait state (T_1 in position B) produce all the maximum output amplitude given by $-g_m V_{st}\Delta T$. Hits coming only when T_1 is connected to C, yield smaller and smaller signals as their time of arrival progressively shifts towards the end of the second integration phase. The resulting trapezoidal weighting function is shown in Fig. 7.16, where for simplicity all the four time intervals (first integration, idle time, second integration and measuring time) have been assumed to be equal. During the measuring interval, T_1 is put back to ground so the noise of the input stage is not relevant in this phase (hence the weighting

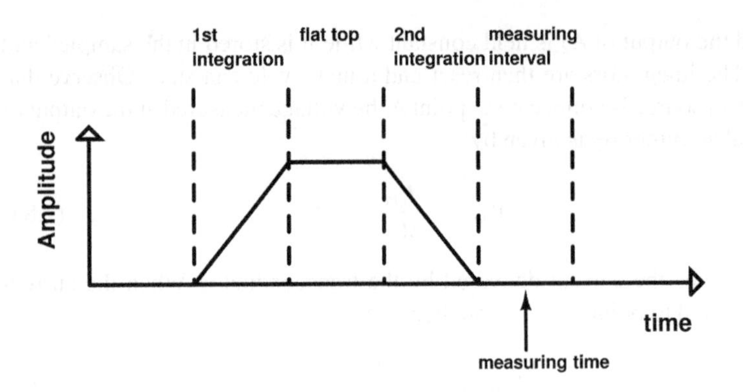

FIGURE 7.16 Weighting function of the circuit of Fig. 7.15.

function is zero). Of course, the signal should arrive when the circuit is cycling in the wait state, for which the weighing function has its flat top and the maximum possible amplitude is produced. If the events arrive asynchronously on the detector, four independent circuits like the one of Fig. 7.15 can be staggered. The timing of the control signals is organized so that there is always one element (called a "wheel" by the authors) that is in the wait mode and can yield the maximum signal and hence the best signal to noise ratio. Since the outputs of all wheels are sampled, an event arriving randomly in time will determine a stair-case waveform at the output of the sample and hold, with the peak being given by the element that was in the "idle state" when the hit stroke. Note that an event coming before a cycle does not produce any output because it becomes part of the rejected "baseline" even if the feedback capacitor in the charge sensitive amplifier has not been cleared. The CSA can use a continuous reset with a time constant long enough that the signal is kept stable during a full cycle. As an alternative, the CSA saturation can be avoided by a reset pulse, which is issued when the output reaches a preset threshold. More details about the described circuit can be found in [17–19]. Potential drawbacks of time-interleaved structures are the power consumption due to the number of processing elements working in parallel and the mismatch between the different signal processors in the same channel.

Let's now examine another example of a time variant circuit that implements a trapezoidal weighting function [20]. The front-end has been designed in a CMOS 0.13 μm CMOS technology with 1.2 V power supply. The purpose of the ASIC is to read-out DEPFET sensors for applications at the new European X-ray Free Electron Laser (XFEL) [21] in Hamburg. DEPFET are sensors in which a transistor (often a p-type MOSFET) is fabricated in a high resistivity substrate. By suitable doping adjustments, a region underneath the channel is formed which provides a minimum potential for the electrons. During operation, the transistor bulk is depleted by applying a reverse-bias voltage and the impinging radiation forms electrons-hole pairs. The holes are swept away by the electric field, while the electrons are collected in the "pocket", which acts as a second gate terminal (the "internal gate"). The accumulated negative charge induces more holes in the channel, increasing the device current and

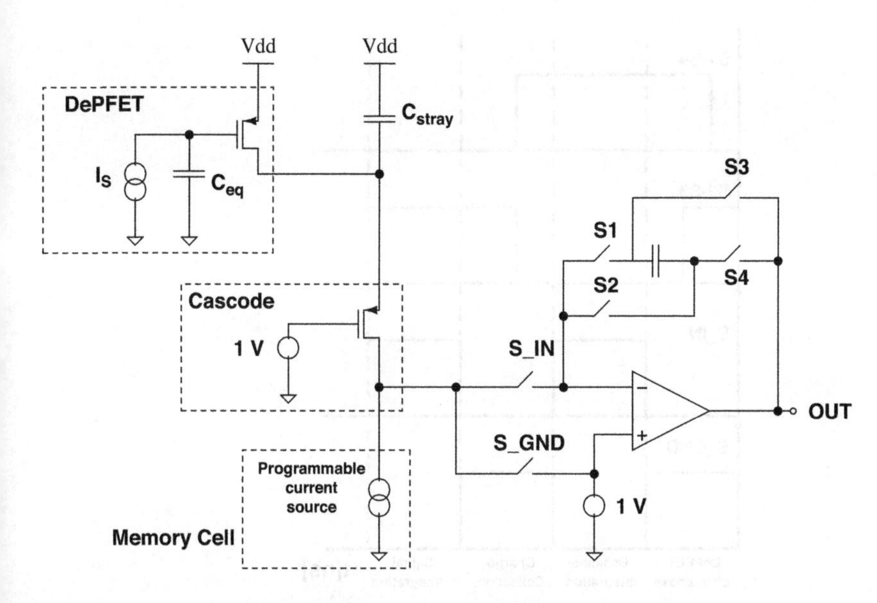

FIGURE 7.17 Flip capacitor integrator. From [20], reprinted with permission.

producing a signal. After the measurement, the electrons are removed by applying a suitable pulse to a dedicated "clear" contact. More details on these sensors, which have shown excellent spectroscopic performance, can be found in [22, 23]. At the XFEL, the photons are grouped in macro bunches 600 μs long, separated by intervals of $\approx$ 100 mS. Within each macro-bunch, X-ray flashes arrive every 200 ns. The known time structure of the beam can be used to regulate the operation of a time variant readout, as shown in Fig. 7.17. A signal can be represented as a fast current pulse appearing at the gate of the DEPFET transistor. This is integrated on C_{eq} and produces a step change in the device current. The 200 ns time budget is divided into two periods of 60 ns and two of 40 ns. First, the sensor is cleared and prepared for the new measurement (60 ns). The baseline is sampled by closing the switch S_{IN} and opening S_{GND}, integrating the current on C_f for 40 ns. The op-amp is then disconnected for another 60 ns. During this time, the X-rays are expected to hit the sensor and the resulting charges accumulate in the internal gate, increasing the DEPFET current, which is temporarily steered to ground. In the same time-frame, switches $S_1 - S_4$ are opened and $S_2 - S_3$ are closed, flipping the capacitor arms around the op-amp. This inversion allows the offset subtraction and the filtering of low frequency noise. Finally, the signal is integrated in the last 40 ns, after which the final value to be measured is ready at the op-amp output. The timing of the circuit is shown in Fig. 7.18. A numerical example helps in understanding the circuit operation better. Consider initially an ideal case in which the current driven by the sensor before the signal is detected matches exactly the one imposed by the current source. In this case, no charge accumulates on C_f during the baseline integration phase and the flipping does

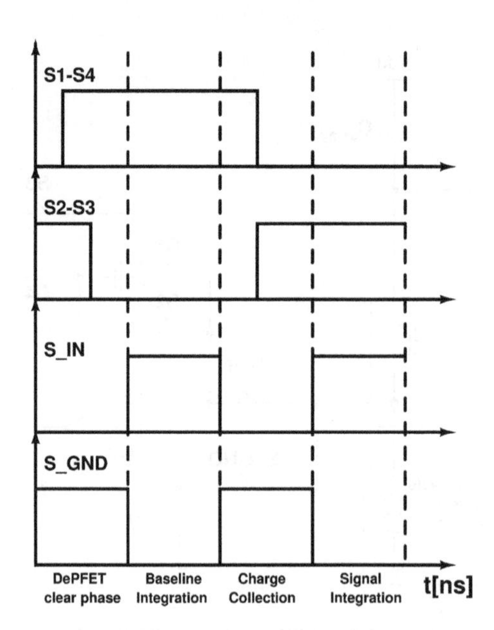

FIGURE 7.18 Timing of the flip capacitor filter. From [20], reprinted with permission.

not produce any effect, since the arms of C_f are equipotential. Suppose that during the charge collection enough electrons have been accumulated in the internal gate to increase the transistor current by 20 μA. When S_{IN} is closed and S_{GND} is opened to start the second integration phase, this current flows into the op-amp node and is integrated on C_f. Assuming for C_f a value of 2 pF, the output will shift downwards by:

$$\Delta V_{out} = \frac{20\ \mu\text{A}}{2\text{pF}} \times 40\ \text{ns} = 0.4\ \text{V} \tag{7.89}$$

going from 1 V to 0.6 V. Let's examine now a case in which during the baseline integration the current delivered by the source is 2 μA bigger than the one provided by the sensor. In this case, the extra current is integrated on C_f rising the op-amp voltage by 40 mV and bringing it to 1.04 V at the end of the first integration. The flipping operation then swaps the capacitor's terminal, so that the output is at 0.96 V before the second integration starts. Even if the signal is always 20 μA, only 18 μA are seen by the integrator, because 2 μA are directly sunk by the current source. Hence, the ΔV_{out} will be 0.36 V. However, the output starts moving downwards from the 0.96 V attained after flipping, so the final value is always 0.6 V, as calculated before.

To derive the noise weighting function, we start as usual by studying the effect of current noise pulses, which produce the same effect as the signal. In this case, a noise pulse originates a step in the sensor current. If the pulse arrives in the middle of the second integration phase, the resulting current variation is integrated for 20 ns, thus the pulse affects the measurement only by one half of its full potential. All the

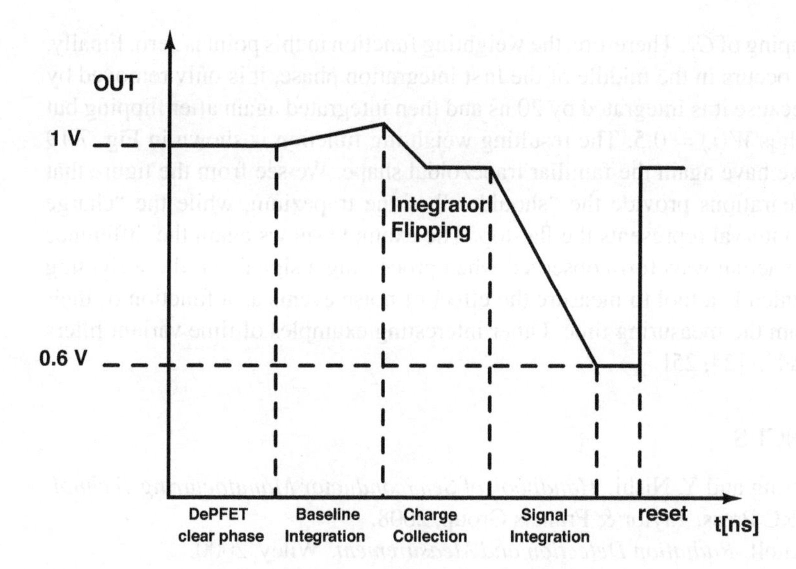

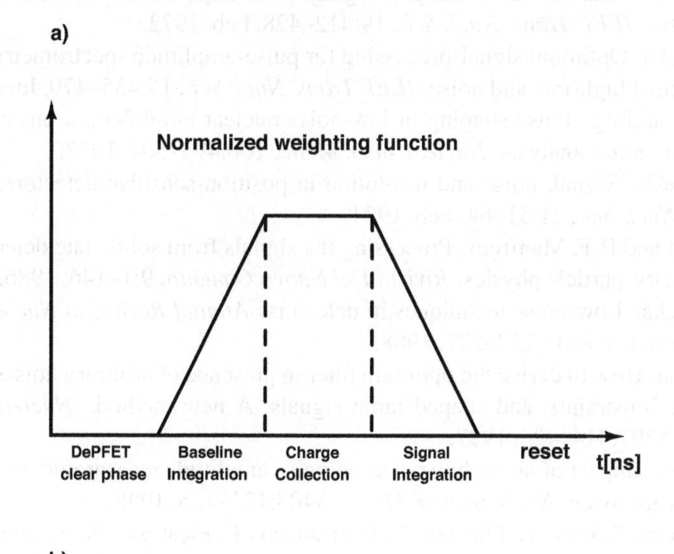

FIGURE 7.19 a): Possible output signal of the flip-capacitor filter described in [20]. b): Circuit weighting function.

noise pulses occurring during the "charge collection phase" are integrated as much as the signal, so they determine the maximum damage to the measurement. Consider now a noise pulse occurring at the very beginning of the first integration. It basically determines a baseline variation, which is canceled by the double integration combined with the flipping of C_f. Therefore, the weighting function in this point is zero. Finally, if the pulse occurs in the middle of the first integration phase, it is only removed by one half, because it is integrated by 20 ns and then integrated again after flipping but for 40 ns, thus $W(t) = 0.5$. The resulting weighting function is shown in Fig. 7.19 b), where we have again the familiar trapezoidal shape. We see from the figure that the two integrations provide the "shoulders" of the trapezium, while the "charge collection" interval represents the flat-top. The example shows again the difference between the actual waveform observed when processing a signal and the weighting function, which is a tool to measure the effect of noise events as a function of their distance from the measuring time. Other interesting examples of time variant filters can be found in [24, 25]

REFERENCES

1. R. Doering and Y. Nishi. *Handbook of Semiconductor Manufacturing Technology.* CRC Press, Taylor & Francis Group, 2008.
2. G. F. Knoll. *Radiation Detection and Measurement.* Wiley, 2000.
3. V. Radeka. Trapezoidal filtering of signals from large Germanium detectors at high rates. *IEEE Trans. Nucl. Sci.*, 19:412–428, Feb. 1972.
4. V. Radeka. Optimum signal-processing for pulse-amplitude spectrometry in the presence of high-rate and noise. *IEEE Trans. Nucl. Sci.*, 15:455–470, June 1968.
5. F. S. Goulding. Pulse-shaping in low-noise nuclear amplifiers: a physical approach to noise analysis. *Nuclear Instr. Meth.*, 100:493–504, 1972.
6. V. Radeka. Signal, noise and resolution in position-sensitive detectors. *IEEE Trans. Nucl. Sci.*, 21:51–64, Feb. 1974.
7. E. Gatti and P. F. Manfredi. Processing the signals from solid-state detectors in elementary-particle physics. *Rivista Del Nuovo Cimento*, 9:1–146, 1986.
8. V. Radeka. Low-noise techniques in detectors. *Annual Review of Nuclear and Particle Science*, 38:217–277, 1988.
9. A. Pullia. How to derive the optimum filter in presence of arbitrary noises, time-domain constraints, and shaped input signals: A new method. *Nuclear Instr. Meth.*, A397:414–425, 1997.
10. A. Pullia. Impact of non-white noises in pulse amplitude measurements: a time-domain approach. *Nuclear Instr. Meth.*, A405:121–125, 1998.
11. D. Gascon, S. Bota, A. Dieguez, L. Garrido, and E. Picatoste. Noise analysis of time variant shapers in frequency domain. *IEEE Trans. Nucl. Sci.*, 58:177–186, Feb. 2011.
12. T. J. Walker, S. Parker, B. Hyams, and S. L. Shapiro. Development of high density readout for silicon strip detectors. *Nucl. Instr. Meth.*, A226:200–2003, Sept. 1984.
13. G. Anzivino et al. First results from a silicon-strip detector with VLSI readout. *Nucl. Instr. Meth.*, A243:153–158, Feb. 1986.
14. P. P. Allport, P. Seller, and M. Tyndel. A low power CMOS VLSI multiplexed amplifier for silicon strip detectors. *Nucl. Instr. Meth.*, A273:630–635, 1988.
15. H. Spieler. *Semiconductor Detector Systems.* Oxford Science Publications, 2005.

16. A. Pullia, C. Fiorini, E. Gatti, A. Longoni, and W. Buttler. A novel VLSI processor for high-rate, high resolution spectroscopy. *Nucl. Instr. Meth.*, A439:385–390, 2000.
17. A. Pullia, C. Fiorini, E. Gatti, A. Longoni, and W. Buttler. ROTOR: the VLSI switched current amplifier for high-rate high-resolution spectroscopy with asynchronous event occurrence. *IEEE Trans. Nucl. Sci.*, 45, Dec. 1998.
18. C. Fiorini, A. Longoni, and W. Buttler. Multichannel implementation of ROTOR amplifier for the readout of silicon drift detectors arrays. *IEEE Trans. Nucl. Sci.*, 49, Aug. 2002.
19. M. Porro, C. Fiorini, and L. Strüder. Theoretical comparison between two different filtering techniques suitable for the VLSI spectroscopic amplifier ROTOR. *Nucl. Instr. Meth.*, A512:179–190, 2003.
20. L. Bombelli, C. Fiorini, S. Facchinetti, M. Porro, and G. De Vita. A fast current readout strategy for the XFEL DEPFET detector. *Nucl. Instr. Meth.*, A624:360–366, 2010.
21. The European XFEL. http://www.xfel.eu/.
22. J. Ulrici et al. Spectroscopic and imaging performance of DEPFET pixel sensors. *Nucl. Instr. Meth.*, A465:247–252, 2001.
23. W. Neeser et al. DEPFET—a pixel device with integrated amplification. *Nucl. Instr. Meth.*, A477:129 - 136, 2002.
24. A. Abusleme, A. Dragone, G. Haller, and B. A. Wooley. BeamCal instrumentation IC: design, implementation, and test results. *IEEE Trans. Nucl. Sci.*, 59:589–596, 2012.
25. A. Dragone et al. eLine100: A front-end ASIC for LCLS detectors in low noise applications. *IEEE Trans. Nucl. Sci.*, 61:1001–1006, 2014.

8 Transistor-Level Front-End Design

In this chapter we focus on the transistor-level implementation of front-end amplifiers, deriving guidelines for topology choice and transistor sizing. Most systems use as the input stage a singled ended cascode amplifier with a transimpedance feedback loop. As it was introduced in the previous chapters, the term "Charge Sensitive Amplifier" (CSA) is rigorously correct only if the feedback time constant is large enough that it does not contribute significantly to the overall signal shaping performed by the front-end. The optimization rules reported here are nevertheless valid also when the feedback resistor has a primary role in the signal processing, i.e. when it would be more appropriate to speak of a "transimpedance configuration" in the strict sense. In the following, we therefore use the word CSA to indicate any input stage based on a high-gain voltage amplifier supplemented with feedback.

The chapter starts with the optimization of the core amplifier because its noise performance determine the ultimate resolution achievable by the full system. We then examine the most common type of feedback networks. Disturbances occurring on the amplifier power supply rails may introduce additional noise that often overshades the one contributed by the transistors and the resistors forming the circuit. The issue of power supply rejection is therefore treated in some length. Following this, the practical implementation of shapers and baseline restorers is presented.

Most radiation sensors provide single-ended signals. For this reason, front-end amplifiers are typically realized with single-ended architectures. Furthermore, single-ended stages offer a better noise figure for a given power budget as far as the intrinsic amplifier noise is considered. Fully differential circuits, on the other hand, are more immune to external interference. They therefore provide a more robust alternative if the front-end is incorporated in a chip embedding complex digital function or when the ASIC must operate in a particularly noisy environment. Examples of fully differential designs are hence shown. The chapter closes with a discussion of the key issues arising in the implementation of current-mode input stages.

8.1 TRANSISTOR LEVEL DESIGN OF CHARGE SENSITIVE AMPLIFIERS

In most circumstances, Charge Sensitive Amplifiers (CSA) employ one of the single-ended cascode schemes studied in chapter 3 and chapter 4. In this section we want to understand how to size the transistors and select their bias points. The problem can be factorized in two parts, addressing separately the design of the input device and the one of the active load.

8.1.1 OPTIMIZATION OF THE INPUT TRANSISTOR

Achieving the best signal-to-noise ratio from the CSA requires a careful choice of the width and length of the input transistor, because a wrong implementation of this device can compromise the whole system performance. It is therefore not surprising that in the literature a lot of attention has been paid to this issue. Sizing the input transistor is not trivial, because its dimensions have a two-fold impact on noise. On the one hand, the larger the device is, the more it contributes to the total input capacitance, C_T, to which both the thermal and flicker components of the ENC are proportional. On the other hand, a bigger transistor offers a larger transconductance and a smaller $1/f$ spectral density. The best balance between these opposite effects should hence be searched. The goal of the optimization procedure is finding the gate width W and length L resulting in the smallest ENC figure, given a certain number of constraints, such as the detector capacitance and its leakage current, the peaking time and the power consumption allocated to the front-end.

The problem finds an easy solution if the transistor works in strong inversion, where the optimization rules emerge out of elementary analytical formulas [1, 2]. Today, unfortunately, this simple optimization procedure is not suitable in most cases. At the current densities normally employed in front-end circuits, modern CMOS transistors operate in fact between weak and moderate inversion, where different recipes must be followed. The study of the classical case remains nevertheless instructive, because it lays down the framework in which other situations, today more recurrent in practice, can be discussed.

We start examining the optimization for flicker noise, which requires less mathematical effort. We have already seen that the ENC due to $1/f$ noise can be expressed as:

$$ENC_f^2 = \frac{K_f}{C_{ox}WL} C_T^2 N_f \tag{8.1}$$

where W and L are the gate width and length, C_{ox} is the gate capacitance per unit area, K_f the flicker noise coefficient and N_f the shaper noise index for flicker noise. The total input capacitance, C_T, can be written as the sum of two components: the detector capacitance C_d and the gate capacitance of the input transistor $C_g = C_{ox}WL$. In C_d we may lump the capacitance of the sensor electrode plus any other capacitance (e.g. stray due to the interconnection) shunting the preamplifier input. Equation (8.1) can then be rewritten as:

$$ENC_f^2 = K_f \frac{(C_d + C_g)^2}{C_g} N_f \tag{8.2}$$

We can now search the condition that minimizes ENC_f^2 by differentiating (8.2) with respect to C_g and equating the result to zero:

$$\frac{dENC_f^2}{dC_g} = N_f K_f \left(\frac{2(C_d + C_g) C_g - (C_d + C_g)^2}{C_g^2} \right) = 0 \rightarrow C_g = C_d \tag{8.3}$$

We can thus conclude that flicker noise is minimized if the gate capacitance of the input transistor is equal to the one of the sensor or, more in general, to the sum of all the capacitances that can be represented in parallel to the input.

We can now study the minimization of the ENC due to thermal noise, for which we assume the following spectral density:

$$v_{nw}^2 = \frac{4k_B T n \gamma \alpha_w}{g_m} \tag{8.4}$$

We remind that in the above equation k_B is the Boltzmann constant, T the absolute temperature, γ the inversion factor, n the body factor, α_w the excess noise factor and g_m the input device transconductance. The ENC due to thermal noise can be written as:

$$ENC_w^2 = 4k_B T n \gamma \alpha_w \frac{(C_d + C_g)^2}{g_m(C_g)} \frac{N_w}{T_p} \tag{8.5}$$

where N_w is the shaper noise index for thermal noise and T_p is the peaking time. We have written g_m as a function of the gate capacitance C_g to emphasize the fact that the latter enters into the noise equation with two opposite effects, which should imply that an optimum value exists. To find this optimum, we differentiate ENC_w^2 with respect to C_g, equating again the result to zero:

$$\frac{dENC_w^2}{dC_g} = \frac{4k_B T n \gamma \alpha_w N_w}{T_p} \left[\frac{2(C_d + C_g) g_m - (C_d + C_g)^2 \frac{dg_m}{dC_g}}{g_m(C_g)^2} \right] = 0 \tag{8.6}$$

The term before the parenthesis becomes irrelevant for the equation, which reduces to:

$$2g_m = (C_d + C_g) \frac{dg_m}{dC_g} \tag{8.7}$$

Note that the above condition is very general, as it does not assume any particular form of $g_m(C_d)$. The solution of the problem is thus contained in the relationship connecting the transconductance to the gate capacitance. We must now define explicitly g_m as a function of C_g. To do so, we need to make a guess on the operating region of the transistor and, as a first step, we assume that the device works in strong inversion. In this case, we can write g_m as:

$$g_m = \sqrt{2\mu C_{ox} \frac{W}{L} I_D} \tag{8.8}$$

where μ is the carrier mobility (electrons or holes) and I_D is the drain-source current. We do not put any restriction on the type of device, which can be either a PMOS or an NMOS, hence the following considerations apply indifferently to both kinds of transistors. In a first approximation, the gate capacitance can be written as $C_g = C_{ox}WL$. In principle, if the device is biased in strong inversion and in saturation, the gate capacitance is $2/3 C_{ox}WL$. However, the capacitance due to the physical overlap between the gate and the source/drain electrodes and to the fringing fields is not negligible and can be up to 1 fF per micron of gate width. Approximating C_g with $C_{ox}WL$ thus provides a reasonable estimate of the total gate capacitance to be considered in the ENC evaluation. By multiplying and dividing by L the quantity under the square-root in equation (8.8) we obtain:

$$g_m = \sqrt{2\mu C_g \frac{I_D}{L^2}} \tag{8.9}$$

Differentiating g_m with respect to C_g we have:

$$\frac{dg_m}{dC_g} = \frac{\mu \frac{I_D}{L^2}}{\sqrt{2\mu C_g \frac{I_D}{L^2}}} \tag{8.10}$$

Inserting now equation (8.10) into (8.7) and solving for C_g we find the following condition:

$$C_g = \frac{1}{3} C_d \tag{8.11}$$

which is the very well known classical rule stating that the thermal noise part of the ENC due to the input transistor is minimized when its gate capacitance is equal to one third of the sensor capacitance, or, more in general, the sum of all the other capacitances shunting the input. One could then deduce the following optimization strategy, assuming that the maximum power consumption and the peaking time are known: the maximum bias current I_D compatible with the power allocated to the input stage is selected, together with the minimum gate length L allowed by the technology in which the circuit is being designed. According to (8.8), both choices contribute in fact to increasing the transconductance. The gate width is finally adjusted to achieve the matching condition defined in (8.11). At this point, the ENC due to flicker noise is calculated. If it is greater than the one given by thermal noise, C_g can be further increased. Depending on the value of K_f and the peaking time, the optimization will result in a W yielding a gate capacitance between $1/3C_d$ and C_d.

Unfortunately, things are not that simple, because increasing W the operating region of the transistor can change. To understand this, we can just rewrite the equation defining the drain-source current of a transistor in strong inversion and in saturation:

$$I_{DS} = \frac{1}{2} \mu C_{ox} \frac{W}{L} (V_{GS} - V_{TH})^2 \tag{8.12}$$

In this particular case, we have written the relationship for the NMOS, but of course the following arguments apply to the PMOS as well. Since the bias current I_{DS} is fixed, increasing W/L reduces the overdrive voltage $V_{GS} - V_{TH}$, eventually driving the transistor in moderate or weak inversion. In particular, in weak inversion the transconductance is given by:

$$g_m = \frac{I_{DS}}{n\phi_T} \tag{8.13}$$

As a result, g_m is not a function of the gate capacitance any longer, but it depends only on the bias current. Therefore, if the transistor works in this region, increasing its gate width too much worsens the noise, because it leads to more gate capacitance without improving the transconductance. In fact, according to (8.13), $dg_m/dC_g = 0$, which, inserted in equation (8.10), yields the optimal value $C_g = 0$, implying that the device should be as small as possible[1].

If the drain current and the detector capacitance are known, it is possible to predict if we are in the regime for which the classical approximation holds [2]. The method proceeds as follows. We suppose a priori to be in the classical case and to have optimized the transistor for minimal thermal noise, hence we assume $C_g = C_d/3$. By multiplying and dividing the second member of equation (8.12) by L we can introduce explicitly the dependence on the gate capacitance:

$$I_{DS} = \frac{1}{2}\mu\frac{C_g}{L^2}(V_{GS} - V_{TH})^2 \tag{8.14}$$

We now express the overdrive voltage as a function of the other parameters, introducing also the condition $C_g = C_d/3$, which yields:

$$V_{GS} - V_{TH} = \sqrt{\frac{2L^2 I_{DS}}{\mu C_g}} = \sqrt{\frac{6L^2 I_{DS}}{\mu C_d}} \tag{8.15}$$

The next step is to require that we are actually in strong inversion, which means that the overdrive voltage must be significantly greater than zero. Since the transition from moderate to strong inversion is not sharp, there is some degree of freedom in the choice of this value. The limit proposed in [2] is to set the boundary at $2n\phi_T$, hence we have:

$$V_{GS} - V_{TH} > 2n\phi_T \rightarrow \sqrt{\frac{6L^2 I_{DS}}{\mu C_d}} > 2n\phi_T \tag{8.16}$$

We can rewrite the above condition as follows:

$$\frac{I_{DS}}{C_d} > \frac{2}{3}\frac{\mu}{L^2}(n\phi_T)^2 \tag{8.17}$$

Given the bias current and the sensor capacitance, equation (8.17) allows us to calculate if the transistor can work in strong inversion and, consequently, if the classical matching condition should be applied. Note that on the right side of the inequality we have mainly constants or technological parameters. The only design variable is the transistor length L which however for a first estimation can be chosen equal to the minimum allowed by the process. As an example, consider the case of an NMOS transistor implemented in a 0.13 μm CMOS technology with $n = 1.5$, $\mu = 400$ cm^2/V·s and operating at room temperature ($\phi_T = 26$ mV). Assuming we use a channel length of 130 nm, the limit for a strong inversion operation requires that the ratio I_{DS}/C_d is greater than $2.4 \cdot 10^9$ V/s. Satisfying this condition for a sensor with 10 pF requires a bias current of 24 mA. This value of current is beyond the power budget of most

[1] Observe that the requirement of minimal transistor area may conflict with the optimization of the flicker noise.

applications, because it would lead to more than 30 mW of power dissipated only in the input branch of the CSA. In a deep submicron technology, the mobility of holes is typically at least four times smaller than the one of the electrons, therefore for a PMOS the limit would lower to 6 mA. Observe that increasing the sensor capacitance for a given power budget shifts the boundary further away from the classical region. In fact, in an attempt to reach the optimal condition $C_g = C_d/3$, one increases the transistor width till the device leaves the strong inversion region and shifts to the moderate or weak inversion, where equation (8.9) is not valid anymore.

The relationship in (8.17) can also be rewritten as:

$$C_d < \frac{3}{2}\frac{L^2}{\mu}\frac{I_{DS}}{(n\phi_T)^2} \tag{8.18}$$

To see if the classical condition holds, the mobility μ and the body factor n are extracted from the model provided by the foundry. The device length is then chosen, according to criteria that will be discussed more extensively in the following pages. However, for a first evaluation, a value equal to the minimum gate length allowed by the process can be assumed. The quantity on the right side of the inequality, which has the unit of a capacitance, can thus be evaluated. If the actual detector capacitance is smaller than the found value, the classical approximation holds, otherwise a different optimization strategy should be applied. Observe that the boundary condition is quite sensitive to the channel length, which enters squared in (8.18).

Example 8.1: Estimation of the input transistor working region.

A CMOS technology has the following parameters: $L_{min} = 0.12\ \mu$m, $\mu_n = 400$ cm^2/V·s, $\mu_p = 100$ cm^2/V·s, $n = 1.5$. If the current that can be spent in the input device is 300 μA, estimate the maximum sensor capacitance for which the device would be sized to work in strong inversion.

* * *

For a first estimate we can assume minimum length transistors. Putting the numbers in (8.18) we obtain for an NMOS input device:

$$\frac{3}{2}\frac{(120\cdot 10^{-9})^2}{0.04}\frac{3\cdot 10^{-4}}{1.52\cdot 10^{-3}} = 1.07\cdot 10^{-13}\ \text{F} = 107\ \text{fF} \tag{8.19}$$

For a PMOS, the only parameter that changes is the mobility, leading to a four times greater value:

$$\frac{3}{2}\frac{(120\cdot 10^{-9})^2}{0.01}\frac{3\cdot 10^{-4}}{1.52\cdot 10^{-3}} = 4.28\cdot 10^{-13}\text{F} = 428\ \text{fF} \tag{8.20}$$

If the sensor we have to read has a capacitance greater than 107 fF for an NMOS input transistor or 428 fF for a PMOS, the inequality given in (8.18) is violated and the device will work in moderate or weak inversion. If we now make a less aggressive choice of the gate length, setting it to 1.5 times the minimum, the limits increase to 240 fF for an NMOS and 960 fF for a PMOS, respectively.

A possible optimization procedure could thus be the following: given the maximum drain current I_D, choose the minimum length allowed by the process. Then, verify (8.18). If the inequality is satisfied, the transistor can work in strong inversion and the condition $C_g = 1/3C_d$ should be met if, given the peaking time and the allocated power budget, the thermal noise is the dominant source. If flicker noise is important, the gate width is increased till the optimum value of the gate capacitance is instead found between $C_d/3$ and C_d. If (8.18) is violated, choose the minimum value of W that put the transistor in weak inversion. Although it sounds at first reasonable, such a strategy is very qualitative. In fact, the transition between strong and weak inversion may span up to two decades of current, hence it is not clear a priori which degree of inversion should be chosen. Furthermore, going from strong to weak inversion the effective gate capacitance changes and a more accurate estimate of its value becomes necessary.

Before discussing the optimization in weak and moderate inversion, it is interesting to examine what happens if the sensor capacitance is very small [2]. In this case, one would reduce the input transistor width, increasing, for a fixed bias current, its overdrive voltage. To keep the device saturated, also the drain-source voltage must be maintained greater than $V_{GS} - V_{TH}$. The electric field across the channel is thereby increased and the transistors may eventually enter into the velocity saturation regime. This typically occurs for fields greater than 1 V/μm, implying that 100 mV applied across a device with 100 nm gate length are sufficient to cause the phenomenon. When this happens, the drain current becomes:

$$I_{DS} = v_{sat} C_{ox} W (V_{GS} - V_{TH}) = v_{sat} \frac{C_g}{L} (V_{GS} - V_{TH}) \tag{8.21}$$

where v_{sat} is the maximum velocity that can be attained by the carriers, approximately equal to 10^7 cm/s. In equation (8.21) we have as usual multiplied and divided the expression of I_{DS} by L to introduce the explicit dependence on C_g. The transconductance and its derivative are thus given by:

$$g_m = \frac{v_{sat}}{L} C_g \rightarrow \frac{dg_m}{dC_g} = \frac{v_{sat}}{L} \tag{8.22}$$

Inserting this relationship into (8.7), we find the optimal condition $C_g = C_d$. Interestingly, in the velocity saturation regime the same value of gate capacitance minimizes both flicker and thermal noise. However, this situation is little more than an academic curiosity. To see why, we can write the velocity saturation as:

$$v_{sat} = \frac{\mu (V_{GS} - V_{TH})}{L} \rightarrow (V_{GS} - V_{TH}) = \frac{L v_{sat}}{\mu} \tag{8.23}$$

With the above equation, we have expressed the overdrive voltage as a function of v_{sat}. We can then reason as before: we assume to be in strong inversion with C_g set equal to $C_d/3$ and we calculate the overdrive voltage, which is given by (8.15). We then require that the resulting value of $(V_{GS} - V_{TH})$ is smaller than the one given in (8.23), that would cause velocity saturation. We can thus write:

$$\sqrt{\frac{6L^2 I_{DS}}{\mu C_d}} < \frac{L v_{sat}}{\mu} \rightarrow C_d < \frac{6\mu I_{DS}}{v_{sat}^2} \tag{8.24}$$

If we now consider a case with $\mu = 400$ cm^2/V·s, $v_{sat} = 10^7$ cm/s, $I_{DS} = 100$ μA, applying (8.24) we find that the velocity saturation condition only occurs if $C_d <$ 2.5 fF. Such a small capacitance hardly occurs even in monolithic CMOS sensors, for which however other considerations, such as the overall power consumption, would likely limit the current in the front-end well below the value assumed in the example.

We can now revert to the size optimization in weak and moderate inversion. To study further the problem, it is important to have equations describing the relevant quantities (in particular the transconductance and the gate capacitance) in a form that is continuous across the different levels of inversion. Therefore, several authors have adopted the EKV model, which provides compact equations satisfying this requirement [3–7]. The EKV-based equations were already introduced in chapter 2. We remind here that, in this approach, the device region of operation is identified by the inversion coefficient I_C defined as:

$$I_C = \frac{I_D}{2n\mu C_{ox}\phi_T^2 \frac{W}{L}} \tag{8.25}$$

The transistor is considered in weak inversion if $I_C < 0.1$ and in strong inversion if $I_C > 10$. For $0.1 < I_C < 10$ the device is in moderate inversion. All the other quantities are then written as functions of I_C. The transconductance becomes:

$$g_m = \frac{I_D}{n\phi_T} \frac{1}{\sqrt{I_C + 0.5\sqrt{I_C} + 1}} \tag{8.26}$$

The inversion factor γ, which is relevant in noise calculations, is given by:

$$\gamma = \frac{1}{2} + \frac{1}{6}\frac{I_C}{I_C + 1} \tag{8.27}$$

The total gate capacitance is:

$$C_g = C_{gs} + C_{gb} \tag{8.28}$$

where C_{gs} is the gate-source capacitance, C_{gb} is the gate-bulk capacitance. The sum of C_{gs} and C_{gb} can also be written as [5]:

$$C_g = C(x) C_{ox} WL \tag{8.29}$$

where the coefficient $C(x)$ is:

$$C(x) = \frac{n - (1 + x)/3}{n} \tag{8.30}$$

The variable x is a function of the inversion coefficient as well:

$$x = \frac{(\sqrt{I_C + 0.25} + 0.5) + 1}{(\sqrt{I_C + 0.25} + 0.5)^2} \quad (8.31)$$

A practical example will now show how the sizing depends on the interplay of several factors.

Example 8.2: Input transistor sizing.

A charge sensitive amplifier is used together with a CR-RC shaper. The maximum current that can be allocated to the CSA input stage is 200 μA. The circuit is implemented in 0.13 μm CMOS process having the following parameters: $t_{ox} = 2.4$ nm, $\mu_n C_{ox} = 400\ \mu A/V^2$, $\mu_p C_{ox} = 100\ \mu A/V^2$, $K_{fN} = 2.4 \cdot 10^{-24}$ J, $K_{fP} = 0.6 \cdot 10^{-24}$ J. The excess noise factor for white noise, α_w, can be assumed equal to one. The system should achieve a noise below 300 electrons rms when used with a sensor that has a capacitance of 5 pF and a leakage current of 1 nA. The total signal duration should be less than 20 μs.

* * *

The technology parameters are the same of the previous example, hence we already know that with the given sensor capacitance both an NMOS and a PMOS transistor will have their optimal bias point outside the strong inversion region. We start assuming to use an NMOS input transistor with a minimum gate length. First, we calculate the gate capacitance per unit area:

$$C_{ox} = \frac{\varepsilon_{ox}}{t_{ox}} = \frac{3.9 \times 8.85 \cdot 10^{-12}}{2.4 \cdot 10^{-9}} = 15\ \mathrm{fF}/\mu\mathrm{m}^2 \quad (8.32)$$

It is then useful to evaluate the current I_M for which a device with the chosen gate length and a gate width of 1 μm will have an inversion coefficient of 1, that puts it at the center of moderate inversion. This is given by:

$$I_M = 2n\mu_n C_{ox}\phi_T^2 \frac{W}{L} = 2 \times 1.5 \times 0.04 \times 1.5 \cdot 10^{-2} \times 6.7 \cdot 10^{-4} \frac{1 \cdot 10^{-6}}{1.3 \cdot 10^{-7}} \approx 9.4\ \mu A \quad (8.33)$$

This equation tell us that if we want $I_C = 1$ for I_{DS}=200 μA, we need a gate width of 20 μm. Using equation (8.26), we can evaluate that device transconductance, which is 3.24 mS. If the transistor works in weak inversion with $I_C = 0.1$, the transconductance would be 4.56 mS. To obtain an inversion coefficient of 0.1 we need to increase the transistor size to 200 μm. A rough estimation of the transistor gate capacitance as $C_g = C_{ox}WL$ yields 390 fF, which is much smaller than the sensor capacitance. To get however a more accurate result, in the following C_g is estimated using equations (8.28)-(8.31) and added to C_d to yield the total input capacitance C_T. Since the device area is rather small, it is worth checking the contribution of the flicker noise before going further with the optimization:

$$ENC_f^2 = \frac{K_{fN}}{C_{ox}WL}(C_d + C_g)^2 N_f \quad (8.34)$$

For a CR-RC shaper, $N_f = 3.57$. Putting in the other numbers, taking the square-root and dividing by the electron charge q to express the ENC in electrons, we get $\text{ENC}_f = 160$ electrons rms. This already covers one half of our noise budget. We can then increase further the gate width to reduce the flicker noise. Note that for the $1/f$ spectral density the relevant quantity is $C_{ox}WL$. However, it is the effective gate capacitance C_g which goes in parallel to the detector capacitance and hence enter in the term $(C_d + C_g)^2 = C_T^2$ which appears in both the flicker and thermal noise expressions. Increasing the gate width by a factor of two, we push the transistor further in weak inversion, reducing the inversion coefficient from 0.1 to 0.05. Using equations (8.28) to (8.31), we can estimate the gate capacitance as $0.36C_{ox}WL = 283$ fF, which is still small with respect to the sensor capacitance. Hence, we can re-evaluate the flicker noise to find 108 electrons rms. We can now optimize the thermal and parallel noise. Using the new inversion coefficient, the transconductance g_m becomes 4.8 mS. The γ factor given by equation (8.27) is 0.508, so we can assume in the calculation the value of 0.5, which ideally characterizes the weak inversion regime. Knowing the leakage current we can find the optimum peaking time as:

$$T_p = \sqrt{\frac{4k_B T n\gamma}{g_m} C_T^2 \frac{1}{2qI_{leak}} \frac{N_w}{N_i}} \tag{8.35}$$

where N_w and N_i are the shaper noise coefficient respectively for thermal and parallel noise. In a CR-RC shaper, the two are identical and equal to 0.923. Putting in our numbers, we get the optimal peaking time value of 478 ns. Since the noise depends on the square-root of the peaking time, a round choice such as 500 ns would not change so much the ENC. Using this value of T_p, we can calculate the ENC due to white series and parallel noise, which are, respectively:

$$ENC_w^2 = \frac{4k_B T n\gamma}{g_m} \frac{C_T^2}{T_p} N_w \tag{8.36}$$

and

$$ENC_i^2 = 2qI_{leak} N_i T_p \tag{8.37}$$

Inserting numbers in the two above equations, we get $ENC_w = 73$ electrons rms and $ENC_i = 75$ electrons rms. Note that we have chosen a peaking time larger than the optimal one, which gives a little bit more weight to the parallel noise and slightly less to the series one. The total rms noise can now be calculated summing in quadrature the three components to get a total noise of 150 electrons rms. The ENC due to the thermal noise is still smaller than the one due to flicker noise. Thus increasing further the gate width would bring a better balance between the two noise sources. However, the noise is already one half of the target value, so the result can be considered already satisfactory.

It is interesting to compare the results if a PMOS is used at the input transistor. Since we have assumed that the flicker noise coefficient of the PMOS is one fourth the one of the NMOS, for the same transistor area the flicker noise will be halved, reducing it to 54 electrons rms for $W = 400\ \mu$m. On the other hand, the mobility of the holes is also one fourth, which means that for the same bias current of 200 μA, the inversion coefficient will be 0.2, shifting the operation towards moderate inversion. In this case, the transconductance is 4.3 mS. Using equations (8.30) and (8.31), we can find the gate capacitance C_g which is 335 fF. Note that since the device works more towards the moderate inversion, its gate capacitance has increased. We can now re-apply equation (8.35), with the new values of the gate capacitance and of the inversion factor

γ (which is now 0.53) to obtain the new optimal peaking time of 522 ns, which is not much different from the 500 ns previously used. With $T_p = 522$ ns we get the thermal and parallel noise contributions at respectively 77.5 and 77.5 rms electrons. If we use the rounded value of 500 ns, we get instead 79 and 76 electrons. The total ENC is then 122.18 rms electrons (522 ns peaking time) or 122.2 rms electrons (for $T_p = 500$ ns), confirming the irrelevant effect of small peaking time adjustments around the optimal value.

The ENC achieved in the previous example is significantly lower than the specified one. On the one hand, keeping some safety factor with respect to the target design values is a good practice, because the noise of the final system is almost always higher than the one achieved in computer simulations. Many reasons in fact (underestimation of parasitics both at the chip and at the system level, inaccuracy in the models, etc.) may contribute to the increase of the total final noise. In the long term, the noise, and in particular its $1/f$ component, can also worsen due to the device aging [8]. It is thus advisable to target in the design phase a noise floor lower than the bare acceptable minimum. It must however also be observed that a small penalty in the noise figure may bring significant advantages to other system parameters, as shown in the next example.

Example 8.3: Redesign of the circuit of example 8.2 for lower power.

Using the data of example 8.2, redesign the system to save at least 50% power in the input device.

* * *

We note that the maximum allowed dead time is 20 μs, so we can use a longer peaking time. Since the noise achieved with the previous sizing was substantially lower than the target one, we can also tolerate a bigger ENC. For instance, if each of the three noise sources contribute with 100 electrons, the total ENC is 173 electrons, still allowing a 40% safety margin. To have a parallel noise of 100 electrons, we can allow a peaking time of:

$$T_p = \frac{ENC_i^2}{2qI_{leak}} = \frac{2.56 \cdot 10^{-34}}{2 \times 1.6 \cdot 10^{-19} \times 1 \cdot 10^{-9}} = 800 \text{ ns} \tag{8.38}$$

Using the new peaking time, we can calculate the transconductance that equalizes the series and parallel noise:

$$g_m = \frac{4k_B T n \gamma}{ENC_w^2} \frac{C_T^2}{T_p} N_w \tag{8.39}$$

where we set ENC$_w$ = 100 electrons, which implies ENC$_w^2 = 2.56 \cdot 10^{-34}$ C^2. We initially use a PMOS transistor with the same aspect ratio $W/L = 400/0.13$ μm employed in the previous example and we assume that the device works deep in weak inversion, (i.e. $I_C = 0$ and $\gamma = 0.5$). With these hypotheses, the transconductance necessary to achieve 100 electrons at 800 ns of peaking time is 1.6 mS, which corresponds to a current $I_{DS} = g_m n \phi_T$ of 62 μA. The actual inversion factor for this current is $I_C = 0.066$, which leads to a γ factor of 0.51, a gate capacitance of 290 fF and an effective transconductance of 1.45 mS. With these values, the thermal noise becomes 105 electrons. The total noise is then 145 electrons rms. These new settings lead to a noise increase of 36 electrons with respect to the case in which the maximum current is used, but the power is reduced by 70%.

Let's now reduce the W of the transistor by one half. If the bias current is not changed, the inversion coefficient is doubled, the transconductance reduces to 1.39 mS and the gate capacitance becomes 185 fF. The reduced gate area brings the flicker noise to 76 electrons, while the thermal noise rises to 106 electrons, mainly because of the loss in transconductance. The total ENC becomes 164 electrons. Therefore, accepting a penalty of 9 electrons in the ENC, one can spare 50% of the input transistor area, which is typically the largest (or one of the largest) component in the CSA.

The above examples show that selecting the maximum allowed power consumption in the input transistor and targeting the lowest possible ENC might not always be the optimal choice from a system point of view. In large multi-channel ASICs and in big detector complexes, a low power consumption is often critical, and design implementations that lead to an adequate rather than to the smallest achievable noise can be at a premium. The saved power may also be invested in extra components (such as on chip regulators, better reference generators, etc.) that can enhance the robustness of the chip and the overall performance of the apparatus. We should also note that we have found adequate dimensions of the input transistor matching or exceeding the specifications, but we might not have performed the best possible optimization.

For a more accurate sizing of the input device, additional effects must be taken into account. First, in a deep submicron technology the choice of the minimum gate length, L_{min}, allowed by the process is usually not recommended. On the one hand, minimum length transistors have degraded output conductance and their use can reduce significantly the DC gain of the amplifier, which is another key design parameter. Furthermore, the following effects on noise have been observed when the channel length approaches L_{min}:

- The excess noise factor for white noise, α_w, increases [9].
- The flicker noise coefficient K_f is enhanced. For NMOS transistors, the value of K_f at minimum gate length can be three to four times bigger than the one at $L = 3L_{min}$[10].

For PMOS transistors, a dependence of K_f on the current density has also been measured, with up to six-fold increase in some process between $V_{GS} - V_{TH} = 0$ V and $V_{GS} - V_{TH} = 0.2$ V [10]. In addition, the deviation of the flicker noise from the ideal $1/f$ power law should also be considered. A more accurate definition of the flicker noise spectral density is given by:

$$v_{nf}^2 = \frac{K_f(L, I_C)}{C_{ox}WL} \frac{1}{f^{\alpha_f}} \tag{8.40}$$

where the exponent α_f is between 0.85 for NMOS and 1.1 for PMOS transistors.

Another important component that we have neglected so far is the gate-drain and gate-source overlap capacitance. This can be close to 1 fF per micron of transistor width and may represent a sizable portion of the total gate capacitance for small devices. The formula of the gate capacitance must hence be extended to include this contribution [6, 9]:

$$C_g(I_C, W, L) = C(x(I_C))C_{ox}WL + 2C_{ov}W = C_{gW}(I_C, L)W \tag{8.41}$$

where $C_{gW} = C(x(I_C))C_{ow}L + 2C_{ov}$ is the gate capacitance per unit of width. Note that the variable x defined in equation (8.31) is a function of the inversion coefficient as well. The overall ENC equation can then be reformulated as:

$$ENC^2 = (C_d + C_{gW}(I_C, L)W)^2 \left[\frac{4k_B T n \gamma(I_C) \alpha_w}{g_m(I_C)} \frac{N_w}{T_p} + \frac{K_f(L, I_C)}{C_{ox}WL} \frac{N_f(\alpha_f)}{T_p^{1-\alpha_f}} \right] + 2qI_{leak}N_i T_p \tag{8.42}$$

where we have introduced the dependence of the flicker noise coefficient on the gate length and on the inversion coefficient. For completeness, the excess noise factor α_w has also been inserted. The optimization proceeds thus as follows. The input capacitance, the detector leakage current, the event rate, the maximum available power consumption and the target ENC are supposed to be known, as these constraints define the fundamental properties of any detection system. On the designer side, an assumption on the type of pulse shaper is necessary to determine the filter noise coefficients. A shaper with one real and two complex-conjugate poles can be a good starting point, as it can be implemented with similar area and power consumption of a CR-RC circuit, but it offers a faster return to the baseline and has more favorable noise coefficients, especially for parallel noise. The peaking time is then determined from the rate requirements and the ENC due to parallel noise is calculated. If at the chosen value of peaking time the parallel noise gives already a small contribution to the total ENC, the peaking time can be considered defined and the optimization can concentrate on the minimization of the series noise. In case the ENC due to parallel noise is too high, the peaking time should be reduced. A possibility can be to start from the peaking time giving a parallel noise equal to ENC/$\sqrt{3}$, assuming that and the end the three noise sources will equally share the ENC budget, and then continue with the series noise optimization. Form equation (8.25) we observe that once the gate length and the drain current are chosen, the current density and the inversion coefficient depends only the transistor width W, which is thus swept till the minimum ENC is found. Equations (8.25)-(8.31) and equation (8.41) can be used to estimate the various functions of the inversion coefficient appearing in (8.42). The procedure can then be iterated for different values of gate lengths. If the value found of the ENC is significantly lower than the target one, a new optimization with a reduced value of the drain current can be performed to improve upon the power consumption. Note also that if a PMOS is used as the input device, due to the dependence of the flicker noise coefficient on the bias current, the optimal value of ENC could be found at a smaller current than the maximum allowed by power dissipation constraints. This is especially true when long shaping times are permitted, thus reducing the white series noise contribution [11]. The inversion coefficient can also be written as:

$$I_C = \frac{I_{DS}}{2n\mu C_{ox}\phi_T^2 \frac{W}{L}} = \frac{L}{2n\mu C_{ox}\phi_T^2} \frac{I_{DS}}{W} = K(L)J_D \tag{8.43}$$

This shows that I_C is proportional to the drain current per unit of gate width through the coefficient $K(L)$, which becomes fixed once the transistor length in chosen. Equation 8.41 can also be written as [9, 11]:

$$ENC^2 = \left[C_d + C_{gW}(J_D,L)W\right]^2 \left[\frac{4k_BTn\gamma(J_D,L)\alpha_w}{g_m(J_D,L)}\frac{N_w}{T_p} + \frac{K_f(J_D,L)}{C_{ox}WL}\frac{N_f(\alpha_f)}{T_p^{1-\alpha_f}}\right] + 2qI_{leak}N_iT_p \tag{8.44}$$

This second formulation can be more practical if one wants to extract $g_m(J_D,L)$ and $C_g(J_D,L)$ from computer simulations or measurements on a limited set of sample devices, as in this case the bias current I_D is changed for a given transistor width to sweep J_D. For a fixed current in the input stage, varying W changes J_D, therefore knowing the functions $g_m(J_D,L)$, $C_{gW}(J_D,L)$ one can perform the optimization as well. The results obtained with numerical calculations are then verified and refined with CAD simulations. Front-end amplifiers contain a relatively small number of transistors and their electrical simulation is very fast on today's computers. Furthermore, modern CAD software offer algorithms that can optimize the transistor sizing on the basis of user-defined constraints. The tool always employs the full set of models provided by the foundry and all the devices in the circuit are taken into account, so critical contributions beyond the input transistor promptly emerge. Direct optimization with the CAD tool can then be more effective in terms of design time. It is nevertheless important that the designer knows well the effect of the major parameters defining the ENC in order to properly guide the CAD through the optimization process. It must also be pointed-out that, sometimes, the noise model provided by the foundry might not be fully accurate because it is extracted with current densities much higher than the one typically used in a front-end. For the best possible noise prediction, the functions $C_gW(J_D,L)$, $\gamma(J_D,L)$, $g_m(J_D,L)$ and $K_f(J_D,L)$ used in (8.44) should be derived from experimental measurements.

For the sake of completeness, we mention that alternative expressions for the inversion coefficient are found in the literature [12]. One can in fact define a specific current $I^* = 2n\mu C_{ox}\phi_T^2$, which is the current necessary to have an inversion coefficient of 1 for a transistor with unitary aspect ratio ($W/L = 1$) in a given process. The specific current is thus a purely technology-dependent parameter. The inversion coefficient is hence defined as:

$$I_C = \frac{LI_{DS}}{WI^*} \tag{8.45}$$

which allows us to clearly separate the different components: process (through I^*), bias (through I_{DS}) and device geometry. The transconductance and the inversion factor can also be expressed as:

$$g_m(I_{DS},W,L) = \frac{I_{DS}}{n\phi_T}\frac{2}{1+\sqrt{1+4\frac{LI_{DS}}{WI^*}}} \tag{8.46}$$

$$\gamma(I_D,W,L) = \frac{1}{1+\frac{LI_{DS}}{WI^*}}\left(\frac{1}{2}+\frac{2}{3}\frac{LI_{DS}}{WI^*}\right) \tag{8.47}$$

Further interesting discussions about the optimal sizing of the input transistor in a CSA can also be found in [13] and [14].

8.1.2 LOAD DESIGN

The design of the active load also presents interesting trade-offs. Consider first a direct cascode as shown in Fig. 8.1. In this case, the full bias current flows both in the input transistor and in the load. The total current noise at the drain is obtained by summing the power spectral densities of M_1 and M_4:

$$i_{nw}^2 = 4k_BTn\gamma_1\alpha_w g_{m1} + 4k_BTn\gamma_4\alpha_w g_{m4} \tag{8.48}$$

Dividing the above equation by g_{m1}^2, one gets the input referred voltage noise, which reads:

$$v_{nw}^2 = 4k_BTn\gamma_1\alpha_w\frac{1}{g_{m1}} + 4k_BTn\gamma_4\alpha_w\frac{g_{m4}}{g_{m1}^2} \tag{8.49}$$

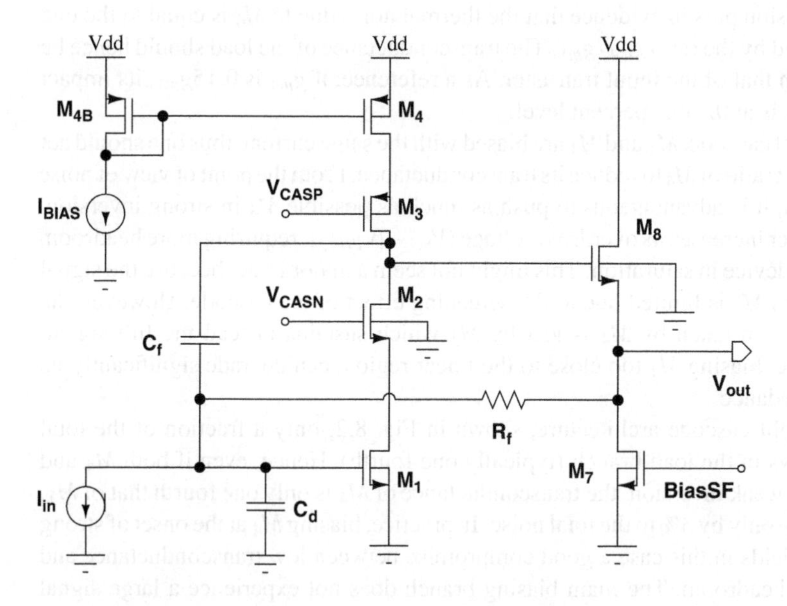

FIGURE 8.1 Charge sensitive amplifier implemented with a direct cascode.

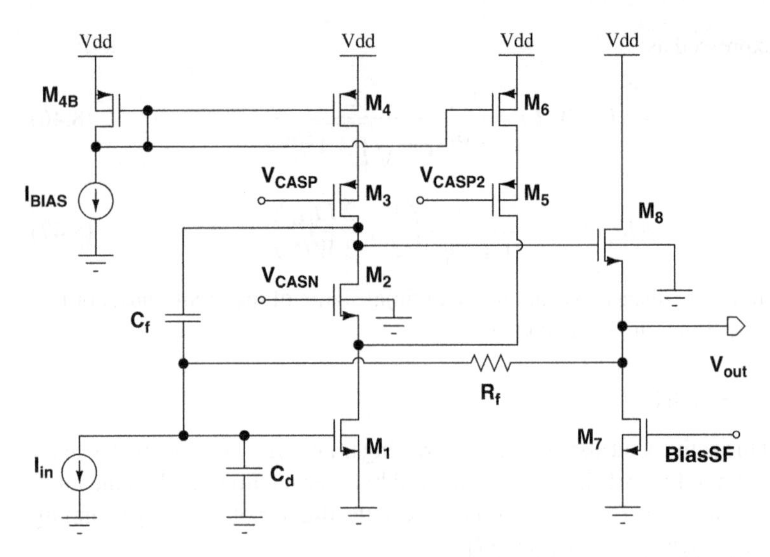

FIGURE 8.2 Charge sensitive amplifier implemented with a direct cascode with split current sources.

The term due to M_4 can be rewritten as:

$$v_{nw4}^2 = 4k_B T n \gamma_4 \alpha_w \frac{1}{g_{m1}} \left(\frac{g_{m4}}{g_{m1}} \right) \tag{8.50}$$

This expression puts in evidence that the thermal noise due to M_4 is equal to the one of M_1, scaled by the ratio g_{m4}/g_{m1}. The transconductance of the load should hence be smaller than that of the input transistor. As a reference, if g_{m4} is $0.15g_{m1}$, its impact on the ENC is at the one percent level.

In a direct cascode, M_1 and M_4 are biased with the same current, thus one should act on the aspect ratio of M_4 to reduce its transconductance. From the point of view of noise suppression, it is advantageous to push as much as possible M_4 in strong inversion. This however increases its overdrive voltage $(V_{SG} - |V_{TH}|)_4$, requiring more headroom to keep the device in saturation. This might not seem a major issue, because the signal swing across M_4 is limited due to the screening effect of the cascode. However, the extra headroom taken by M_4 is lost by M_3 which sustains instead the full signal. Furthermore, biasing M_4 too close to the linear region, can degrade significantly its output impedance.

In the split cascode architecture, shown in Fig. 8.2, only a fraction of the total current flows in the load branch (typically one fourth). Hence, even if both M_4 and M_1 work in weak inversion, the transconductance of M_4 is only one fourth that of M_1, contributing only by 3% to the total noise. In practice, biasing M_4 at the onset of strong inversion yields in this case a good compromise between low transconductance and reasonable headroom. The main biasing branch does not experience a large signal swing. Therefore, the current source M_6 can be put deep in strong inversion. One should however pay attention to the aforementioned increase of α_w and K_f with the inversion level. Note that in the split bias cascode, M_6 may take up the role of second most important noise source, as it is biased at a greater current than M_4. To refer the flicker noise contribution of the load to the preamplifier input we need first to convert it into a current through the load transconductance and then divide by g_{m1}^2. For instance, in case of M_4 we have:

$$v_{n4f} = \frac{K_f}{C_{ox}W_4L_4}\left(\frac{g_{m4}}{g_{m1}}\right)^2 \tag{8.51}$$

A similar equation can be written for M_6.

It is worth noting that the cascode transistors give little contribution to the noise because their transconductance is degenerated by the output resistance of the load. For instance, the equivalent transconductance of M_3 is given by:

$$g_{m3,eq} \approx \frac{g_{m3}}{1+g_{m3}r_{04}} \tag{8.52}$$

Interestingly, the source degeneration technique could be applied also to the load transistors M_4 and M_6 to further suppress their transconductance. In some circuits implemented in very deep submicron technologies this principle has been applied systematically to all current sources to decrease their transconductance and increase their output impedance [15]. However, the insertion of a source degeneration resistor requires an additional voltage drop, thus reducing the useful output swing. One can easily calculate the value of R_S and the associated voltage drop necessary to achieve a given degeneration factor. The equivalent transconductance in case of source degeneration can be written as:

$$g_{m,eq} = \frac{g_m}{1+g_mR_S} = \frac{g_m}{1+K} \rightarrow R_S = \frac{K}{g_m} \tag{8.53}$$

Using equation (8.26), we have:

$$R_S = \frac{K}{g_m} = \frac{Kn\phi_T\sqrt{I_C+0.5\sqrt{I_C}+1}}{I_{DS}} \tag{8.54}$$

Thus, the voltage drop across R_S depends only on fundamental or technology constants and the device inversion coefficient:

$$R_SI_{DS} = Kn\phi_T\sqrt{I_C+0.5\sqrt{I_C}+1} \tag{8.55}$$

As an example, we apply this technique to M_6, as shown in Fig. 8.3, assuming that the bias current is 150 μA and the device works at the center of the moderate inversion region, i.e. $I_C = 1$. The choice of $K = 5$ achieves a degeneration factor of 6, requiring a voltage drop across R_S of 0.3 V and a value of R_S given by $R_S = 0.3/I_{DS} = 2$ kΩ. The equivalent transconductance of M_6 is thus 2.43 mS without degeneration and 0.4 mS with degeneration.

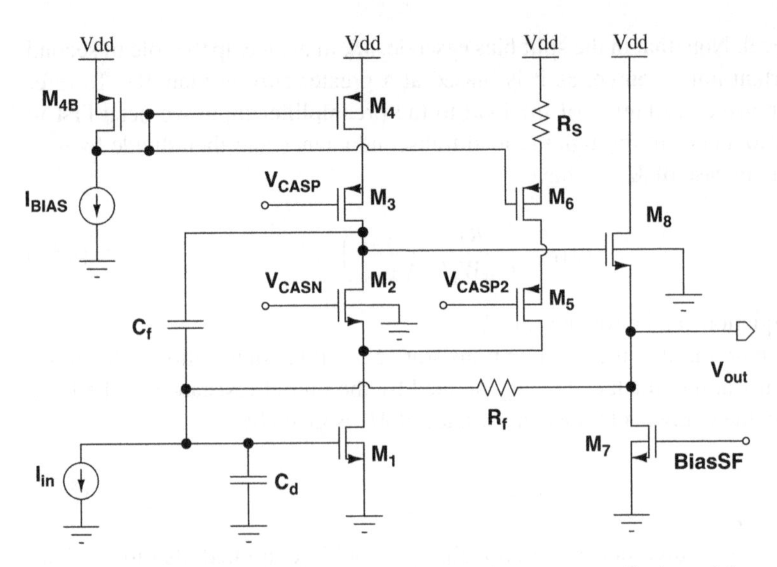

FIGURE 8.3 Use of a degeneration resistor to reduce the noise contribution of the current source in a CSA.

8.1.3 CHOICE OF INPUT TRANSISTOR TYPE

In many charge sensitive amplifiers, a PMOS is used as the input device. Historically, one of the reasons for this preference is the lower flicker noise shown by PMOS transistors. The advantage of PMOS versus NMOS in terms of $1/f$ noise is confirmed by measurements till the 90 nm generation [10], while the two types of devices have been found to have almost equivalent performance in a 65 nm process [16]. In some technologies, PMOS transistors show also a significant dependence of the flicker noise coefficient on the drain current density [10]. The advantage of PMOS in terms of flicker noise can be fully exploited only when long shaping times reduce significantly the contribution of thermal noise.

Due to the higher mobility, NMOS transistors offer better transconductance for the same power budget. Note that once the device enters in weak inversion, g_m depends only on the bias current and it is thereby the same for PMOS and NMOS. However, for the same current, it is easier to make an NMOS work in weak inversion and a PMOS in strong inversion. This is because the inversion coefficient I_C is inversely proportional to the mobility, hence for the same current, gate width and gate length, I_C is $3-4$ times larger for a PMOS than for an NMOS. As far as thermal noise is concerned, the large ratio between μ_n and μ_p observed in modern technology ($\mu_n/\mu_p > 4$)) favors the NMOS as the input transistor and the PMOS as the load device.

In technologies above the 0.25 μm generation only PMOS transistors can usually be put in an isolated nwell, shielding them from the noise injected into the substrate by digital circuits embedded on the same chip. In modern CMOS technologies this aspect becomes less relevant, because the availability of deep nwells allows an effective isolation also of NMOS transistors.

Another aspect that must be taken into account is the expected signal polarity. With a PMOS, the input DC node voltage of the CSA is set to V_{DD} minus the source-gate voltage of the device, whereas for an NMOS the input voltage is found one gate-source voltage above ground. If a simple resistive feedback is used, these voltages are also transferred to the preamplifier output. Therefore, a circuit with a PMOS input transistor has a fairly high output DC level (typically 0.3 to 0.6 V below V_{DD}) and is thus more suitable to process currents entering into the preamplifier, which determine an output signal moving downwards from the baseline. The opposite is true for a circuit with an NMOS input transistor. The DC levels can however also be adjusted with suitable feedback systems.

These considerations show that several factors must be pondered in selecting the type of input device. Many of these factors (the carrier mobility, the flicker noise coefficient and its dependence on the transistor bias) are strongly process dependent, so the choice must be carefully made even in case an already silicon proven design is ported to a new technology.

8.2 PASSIVE FEEDBACK NETWORKS

In simplified schematics, feedback networks of CSA and shapers are represented as passive components. Feedback capacitors are in general implemented with highly linear metal-to-metal or poly-poly capacitors. Passive resistors are attractive for their linearity, but area and parasitic capacitance considerations limit their use to values below 1 MΩ.

8.2.1 INTEGRATED CAPACITORS

Due to the thin gate oxide, the highest capacitance density is obtained with MOS structures. However, the capacitor value changes significantly with the applied voltage, leading to undesirable non-linearities. For this reason, fully passive components are in general preferred in the implementation of front-end amplifiers. In the following, the most common structures available in modern CMOS technologies are briefly reviewed. The reader can find more detailed descriptions in standard microelectronics textbooks such as [17]. Till the 0.35 μm technology node, highly-linear capacitors were usually implemented by sandwiching silicon dioxide between two polysilicon layers. Poly-poly capacitors have relatively high capacitance density. However, the polysilicon has a non-negligible resistance and the proximity of the lower polysilicon layer to the chip substrate increases the parasitic capacitance associated to the bottom plate. These features are undesirable in very high frequency (RF) applications [18], which form an important fraction of the market of today's analog circuits. Therefore, starting from the 0.25 μm process generation, poly-poly capacitors have been replaced with metal-insulator-metal (MiM-cap) ones. The capacitor plates are made of metals, which are less resistive than polysilicon. To reduce the coupling to the substrate, the metal closest to the substrate is avoided. An oxide with a typical thickness of 30 – 50 nm provides the dielectric, so that the capacitance density is between 0.7

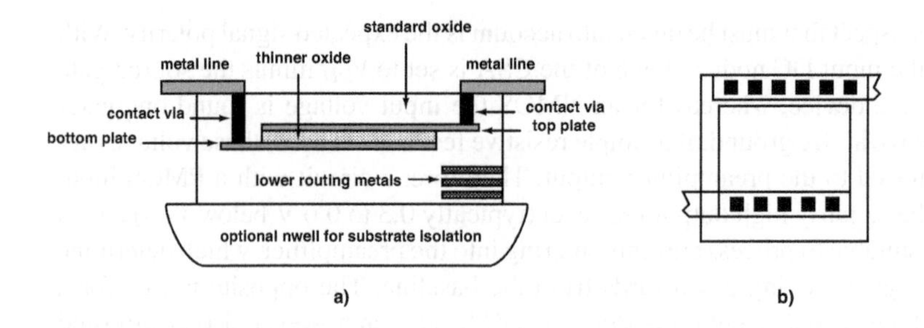

FIGURE 8.4 Front and top view of a metal-insulator-metal capacitor.

and 2 fF/μm^2. Fig. 8.4 shows the section (in a) and the top view (in b) of such a capacitor. MiM-cap structures require dedicated processing steps and a few extra masks, increasing the cost of the IC fabrication beyond the bare minimum. In applications where only small capacitors are needed or when the area is not the primary concern (such as in integrated circuits with only a small number of channels), the natural parasitic capacitance between the different routing metals can be exploited. These capacitors use as dielectric the oxide that isolates the different interconnection layers, which has a thickness in the order of $0.7 - 0.8$ μm. Therefore their capacitive density can be as low as 0.05 fF/μm^2 for a simple two plate capacitor. The availability of several metal layers offers nevertheless the opportunity of increasing the capacitance density by using more levels and alternating the plates in the vertical direction, as shown in Fig. 8.5, so that a density of $0.2 - 0.3$ fF/μm^2 becomes possible. In scaling CMOS technologies, also the density of the interconnections must be increased in order to take full benefit of the shrunk transistor dimensions. As a result, the width of the wiring metals is reduced, as well as their pitch. The resistance of a metal bar is given by:

$$R = \rho\frac{L}{A} = \rho\frac{L}{Wt} \tag{8.56}$$

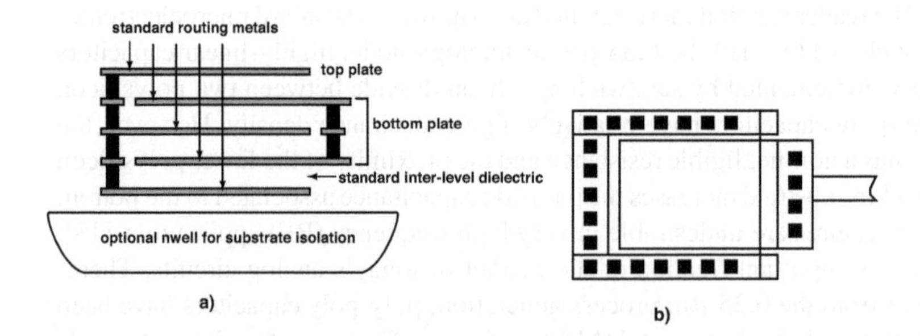

FIGURE 8.5 Front and top view of a metal-oxide-metal capacitor. The plates of the capacitor are formed by the same metals used for the interconnections and the dielectric is provided by the oxide isolating the different routing layers.

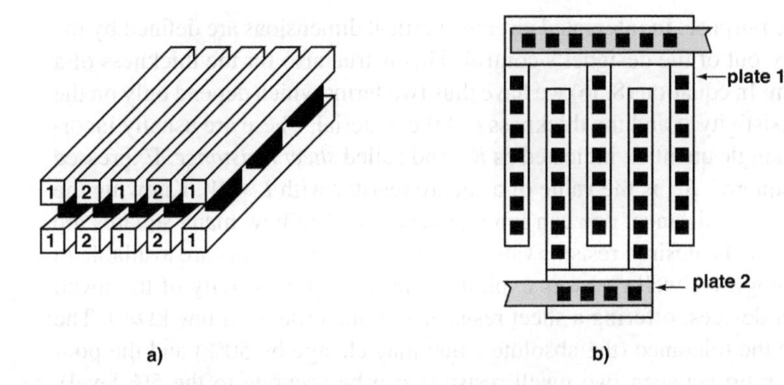

FIGURE 8.6 Vertical parallel plate capacitors. The two arms are respectively indicated by the numbers “1” and “2”. The structure exploits the parasitic capacitance between routing metals located on the same layer.

where W is the width and t is the vertical thickness of the line. If W is reduced to allow for a greater density, t must also be increased to maintain the line resistance within acceptable limits. The horizontal spacing between two metal lines can be less than 300 nm, therefore the dielectric thickness in the lateral direction is smaller than in the vertical one. The capacitance between adjacent tracks can thus be exploited to realize highly linear and fairly dense capacitors. Fig. 8.6 shows the concept of such a device, called vertical parallel plate capacitor. To increase the capacitance, more metal strips are superimposed, shorted by vias through the inter-level oxide and arranged in a fingered pattern. With this approach, a capacitance per unit area equal or even greater than the one offered by MiM-cap is achieved, with the additional benefit that no special masks are necessary for fabrication. A drawback of vertical capacitors is that there is not a “top” and bottom “plate” and both arms are equally affected by the parasitic capacitance towards the substrate. In parallel-plate structures, like the ones of Fig. 8.4 and Fig. 8.5 one can instead have a “privileged” plate in which this parasitic coupling is minimized. In many situations, the two points between which the capacitor is connected are not equally sensitive to the parasitics affecting the capacitor arms. Therefore, the plate with less parasitic capacitance, if any, must be connected to the more critical node.

All metal-metal capacitors have a weak dependence on both voltage and temperature, in the order of 20-40 parts per million per Volt (ppm/V). The accuracy on the absolute value is at the 10% level, while the accuracy of the ratio between the values of two carefully laid-out devices is typically better than 0.5%.

8.2.2 PASSIVE RESISTORS

A resistor to be used in the feedback path of a charge sensitive amplifier or shaper should be compact and have a weak dependence versus temperature and applied voltage. For these reasons, not all the resistors that are offered in a CMOS process can be adequate to the purpose. In integrated circuits, vertical dimensions are defined by the process and are out of the designer's control. This is true also for the thickness of a conductive film. In equation (8.56) we have thus two terms which depend only on the process, the resistivity ρ and the thickness t of the material. These are usually incorporated into a single quantity, indicated as $R_{\square}$ and called *sheet resistance*. Expressed in "Ohm per square", $R_{\square}$ is the value of a square resistor with $L = W$. Knowing the value of the sheet resistance, one can immediately calculate how many squares are needed to achieve the desired resistor value. Several type of resistors are available in CMOS technologies. Nwell resistors exploit the fairly high resistivity of the nwell hosting PMOS devices, offering a sheet resistance in the order of a few kΩ/□. The drawbacks are the tolerance (the absolute value may change by 50%) and the poor matching (the ratio between two nwell resistors can be accurate to the 5% level). The nwell forms a junction with the substrate, which is always kept reverse-biased. Since the nwell is lightly doped, the depletion region associated to this junction extends significantly also into the nwell itself. When the voltage at the resistor terminals changes, the size of the depletion region is modulated, altering the effective section of the resistor and thus its value. This phenomenon is at the origin of the large voltage coefficient associated to nwell resistors, whose value can change with the applied voltage by as much as 10.000 ppm/V. Nwell resistors are hence only suitable for those applications in which a big resistor in needed, but its precision and linearity are not an issue. Diffusion resistor with a smaller sheet resistance (20 – 100 Ω/□), but with much better tolerance (10%) and matching accuracy (better than 1%) are also available. The smaller sheet resistance implies a higher doping. Therefore, the depletion regions extend mainly in the substrate and the voltage coefficient is thus reduced to about 200 ppm/V. Both diffusion and nwell resistors suffer from high parasitic capacitance to the substrate which, being due to a reverse-biased junction, changes with the applied voltage. In mixed-signal systems, this capacitance can favor the pick-up of substrate noise injected by digital gates fabricated on the same chip.

A better alternative is offered by poly-silicon resistors, which have a smaller parasitic to the substrate that is also constant with the voltage. The resistivity of polysilicon can be altered by doping. Several flavors of polysilicon resistors can thus be found in the same technology, with a sheet resistance that may range from 20Ω/□ (standard polysilicon used as the base material for transistor gates) to a few kΩ/□ of high resistivity poly. High values of $R_{\square}$ allow for more compact devices, but they are in general less accurate. Medium resistivity polysilicon, with $R_{\square}$ between 100 and 200 Ω/□ results usually in a reasonable compromise between accuracy and area. Large resistors are obtained by connecting in series a suitable number of smaller bars, as shown in Fig. 8.7. Care must also be paid to the resistance of the contact vias, which might not be negligible. In designing a resistor, two aspects must be taken into account:

- The design rules may permit the design of very narrow and compact resistors. For instance, the minimum width of a polysilicon resistor could be similar to the minimum gate width of a transistor. However, both the absolute accuracy and the matching of such components can be severely degraded. Therefore, the minimum acceptable width can be significantly higher than the limit

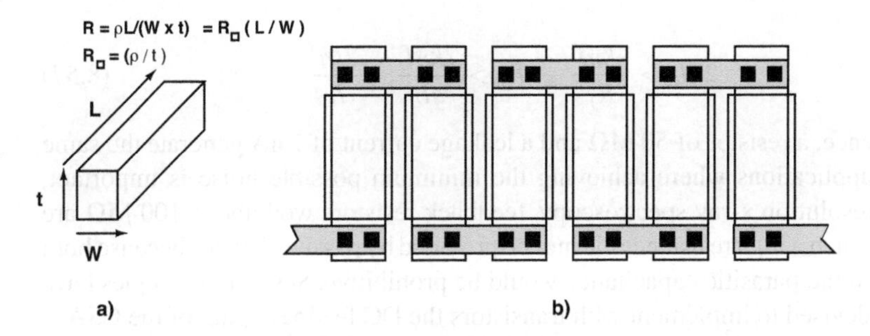

FIGURE 8.7 a). Illustration of the concept of sheet resistance. b) Typical layout of a large resistor, obtained by connecting in series a suitable number of individual bars.

imposed by process resolution.

- If the resistor has to sustain a significant amount of current, electromigration rules become a concern and may also enforce the use of a higher width. Polysilicon resistors are more sensitive to electromigration than diffusion ones, which could be preferred in case large amounts of current must be accommodated.

In designing a resistor, first the appropriate width W is chosen on the basis of the above considerations, and then the length L is obtained from 8.56. Due to the parasitic capacitance, an integrated resistor can not be modeled as an ideal component, but as a distributed RC network, as shown in Fig. 8.8. When inserted in the amplifier feedback loop, the parasitic capacitance associated to the resistor may degrade significantly the phase margin. Due to this effect, passive resistors usually have values below 500 kΩ and are mostly used in the feedback network of shapers, where it is necessary to maintain the linearity over a large signal swing. However, CSA designs employing passive resistors above 1 MΩ have also been reported [19, 20].

8.3 ACTIVE FEEDBACK NETWORKS

The ultimate noise performance in a system can be achieved if series noise is entirely dominated by the input transistor and the major contribution to parallel noise is due to the sensor. Therefore, the noise introduced by the feedback resistor of the CSA should be smaller than the one caused by the detector leakage current. This implies

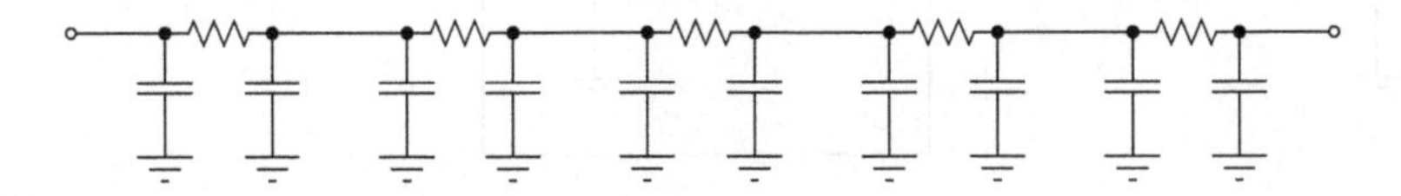

FIGURE 8.8 Integrated resistor with parasitics.

that:

$$2qI_L > \frac{4k_BT}{R_f} \to R_f > \frac{2k_BT}{qI_L} = \frac{2\phi_T}{I_L} \tag{8.57}$$

As a reference, a resistor of 50 MΩ and a leakage current of 1 nA generate the same noise. In applications where achieving the minimum possible noise is important, like high-resolution x-ray spectroscopy, feedback resistors well above 100 MΩ are necessary. Such a high resistance can not be provided by passive devices because both the area and the parasitic capacitance would be prohibitive. Several techniques have thus been devised to implement with transistors the DC feedback path of the CSA.

8.3.1 FEEDBACK TRANSISTORS IN LINEAR REGION

A resistor can be mimicked by a transistor working in the linear region. The technique can be better understood with the help of the circuit shown in Fig. 8.9. In this case, the main amplifier employs a cascode topology with split current source and NMOS input transistor. The feedback device is an NMOS as well. Consider first a device biased in strong inversion and working in the linear region, whose characteristics can be written as:

$$I_{DS} = \mu_n C_{ox} \frac{W}{L} \left[(V_{GS} - V_{TH}) V_{DS} - \frac{{V_{DS}}^2}{2} \right] \tag{8.58}$$

If the drain-source voltage is small enough, equation (8.58) can be approximated as:

$$I_{DS} = \mu_n C_{ox} \frac{W}{L} (V_{GS} - V_{TH}) V_{DS} \to R_{ON} = \frac{V_{DS}}{I_{DS}} = \frac{1}{\mu_n C_{ox} \frac{W}{L} (V_{GS} - V_{TH})} \tag{8.59}$$

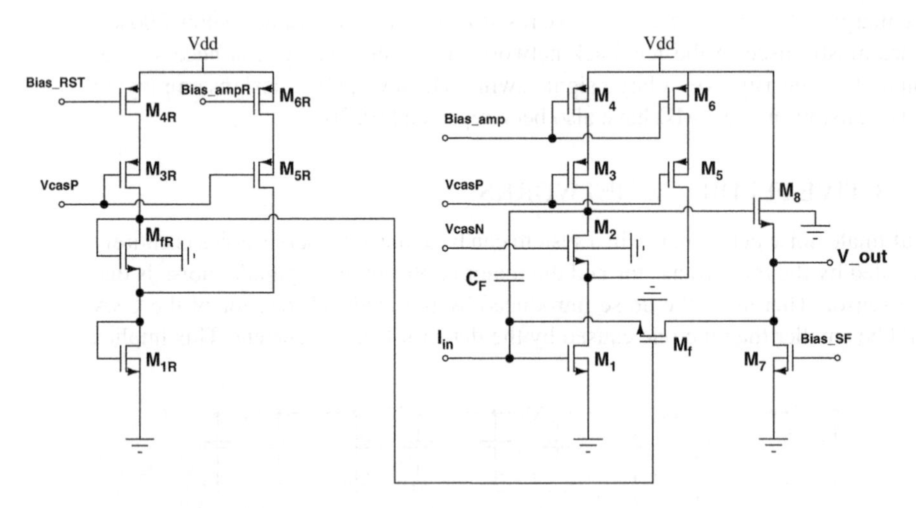

FIGURE 8.9 Charge sensitive amplifiers using transistors biased in the linear region to discharge the feedback capacitor. The replica bias circuit is necessary to reduce the impact of process spreads on the effective value of the equivalent resistance.

The equivalent resistance is hence defined by the device aspect ratio and the overdrive voltage. If no current flows in the input node, the current in M_f must be zero, as the gate of M_1 has almost infinite impedance. From equation (8.58) we see that also the drain-source voltage of M_f must be zero and the DC output voltage is equal to the gate voltage of M_1. Suppose now that a pulse containing a total charge Q_{in} is presented to the CSA input. If the amplifier is fast enough, the pulse is fully integrated on C_f, giving a step of amplitude Q_{in}/C_f and M_f only defines the reset time of the feedback capacitor. If the signal current is pulled from the input node towards ground, the output voltage rises and the terminal of M_f connected to the input acts as the transistor source. If the drain-source voltage of M_f is small, equation (8.59) describes the transistor behavior with good approximation and C_f is discharged with the usual exponential tail. If the signal is increased, the quadratic term in equation (8.58) gives a non-negligible contribution. When the input pulse is so large that $V_{DSf} \geq (V_{GS} - V_{TH})_f$ the transistor enters into the saturation region and its current is primarily defined by $(V_{GS} - V_{TH})_f$. In this condition, C_f is discharged by a constant current. Consider now the case in which a DC leakage current is instead sunk from the input. If the leakage current is so big that M_f is pushed in saturation, it provides only the maximum current given by:

$$I_{MAX} = \frac{1}{2}\mu_n C_{ox}\left(\frac{W}{L}\right)_f (V_{GS} - V_{TH})_f^2 \tag{8.60}$$

The leakage current in excess is then integrated on C_f and the output node rises till the CSA saturates. Therefore, the circuit can only accept values of leakage current small enough to allow M_f to work in the linear region. If we have instead a current entering into the input node, the output swings downwards and the terminal of M_f connected to the output acts as the source. The input signal thus controls directly the gate-source voltage of M_f because when the output swings, $V_{GS} - V_{TH}$ is increased and the equivalent resistance of M_f is reduced. The same considerations apply to a continuous leakage current entering into the input node. The greater the current, the lower the output node is pulled-down, increasing $V_{GS} - V_{TH}$ and decreasing R_{ON}, thus desensitizing the circuit to the effect of the leakage current. In this sense, it is said that the circuit "self-adapts" to the leakage current. The amount of DC current that can be accommodated by M_f is not limited by the constraint that the device must work in the linear region. To increase the value of the equivalent resistor, $V_{GS} - V_{TH}$ must be reduced and M_f eventually enters in weak inversion. Biasing the transistor in this region, the equivalent resistance is increased, but the range over which the device can be maintained in the linear region is smaller, since $V_{DS,sat}$ in weak inversion is typically 125 mV.

One of the issues in using a transistor as a resistor is keeping the value of the equivalent resistance fairly constant against the change in process parameters. The value of V_{GS} must thus be defined so that the difference $V_{GS} - V_{TH}$ remains constant in case the threshold voltage changes. This can be achieved through the use of replica bias circuits [21–23] as shown in Fig. 8.9. Transistor M_{1R} has the same dimensions of the input device and it is biased with the same current, so the gate-source voltage of

M_1 and M_{1R} are also identical[2]. Transistor M_{fR} is a scaled-up replica of the feedback transistor, hence $(W/L)_{M_{fR}} = k(W/L)_{M_f}$, with $k > 1$. If M_{fR} works in saturation, its gate-source voltage is defined as:

$$(V_{GS} - V_{TH})_{M_{fR}} = \sqrt{\frac{2L_f I_{RST}}{\mu_n C_{ox} k W_f}} \tag{8.61}$$

where I_{RST} is the bias current flowing in M_{fR} and W_f and L_f are respectively the gate width and gate length of M_f. Owing to the replica scheme, the gate-source voltage of M_{fR} and M_f are approximately equal, so the value of the overdrive voltage given in equation (8.61) can be replaced in (8.59), yielding:

$$R_{ON,M_f} = R_f = \sqrt{\frac{k}{2\mu_n C_{ox} \left(\frac{W}{L}\right)_f I_{RST}}} \tag{8.62}$$

The above equation shows that the value of the equivalent resistance depends only on the bias current I_{RST} and on the ratio between the transistor widths of M_{fR} and M_f. The device length is kept usually the same to improve the matching between the two devices and only the width is changed. Let's now consider the case in which M_{fR} and M_f work in weak inversion. In this case the transistor characteristics can be written as:

$$I_{DS} = 2n\mu_n C_{ox} \frac{W}{L} \phi_T^2 e^{\frac{V_{GS}-V_{TH}}{n\phi_T}} \left(1 - e^{-\frac{V_{DS}}{\phi_T}}\right) \tag{8.63}$$

If V_{DS} is very small, we can replace the exponential in the parenthesis with a first order Taylor series:

$$e^{-\frac{V_{DS}}{\phi_T}} = 1 - \frac{V_{DS}}{\phi_T} \tag{8.64}$$

which, inserted into (8.63), gives:

$$I_{DS} = 2n\mu_n C_{ox} \frac{W}{L} \phi_T^2 e^{\frac{V_{GS}-V_{TH}}{n\phi_T}} \frac{V_{DS}}{\phi_T} \tag{8.65}$$

Therefore, the equivalent resistance in the linear region and in weak inversion can be written as:

$$R_{ON,wi} = \frac{1}{2n\mu_n C_{ox} \frac{W}{L} \phi_T e^{\frac{V_{GS}-V_{TH}}{n\phi_T}}} \tag{8.66}$$

The replica bias transistor, M_{fR}, is diode connected. If we assume that it works in saturation, we can write:

$$I_{RST} = 2n\mu_n C_{ox} k \frac{W}{L} \phi_T^2 e^{\frac{V_{GS}-V_{TH}}{n\phi_T}} \rightarrow e^{\frac{V_{GS}-V_{TH}}{\phi_T}} = \frac{I_{RST}}{2n\mu_n C_{ox} k \frac{W}{L} \phi_T^2} \tag{8.67}$$

[2] To have the same gate-source voltage, it is actually sufficient that both the aspect ratio and the bias current of M_{1R} are scaled, so that the current density in M_{1R} and M_1 is the same.

Replacing the exponential term in (8.66), we can express the equivalent resistance of the feedback transistor M_9 as:

$$R_{ON,M_f} = R_f = \frac{k\phi_T}{I_{RST}} \tag{8.68}$$

The equivalent resistance scales now linearly with the ratio k between the widths of the feedback and of the replica transistors and the bias current I_{RST}.

Example 8.4: Calculation of the equivalent feedback resistance.

In the circuit of Fig. 8.9, $(W/L)_{M_f}$ is 0.5/100 and the ratio k between the widths of M_f and M_{fR} is 100. Calculate the equivalent feedback resistance of M_f when M_{fR} is biased at the onset of strong and weak inversion. Use the following device parameters: $\mu_n C_{ox} = 250\ \mu\text{A/V}^2$, $V_{TH} = 0.5$ V, $\gamma = 0.5\ \text{V}^{1/2}$.

* * *

For M_{fR}, the current for which the inversion coefficient is one is given by:

$$I_{wi,si} = 2n\mu_n C_{ox}\frac{W}{L}\phi_T^2 = 200 \text{ nA} \tag{8.69}$$

The device can be considered in strong inversion if $I_C = 10$, which requires us to set the bias current I_{RST} to 2 μA. Applying equation (8.62), we get $R_f = 4.47$ MΩ. The onset of weak inversion is at $I_C = 0.1$, which implies $I_{RST} = 20$ nA. From equation (8.68) we can calculate that R_{ON,M_f} and thus the equivalent small signal feedback resistance of the CSA is 130 MΩ. Note that both M_f and M_{fR} are equally affected by the body effect because V_{SB} is identical for both devices. Assuming that the gate-source voltage of M_1 and M_{1R} is 0.5 V, we have that the threshold voltage of M_f and M_{fR} is:

$$V_{TH} = V_{TH0} + \gamma\left(\sqrt{2\phi_F + V_{SB}} - \sqrt{V_{SB}}\right) \approx 0.64 \text{ V} \tag{8.70}$$

In strong inversion, the gate-source voltage of M_f and M_{fR} is:

$$(V_{GS} - V_{TH})_{M_{fR}} = \frac{2LI_{RST}}{\mu_n C_{ox} kW} = 180 \text{ mV} \rightarrow V_{GS}(M_{fR}) = 820 \text{ mV} \tag{8.71}$$

The gate voltage of M_{fR} is therefore at 1.32 V. For the weak inversion case, the gate-source voltage can be found by solving the following equation:

$$2n\mu_n C_{ox}\frac{kW}{L}e^{\frac{V_{GS}-V_{TH}}{n\phi_T}} = 20 \text{ nA} \tag{8.72}$$

which yields $V_{GS}(M_{fR}) = 0.56$ V.

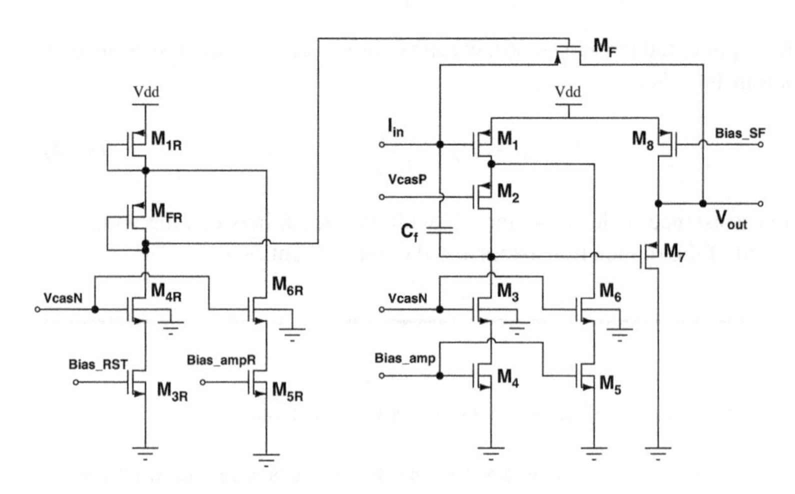

FIGURE 8.10 PMOS input CSA with transistor in linear region in the feedback path.

A version of the circuit of Fig. 8.9 implemented with PMOS input transistor and PMOS feedback is shown in Fig. 8.10. Observe that with direct cascode topologies it can be problematic to have the input and the feedback transistor of complementary types. For instance, if in Fig. 8.9 we replace M_f with a PMOS, its gate voltage must be kept below the gate voltage of M_1. In a deep submicron technology the gate-source voltage of the input device can be as low as 300 mV and within this headroom one should accommodate at least one diode connected transistor and one current source. When it is desirable to have the input and the feedback device of complementary types folded cascode architectures like the one described in [21] and [24] are more suitable. The value of feedback resistance found with equations (8.62) and (8.68) are valid only at the equilibrium, as the input signal can change significantly the operating region of the transistor. When the bandwidth limitation of the CSA is taken into account, the output peak of the CSA can be written as:

$$V_{out,peak} = \frac{Q_{in}}{C_f}\left(\frac{\tau_f}{\tau_r}\right)^{\frac{\tau_r}{\tau_r-\tau_f}} \tag{8.73}$$

where τ_r is the time constant associated to the bandwidth limitation of the core amplifier and τ_f is the feedback time constant. If $\tau_f \gg \tau_r$, the output reaches the peak given by Q_{in}/C_f and non-linearities in the feedback resistance just cause different return time to the baseline for different signals. If, on the other hand, τ_r and τ_f are comparable, the non-linearity affects also the peak of the CSA output voltage. This problem can be however circumvented if a pole-zero cancellation technique is applied, as shown in Fig. 8.11. Here, M_z is built by connecting in parallel m copies of M_f, with $m = C_z/C_f$. If the gate-source and drain-source voltages of M_z and M_f are identical, the current in M_z is m times the one in M_f, regardless of their operating region. The linearity of the system is thus maintained, in the sense that the current fed to the second stage is a replica of the input current, scaled-up by a factor m. Observe

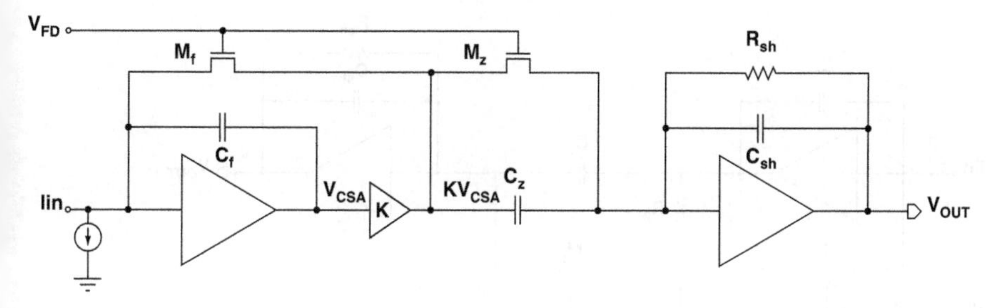

FIGURE 8.11 CSA feedback with reset transistor working in the linear region and pole-zero cancellation.

that if no other measure is taken, also the leakage current is amplified by m. Suppose that the output buffer of the CSA has a gain $k < 1$ (as it happens when it is implemented with a source follower affected by the body effect). Assuming for simplicity that the input is an ideal virtual ground, we can write:

$$I_{in} - V_{CSA}sC_f - \frac{kV_{CSA}}{R_f} = 0 \rightarrow \frac{V_{CSA}}{I_{in}} = \frac{R_f}{k\left(1 + sC_f\frac{R_f}{k}\right)} \tag{8.74}$$

where R_f and R_z are, respectively, the equivalent small signal resistance of M_f and M_z. The current flowing into the second stage is given by:

$$I_{out} = \frac{R_f}{k\left(1 + sC_f\frac{R_f}{k}\right)} \frac{(1 + sR_zC_z)}{R_z} \tag{8.75}$$

The pole-zero cancellation is achieved if:

$$C_f\frac{R_f}{k} = C_zR_z \tag{8.76}$$

If M_z is made of m identical copies of M_f connected in parallel, $R_z = R_f/m$, and C_z must thus be chosen equal to $C_f \cdot (m/k)$. To increase further the value of the equivalent feedback resistance a T-network can be introduced [25]. Shown in Fig. 8.12, the scheme makes use of a voltage divider that scales the output voltage before passing it to the feedback transistor. The circuit analysis reveals that the equivalent feedback resistor is given by:

$$R_{feq} = R_2 + R_f\left(1 + \frac{R_2}{R_1}\right) \tag{8.77}$$

where R_f is the equivalent resistance of the feedback device, which get boosted by the coefficient $(1 + R_2/R_1)$. The technique reduces also the swing across the non-linear feedback element, improving the overall linearity.

From a practical point of view, the limit in the accuracy of the replica bias schemes resides in the matching between the transistors in the main amplifier and in the replica circuit, which entails also a trade-off between accuracy and power dissipation. In a multi-channel system, for better accuracy each channel should have its replica circuit,

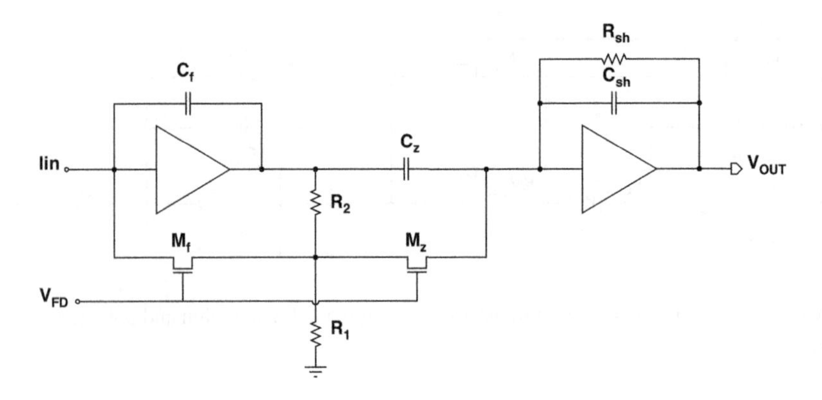

FIGURE 8.12 CSA feedback with T-network to increase the equivalent feedback resistance.

but this increases the power required by the CSA. The replica circuit is thus normally shared among several channels. In very large ICs, having the replica circuit far from the final destination worsens the performance because of long distance gradients affecting the transistor parameters and voltage drops on the power supply and ground lines. Instantiating one replica circuits every few channels (e.g. eight or sixteen) may hence provide a good compromise.

8.3.2 TRANSCONDUCTANCE FEEDBACK

Consider the circuit of Fig. 8.13 in which a transconductance amplifier is connected in feedback of a voltage amplifier. If the voltage amplifier has infinite input impedance and $C_f = 0$, all the current I_{in} must be provided by the transconductor. We therefore

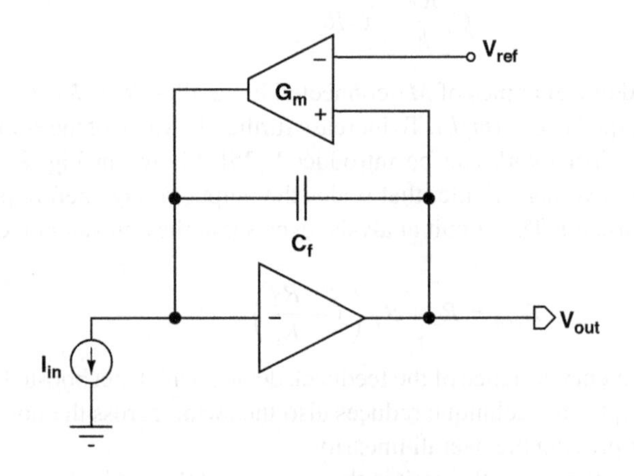

FIGURE 8.13 Concept of transconductance feedback.

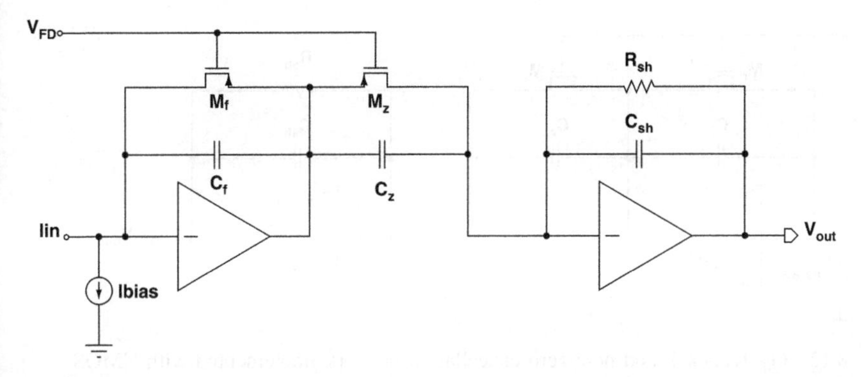

FIGURE 8.14 G_m feedback and pole-zero cancellation network implemented with PMOS transistors.

have:

$$I_{in} = I_{fb} = G_m V_{out} \rightarrow \frac{V_{out}}{I_{in}} = \frac{1}{G_m} \tag{8.78}$$

The trasconductor thus provides and equivalent resistance of value $1/G_m$ that can be used to discharge C_f, establishing also a DC feedback around the core amplifier. Note that if the transconductor has a differential input, the output of the amplifier locks the reference voltage V_{ref}. In this way, the input and output DC voltages are decoupled and the latter can be adjusted to optimize the dynamic range or the coupling to the second stage. In principle, any transconductance amplifier can be used as the feedback element. A complex building block will however introduce additional poles in the loop, making the stability an issue. The simplest transconductance feedback can be realized with a single MOS transistor working in saturation [26]. Shown in Fig. 8.14, the circuit works as follows. The feedback transistor is biased with a DC current, that can be provided either by the detector leakage current or by a dedicated biasing circuit. In case a PMOS transistor is used, the current must be sunk from the input, so that the output node rises above the amplifier input voltage and becomes the source of the feedback device. The source voltage "follows" the one applied to the gate. Therefore, the output voltage can be controlled by acting on the gate bias of M_f. The only critical point is that the output voltage is high enough to keep the feedback transistor in saturation, so the value of the bias voltage V_{FD} is not critical and can be generated by simple circuits. The equivalent feedback resistance is in fact only determined by the bias current of M_f and by its dimensions and not by its gate voltage. Transistors M_f and M_z operate with the same source-gate voltage, therefore the current in M_z is a scaled version of the one in M_f and can be used to provide a precise pole-zero cancellation. For this to be effective, the two transistors must have the same length and only the widths must be changed. Additionally, they must operate with the same source-drain voltage, hence the DC voltage at the input of the first and of the second stages must be approximately equal. Fig. 8.15 shows the scheme employing NMOS transistors, which is suitable for currents entering into the input node. Note that, with direct cascode topologies, the circuit of Fig. 8.14 is more

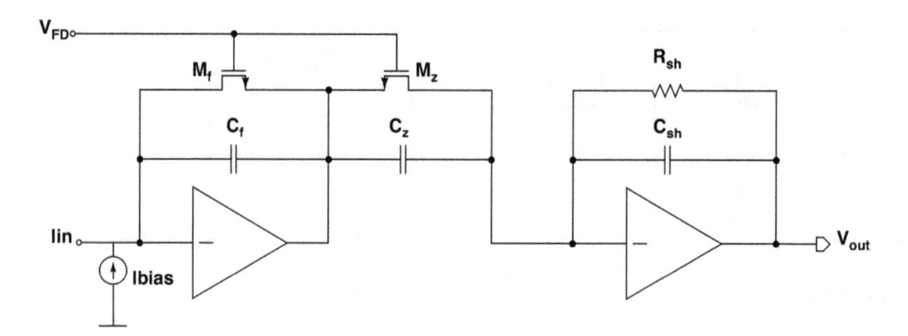

FIGURE 8.15 G_m feedback and pole-zero cancellation network implemented with NMOS transistors.

appropriate for core amplifiers based with an NMOS input device, while the one of Fig. 8.15 is better suited to a stage with a PMOS input transistor.

Example 8.5: Analysis of circuits with g_m feedback implemented with single MOS transistors.

In the circuit of Fig. 8.14 the DC bias current is 1 nA and is provided by an MOS current source. Size M_f and M_z so that the gain in the first stage is 40 and the noise contribution of the feedback network is less than 100 electrons for a peaking time of 150 ns. The circuit is implemented in a 0.25 μm CMOS technology with the following device parameters: $\mu_n C_{ox} = 250 \mu$A/V^2, $\mu_p C_{ox} = 50 \mu$A/V^2, $V_{THN} = |V_{THP}| = 0.5$ V.

* * *

If we bias the feedback transistor with a current source, we have two parallel noise sources, one due to M_f and one due to the transistor that implements the current source. The first step is to calculate the maximum noise power spectral density that can be tolerated. To allow for some safety margin, we fix the limit of the noise introduced by the feedback network to 50 electrons rms. The noise spectral density is then calculated from:

$$ENC^2 = S_i N_i T_p \tag{8.79}$$

where S_i is the noise spectral density, N_i the shaper noise index for parallel noise and T_p the peaking time. For simplicity, we can assume $N_i = 1$. Note that in the above expression ENC^2 must be expressed in Coulomb^2, and it is equal to $6.4 \cdot 10^{-35}$ C^2. Divided by the peaking time of 150 ns, this yields a total noise spectral density of $4.3 \cdot 10^{-28}$ A^2/Hz. The noise budget must be shared by the feedback transistor and its biasing current source. Assuming that they work in strong inversion, we have:

$$4k_BT\frac{2}{3}g_{mB} + 4k_BT\frac{2}{3}g_{mf} = 4.3 \cdot 10^{-28} \text{ A}^2/\text{Hz} \tag{8.80}$$

where g_{mB} is the transconductance of the biasing transistor and g_{mf} the one of the feedback device. If we choose $g_{mB} = g_{mf}$, we get $g_{mB} = g_{mf} = 20$ nS. For the transconductance we can use the expression:

$$g_m = \frac{I_{DS}}{n\phi_T} \frac{1}{\sqrt{I_C + 0.5\sqrt{I_C} + 1}} \tag{8.81}$$

With $n = 1.3$ and $I_{DS} = 1$ nA, the term $I_{DS}/n\phi_T$ is 30 nS. An inversion coefficient of 1.5 would reduce the transconductance by a factor of two, bringing it below the target value. The W/L ratio can now be obtained from:

$$I_C = \frac{I_{DS}}{2n\mu C_{ox}\frac{W}{L}\phi_T^2} \rightarrow \frac{W}{L} = \frac{I_{DS}}{2n\mu C_{ox} I_C \phi_T^2} \tag{8.82}$$

Putting in the parameter values for the PMOS transistor, we find $(W/L) = 7.6 \cdot 10^{-3}$. Assuming a width of 0.4 μm, the length must be 50 μm. M_z will then be made by 40 identical copies of M_f. Due to the much higher transconductance parameter of NMOS transistors, a length of 250 μm is necessary for the transistor implemented NMOS current source. Usually, design rules enforce a maximum gate area and maximum gate width and length when the gate is designed as a single piece of polysilicon. In this case, more transistors can be connected in series to form a longer device. With a g_{mf} of 20 nS, the equivalent feedback resistance is $R_f = 1/g_{mf} = 50$ MΩ.

It is interesting to recalculate the transistor dimensions in case the noise target is chosen without safety factors and a shaper with a more favorable noise coefficient for parallel noise is used (e.g. $N_i = 0.6$). Repeating the calculations, one finds that the maximum acceptable transconductance is 150 nS, well above the value of 30 nS which results from (8.81) for $I_C = 0$. Therefore, the feedback device can work also in weak inversion and much shorter transistors can be used. In principle, even a length as short as 1 μm could be used. However, decreasing too much the transistor area worsens the matching between M_f and M_z. Choosing, for instance, $L = 10$ μm can be a good starting point, and the result must then be refined with computer simulations.

The example shows that while taking safety margins in a design is in general a good policy, one should be careful to not exceed and make sure the given specifications contain safety factors already. The reader can repeat as an exercise the calculations for the circuit of Fig 8.15.

Biasing the circuits of Fig. 8.14 and Fig. 8.15 with an on-chip current source increases the noise, but makes the bias point well defined also in case the circuit is not connected to a detector. In fact, if the input current is zero, the transistors go into the linear region and the equivalent resistance is defined by the gate-source voltage. If a replica bias scheme is not used, the equivalent resistance may change significantly from channel to channel and might become also too high to provide sufficient DC feedback. When a circuit is tightly optimized for a given sensor, once must make sure that it works also when it is not connected to the detector, to allow for independent electrical tests at the bench. To minimize noise, one can reduce the bias current, but too small currents are also difficult to control with acceptable matching. In spectroscopy applications, when minimum noise is paramount, sensors with very small leakage current (in the pA range) are necessary. In this case, schemes like the one of Fig. 8.12, with the feedback device working in weak inversion may provide better results. The circuits of Fig. 8.14 and Fig. 8.15 are instead more practical when ultimate reduction of parallel noise is not necessary. Fig. 8.16 shows another approach to the transconductance feedback, implemented by mean of current mirrors [11, 27]. The circuit in Fig. 8.16 a) is suitable to process currents sunk from the CSA input, while the one of Fig. 8.16 b) is its counterpart adequate for opposite polarity signals. The CSA output voltage is converted into a current by M_{fb2}, which is then mirrored towards the input, scaled by the ratio between the widths of M_{fb1} and M_f. The same current is also mirrored by M_z towards the output node. As usual with current mirrors, the length is the same for all transistors, while the ratio between the width of M_z and the one of M_f is identical to the ratio between C_z and C_f. The equivalent feedback resistance is given by:

$$R_f = \frac{1}{g_{mfb2}} \frac{(W/L)_f}{(W/L)_{fb1}} \tag{8.83}$$

In the circuits of Fig. 8.16, M_f and M_z could also be cascoded, boosting their output impedance. In this way it becomes less critical that the voltage at the drain of M_f and the one at the drain of M_z is the same. The circuits of Fig. 8.16 can also be cascaded, since the output polarity of one version is compatible with the input polarity of the other. In this way, a total current gain of $M \times N$ is achieved, where M and N are the mirror scale factors used in the first and second stage. The feedback loops are strictly unipolar, because a current of polarity opposite to the expected one and larger than the DC bias current breaks the loop, sending the circuit in a open-loop state. Note also that a minimum DC bias current could be necessary to activate the loop and that the gate-source voltage of M_{fb2} must be large enough to allow a proper biasing of the output stage of the CSA. More complicated transconductance feedback are also possible, although rarely used in the implementation of charge sensitive amplifiers for radiation detectors. An interesting overview of these topologies can be found in [28] and [29].

8.3.3 FEEDBACK WITH DC CURRENT COMPENSATION

All the feedback circuits described in the previous subsection are sensitive to the DC leakage current coming from the sensor, which is amplified with the same gain as the signal. In some of the topologies a minimum current is even necessary to activate the feedback loop. The amplified leakage current is then passed to the following stages. Usually, in some point of the chain the leakage current must be removed, either by introducing AC coupling or by other suitable methods, to avoid that the baseline of the front-end is affected. Originally proposed by F. Krummenacher in [30], the circuits of Fig. 8.17 allow to compensate the leakage current already at the CSA level. Consider first the configuration of Fig. 8.17 a), which is suitable if the DC current flows inside the preamplifier. The differential pair is biased with a current I_{FEED} through M_1, which works as a current source. Transistor M_5 works also as a current source, sinking the current $I_{FEED}/2$. If I_{dect} is zero, the remaining half of the feedback current flows in

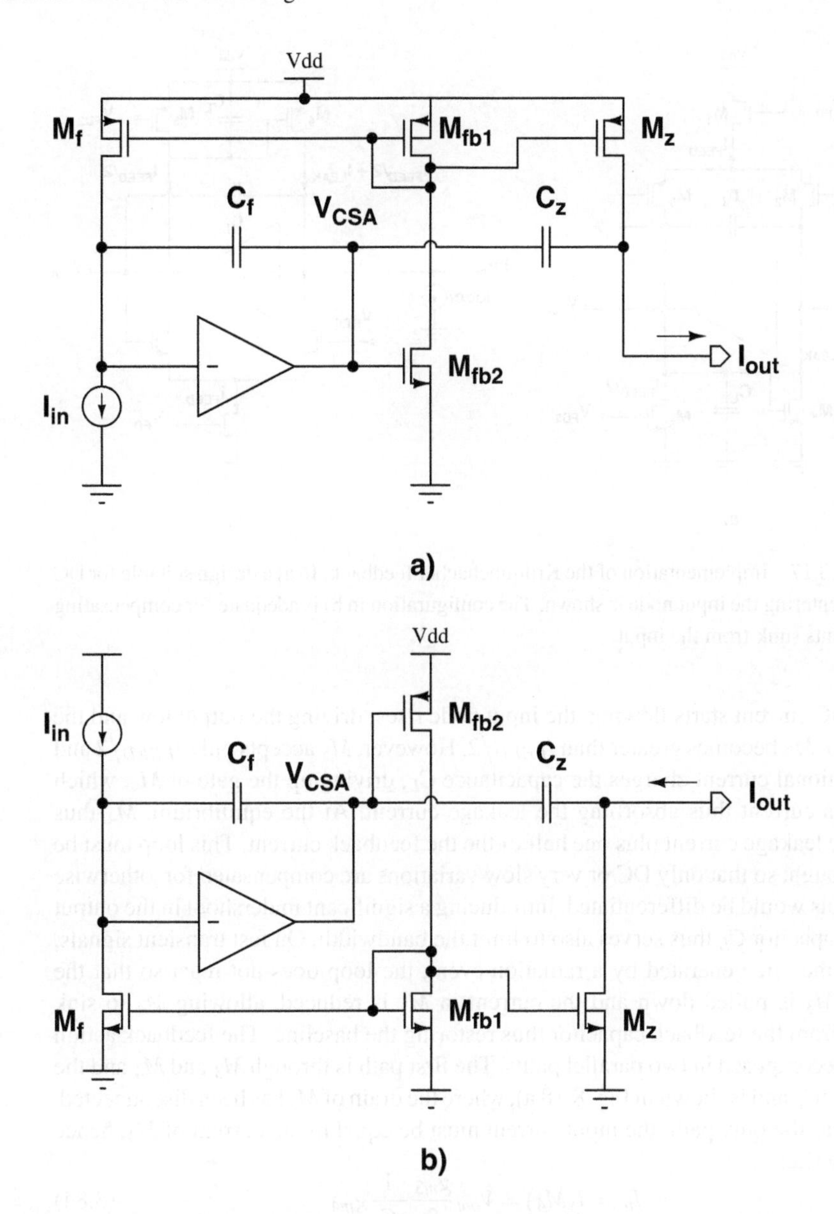

FIGURE 8.16 G_m feedback implemented with current mirrors.

M_2. The devices in the differential pair, M_2 and M_3 must have the same source-gate voltage, thus the DC level of the output is locked to the reference voltage V_{REF}. If

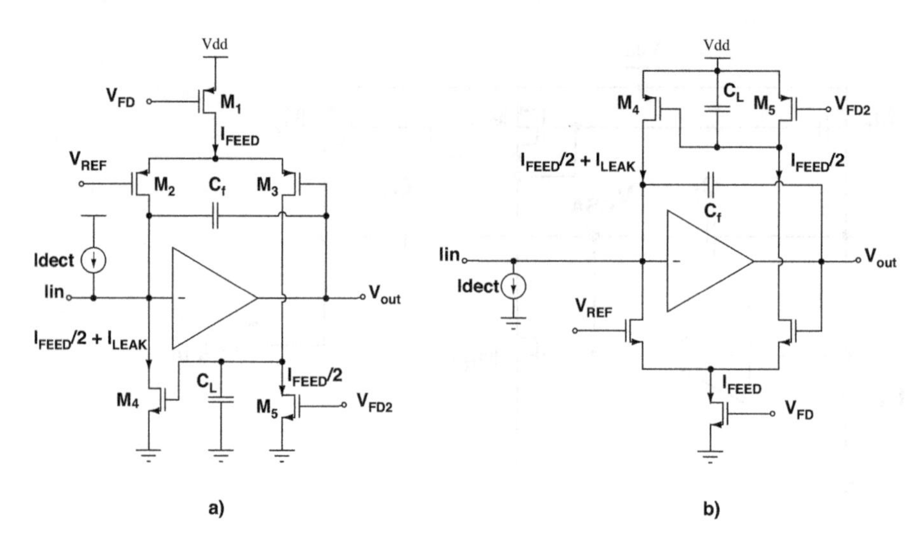

FIGURE 8.17 Implementation of the Krummenacher feedback. In a) a design suitable for DC currents entering the input node is shown. The configuration in b) is adequate for compensating DC currents sunk from the input.

now a DC current starts flowing, the input node rises, driving the output low and the current in M_3 becomes greater than $I_{FEED}/2$. However, M_5 accepts only $I_{FEED}/2$ and the additional current charges the capacitance C_L, driving up the gate of M_4, which increases current thus absorbing the leakage current. At the equilibrium, M_4 thus sinks the leakage current plus one half of the the feedback current. This loop must be slow enough, so that only DC or very slow variations are compensated for, otherwise the signals would be differentiated, introducing a significant undershoot in the output pulse. Capacitor C_L thus serves also to limit the bandwidth. On fast transient signals, such as the one generated by a radiation event, the loop does not react so that the gate of M_3 is pulled down and the current in M_2 is reduced, allowing M_4 to sink current from the feedback capacitor thus restoring the baseline. The feedback action can be decomposed in two parallel paths. The first path is through M_3 and M_4 and the capacitor C_L and is shown in Fig. 8.18 a), where the drain of M_2 has been disconnected. If this was the only path, the input current must be equal to the current of M_4, hence we can write:

$$I_{in} = I(M_4) = V_{out} \frac{g_{m3}}{2} \frac{1}{sC_L} g_{m4} \tag{8.84}$$

Note that the differential pair is driven in single-ended mode, therefore only one-half of the transconductance of M_3 is exploited. The equivalent feedback resistance can then be expressed as:

$$\frac{V_{out}}{I_{in}} = \frac{2sC_L}{g_{m3}g_{m4}} \tag{8.85}$$

From the above equation we see that the equivalent feedback impedance has an inductive behavior, it is zero at DC ($s = 0$) and then rises with frequency. This is

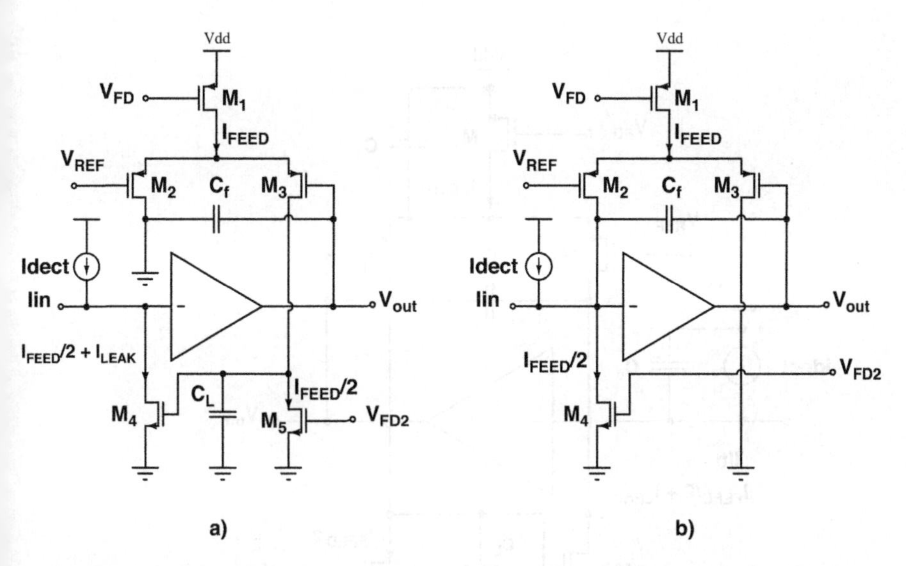

FIGURE 8.18 Decomposition of the circuit of Fig. 8.17 in two independent feedback paths. The path in a) dominates at low frequency and suppresses the effect of the input DC current. The path in b) discharges the feedback capacitor after a fast signal has been detected.

what is needed to minimize the circuit sensitivity to DC input currents. The second feedback path is through the differential pair and is shown in Fig. 8.18 b). Here, the purpose of M_4 is just to provide the bias current at the equilibrium, but since its gate is connected to a fixed DC point its current can not change. Therefore, any variation of the input current must be accommodated by M_2 and the corresponding feedback resistance is given by:

$$\frac{V_{out}}{I_{in}} = \frac{2}{g_{m3}} \tag{8.86}$$

When the two paths act simultaneously, the former guarantees low gain for DC currents, while the latter provides a discharge path to the feedback capacitor for fast signals. Fig. 8.19 reports the circuit of Fig. 8.17 a) redesigned to include all the capacitances that influence the loop stability. Note that C_1 is referenced to V_{DD}, but for the purpose of our discussion it can be referenced to any small signal ground. We suppose in addition that the core amplifier has a first order transfer function, characterized by a DC gain A_0 and a pole with a time constant τ_A, associated to the angular frequency $\omega_A = 1/\tau_A$. The detailed small signal analysis of the circuit reveals that the loop contains four major poles. The pole at highest frequency is due to the intrinsic bandwidth limitation of the core amplifier and is given by:

$$\omega_4 = \frac{C_f}{C_d} A_0 \omega_A \tag{8.87}$$

The parasitic capacitance seen from the source of M_2 and M_3, represented with C_1 in Fig. 8.19, yields a pole located at intermediate frequencies that can be written as:

$$\omega_3 = \frac{g_{m3}}{C_1} \tag{8.88}$$

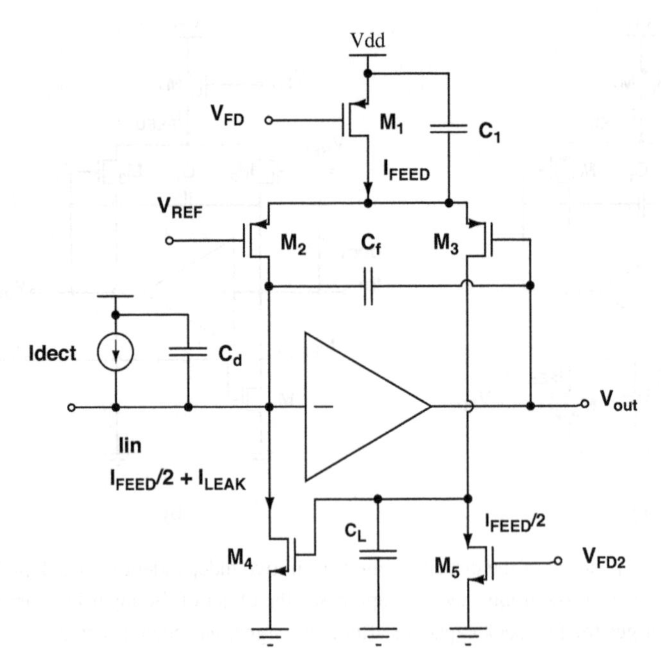

FIGURE 8.19 Circuit showing the capacitances relevant for the stability of the Krummenacher feedback.

There are then two poles at lower frequencies associated respectively to C_f and C_L. The pole associated to C_f is found at:

$$\omega_2 = \frac{g_{m3}}{\frac{C_d}{A_0} + C_f} \approx \frac{g_{m3}}{C_f} \tag{8.89}$$

where the rightmost approximation holds if the open loop gain of the core amplifier is sufficiently high. The inverse of g_{m3} can be considered the equivalent feedback resistance provided by the Krummenacher circuit, which discharges the feedback capacitor. Therefore the quantity C_f/g_{m3} is the time constant of the CSA feedback loop, analogous to the term $R_f C_f$ found in case R_f is a passive component. Finally, the pole at lowest frequency is given by:

$$\omega_1 = \frac{g_{m4}}{C_L} \tag{8.90}$$

To grant the stability of the loop, the four poles must be well separated and, in general, the condition $\omega_1 \ll \omega_2 \ll \omega_3 \ll \omega_4$ must be satisfied. It must be observed that the position of the lowest frequency pole depends on the transconductance of M_4, which is the transistor that absorbs the sensor leakage, diverting it from the main signal path. It is therefore necessary to check the circuit stability for the maximum value of leakage current that is expected during the system lifetime. The design of a stable Krummenacher feedback entails therefore the following trade-offs:

- To make ω_1 dominant, C_L should be large. An adequate damping of the loop may require a big value of C_L and thus a significant circuit area.
- To improve stability, the parasitic capacitance C_1 must be minimized, therefore M_1, M_2 and M_3 should be made small. This is in contrast to having a good uniformity of the feedback currents between different channels and to achieving low offset in the differential pair. In fact, due to the mismatch between M_2 and M_3, the DC output voltage locks at $V_{ref} \pm V_{os}$, originating a spread in the CSA baseline from channel to channel and a non-uniformity between the times of return to the baseline after a signal is detected.
- To separate the second from the third pole, C_f can be made larger, which pushes also ω_4 to higher frequencies, improving the phase margin. However, increasing C_f reduces also the CSA closed-loop gain.

An important point to note about the Krummenacher feedback is that the circuit operation on the signal can be divided into two regions. When the signal is small, the linear models holds and the feedback mimics a resistor of value proportional to $1/g_{m3}$. If the signal is big enough, the differential pair becomes completely imbalanced and the feedback capacitor is discharged or recharged to the baseline with a constant current given by $I_{FEED}/2$.

On fast signals, the circuit handles equivalently well either signal polarity, regardless if the differential pair is PMOS or NMOS based. However, this is not true for the leakage current. To understand better this point, consider, for instance, the circuit of Fig. 8.19. Here, the leakage current which is pushed into the circuit is absorbed by M_4, which rises its gate voltage. If the gate voltage of M_4 rises too much, it might eventually push M_3 in linear region. Therefore, the maximum leakage that can be absorbed is ultimately limited by the aspect ratio of M_4 and it is much bigger than I_{FEED}. If, on the other hand, the polarity of the sensor is reversed and the leakage current is sunk from the circuit, the only thing that the feedback can do to keep the differential pair balanced is to cut M_4 off. Therefore, in this case the maximum leakage current that can be sunk from the circuit without unbalancing the differential pair is $I_{FEED}/2$. The arguments are of course specular for the NMOS based circuit of Fig. 8.17 b).

The scheme of Fig. 8.17 can also be used to compensate only the DC current, as shown in Fig, 8.20. In these circuits, two fully independent feedback paths are provided. The fast one, formed by R_f and C_f processes the fast signals induced by radiation quanta. The second path, formed by the operational amplifiers and the compensation transistors M_C takes care of sourcing or sinking the sensor leakage current. The input terminals of the OTA are connected respectively to the input and output of the CSA. In absence of leakage, no current flows in R_f and the two terminals are equipotential. When a DC current is present in the circuit of Fig. 8.20 a), the output voltage of the CSA tends to rise, pulling down the OTA output, which increases the current in M_C thus compensating the leakage current. The circuit of Fig. 8.20 b) works in the same way for DC currents entering into the CSA. The bandwidth of the loop must be very low, in order to avoid cutting also the signals. Large bandwidth loops can also lead to instability. The feedback resistor R_f can of course also be implemented with a suitably biased transistor.

8.3.4 CONSTANT CURRENT FEEDBACK

Discharging the feedback capacitor with a constant current allows us to perform a rudimentary analog to digital conversion through the method of Time over Threshold (ToT). As we have seen above, the Krummenacher feedback provides a constant current discharge when the signal at the CSA output is big enough to fully unbalance the differential pair. Fig. 8.21 shows two other circuits that can also be used to discharge the feedback capacitor at a constant rate. The feedback current is fed to a diode-connected transistor, which has its source tied to the CSA output. If $I_{dect} = 0$, M_f is forced in the linear region. For sufficiently large value of the output signal, M_f goes in saturation and forms a current mirror together with M_{fb}. The feedback current, eventually scaled by the ratio between the widths of the two devices, is mirrored from M_{fb} to M_f, discharging the feedback capacitor at a constant pace. Many versions of the constant current discharge concept have been implemented, often mated to additional circuitry for leakage compensation [31–33].

8.4 POWER SUPPLY REJECTION

Design robustness against power supply fluctuations attracts often less attention than the reduction of the ENC caused by the intrinsic noise sources of the amplifier-sensor system. Achieving an adequate Power Supply Rejection Ratio (PSRR) is nevertheless paramount. Circuits showing satisfactory results in electrical tests at the bench sometimes exhibit degraded performance when operated in a more realistic environment, where the quality of the power supply and ground lines is worse than in the cleaner laboratory set-up. Power supply rejection is a primary figure of merit for any analog circuit, but we focus here on its optimization in front-end amplifiers, where the weak

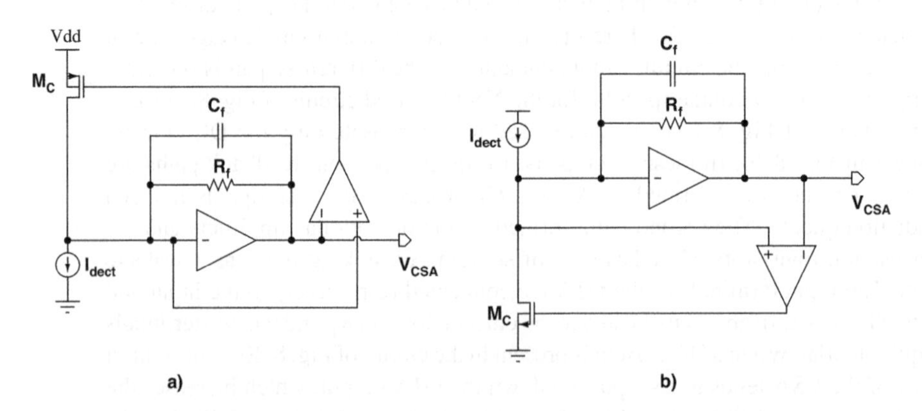

FIGURE 8.20 Two independent circuits can be used to compensate the sensor leakage current and to discharge the feedback capacitor on fast signals.

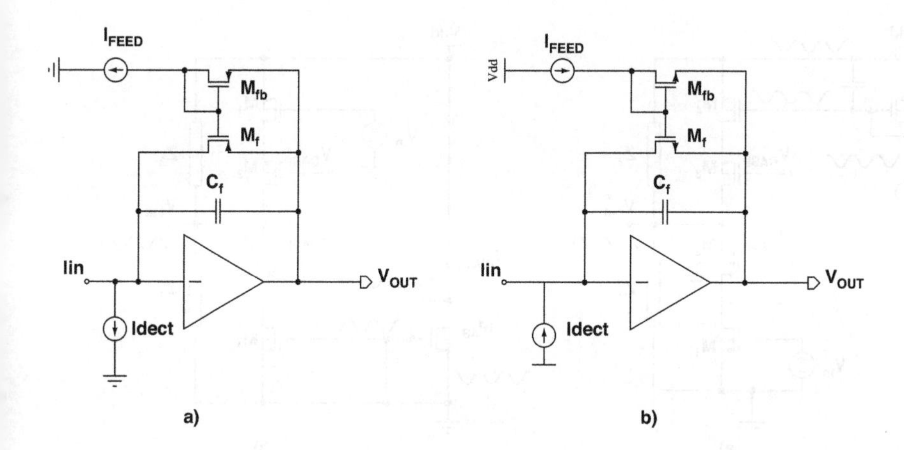

FIGURE 8.21 Circuits providing a constant current discharge of the feedback capacitor.

detector signals are first processed. Overlooking the power supply performance of this stage can lead to very noisy systems that do not meet the application goals. The effect of the power supply noise on a single-ended charge sensitive amplifier is different if the disturbance occurs on the rail connected to the load or on the one connected to the source of the input transistor. Most modern front-end ASICs are designed for a single supply rail, i.e. to be powered between V_{DD} and ground. We thus study primarily this case in the following. In the configuration with NMOS input transistor and PMOS loads, the source of the input device is connected to ground and the source of the loads is tied to V_{DD}, while the opposite is true for the complementary topology. We start considering the effect of perturbations occurring on the rail feeding the loads, first examining the response of the voltage amplifier employed as the core stage. The feedback network is therefore not included and the circuit is driven directly by an ideal voltage source. We use as a reference the stage with NMOS input transistor, so the node of interest is V_{DD}. For the PMOS-based architecture, the same considerations apply to disturbances arising on the ground line. The symmetry between the two situations is shown in Fig. 8.22. In the figure, the source followers are omitted for brevity, as they are not central to this discussion.

Consider a noise affecting V_{DD} in the circuit of Fig. 8.22 a). If the bias generator is ideal, the current in the diode-connected transistor M_{4B} is fixed independently of V_{DD}. Therefore, the power supply modulation is transferred also to the gate of the device, because the source-gate voltage of M_{4B} can not change. As a result, the noise on the power supply can not affect the bias current of the stage. For this to happen in practice, the reference generator has to fulfill two conditions: it must be itself as insensitive as possible to V_{DD} variations and have a very high output impedance. This is the reason why a simple resistor to ground, although very practical in small test circuits, is not recommended in the final implementation. Even a well designed current source will not have a hermetic immunity to power supply noise. The rejection of high frequency components can be improved by connecting filtering capacitors in

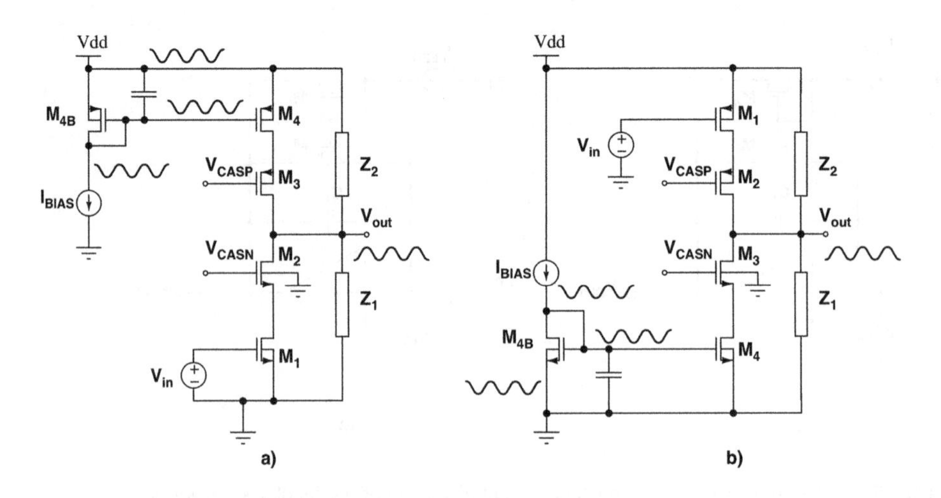

FIGURE 8.22 Effect of power supply noise in open-loop cascode amplifiers.

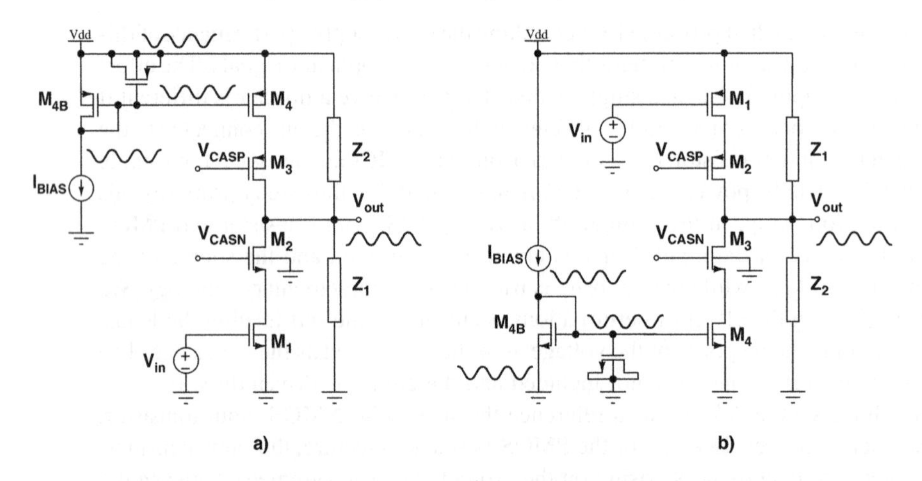

FIGURE 8.23 A transistor with the source and drain shorted together can be used as a decoupling capacitor to reduce the power supply noise and to filter-out the thermal noise from the diode-connected reference device.

parallel to the diode-connected transistor, as shown in Fig. 8.22. The idea is that when the frequency increases, the impedance of the filtering capacitor is reduced, shorting the gate and the source of the transistor and blocking the change in the bias current. In modern CMOS technologies, the gate capacitance per unit area can be between 10 and 20 fF/μm^2 and the filtering units can be implemented with transistors with the source and drain short-circuited, as shown in Fig. 8.23. In many applications, the bias line is common to many channels, hence putting one decoupling capacitor per channel it is possible to reach a total value above hundreds of pF. With this approach, one should make sure that the gate leakage of so many devices connected in parallel does not affect the bias point. This is mostly relevant for very deep submicron technologies (130 nm and below) if the controlled current is very small. A typical example can be the front-end for hybrid pixel sensors, where thousands of front-end cells are incorporated on the same ASIC and currents below the μA must be provided. Additionally, one should check that the integrated decoupling capacitors do not form resonant *LC* circuits with the inductance of the bonding wires.

Direct modulation of the bias current is not the only mechanism through which noise can propagate from the power supply to the circuit output. Another path is through the output impedance of the input stage and the load, which are indicated explicitly in Fig. 8.22 respectively as Z_1 and Z_2. The gain from V_{DD} to the output is calculated with the voltage divider rule:

$$V_{out,VDD} = \Delta V_{DD} \frac{Z_1}{Z_1 + Z_2} \rightarrow A_{V_{DD}} = \frac{Z_1}{Z_1 + Z_2} \tag{8.91}$$

The power supply rejection ratio is thus defined as the ratio between the gain A_V from input to output and the gain $A_{V_{DD}}$ from the power rail to the output. In an open-loop cascode amplifier A_V is given by:

$$A_V = g_{m1} (Z_1 // Z_2) = \frac{g_{m1} Z_1 Z_2}{Z_1 + Z_2} \tag{8.92}$$

Hence the PSSR is:

$$\left| \frac{A_V}{A_{V_{DD}}} \right| = g_{m1} Z_2 \tag{8.93}$$

The measure of the PSSR is usually given in decibels. One could surmise that the PSSR of the circuits of Fig. 8.22 is very good, since the cascode impedance is typically high. However, this is only true at low frequency, because the reactive components of Z_2 (and Z_1) drop their value as the frequency is increased. It is thereby important to check the *PSSR* in the full frequency spectrum.

In a CSA, the cascode amplifier is employed in a closed-loop configuration. It is thus interesting to study how the PSSR changes when feedback is applied. To do so, we first examine the most general case and we then apply the result to the CSA. Before going further, it is useful to rewrite the fundamental concepts and equations of a feedback system. The output signal Y_{out} is given by the open-loop gain of the amplifier multiplied by the error signal X_ε:

$$Y_{out} = A X_\varepsilon \tag{8.94}$$

where $X_\varepsilon = X_{in} - X_{fb} = X_{in} - k_F Y_{out}$, i.e. the error X_ε is the difference between the input and the feedback signal, generated by the feedback network through a suitable processing of the output. We can now apply the superposition theorem and write the output as the sum of two contributions that can be evaluated separately. One is the response to a given input signal measured without noise on the power supply and the other is the noise N_{out}, observed by inducing a perturbation on the power rail without any input signal. Consider first an open loop voltage amplifier with gain A in which the noise on the power supply induces a perturbation of amplitude N_{out} at the output. The effect can be modeled with a circuit in which the power supply is clean and a signal of amplitude N_{out}/A is applied to the input. Let's now apply a feedback network that samples the output signal Y_{Nout} and generates a feedback given by $k_F Y_{Nout}$. Writing the feedback equation we have:

$$Y_{Nout} = AX_{\varepsilon} = A\left(\frac{N_{out}}{A} - k_F Y_{Nout}\right) \rightarrow Y_{Nout} = \frac{N_{out}}{1+k_F A} \tag{8.95}$$

Observe that Y_{Nout} is the signal measured at the output due to the power supply noise after feedback has been applied. Equation (8.95) reveals therefore an interesting point: if feedback is employed, the power-supply induced noise N_{out} observed at the output of a open-loop amplifier is suppressed by a factor equal to the loop gain. If there is no noise on the power supply and an input signal X_{in} is fed to the circuit, the output Y_{Sout} due to the input alone is given by:

$$Y_{Sout} = \frac{A}{1+k_F A} X_{in} \tag{8.96}$$

Summing (8.95) and (8.96) we can write the global output signal Y_{out} measured when both the input stimulus and the noise on the power supply are present:

$$Y_{out} = \frac{A}{1+k_F A} X_{in} + \frac{N_{out}}{1+k_F A} \tag{8.97}$$

We can now apply these concepts to the particular case of the charge sensitive amplifier. A CSA is a form of transimpedance amplifier and can be seen as made by an open-loop transimpedance amplifier with gain $A_V(s)Z_F(s)$ and a feedback network with transfer function $k_F = 1/Z_F(s)$. The impedance $Z_F(s)$ is the parallel of the feedback capacitance and resistance, while $A_V(s)$ is the open-loop voltage gain of the cascode stage. The loop gain is:

$$L(s) = A_V(s)Z_F(s)\frac{1}{Z_F(s)} = A_V(s) \tag{8.98}$$

Hence, equation (8.97) can be rewritten as:

$$V_{out} = \frac{A_V(s)}{1+A_V(s)} Z_F(s) I_{in} + \frac{N_{out}}{1+A_V(s)} \tag{8.99}$$

Equations (8.93) and (8.99) show that to obtain a good power supply rejection in a CSA the load impedance and the open-loop gain of the core amplifier should be maximized.

Let's now examine the effects of fluctuations affecting the source voltage of the input transistor and the sensor power supply. In this case, the detector and its biasing network play a primary role in determining the amount of noise which is coupled to the amplifier output. To understand why, we study the circuit of Fig. 8.24. Here the sensor is represented as a semiconductor diode, however the following considerations apply to any type of sensor presenting a capacitive load to the front-end. In most applications, a sensor bias voltage of tens to hundreds of Volts is applied to create an electric field that allows a prompt collection of the radiation-induced charges. In case of semiconductor sensors, the high voltage is also necessary to reverse-bias the junction, depleting the device of mobile charges. In the small signal equivalent model of the circuit of Fig. 8.24 the high voltage HV, the ground and the amplifier power supply are replaced by signal grounds. The noise affecting each of these lines can be represented as a voltage signal source connected in series to the DC supply. For simplicity, the output buffer and the feedback resistor are omitted in the small signal model and R_f is supposed to be infinite. We consider now three different cases. The first one is when the noise affects the sensor power supply, while the reference of the input transistor is clean. This situation is represented in Fig. 8.25. We can write the equations for the input and output node as:

$$\begin{cases} (V_1 - V_{nHV})\, sC_d + (V_1 - V_{out})\, sC_f = 0 \\ g_{m1}V_1 + (V_{out} - V_1)\, sC_f + \frac{V_{out}}{R_L} + V_{out}sC_L = 0 \end{cases} \tag{8.100}$$

where V_{nHV} is the noise on the sensor power supply. We now solve the system of

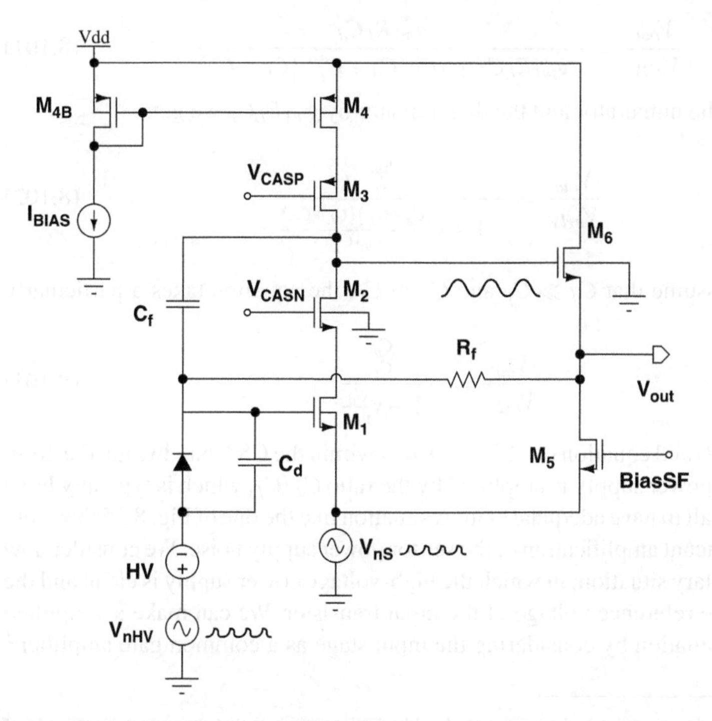

FIGURE 8.24 A charge sensitive amplifier with noise affecting the detector power supply and the reference voltage of the input transistor. The figure shows an NMOS input stage, where the critical voltage in a single-rail implementation is the ground.

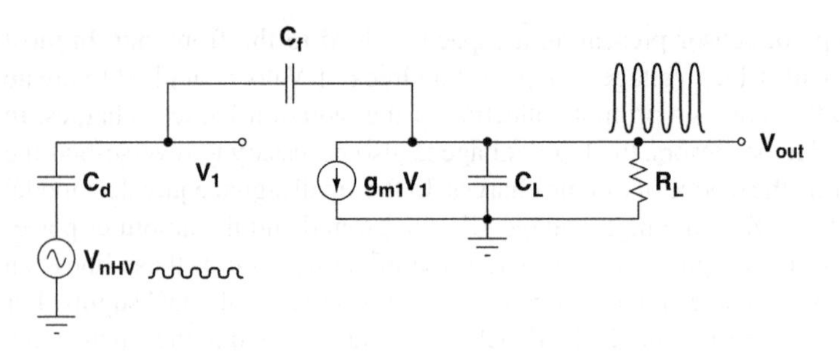

FIGURE 8.25 Small signal model to calculate the propagation at the output of noise corrupting the sensor power supply.

equations under two hypotheses:

- The zero in the numerator introduced by C_f is neglected;
- The quantity $g_{m1}R_LC_f$ is much bigger than $C_d + C_f$ which is thus neglected as well.

With the above conditions, the gain from the detector power supply to the output becomes:

$$\frac{V_{out}}{V_{nHV}} = -\frac{g_{m1}R_LC_d}{g_{m1}R_LC_f + sR_L\left(C_d + C_f\right)\left(C_L + C_f\right)} \tag{8.101}$$

Dividing both the numerator and the denominator by $g_{m1}R_LC_f$ we get:

$$\frac{V_{out}}{V_{nHV}} = -\frac{\frac{C_d}{C_f}}{1 + s\frac{(C_d+C_f)(C_L+C_f)}{g_{m1}C_f}} \tag{8.102}$$

If we further assume that $C_d \gg C_f$ and $C_f \gg C_L$, the equation takes a particularly simple form:

$$\frac{V_{out}}{V_{nHV}} = -\frac{\frac{C_d}{C_f}}{1 + s\frac{C_d}{g_{m1}}} \tag{8.103}$$

Equation (8.102) and equation (8.103) show that within the CSA bandwidth, the noise on the detector power supply is amplified by the ratio C_d/C_f, which is typically large as C_f is kept small to have adequate gain. A situation like the one of Fig. 8.25 therefore leads to a significant amplification of the sensor power supply noise. We consider now the complementary situation, in which the high-voltage power supply is clean and the noise affects the reference voltage of the input transistor. We can make a simplified model of this situation by considering the input stage as a common gate amplifier.[3]

[3]The model is only approximate because one should take into account that the output impedance of the input cascode is not infinite. However, the direct transmission of the ground noise through the output impedance of the input cascode is small in comparison to the one due to the direct modulation of the source of the input device.

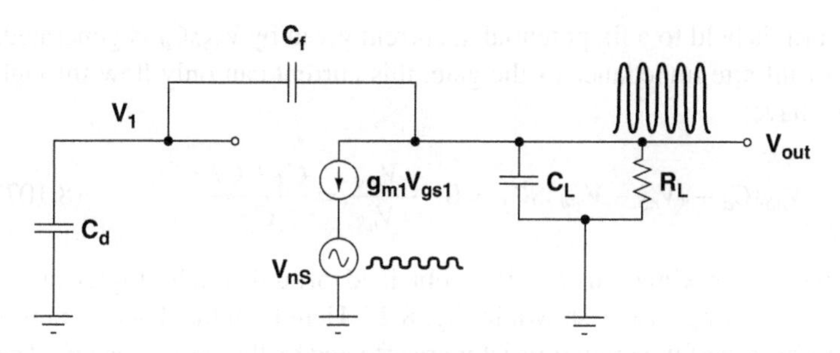

FIGURE 8.26 Effect of bouncing on the reference line of the input transistor in a charge sensitive amplifier. The reference line is the one to which the source of the input device is connected.

The corresponding circuit is reported in Fig. 8.26. Note that the source of the input transistor is not anymore at ground. The current generated by the input device must be therefore written as $g_{m1}V_{gs1}$ where $V_{gs1} = V_1 - V_{nS}$ and V_{nS} is the noise at the device source. The nodal equations now read:

$$\begin{cases} V_1 sC_d + (V_1 - V_{out})\, sC_f = 0 \\ g_{m1}(V_1 - V_{nS}) + \frac{V_{out}}{R_L} + V_{out} sC_L + (V_{out} - V_1)\, sC_f = 0 \end{cases} \tag{8.104}$$

Applying the same simplifying assumptions as before, we have:

$$\frac{V_{out}}{V_{nS}} = \frac{\frac{C_d + C_f}{C_f}}{1 + s\frac{(C_d+C_f)(C_L+C_f)}{g_{m1}C_f}} \approx \frac{\frac{C_d}{C_f}}{1 + s\frac{C_d}{g_{m1}}} \tag{8.105}$$

We therefore see that in case the reference voltage of the input transistor is noisy and the detector power supply is clean, the noise affecting the source is amplified again by the ratio C_d/C_f, the only difference being the sign of the amplification. Also in this case, a small noise affecting the reference potential can be seen greatly magnified at the circuit output. For the circuit of Fig. 8.26 it is interesting to calculate in addition the relationship between V_1 and V_{nS}, which is:

$$\frac{V_1}{V_{nS}} = \frac{1}{1 + s\frac{(C_d+C_f)(C_L+C_f)}{g_{m1}C_f}} \approx \frac{1}{1 + s\frac{C_d}{g_{m1}}} \tag{8.106}$$

The equation shows that within the CSA bandwidth, the noise on the source voltage is transferred with unitary gain on the transistor gate. This results has a simple interpretation. The input transistor is biased by a current source load, designed to have an impedance as high as possible, therefore the current in the input device must be kept constant. If the source potential is moved, the gate must follow to prevent variations of the gate source voltage. The noise V_{ns} is then transferred directly on the gate of the input transistor, which is also connected to one arm of the sensor capacitance

C_d. If the other arm is held to a fix potential, a current given by $V_{nS}sC_d$ is generated. Given the almost infinite impedance of the gate, this current can only flow through C_f, therefore we have:

$$V_{nS}sC_d + (V_{nS} - V_{out})\,sC_f = 0 \rightarrow \frac{V_{out}}{V_{nS}} = \frac{C_d + C_f}{C_f} \tag{8.107}$$

which is the in-band approximation of (8.105), obtained through much simpler calculations. The last interesting case is shown in Fig. 8.27. Here both the detector power supply and the reference of the input transistor are affected by the same noise V_n. The corresponding nodal equations are:

$$\begin{cases} (V_1 - V_n)\,sC_d + (V_1 - V_{out})\,sC_f = 0 \\ g_{m1}\,(V_1 - V_n) + \frac{V_{out}}{R_L} + V_{out}sC_L + (V_{out} - V_1)\,sC_f = 0 \end{cases} \tag{8.108}$$

Solving the above system for V_{out}/V_n, we get:

$$\frac{V_{out}}{V_n} = \frac{1}{1 + s\frac{(C_d+C_f)(C_L+C_f)}{g_{m1}C_f}} \approx \frac{1}{1 + s\frac{C_d}{g_{m1}}} \tag{8.109}$$

We have thus obtained the lowest amplification of the power supply noise among the three considered cases. This shows that the detector power supply and the reference voltage of the input transistor should ideally be the same line, in which case the noise affecting the rail is transferred to the amplifier input only with unity gain. In many cases, however, the sensor power supply voltage needs to be several hundreds of Volts and it can not be tied to the amplifier ground. However, what is important is that the two fluctuate together. This can be achieved with the circuit of Fig. 8.28. The sensor is connected to the high voltage source through a resistor. A capacitor to ground creates a high-frequency short circuit between the power supply and the CSA ground, so that above the cut-off frequency of the filter a noise pulse affecting one line will be transferred also to the other. The filter is effective for frequencies above $1/(2\pi C_{HV}R_{HV})$, therefore both R_{HV} and C_{HV} should be large. These components must also be suitable to withstand high voltages.

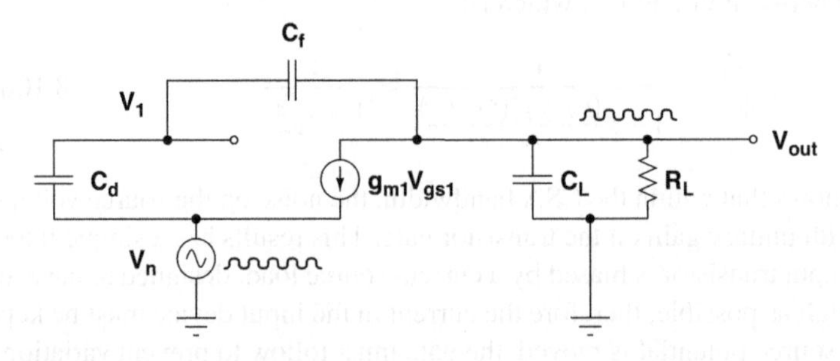

FIGURE 8.27 Effect of noise affecting equally the reference of the input transistor and the sensor power supply line.

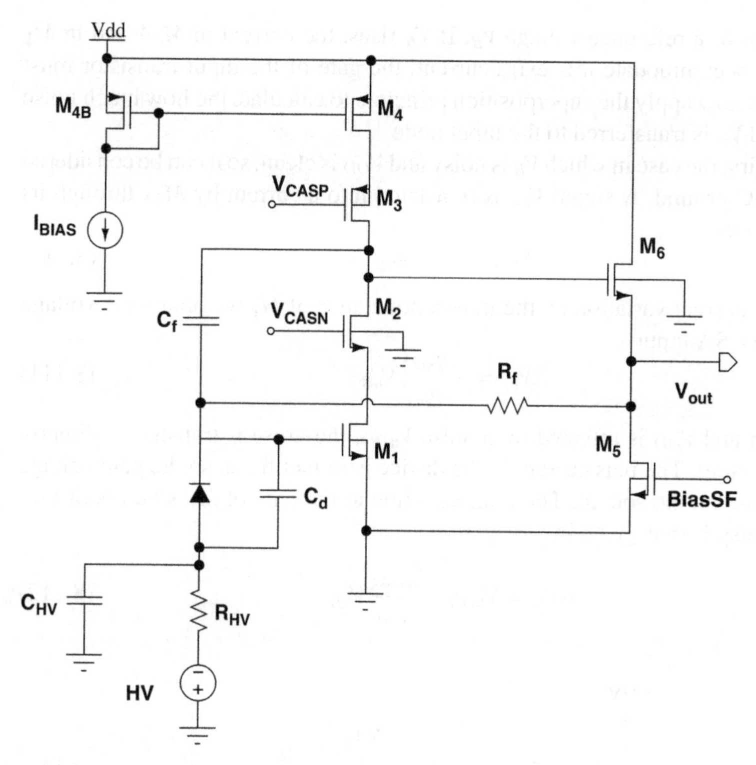

FIGURE 8.28 Decoupling of the sensor high-voltage power supply in case the source of the input transistor of the CSA is referenced to ground.

The circuit of Fig. 8.28 satisfies the widespread intuition that, to solve noise problems in a circuit, decoupling capacitors between the critical line and ground should be used. However, it is also necessary to place the decoupling capacitors towards the appropriate reference voltage. In fact, we have based our discussion on a stage with NMOS input transistor. However, if we use a PMOS input stage, the reference voltage of the input device is V_{DD}. All the considerations made above are still valid, provided that the detector power supply and V_{DD} are made to bounce simultaneously. The decoupling scheme for a PMOS input device is shown in Fig. 8.29, in which an extra capacitor C_S, referenced to ground, is also shown. This capacitor models any parasitic capacitance (such as stray capacitance due to the interconnections, capacitance of protection diodes on the input pad, etc.) that is located between the input and ground. Such a capacitance is small but not negligible (typically between 0.5 and 1 pF) and can still provide the path for a significant magnification of the noise on V_{DD}, because in many cases the feedback capacitance is smaller (value of C_f between 50 fF and 500 fF are rather common). Fig. 8.30 shows also an interesting solution to circumvent the problem [34]. In this case, a cascode with split bias current source is employed. The gate of the NMOS transistor providing most of the bias current is directly tied to a reference voltage V_B. If V_B rises, the current in M_4A and in M_1 increases. To accommodate this extra current, the gate of the input transistor must lower. We can now apply the superposition principle to calculate the how much noise from V_{DD} and V_B is transferred to the input node.

Consider first the case in which V_B is noisy and V_{DD} is clean, so it can be considered as an ideal AC ground. A signal V_{nB} is translated into a current by M_{4A} through its transconductance:

$$\Delta I_{4A} = g_{m4A} V_{nB} \tag{8.110}$$

Dividing this current variation by the transconductance of M_1 we obtain the voltage change at the CSA input:

$$\Delta V_{in} = -\frac{g_{m4A}}{g_{m1}} V_{nB} \tag{8.111}$$

If V_B is clean and V_{DD} is affected by a noise V_{nDD}, the noise is transferred directly to the circuit input. The bias current in the device is in fact fixed, so the gate voltage follows the one on the source. The total variation at the gate of M_1 when both V_{DD} and V_B are noisy is thus given by:

$$\Delta V_{in} = V_{nDD} - \frac{g_{m4A}}{g_{m1}} V_{nB} \tag{8.112}$$

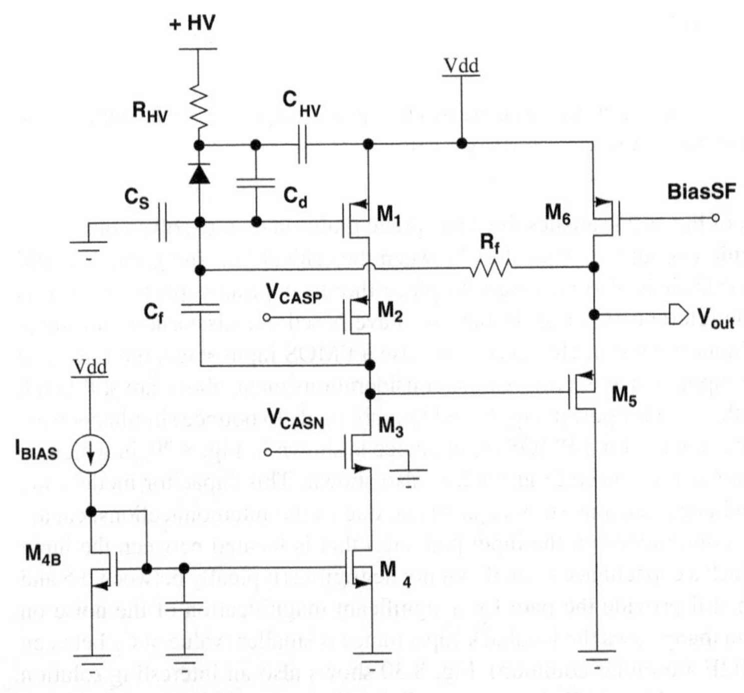

FIGURE 8.29 Decoupling of the sensor high-voltage power supply in case the source of the input transistor of the CSA is referenced to V_{DD}.

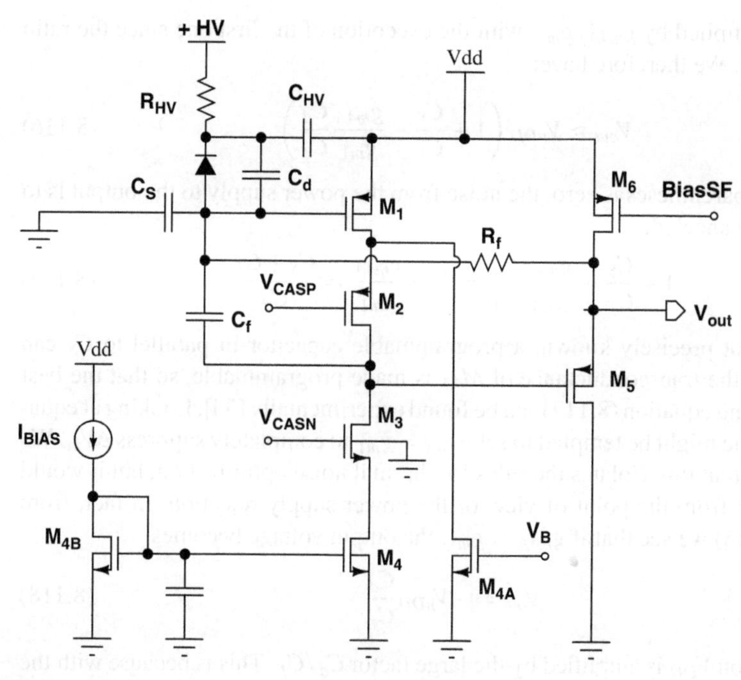

FIGURE 8.30 PMOS input CSA with split current source and stray capacitance. See text for discussion.

Let's now modify the circuit as shown in Fig. 8.31 a), where the gate of the current source M_{4A} has been tied to V_{DD}. Equation (8.112) can be rewritten as:

$$\Delta V_{in} = V_{nDD} - \frac{g_{m4A}}{g_{m1}} V_{nDD} \tag{8.113}$$

We can use the model of Fig. 8.31 b) to make a first order calculation of how much noise is observed at the circuit output. Writing the nodal equation we have:

$$\begin{gathered}\left[V_{nDD}\left(1-\frac{g_{m4A}}{g_{m1}}\right)-V_{nDD}\right]sC_d+\left[V_{nDD}\left(1-\frac{g_{m4A}}{g_{m1}}\right)\right]sC_S+ \\ +\left[V_{nDD}\left(1-\frac{g_{m4A}}{g_{m1}}\right)-V_{out}\right]sC_f=0\end{gathered} \tag{8.114}$$

Solving for V_{out}, we get:

$$V_{out} = -V_{nDD}\frac{g_{m4A}}{g_{m1}}\frac{C_d}{C_f} + V_{nDD}\frac{C_S}{C_f} - V_{nDD}\frac{g_{m4A}}{g_{m1}}\frac{C_S}{C_f} + V_{nDD} - V_{nDD}\frac{g_{m4A}}{g_{m1}} \tag{8.115}$$

From previous considerations on thermal noise, we already know that the transconductance of the load current source must be much smaller than the one of the input transistor. To the first order, we can thus neglect in the previous equation the terms which are multiplied by g_{m4A}/g_{m1}, with the exception of the first one since the ratio C_d/C_f is large. We therefore have:

$$V_{out} \approx V_{nDD}\left(1+\frac{C_S}{C_f}-\frac{g_{m4A}}{g_{m1}}\frac{C_d}{C_f}\right) \tag{8.116}$$

If the term in parentheses is zero, the noise from the power supply to the output is to the first order canceled:

$$1+\frac{C_S}{C_f}-\frac{g_{m4A}}{g_{m1}}\frac{C_d}{C_f}=0 \rightarrow \frac{g_{m4A}}{g_{m1}}=\frac{C_S+C_f}{C_d} \tag{8.117}$$

Since C_S is not precisely known, a programmable capacitor in parallel to C_S can be added and the transconductance of M_{4A} is made programmable, so that the best values satisfying equation (8.117) can be found experimentally [34]. Looking at equation (8.113) one might be tempted to set $g_{m4A}=g_{m1}$ to completely suppress ΔV_{in}. We already know that this violates the rules for thermal noise optimization, but it would be wrong also from the point of view of the power supply rejection. In fact, from equation (8.115) we see that if $g_{m4A}=g_{m1}$, the output voltage becomes:

$$V_{out}=-V_{nDD}\frac{C_d}{C_f} \tag{8.118}$$

and the noise on V_{DD} is amplified by the large factor C_d/C_f. This is because with the choice $g_{m4}=g_{m1}$ we make the amplifier input a perfect virtual ground from the point of view of power supply variation and since V_{DD} is bouncing the charge $V_{nDD}C_d$ is injected into the system. It must finally be observed that nothing would prevent the circuit of Fig. 8.30 to be powered between zero and a negative potential. In this case, the source of the input transistor would be referenced to ground. However, systems with negative power supply are by far less common, so the interface between the front-end and the outside world becomes more problematic, requiring AC coupling.

In circuit simulations, the PSSR should be studied with two techniques. The first is through the small signal analysis, in which the transfer function from the power supply to the output is calculated. The second is by transient simulations, in which an abrupt change is introduced on the power supply voltage source and the output of the circuit is inspected. A voltage step with a fast rise time contains in fact many high-order harmonics and through a time domain simulations instability problems that might not be trivial to indentify only with the AC analysis can be identified. Critical frequencies are 50 – 120 Hz, which corresponds to the residual ripple stemming from the power distribution network. At those frequencies the impedance of the reactive element inside the circuit is still very high and the PSSR is usually very good. However, today systems make large use of switching power supplies, because they are both more power efficient and cost effective. With these regulators, the residual ripple on the power supply is at high frequencies (beyond 100 kHz), where the PSSR is smaller. Therefore, particular attention must be paid to provide an adequate protection to the CSA power supply line. This is often achieved interposing a linear regulator between the switching power supply and the chip and mounting decoupling capacitors as close as possible to the ASIC pins.

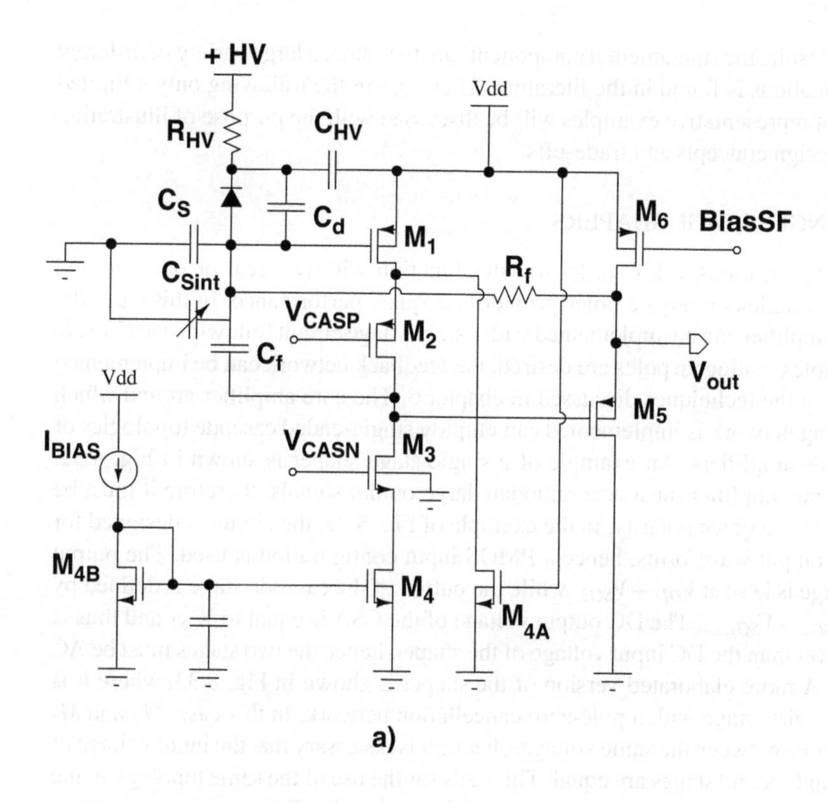

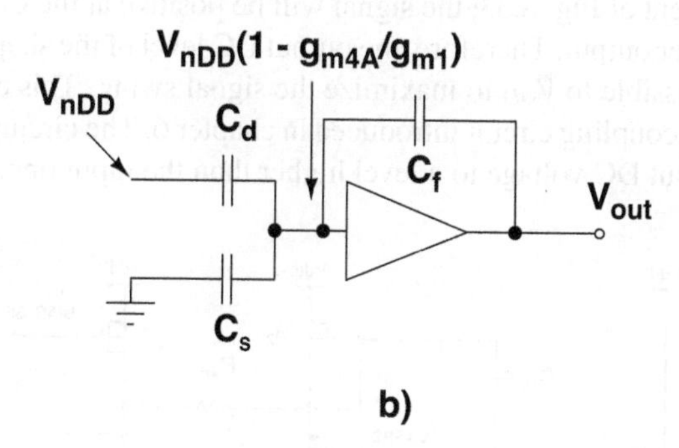

FIGURE 8.31 Current source connection for best PSSR with stray capacitance for a PMOS-input CSA (in a). In b) a simplified circuit to calculate the propagation of noise from the power supply to the output is shown.

8.5 IMPLEMENTATION OF SHAPING AMPLIFIERS

The design of shaping amplifiers entails the developments of four major building blocks: the core amplifier, the high frequency feedback network, the output stage and possibly the baseline holder/baseline restorer, that keeps the DC levels under control. Despite the fundamental components are the same, a large variety of different implementations is found in the literature. Therefore in the following only a limited number of representative examples will be discussed with the purpose of illustrating the key design concepts and trade-offs.

8.5.1 SINGLE STAGE SHAPERS

In many applications, a low order transfer function with two real or one real and a pair of complex conjugate poles provides adequate performance. In this case, the shaping amplifier can be implemented with a single stage circuit following the CSA. In case complex conjugate poles are desired, the feedback network can be implemented with one of the techniques discussed in chapter 6. The core amplifier around which the shaping network is implemented can employ single-ended cascode topologies or operational amplifiers. An example of a single stage shaper is shown in Fig. 8.32. The shaping amplifier must accommodate large output signals, therefore it must be optimized for a given polarity. In the example of Fig. 8.32, the circuit is designed for negative output waveforms, hence a PMOS input configuration is used. The output DC voltage is kept at $V_{DD} - V_{SG2}$, while the output of the cascode stage is defined by $V_{DD} - V_{SG2} - V_{SG_{MF2}}$. The DC output voltage of the CSA is equal to V_{GS1} and thus is much lower than the DC input voltage of the shaper, hence the two stages must be AC coupled. A more elaborated version of the shaper is shown in Fig. 8.33, where it is mated to a first stage with a pole-zero cancellation network. In this case, M_f and M_z must operate between the same voltages, hence it is necessary that the input voltage of the first and second stages are equal. This calls for the use of the same topology in the two stages. With the arrangement of Fig. 8.33, the signal will be positive at the CSA output and negative at the shaper output. Therefore, the output DC level of the shaper must be brought as close as possible to V_{DD} to maximize the signal swing. This can be obtained with the active AC coupling circuit introduced in chapter 6. The circuit is suitable for regulating the output DC voltage to a level higher than the input one. In

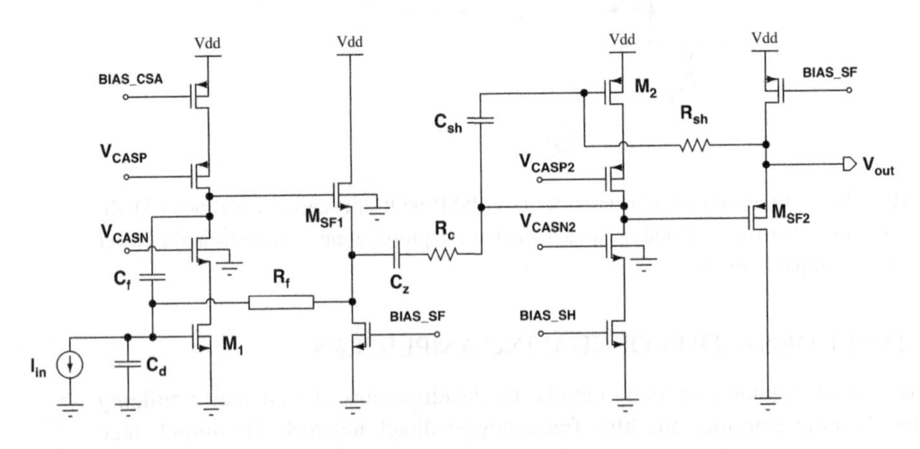

FIGURE 8.32 Two stage preamp-shaper with AC coupling.

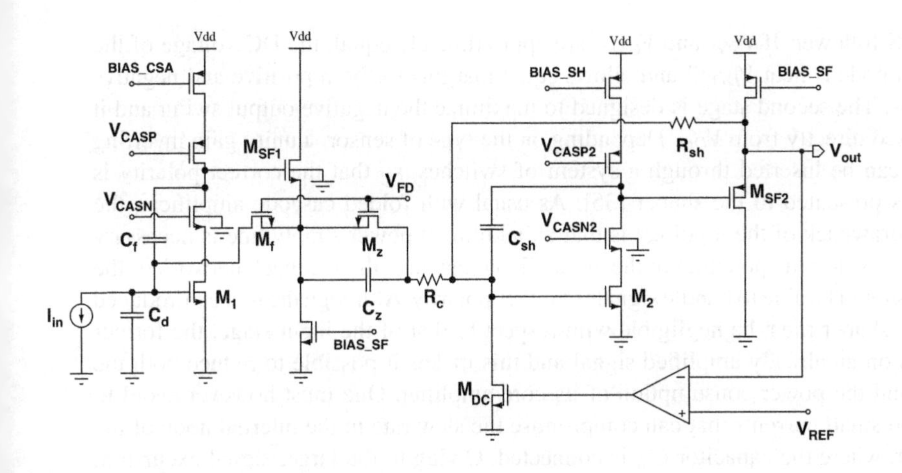

FIGURE 8.33 Two stage preamp-shaper with active AC coupling.

fact, if the "minus" terminal of the differential stage is lower than the "plus" one, its output is driven high, increasing the gate voltage of M_{DC} and thus its current, which, flowing in the feedback resistor, rises the output node voltage. A negative feedback mechanism is thus established that makes the output DC level track the reference voltage V_{REF}. The active AC coupling can also handle the DC current coming from the CSA. In designing the stage, care must be paid to the extra parallel noise introduced by M_{DC}. Fig. 8.34 shows another implementation of a CSA followed by a single stage shaping amplifier. In this case, the input stage employs a folded cascode topology. Note that the input transistor is biased by a dedicated power supply, that should be lower than V_{DD}. The gate voltage of the cascode transistor, V_{CASN1}, is regulated to provide the minimum headroom that allows the main biasing current source to work properly. The power supply of the input transistor is set halfway between ground and V_{DD}. The output node of the CSA is found at $V_{DDpre} - V_{SG1}$, while before the buffer the DC voltage is $V_{DDpre} - V_{SG1} + V_{GSF1}$ where V_{GSF1} is the gate-source voltage of the

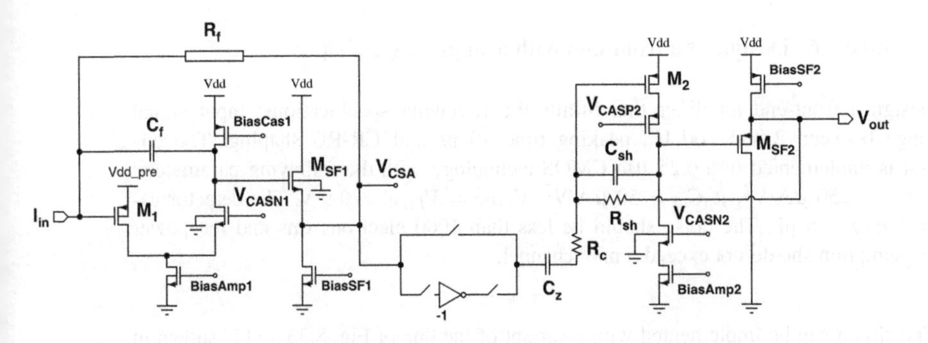

FIGURE 8.34 Two stage preamp-shaper suitable for both input signal polarities.

NMOS follower. If V_{SG1} and V_{GSF1} are approximately equal, the DC voltage of the output node is near $V_{DD}/2$ and allows equal margins for both positive and negative signals. The second stage is designed to maximize the negative output swing and it is biased directly from V_{DD}. Depending on the type of sensor, a unity gain inverting stage can be inserted through a system of switches, so that the correct polarity is always presented to the shaper [35]. As usual with folded cascode amplifiers, the main drawback of the topology is that an additional power supply line is necessary to fully exploit the potential of the circuit. Note that also the feedback network of the CSA should be able to handle signals of either polarity. Although the noise introduced by the shaper must be negligible with respect to that of the input stage, the former works on an already amplified signal and this makes it possible to reduce both the area and the power consumption of its core amplifier. One must however avoid to use too small currents that can compromise the slew rate in the internal node of the shaper, where the capacitor C_{sh} is connected. Owing to the larger signal excursion, it can be beneficial to increase also the channel length of the input and the cascode transistors in order to reduce the distortion due to the non-linear output conductance of the devices. In many circumstances, a channel length 3-4 times the minimum allowed by the process is an adequate choice. Despite their simplicity, these types of shapers cover a broad range of applications and have been adopted in many ASICs [35–40].

The core amplifiers can also be implemented with Operational Transconductance Amplifiers (OTAs). The use of OTAs provide an additional input pin that can serve to adjust the DC levels [41, 42]. In principle, the OTA can be employed in the inverting and non-inverting configuration. The latter has the disadvantage that the excursion at the op-amp input is wider, possibly requiring the use of rail-to-rail input stages [43] when large signals have to be handled. One of the key advantages of OTAs is the possibility of having rail-to-rail output stages, that can also be designed in class AB configuration to optimize the power consumption. Many examples of OTA-based shapers can be found in the literature [1, 44–47]. The following example describes a front-end amplifier with large output swing exploiting an OTA in the shaping stage.

Example 8.6: Design of a front-end with a single stage shaper.

Design a front-end amplifier that fulfills the following specifications: input signal range between 2 and 200 fC, peaking time 20 ns and CR-RC shaping. The circuit is implemented in a 0.25 μm CMOS technology with the following parameters: $\mu_n C_{ox} = 250\ \mu A/V^2$, $\mu_p C_{ox} = 50\ \mu A/V^2$, $V_{THN} = |V_{THP}| = 0.5$ V. The detector capacitance is 5 pF. The noise should be less than 1000 electrons rms and the power consumption should not exceed 5 mW/channel.

* * *

The circuit can be implemented with a variant of the one of Fig. 8.33 and is shown in Fig. 8.35. The large dynamic range calls for a shaper with a rail-to-rail output stage, thus an OTA is the preferred choice for the shaper core amplifier. The short peaking

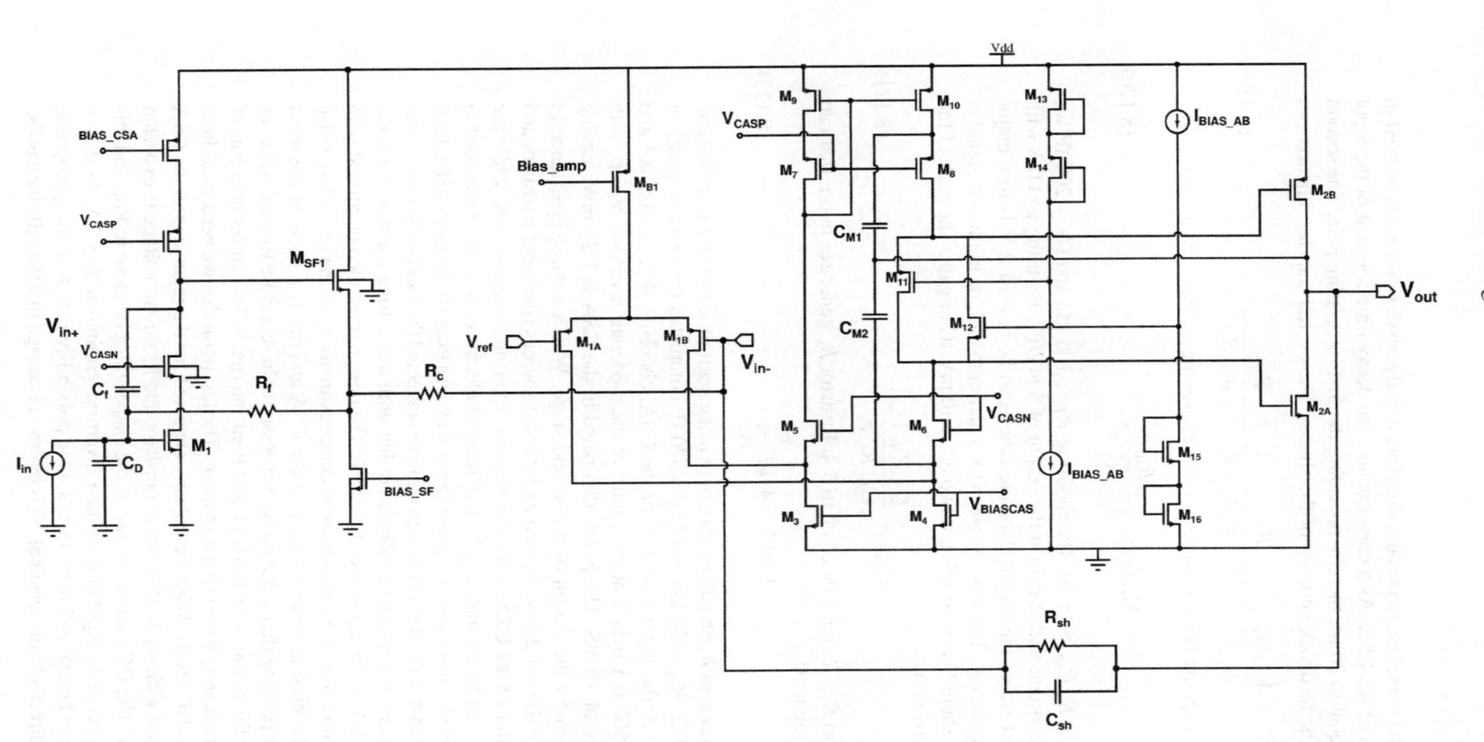

FIGURE 8.35 Single stage shaper implemented with a class AB OTA.

time makes parallel noise less important, therefore a fairly small resistor can be used in the feedback path of the CSA. As a consequence, the decay time constant of the input stage is chosen equal to 20 ns, thereby providing the first integrating pole. The second pole is given by the feedback network of the shaper. The overall transfer function can be written as:

$$\frac{V_{out}(s)}{I_{in}(s)} = \frac{R_f}{(1+sC_fR_f)} \frac{1}{R_c} \frac{R_{sh}}{(1+sC_{sh}R_{sh})} \tag{8.119}$$

The time domain response for a input delta pulse conveying a charge Q_{in} is given by:

$$V_{out}(t) = \frac{Q_{in}}{C_f} \frac{R_{sh}}{R_c} \frac{t}{\tau} e^{-\frac{t}{\tau}} \tag{8.120}$$

where $\tau = R_fC_f = R_{sh}C_{sh} = 20$ ns. By choosing $R_f = 100$ kΩ and $C_f = 200$ fF we get the first time constant with a gain at the output of 5 mV/fC. By using an OTA with a rail-to-rail output stage, the output dynamic range can be at least 2 V before major distortions start appearing. The ratio between the maximum and the minimum signal is 100, therefore we should have 20 mV per two fC or 10 mV/fC of gain. The peak of the CR-RC shaper is given by:

$$V_{peak} = \frac{Q_{in}}{C_f} \frac{R_{sh}}{R_c} \frac{1}{e} \tag{8.121}$$

Therefore, the ratio R_{sh}/R_c must be equal to 5.4. Resistor R_c generates noise, that can be referred to the input as:

$$I_n(R_c) = 4k_BT\frac{R_c}{R_f^2} \tag{8.122}$$

therefore a small value of R_c should be chosen to make negligible its noise contribution. Choosing $R_c = 5$kΩ, $R_{sh} = 54$ kΩ and $C_{sh} = 370$ fF completes the first cut sizing of the components. For the input transistor of the CSA, choosing $W/L = 400/0.3$ and a bias current of 500 μA puts it at the center of the moderate inversion region, with a transconductance of 10 mS. The power dissipated in the CSA is 1.25 mW, leaving a significant margin for the design of the second stage. With a sensor capacitance of 5 pF and a peaking time of 20 ns, the two major noise sources (the input transistor and the feedback resistor of the CSA) contributes about 300 rms electrons each, while the contribution of R_c can be estimated to be less than 80 electrons that, when summed in quadrature with the two main sources, contributes with less than 10 electrons to the total noise budget. Note that in the first cut design we have neglected the bandwidth limitations of the core amplifiers, but it can be calculated that with a rise time constant 1 ns, the peak gain of the CSA is already reduced by 15%. The design must then be refined with computer simulations, that allow also more accurate estimations of the secondary noise sources, such as the biasing current source of the CSA and the $1/f$ noise of the input transistor. At short peaking times, the power necessary to have a close to ideal behavior from the core amplifiers might be too high and their internal poles can become part of and contribute significantly to the signal shaping. The last point that we need to address is the setting of the DC levels. With the polarity expected for the input signal, the CSA is naturally designed with an NMOS input transistor and its output voltage is expected to be around 0.6 V. The DC output voltage of the shaper must be close to V_{DD} and this can be achieved by properly regulating the non-inverting terminal of the OTA. As shown in Fig. 8.36, this can be realized in two ways. One possibility is to feed an appropriate reference voltage directly to the terminal. This has the advantage that the full chain is DC

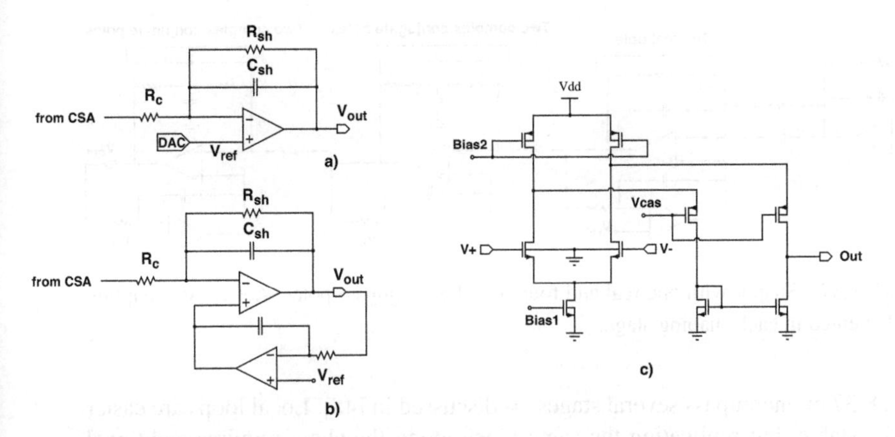

FIGURE 8.36 DC (in a) and active AC coupling between preamp and shaper. The circuit in Fig. c) can be used to implement the integrator in the active AC coupling.

coupled, avoiding the risk of baseline drifts at high rates. On the other hand, the system will be also sensitive to the sensor leakage current and to process fluctuation affecting either the CSA and the reference voltage, which must then be provided through a digital to analog converter, so that fine regulations are possible. The other method is to employ a feedback network providing the active AC coupling, as shown in Fig. 8.36 b). The differential stage is configured as an integrator, so that fast transient signals are filtered-out and only slow changes are compensating for. Observe that the OTA implementing the integrator should work with a high common mode voltage, as the reference voltage is close to V_{DD}. Since the output voltage will be lower than 1 V, the circuit can be implemented with the folded topology of Fig. 8.36 c).

8.5.2 MULTI-STAGE SHAPERS

Higher order transfer functions are realized by combining a given number of low order cells. For real poles, assuming that each stage contributes with one time constant, n stages are needed to have n real coincident poles. For n complex conjugate poles, the number of stages is given by $n/2$, as the poles come as complex conjugate pairs. Higher order shapers can be divided in cascaded and nested feedback topologies. In a cascaded topology, each basic cell is independent of the others and has its dedicated network to generate the poles. Figure 8.37 shows as an example a shaper realizing one real and four complex conjugate poles. As anticipated in the previous example, the DC coupling between the stages can be handled in different ways. If OTAs are used, an adequate reference voltage can be provided to set the output of each stage at the desired DC level [41, 42]. In this case, one has to consider that if the DC gain of the stage is large, the output would be very sensitive to the value of the reference voltage, which must be controlled with adequate accuracy. The other alternative is to have AC coupling between the stages. When active loops are used, these can be local, as shown

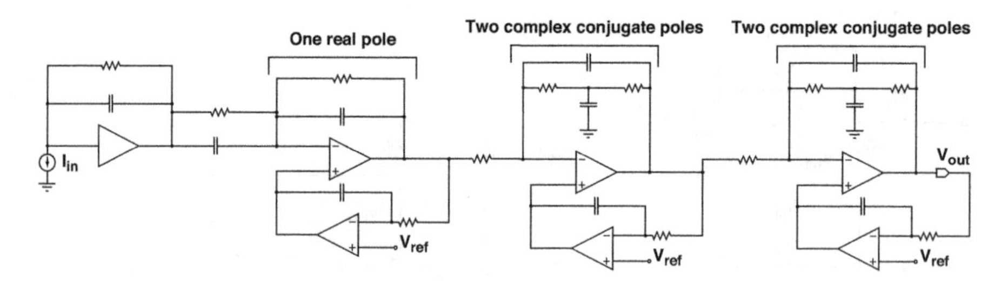

FIGURE 8.37 Shaper with one real and four complex conjugate poles. Active AC coupling is implemented in each shaping stage.

in Fig. 8.37 or encompass several stages, as discussed in [48]. Local loops are easier to make stable, but replicating the same block along the chain requires additional area. The shaper should be optimized so that its contribution to the overall front-end noise is negligible. A detailed discussion of the optimization of shapers for minimum noise can be found in [49]. In the paper it is shown that the best results are achieved using the concept of Delayed Dissipative Feedback (DDF), which consists in taking the feedback of each stage from a point as far as possible down in the processing chain. The concept is illustrated in Fig. 8.38, which compares three realizations of a shaper with one real and two complex conjugate poles. The components are sized for a peaking time of 500 ns and a gain of 10 mV/fC [49]. The first circuit in the top part of the figure is the standard implementation. One stage with its local feedback provides the real pole, while the second shaping unit realizes the two complex conjugate ones. The circuit in Fig. 8.38 b) shows the application of the DDF concept. Note that the feedback of the first stage is taken from the output of the chain, so it is delayed with respect to signal across the capacitor. The circuits of Fig. 8.38 a) and b) are designed for the same gain and signal-to-noise ratio. Note that the resistor values are increased (in particular the one of the resistor implementing the DDF feedback), but the capacitor values are significantly reduced. In particular, the coupling capacitor between the CSA and the shaper scales from $106C_f$ to $28C_f$, thus allowing for a significant saving in circuit area (almost 70%) for the same dynamic range. The implementation of Fig. 8.38 c) has the same area of the one in Fig. 8.38 c), but the rms noise is reduced to one half. Note that the DDF component must provide negative feedback. On the other hand, a capacitor is also connected between the second and the third shaping stage. This positive feedback is necessary to obtain complex conjugate poles in the transfer function. The DDF feedback concept can be implemented with shapers of any order, but its advantages over cascading configuration reduces as the order of the filter is increased. The reader is referred to the original paper for the full details on the technique and the comparison with other more traditional methods.

8.5.3 ACTIVE SHAPING NETWORKS

Shaping amplifiers usually handle large signals. The use of passive resistors in their feedback network is therefore common, especially in applications where a good

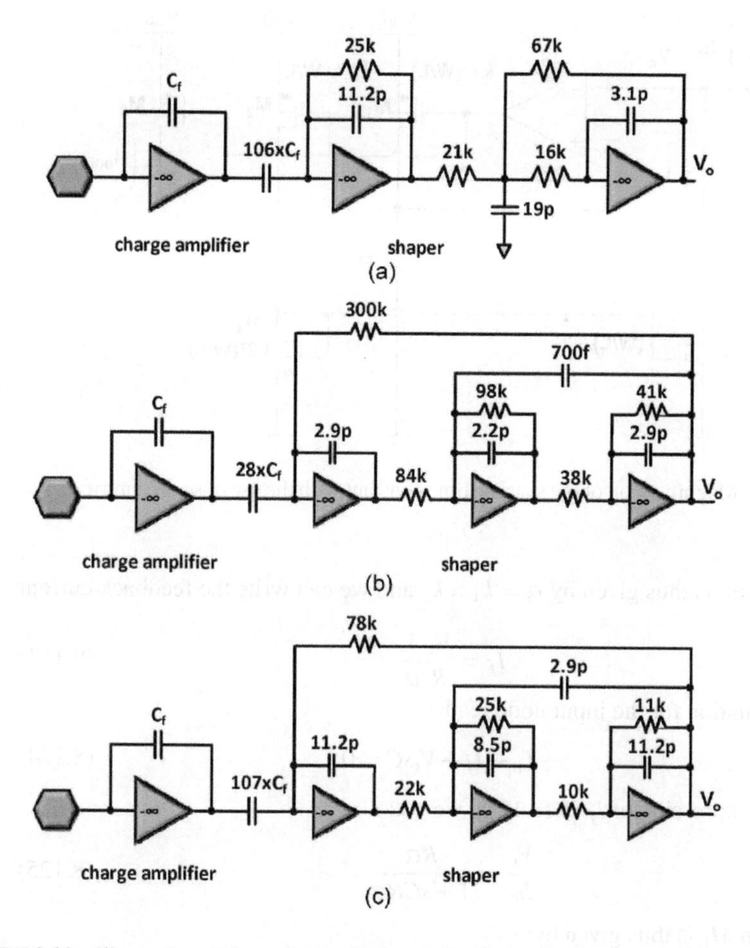

FIGURE 8.38 Illustration of the concept of delayed dissipative feedback. See text for details. © [2011] IEEE. From [49], reprinted with permission.

linearity must be maintained over a wide dynamic range. Active feedback networks are however also used in shapers and they can be particularly advantageous if long shaping times are necessary. In fact, using a transconductor as the feedback element the equivalent resistance can be controlled by regulating its bias current. Source degeneration and other techniques can also be employed to improved upon linearity. In many topologies, the filter time constant is increased by introducing a current mirror in the feedback path. The concept can be understood with the help of Fig. 8.39. Observe that due to the action of the differential amplifier, the voltage V_x is copied on its positive terminal, originating a current given by V_x/R. Since the aspect ratio of M_2 is k_1 smaller than the one of M_1, the current is reduced by the same factor. The current is further scaled down by k_2 before being presented to the input node. The

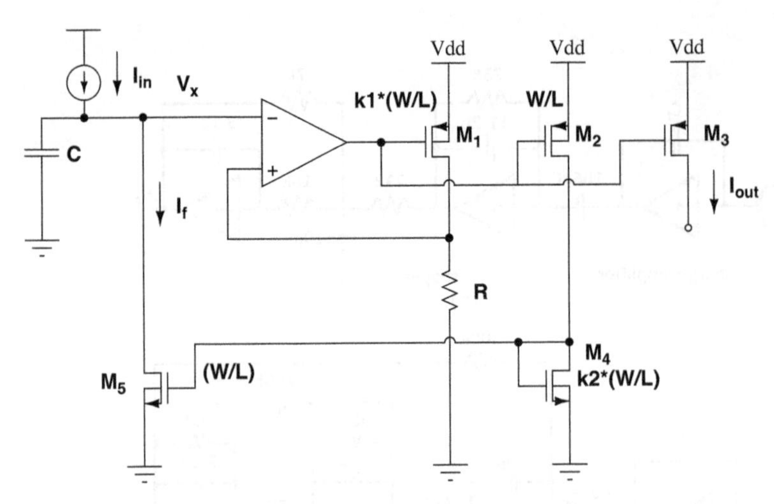

FIGURE 8.39 Magnification of the shaping time constant with the use of scaling mirrors. See text for details.

total attenuation is thus given by $\alpha = k_1 \times k_2$ and we can write the feedback current as:

$$I_f = \frac{V_x}{R}\frac{1}{\alpha} \tag{8.123}$$

The nodal equation for the input node reads:

$$-I_{in} + I_f + V_x sC = 0 \tag{8.124}$$

Inserting the expression of I_f in (8.124), we get:

$$\frac{V_x}{I_{in}} = \frac{R\alpha}{1 + sCR\alpha} \tag{8.125}$$

The current in M_1 is thus given by:

$$I(M_1) = \frac{I_{in}\alpha}{1 + sCR\alpha} \tag{8.126}$$

We thus see that in the definition of the time constant the physical value of the resistor R is multiplied by the scaling factor α. Since transistors are smaller than passive resistors, the scaling technique allows to achieve large time constants in a relatively compact space. The area of the circuit can be further reduced by replacing the differential amplifier with a simple source follower, as shown in Fig. 8.40 although in CMOS the linearity is affected. Better results are obtained in BiCMOS technologies where M_1 can be implemented with a bipolar transistor [50]. It is interesting to compare the noise generated by resistor R in Fig. 8.39-8.40 with the one of the feedback resistor of the cell implementing the real pole in Fig. 8.37. When referred to the input, the noise of the feedback resistor in Fig. 8.37 is given by:

$$i_n^2 = \frac{4k_BT}{R} \tag{8.127}$$

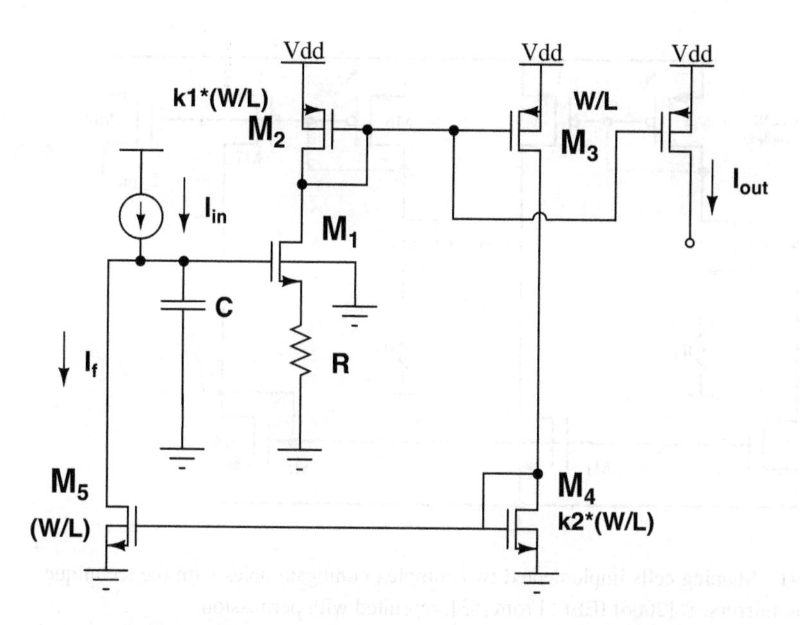

FIGURE 8.40 A compact scaling mirror circuit can be obtained by using source followers.

To achieve the same time constant with a scaling factor α in the mirrors, the resistors of Fig. 8.39 and Fig. 8.40 are designed α times smaller than the one of Fig. 8.37 and their current noise is:

$$i_n(R)^2 = \frac{4k_B T\alpha}{R} \tag{8.128}$$

However, when referred to the input of the shaping cell, this noise is divided by α^2, so the input referred noise becomes:

$$i_n^2 = \frac{4k_B T}{\alpha R} \tag{8.129}$$

This suggests that shaping cells applying the scaling technique offer better noise performance while allowing a smaller feedback resistor. However, the additional transistors in the feedback chain contribute with their own noise, thus mitigating this advantage. Furthermore, the current change might be significant with respect to the one at the quiescent point, therefore also non-stationary noise components must be considered [49].

Note that the current in the output transistor M_3 has the same sense of the input current and that current of either polarity can anyway be achieved with this technique. Therefore, more cells can be easily cascaded to implement the required filter order. By cascading two cells and taking the feedback from the output of the second towards the input of the first one complex conjugate poles are also obtained [51, 52]. A CMOS implementation of this concept is shown in Fig. 8.41 [53]. Example of shapers that make use of the scaling mirror technique can be found in [54–56].

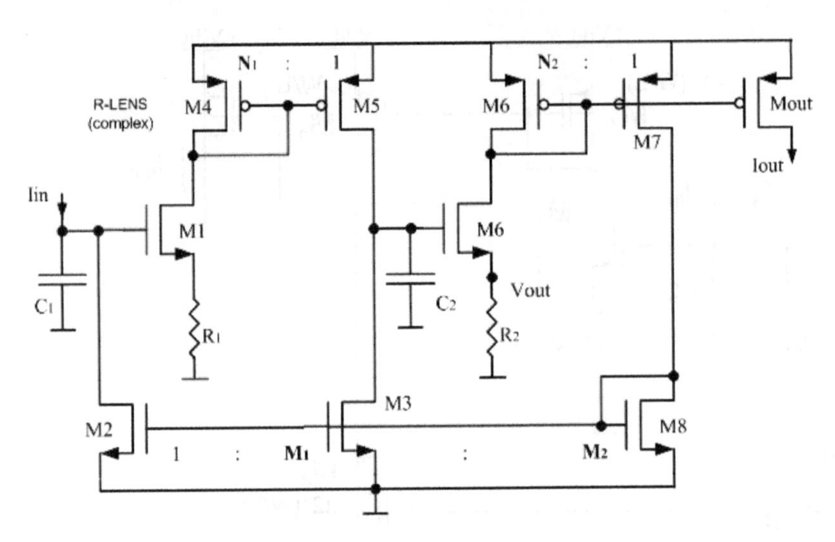

FIGURE 8.41 Shaping cells implemented two complex conjugate poles with the technique of the scaling mirrors. © [2006] IEEE. From [53], reprinted with permission.

8.5.4 OTA-C SHAPERS

Time constants in shapers can also be implemented using OTA as the resistive element [57–60]. The concept is explained in Fig. 8.42 for the case of a simple differential pair. If $R_L = r_{02}//r_{04}$ is the load resistor, the open loop output resistance of the circuit is R_L and the DC gain is given by $A_v = g_{m1}R_L$. When the circuit is closed in a unity feedback loop (see Fig. 8.42 b), the output resistance becomes:

$$r_{out,CL} = \frac{R_L}{1 + g_{m1}R_L} \approx \frac{1}{g_{m1}} \tag{8.130}$$

The Thevenin equivalent of the circuit is shown in Fig. 8.42 c), where one can see that the circuit behaves as a low pass filter with a time constant given by:

$$\tau_{OTA} = \frac{C_L}{g_{m1}} \tag{8.131}$$

By adjusting the value of the load capacitor and of the transconductance of the driving circuit a wide range of time constants can be realized. As reported in Fig. 8.43, the unity gain cell can be used as the load of another OTA, producing a DC gain given by g_{m1}/g_{1mL}, where g_{1mL} is the transconductance of the cell used as the load. In OTA-C filter design, the operational amplifier should mimic as much as possible an ideal transconductor, therefore structures with high output impedance are desired. Practical examples of shapers based on the OTA-C technique can be found in [61–63].

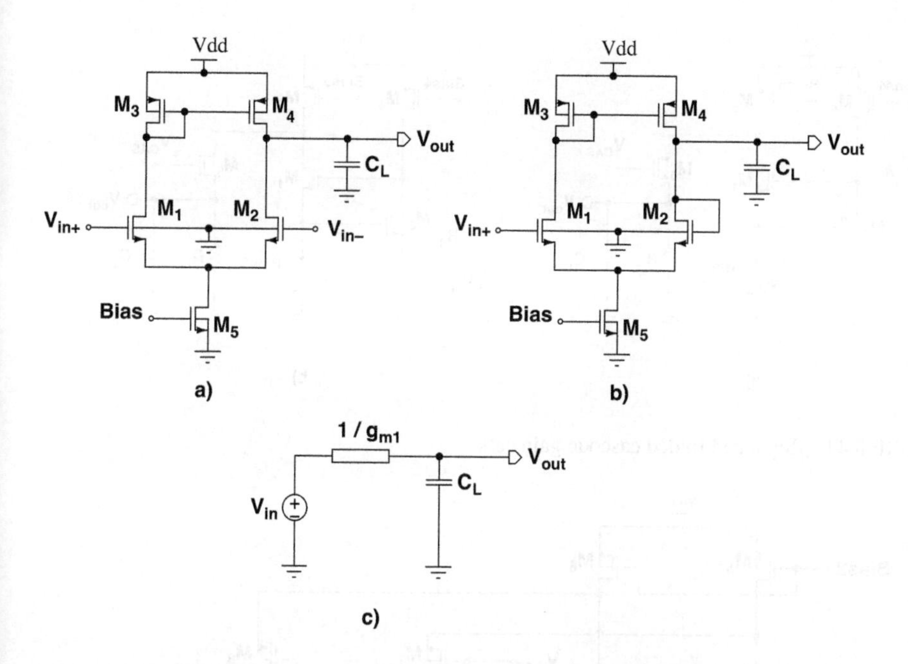

FIGURE 8.42 Implementation of a first order low pass filter with a transconductance stage.

8.5.5 COMPACT SHAPING CELLS

In applications where linearity does not need to be kept over a large range, such as in binary circuits, simpler shaping cells can also be used. Fig. 8.44 shows as an example an amplifying cell based on a regulated common gate input transistor and a folded cascode output stage. The transconductance of the input transistor, M_1 is boosted by the local regulating loop provided by the common source amplifier formed by M_2 and M_4, thus allowing for a low input impedance of the circuit. The current delivered by

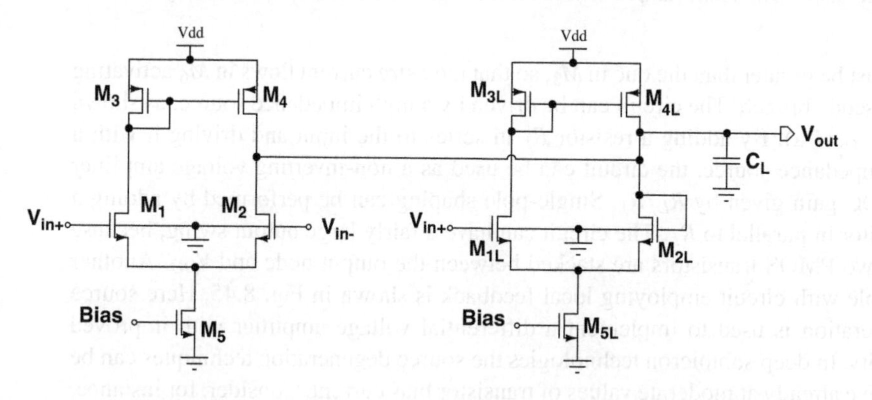

FIGURE 8.43 First order low pass filter with DC gain implemented with the OTA-C technique.

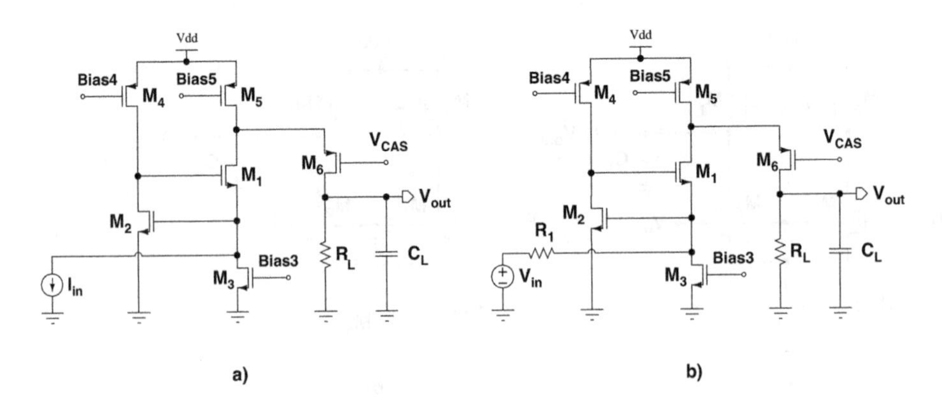

FIGURE 8.44 Regulated folded cascode gain cell.

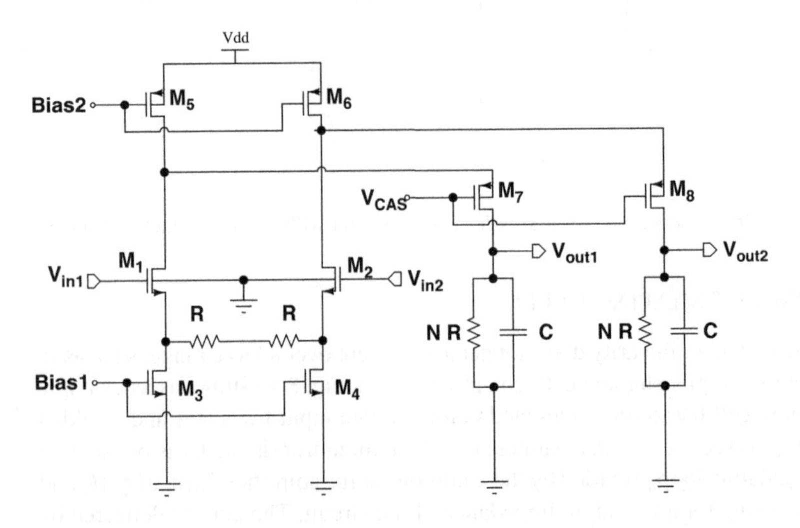

FIGURE 8.45 Differential amplifier with source degeneration to improve linearity.

M_5 must be greater than the one in M_3, so that the extra current flows in M_6 activating the cascode branch. The circuit can be driven by a high impedance source, as shown in Fig. 8.44 a). By adding a resistor R_1 in series to the input and driving it with a low impedance source, the circuit can be used as a non-inverting voltage amplifier with DC gain given by R_L/R_1. Single-pole shaping can be performed by adding a capacitor in parallel to R_L. The circuit can have a fairly large output swing, because only two PMOS transistors are stacked between the output node and V_{DD}. Another example with circuit employing local feedback is shown in Fig. 8.45. Here source degeneration is used to implement a differential voltage amplifier with improved linearity. In deep submicron technologies the source degeneration techniques can be effective already at moderate values of transistor bias current. Consider, for instance, a 0.13 μm technology with $\mu_n C_{ox} = 350$ μA/V^2. A transistor with $W/L = 50/0.2$

biased with 50 μA has an inversion coefficient of 0.33 and a transconductance of about 1.1 mS. A resistor of 10 kΩ provides already a degeneration factor of 10. In case of source degeneration, the differential DC gain of the stage is approximately equal to the ratio N between the values of the load and degeneration resistors. The stage can be particularly useful when single-ended to differential conversion is needed. An interesting technique to exploit local feedback to design differential amplifiers with high linearity is described in [64]. Note that, in general, the load resistors of the circuits in Fig. 8.44 and Fig. 8.45 are not small (20-50 kΩ can be typical values) and buffering of the output voltage may become necessary. Further examples of front-end amplifiers based on compact shaping cells can be found in [15, 65, 66].

8.6 BASELINE HOLDER DESIGN

Baseline holders [67] allow the locking of the baseline to a given reference voltage, avoiding at the same time the rate dependent drifts typical of linear AC coupling systems. The theory beyond BLH circuits was introduced in chapter 6, hence we just summarize here the key design criteria. A BLH must incorporate some form of non-linearity to suppress the fast signals, generated when a quantum of radiation impinges on the detector. For stability reasons, the BLH needs also a strong dominant pole having a time constant $\tau_l \gg G_0 \tau_{sh}$, where G_0 is the DC gain of the loop including both the shaper and the BLH and τ_{sh} is the shaping time constant. To make our treatment more concrete, we discuss now the design of practical circuit. The full front-end is shown in Fig. 8.46 and we use a 0.25 μm CMOS technology with a power supply of 2.5 V. The first stage of the front-end is formed by a preamplifier with pole-zero cancellation and current mirror feedback and it is designed to accommodate current pulses with negative polarity. At the output of the pole-zero stage the signal is thus "pushed" into the shaper, for which a simple CR-RC implementation with two identical time constants is chosen.

The shaper is designed for a gain of 1 mV/fC and a peaking time of 100 ns, obtained with $R_{sh} = 100$ kΩ, $C_{sh} = 1$ pF and $R_c = 36.8$ kΩ. We concentrate now on the optimal setting of the DC levels to achieve an adequate dynamic range in each stage. At the output of the first shaping block, the signal is swinging downwards, therefore the DC voltage of this point should be kept as high as possible. The opposite is true for the global output, where the DC quiescent point should be set reasonably close to ground. For the time being, we assume that this voltage is chosen to 0.4 V. A possible transistor level implementation of the shaper is shown in Fig. 8.47, in which the relevant DC levels inside the circuit are also reported. Both core amplifiers, A_1 and A_2 are implemented as PMOS input cascodes. It is assumed that the circuit is sized so that the voltage at the gate of the input transistor is 1.8 V for both stages. If no current flows in the feedback resistor of A_2 also its output is at the same voltage. To bring it down to 0.4 V we need a current in the feedback resistor of $(1.8-0.4)/(100000\,\Omega) = 14\,\mu$A. This current must be provided by A_1 through the coupling resistor R_c. The voltage drop across R_c is thus $36800\,\Omega \times 14\,\mu$A $= 0.515$ V. The output of A_1 must then be at $1.8+0.515 = 2.315$ V. The drain-source voltage of the transistor biasing the output source follower of A_1 is 185 mV, which is barely sufficient to saturate it. Note

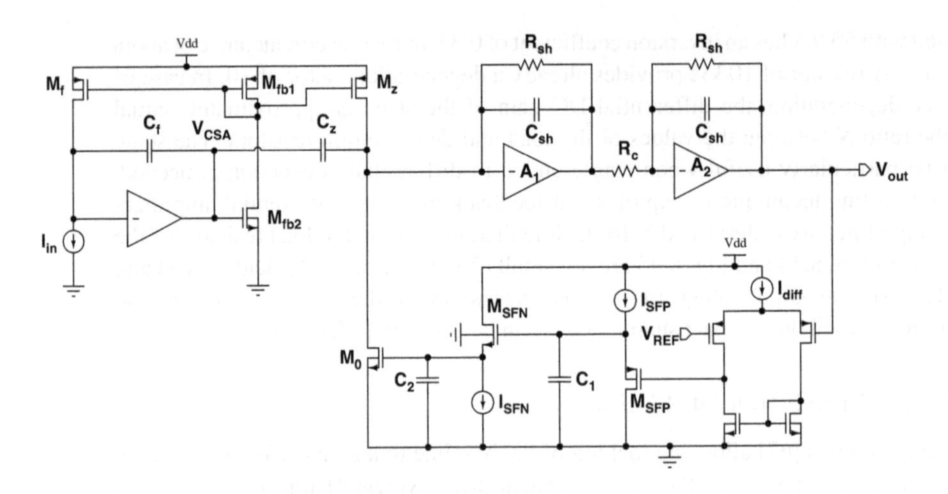

FIGURE 8.46 Full front-end chain with baseline holder.

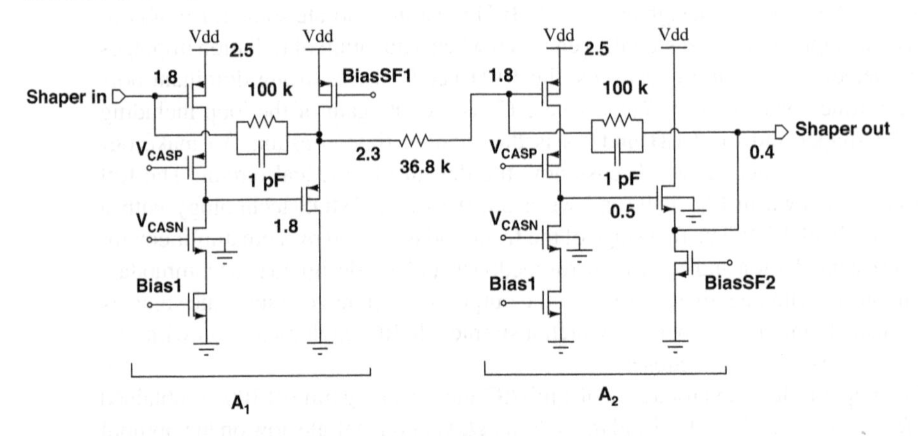

FIGURE 8.47 Schematic level design of the shaper used in Fig. 8.46. The relevant DC levels are also indicated in the figure. See the text for more details.

that the current to R_c must be provided by this transistor, therefore its bias current should be large enough. The voltage at the internal cascode node is scaled down by the gate-source voltage of the follower, and it is again around 1.8 V, that is sufficient to keep the input transistor and its cascode in saturation. The input DC voltage of A_1 is 1.8 V, hence to bring its output to 2.3 V a current of 5.15 μA must be absorbed from its input. This current must be provided by the baseline holder. Note that the source follower at the output of the second stage is implemented with an NMOS transistor. This is necessary because a PMOS transistor would shift downwards by its gate-source voltage the potential of the high impedance node, squeezing the load transistor and its cascode in the linear region. On the other hand, a standard NMOS device would bring the node up by $0.6 - 0.7$ V with respect to the output, compromising the dynamic range. Here it is assumed that this source follower is implemented with a "native threshold" or "zero threshold" transistor, a device which has a threshold close to zero and which has become standard in CMOS technologies from the quarter micron generation.

Let's now come back to the baseline holder. The input stage must compare the baseline value with the reference voltage, hence a differential amplifier is needed. This stage must handle low DC voltages at its input, which calls for a configuration with PMOS input transistors. A non-linearity that clips the standard fast signals must be introduced and it is implemented with PMOS source follower M_{SFP}. Observe that the signal on the gate of M_{SFP} has the same polarity of the output. Therefore, when the latter swings, the gate of M_{SFP} is pulled up, switching M_{SFP} off. The current available to charge C_1 is thus limited by the current source I_{SFP}. For instance, if I_{SFP} is chosen to be 20 nA and C_1 is 1 pF, the positive slew rate is 20000 V/s, which means that the voltage on the capacitor can move only by 2 mV in a 100 ns time frame, which corresponds to the peaking time. The signal is then low pass-filtered by the second source follower. This would also act as a slew rate limiter in case a signal of the "wrong" polarity is presented to the circuit, a situation occurring if the front-end chain is tested injecting voltage steps through a calibration capacitor connected in series to its input. We can now evaluate the DC gain of the shaper alone and of the full loop. In absence of the BLH, a DC current presented to the shaper input is translated to an output voltage with a gain given by:

$$T_{DC} = R_{sh}\frac{R_{sh}}{R_c} \approx 270\ \text{k}\Omega \tag{8.132}$$

As a reference, a change in the input DC current by 1 μA produces an output DC shift of 270 mV. The total loop gain can be evaluated by disconnecting the BLH from the shaper, injecting a voltage at its input and measuring the response at the shaper output. A voltage applied to the BLH input is first multiplied by the gain of the differential amplifier, then is converted to a current by M_0 and finally re-transformed to a voltage by the DC transimpedance gain of the shaper. Observe that M_0 must provide the 5.15 μA current needed for the appropriate setting of the DC levels in the shaper plus any current delivered by the preamplifier. As a first approximation we can neglect the latter and evaluate the transconductance of M_0 assuming that it works in weak inversion, which leads to:

$$g_{m0} = \frac{I_{DS0}}{n\phi_T} \approx 160\ \mu\text{S} \tag{8.133}$$

With a gain of 100 in the differential cell the total DC gain of the loop is:

$$G_{L0} = A_d g_{m0} T_{DC} \approx 4500 \tag{8.134}$$

When the BLH is connected to the shaper, the shaper DC gain is reduced by the loop gain and becomes about 60 Ohm, implying that a change in 1 μA in the input current determines a shift of only 60 μV in the output baseline. The dominant time constant of the BLH must be greater than $G_{L0}\tau_{sh} = 450\ \mu$s to ensure frequency stability. By

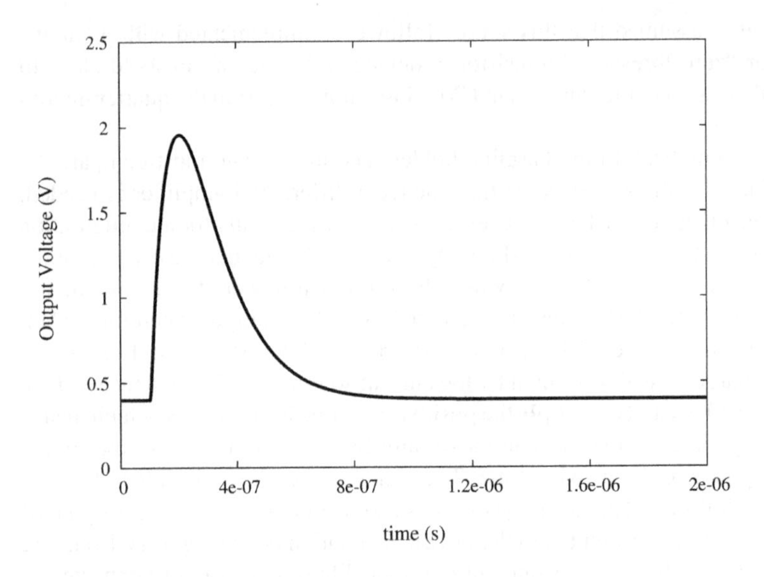

FIGURE 8.48 Response of a front-end with BLH to a full scale signal.

choosing $C_2 = 10$ pF and $I_{SFN} = 50$ pA, one gets a time constant given by C_2/g_{mfn} of 6.6 ms, which satisfies the stability condition. Note that this time constant should be also greater of the non-dominant time constant of the BLH, given by C_1/g_{mfp}, which is 1.6 μs. Both g_{mfn} and g_{mfp} are calculated assuming the source followers are in weak inversion. Circuit containing very long time constants should be simulated for an adequate time duration to verify the circuit stability. To illustrate this point, Fig. 8.48 shows the response of the full front-end to large signal using the value of the biasing currents in the BLH discussed so far. Fig. 8.49 reports the circuit response simulated for the same time scale of 2 μs, changing the current in the NMOS follower from 50 pA to 10 nA. The signals of Fig. 8.49 and Fig. 8.48 are seemingly equivalent, suggesting the idea that the circuit can work also with a much larger bias current in the bandwidth limiting stage. This is in principle desirable, because larger currents are easier to generate and reproduce uniformly across chip. Fig. 8.50 shows the same simulation extended to a time frame of 60 μs and makes visible large oscillations of the baseline. This confirms the necessity of keeping the low pass filtering constant big enough and that, in this type of circuits, simulations lasting only as much as the pulse of interest can be misleading. Finally, Fig. 8.51 reports the response of the properly damped circuit to a train of pulses spaced by 2 μs and shows that there is no significant baseline drift. The compensation of this type of baseline holding circuits requires the generation of very small currents. The uniformity of this current across the chip is an issue, however they do not need to be very precise, as a change by a factor of two is usually tolerated. Finally we should observe that if the current in M_0 increases, so does its transconductance and thus the loop gain. This, on the one hand, tends to further suppress the impact of the additional current, but on the other makes

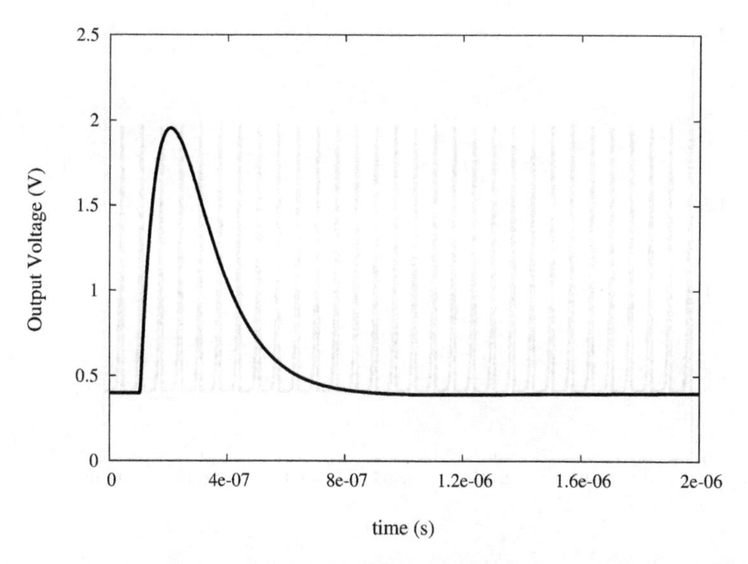

FIGURE 8.49 Response of an underdamped BLH loop over a scale comparable with the pulse duration.

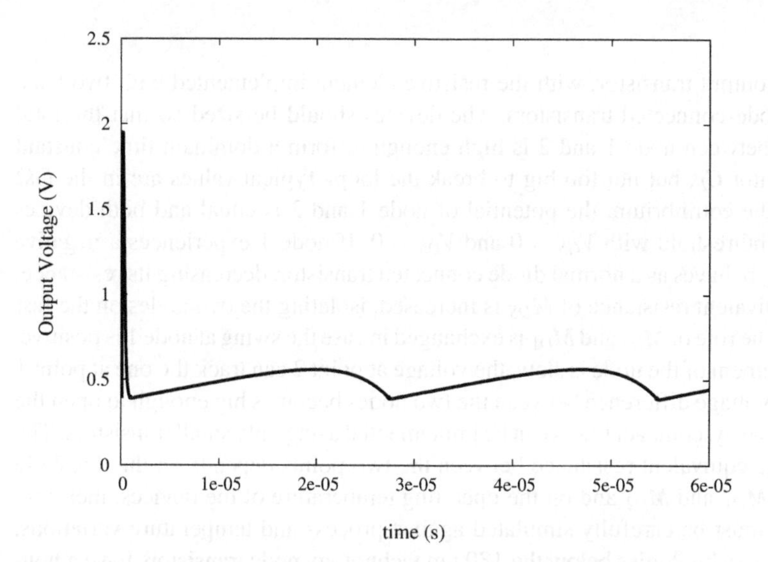

FIGURE 8.50 Response of an underdamped BLH loop simulated for an appropriate time scale.

the compensation more difficult. Therefore, the stability of these circuits must be checked for the maximum DC current they are expected to compensate. Alternative schemes of implementing baseline holding circuits are shown in Fig. 8.52 [68] and Fig. 8.53 [45]. In the circuit of Fig. 8.52 a non-linear low pass filter is introduced

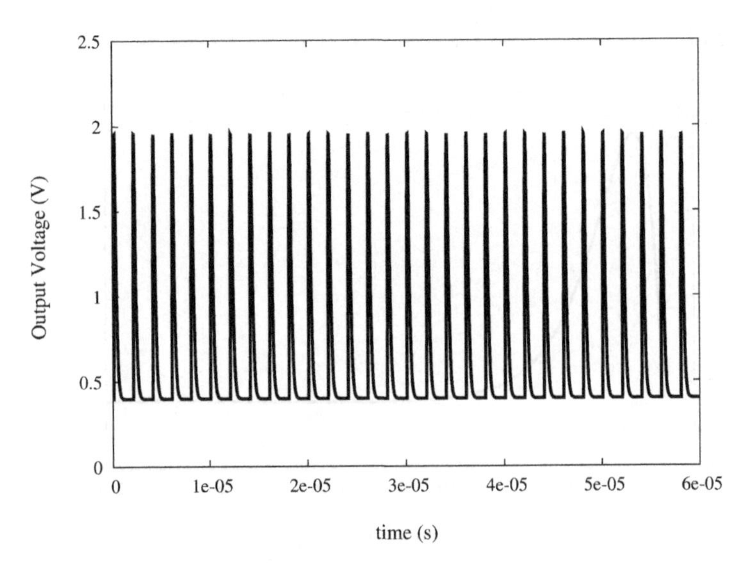

FIGURE 8.51 Response of the BLH of Fig. 8.46 to a train of pulses spaced by 2 μs. The loop is properly damped by starving the NMOS source follower.

before the output transistor, with the resistive element implemented with two back to back diode-connected transistors. The devices should be sized so that the total resistance between node 1 and 2 is high enough to form a dominant time constant with capacitor C_L, but not too big to break the loop. Typical values are in the GΩ range. At the equilibrium, the potential of node 1 and 2 is equal and both devices work in subthreshold with $V_{GS} = 0$ and $V_{DS} = 0$. If node 1 experiences a negative swing, M_{D2} behaves as a normal diode connected transistor, decreasing its resistance, but the equivalent resistance of M_{DS} is increased, isolating the two nodes on the fast transient. The role of M_{D2} and M_{D1} is exchanged in case the swing at node 1 is positive. If the movement of the node is slow, the voltage at point 2 can track the one at point 1 before the voltage difference between the two nodes becomes big enough to open the loop. In this way, compact filters can be implemented using only small transistors. The value of the equivalent resistance between the two points depends on the threshold voltage of M_{D1} and M_{D2} and on the operating temperature of the devices, therefore the circuit must be carefully simulated against process and temperature variations. However, in technologies below the 180 nm technology node transistors have a non-negligible conductance even at $V_{GS} = 0$ and the gate width and length of M_{D2} and M_{D1} can be chosen in a fairly broad range. In the circuits of Fig. 8.53 the slew rate limiting source follower is embedded in a differential cell which then works in closed loop configuration, thus providing high bandwidth. All the poles due to this stage are located at high frequencies and do not interfere with the loop stability. The high output impedance of the differential cascode amplifier coupled to the load capacitor provides the dominant low pass filtering time constant. Both circuits of Fig. 8.52 and

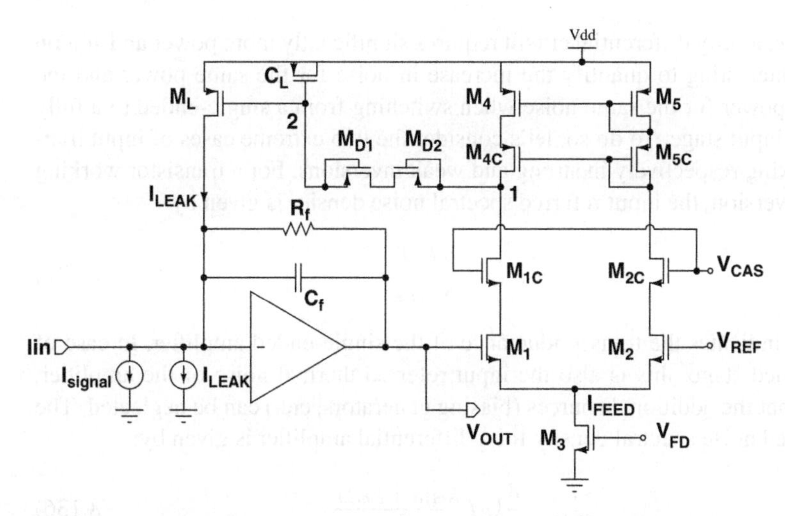

FIGURE 8.52 Baseline holder with back-to-back diode connected transistors to implement an equivalent resistor in the GΩ range.

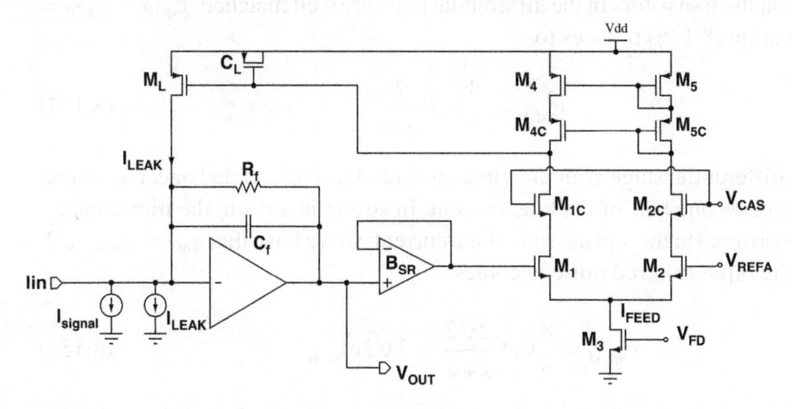

FIGURE 8.53 Baseline holder with a high speed, slew rate limited buffer. The voltage follower B_{SR} has a very small slew rate but a high bandwidth, so that it clips the fast signals without introducing a low-frequency pole. The active AC coupling is realized by the differential amplifier in combination with C_L and M_L.

Fig. 8.53 do not require the control of very small currents, as in both cases the bias current of the differential stage can be in the order of several hundreds nA.

8.7 FULLY DIFFERENTIAL FRONT-END

Fully differential topologies are the baseline choice for most signal processing applications as they allow for a better rejection of interference either generated on chip or off chip. In front-end for radiations sensors, fully differential architectures are not as popular as in other electronics domains. One reason is that, for the same noise performance, a fully differential circuit requires significantly more power and silicon area. It is interesting to quantify the increase in noise for the same power and the increase in power for the same noise when switching from a single-ended to a fully differential input stage. To do so, let's consider the two extreme cases of input transistors working respectively in strong and weak inversions. For a transistor working in strong inversion, the input referred spectral noise density is given by:

$$v_{nse,si}^2 = \frac{8}{3}\frac{k_B T}{g_{mse}} \tag{8.135}$$

where g_{mse} indicates the transconductance of the single-ended amplifier. In case of a single-ended stage, this is also the input referred thermal noise of the amplifier, providing that the additional sources (biasing generators, etc.) can be neglected. The input referred noise spectral density for a differential amplifier is given by:

$$v_{nd,si}^2 = \frac{8}{3}k_B T \frac{g_{m1d} + g_{m2d}}{g_{m1d}^2} \tag{8.136}$$

Assuming that the transistors in the differential pairs are well matched, $g_{m1d} = g_{m2d} = g_{md}$ and equation (8.136) reduces to

$$v_{nd,si}^2 = \frac{8}{3}k_B T \frac{2}{g_{md}} \tag{8.137}$$

If we bias a differential stage with the same current of a single-ended one, each input transistor receives one half of the total current. In strong inversion, the transconductance is proportional to the square-root of the current, so we have that $g_{md} = g_{mse}/\sqrt{2}$. Therefore, the input referred noise becomes:

$$v_{nd,si}^2 = \frac{8}{3}k_B T \frac{2\sqrt{2}}{g_{mse}} = 2\sqrt{2}v_{nse,si}^2 \tag{8.138}$$

We can thus conclude that for the same total power, the input-referred spectral noise density for a differential stage is $2\sqrt{2}$ times the one of the single-ended counterpart and the ENC is $\sqrt{2\sqrt{2}}$ times bigger. In weak inversion, the transconductance scales linearly with the current, hence we have:

$$v_{ns,wi}^2 = 2k_B T \frac{1}{g_{mse}} \tag{8.139}$$

For the differential pair, we can write:

$$v_{nd,wi}^2 = 2k_B T \frac{2}{g_{md}} = 2k_B T \frac{4}{g_{mse}} = 4v_{nse,wi}^2 \tag{8.140}$$

In other words, in weak inversion, a differential stage biased at the same total power of a single-ended one has a spectral noise density four times bigger and an ENC two times greater.

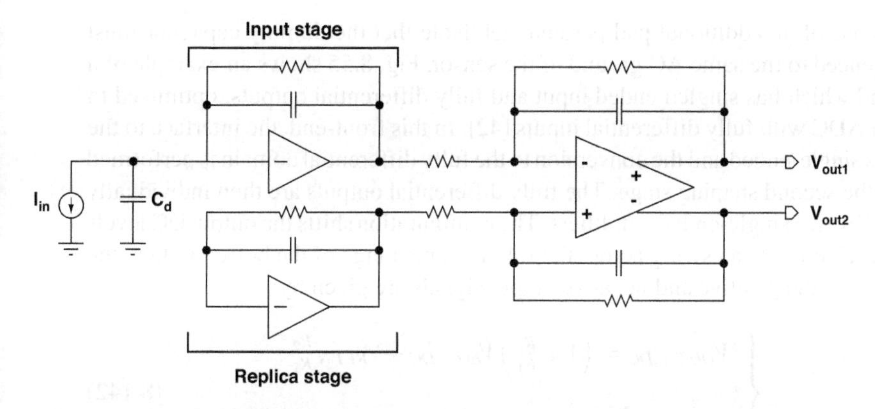

FIGURE 8.54 Fully differential shaper with replica bias to define a reference for the fully differential amplifier. The common mode feedback necessary in the second stage is not shown.

If we need to preserve in a differential stage the same noise of the single-ended counterpart, we must have identical spectral noise densities in both cases, which means:

$$g_{md} = 2g_{mse} \tag{8.141}$$

In strong inversion, the transconductance is proportional to the square-root of the current, hence we need a four times bigger current in the input transistor. However, a differential pair has two transistors, hence the tail current must be eight times the one of the single-ended stage. In weak inversion, the transconductance is proportional to the current, so we need a total current four times as big. In the differential pair, the transistors need also to be larger to have the same impact of the flicker noise, since both transistors equally contribute. When achieving minimum noise at lowest possible power is paramount, single-ended solutions are mandatory. When a higher noise can be accepted, a differential topology can be considered.

The second reason that discourages the use of fully differential front-ends is that multi-channel radiation sensors are normally single-ended devices, in which a reference terminal is common to several channels. Therefore the connection between the sensor and the front-end input stage is implemented with "pseudo-differential" schemes, as shown Fig. 8.54. In the figure, a replica of the input stage is used to provide a suitable reference voltage to the first differential stage in the chain, which makes a conversion of the signal from single-ended to fully differential. The input of the replica stage can be left floating. In this case, the impedance seen by the two input stages is very different and any noise affecting the detector ground is amplified as in a standard single-ended system. The replica bias thus serves as a mere reference generator. A second option is to add a capacitor between the input of the replica bias and ground. If the detector capacitance is small, this capacitor can be integrated on chip. In case the sensor capacitance is large, the input of the replica stage is connected as well to an input pad and the dummy capacitor becomes an external component. This option provides the opportunity for better symmetry between the two inputs at the expense of an additional pad per channel. Note that the dummy capacitor must be referenced to the same AC ground of the sensor. Fig. 8.55 shows an example of a front-end which has singled ended input and fully differential outputs, optimized to drive an ADC with fully differential inputs [42]. In this front-end, the interface to the sensor is single-ended and the conversion to the fully differential domain is performed only in the second shaping stage. The fully differential outputs are then individually buffered by two single-ended amplifiers. The configuration shifts the output DC levels so that each output can swing across the full dynamic range of the buffer. In fact, the DC levels of the positive and negative output signals are given by:

$$\begin{cases} V_{OUT+,DC} = \left(1+\frac{R_2}{R_1}\right) V_{out+,DC} - V_{REFN}\frac{R_2}{R_1} \\ V_{OUT-,DC} = \left(1+\frac{R_2}{R_1}\right) V_{out-,DC} + V_{REFP}\frac{R_2}{R_1} \end{cases} \tag{8.142}$$

To achieve the maximum output range, the DC levels of the positive and negative output are thus positioned close to the negative and positive rails. Fig. 8.56 shows the core amplifier employed in the fully differential stage, based on the folded cascode topology. Observe that the common mode voltage is sensed by a low-gain differential amplifier with diode connected loads, as most of the gain necessary to the common mode feedback loop is generated inside the cascode stage of the main amplifier.

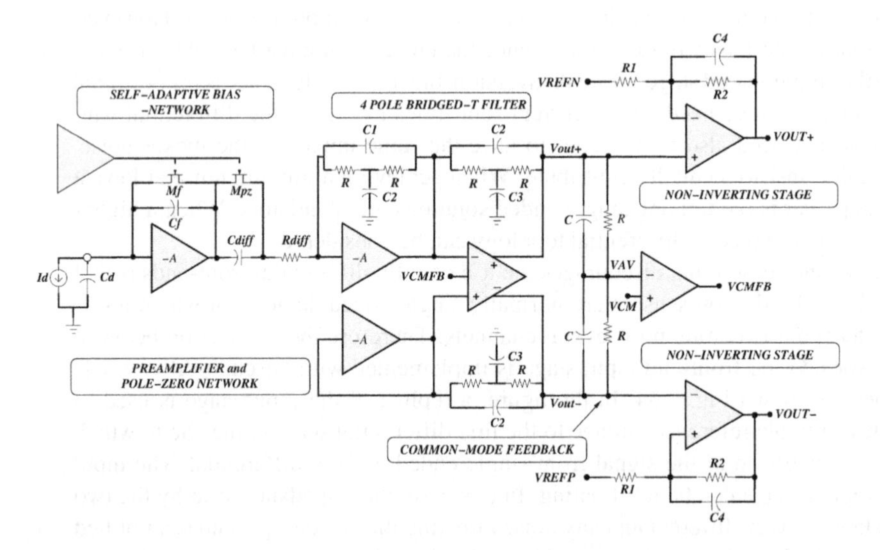

FIGURE 8.55 Front-end amplifier with fully differential output designed for the Time Projection Chamber (TPC) of the ALICE detector at CERN. From [42], reprinted with permission.

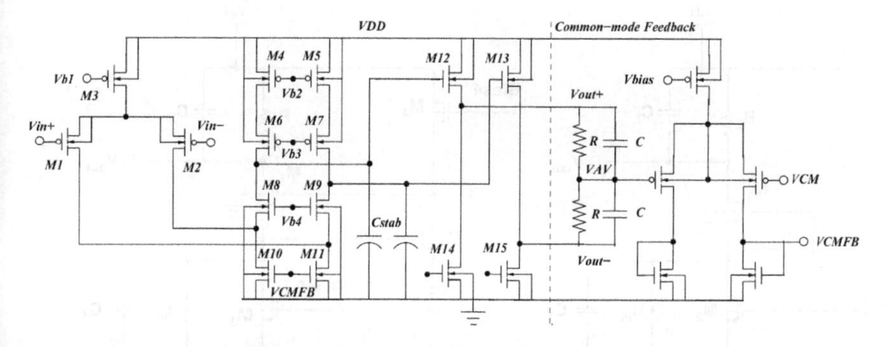

FIGURE 8.56 Detail of the fully differential folded cascode CMOS OTA used in the circuit of Fig. 8.55. From [42], reprinted with permission.

8.8 IMPLEMENTATION OF CURRENT-MODE FRONT-ENDS

Current-mode front-ends are particularly useful to perform fast readout of sensors that have a large capacitance. As we have studied in chapter 3 and chapter 4, the simplest current-mode configuration is the open-loop common gate amplifier, illustrated in Fig. 8.57 a). In this case, the input transistor acts as current buffer, providing low input impedance to the source and delivering to the load resistor R_L the same current on a higher output impedance. In chapter 4 we have also seen that, neglecting the bulk effect, the transfer function of the circuit can be approximated as:

$$T(s) = \frac{V_{out}(s)}{I_{in}(s)} = \frac{R_L}{\left(1+s\frac{C_T}{g_{m1}}\right)(1+sR_LC_L)} = \frac{R_L}{(1+s\tau_{in})(1+s\tau_L)} \tag{8.143}$$

where g_{m1} is the transconductance of the input transistor, C_T is the total input capacitance while C_L is the output capacitance. Depending on the application, C_L can just represent the unavoidable parasitic capacitance seen from the output node or it can be a physical component added on purpose to provide filtering functions. Equation (8.143) shows that the circuit has two real poles, one associated to the input node, with a time constant given by $\tau_{in} = C_T/g_{m1}$, and the other corresponding to the output node, whose time constant is $\tau_L = R_LC_L$. Before delving into the noise analysis of the common gate amplifier, we study the circuit response when it is driven by a voltage generator connected between the source of the input transistor and the drain of the biasing device. The corresponding scheme is reported in Fig. 8.58. The knowledge of the transfer function of this circuit is in fact necessary to compute the contribution of the series noise term. With reference to the small signal equivalent circuit shown

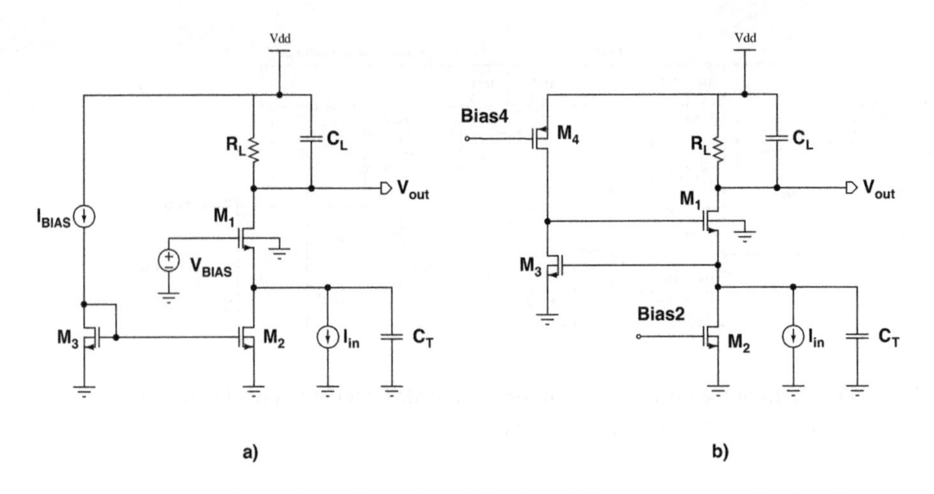

FIGURE 8.57 Common gate and regulated cascode amplifiers.

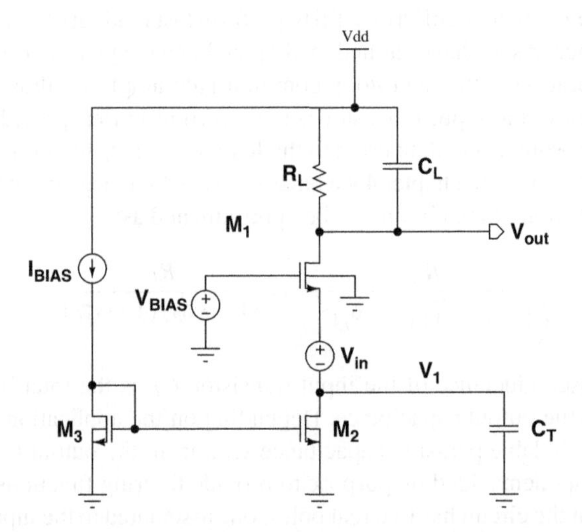

FIGURE 8.58 Common gate amplifier driven by a voltage signal inserted between the source of the input device and the drain of the biasing transistor.

in Fig. 8.59, we can write the following nodal equations:

$$\begin{cases} g_{m1}V_{gs1} + \frac{V_{out}}{R_L} + V_{out}sC_L = 0 \\ -g_{m1}V_{gs1} + \frac{V_1}{r_{02}} + V_1 sC_T = 0 \\ V_{gs1} = 0 - (V_{in} + V_1) \end{cases} \tag{8.144}$$

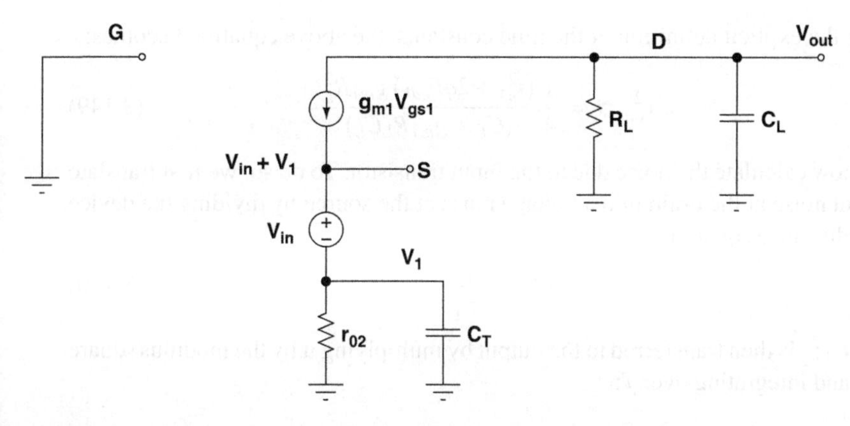

FIGURE 8.59 Small signal equivalent model to calculate the transfer function of the circuit of Fig. 8.58.

Replacing in the first two equations of the above set the explicit definition of V_{gs1} and solving for V_{out}, we find:

$$V_{out}\left(1+sC_LR_L\right) = \frac{g_{m1}R_L\left(1+sC_Tr_{02}\right)V_{in}}{\left(1+g_{m1}r_{02}+sC_Tr_{02}\right)} \tag{8.145}$$

Assuming that $g_{m1}r_{02} \gg 1$, the transfer function can be finally written as:

$$\frac{V_{out}(s)}{V_{in}(s)} = T_v(s) = \frac{R_L\left(1+s\tau_z\right)}{r_{02}\left(1+s\tau_{in}\right)\left(1+s\tau_L\right)} \tag{8.146}$$

where we have put $\tau_z = r_{02}C_T$, $\tau_{in} = C_T/g_{m1}$ and $\tau_L = R_LC_L$. The result can be interpreted as follows: the voltage V_{in} first originates a current through the impedance stemming from the parallel combination of the input capacitance C_T and the output resistance of the current source, r_{02}. This input current is then processed by $T(s)$.

The simple common gate amplifier has three major noise sources, due to the input transistor, the bias transistor and the load resistor. Current-mode circuits are suitable for short peaking time and thermal noise is expected to be dominating, hence we do not consider in the following the impact of $1/f$ noise. The noise generated by the current source, M_2 is given by $4k_BT\gamma g_{m2}$ and is processed by the same transfer function as the signal. To this thermal noise source, we should also add the shot noise generated by the sensor leakage current. As usual, the noise spectral density is multiplied by the square of the transfer function modulus and integrated over the full frequency spectrum, thus we have:

$$< v_{out}^2 >_i = \left(i_{n2}^2 + 2qI_{leak}\right)R_L^2\int_0^\infty \left|\frac{1}{\left(1+jf/f_{in}\right)\left(1+jf/f_L\right)}\right|^2 df \tag{8.147}$$

Solving the integral, we obtain:

$$< v_{out}^2 >_i = \frac{1}{4}\left(i_{n2}^2 + 2qI_{leak}\right)\frac{R_L^2}{\tau_{in}+\tau_L} \tag{8.148}$$

Inserting the explicit definition of the time constants, the above equation becomes:

$$< v_{out}^2 >_i = \frac{1}{4}\frac{\left(i_{n2}^2 + 2qI_{leak}\right) g_{m1}R_L^2}{\left(C_T + g_{m1}R_LC_L\right)} \tag{8.149}$$

We can now calculate the noise due to the input transistor. To do so, we first translate its current noise at the drain into a voltage noise at the source by dividing the device transconductance squared:

$$v_{n1}^2 = \frac{i_{n1}^2}{g_{m1}^2} \tag{8.150}$$

The noise v_{n1}^2 is then transferred to the output by multiplying it by the modulus square of $T_v(s)$ and integrating over f:

$$< v_{out}^2 >_{w1} = \frac{i_{n1}^2 R_L^2}{g_{m1}^2}\int_0^\infty \left|\frac{1+jf/f_z}{(1+jf/f_{in})(1+jf/f_L)}\right|^2 df \tag{8.151}$$

In this case, the solution of the integral yields:

$$< v_{out}^2 >_{w1} = \frac{1}{4} i_{n1}^2 \frac{R_L^2}{g_{m1}^2 r_{02}^2}\frac{1}{\tau_{in}+\tau_L}\left(1+\frac{\tau_z^2}{\tau_{in}\tau_L}\right) \tag{8.152}$$

If the condition:

$$\frac{\tau_z^2}{\tau_{in}\tau_L} \gg 1 \tag{8.153}$$

is verified, equation (8.152) assumes the following form:

$$< v_{out}^2 >_{w1} = \frac{1}{4}\frac{i_{n1}^2 R_L C_T}{C_L\left(C_T + g_{m1}R_LC_L\right)} \tag{8.154}$$

The current noise spectral density of M_1 can be written as $i_{n1}^2 = 4k_BT\gamma g_{m1}$. If we further suppose that $g_{m1}R_LC_L \gg C_T$ and insert the explicit expression of i_{n1}^2 into (8.154), we get:

$$< v_{out}^2 >_{w1} = \frac{k_BT\gamma C_T}{C_L^2} \tag{8.155}$$

Interestingly, the effect of the noise due to M_1 depends only on the ratio between the input and the load capacitance and the inversion coefficient of the device, but not on its transconductance. Finally, the load resistor generates a current noise given by $i_{nL}^2 = 4k_BT/R_L$ and its contribution to the voltage node is given by:

$$< v_{out}^2 >_{wL} = 4k_BTR_L\int_0^\infty \left|\frac{R_L^2}{1+jf/f_L}\right|^2 df = \frac{k_BT}{C_L} \tag{8.156}$$

The total output noise is finally obtained by summing up the three contributions. In case $\tau_{in} \neq \tau_L$, the peak output voltage due to a charge of one electron delivered in the form of a Dirac-delta is:

$$V_{q,out} = \frac{q}{C_L}\left(\frac{\tau_L}{\tau_{in}}\right)^{\frac{\tau_{in}}{\tau_{in}-\tau_L}} \tag{8.157}$$

whereas if $\tau_{in} = \tau_L$, we have:

$$V_{q,out} = \frac{q}{C_L} \frac{1}{e} \tag{8.158}$$

Dividing the output voltage noise by $V_{q,out}$ one obtains the ENC noise in rms electrons. In simple common gate amplifiers, the noise contribution of the biasing current source can be significant. The g_m-boosting technique, shown in Fig. 8.57 b), can thus be used to have the same transconductance by lowering the current in the input transistor. In this way, also the current in M_2 is reduced, thus minimizing its impact on the overall noise. In alternative, the boosting technique can be used to obtain a larger equivalent transconductance and hence a lower input impedance for the same bias current. In the following, we neglect for simplicity the bandwidth limitation of the boosting amplifier and we suppose that it provides a gain A constant over frequency. In this approximation, the equivalent transconductance of g_{m1} becomes Ag_{m1}. The transfer functions of the circuit $T(s)$ and $T_v(s)$ can still be approximated by equation (8.143) and (8.146), with τ_{in} now defined as:

$$\tau_{in} = \frac{C_T}{g_{m1}A} \tag{8.159}$$

The noise contribution of the input transistor hence becomes:

$$< v_{out}^2 >_{w1} = \frac{1}{4} i_{n1}^2 \frac{R_L^2}{A^2 g_{m1}^2 r_{02}^2} \frac{1}{\tau_{in} + \tau_L} \left(1 + \frac{\tau_z^2}{\tau_{in}\tau_L}\right) \tag{8.160}$$

In addition to the previous noise sources, we must consider also the noise introduced by the auxiliary common-source boosting amplifier. The noise at the drain of T_3 is i_{n3}^2 and it can be transformed into a voltage source in series with the gate of value $v_{n3,in}^2 = i_{n3}^2 / g_{m3}^2$. This noise is the transferred to the output by $T_v(s)$, so we have:

$$< v_{out}^2 >_{w3} = \frac{1}{4} i_{n3}^2 \frac{R_L^2}{g_{m3}^2 r_{02}^2} \frac{1}{\tau_{in} + \tau_L} \left(1 + \frac{\tau_z^2}{\tau_{in}\tau_L}\right) \tag{8.161}$$

The loop is designed so that $Ag_{m1} \gg g_{m3}$, hence it is M_3 which now gives the dominant noise contribution. In principle, one should also consider the noise of the load transistor M_4, that is optimized with the same rules followed for the design of standard common source amplifiers. If the input DC voltage has to be programmable, the common source boosting amplifier is replaced by a differential amplifier, as shown in Fig. 8.60. In this case, the series noise is dominated by the transistors in the differential pair. As anticipated in chapter 3, a common gate amplifier is often loaded with a diode connected transistor, which forms the input stage of a current mirror. In this way, the input current can be fanned-out to multiple destinations with different gains, as shown in Fig. 8.61. The gain can easily be made programmable by having more transistors in the output branches that can be inserted in or excluded from the signal path by switches. Note that in this case, dummy transistors must be used in order to have a good match between the reference and the output branches. In some cases, the output signal is fed to a current-mode discriminator or converted back to a voltage

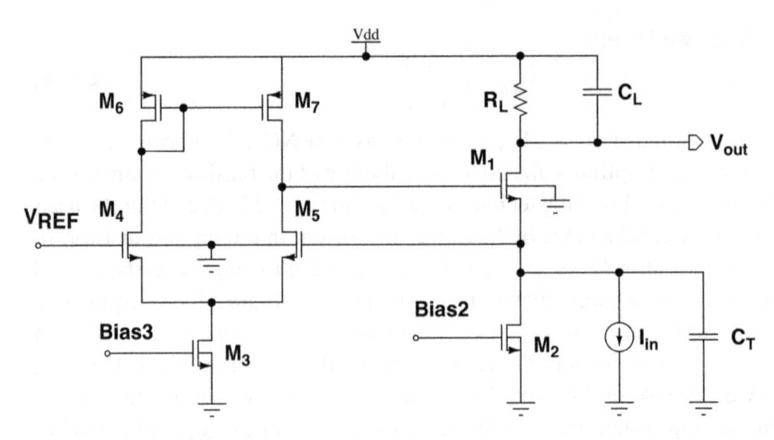

FIGURE 8.60 The common source gain boosting amplifier can be replaced with a differential cell. In this case, the input DC voltage is locked to the reference voltage V_{REF}.

through a simple terminating resistor connected between drain of the output transistor and ground. If more elaborated signal processing is necessary, standard shaping amplifiers can be employed. A detailed treatment of the common gate and regulated cascode configurations can be found in [69]. Examples of current-mode front ends are provided by [70–76].

Finally, it must be observed current-mode front-ends can also be designed with fully differential architectures [73–77]. Fig. 8.62 shows a possible implementation, based on regulated cascode in which the regulation loop is implemented with a fully differential amplifier. In this case, the DC input levels are defined by the common mode feedback circuit of the differential amplifier, not shown in the figure. If both sensor terminals are accessible, a true fully differential readout can be performed

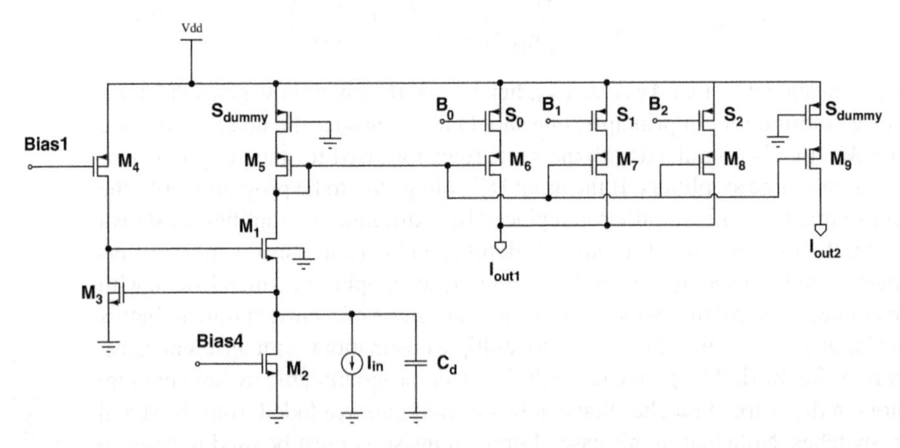

FIGURE 8.61 Regulated cascode amplifier with multiple output branches and programmable output current.

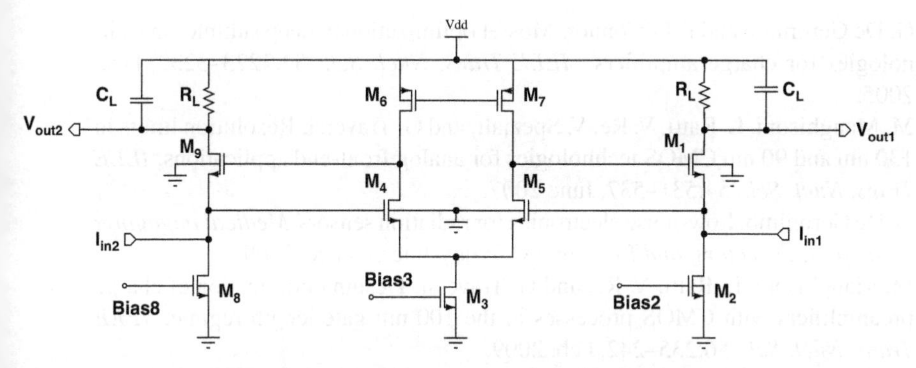

FIGURE 8.62 Example of fully differential current-mode front-end. The common mode feedback network is not shown in the figure.

by connecting the sensor across the amplifier input ports, as proposed in [78] for the readout of individual silicon photomultipliers pixels. The drawback is that the number of input connections is doubled. More considerations on the single-ended versus differential connection schemes in capacitive sensors can be found in [79].

REFERENCES

1. W. M. C. Sansen and Z. Y. Chang. Limits of low noise performance of detector readout front ends in CMOS technology. *IEEE Trans. Circ. Syst.*, 37:1375–1382, 1990.
2. P. O'Connor and G. De Geronimo. Prospects for charge sensitive amplifiers in scaled CMOS. *Nucl. Instr. Meth.*, A480:713–725, 2002.
3. C. C. Enz, F. Krummenacher, and E. A. Vittoz. An analytical MOS transistor model valid in all regions of operation and dedicated to low-voltage and low-current applications. *Analog Integrated Circuits and Signal Processing*, 8:83–114, 1995.
4. C. C. Enz. A short story of the EKV MOS transistor model. *IEEE Solid-state Circuits Newsletter*, 13:24–30, 2008.
5. D. M. Binkley. *Tradeoffs and Optimization in Analog CMOS Design.* Wiley, 2008.
6. P. Grybos, M. Idzik, and P. Maj. Noise optimization of charge amplifiers with MOS input transistors operating in moderate inversion region for short peaking times. *IEEE Trans. Nucl. Sci.*, 54:555–560, 2007.
7. M. A. Karagounis. *Analog Integrated CMOS Circuits for the Readout and Powering of Highly Segmented Detectors in Particle Physics Applications.* PhD thesis, Fernuniversität in Hagen, 2010.
8. S. More. *Aging Degradation and Countermeasures in Deep-submicrometer Analog and Mixed Signal Integrated Circuits.* PhD thesis, Lehrstuhl für Technische Elektronik der Technischen Universität München, 2012.

9. G. De Geronimo and P. O'Connor. Mosfet optimization in deep submicron technologies for charge amplifiers. *IEEE Trans. Nucl. Sci.*, 52:3223–3232, Dec. 2005.
10. M. Manghisoni, L. Ratti, V. Re, V. Speziali, and G. Traversi. Resolution limits in 130 nm and 90 nm CMOS technologies for analog front-end applications. *IEEE Trans. Nucl. Sci.*, 54:531–537, June 2007.
11. G. De Geronimo. Low-noise electronics for radiation sensors. *Medical Imgaging: Principles, Detectors, and Electronics.*, Wiley-Interscience, 2009.
12. M. Manghisoni, L. Ratti, V. Re, and G. Traversi. Design optimization of charge preamplifiers with CMOS processes in the 100 nm gate length regime. *IEEE Trans. Nucl. Sci.*, 56:235–242, Feb. 2009.
13. E. Alvarez, D. Avila, H. Campillo, A. Dragone, and A. Abusleme. Noise in charge amplifiers - A g_m/I_D design approach. *IEEE Trans. Nucl. Sci.*, 59:2457–2462, Oct. 2012.
14. H. Spieler. *Semiconductor Detector Systems.* Oxford Science Publications, 2005.
15. J. Kaplon and M. Noy. Front end electronics for silicon strip detectors in 90nm CMOS technology: advantages and challenges. *Journal of Instrumentation*, 5, 2010.
16. M. Manghisoni, L. Ratti, V. Re, V. Speziali, and G. Traversi. Survey of noise performances and scaling effects in deep submicrometer CMOS devices from different foundries. *IEEE Trans. Nucl. Sci.*, 54:531–537, June 2007.
17. P. E. Allen and D. R. Holberg. *CMOS Analog Circuit Design*. 2nd edition, Oxford University Press, 2002.
18. J. S. Dunn et al. Foundation of RF CMOS and SiGe BiCMOS technologies. *IBM J. Res. & Dev.*, 47:101–138, March/May 2003.
19. N. Randazzo, G. V. Russo, D. Lo Presti, S. Panebianco, C. Petta, and S. Reito. A four-channel, low-power CMOS charge preamplifier for silicon detectors with medium value of capacitance. *IEEE Trans. Nucl. Sci.*, 44:31–35, 1997.
20. M. Idzik, S. Kulis, and D. Przyborowski. Development of front-end electronics for the luminosity detector at ILC. *Nucl. Instr. Meth.*, A608:169–174, 2009.
21. G. Gramegna, P. O'Connor, P. Rehak, and S. Hart. CMOS preamplifier for low-capacitance detectors. *Nucl. Instr. Meth.*, A390:241–250, 1997.
22. G. Gramegna, P. O'Connor, P. Rehak, and S. Hart. Low-noise CMOS preamplifier-shaper for silicon drift detectors. *IEEE Trans. Nucl. Sci.*, 44:385–388, 1997.
23. K. T. Z. Oo, E. Mandelli, and W. W. Moses. A high-speed low-noise 16 channel CSA with automatic leakage compensation in 0.35 μm CMOS process for APD-based PET detectors. *IEEE Trans. Nucl. Sci.*, 54:444–453, 2007.
24. P. Grybos and R. Szczygiel. Pole-Zero cancellation circuit with pulse pile-up tracking system for low noise charge sensitive amplifiers. *IEEE Trans. Nucl. Sci.*, 55:583–590, 2008.
25. P. O'Connor, G. Gramegna, P. Rehak, F Corsi, and C. Marzocca. Ultra low noise CMOS preamplifier-shaper for X-ray spectroscopy. *Nucl. Instr. Meth.*, A409:315–321, 1998.

26. G. De Geronimo and P. O'Connor. A CMOS detector leakage current self-adaptable continuous reset system: Theoretical analysis. *Nucl. Instr. Meth.*, A421:322–333, 1999.
27. G. De Geronimo et al. Front-end ASIC for a GEM based time projection chamber. *IEEE Trans. Nucl. Sci.*, 51:1312–1317, 2004.
28. A. Veeravalli, E. Sanchez-Sinencio, and J. Silva-Martinez. Transconductance amplifier structures with very small transconductances: a comparative design approach. *IEEE J. Solid-state Circ.*, 37:770–775, 2002.
29. I. Pachnis, A. Demosthenous, and N. Donaldson. Comparison of transconductance reduction techniques for the design of a very large time-constant CMOS integrator. In *ICES 2006*, 2006.
30. F. Krummenacher. Pixel detectors with local intelligence: an IC designer point of view. *Nucl. Instr. Meth.*, A305:527–532, 1991.
31. L. Blanquart, A. Mekkaoui, V. Bonzom, and P. Delpierre. Pixel analog cells prototypes for ATLAS in DMILL technology. *Nucl. Instr. Meth.*, A395:313–317, 1997.
32. L. Blanquart et al. Analog front-end cell designed in a commercial 0.25 μm process for the ATLAS pixel detector at LHC. *IEEE Trans. Nucl. Sci.*, 49:1778–1782, 2002.
33. I. Peric et al. The FEI3 readout chip for the ATLAS pixel detector. *Nucl. Instr. Meth.*, A565:178–187, 2006.
34. T. Zimmermann. The Nova APD readout chip. *Sixth International Meeting on Front-End Electronics*, Perugia, May 2006.
35. L. L. Jones et al. The APV25 deep submicron readout chip for CMS detectors. *Proceedings of the 5th Workshop on Electronics for LHC Experiments, Snowmass, Sept. 1999*, CERN/LHCC/1999/009:162–166, 1999.
36. O. Toker, S. Masciocchi, E. Nygård, A. Rudge, and P. Weilhammer. VIKING, a CMOS low noise monolithic 128 channel frontend for Si-strip detector readout. *Nucl. Instr. Meth.*, A340:572–579, 1994.
37. M. J. French et al. Design and results from the APV25, a deep sub-micron CMOS front-end chip for the CMS tracker. *Nucl. Instr. Meth.*, A466:359–365, 2001.
38. F. F. Khalid, L. L. Jones, R. Stephenson, and J. D. Lipp. A programmable analogue front-end ASIC for gas microstrip detectors having a wide range of input capacitance. In *IEEE NSS-MIC Conference Records*, pages 860–864, 2006.
39. N. Ollivier-Henry et al. A front-end readout mixed chip for high-efficiency small animal PET imaging. *Nucl. Instr. Meth.*, A571:312–316, 2007.
40. S Löchner. *Development, Optimisation and Characterisation of a Radiation Hard Mixed-Signal Readout Chip for LHCb.* PhD thesis, Ruperto-Carola University of Heidelberg, 2006.
41. P. Wieczorek and H. Flemming. Low noise preamplifier ASIC for the PANDA EMC. In *IEEE NSS-MIC Conference Records*, pages 1319–1322, 2010.
42. H.K. Soltveit et al. The preamplifier shaper for the ALICE TPC detector. *Nucl. Instr. Meth.*, A676:106–119, 2012.

43. J. Hijsing. *Operational Amplifiers: Theory and Design*. 2nd edition, Springer, 2011.
44. J.-F. Pratte, C. M. Pepin, D. Rouleau, O. Menard, J. Mouine, and R. Lecomte. Design of a fast-shaping amplifier for PET/CT APD detectors with depth-of-interaction. *IEEE Trans. Nucl. Sci.*, 49:2448–2454, 2002.
45. M. Chiosso, O. Cobanoglu, G. Mazza, D. Panzieri, and A. Rivetti. A fast binary front-end ASIC for the RICH detector of the COMPASS experiment at CERN. In *IEEE NSS-MIC Conference Records*, pages 1495–1500, 2008.
46. A. Kandasamy, E. O'Brien, P. O'Connor, and W. von Achen. A monolithic preamplifier-shaper for measurement of energy loss and transition radiation. *IEEE Trans. Nucl. Sci.*, 46:150–155, 1999.
47. K. Barish et al. Front-end electronics for PHENIX time expansion chamber. *IEEE Trans. Nucl. Sci.*, 49:1141–1146, 2002.
48. G. De Geronimo, P. O'Connor, and J. Grosholz. A generation of CMOS readout ASICs for CZT detectors. *IEEE Trans. Nucl. Sci.*, 47:1857–1867, 2000.
49. G. De Geronimo and S. Li. Shaper design in CMOS for high dynamic range. *IEEE Trans. Nucl. Sci.*, 58:2382–2390, 2011.
50. G. Bertuccio, P. Gallina, and M. Sampietro. 'R-lens filter': An (RC)n current-mode low pass filter. *Electronics Letters*, 35, 1999.
51. S. Buzzetti, C. Guazzoni, and A. Longoni. Theoretical analysis and experimental characterization of a novel VLSI current-mode shaping cell for high-resolution spectroscopy. *IEEE Trans. Nucl. Sci.*, 51:1343–1348, 2004.
52. S. Buzzetti, C. Guazzoni, and A. Longoni. Multichannel current-mode spectroscopy amplifier in BiCMOS technology with selectable shaping time. *IEEE Trans. Nucl. Sci.*, 52:1617–1623, 2005.
53. J. L. Britton et al. PATARA: solid-state neutron detector readout electronics with pole-zero and complex shaping and gated baseline restorer for the SNS. In *IEEE NSS-MIC Conference Records*, pages 27–31, 2006.
54. C. Fiorini and M. Porro. Integrated RC cell for time-invariant shaping amplifiers. *IEEE Trans. Nucl. Sci.*, 51:1953–1960, 2004.
55. C. Fiorini and M. Porro. DRAGO chip: a low-noise CMOS preamplifier shaper for silicon detectors with integrated front-end JFET. *IEEE Trans. Nucl. Sci.*, 52:1647–1653, 2005.
56. X. Yun, M. Stanacevic, and S. Luryi. Low-power amplifier for readout interface of semiconductor scintillator. *IEEE Trans. Nucl. Sci.*, 58:2129–2136, 2011.
57. R. L. Geiger and E. Sánchez-Sinencio. Active filter design using operational transconductance amplifiers: a tutorial. *IEEE Circuits and Devices Magazine*, 1:20–31, 1985.
58. E. Sanchez-Sinencio and J. Silva-Martinez. CMOS transconductance amplifiers, architectures and active filters: a tutorial. *IEEE Proceedings-Circuits Devices and Systems*, 147, 2000.
59. T. Noulis, C. Deradonis, S. Siskos, and G. Sarrabayrouse. Detailed study of particle detectors OTA-based CMOS semi-Gaussian shapers. *Nucl. Instr. Meth.*, A583:469–478, 2007.

60. T. Noulis, C. Deradonis, and S. Siskos. Design guidelines and comparison of detector readout front end integrated SG shapers using transconductance circuits. *International Journal of Electronics*, 94:943–959, 2007.
61. T. Noulis, C. Deradonis, S. Siskos, and G. Sarrabayrouse. Particle detector tunable monolithic semi-gaussian shaping filter based on transconductance amplifiers. *Nucl. Instr. Meth.*, A589:330–337, 2008.
62. B. Krieger, I. Kipnis, and B. A. Ludewigt. XPS: a multi-channel preamplifier-shaper IC for X-ray spectroscopy. *IEEE Trans. Nucl. Sci.*, 45:732–734, 1998.
63. T. Noulis, S. Siskos, G. Sarrabayrouse, and L. Bary. Advanced low-noise X-ray readout ASIC for radiation sensor interfaces. *IEEE Trans. Circ. and Syst. I*, 55:1854–1862, 2008.
64. J. J. F. Rijns. CMOS low-distortion high-frequency variable-gain amplifier. *IEEE J. Solid-state Circ.*, 31:1029–1034, 1996.
65. J. Kaplon and W. Dabrowski. Fast CMOS binary front end for silicon strip detectors at LHC experiments. *IEEE Trans. Nucl. Sci.*, 52:2713–2720, 2005.
66. E. Martin et al. The 5ns peaking time transimpedance front end amplifier for the silicon pixel detector in the NA62 gigatracker. In *IEEE NSS-MIC Conference Records*, pages 381–388, 2009.
67. P. O'Connor, G. De Geronimo, and C. Grosholtz. A CMOS baseline holder (BLH) for readout ASICs. *IEEE Trans. Nucl. Sci.*, 47:818–822, June 2000.
68. T. Kugathasan, G. Mazza, A. Rivetti, and L. Toscano. A 15 μW 12-bit dynamic range charge measuring front-end in 0.13 μm CMOS. In *IEEE NSS-MIC Conference Records*, pages 1667–1673, 2010.
69. L. B. Oliveira, C. M. Leitao, and M. Medeiros Silva. Noise performance of regulated cascode transimpedance amplifiers for radiation detectors. *IEEE Trans. Circ. Sys.-I*, 59:1841–1848, 2012.
70. F. Anghinolfi, P. Aspell, M. Campbell, E. H. M. Heijne, P. Jarron, G. Meddeler, and J. C. Santiard. ICON, a current-mode preamplifier in CMOS technology for use with high rate particle detectors. *IEEE Trans. Nucl. Sci.*, 40:271–274, 1993.
71. D. Moraes et al. CARIOCA: CMOS fast binary front-end for sensor interface using a novel current-mode feedback technique. In *IEEE International Symposium on Circuits and Systems*, pages 360–363, 2001.
72. F. Corsi, M. Foresta, C. Marzocca, G. Matarrese, and A. Del Guerra. CMOS analog front-end channel for silicon photo-multipliers. *Nucl. Instr. Meth.*, A617:319–320, 2010.
73. F. Anghinolfi et al. NINO: an ultra-fast and low-power front-end amplifier/discriminator ASIC designed for the multigap resistive plate chamber. *Nucl. Instr. Meth.*, A533:183–187, 2004.
74. N. Ollivier-Henry et al. Design and characteristics of a multichannel front-end ASIC using current-mode CSA for small-animal PET imaging . *IEEE Trans. Biomed. Cir. and Sys.*, 5.
75. P. Carniti, M. De Matteis, A. Giachero, C. Gotti, M. Mainoa, and G. Pessina. CLARO-CMOS, a very low power ASIC for fast photon counting with pixellated photodetectors. *JINST*, pages 1–24, 2012.

76. D. Gascon et al. Low noise front end ASIC with current-mode active cooled termination for the upgrade of the LHCb calorimeter. *IEEE Trans. Nucl. Sci.*, 59:2471–2478, 2012.
77. Y. Dong and K. W. Martin. A high-speed fully-integrated POF receiver with large-area photo detectors in 65 nm CMOS. *IEEE J. Solid-state Circ.*, 47:2080–2092, 2012.
78. F. Powolny et al. Time based readout of a silicon photomultiplier (SiPM) for time of flight positron emission tomography (TOF-PET). *IEEE Trans. Nucl. Sci.*, 58:1212–1219, 2009.
79. M. D. Rolo. *Integrated Circuit Design for Time-of-Flight PET*. PhD thesis, University of Turin, 2014.

9 Discriminators

In the field of radiation detectors, current and voltage comparators are often called discriminators to emphasize their function of separating the signal from the unwanted background. Fast and high accuracy comparators can be built using positive feedback circuits, periodically enabled and reset by a clock. These synchronous comparators are systematically employed in the implementation of Analog to Digital Converters (ADC) and in many other analog systems and are the baseline choice for high performance design. The literature on comparators is thus mostly focused on clocked topologies.

In applications involving radiation sensors, the arrival time of the event is often unknown. Employing a synchronous discriminator would hence require sampling the output of the front-end amplifier at a fairly high speed, increasing the power consumption and demanding the routing of high speed digital control lines in close proximity of very sensitive analog blocks. Even in those applications in which particles arrive at known times, such as in colliding beam experiments, the simultaneous triggering of many digital cells can become an issue, especially in chips with many channels. For these reasons, asynchronous topologies are in general preferred. The picture may change with ultra-deep submicron technologies (65 nm and beyond), in which the supply voltage is reduced to 1 V or less and the parasitic capacitance becomes very small, allowing fast switching at very low power. The discriminator plays a particular role in those applications requiring the precise tagging of the particle arrival time. Special techniques have consequently been developed for these high performance "timing discriminators".

In the first part of the chapter, the basic features of comparators are discussed and a few general purpose topologies are presented. In the second part, the key factors affecting the accuracy of a timing measurement are reviewed and the design of timing discriminators is treated. High speed, positive feedback comparators are then discussed. At the end of the chapter, offset and offset compensation techniques are finally reviewed.

9.1 BASIC DISCRIMINATOR PROPERTIES

The discriminator task is to generate a logic signal that distinguishes if the front-end output is below or above a given threshold. This information is then passed to a digital control machine that takes the appropriate decisions on the following signal processing and/or data transmission steps. The discriminator output has therefore two possible states, one corresponding to the logic level "zero" (0 V in a single supply system) and the other to the logic level "one" (in general equal to V_{DD}), and switches instantaneously between the two states when the threshold is crossed. This is equivalent to asserting that the ideal comparator has infinite gain and bandwidth,

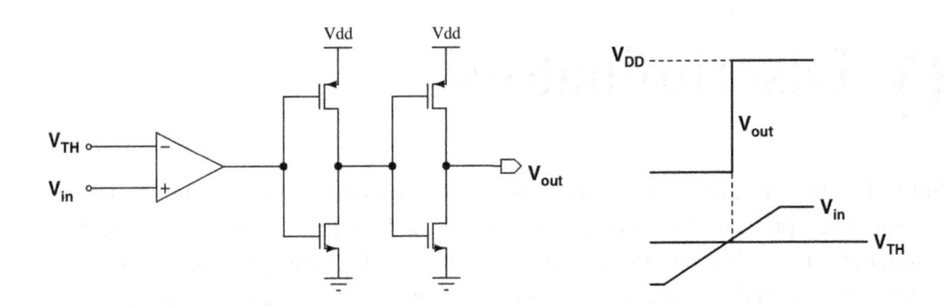

FIGURE 9.1 Conceptual scheme of a voltage discriminator with its idealized response.

which is of course an abstraction. The effects of gain and speed limitations in practical circuits must thus be considered.

9.1.1 DISCRIMINATOR GAIN AND SPEED REQUIREMENTS

As shown in Fig. 9.1, a discriminator basically consists of a high gain differential voltage amplifier followed by a chain of CMOS digital inverters. The output gates serve two purposes: on the one hand they ensure that the comparator delivers to the following digital stage a well defined logic signal. On the other, they provide adequate buffering power to drive the output load. The number of inverters shown in the figure is only indicative and the output stage needs to be optimize according to the load the circuit is expected to drive. In case the comparator output needs to be fanned-out to multiple destinations, a buffer with more stages than that shown in Fig. 9.1 can be required. If the output load is small, oversizing the buffer is unnecessary and leads to additional power consumption and to a higher risk of coupling digital noise into the front-end. CMOS logic gates are described in detail in textbooks on digital integrated circuits, such as [1]. For the purpose of our discussion, it is sufficient to recall that the inverter output flips when the input crosses a built-in threshold which depends on the relative sizing of the NMOS and PMOS transistor and it is typically set to $V_{DD}/2$. The inverter interprets safely as a "zero" any DC level which is no higher than $0.2 \cdot V_{DD}$ and as a "one" any level which is greater than $0.8 \cdot V_{DD}$. When $V_{in} \ll V_{TH}$ the output of the differential stage is in proximity of the negative rail and produces a voltage V_{OL} that, from the above discussion, must be smaller than $0.2 \cdot V_{DD}$. When $V_{in} \gg V_{TH}$, the differential stage output is close to V_{DD} and must deliver an output voltage V_{OH} greater than $0.8 \cdot V_{DD}$. Delivering already clear logic signals to the first inverter is necessary because otherwise the gate might flip on noise, introducing spurious events. Furthermore, both the PMOS and the NMOS transistor could be simultaneous in conduction, increasing significantly the static power consumption of the circuit. The difference $V_{OH} - V_{OL} = \Delta V_{O,HL}$ divided by the DC gain A_{v0} of the differential amplifier yields the minimum voltage difference ΔV_{in} that must be applied across the comparator inputs to determine a flip in the output voltage, thus defining the resolution. However, additional factors limit the circuit ability to detect small signals in practical applications. A comparator should in fact provide an answer within acceptable times, hence also speed considerations play a critical role in its design.

The speed performance of a discriminator are constrained by the small signal bandwidth of the circuit and by its slew-rate. To study the former, we assume that the response of the differential amplifier can be modeled with a single pole transfer function:

$$A_v(s) = \frac{A_{v0}}{(1+s\tau_p)} \tag{9.1}$$

where τ_p is the time constant associated to the pole. If a rectangular voltage step of amplitude V_{in} is applied to the input, the output voltage is given by:

$$V_{out} = V_{in}A_{v0}\left(1-e^{-\frac{t}{\tau_p}}\right) \tag{9.2}$$

Due to the bandwidth limitation, the output does not flip instantaneously, but it will change following (9.2). In order to make the inverter flip, the output must swing by at least $\Delta V_{O,HL}/2$, crossing the inverter threshold point. Let's now normalize the input signal on the minimum difference that the circuit is able to detect. From the above discussion we have:

$$V_{in,min} = \frac{\Delta V_{O,HL}}{A_{v0}} \tag{9.3}$$

We can also write:

$$\frac{V_{in}}{V_{in,min}} = \alpha \rightarrow V_{in} = \alpha\frac{\Delta V_{O,HL}}{A_{v0}} \tag{9.4}$$

By replacing the above expression of V_{in} in (9.2) and requiring that the output signal reaches the minimum value to make the comparator flip, equal to $\Delta V_{OHL}/2$, we obtain:

$$\frac{1}{2} = \alpha\left(1-e^{-\frac{t_d}{\tau_p}}\right) \tag{9.5}$$

In the above equation, t_d is the time the output of the comparator needs to move by $\Delta V_{O,HL}/2$ due to its limited bandwidth and is a metric of the propagation delay of the circuit. By solving equation (9.5) with respect to t_d, one can express the propagation delay as a function of the pole time constant and the ratio α between the actual and the minimum detectable signal:

$$t_d = \tau_p \ln\left(\frac{2\alpha}{2\alpha-1}\right) \tag{9.6}$$

As a reference, the propagation delay is $0.693\tau_p$ for a signal equal to the minimum one ($\alpha = 1$) and reduces to 10% of τ_p for $\alpha = 5$. Fig. 9.2 shows the propagation delay as a function of α for a comparator with a single pole time constant of 500 ns, corresponding to a bandwidth of 318 kHz. The decrease in the propagation delay is steep till $\alpha \approx 5$, then it smooths out, reaching asymptotically zero as α tends to infinity. It must be pointed out that the estimation of the propagation delay with the linear model is only indicative, because the comparator is an open-loop high gain amplifier and even small values of the input signal can push the circuit in an operating region different from the one used to calculate the small signal equivalent circuit,

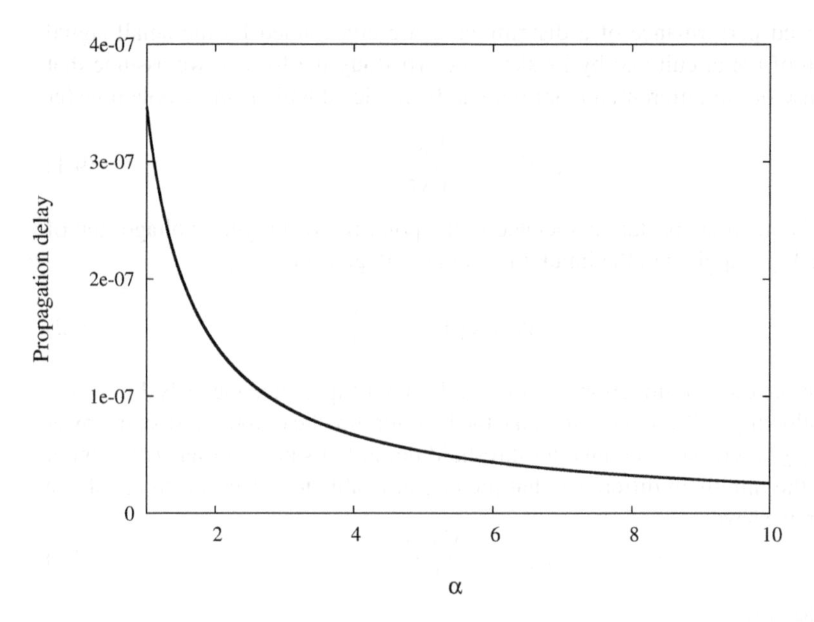

FIGURE 9.2 Propagation delay of a single pole discriminator as a function of the ratio α between the actual and the minimum signal. A time constant $\tau_p = 500$ ns is assumed for the pole.

which assumes that the two inputs are equipotential. When the input signal is large enough, the comparator can become fully unbalanced and the capacitance is charged or discharged with the maximum available current. In this case, the circuit speed is limited by its slew rate and the output changes at a constant pace.

9.1.2 DISCRIMINATOR THRESHOLD SETTING

Consider now the case in which one input of the comparator is connected to the front-end output and the other is held fixed at a voltage equivalent to the front-end baseline. Due to the noise, the baseline level is crossed many times, resulting in a series of pulses at the discriminator output. To avoid this effect, the comparator must be unbalanced, putting the threshold at a different value from the baseline.

The problem of calculating the frequency of induced events occurring when a given threshold level is crossed by noise pulses was mathematically studied by Rice in well known works published in 1944-1945 [2, 3]. The analytical derivation is quite elaborate, so we state here without proof the two most important results. The first one is that if no threshold is applied, the frequency of zero crossings due to noise f_{n0} is given by:

$$f_{n0}^2 = 4\frac{\int_0^\infty f^2 H(f)^2 df}{\int_0^\infty H(f)^2 df} \tag{9.7}$$

In other words, f_{n0} gives the number of times per second a given baseline level is crossed in either direction under the effect of noise. If the noisy level is applied to a comparator with threshold equal to the baseline (or, equivalently, "zero threshold"), the frequency of rectangular pulses at the comparator output is $f_{n0}/2$, because one zero crossing is associated to the leading edge of the output pulse and the other to the trailing edge, hence two zero crossings are necessary to yield a rectangular pulse at the discriminator output.

The second important formula yields the frequency of noise hits as a function of f_{n0} and of the threshold to noise ratio:

$$f_n = \frac{1}{2} f_{n0} e^{-\frac{S_{TH}^2}{2S_n^2}} = \frac{1}{2} f_{n0} e^{-\frac{V_{TH}^2}{2V_n^2}} = \frac{1}{2} f_{n0} e^{-\frac{Q_{TH}^2}{2Q_n^2}} \tag{9.8}$$

The threshold and the noise can be expressed in the appropriate units (voltage or current) at the front-end output or referred to its input and given in terms of equivalent charges. As an example, if the front-end transfer function can be modeled with a low pass filter with two real coincident poles, equation (9.7) becomes:

$$f_{n0}^2 = 4\frac{\int_0^\infty f^2 \left|\left(\frac{1}{1+j2\pi f\tau}\right)^2\right|^2 df}{\int_0^\infty \left|\left(\frac{1}{1+j2\pi f\tau}\right)^2\right|^2 df} = \frac{1}{\pi^2\tau^2} \tag{9.9}$$

The frequency of noise hits at zero threshold is thus given by $f_{n0}/2 = 1/2\pi\tau$ and, for a voltage discriminator, the frequency as a function of the threshold is:

$$f_n = \frac{1}{2\pi\tau} e^{-\frac{V_{TH}^2}{2V_n^2}} \tag{9.10}$$

Suppose, for instance, to have a second order system with $\tau = 20$ ns. The expected frequency of noise-induced pulses if no threshold is applied is given by $1/2\pi\tau = 7.96$ MHz. Observe that this is true in the idealized case that the discriminator has infinite gain and bandwidth. The fact that the comparator needs a minimum input signal to flip may limit the amount of noise pulses. Furthermore, the discriminator finite bandwidth should also be considered in the evaluation of f_{n0}. Equation (9.10) provides nevertheless a useful metric to estimate the threshold necessary to limit the spurious induced-noise hits below a given amount. For instance, if we apply to our system a threshold four times bigger than the noise, the frequency of spurious hits becomes:

$$f_n = \frac{1}{2} f_{n0} e^{-8} = 7.96 \cdot 10^6 \times e^{-8} = 2670 \text{ Hz} \tag{9.11}$$

whereas for a threshold eight times bigger than the noise the spurious hits are suppressed by a factor e^{-32} with respect to the zero-threshold case, reducing f_n to $1 \cdot 10^{-7}$. This is equivalent to saying that in a hypothetical system with ten millions identical channels, only one channel per second will fire because of noise. The number of noise hits in a given time frame, $f_n\Delta t$, expressed in percentage, yields the noise occupancy. The appropriate choice of the time interval Δt over which the occupancy is calculated depends on the particular application. For instance, in a colliding beam experiment, where events happen at regular intervals, Δt can be the time between two successive beam crossings. In a fully asynchronous system continuously sensitive Δt can be the unit time interval.

Equation (9.11) shows that there is a trade-off between the sensitivity of the system and the fraction of spurious hits which are embarked. Even if noise induced events can often be rejected by further analysis steps, they need to be processed as good events by the front-end electronics, eventually compromising its data transmission capability. Furthermore, in ASIC with a big number of channels, a high noise hit rate may increase the interference on chip caused by the simultaneous switching of many comparators, triggering additional channels and leading to a positive feedback that makes the chip unusable. Inverting (9.11), one obtains an interesting relationship:

$$\frac{V_{TH}}{V_n} = \sqrt{-2\ln f_n 2\pi\tau} \tag{9.12}$$

From the above equation, we see that a given threshold to noise ratio stabilizes the product $f_n\tau$, i.e the product between the frequency of noise hits and the system bandwidth. Therefore if the bandwidth is increased by reducing τ, also f_n increases, implying that for the same threshold to noise ratio, faster systems will have a higher frequency of fake hits and thus a higher noise occupancy.

In our discussion so far we have assumed that the comparator has zero offset. This is never the case, because of the mismatch existing between the nominally identical transistors that form the decision circuit. An offset can be represented as an additional voltage source in series to the threshold. Since its polarity is a priori unknown, it can either increase or decrease the effective threshold. In a multi-channel ASIC, each comparator will flip around its effective threshold. If the same nominal threshold is set for the full system, the highest value that guarantees an acceptable fake hit rate in the worst channel must be chosen, thus reducing the sensitivity in the other ones. The effect of channel-to-channel threshold variation thus becomes equivalent to that of an additional random noise source, that in many cases is higher than the intrinsic front-end noise. To keep the threshold adequately uniform, offset correction techniques are often mandatory and they will be dealt with in a separate section at the end of the chapter.

9.2 GENERAL PURPOSE VOLTAGE DISCRIMINATORS

In case the speed requirements are not very severe, voltage discriminators can be designed as open-loop operational amplifiers, removing the compensation capacitors that reduce the circuit bandwidth. Many different architectures are thus possible, but in this section we focus our attention on two topologies that are widely adopted.

9.2.1 TWO STAGE DISCRIMINATORS

In the simplest implementation, the amplifier driving the inverters could consist of a single differential cell with active load. Due to the low intrinsic gain of transistors in

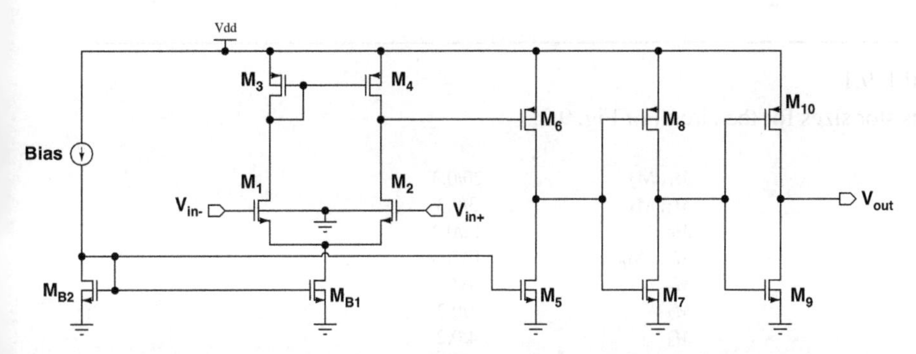

FIGURE 9.3 Simple discriminator formed by a two stage CMOS OTA followed by a digital buffer.

deep submicron CMOS technologies, this straightforward approach often results in a poor resolution, demanding the use of additional gain stages before the digital gates. A two stage circuit similar to the Miller OTA can then be used, as shown in Fig. 9.3. The gain provided by the circuit must be sufficient to guarantee a transition from $0.2 \cdot V_{DD}$ to $0.8\ V_{DD}$ for the minimum signal of interest. This corresponds to about 1.5 V for a 0.25 μm technology powered at 2.5 V and 0.76 V for a 130 nm process, normally running at 1.2 V. Observe that because of the closer proximity of the logic levels at lower power supply, a smaller gain is required for the same resolution in the more scaled technology. This makes less severe the reduction of the amplifier gain caused by the worsening of the device output conductance associated to the smaller feature size. The gain of the first and second stage is respectively given by:

$$A_{V01} = g_{m1}\left(r_{02}//r_{04}\right) = \frac{g_{m1}}{g_{ds2}+g_{ds4}} \tag{9.13}$$

$$A_{V02} = g_{m6}\left(r_{06}//r_{05}\right) = \frac{g_{m6}}{g_{ds6}+g_{ds5}} \tag{9.14}$$

Each stage contributes with its own dominant time constant to the speed limitation, so the overall transfer function can be written as:

$$A(s) = \frac{A_{V01}A_{V02}}{(1+s\tau_1)(1+s\tau_2)} \tag{9.15}$$

The strongest cut to the bandwidth is usually introduced by the first stage, because the Miller multiplication increases the effective value of the gate-drain capacitance of M_6. Hence, the use of the simple single pole model often yields reasonable predictions. A detailed analysis of the comparator speed performance employing a two-pole model can be found in [4].

Let's now consider the typical performance that can be achieved by the two-stage comparator in a 130 nm process, using the device sizes given in table 9.1. A length of 0.3 μm is chosen for the transistors in the two gain stages in order to avoid a too strong degradation of the output conductance. The width of the input devices if fairly large to reduce the impact of threshold voltage mismatch. We examine the

TABLE 9.1
Transistor sizes for the circuit of Fig. 9.3.

M_1, M_2	20/0.3
M_3, M_4	3/0.3
M_6	12/0.3
M_{B1}, M_{B2}	10/0.8
M_5	20/0.8
M_7	1/0.2
M_8	4/0.2
M_9	2.7/0.2
M_{10}	10.8/0.2

circuit behavior with two different bias currents: 2 μA and 20 μA. The former value is typical of applications in which very low power consumption is mandatory, such as front-end ASICs for hybrid pixel detectors, where thousands of channels are packed together on the same chip. The latter still allows a power dissipation that makes the comparator contribution negligible in most applications. The small signal gain is calculated around the equilibrium point, i.e. for $V_{in+} = V_{in-}$. A common mode input value of 0.6 V is chosen, which easily keeps all transistors in saturation. In a typical 130 nm process, $\mu_n C_{ox}$ can be assumed to be 350 μA/V^2 and $\mu_p C_{ox}$ 90 μA/V^2. For the input transistors, the value of the inversion coefficient evaluated for a bias current of 1μA is:

$$I_C = \frac{I_{DS}}{2n\mu_n C_{ox}\phi_T^2 \frac{W}{L}} = 0.024 \tag{9.16}$$

We can thus assume that the input devices work in weak inversion, with a transconductance given by $I_{DS}/n\phi_T$ equal to 30 μS. For the PMOS transistor M_6, the nominal bias current at the equilibrium is 4 μA and the inversion coefficient is 0.61, bringing the device close to moderate inversion. Its transconductance is hence better estimated with the more general formula:

$$g_m = \frac{I_{DS}}{n\phi_T} \frac{1}{\sqrt{I_C + 0.5\sqrt{I_C} + 1}} \tag{9.17}$$

yielding a transconductance of 65 μS. The list of the relevant operating points obtained through a computer simulation is reported in table 9.2. Inserting the values of table 9.2 in (9.13) and in (9.14), one gets $A_{V1} = 18.62$ and $A_{V2} = 34.7$, leading to a total gain of 646 or 56.2 dB, sufficient to make the comparator flip when the input voltage changes by ± 1 mV around the threshold point.

For a total bias current of 20 μA, the nominal current at the equilibrium is 10 μA in M_1-M_2 and 40 μA in M_6. The inversion coefficient is 0.25 for the input transistors and 6.5 in M_6, while the transconductance calculated with (9.17) is 248 μS for M_1-M_2 and 400 μS for M_6. The values obtained in simulation are reported in table 9.3. Using

TABLE 9.2
DC operating points for the circuit of Fig. 9.3 with a bias current of 2 μA.

Device	g_m	g_{ds}
M_1, M_2	28 μS	0.78μS
M_4		0.72 μS
M_6	73 μS	1μS
M_5		1.1 μS

TABLE 9.3
DC operating points for the circuit of Fig. 9.3 with a bias current of 20 μA.

Device	g_m	g_{ds}
M_1, M_2	248 μS	6.2μS
M_4		4 μS
M_6	432 μS	13μS
M_5		6.5 μS

these numbers, one can compute a total gain of 538 V/V. Observe that increasing the bias current slightly reduces the gain, but it leads to an almost tenfold increase in the bandwidth. Fig. 9.4 reports the comparator delay as a function of the overdrive voltage for the two bias conditions. In the simulation, a threshold of 10 mV is applied and a rectangular voltage pulse with a duration of 100 ns is fed to the input. Observe the much faster speed at the higher bias current and the steep increase of the delay at small overdrive voltages. The gain and speed performance of the comparator are mainly limited by M_6, which has a fairly big output conductance and whose gate-drain capacitance is multiplied by the Miller effect. By cascoding M_6, as shown in Fig. 9.5, the gain of the circuit can be significantly improved, without compromising the speed. In fact, the gain of the second stage is now given by $g_{m6}r_{05}$, because the output impedance of the cascode formed by M_6 and M_{6C} is much bigger than r_{05} and can be neglected. This allows increasing the gain of the second stage to ≈ 67 and the total gain to more than 1500 or 64 dB. Fig. 9.6 reports the delay measured with the output unloaded and connected to a 1 pF capacitor, which introduces a marginal degradation in the speed. The comparator is hence capable of driving small on chip loads without the need of additional buffering. Examples of discriminators based on low-power, high gain stages can be found in [5–10].

9.2.2 DISCRIMINATOR WITH CROSS-COUPLED LOADS

Another very common two stage topology is shown in Fig. 9.7 [4, 11]. The circuit is derived from the symmetrical two stage OTA with the addition of the cross-coupled

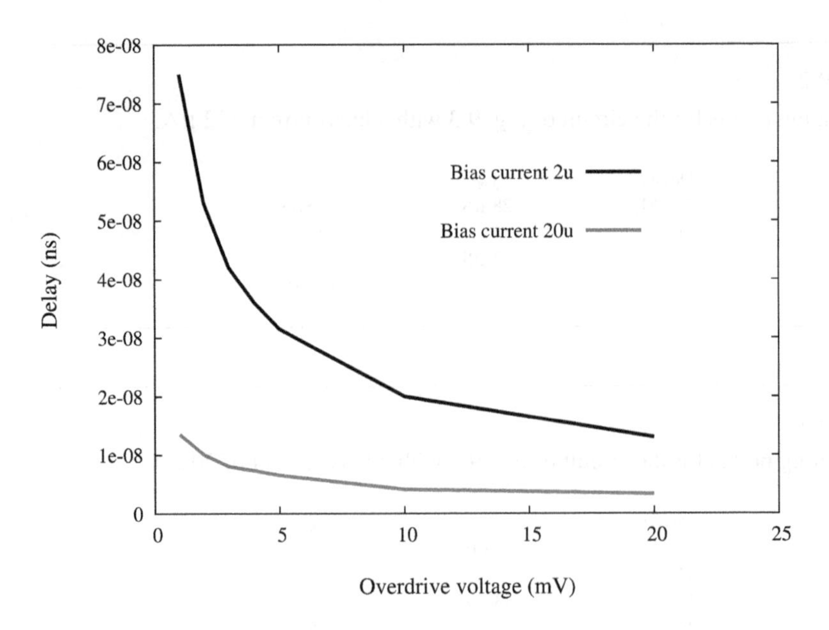

FIGURE 9.4 Delay of the comparator of Fig. 9.3 with a bias current of 2 and 20 μA.

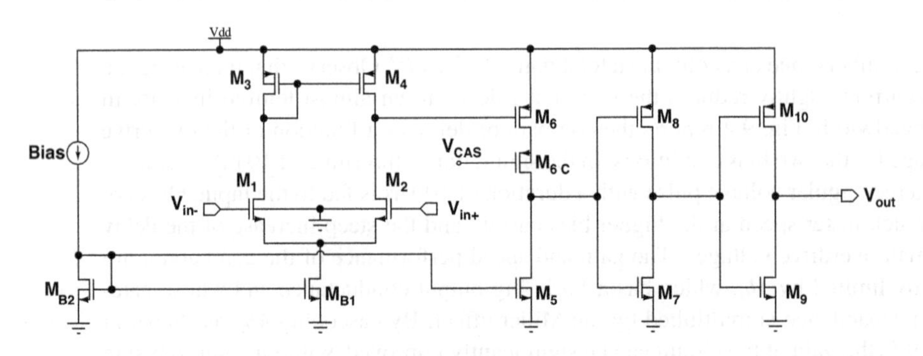

FIGURE 9.5 Two stage comparator with cascode in the second stage.

pair formed by M_5 and M_6. Consider first the case in which the aspect ratio of M_5 and M_6 is equal to the one of M_3 and M_4. When the circuit is driven by a differential signal, i.e. $V_{in1} = -V_{in2}$ the two outputs V_{01} and V_{02} have equal magnitude and opposite phase. Suppose for instance, that V_{in+} increases, so that V_{01} changes by $-\Delta V$ and V_{02} by $+\Delta V$. The lowering of the voltage at the drain of M_1 increases the current in the diode connected transistor M_3 by $g_{m3}\Delta V$. However, M_6 reduces its current by the same amount, because its gate, connected to V_{02} experiences a positive excursion of $+\Delta V$. Therefore M_6 behaves as negative resistance, that cancels out the effect of the diode connected device. This point can be better understood with the help of the small signal equivalent circuit of the left part of the stage, reported in Fig. 9.8. The model is

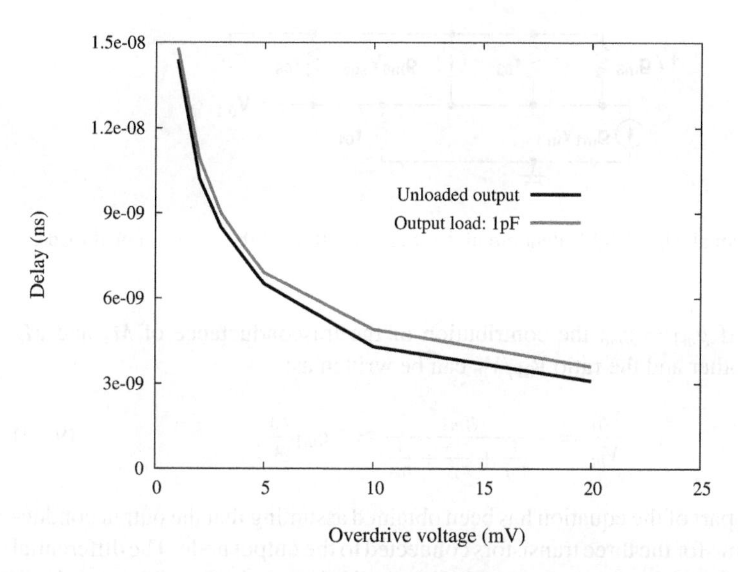

FIGURE 9.6 Delay of the comparator of Fig. 9.5 with the output unloaded and connected to a capacitor of 1 pF.

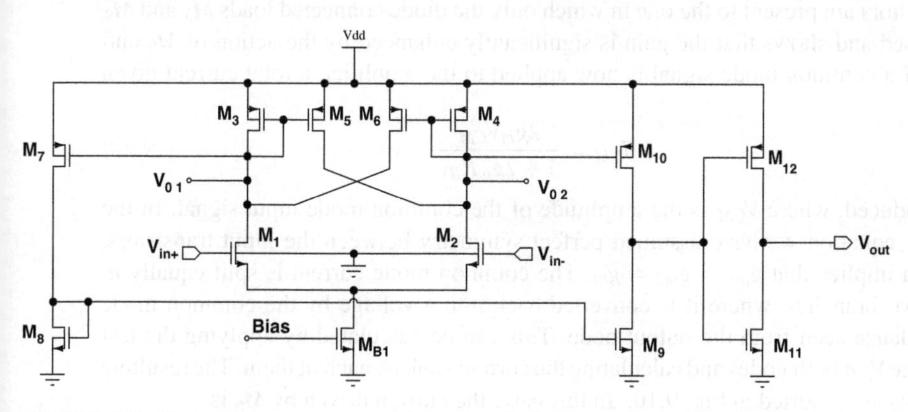

FIGURE 9.7 Two stage comparator with symmetrical input stage and cross-coupled active loads.

derived considering that for a pure differential mode signal the source node of M_1 and M_2 behaves as a virtual ground, allowing the splitting of the circuit in two identical parts. The current in M_6 increases if its source gate voltage increases. In case of a differential signal we have that $V_{sg6} = 0 - V_{02} = +V_{01}$, because for a pure differential mode signal the two outputs swing by the same amount in opposite directions. We can thus write:

$$g_{m1}V_{in} + V_{01}\left(\frac{1}{r_{01}} + \frac{1}{r_{03}} + \frac{1}{r_{06}}\right) + V_{01}g_{m3} - V_{01}g_{m6} = 0 \qquad (9.18)$$

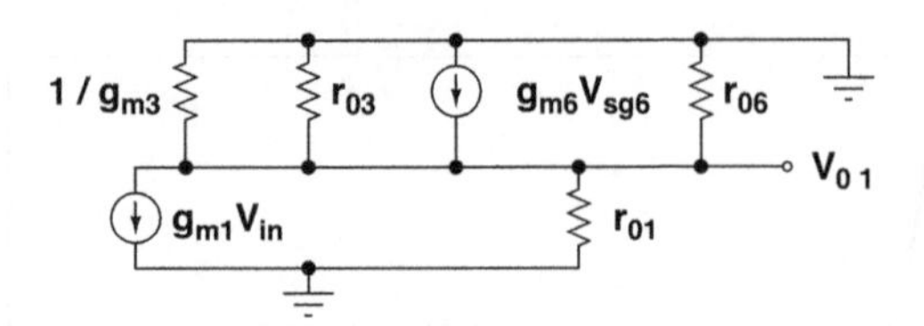

FIGURE 9.8 Small signal equivalent circuit to study the effect of the cross-coupled loads.

Observe that if $g_{m3} = g_{m6}$ the contribution of the trasconductance of M_3 and M_6 cancels each other and the ratio V_{01}/V_{in} can be written as:

$$\frac{V_{01}}{V_{in}} = -\frac{g_{m1}}{\frac{1}{r_{01}} + \frac{1}{r_{03}} + \frac{1}{r_{06}}} \approx -g_{m1}\frac{r_0}{3} \tag{9.19}$$

where the last part of the equation has been obtained assuming that the output conductance is the same for the three transistors connected to the output node. The differential gain is almost the same as the one provided by the single ended stage with active load. Fig. 9.9 compares with an AC simulation the gain in case the cross-coupled PMOS transistors are present to the one in which only the diode connected loads M_3 and M_4 are used and shows that the gain is significantly enhanced by the action of M_5 and M_6. If a common mode signal is now applied to the amplifier, a total current given by:

$$I_{CM} = \frac{2g_m V_{CM}}{1 + 2g_m R_{B1}} \tag{9.20}$$

is produced, where V_{CM} is the amplitude of the common mode input signal. In the above equation we have assumed perfect symmetry between the input transistors, which implies that $g_{m1} = g_{m2} = g_m$. The common mode current is split equally in the two branches, where it is converted back into a voltage by the common mode impedance seen from the output node. This can be calculated by applying the test voltage V_t at both nodes and calculating the current sunk by each of them. The resulting circuits are reported in Fig. 9.10. In this case, the current driven by M_6 is:

$$I(M_6) = g_{m6}V_{sg6} = g_{m6}(0 - V_t) \tag{9.21}$$

The nodal equation at the drain of M_6 can thus be written as:

$$V_t\left(\frac{1}{r_{03}} + \frac{1}{r_{06}}\right) + V_t g_{m3} + V_t g_{m6} = I_t \tag{9.22}$$

The contribution of the output conductance can now be neglected and equation (9.22) reduces to:

$$\frac{V_t}{I_t} = \frac{1}{g_{m3} + g_{m6}} \tag{9.23}$$

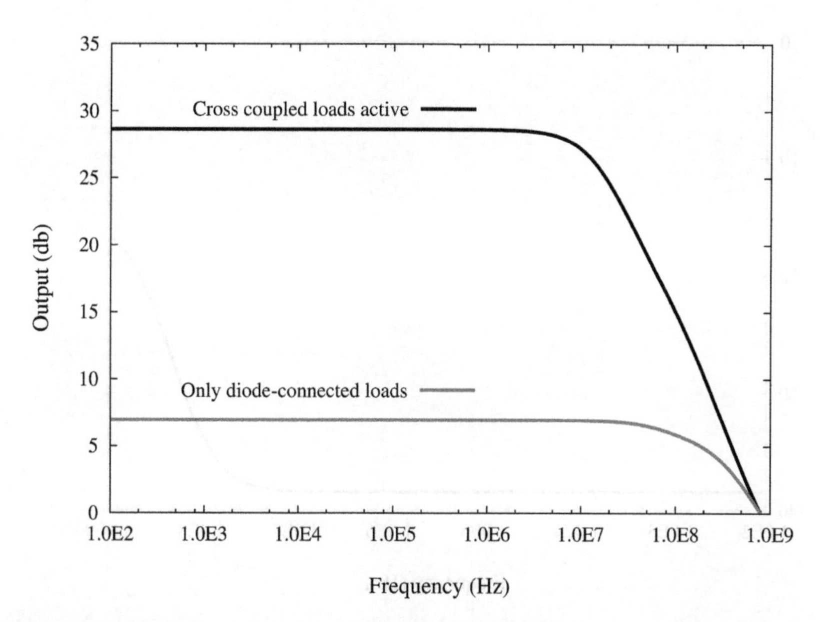

FIGURE 9.9 Differential gain of the first stage of the comparator in Fig. 9.7 with and without transistors M_5 and M_6.

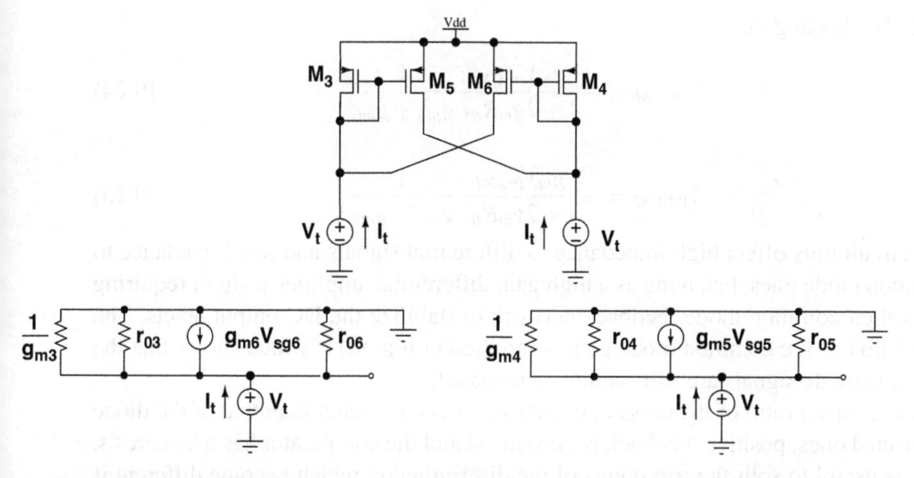

FIGURE 9.10 Circuits for calculating the common mode impedance seen from the output nodes of the differential amplifier with cross-coupled loads.

which shows that the common mode impedance is small. The common mode output voltage is finally obtained by multiplying the common mode current of each branch by the common mode impedance seen at the drain node of the input device given

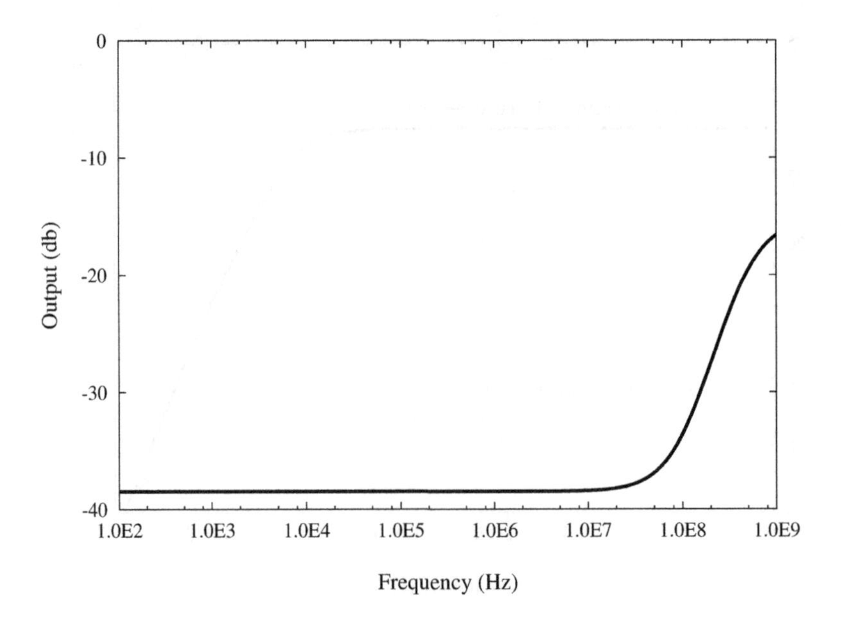

FIGURE 9.11 Common mode gain of the first stage of the comparator in Fig. 9.7.

by (9.21), leading to:

$$V_{01,CM} = -\frac{g_m V_{in,CM}}{1+2g_m R_{B1}} \frac{1}{g_{m3}+g_{m6}} \tag{9.24}$$

and

$$V_{02,CM} = -\frac{g_m V_{in,CM}}{1+2g_m R_{B1}} \frac{1}{g_{m4}+g_{m5}} \tag{9.25}$$

The circuit thus offers high impedance to differential signals and low impedance to common mode ones, behaving as a high gain differential amplifier without requiring an explicit common mode feedback network to stabilize the DC output levels. The simulation of the common mode gain is reported in Fig. 9.11, which shows that the common mode signals are substantially attenuated.

If the aspect ratio of the cross-coupled transistors is greater than that of the diode connected ones, positive feedback is introduced and the comparator has a hysteresis. This is useful to split the trip points of the discriminator, which become different if the input signal is heading positive or negative, thus avoiding the risk of multiple commutations due to noise when the signal is in proximity of the threshold. The hysteresis mechanism can be understood by referring to Fig. 9.7. Suppose that initially V_{in+} is much smaller than V_{in-}, so that M_1 is off and all the bias current is taken by M_2. M_6 tries to mirror the current in M_4 towards V_{01}, but since M_1 is off it can not sink any current, hence M_6 is driven into the linear region, pulling V_{01} to V_{DD}. If we now increase V_{in+}, M_1 start sinking current. When M_1 turns on, the voltage V_{01} is high, so M_1 is in saturation and M_3 is off. M_1 sinks its current from M_6, which is in the linear region and behaves as a resistor. As the current in M_1 increases, the current in M_2 decreases and the resistance of M_6 increases. When the current required by M_1 becomes greater than the one which can be provided by M_6, M_1 is driven in linear region and V_{01} decreases sharply, making the comparator quickly flip. All the bias current is now taken by M_1 and mirrored by M_5. However, M_2 can not receive the current from M_5, which is driven in the linear region, pulling V_{02} to V_{DD}. At the switching point, the following set of conditions must be verified:

$$\begin{cases} I_1 = I_6 \\ I_4 = I_2 \\ I_6 = kI_4 = kI_2 \\ I_1 + I_2 = I_{B1} \end{cases} \tag{9.26}$$

Combining the above relationships, we can write I_1 and I_2 as a function of the bias current and of the parameter k:

$$\begin{cases} I_1 = \frac{k}{1+k} I_{B1} \\ I_2 = \frac{1}{1+k} I_{B1} \end{cases} \tag{9.27}$$

The current in M_1 is greater than the one in M_2, so at the trip point V_{GS1} must be greater than V_{GS2} and the comparator flips for $V_{GS1} - V_{GS2} = \Delta V > 0$. All the quantities in equation (9.27) are known, so V_{GS1} and V_{GS2} can be computed. When the voltage at the gate of M_1 swings back towards a low value, the role of M_1 and M_2 are exchanged, so the comparator switches again when:

$$\begin{cases} I_2 = \frac{k}{1+k} I_{B1} \\ I_1 = \frac{1}{1+k} I_{B1} \end{cases} \tag{9.28}$$

Therefore, the circuit flips in the negative direction for $V_{GS1} - V_{GS2} = -\Delta V$. A numerical example can help in clarifying the hysteresis concept.

Example 9.1: Calculation of the switching points in a comparator with hysteresis

The comparator of Fig. 9.7 is implemented in a 130 nm CMOS technology with $\mu_n C_{ox} =$ 350 μA/V^2, $\mu_p C_{ox} = 90$ μA/V^2, $V_{THN} = |V_{THP}| = 0.28$ V. Calculate the switching points of the comparator. The transistor sizes are given in table 9.4. The bias current of the differential pair is 30 μA and the threshold voltage applied to the gate of M_2 is 0.6 V

TABLE 9.4
Device size of the circuit of Fig. 9.7.

Device	Aspect ratio
M_1, M_2	20/0.3
M_3, M_4	3/0.3
M_5, M_6	6/0.3
M_7, M_{10}	12/0.3
M_8, M_9	4/0.3
M_{11}	2/0.3
M_{12}	8/0.3

* * *

The aspect ratio of $M_4 - M_6$ is twice than that of $M_3 - M_4$, hence $k = 2$. When the input moves from below to above the threshold, equation (9.27) must be used. The comparator thus flips when I_1 becomes equal to 20 μA and I_2 is 10 μA. Observe that without the hysteresis, the discriminator switches when the input crosses the threshold, i.e. around the equilibrium point where $I_1 = I_2 = 15$ μA. Knowing the currents in M_1 and M_2 we can calculate their gate-source voltages. For a more accurate result, the general expression of I_{DS} is used, which reads:

$$I_{DS} = 2n\mu C_{ox}\frac{W}{L}\phi_T^2\left[ln\left(1+e^{\frac{V_{GS}-V_{TH}}{2n\phi_T}}\right)\right]^2 \quad (9.29)$$

By solving the above equation, one finds that $V_{GS1} \approx V_{TH} = 0.28$ V and $V_{GS2} = 0.246$ V. The gate of M_2 is held fixed at 0.6 V. Hence the value of V_{th+} at which the flipping occurs is:

$$V_{th+} = 0.6 - 0.246 + 0.28 = 0.634 \text{ V} \quad (9.30)$$

When V_{in+} goes back to the baseline, the comparator flips again when the current in M_2 becomes equal to the one in M_5. The role of M_1 and M_2 are exchanged and the switching occurs when $I_2 = 20$ μA and $I_1 = 10$ μA, implying that the gate source voltage of M_1 is now 0.246 V and the one of M_2 is 0.28 V. The negative crossing point is thus given by:

$$V_{th-} = 0.6 - 0.280 + 0.246 = 0.566 \text{ V} \quad (9.31)$$

The hysteresis hence determines a splitting of the threshold levels by a symmetrical amount ΔV around the nominal point. Fig. 9.12 shows a simulation where the comparator input is fed with a linear ramp, which allows us to see the different effective thresholds on the rising and trailing edge. Observe that to find accurately the trip points in simulation requires a DC sweep. In fact, in a transient analysis, the comparator fires after the threshold has been crossed because of the propagation delay, so determining the trip point just by finding on the plot the intersection between the two waveforms in Fig. 9.12 leads to incorrect results, unless the slew rate of the ramps is adequately low.

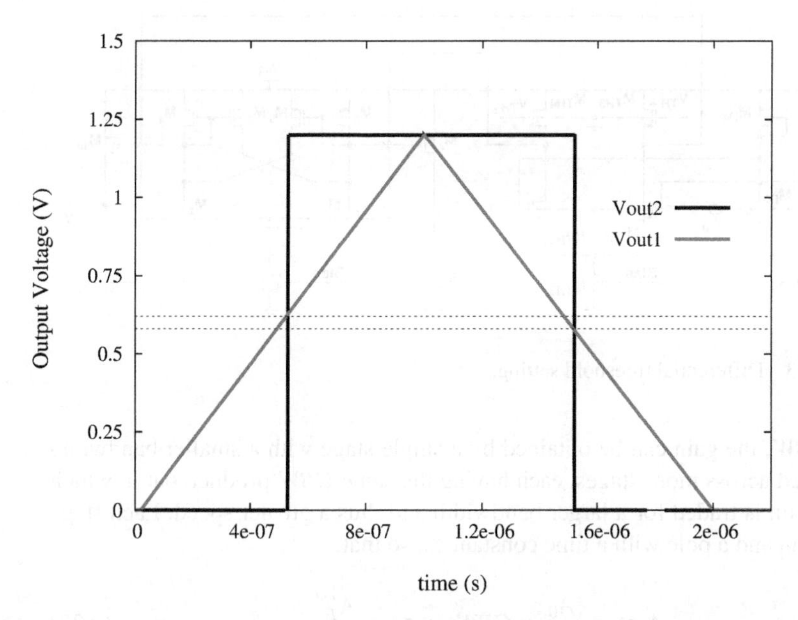

FIGURE 9.12 Example of time domain response of the circuit of Fig. 9.7 when hysteresis is applied.

An important point is that when hysteresis is used, the baseline of the circuit driving the comparator must be below the lower threshold, otherwise the comparator does not return to the zero state when the signal is over, but keeps memory of the previous transition. To minimize the power consumption, the comparator should be used so that at the quiescent state the gate of M_1 is lower than the one of M_2. In fact, the current of M_1 is mirrored by M_3 and M_7 and fed to the diode connected load M_8. Hence, if at the quiescent point M_1 drives current, so does the branch formed by M_7 and M_8, increasing significantly the power consumption of the circuit.

To improve the noise immunity, the threshold can be provided with a differential scheme, as shown in Fig. 9.13. The single ended signal coming from the front-end is converted into a differential one by the low gain differential amplifier. The second stage is formed only by NMOS transistors and has approximately unity gain. M_{TH3} and M_{TH4} behave as source followers, and set the threshold to the discriminator. Design examples of discriminators employing differential stages with cross-coupled loads can be found, for instance in [12, 13].

9.3 HIGH SPEED DISCRIMINATORS

In the previous section we have seen that to achieve enough gain more stages need to be cascaded. In case of first order systems, it is easy to calculate the optimal gain per stage that reaches the desired resolution with the maximum speed. Let's call A_T the total gain which is necessary in the comparator. For a fixed gain-bandwidth

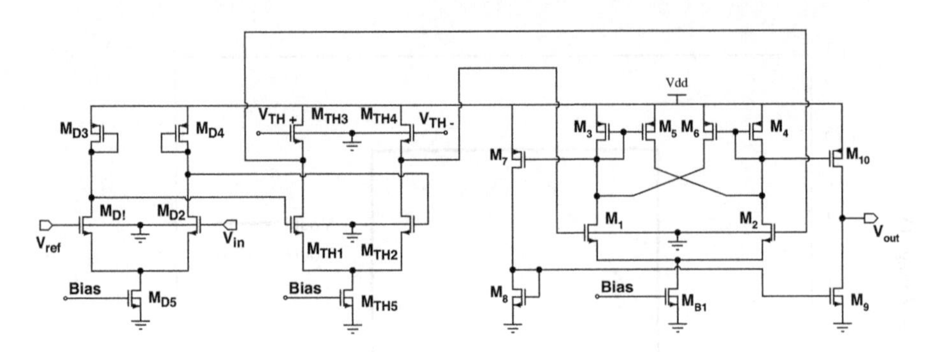

FIGURE 9.13 Differential threshold setting.

product *GBW*, the gain can be obtained by a single stage with a smaller bandwidth or distributed across more stages, each having the same *GBW* product, but in which a smaller gain is traded for a larger bandwidth and thus a greater speed. Each stage has a gain A_0 and a pole with a time constant τ_0, so that:

$$A_0\omega_0 = \frac{A_0}{\tau_0} = GBW \rightarrow \tau_0 = \frac{A_T^{1/n}}{GBW} \tag{9.32}$$

The total gain of the comparator is the product of the gains of the cascaded stages, hence we can write:

$$A_T = A_0^n \rightarrow A_0 = A_T^{1/n} \tag{9.33}$$

In the Laplace domain, the response of each stage to a step input signal of amplitude K can be written as:

$$A_0(s) = \frac{K}{s}\frac{A_0}{(1+s\tau_0)} \tag{9.34}$$

while the one of the full chain is:

$$A_T(s) = \frac{K}{s}\frac{A_0^n}{(1+s\tau_0)^n} \tag{9.35}$$

The rise time of n cascaded stages can be approximated by the following formula:

$$\tau_r = \sqrt{\tau_1^2 + \tau_2^2 + \cdots + \tau_n^2} \tag{9.36}$$

If the individual rise times are all equal to τ_0 we can write:

$$\tau_r = \sqrt{n}\tau_0 \tag{9.37}$$

Inserting into the above equation the expression of τ_0 as a function of the total gain and of the gain-bandwidth of the single stage, we obtain:

$$\tau_r = \sqrt{n}\tau_0 = \frac{\sqrt{n}A_T^{\frac{1}{n}}}{GBW} \tag{9.38}$$

where, using (9.32) τ_0 has been expressed in terms of the total gain and of the gain-bandwidth product. Differentiating (9.38) with respect to n, yields:

$$\frac{d\tau_r}{dn} = \frac{1}{GBW}\left(\frac{A_T^{\frac{1}{n}}}{2\sqrt{n}} - \frac{A_T^{\frac{1}{n}}\ln A_T}{n^{\frac{3}{2}}}\right) \tag{9.39}$$

which, solving for $d\tau_r/dn = 0$ to find the minimum, gives the optimal number of stages resulting in the best speed performance:

$$n = 2\ln A_T \tag{9.40}$$

For instance, for a given gain-bandwidth per stage and $A_T = 1000$, the optimal speed is achieved using 14 stages with gain 1.63. Therefore, high speed architectures implies the cascading of low-gain stages with adequate gain-bandwidth product, leading to a large power consumption. To limit the power dissipation, a compromise is usually accepted employing three to five stages, each with a gain between five and ten. Fig. 9.14 shows the typical architecture of a high speed comparator. The high speed, low gain blocks normally use a fully differential topology and are followed by a final high gain module that performs also the differential to single ended conversion. Fig. 9.15 shows three possible implementations of the differential amplifiers. The baseline topology, reported in a), consists of a differential pair with active diode-connected loads and gain given by g_{m1}/g_{m3}. In deep sumbicron technology and at low current densities, both NMOS and PMOS transistors operate close to moderate or weak inversion. Achieving an adequate gain might require to increase significantly the length of the PMOS devices, increasing their size and thus the capacitance at the output node. Therefore, additional current sources are added so that the current that flows in the loads is smaller than the one biasing the input differential pair. The ratio g_{m1}/g_{m4} can hence be easily controlled and optimized. The input transistors can also be cascoded to reduce the impact of the Miller effect on their gate-drain capacitance, as shown in Fig. 9.15 b). When high resistance polysilicon film are available, it can be advantageous to implement the loads as passive resistors, that might offer a smaller parasitic capacitance. To optimize the output DC node voltage, current sources can

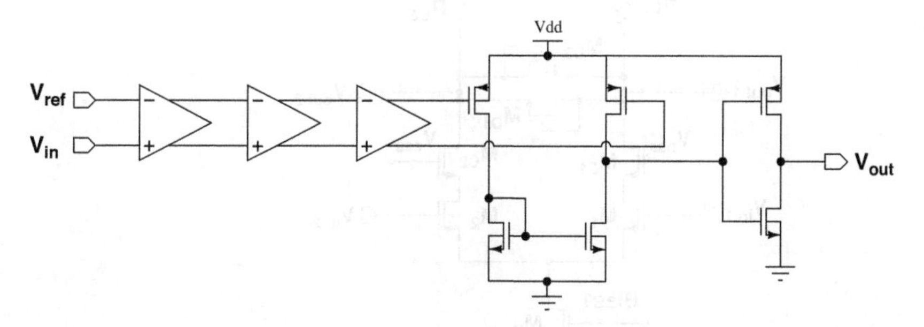

FIGURE 9.14 To maximize the speed, a discriminator must be implemented cascading low-gain cells.

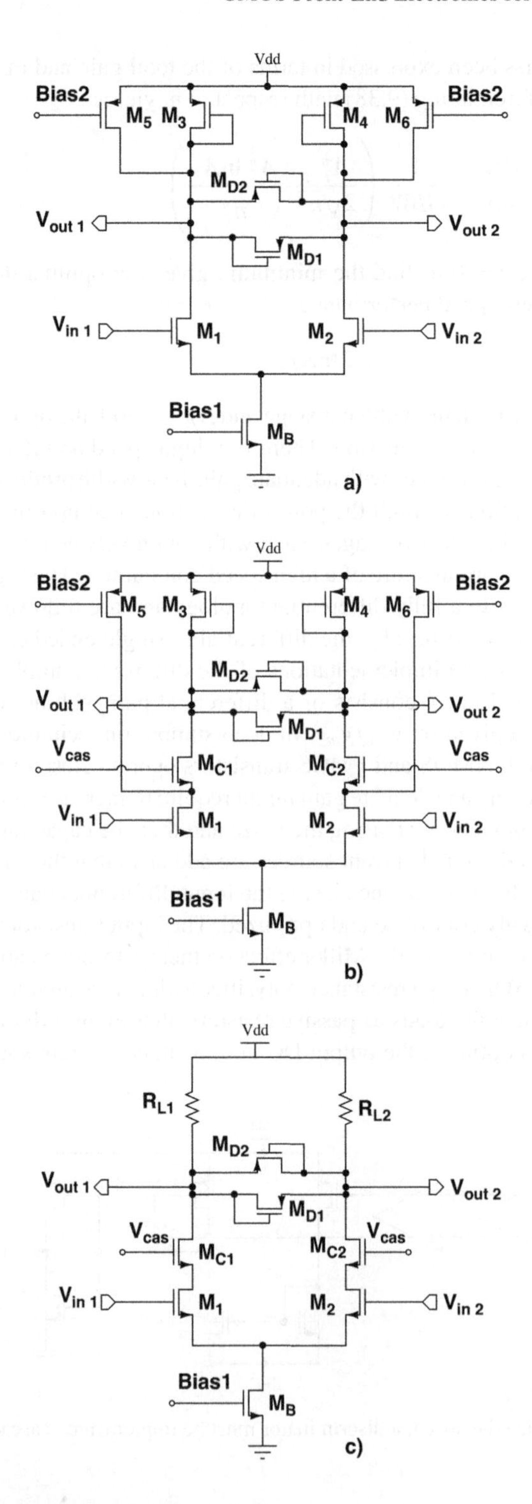

FIGURE 9.15 Gain cells for fast discriminators.

be used to regulate the current in the loads, with the same scheme of Fig. 9.15 a) and b). In all the cells two diode connected transistors are inserted between the two arms of the differential pair. The purpose is to limit the maximum voltage swing, in order to have a fast recovery in case of large input signals. High speed discriminators implemented with the multi-stage approach are described in [14–18]. The design of low-gain differential cells is further discussed in Appendix 1.

9.4 TIMING DISCRIMINATORS

Timing discriminators are comparators optimized to yield the best accuracy in capturing the time of occurrence of an event. The performance of a timing system depend on the interplay between the sensor, the front-end amplifier and the discriminator, hence timing measurements can be corrupted by several error sources. The most important are random noise in the front-end electronics, rise time variations of the detector pulse and time walk. We briefly examine these three effects before discussing the design of discriminators optimized for timing applications.

9.4.1 JITTER AND RISE TIME VARIATIONS

Jitter was already discussed in chapter 1. Here we just remember that if a signal with constant amplitude but affected by noise is presented to a discriminator and many iterations are run, the discriminator transition time wanders randomly back and forth around an average value. The sigma of the distribution of the comparator flipping times is related to the one of the noise affecting the input signal by the following relationship:

$$\sigma_t = \frac{\sigma_V}{\frac{dV}{dt}} \tag{9.41}$$

where σ_V and dV/dt are, respectively, the sigma of the noise and the slope of the signal fed to the discriminator. If we assume that the input signal is a linear ramp, we can approximate its slope as V_{peak}/t_r, where V_{peak} is the peak voltage and t_r the rise time. The jitter can thus be expressed as:

$$\sigma_t = \frac{\sigma_V t_r}{V_{peak}} \tag{9.42}$$

For the same noise and rise time, the jitter thus improves with the signal amplitude. In chapter 1 we have also seen that faster systems offer better timing performance and that the integration time of the front-end electronics should match the charge collection time of the sensor.

Rise time variations inside the sensor change the slope of the output signal and thus introduce an additional source of inaccuracy, even if the total collected charge is the same. The greatest spread in rise time for the same deposited charge occurs if all the charge is released in a punctual interaction. Depending on whether the interaction takes place, the charge takes different times to reach the electrodes, thus originating a variation in the sensor signal shape because both the collection time and the induced current change from event to event. A smaller spread is observed for fast charged particles that cross the full sensor, loosing a constant energy per unit of length. However, also in this case, the energy deposition along the particle track may not be perfectly homogeneous. More important, the electric field is not completely uniform inside the sensor, so the speed of the charge may change whether the crossing occurs at the center or close to the border of the sensing element. The ideal timing detector should thus produce large signals and have a short collection time together with small capacitance, in order to allow high intrinsic speed and low noise in the front-end electronics with a reasonable power consumption.

9.4.2 TIME WALK

Time walk occurs when signals of identical shape and different amplitudes cross a threshold. In fact, since the signals have all the same shape, they reach the peak amplitude at equal times, hence larger signals have a greater slope and cross the threshold earlier than smaller ones. Known as time walk, this effect originates a *deterministic* error in the timing measurement which can be in principle calibrated out if the amplitude of the signals is also measured. Observe that unlike the comparator delay, which is not present in case of ideal components with infinite bandwidth, time walk arises also if the chain is entirely implemented with ideal blocks, because it has its roots in the shape of the signals which are fed to the discriminator. If a unipolar shaper is connected to an ideal discriminator, the longest time walk corresponds to the peaking time and is produced by signals that barely cross the threshold. The time walk then reduces with $\approx 1/V_{out}$ as the signal amplitude increases.

9.4.3 LEADING EDGE DISCRIMINATORS

Good timing performance can be achieved with simple leading edge discriminators, where the timing information is extracted directly from the first transition of the discriminator output pulse. To minimize the contribution of the internal delay of the comparator, a fast topology with cascaded low gain stages is usually preferred. The solution is optimally suited for sensors delivering large signals, such as silicon photomultipliers, micro-channel plates and standard vacuum-tube photomultipliers, where the minimum signal of interest substantially exceeds the electronics noise. In this case, the threshold can be put very close to the signal origin, minimizing simultaneously time walk and the impact of sensor rise time variations. It is interesting to observe that with this method a low order shaping filter is preferable in the front-end. In fact, the time response of a generic CR-RC^n shaper is given by:

$$V_{out}(t) = K \left(\frac{t}{\tau}\right)^n e^{-\frac{t}{\tau}} \tag{9.43}$$

where K is the gain factor, which is unessential in this discussion. The slope of the output signal is thus given by:

$$\frac{dV_{out}}{dt} = \frac{n}{\tau}\left(\frac{t}{\tau}\right)^{n-1} e^{-\frac{t}{\tau}} - \left(\frac{t}{\tau}\right)^n \frac{1}{\tau} e^{-\frac{t}{\tau}} \tag{9.44}$$

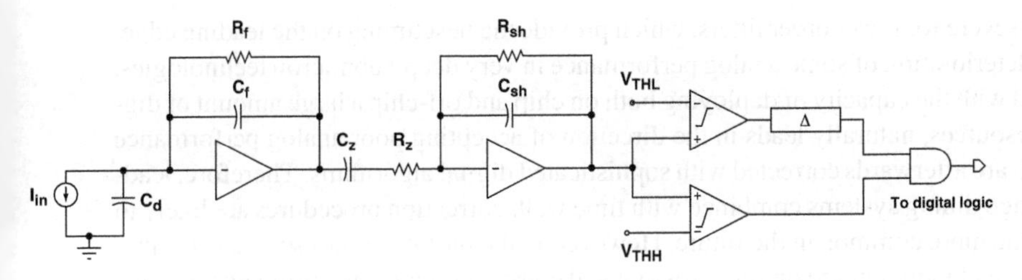

FIGURE 9.16 Front-end using two leading edge discriminators with different thresholds to achieve precise timing.

For the CR-RC shaper ($n = 1$) the slope in the origin is $\frac{1}{\tau}$, whereas for all the other configurations it is zero. Higher order shapers have thus a smoother "start-up" which increases the jitter, particularly if the threshold is put very close to the signal origin. An example of a leading edge discriminator that achieves a time resolution of 277 ps FWHM with Silicon PM is reported in [19].

If the threshold is not significantly smaller that the minimum signal, time walk becomes an issue. In this case one possibility to increase the time resolution is to implement a dual threshold scheme, as shown in Fig. 9.16. The output of the front-end, here represented as a simple CR-RC filter, is fed simultaneously to two discriminators. One discriminator works with a low threshold, V_{THL} and thus has a substantial number of fake hits due to noise. Its output is gated by a second comparator, that employs a higher threshold, V_{THH}. Only signals that trigger both discriminators can survive after the AND gate, delivering a valid output pulse. The circuit with lower threshold fires first. Therefore its output is delayed by a digital delay line (indicated as Δ in the figure) so that it arrives at the AND gate after the signal generated by the high-threshold discriminator. In this way, the transition at the output of the AND is determined by the circuit with lower threshold, which has better timing performance.

Another method to obtain precise timing with leading edge discriminators is to correct the time walk by measuring the signal amplitude. This can be done by registering also the trailing edge of the discriminator pulse and exploiting the Time over Threshold (ToT) technique. Digitization with ADCs is also possible, but the ToT method allows us to infer the amplitude information by re-using the same hardware which is already necessary to perform the timing measurement. This approach has been demonstrated in a front-end for hybrid pixel detectors, that has achieved better than 200 ps rms resolution in beam tests with minimum ionizing particles delivering signals in the $1 - 10$ fC range [20].

High accuracy systems require a precise calibration of the time walk versus amplitude relationship, which is non-linear and may need several calibration points per each curve. When extended to many channels, the procedure can become cumbersome, necessitating long calibration runs. If the ToT method is employed, it must also be taken into account that the amplitude resolution is limited by the jitter on the trailing edge which is substantially larger than the one affecting the leading edge because the return to the baseline is significantly slower than the rise to the peak. The effect is more severe for lower order filters, which provide the best timing on the leading edge. The deterioration of some analog performance in very deep-submicron technologies, mated with the capacity of deploying both on chip and off-chip a large amount of digital resources, naturally leads in the direction of accepting poor analog performance which are afterwards corrected with sophisticated digital algorithms. Therefore, leading edge timing systems combined with time walk correction procedures are likely to become more common in the future. However, in the past it was not so and walk-free analog techniques based on zero crossing and constant fraction discriminators needed to be applied. Such methods are still in widespread use today because they can offer satisfactory performance without complex off-line calibrations.

9.5 ZERO CROSSING DISCRIMINATORS

In a linear front-end, pulses of different amplitudes reach the peak at the same time. Therefore, extracting the timing information from the peak position yields in principle a walk-free measurement. In the following, we illustrate the method starting from a CR-RC shaper, but the description can be generalized to higher order filters. The impulse response of a CR-RC shaper with peaking time τ is:

$$V_{out,CR-RC}(t) = \frac{Q_{in}}{C_f}\left(\frac{t}{\tau}\right)e^{-\frac{t}{\tau}} \tag{9.45}$$

As studied in chapter 6, when the output of a CR-RC shaper is further processed by a high pass filter, a CR^2-RC system is obtained with an impulse response given by:

$$V_{out,CR2-RC}(t) = \frac{Q_{in}}{C_f}\left[\frac{t}{\tau} - \frac{1}{2}\left(\frac{t}{\tau}\right)^2\right]e^{-\frac{t}{\tau}} \tag{9.46}$$

The waveform in (9.45) has a bipolar shape, crosses the zero at $t = 2\tau$ and then changes its sign. We must remember here that the zero-crossing does not coincide with the peak of the unipolar waveform, because the high pass filter does not perform an *exact* mathematical derivative of the unipolar signal. The crossing time is nevertheless independent of the signal amplitude and thus provides a timing measurement which is not affected by time walk. Triggering on the zero crossing increases however the jitter. To understand this point we need to compare the slopes of the CR-RC and CR^2-RC filters. If we trigger directly on the output of the CR-RC stage, we need to put a threshold before the peaking time and ideally as close as possible to the baseline. As a working hypothesis, we can assume that the threshold is set at 25% of the peak amplitude, which occurs for $t \approx 0.1\tau$. Taking the derivative of (9.45) we get:

$$\frac{dV_{out,CR-RC}}{dt} = \frac{Q_{in}}{C_f}\frac{1}{\tau}e^{-\frac{t}{\tau}}\left(1-\frac{t}{\tau}\right) \tag{9.47}$$

The slope in the crossing point is thus:

$$\left.\frac{dV_{out,CR-RC}}{dt}\right|_{t=0.1\tau} = \frac{Q_{in}}{C_f}\frac{0.814}{\tau} \tag{9.48}$$

The slope of CR^2-RC signal is:

$$\frac{dV_{out,CR2-RC}}{dt} = \frac{Q_{in}}{C_f}\left(\frac{1}{\tau} - \frac{t}{\tau^2}\right)e^{-\frac{t}{\tau}} - \left[\frac{t}{\tau} - \frac{1}{2}\left(\frac{t}{\tau}\right)^2\right]\frac{1}{\tau}e^{-\frac{t}{\tau}} \tag{9.49}$$

The slope must be evaluated in the zero-crossing point, which occurs at $t = 2\tau$, yielding:

$$\left.\frac{dV_{out,CR2-RC}}{dt}\right|_{t=2\tau} = \frac{Q_{in}}{C_f}\frac{1}{\tau e^2} \tag{9.50}$$

Dividing (9.49) by (9.50), we get that the ratio between the slopes is $0.814 \cdot e^2 \approx 6$. In other words, in this particular case the slope of the zero crossing signal is 6 times smaller than that of the unipolar one, leading to a substantial jitter increase. Therefore, zero-crossing timing is advantageous only if time walk is the dominant source of uncertainty.

The building block implementation of the zero crossing system is similar to the one of the double threshold leading edge circuit and requires two comparators, one for arming and the other one that provides the accurate timing. Observe that a zero crossing discriminator in practice determines the instant in which the difference between the voltages applied to its input changes sign. Therefore, the inputs of the zero crossing detectors must be kept equipotential, but they do not need to be at zero. In practice, in single rail systems, the inputs are set to a convenient voltage that guarantees an adequate biasing of the input differential pair of the zero-crossing detector. This can be achieved as shown in Fig. 9.17, where a low impedance voltage source provides a reference voltage V_{ZC} that sets the comparator inputs to the appropriate common mode level. If the bipolar signal is generated before the discriminator, the value of the resistance must be high enough, so that high pass filter in front of the discriminator serves only to isolate the DC component. In this way, the signal current flowing to the common reference voltage can be kept small, which relaxes the design of the circuit that has to provide it.

Note that in the scheme of Fig. 9.17 no delay line is present, because the arming comparator fires on the leading edge of the signal, which occurs well before the zero

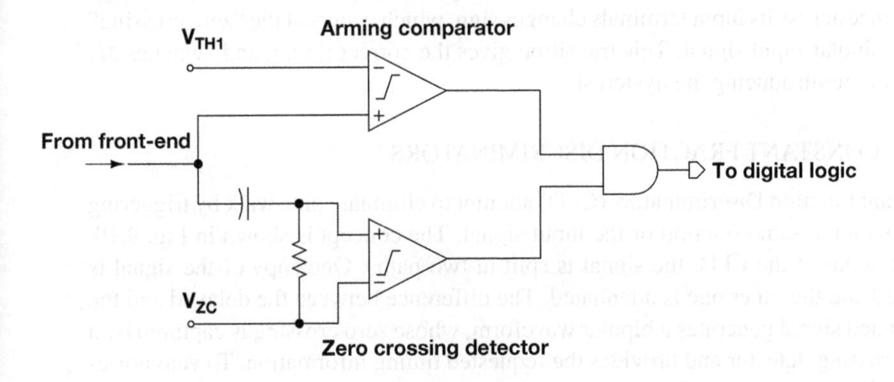

FIGURE 9.17 Principle of a zero crossing detector.

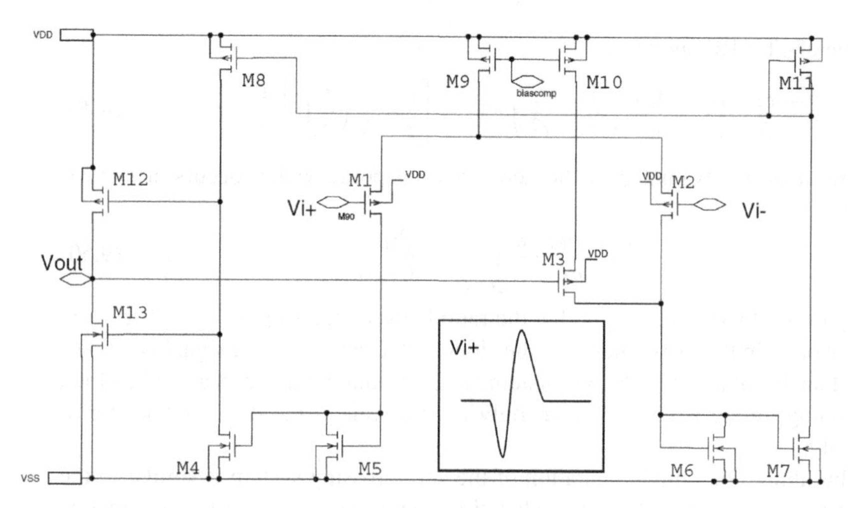

FIGURE 9.18 Zero crossing comparator with dynamic hysteresis. © [2004] IEEE. From [21], reprinted with permission.

crossing. The gating of the zero crossing signals prevents the fake hits generated by the noise in the zero crossing comparator from propagating to the digital logic. Nevertheless, if no additional measure is taken, the output of this circuit flips continuously. In multi-channels system this can be a severe source of noise that can compromise the performance of the full ASIC. The issue can be addressed by adding an asymmetric hysteresis to the zero crossing comparator. The concept is illustrated in Fig. 9.18 [21]. The signal driving the comparator is shown in the inset. Here the circuit is designed so that at the equilibrium V_{out} is zero and M_3 is on. Transistor M_{10} thus feeds to M_6 an additional current which imbalances the differential pair, limiting the number of noise commutations. The bipolar signal goes negative before the zero-crossing. When the trailing edge of the bipolar signal is big enough, V_{out} swings from 0 to V_{DD} and M_3 is disabled cutting the extra current off. The symmetry between the two arms of the circuit is thus fully restored and the comparator flips again when the voltage difference across its input terminals changes sign, which occurs at the "zero crossing" of the bipolar input signal. This transition gives the correct timing and switches M_3 on again, reintroducing the hysteresis.

9.5.1 CONSTANT FRACTION DISCRIMINATORS

Constant Fraction Discriminators (CFD) attempt to eliminate time walk by triggering always on the same fraction of the input signal. The concept is shown in Fig. 9.19. At the input of the CFD, the signal is split in two paths. One copy of the signal is delayed and the other one is attenuated. The difference between the delayed and the attenuated signal generates a bipolar waveform, whose zero crossing is captured by a zero crossing detector and provides the requested timing information. To veto noise-induced hits, the output of the zero crossing circuit is gated by a leading edge arming

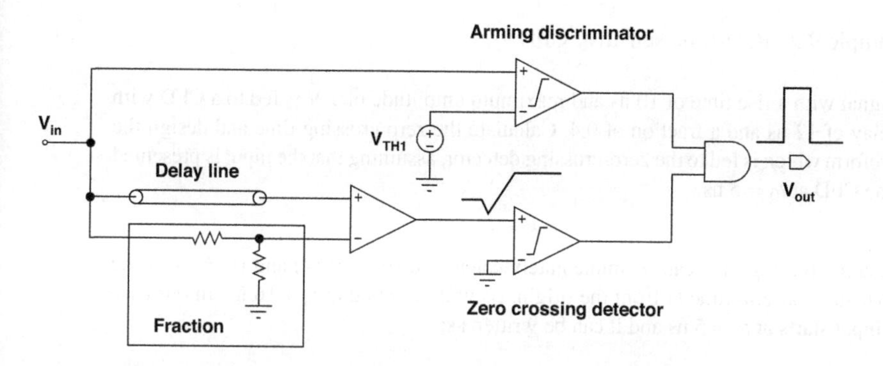

FIGURE 9.19 Principle of the Constant Fraction Discriminator (CFD).

discriminator, to which the input signal is directly presented. To understand how the circuit works, let's assume that the input is a voltage step of amplitude V_M and linear rise time t_r. For simplicity, we assume also that the signal starts at $t = 0$. Between 0 and t_r the input signal is thus given by:

$$V(t) = \frac{V_M}{t_r} t \tag{9.51}$$

while for t greater than t_r it is constant and equal to V_M. The delayed copy can be written as:

$$V_d(t) = \frac{V_M}{t_r}(t - t_d) \tag{9.52}$$

where t_d is the delay time. The above expression is valid between t_d and $t_d + t_r$, while $V_d(t)$ is zero before t_d and equal to V_M after $t_d + t_r$. If the delay time is greater than the rise time, the delayed copy starts when the attenuated one has already reached its maximum, given by fV_M, where f is the fraction by which the input is attenuated ($f < 1$). Between t_d and $t_d + t_r$, the difference between the delayed and the attenuated copy can be expressed as:

$$V_{zc}(t) = \frac{V_M}{t_r}(t - t_d) - fV_M \tag{9.53}$$

The zero crossing time is found by imposing that $V_{zc}(t)$ is zero and is:

$$t_{zc} = t_d + ft_r \tag{9.54}$$

Observe that t_{zc} depends on the delay, the fraction and the rise time of the input signal. The system is therefore sensitive to rise time variations affecting the input. Note also that t_{zc} as given in (9.54) is measured from the input signal start.

Example 9.2: Rise-time sensitive CFD

A signal with a rise time of 10 ns and maximum amplitude of 1 V is fed to a CFD with a delay of 12 ns and a fraction of 0.4. Calculate the zero crossing time and design the waveform which is fed to the zero-crossing detector, assuming that the input is presented to the CFD at $t_0 = 5$ ns.

* * *

The zero crossing time can be immediately calculated from (9.54) and is 16 ns. If the signal starts after a time t_0 from the origin, t_{zc} will be found at $t_0 + 16$ ns. In our case the input starts at $t_0 = 5$ ns and it can be written as:

$$V_{in}(t) = \frac{1}{10^{-8}} \cdot (t-5\cdot 10^{-9}) \cdot u(t-5\cdot 10^{-9}) - \frac{1}{10^{-8}} \cdot (t-1.5\cdot 10^{-8}) \cdot u(t-1.5\cdot 10^{-8}) \quad (9.55)$$

The unit-step function is necessary to have no signal before 5 ns while the second term is subtracted after 15 ns to flatten-out the waveform at 1 V after the full rise time of 10 ns has elapsed. The input signal is shown in the top part of Fig. 9.20. The attenuated signal is obtained by multiplying (9.55) by the fraction of 0.4 and is reported in the central part of the same figure. The delayed minus the attenuated signal is thus given by:

$$\begin{aligned} V_{zc}(t) \quad = \quad & \frac{1}{10^{-8}} \cdot (t-1.7\cdot 10^{-8}) \cdot u(t-1.7\cdot 10^{-8}) \\ & -\frac{1}{10^{-8}} \cdot (t-2.7\cdot 10^{-8}) \cdot u(t-2.7\cdot 10^{-8}) - 0.4V_{in}(t) \end{aligned} \quad (9.56)$$

and is plotted in the bottom part of Fig. 9.20. Observe that between 5 and 15 ns only the attenuated signal is active and the bipolar waveform becomes negative. After 12ns the delayed signal arrives and the waveform starts swinging back, crossing the 0 axis at $t = 21$ ns, i.e. 16 ns after the main signal starts, as predicted by (9.54).

Note that, in most cases, the circuit will be powered between ground and a positive rail, therefore all the signals will be shifted upwards by the offset necessary to keep the transistors properly biased. The "zero crossing time" must thus be interpreted as the time at which a generic baseline level (not necessarily the ground) is crossed and the potential difference at the input of the zero crossing detector changes its sign. The CFD with $t_d > t_r$ compensates only for amplitude variations and in this respect it is similar to the zero crossing method based on the derivative of a unipolar pulse. By choosing a delay shorter than the signal rise time, the rise-time variations of the input signal can also be rejected. If t_d is smaller than t_r, the bipolar waveform in fact becomes:

$$V_{zc}(t) = \frac{V_M}{t_r}(t-t_d) - f\frac{V_M}{t_r}t \quad (9.57)$$

The zero crossing time is found by solving for $V_{zc}(t) = 0$ and is:

$$t_{zc} = \frac{t_d}{1-f} \quad (9.58)$$

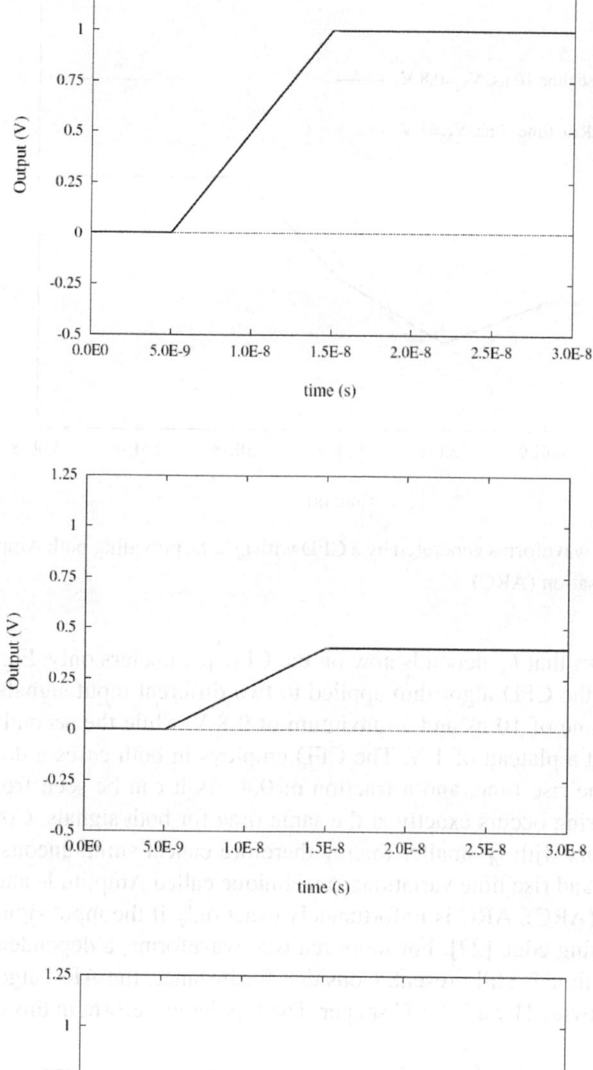

FIGURE 9.20 Waveforms for the circuit discussed in the example. From top to bottom: input signal, attenuated input signal and bipolar waveform used for timing.

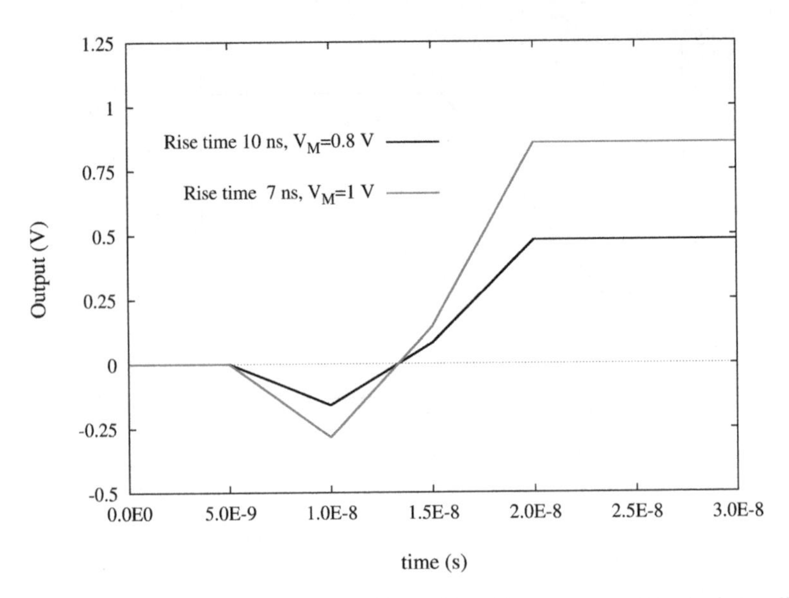

FIGURE 9.21 Output waveforms generated by a CFD with $t_d < t_r$, providing both Amplitude and Rise time Compensation (ARC).

Equation (9.58) shows that t_{zc} depends now on the CFD parameters only. Fig. 9.21 reports the result of the CFD algorithm applied to two different input signals. The first one has a rise time of 10 ns and a maximum of 0.8 V, while the second has a rise time of 7 ns and a plateau of 1 V. The CFD employs in both cases a delay of 5 ns, smaller than the rise time, and a fraction of 0.4. As it can be seen from the figure, the zero-crossing occurs exactly at the same time for both signals. Constant fraction discriminators with t_d smaller than t_r therefore cancel simultaneously the effects of amplitude and rise time variations, a technique called Amplitude and Rise time Compensation (ARC). ARC is unfortunately exact only if the input signal is a step with a linear rising edge [22]. For more realistic waveforms, a dependence on the input signal rise time is still present. Consider, for instance, the ARC algorithm applied to a pulse provided by a CR-RC shaper. The bipolar waveform in this case is given by:

$$V_{zc}(t) = \left(\frac{t-t_d}{\tau}\right) e^{-\left(\frac{t-t_d}{\tau}\right)} - f\left(\frac{t}{\tau}\right) e^{-\frac{t}{\tau}} \tag{9.59}$$

The zero crossing time is thus given by:

$$t_{zc} = \frac{t_d e^{\frac{t_d}{\tau}}}{e^{\frac{t_d}{\tau}} - f} \tag{9.60}$$

From the above equation, we see that the time of the zero crossing still depends on a term that contains the peaking time of the input signal. Ideally, the dependence is canceled if $f = 0$ or $t_d = 0$, suggesting that in practice small delay and fractions

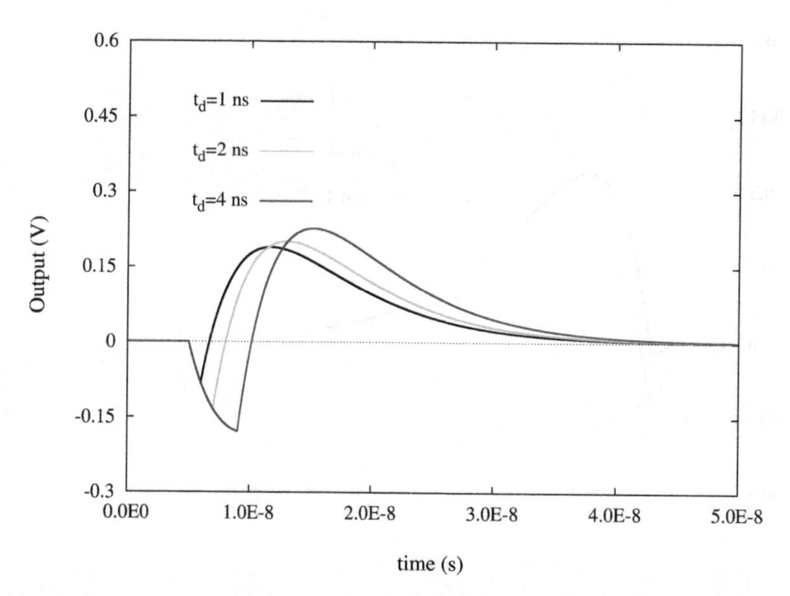

FIGURE 9.22 Output waveforms generated by a CFD algorithm applied to the pulse provided by a CR-RC shaper with 5 ns peaking time. The fraction is set at 0.5 and the delay is varied.

should be used. Fig. 9.22 shows the bipolar waveform obtained applying (9.59) to a CR-RC signal with 5 ns peaking time. The fraction is kept constant to 0.5 and three different delays are considered (1, 2 and 4 ns). In Fig. 9.23 the delay is kept constant at 2 ns, while fractions of 0.1, 0.25 and 0.5 are employed We see from the plots that, as predicted by (9.60), increasing the delay or the fraction shifts the zero-crossing time forward. The figures show however also a more interesting trend. Decreasing t_d and f the portion of the signal the precedes the zero crossing becomes smaller. This portion of the bipolar signal is called "under-drive" (or "over-drive" in case the polarity is reversed) and must be well above the noise to ensure a clean detection of the zero crossing. There is hence a trade-off between the rejection of rise-time variations (which calls from small delays and/or fraction) and the capability of the circuit to detect small signal, for which the under-drive could be over-shaded by noise. It must be pointed-out that CR-RC like shapers are not the most appropriate choice as far as timing is concerned. In fact, in these shapers the differentiation and the integration time constants are chosen to be equal. However, the differentiation time constant also cuts the signal amplitude. Therefore, while the integration time constant should be set approximately equal to the sensor collection time, in a timing system the differentiation time constant should be chosen as long as permitted by rate considerations [23].

The circuits described so far are based on ideal delay lines that, in discrete circuits, can be well approximated by transmission lines and coaxial cables. In integrated circuits, the delay element must be realized with appropriate filters, which however degrade the pulse slope and deteriorate the jitter. The optimization of the delay

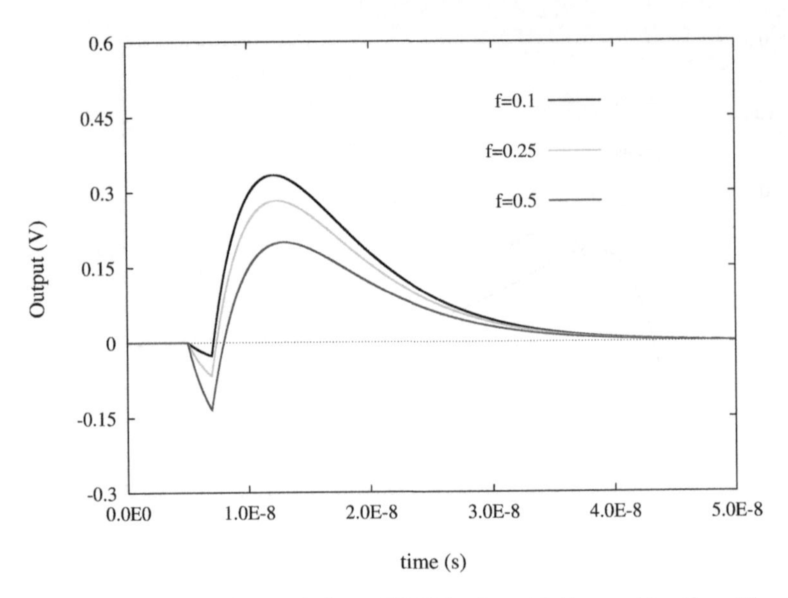

FIGURE 9.23 CFD algorithm applied to a CR-RC pulse with 5 ns peaking time. The delay is kept to 2 ns and the fraction is changed.

function is thus a central issue in the design of constant fraction discriminators in CMOS technologies. Some implementation mimic the delay line on chip with a distributed RC line, obtained by drawing a serpentine of polysilicon or metal over a conductive backplane [24]. The main drawback is that fairly long lines are needed to achieve delays in the range of 1-5 ns. In most integrated implementations, the delay is thus created by processing the signal with RC filters. Fig. 9.24 shows a constant fraction discriminator in which the signal is delayed with a single pole low pass filter. If $t_d = RC$ is the time constant of the low pass filter and f is the fraction, the transfer

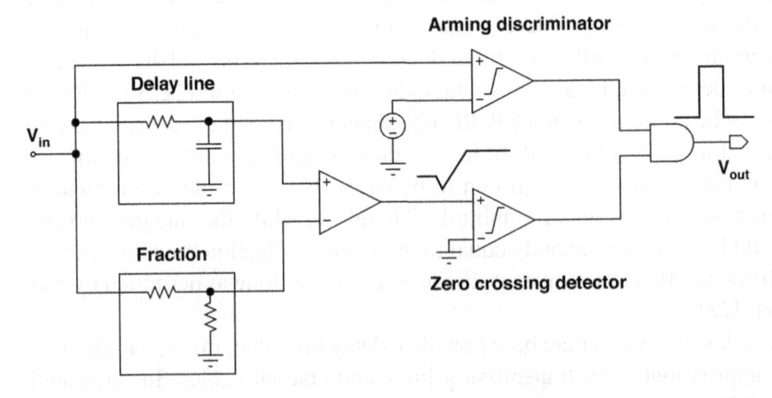

FIGURE 9.24 Constant fraction discriminator in which the delay is implemented through a single pole RC low pass filter.

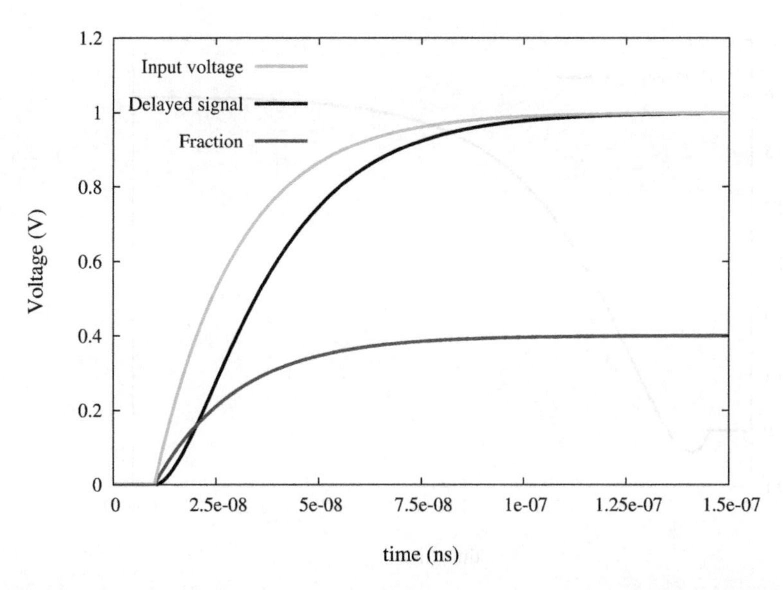

FIGURE 9.25 Input signal, delayed signal and fraction of a CFD with a single-pole low pass filter.

function of the circuit can be written as:

$$V_{zc} = \frac{V_{in}}{1 + st_d} - fV_{in} \tag{9.61}$$

The above equation can be easily rearranged as following:

$$V_{zc}(s) = V_{in} \frac{(1-f)\left[1 - \frac{fst_d}{1-f}\right]}{1 + st_d} \tag{9.62}$$

Fig. 9.25 reports the input signal, the delayed and the attenuated signal for a first order low pass filter CFD. For simplicity, it is assumed that the input is a voltage step filtered by a single time constant, i.e. V_{in} has the form:

$$V_{in} = \frac{V_0}{s} \frac{1}{(1 + s\tau_{in})} \tag{9.63}$$

where V_0 has been chosen equal to one. The zero crossing signal is shown in Fig. 9.26. In the plots, it has been assumed that τ_{in} is 20 ns, $t_d = 10$ ns and the fraction f is fixed to 0.4. Integrated delay lines for CFDs have been more frequently implemented with high pass filters [25]. The resulting circuit is shown in Fig. 9.27. Note that delay and fraction have been exchanged for a reason that becomes clear if one analyzes the transfer function of the circuit. If we call f' the fraction of the CFD with high pass filter delay, we can write:

$$V_{zc}(s) = f'V_{in} - \frac{V_{in}st_d}{(1 + st_d)} \tag{9.64}$$

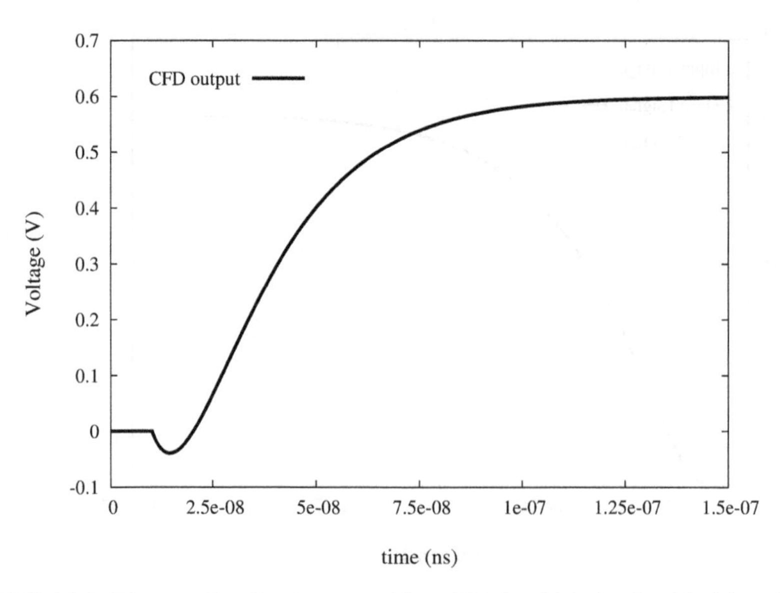

FIGURE 9.26 Zero crossing signal generated by a CFD in which the signal is delayed by a single pole low pass filter.

The above equation can be put in a more effective form, rewriting it as:

$$V_{zc}(s) = V_{in} f' \frac{\left[1 - \frac{1-f'}{f'} st_d\right]}{(1+st_d)} \tag{9.65}$$

From (9.65) we see that transfer function can be made coincident with the one described by (9.62) under the following conditions:

- The fraction f' is chosen equal to $1-f$.
- The time constants of the filters are equal.
- In the CFD based on the high pass filter, the delay is subtracted from the fraction in order to keep unaltered the polarity of the zero-crossing waveform.

The delay and the fraction of the high pass filter based CFD are reported in Fig. 9.28. The zero crossing waveform is shown in Fig. 9.29. The delayed signal is completely different from the one produced by the low pass filter, but, as predicted by (9.65) the bipolar signal is the same, provided that the three conditions stated above are met. A fraction of 0.6 and a time constant of 10 ns have thus been chosen to produce the plot of Fig. 9.28 and Fig. 9.29.

CFD based on passive filters are discussed in detail in [26], where it is shown that it is preferable to implement the delay filter by cascading more stages each with a smaller time constant rather than trying to achieve a given delay with a single large time constant. Higher order-filters have in fact a better slope-to-delay ratio, implying that for a given delay they filter and thus slow-down less the input signal. Fig. 9.30

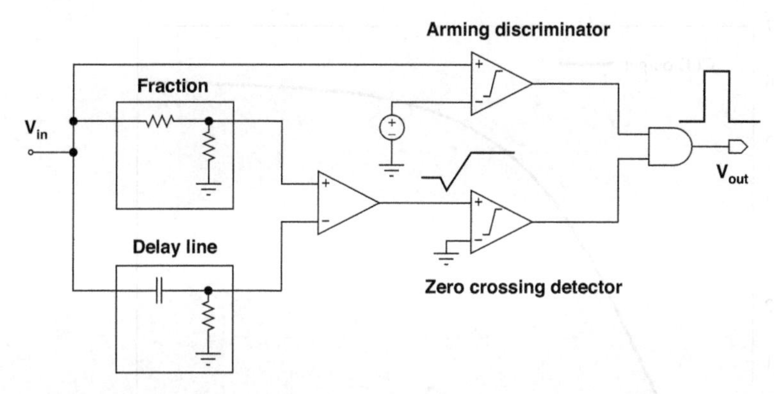

FIGURE 9.27 Constant fraction discriminator in which the delay is obtained with a first order high pass filter.

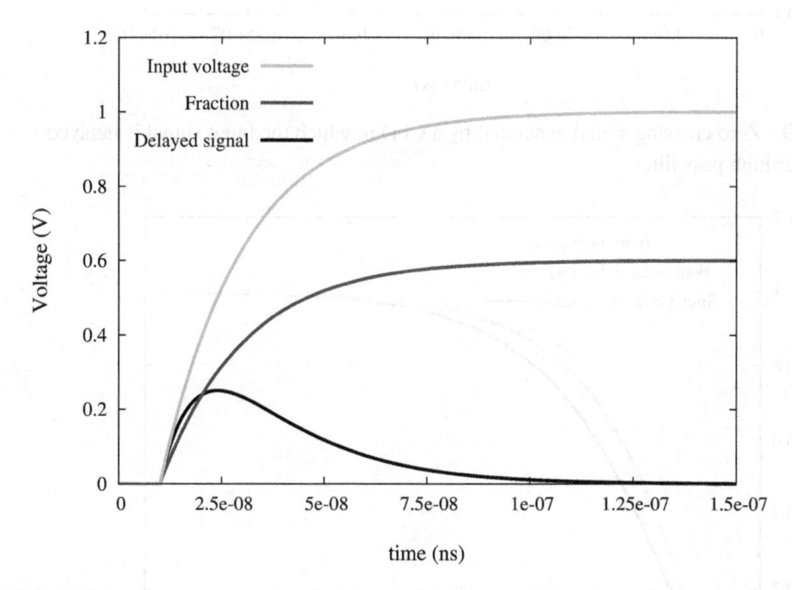

FIGURE 9.28 Input, attenuated and delayed signals in a CFD with a first order high pass filter.

shows as an example the delay obtained with a single pole low pass filter and a fourth order filter, optimized to produce the same initial delay. The corresponding zero-crossing waveforms are shown in Fig. 9.31, where it is seen that the CFD employing the higher order filter has a better underdrive and a greater slope at the zero crossing point, thus offering better jitter performance and more robust signal detection. As discussed in [26], the CFD shaping deteriorates however the jitter at the zero crossing point. The penalty in comparison to an ideal delay line is more severe if the input signal is a voltage step filtered by a single time constant, as the one assumed in our discussion, while better results are achieved if the input step voltage is already fil-

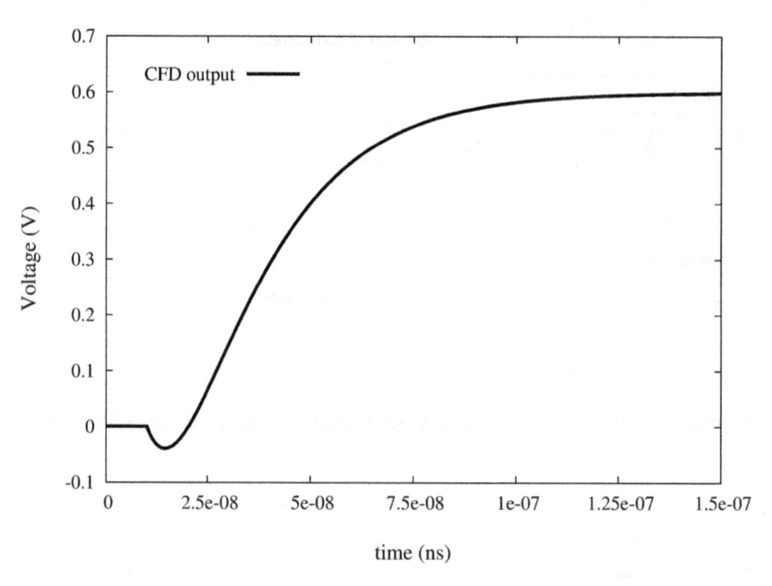

FIGURE 9.29 Zero crossing signal generated by a CFD in which the input signal is delayed by a first order high pass filter.

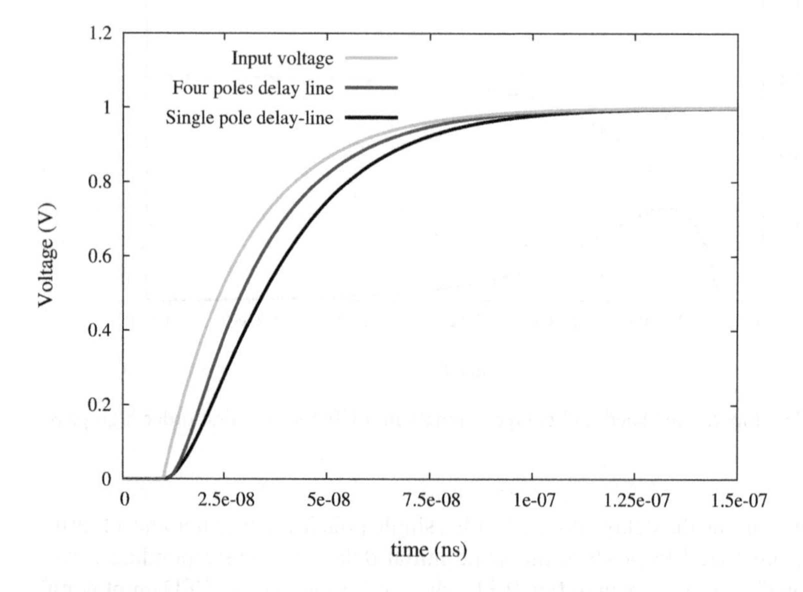

FIGURE 9.30 Delay signals generated by a first and fourth order low pass filters.

tered by two poles [26]. Using cascaded RC filters, compact CFD can be achieved and circuits fitting in 160 μm$\times$80 μm have been reported [27, 28]. Constant fraction discriminators rely on the linearity of the signal processing, therefore additional errors are introduced by distortions determined by saturation or slew rate limitations

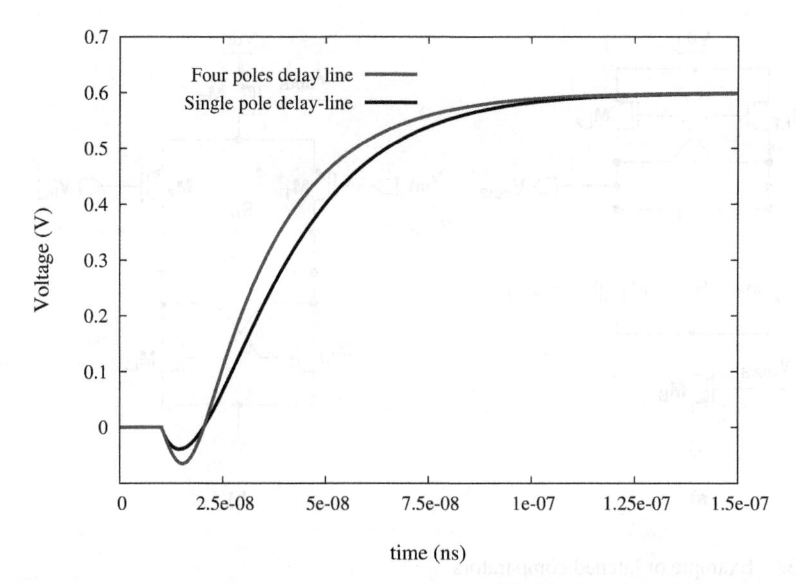

FIGURE 9.31 Zero crossing signals with delays implemented with single pole and four poles low pass filters.

occurring in the CFD circuitry. The offset in the zero-crossing detector must also be carefully considered, because it can shift the triggering point far from the zero crossing of the bipolar waveform.

9.6 LATCHED COMPARATORS

Latched comparators employ positive feedback to regenerate quickly a small voltage difference to full digital levels. Fig. 9.32 shows CMOS differential pairs loaded by cross-coupled transistors. The circuits operate between two states: in the reset mode, the switch S_R is closed and the positive feedback is disabled. The non-zero resistance of the reset switch allows however to establish a small voltage difference between the output nodes in case a differential signal is applied to the inputs. When S_R is open, the positive feedback is enabled and amplifies quickly the initial imbalance, generating a signal large enough to drive the following logic. The circuit must then be reset before a new event can be processed. Fig. 9.33 reports the small signal equivalent circuit of the amplifier with cross-coupled load. In the figure, C_L is the parasitic capacitance that loads the output node, R_S is the resistance of the switch S_R, g_{mL} is the transconductance of the transistors forming the latch and g_{m1} is the transconductance of the input transistor. For simplicity, the output resistance of all transistors is neglected. Assuming that the circuit is driven by a pure differential mode

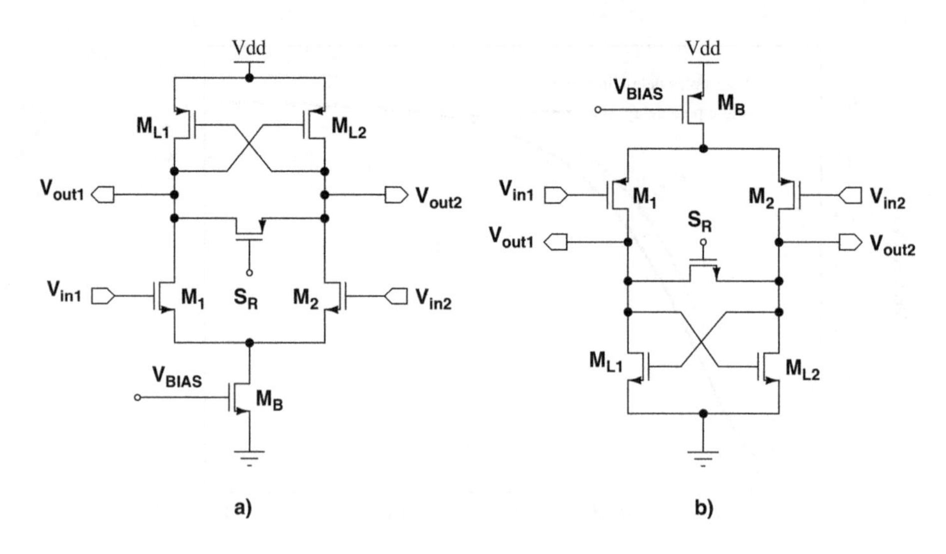

FIGURE 9.32 Example of latched comparators.

signal, we can write the following set of nodal equations:

$$\begin{cases} g_{mL}V_2 + V_1 sC_L + \frac{V_1 - V_2}{R_S} = g_{m1}\frac{V_{in}}{2} \\ g_{mL}V_1 + V_2 sC_L + \frac{V_2 - V_1}{R_S} = -g_{m1}\frac{V_{in}}{2} \end{cases} \tag{9.66}$$

Solving for $V_1 - V_2$, we obtain:

$$V_1 - V_2 = \frac{g_{m1}R_S}{2 + sC_L R_S - g_{mL}R_S} V_{in} \tag{9.67}$$

The above equation can be further rearranged as:

$$A_{vL} = \frac{V_1 - V_2}{V_{in}} = \frac{g_{m1}R_S}{(2 - g_{mL}R_S)} \frac{1}{\left(1 + \frac{sC_L R_S}{2 - g_{mL}R_S}\right)} \tag{9.68}$$

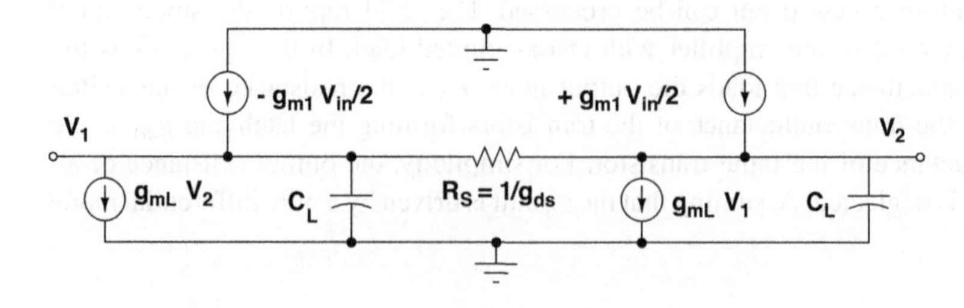

FIGURE 9.33 Equivalent circuit of a latched comparator.

Using now the fact that $R_S = 1/g_{ds}$, we finally get:

$$A_{vL} = \frac{g_{m1}}{(2g_{ds} - g_{mL})} \frac{1}{\left(1 + s\frac{C_L}{2g_{ds} - g_{mL}}\right)} \tag{9.69}$$

Multiplying the above expression by K/s and taking the Inverse Laplace Transform, we obtain the time domain response when the circuit is driven by a differential voltage step of amplitude K:

$$V_{out}(t) = \frac{Kg_{m1}}{(2g_{ds} - g_{mL})} \left(1 - e^{-t\frac{2g_{ds} - g_{mL}}{C_L}}\right) \tag{9.70}$$

Equation (9.70) shows an interesting feature. If the resistance between the two output nodes is low enough, the time constant appearing in the exponential is negative and the circuit behaves as an amplifier with a first order response, whose gain is given by:

$$A_{vL0} = \frac{g_{m1}}{(2g_{ds} - g_{mL})} \tag{9.71}$$

The conductivity between the two arms of the differential pair is granted by the reset switch, therefore to obtain a negative time constant in the exponential we need to have:

$$2g_{ds} - g_{mL} > 0 \rightarrow g_{ds} > g_{mL}/2 \tag{9.72}$$

If g_{ds} is smaller than $g_{mL}/2$, the time constant in the exponential becomes positive and the outputs diverge. When the switch is open, its conductance becomes zero and the circuit quickly amplifies the initial imbalance between the gates of the cross-coupled transistors with a time constant given by C_L/g_{mL}. There is therefore an interesting trade-off in the design of the reset switch. In fact, the condition given by (9.72) must be fulfilled in order to allow a proper reset of the stage. On the other hand, if the conductance of the closed switch is too high, the reset-mode gain of the circuit becomes too low, so only a very little voltage difference can be established between the outputs at the beginning of the regeneration phase and the sensitivity of the circuits is reduced. Fig. (9.34) shows a more complete implementation of a latched comparator. In this circuit, two reset switches are connected in parallel with the load. The inverters at the output speed-up the generation of full swing digital CMOS signals. Observe that an additional switch is placed in series with the tail current source biasing the differential pair. In this way, no static power is dissipated when the circuit is in the reset state. From the point of view of the power efficiency, latched comparators can be distinguished between static or class A circuits, class AB and dynamic only. Class A circuits have a permanent conductive path between the power supply rails and dissipate always static power consumption. An example is shown in Fig. 9.35. Class AB circuits dissipate static power only when the reset is disabled, as the circuit reported in Fig. 9.34. Finally, dynamic comparators dissipate power only during the regenerative phase, i.e. during the transistion from one logic level to the other. An example of a dynamic comparator is shown in Fig. 9.36. During the reset phase, both outputs are pulled to V_{DD} and the differential pair is off. In the evaluation phase, the

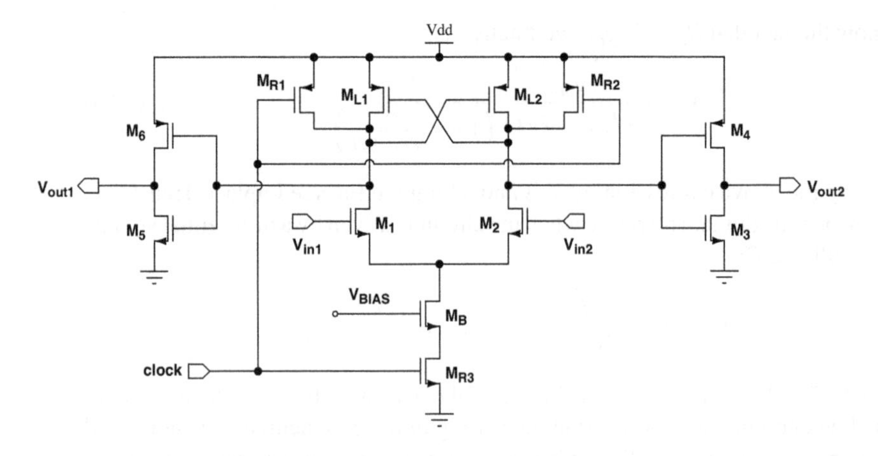

FIGURE 9.34 Example of a complete latched comparator.

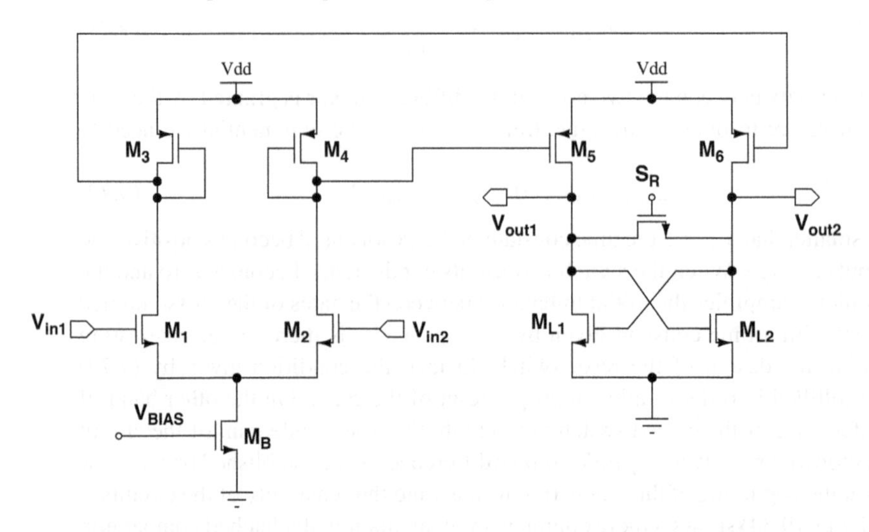

FIGURE 9.35 Class A latch with preamplifier.

differential pair is enabled and M_1 and M_2 start working in saturation, since initially they have the drain at V_{DD}. If V_{in1} is greater than V_{in2}, V_{out1} is pulled down faster than V_{out2}, starting the positive feedback that will drive V_{out1} low and V_{out2} high. At the end of the comparison, in each inverter forming the latch one transistor will be off, cutting the DC path between power supply and ground. As a digital gate, the circuit dissipates only during transitions. The discussed architectures formed only a small sample of the many different configurations of latched comparators which are described in the literature. Additional examples can be found in [29–33].

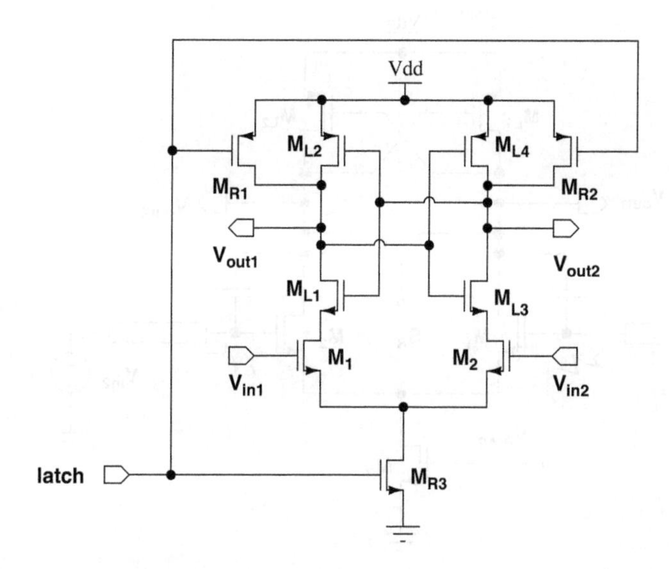

FIGURE 9.36 Dynamic comparator.

A major issue with latched comparators is the kick-back noise[1], illustrated in Fig. 9.37 [34, 35]. The large voltage swings at the comparator outputs couple to the input through the parasitic capacitance of the input transistors. Since the source driving the comparator does not have zero output impedance, this injects a disturbance on the input signals, which in turn can deteriorate the comparator accuracy. A straightforward solution to reduce the impact of kick-back noise is to increase the isolation between the comparator and its driving circuit, introducing a further stage in between. Shown in Fig. 9.35, the technique consists of using a preamplifier with moderate gain and fairly low output impedance before the stage with positive feedback. The use of preamplifiers has the disadvantage that it slows down the circuit and increases the power consumption. It is therefore interesting to reduce the kick-back noise while maintaining a single stage dynamic latch, which offers the best performance in terms of power consumption and speed. A possible implementation is shown in Fig. 9.38 [36]. The input signal is applied to M_{11}-M_{12} whose source is kept at ground, therefore the kick-back can only pass through the drain-gate capacitance. When the reset signal is low, the comparator is disabled and both outputs are pulled up to V_{DD}. The outputs are buffered by inverters and fed to an XNOR gate. In reset mode, the two outputs are equal and the XNOR gate produces a logic high signal, which enables M_9 and M_{10}. When the reset goes high, the regeneration starts and the outputs become different.

---

[1] Kick-back noise is also observed in asynchronous comparators.

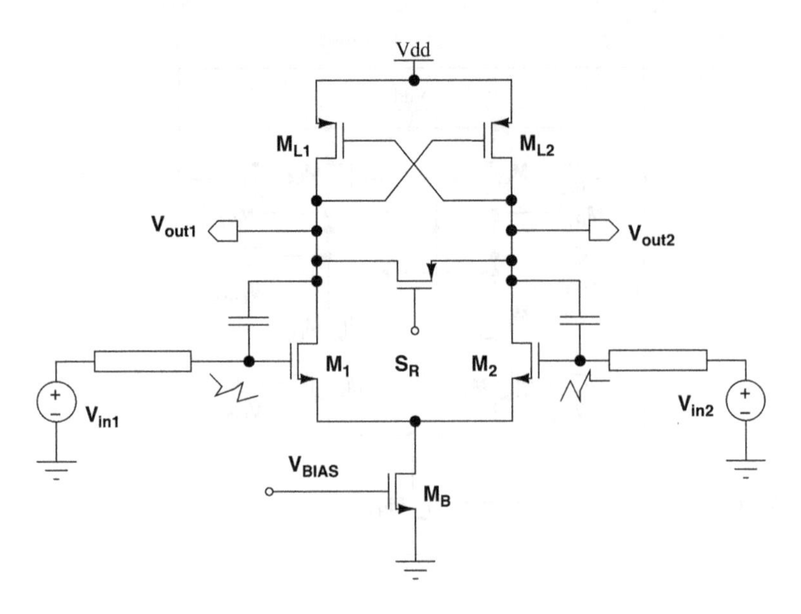

FIGURE 9.37 Kick-back noise in a latched comparator.

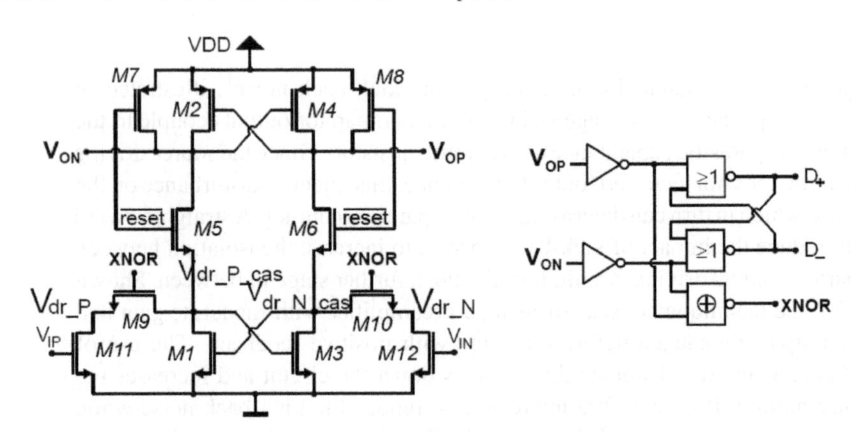

FIGURE 9.38 Dynamic latch with reduced kick-back noise. © [2013] IEEE. From [36], with permission.

The inverters amplifies the latch output signal, so that the XNOR gate flips to zero, disabling M_9 and M_{10} and disconnecting the latch from the input transistors. The key point is that M_9 and M_{10} are switched off before the latch outputs reach the full swing, reducing in proportion the kick-back towards the input.

The fast response and low power consumption of dynamic latches is paid with an additional noise on the power supply rail, due to the switch-on and switch-off transients. In systems having many channels, it is preferable to keep the power supply current as constant as possible. When power consumption is not the primary concern.

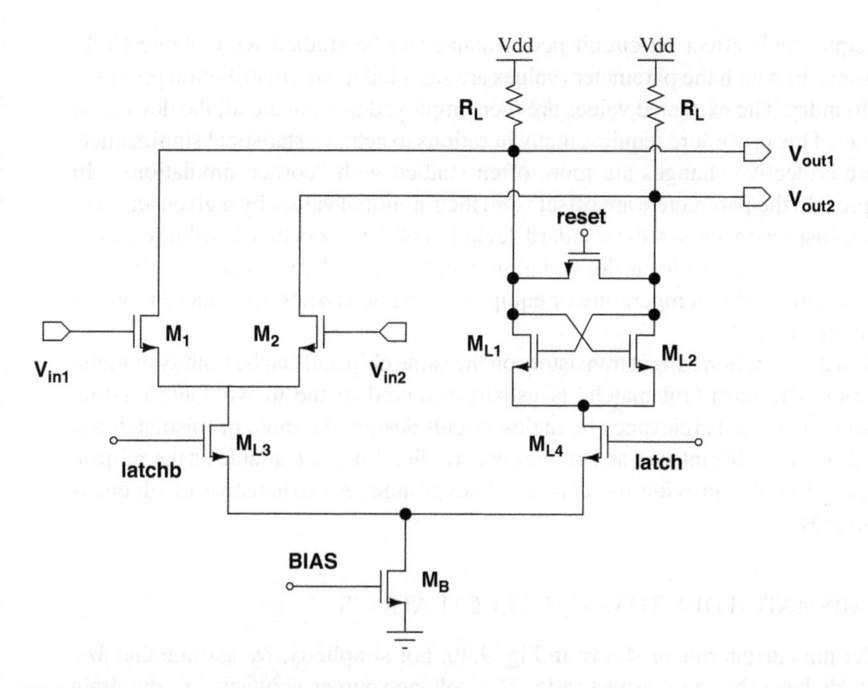

FIGURE 9.39 Current-Mode Logic (CML) latch.

Current-Mode Logic (CML) circuits can be used. Shown in Fig. 9.39, this type of comparator steers the current between the input differential pair in the reset phase and the latch in the decision phase, keeping thus constant the current drained from the power supply [37].

In front-end electronics ASIC, latched comparators are primarily used in analog to digital converters, while they are seldom found as discriminators immediately after the front-end amplifiers, because their switching noise can affect the performance of the latter. The reduction of feature size and voltage swing in deep submicron CMOS technologies is however making them attractive also in this area, at least for those applications in which the possible arrival time of an event is known and can be used to trigger the latch.

9.7 MISMATCH IN CMOS TRANSISTORS

As we have already anticipated, discriminators are affected by offset, arising from the non-uniformity between the characteristics of their transistors. In practice, identically designed transistors show different behaviors once they are fabricated on silicon. Variations of device properties can be classified in within-die (WID), die-to die (D2D), wafer-to-wafer (W2W) and lot-to-lot (L2L) variations.

Fluctuations affecting all transistors of a chip in the same way are referred to as "systematic". It must be pointed out, however, that such shifts are not predictable as they change randomly from one wafer to the next or one production lot to the other.

In principle, their effect on circuit performance can be studied with Monte Carlo simulations, in which the parameters values are sampled from a distribution provided by the foundry. The extracted values are then employed to simulate all the devices in the circuit. This procedure requires many iterations to achieve statistical significance, therefore collective changes are more often studied with "corner simulations". In this approach, the parameters are offset from their nominal values by a given amount, usually measured in units of the standard deviation of the associated distribution. Die-to-die, wafer-to-wafer or lot-to-lot variations can be caused by mechanical polishing, changes in processing temperature or equipment characteristics, tolerances in wafer positioning, etc. [38].

Within die variations affect transistors on the same chip and can be both systematic or random. The term "mismatch" is usually reserved to the unavoidable random component. Due to its relevance in analog circuit design, the topic of mismatch has attracted considerable interest and comprehensive literature is available on the subject. References [39–42] provide the classical background. An exhaustive treatment is found in [43].

9.7.1 MISMATCH DUE TO LOCAL FLUCTUATIONS

Consider the current mirror shown in Fig. 9.40. For simplicity, we assume that M_0, M_1 and M_2 have the same aspect ratio. The voltage sources connected to the drain of the output devices M_1 and M_2 keep the V_{DS} identical to the one of the reference transistor M_0. We therefore expect that the currents in M_1 and M_2 are exactly equal. When many such circuits are produced on the same chip and the difference $I_{DS1} - I_{DS2}$ between the currents in M_1 and M_2 is measured for each mirror, a Gaussian distribution around the expected zero mean value is obtained, revealing a non-uniform behavior of identically designed and laid-out transistors. The spread can be measured giving the standard deviation of the distribution, $\sigma(\Delta I_{DS})$. To understand where these variations

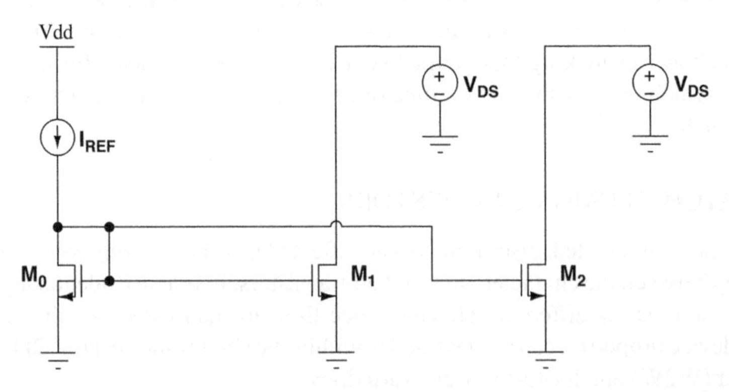

FIGURE 9.40 The currents of two nominally identical transistors are different due the unavoidable imperfections in the fabrication process.

arise, let's write again the relationship of the drain-source-current as a function of the terminal voltages:

$$I_{DS} = \frac{1}{2}\mu_n C_{ox}\frac{W}{L}(V_{GS} - V_{TH})^2 \tag{9.73}$$

We suppose here that the devices work in strong inversion, but the general considerations are valid also for weak and moderate inversion. We see that, for the same V_{GS}, the current can change if there is a difference in the threshold voltage and/or in the transconductance parameter $\beta = \mu_n C_{ox}\frac{W}{L}$. To relate the threshold voltage variations to more fundamental quantities, we remember that V_{TH} is proportional to the charge per unit area in the depletion layer divided by the gate oxide capacitance per unit area, C_{ox}:

$$V_{TH} \propto \frac{Q_B}{C_{ox}} \rightarrow \Delta V_{TH} \propto \frac{\Delta Q_B}{C_{ox}}. \tag{9.74}$$

To evaluate ΔQ_B, we first observe that the total number of active ions in the transistor channel is given by:

$$N_t = N_B W L t_d \tag{9.75}$$

where N_B is the doping concentration in the bulk, t_d is the depth of the depletion layer and WLt_d is the total depleted volume. The depletion layer depth t_d can be expressed as:

$$t_d = \sqrt{\frac{2\varepsilon_{Si}}{qN_B}\psi_s} \tag{9.76}$$

where ψ_s is the potential at the surface of the transistor channel. The standard deviation of the total number of ions due to random fluctuations is given by $\sqrt{N_t}$. Inserting (9.76) in (9.75) we have:

$$N_t = WL\sqrt{\frac{2\varepsilon_{Si}N_B}{q}\psi_s} \rightarrow \sqrt{N_t} = \sqrt{WL}\left(\frac{2\varepsilon_{Si}N_B\psi_s}{q}\right)^{\frac{1}{4}} \tag{9.77}$$

The total charge variation is given by $q\sqrt{N_t}$ and the charge variation per unit area is:

$$\sigma(Q_B) = \frac{\left(2q^3\varepsilon_{Si}N_B\psi_s\right)^{\frac{1}{4}}}{\sqrt{WL}} \tag{9.78}$$

Dividing $\sigma(Q_B)$ by $C_{ox} = \varepsilon_{ox}/t_{ox}$ we obtain the standard deviation on the threshold voltage of a single transistor, $\sigma(V_{TH})$. This means that if we built many identical transistors and measure their threshold, the measurements will follow a Gaussian distribution with a standard deviation $\sigma(V_{TH})$. In other words, 68% of the devices will have a threshold ranging from $V_{TH,average} + \sigma(V_{TH})$ and $V_{TH,average} - \sigma(V_{TH})$. In matching theory, we are interested in the *differences* between transistor parameters. These will follow also a Gaussian distribution, with a standard deviation $\sqrt{2}$ bigger than the one characterizing the individual device. We therefore have:

$$\sigma(\Delta V_{TH}) = \sqrt{2}\frac{1}{\sqrt{WL}}\frac{t_{ox}}{\varepsilon_{ox}}\left(2q^3\varepsilon_{Si}N_B\psi_s\right)^{\frac{1}{4}} = \frac{A_{VTH0}}{\sqrt{WL}} \tag{9.79}$$

where we have introduced the technological constant A_{VTH0}, measured in mV$\cdot\mu$m and given by:

$$A_{VTH0} = \sqrt{2\sqrt{2}\frac{t_{ox}}{\varepsilon_{ox}}}\left(q^3\varepsilon_{Si}N_B\psi_s\right)^{\frac{1}{4}} \tag{9.80}$$

The use of V_{TH0} in the subscript indicates that the quantity is calculated at $V_{SB} = 0$. Remember now that the dependence of threshold voltage on V_{SB} is given by:

$$V_{TH} = \frac{\sqrt{2q\varepsilon_{si}N_B}}{C_{ox}}\left(\sqrt{V_{SB}+2\phi_F}-\sqrt{2\phi_F}\right) = \gamma\left(\sqrt{V_{SB}+2\phi_F}-\sqrt{2\phi_F}\right) \tag{9.81}$$

A variation in N_B therefore implies also a change in γ, adding an extra component to the threshold voltage mismatch when $V_{SB} \neq 0$. This additional contribution is usually negligible for $V_{SB} \ll 2\phi_F \approx 0.7\ V$, a situation often recurring in analog design. Furthermore, the triple-well structure available in deep submicron technologies allows us to keep $V_{SB} = 0$ for NMOS transistors also in circuit topologies, like source followers and differential pairs, for which this was not possible in older technology nodes. Moving from one process generation to the next, N_B increases and t_{ox} decreases. Since the dependence of A_{VTH} in N_B is weaker than the one on t_{ox} it is expected that, for the *same* gate area, the matching performance improves by implementing a given circuit in a more advanced process. This is confirmed by the data reported in the literature, which shows that A_{VTH0} scales from 12.2 mV$\cdot\mu$m in 0.7 μm processes with 18 nm gate oxide to the 3.5 mV$\cdot\mu$m reported for 0.12 μm technology with a 2 nm oxide.

The other mismatch source comes from the transconductance coefficient $\beta = \mu_n C_{ox}\frac{W}{L}$. This can change because of variation in μC_{ox} and by irregularities in the gate shape (edge effects). Fluctuations of μC_{ox} may stem from change in the gate thickness and/or carrier mobility. We can think of the full device as made of elementary structures having gate width and length given by W_{el} and L_{el}. The number of cells needed to form the complete transistor is given by $n = (WL)/(W_{el}L_{el})$. Each elementary unit has its own value of gate capacitance and mobility, $C_{ox,el}$ and μ_{el}. We suppose that the values of $C_{ox,el}$ and μ_{el} follow a Gaussian distribution, with standard deviation given by $\sigma(C_{ox})$ and $\sigma(\mu)$, respectively. The oxide capacitance per unit area, C_{ox} and the mobility μ of the full transistor are obtained by averaging the corresponding parameters of its elementary constituents. The relative error (at one sigma level) on the total value of C_{ox} and μ is then given by $\sigma(C_{ox})/\sqrt{n}$ and $\sigma(\mu)/\sqrt{n}$. If we consider again the distribution of the differences in β between couples of nominally identical transistors, as done for the threshold voltage, we have

$$\frac{\sigma(\Delta\beta)}{\beta}\Big|_{\mu C_{ox}} = \frac{A_{\mu C_{ox}}}{\sqrt{WL}} \tag{9.82}$$

where $A_{\mu C_{ox}}$ is a constant characterizing a particular process.

When a transistor is fabricated on silicon, its gate edges are not perfectly sharp, but exhibit an irregular pattern, which determines variations both in the effective width and length from device to device. To study the problem for the device length, we can again divide a transistor of a given width W and length L in many elementary devices connected in parallel, each with W equal to W_{el} and with its own length.

As usual, we suppose that the values of L of the elementary transistors follow a Gaussian distribution with standard deviation $\sigma(L)$. The number of elementary cells is given by $n = W/W_{el}$. The total length is obtained by averaging the lengths of the n basic units and its standard deviation is given by $\sigma(L)/\sqrt{n}$. The standard deviation of the differences ΔL between the lengths of two nominally identical devices will be $\sqrt{2}\sigma(L)$:

$$\frac{\sigma(\Delta L)}{L} \propto \frac{1}{\sqrt{n}} \propto \sqrt{\frac{W_{el}}{W}\frac{1}{L}} = \frac{k_L}{L\sqrt{W}} \tag{9.83}$$

where we have grouped in k_L all the proportionality constants. Equation (9.83) shows that the mismatch affecting the transistor length is inversely proportional to the square-root of the device width. With a similar reasoning, one can deduce the relative error on the gate width:

$$\frac{\sigma(\Delta W)}{W} = \frac{k_W}{W\sqrt{L}} \tag{9.84}$$

The constants k_W and k_L depend on the particular technology. Assuming that the variation in width and length are uncorrelated, we have:

$$\left(\frac{\sigma(\Delta(W/L))}{W/L}\right)^2 = \left(\frac{\Delta W}{W}\right)^2 + \left(\frac{\Delta L}{L}\right)^2 = \frac{1}{WL}\left(\frac{k_W{}^2}{W} + \frac{k_L{}^2}{L}\right) \tag{9.85}$$

From the above equation, we see that the mismatch due to edge effects is reduced by increasing the gate area and avoiding both short and narrow channel devices.

The total transconductance mismatch is usually written as:

$$\frac{\sigma(\Delta\beta)}{\beta} = \left[\frac{1}{WL}\left(\sigma(\Delta|_{\mu C_{ox}})^2 + \frac{k_W^2}{W} + \frac{k_L^2}{L}\right)\right]^{1/2} \tag{9.86}$$

It is customary to report the β parameter mismatch as:

$$\frac{\sigma(\Delta\beta)}{\beta} = \frac{A_\beta}{\sqrt{WL}} \tag{9.87}$$

with A_β measured in %$\cdot$ μm. A_β lies in the 1-2%$\cdot$ μm range for many different processes. Experimental measurements show that in most cases the threshold voltage is the dominant mismatch source.

9.7.2 MISMATCH CALCULATIONS IN CIRCUITS

To calculate the effect of mismatch in circuits, we can use a small signal approach. Consider the current mirror shown in Fig. 9.40. The gate-source voltage is fixed by the reference bias circuit. Since the statistic variations on V_{TH} and β are supposed to be uncorrelated, we can evaluate them separately and sum the results in quadrature. To study the contribution of β fluctuations, we can generically write the current as $I_{DS} = \beta f(V_{GS} - V_{TH})$ and suppose that V_{TH} is the same for both devices. A discrepancy $\Delta\beta$ in the transconductance parameter translates into a current difference expressed

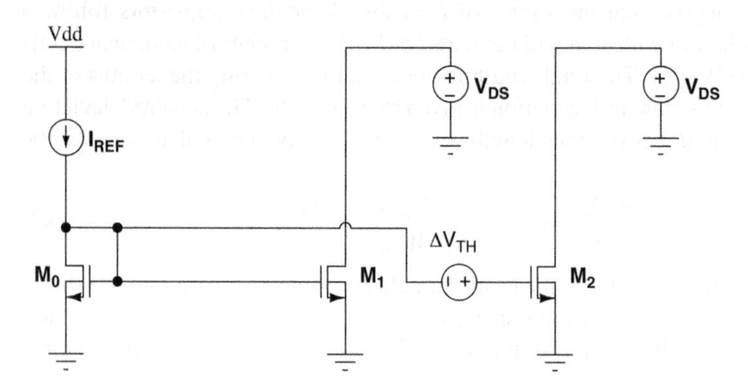

FIGURE 9.41 Study of the effect of the threshold voltage mismatch in current mirrors.

by $\sigma(\Delta I_{DS})$=$\sigma(\Delta\beta)f(V_{GS}-V_{TH})$. The relative current variation at 1σ level is hence given by:

$$\frac{\sigma(\Delta I_{DS})}{I_{DS}} = \frac{\sigma(\Delta\beta)}{\beta} \tag{9.88}$$

We now consider the effect of a threshold difference $\sigma(\Delta V_{TH})$. This can be represented as a voltage source connected between the transistor gates. Since the offset is random, the polarity of the source in arbitrary. With the arrangement of Fig. 9.41, M_2 drives an additional current given by $\sigma(\Delta I_{DS}) = g_m\sigma(\Delta V_{TH})$. The relative current variation due to the threshold mismatch alone is $\sigma(\Delta I_{DS})/I_{DS} = (g_m/I_{DS})\sigma(\Delta V_{TH})$ and the total relative mismatch is:

$$\frac{\sigma(\Delta I_{DS})}{I_{DS}} = \sqrt{\left(\frac{\sigma(\Delta\beta)}{\beta}\right)^2 + \left(\frac{g_m}{I_{DS}}\sigma(\Delta V_{TH})\right)^2} \tag{9.89}$$

The equation shows that to minimize the mismatch in the drain currents, a low value of g_m/I_{DS} is needed, i.e. the devices must work as deeply as possible in strong inversion.

Let's now consider the differential pair of Fig. 9.42 a). Here we impose to both transistors the same gate-source voltage and we measure a difference ΔI_{DS} between their drain currents. Using equation (9.89) we can write its standard deviation $(\Delta I_{DS})^2$ as:

$$(\sigma(\Delta I_{DS}))^2 = \left(I_{DS}\frac{\sigma(\Delta\beta)}{\beta}\right)^2 + (g_m\sigma(\Delta V_{TH}))^2 \tag{9.90}$$

We now want to calculate the V_{GS} that we need to apply between the inputs to rebalance the circuit. To do so we can divide equation (9.90) by g_m^2 and take the square-root:

$$\sigma(\Delta V_{GS}) = \sqrt{(\sigma(\Delta V_{TH}))^2 + \left(\frac{1}{g_m/I_{DS}}\frac{\sigma(\Delta\beta)}{\beta}\right)^2} \tag{9.91}$$

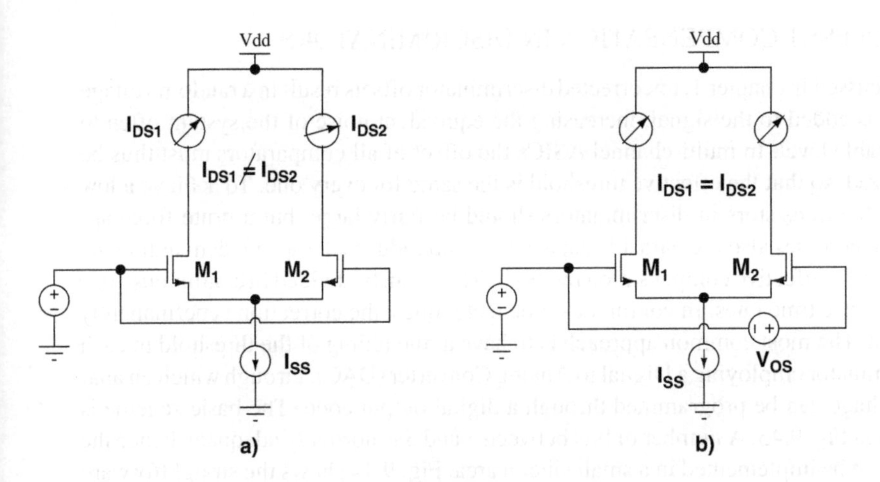

FIGURE 9.42 When the two inputs of a differential pair are equipotential, a difference between the drain current of M_1 and M_2 is observed. The voltage V_{OS} is the voltage necessary to re-balance the circuit, so that $I_{DS1} = I_{DS2}$.

Note that $\sigma(\Delta V_{GS})$ is minimized if g_m/I_{DS} is big, i.e. if the transistors work in weak inversion. If the term related to $\Delta\beta$ can be neglected, we have $\sigma(\Delta V_{GS}) \approx \sigma(\Delta V_{TH})$. The amount of voltage that we have to apply to the input of the differential amplifier to equalize the output currents is the input referred offset of the circuit. In practical cases, the output current is converted back to a voltage by a load. In this case, when the two inputs of the circuit are equal, one would observe a deviation of the output voltage from its nominal equilibrium point, i.e. an output referred offset. The input referred offset is obtained dividing the output offset by the amplifier gain. If we now measure the offsets of many identical pairs, they will follow a Gaussian distribution with standard deviation given by (9.91). Note that only 68% of the pairs will have an offset laying within $\pm\, 1\sigma$ from the average value, while 32% will exceed it. The same argument can be applied to the current spread in mirrors. While the standard deviation provides a useful metric to quantify the matching performance, designing for one sigma only is not a good practice, as large number of circuits will fail in meeting the specs.

Gradients in oxide thickness and doping concentrations determine parameter variations between nominally identical transistors located far from each other on the same die. This is in general a relevant issue for front-end ASICs, which are usually made of many identical channels operating in parallel. To spare area and power, common components such as the biasing networks are often shared. In many situations, a reference voltage is generated by a common block and propagated across the full chip. If there is gradient in transistors parameters, each channel will be biased at different currents, leading to performance discrepancies and, potentially, also to catastrophic failures.

9.8 OFFSET COMPENSATION IN DISCRIMINATORS

As discussed in chapter 1, uncorrected discriminator offsets result in a random voltage which is added to the signal, increasing the equivalent noise of the system often to intolerable level. In multi-channel ASICs the offset of all comparators must thus be equalized, so that the effective threshold is the same for every one. To achieve a low offset, the transistors in discriminators should be fairly large, but a brute force approach increases also the parasitic capacitance and reduces the speed, demanding the use of more effective compensation methods. These can be divided in continuous-time and discrete time ones. In continuous time techniques, the correction is permanently applied. The most common approach is to have a fine tuning of the threshold in each discriminator employing a Digital to Analog Converter (DAC), through which an analog voltage can be programmed through a digital output code. The basic scheme is shown in Fig. 9.43. A number of bits between 3 and 5 is normally adequate, hence the DAC can be implemented in a small silicon area. Fig. 9.44 shows the straightforward realization of a threshold tuning DAC, consisting of an array of binary weighted current mirrors. The current of each bit cell can be routed through a switch to the output node, where it is converted to a voltage by a resistor and filtered with a capacitor. The offset to be compensated has typically a value of 20-50 mV peak-to-peak, which defines the full scale range of the DAC. This fluctuation is superimposed on a baseline that, depending on the circuit implementation, can be of several hundreds of milliVolts. The baseline must then be set independently by an additional current, provided by M_{BL}. When power consumption is critical, this approach is not optimal, as it would require large terminating resistors. A better alternative is shown in Fig. 9.45. Here, the current in each bit cell can be steered between to outputs. This current is fed to a circuit that converts it to a voltage. For instance, in Fig. 9.45 this current is added or subtracted from the diode-connected loads. The generated voltage difference is reasonably linear and is fed to the following stage of the comparator. Using the S-curve method described in the first chapter, the threshold of each comparator is measured and the appropriate correction code is calculated and applied. Each channel thus runs with his own threshold, so that the switching points of all channels are equalized, and the comparators flip for the same value of the input signal. DAC-based compensation methods require also a local digital memory to store the chosen bit pattern for the converter.

An alternative method to reduce the threshold spread is shown in Fig. 9.46. The comparator output is low pass filtered before being applied to the compensation circuit,

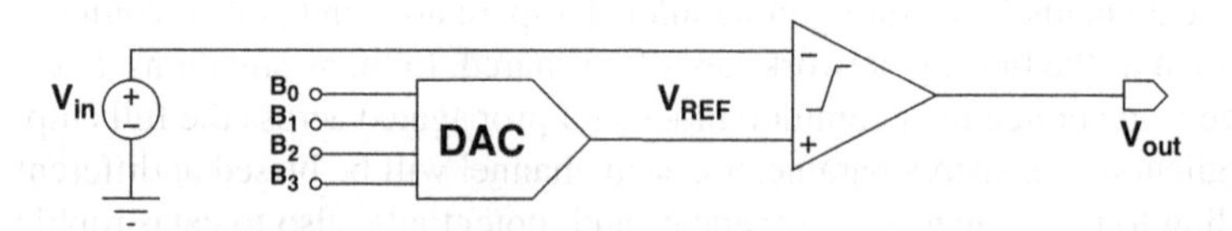

FIGURE 9.43 Discriminator offset compensation with a tuning Digital to Analog Converter (DAC).

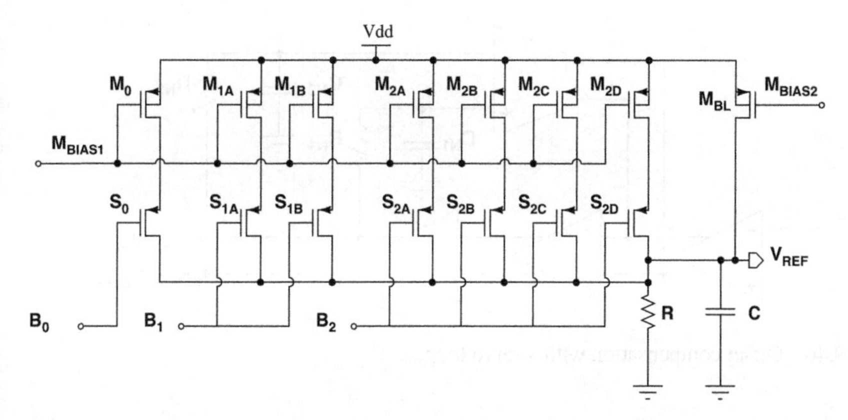

FIGURE 9.44 Example of binary weighted Digital to Analog Converter (DAC) based on current mirrors.

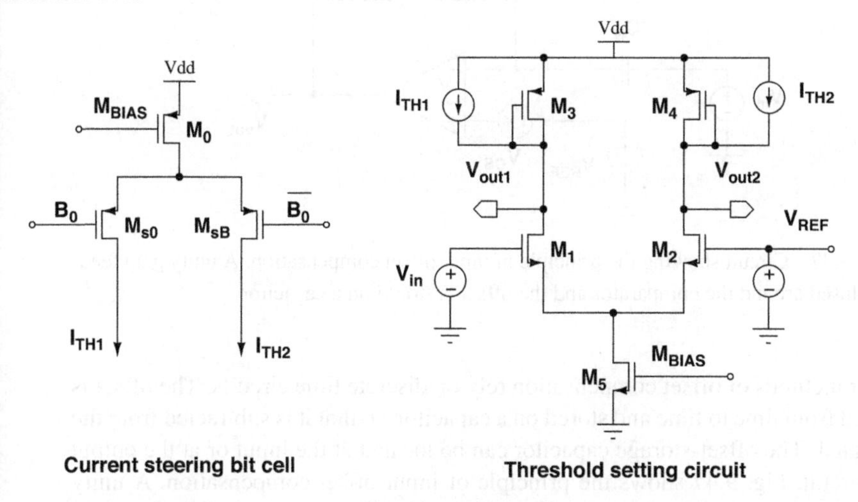

FIGURE 9.45 Current-steering DAC and threshold setting circuit to control the comparator threshold.

in order that only slow variations are presented to the latter. The compensation circuit consists of a transconductance differential amplifier, which generates an output current that minimizes the difference between its input voltages. The residual output offset will be given by the offset of the differential pair and will be seen at the input divided by the gain of the chain. This circuit is particularly interesting for fast, multi-stage comparators, where the input transistors of the cascaded differential cells are kept small to maximize the speed. The compensation circuit, on the other hand, is slow, therefore its input transistors can be made larger to minimize their offset. The principle is similar to the baseline restorer circuits discussed in chapter 5 and chapter 7 and implements a form of active AC coupling, thus rate-dependent threshold variations must be carefully watched.

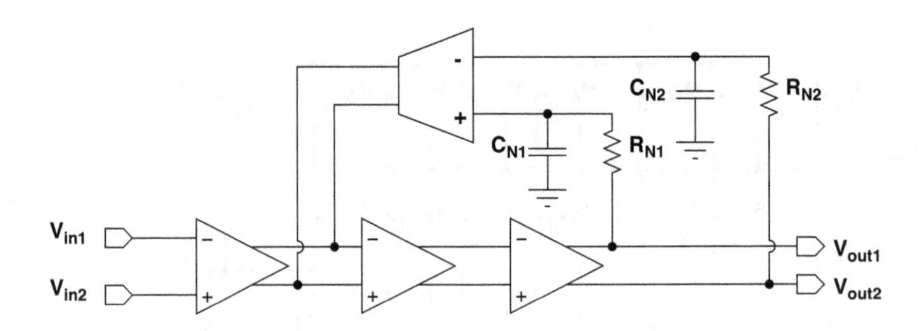

FIGURE 9.46 Offset compensation with a servo loop.

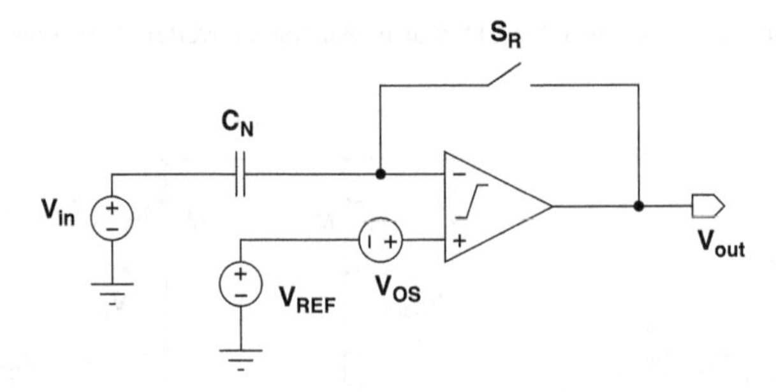

FIGURE 9.47 Circuit showing the principle of input offset compensation. A unity gain feedback is closed around the comparator and the offset is stored on a capacitor.

Other methods of offset compensation rely on discrete time circuits. The offset is measured from time to time and stored on a capacitor, so that it is subtracted from the input signal. The offset-storage capacitor can be located at the input or at the output of the circuit. Fig. 9.47 shows the principle of input offset compensation. A unity gain feedback loop is closed around the stage whose offset must be compensated. In absence of offset and if the gain of the amplifier is high enough, when S_R is closed the output would be approximately equal to V_{REF}. Due to offset, the DC level will be actually different by a quantity ΔV_{out} that can be related to the input offset with a small signal model. In the figure, we have assumed that the offset is added to the reference voltage, but the offset polarity is of course inessential for this discussion. Considering only the small signal contributions, we can write.

$$\Delta V_{out} = \frac{A}{1+A} V_{os} \tag{9.92}$$

where A is the open-loop gain of the amplifier. The signal given by (9.92) is stored on capacitor C_N. The difference between the two input terminals is given by:

$$V_{os} - \frac{A}{1+A} V_{os} = \frac{V_{os}}{1+A} \tag{9.93}$$

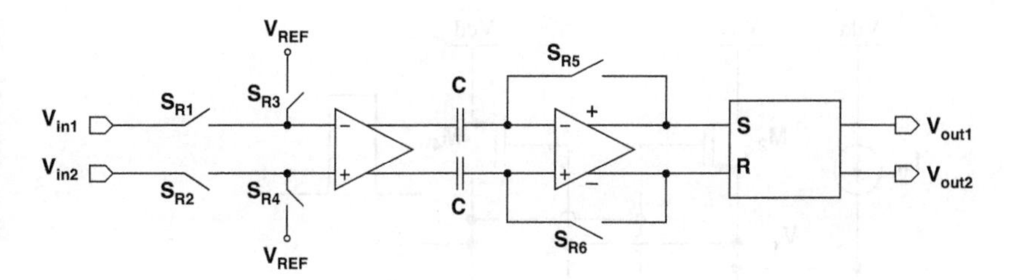

FIGURE 9.48 The offset can be compensated by shorting the amplifier input and storing the output voltage on the coupling capacitors (output offset compensation). In the circuit in the figure output compensation is applied in the first stage and input offset compensation in the second one.

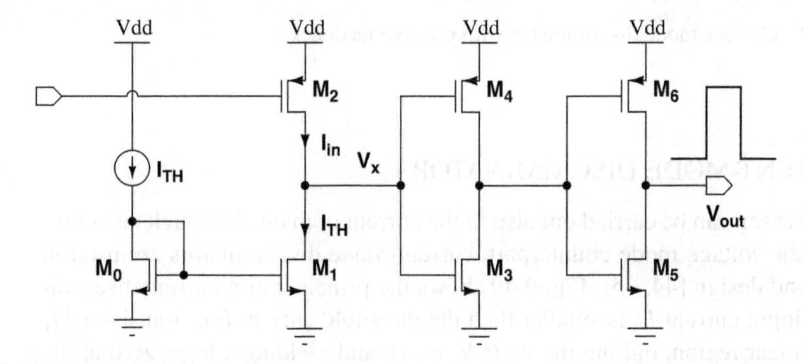

FIGURE 9.49 Principle of a current-mode discriminator. The circuit relies on the fact that when two transistors are connected in series at their drains, the device that has to drive a larger current is pushed in the linear region.

The offset voltage is thus copied on capacitor C_N with an error given by (9.93), which represents the residual offset after compensation. The alternative method is the output offset compensation. Fig. 9.48 shows a comparator in which both techniques are applied simultaneously. In fact, in the second stage the input offset compensation just described is adopted. The residual voltage of the second stage is further divided by the gain of the first one. During offset compensation, the comparator inputs are shorted and the output DC levels of the first stage are stored on the coupling capacitors. In this way, the offset contribution of the first stage is completely suppressed. Three important observations must be done on the dynamic offset compensation techniques:

- With output offset compensation, the gain of the stage must be low enough so that the circuit does not saturate under the effect of its own offset when the inputs are shorted.
- In case the input compensation is applied, the circuit must be stable when the feedback switches are closed.
- The offset compensation must be periodically refreshed.

Due to the random arrival time of the events, in front-end electronics offset compensation with local tuning DACs is very often the preferred choice.

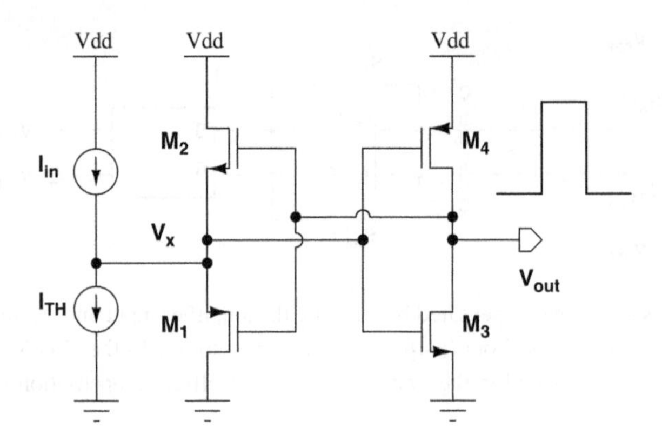

FIGURE 9.50 Current-mode discriminator with positive feedback.

9.9 CURRENT-MODE DISCRIMINATORS

Signal comparison can be carried out also in the current domain. Although less common than their voltage mode counterpart, current-mode discriminators are used in some front-end design [44, 45]. Fig. 9.49 shows the principle of a current discriminator. If the input current I_{in} is smaller than the threshold current I_{TH}, transistor M_1 goes in the linear region, pulling the node V_x down and yielding a logic zero at the output. When I_{in} becomes greater than I_{TH}, it is M_2 which is driven in the linear region, bringing V_x up and making the discriminator output flip. The comparator speed can be improved by using positive feedback [46]. This is shown in Fig. 9.50, which reports a very common topology employed for current comparison. When the threshold current is bigger than the input signal, node V_x is pulled down. The variation on V_x is amplified by the inverter, whose output is applied to the gates of M_1 and M_2. As a consequence, M_2 is quickly switched on and M_2 is switched off. When the signal becomes greater than I_{TH}, V_x starts rising and the change is again amplified by the inverter, so that M_1 is turned on and M_2 off, making the circuit flip.

REFERENCES

1. J. M. Rabaey, A. Chandrakasan, and B. NiKolic. *Digital Integrated Circuits: A Design Perspective.* 2nd edition, Prentice Hall, 2003.
2. S. O. Rice. Mathematical analysis of random noise. *Bell System Technical Journal*, 23:282–332, 1944.
3. S. O. Rice. Mathematical analysis of random noise. *Bell System Technical Journal*, 24:46–156, 1945.

4. Ph. E. Allen and D. R. Holberg. *CMOS Analog Circuit Design*. 2nd edition, Oxford University Press, 2002.
5. W. L. Bryan et al. TGLD: a 16-channel charge readout chip for the PHENIX pad chamber detector subsystem at RHIC. *IEEE Trans. Nucl. Sci.*, 45:754–757, 1998.
6. J. F. Pratte et al. The RatCAP front-end ASIC. In *IEEE NSS-MIC Conference Records*, pages 19–25, 2007.
7. T. Kugathasan, G. Mazza, A. Rivetti, and L. Toscano. A 15 μW 12-bit dynamic range charge measuring front-end in 0.13 μm CMOS. In *IEEE NSS-MIC Conference Records*, pages 1667–1673, 2010.
8. FEI-4 Collaboration. *The FE-I4A Integrated Circuit Guide-version 11.3*.
9. M. Menouni et al. Low power discriminator for the ATLAS pixel chip. In *TWEPP Conference Records*, pages 606–609, 2009.
10. A. Rhouni, J. D. Techer, G. Sou, and M. Berthomier. Design and characterization of a high dynamic range and ultra low power 16-Channel ASIC for an innovative 3D imaging space plasma analyzer. *IEEE Trans. Nucl. Sci.*, 59:2580–2586, 2012.
11. K. Bult and A. Buchwald. An embedded 240 mW 10-b 50MS/s CMOS ADC in 1 mm^2. *IEEE J. Solid-state Circ.*, 32:1887–1895, 1997.
12. P. Grybos et al. RX64DTH–a fully integrated 64-channel ASIC for a digital X-ray imaging system with energy window selection. *IEEE Trans. Nucl. Sci.*, 52:839–846, 2005.
13. J. Kaplon and M. Noy. Front-end electronics for SLHC semiconductor trackers in CMOS 90 nm and 130 nm processes. *IEEE Trans. Nucl. Sci.*, 59:1611–1620, 2012.
14. M. L. Simpson, C. L. Britton, A. L. Wintenberg, and G. R. Young. An integrated CMOS time interval measurement system with subnanosecond resolution for the WA98 calorimeter. *IEEE J. Solid-state Circ.*, 32:198–205, 1997.
15. F. Anghinolfi et al. NINO: an ultra-fast and low-power front-end amplifier/discriminator ASIC designed for the multigap resistive plate chamber. *Nucl. Instr. Meth.*, A533:183–187, 2004.
16. M. Ciobanu, N. Herrmann, K. D. Hildenbrand, M. Kis, and A. Schuttauf. PADI, a fast preamplifier-discriminator for Time-of-Flight measurements. In *IEEE NSS-MIC Conference Records*, pages 2018–2024, 2008.
17. M. Chiosso, O. Cobanoglu, G. Mazza, D. Panzieri, and A. Rivetti. A fast binary front-end ASIC for the RICH detector of the COMPASS experiment at CERN. In *IEEE NSS-MIC Conference Records*, pages 1495–1500, 2008.
18. J. Nissinen, I. Nissinen, and J. Kostamovaara. Integrated receiver including both receiver channel and TDC for a pulsed Time-of-Flight laser rangefinder with cm-level accuracy. *IEEE J. Solid-state Circ.*, 44:1486–1497, 2009.
19. F. Powolny et al. Time based readout of a silicon photomultiplier (SiPM) for time of flight positron emission tomography (TOF-PET). *IEEE Trans. Nucl. Sci.*, 58:1212–1219, 2009.
20. E. Martin et al. Review of results for the NA62 Gigatracker read-out prototype. *Journal of Instrumentation*, 7, 2012.

21. J. F. Pratte et al. Front-end electronics for the RatCAP mobile animal PET scanner: timing discriminator and 32 line address priority serial encoder. In *IEEE NSS-MIC Conference Records*, pages 2211–2215, 2004.
22. R. L. Chase. Pulse timing system for use with gamma-rays on Ge(Li) detectors. *Review of Scientific Instruments*, 39:1318–1326, 1968.
23. H. Spieler. *Semiconductor Detector Systems*. Oxford Science Publications, 2005.
24. M. L. Simpson, G. R. Young, R. G. Jackson, and M. Xu. A monolithic, constant-fraction discriminator using distributed RC delay line shaping. *IEEE Trans. Nucl. Sci.*, 43:1695–1699, 1996.
25. C. H. Nowlin. Low-noise lumped-element timing filters with rise-time invariant crossover times. *Review of Scientific Instruments*, 63:2322–2326, 1992.
26. D. M. Binkley. Performance of non-delay-line constant-fraction discriminator timing circuits. *IEEE Trans. Nucl. Sci.*, 41:1169–1175, 1994.
27. S. Martoiu, A. Rivetti, G. Mazza, F. Osmic, and F. Marchetto. A low power front-end prototype for silicon pixel detectors with 100ps time resolution. In *IEEE NSS-MIC Conference Records*, pages 2958–2961, 2008.
28. S. Garbolino, S. Martoiu, and A. Rivetti. Implementation of Constant-Fraction-Discriminators (CFD) in sub-micron CMOS technologies. In *IEEE NSS-MIC Conference Records*, pages 1530–1535, 2011.
29. H. Jeon and Y. Kim. A CMOS low-power low-offset and high-speed fully dynamic latched comparator. *IEEE International SoC Conference Records*, pages 285–288, 2010.
30. B. Chen, J. Wang, and C. Tsai. A 3GHz, 22-ps/dec dynamic comparator using negative resistance combined with input pair. In *IEEE Asia-Pacific Conference on Circuits and Systems*, 2010.
31. S. Chin, C. Hsieh, C. Chiu, and H. Tsai. A new rail-to-rail comparator with adaptive power control for low power SAR ADCs in biomedical application. In *IEEE International Symposium on Circuits and Systems*, pages 1575–1578, 2010.
32. B. Goll and H. Zimmermann. A comparator with reduced delay time in 65-nm CMOS for supply voltages down to 0.65 V. *IEEE Trans. on Circ. and Syst. II-express Briefs*, 56:810–814, 2009.
33. S. Lan, C. Yuan, Y. Y. H. Lam, and L. Siek. An ultra low-power rail-to-rail comparator for ADC designs. In *Midwest Symposium on Circuits and Systems*, pages 1–4, 2011.
34. P. M. Figueiredo and J. C. Vital. Kickback noise reduction techniques for CMOS latched comparators. *IEEE Trans. on Circ. and Syst. II-express Briefs*, 53:541–545, 2006.
35. G. A. Fahmy, R. K. Pokharel, H. Kanaya, and K. Yoshida. A 1.2 V 246 μW CMOS latched comparator with neutralization technique for reducing kickback noise. In *TENCON, IEEE Region 10 International Conference*, pages 1162–1165, 2010.
36. Y. Huang, H. Schleifer, and D. Killat. Design and analysis of novel dynamic latched comparator with reduced kickback noise for high-speed ADCs. In *2013*

IEEE European Conference on Circuit Theory and Design (ECCTD), 2013.
37. V. Singh, N. Krishnapura, S. Pavan, B. Vigraham, N. Nigania, and D. Behera. A 16 MHz BW 75 dB DR CT $\Sigma - \Delta$ ADC compensated for more than one cycle excess loop delay. *IEEE J. Solid-state Circ.*, pages 1884–1895, 2012.
38. K. A. Bowmann, S.G. Duvall, and J.D. Meindl. Fluctuations on the maximum clock frequency distribution for Gigascale integration. *IEEE J. Solid-state Circ.*, 37:183–190, 2002.
39. J. B. Shyu, G. B. Temes, and K. Yao. Random errors in MOS capacitors. *IEEE J. Solid-state Circ.*, SC-17:1070–1075, 1982.
40. J. B. Shyu, G. B. Temes, and F. Krummenacher. Random error effects in matched MOS capacitors and current sources. *IEEE J. Solid-state Circ.*, 19:948–956, 1984.
41. K. R. Lakshmikumar, R. A. Hadaway, and M. A. Copeland. Characterization and modeling of mismatch in MOS transistors for precision analog design. *IEEE J. Solid-state Circ.*, 21:1057–1066, 1986.
42. M. J. M. Pelgrom, A. C. J. Duinmaijer, and A. P. G. Welbers. Matching properties of MOS transistors. *IEEE J. Solid-state Circ.*, 24:1433–1439, 1989.
43. P. R. Kinget. Device mismatch and tradeoffs in the design of analog circuits. *IEEE J. Solid-state Circ.*, 40:1212–1224, 2005.
44. F. Corsi, M. Foresta, C. Marzocca, G. Matarrese, and A. Del Guerra. Experimental results from an analog front-end channel for silicon photomultiplier detectors. In *IEEE NSS-MIC Conference Records*, pages 2010–2014, 2008.
45. W. Shen and H.-C. Schultz-Coulon. STIC — A current-mode constant fraction discriminator for positron emission tomography using SiPM (MPPC). In *IEEE NSS-MIC Conference Records*, pages 364–367, 2009.
46. H. Traff. Novel approach to high speed CMOS current comparators. *Electronics Letters*, 28, 1992.

10 Data Converters

In the first generations of integrated front-ends, it was common to implement on the ASIC only the indispensable analog functions and to digitize the data off-chip with standard components. The capability of modern CMOS technologies to fabricate complex digital circuits in small areas and with low power consumption makes it preferable today to embed data converters directly on the front-end chip. In the readout of radiation detectors, three types of data converters are of interest: Digital to Analog Converters (DAC), Analog to Digital Converters (ADC) and Time to Digital Converters (TDC). Digital to Analog Converters are primarily used for slow control functions, such as setting bias currents and voltages or discriminator thresholds. The required resolution is in most cases from a few bits to 7-8 bits and the operation speed is typically modest. Their design does not pose specific challenges and it is therefore not discussed in detail here. The reader is referred to standard microelectronics textbooks for the topic. The design of ADC and TDC presents instead peculiar features. The required resolution is often moderate (less than 10 bits), but the large number of converters that must be accommodated on chip and the low power consumption allowed for their operation are rarely found in other applications. Therefore, these components deserve a specific treatment. In the first part of the chapter, we review the key ADC performance metrics, emphasizing the aspects that are more relevant for the readout of radiation sensors. Since ADCs are extremely common components, many different architectures have been reported in the literature. In the second section, we focus our attention on two of the most useful topologies for radiation sensor front-ends, namely the Wilkinson and the the Successive Approximation Register (SAR). Descriptions of other ADC architectures can be found in specialized books, such as [1–4]. The second part of the chapter is devoted to TDCs. Also for this building block a lot of different variants exist. An exhaustive review is available in [5]. Here, after a general introduction, we describe a few schemes which are of greater interest to the field of radiation detection.

10.1 BASIC ADC PROPERTIES

An ADC associates an analogue input voltage to a digital output code. In the converter, a reference voltage V_{REF} is partitioned in 2^N levels, against which the input voltage is compared. The total number of bits in the binary output code defines the resolution of the ADC, while the quantity $V_{REF}/2^N$ identifies the Least Significant Bit (LSB). A 4 bits ADC can thus encode $2^4 = 16$ different analog levels, while a 10 bit ADC can represent $2^{10} = 1024$ analog steps. For an N bit ADC, the digital output word can be written as:

$$B_{out} = B_{N-1}B_{N-2}\cdots B_0 \tag{10.1}$$

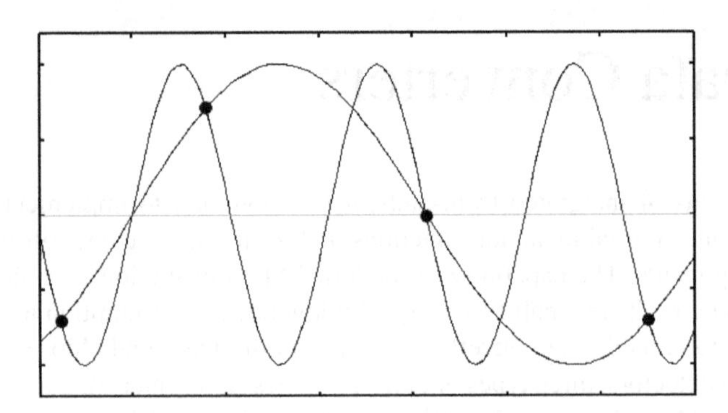

FIGURE 10.1 Intuitive illustration of the aliasing phenomenon. A sinusoid of frequency f_{in} is sampled at a frequency smaller than $2 \cdot f_{in}$. When the signal is recovered from its samples, a sinusoid at a lower frequency is also introduced, creating an ambiguity in the signal reconstruction.

where B_{N-1} is the Most Significant Bit (MSB) and B_0 is the LSB. The analogue value can be reconstructed from the digital code as:

$$V_{in} = B_{N-1}\frac{V_{REF}}{2} + B_{N-2}\frac{V_{REF}}{4} + \cdots + B_0\frac{V_{REF}}{2^N} \tag{10.2}$$

The number of conversions per second defines the "sampling rate" of the converter, typically measured in Mega-Samples per second (MS/s). From this point of view, ADCs are classified in two categories: Nyquist rate and oversampling converters. Nyquist rate ADCs are designed to process input signals containing a maximum frequency that can be up to one-half the ADC sampling frequency. The Nyquist-Shannon theorem states in fact that a signal can be reconstructed from its samples if the sampling frequency is at least twice the maximum frequency contained in the signal itself. If the signal is under-sampled, the phenomenon of aliasing is introduced, as shown in Fig. 10.1. Aliasing can be avoided by low pass filtering the signals before presenting them to the converter, so that the Nyquist criterion is fulfilled. In oversampling converters, the sampling frequency is much higher than the one of the input signal (typical oversampling ratios are 64 or 128) and the redundant information is used to achieve, through digital signal processing, a very good accuracy. Front-end for radiation detectors usually demand moderate resolution and fast conversion speed, therefore Nyquist rate ADCs are preferred. The front-end amplifiers have normally a modest bandwidth and in most cases an ad-hoc anti-aliasing filter is not necessary.

Fig. 10.2 shows the ideal characteristics of a 3 bit ADC under the hypothesis that the input signal, also shown in the figure, is a ramp. In this case, the ADC characteristics have a staircase shape, where the ladder (bins) are equally spaced by $V_{REF}/2^N$. All analog inputs falling within the same bin generate the same digital output. For instance, all values below $V_{REF}/8$ are associated to the code 000, all values between $V_{REF}/8$ and $V_{REF}/4$ are mapped on the code 001, and so on. Therefore, there is an intrinsic error in the conversion process. The error made depends on the difference between

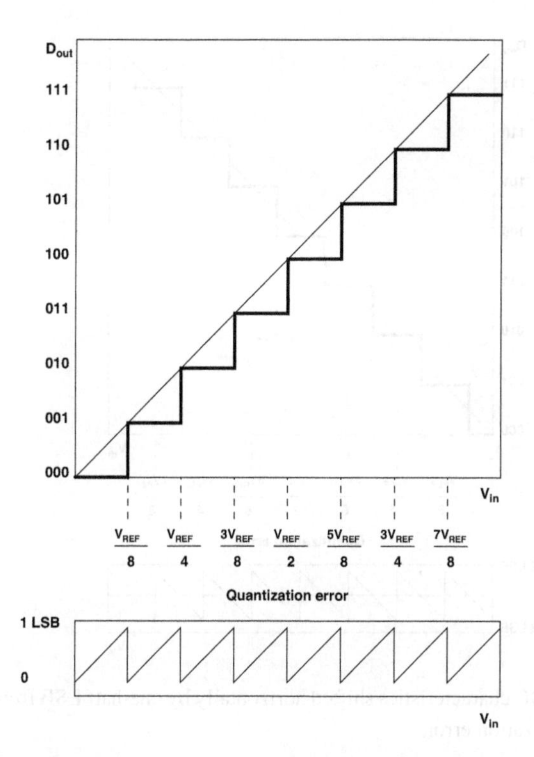

FIGURE 10.2 Ideal characteristics of a 3 bit ADC. The bottom part of the plot shows the quantization error as a function of the input signal.

the input signal and the voltage that determines a given code transition. For instance, starting from the origin the error is zero and it increases up to a maximum amplitude of 1 LSB when the input voltage is just about to reach $V_{REF}/8$, then it goes back to zero when the latter is reached and starts re-growing till the next transition is made and so on, generating the characteristic saw-tooth pattern sketched in the bottom of Fig. 10.2. Fig. 10.3 shows a slightly different characteristic, which is shifted by one half LSB from the origin. In this way, the value conventionally attributed to each code falls in the middle of the bin and the quantization error is symmetrical. In principle, all analog inputs have the same probability of falling in any point of a bin. The probability density function is thus uniform and inversely proportional to the bin width. The rms of the quantization error can hence be calculated as:

$$V_{rms}^2 = \int_{-LSB/2}^{LSB/2} x^2 \frac{1}{LSB} dx \rightarrow V_{rms} = \frac{LSB}{\sqrt{12}} \tag{10.3}$$

For instance, for an ideal 10 bit ADC with a full scale range of 1 V, the amplitude of the LSB is 976 μV and the rms quantization error is 282 μV. The Signal to Noise Ratio (SNR) for an ideal quantizer is the ratio between the power of the input signal and the one of the quantization error. This figure is usually defined assuming as input a sinusoid with peak-to-peak corresponding to the full scale range V_{REF} and amplitude

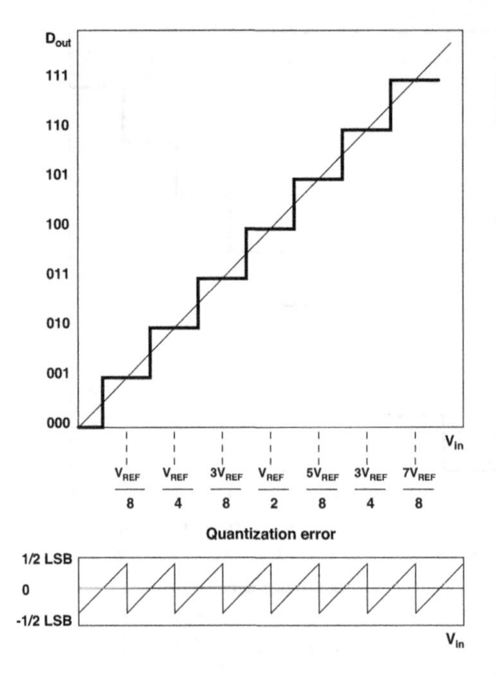

FIGURE 10.3 Ideal ADC characteristics shifted horizontally by one-half LSB from the origin to symmetrize the quantization error.

equal to $V_{REF}/2$. The SNR can be written in dB as:

$$SNR = 20\log\frac{\frac{V_{REF}}{2\sqrt{2}}}{\frac{V_{REF}}{2^N\sqrt{12}}} \tag{10.4}$$

The above equation can be easily simplified to:

$$SNR = 20\log\sqrt{\frac{3}{2}} + 20N\log 2 = 6.02N + 1.76\text{ dB} \tag{10.5}$$

If the signal-to-noise ratio is known, the number of bits of the ADC can thus be inferred by:

$$N = \frac{SNR - 1.76}{6.02} \tag{10.6}$$

The quantization error will sum-up in quadrature with the uncorrelated noise already contained in the signal which is presented to the ADC. In a front-end system, the noise should be dominated by the input stage of the front-end amplifier. If the value of the LSB is chosen equal to the rms of the front-end noise, the rms due to the quantization error is $\sqrt{12} \approx 3.46$ times smaller and when summed in quadrature it increments the overall noise only by 4%. Increasing further the ADC resolution allows a more precise characterization of the background and further reduces the impact of the quantization error. However, the design complexity, the power and the area of an ADC increases exponentially with the number of bits, and overstating the ADC resolution can result in complex and power-hungry circuits. At the system design level, it is therefore important to understand in detail the required ADC resolution by studying the deterioration of the system performance as the ADC number of bits is reduced.

The maximum SNR defined in (10.4) is the one obtained from an ideal quantizer which is only limited by the unavoidable quantization error intrinsic to the conversion process. In practice, several additional effects can easily degrade the resolution of a real ADC below this theoretical limit. We now examine the most common non-idealities which are found in Analog-to-Digital Converters, starting from the ones that can be already identified from a characteristics obtained through very slow changes of the input voltage.

10.1.1 ADC STATIC PERFORMANCE

Fig. 10.4 shows the offset error. The staircase begins sooner or, as shown in the figure, later than what it should do, with the results that either the smaller or the higher signals are clipped. The offset can easily be counteracted, for instance, by adjusting the baseline level of the front-end through a Digital to Analog Converter. The gain error changes the values of the LSB, making it smaller or larger. If the LSB is smaller, as shown in Fig. 10.5, the resolution is increased, but the circuit saturates earlier on large signals. If the LSB becomes larger, the resolution becomes coarser

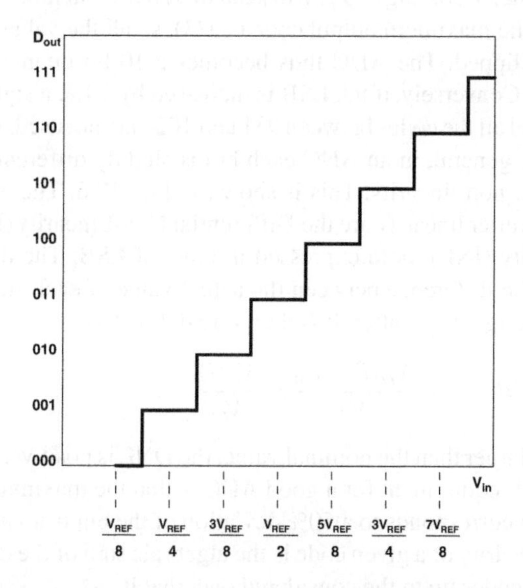

FIGURE 10.4 Effect of offset in ADC. The offset shifts horizontally the converter characteristics. Depending on the sign of the offset, small signals can be completely cut-off or large signals close to the full scale range are clipped.

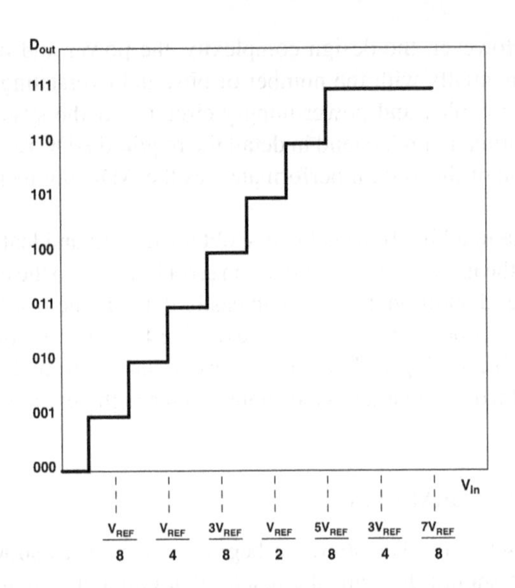

FIGURE 10.5 Example of ADC gain error, showing the case in which the amplitude of the LSB is reduced. This leads to a premature saturation of the circuit.

and the codes in the output word are not optimally used. Suppose, for instance, that in a 10 bit ADC that works with a full scale range of 1 V the size of the LSB is reduced by 10% of its theoretical value, becoming 879 μV instead of 976 μV. An input voltage of 0.9 V already produces the maximum output code (1023), so all the values between 0.9 and 1 V would be clipped. The ADC thus becomes a 10 bit quantizer with a full scale range of 0.9 V. Conversely, if the LSB is increased by 10%, a signal of 1 V produces the code 931 and all the codes between 931 and 1023 are not used, increasing the quantization error. In general, in an ADC each bin is slightly different from the others, introducing a true non-linearity. This is shown in Fig. 10.6. The parameters that characterize the converter linearity are the Differential Non-Linearity (DNL) and the Integral Non-Linearity (INL), both expressed in units of LSB. The differential non-linearity measures the difference between the actual value of each step and the theoretical one, normalizing on the latter. It is therefore defined as:

$$DNL(i) = \frac{V_m(i) - V_{th}}{V_{th}} = \frac{V_m(i)}{V_{th}} - 1 \tag{10.7}$$

If the actual bin width is larger than the nominal value, the DNL is positive, otherwise it is negative. The typical requirement for a good ADC is that the maximum DNL is less than 0.5 LSB, which corresponds to a 50% deviation of the bin from its nominal size. The integral non-linearity of a given code is the algebraic sum of the differential non-linearities of all the codes up to the considered one, that is:

$$INL(i) = \sum_{k=0}^{k=i} DNL(k) \tag{10.8}$$

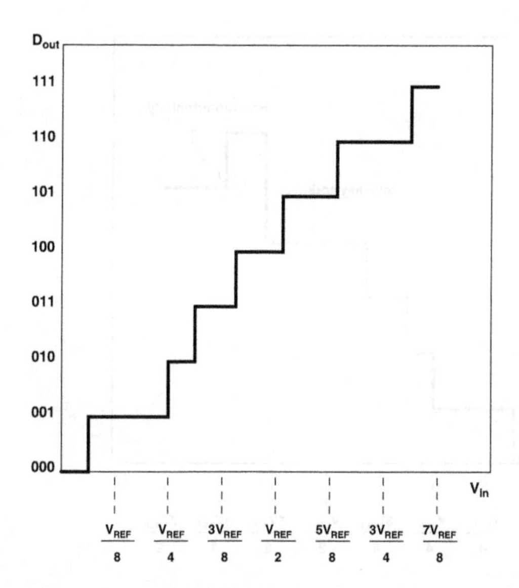

FIGURE 10.6 The non-uniform deviation of the amplitude of each step from its nominal value produces a non-linearity.

Like the DNL, also the INL is usually specified to be less than half an LSB. In applications with radiation sensors several factors limit the linearity of a measurement. Signal fluctuations inside a sensor and distortion in the front-end amplifier often lead to an overall non-linearity of a few per cent of the full scale range. The resolution of the ADC is defined by the ratio between the maximum signal of interest and the noise floor, but often it is not necessary to maintain the INL below the 0.5 LSB limit. When very high dynamic range is necessary, converters with bi-linear or multilinear transfer functions are sometimes employed. The number of bits in the converter output word is fixed, but the full scale range is adapted to the signal to be digitized, so that in each range the quantization error is kept approximately constant. A DNL of 0.5 LSB is acceptable in most applications except a few, like nuclear spectroscopy, where the energy levels of a nucleus need to be classified with the best possible accuracy. In this case, a much better DNL performance is desired, with deviations in the order of a few per cent from the theoretical value, that is equivalent to a DNL of 0.01-0.02 LSB.

Fig. 10.7 shows the characteristics of an ADC with two serious pathologies. First, the DNL of the fourth code is greater than one LSB, and the code incorporates the next one, which never appears in the ADC output stream, a condition known as "missing code". Furthermore, the characteristics are non-monotonic because for the last step the output code decreases when the analog input is increased. Although the two defects are shown simultaneously in the picture, the fact that an ADC has missing codes does not imply that it is also non-monotonic.

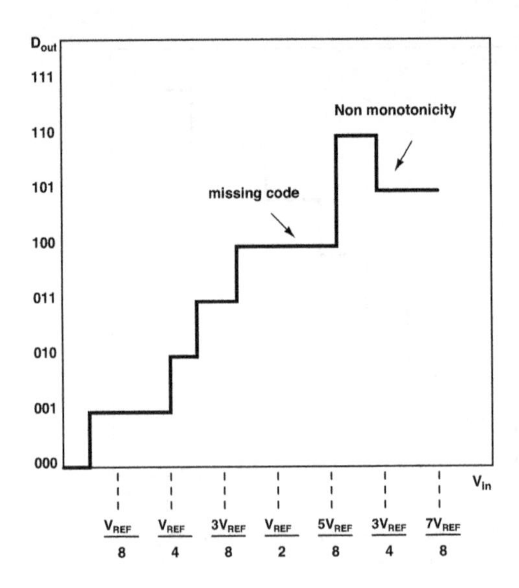

FIGURE 10.7 Example of non-monotonic ADC transfer function with missing codes.

10.1.2 ADC DYNAMIC PERFORMANCE

The performance of an ADC can deteriorate quickly as the frequency of the input signal increases. Slew rate and bandwidth limitations in the converter components, settling time problems and ringing on the reference voltages are progressively exacerbated as the input signal becomes faster. It is therefore necessary to study the converter performance with appropriate dynamic input signals. A linear voltage ramp, although immediately yields at the output the more intuitive staircase curve, it is not a suitable signal for precise ADC characterization. In fact, fast voltage ramps are difficult to generate and often their linearity is no better than 1%, which is insufficient to test even a 7 bit converter. Though less intuitive, a sinusoidal signal is a much better choice for the following reasons:

- Sinusoidal signal generators with very good spectral purity and the second harmonic located more than 95 dB below the fundamental are easily available.
- A sinusoid is easy to treat mathematically and the data analysis and interpretation is simplified.
- A sinusoid is bidirectional and points out problem depending on the sign of the slope of the input signal.

For a complete characterization of the converter it is important to use full scale signals, so that all the possible codes are explored. In fact, the performance measured with small signals can not be reliably extended to the full scale range, because slew rate limitations have a different impact on small and large signals.

The DNL and the INL can be measured with the histogram testing [6]. A large number of samples is collected, counting the number of times a given code is found. This number, indicated as N_m, is compared to the one expected from theoretical calculations, N_{th}. The DNL is then found as:

$$DNL = \frac{N_m}{N_{th}} - 1 \tag{10.9}$$

Larger bins in the ADC characteristics result in more populated codes, while narrower bins are under-filled. Missing codes are easily detected as “holes” in the distribution. For a sinusoidal wave of the form $V(t) = A \sin \omega t$, it can be proven that the probability that the voltage lies between V and $V + dV$ is given by the following probability density function:

$$p(V) = \frac{1}{\pi \sqrt{A^2 - V^2}} \tag{10.10}$$

The probability that a sample is found between V_a and V_b is thus given by:

$$P(V_a < V < V_b) = \int_{V_a}^{V_b} p(V) dV = \frac{1}{\pi}\left(\arcsin \frac{V_b}{A} - \arcsin \frac{V_a}{A} \right) \tag{10.11}$$

Once the INL is measured, the DNL is calculated by applying (10.8).

By processing the data samples obtained with a sinusoidal input, further information on the converter performance can be obtained. Applying in fact the Discrete Fourier Transform (DFT), the ADC behavior can be studied in the frequency domain. For a set of n samples ($n = 0, 1, \cdots N-1$), taken every T_S seconds, the DFT is defined as:

$$X_k = \sum_{n=0}^{N-1} x(nT_S) e^{-j2\pi k \frac{n}{N}} \tag{10.12}$$

In practice, the more efficient Fast Fourier Transform (FFT) is employed to reduce the total number of computations from the N^2 required by the DFT defined in (10.12) to $N \log_2 N$. If f_S is the sampling frequency, the number of samples N collected during a time T is given by:

$$N = T f_S \tag{10.13}$$

In this window, the input signal should also make an integer number of cycles M:

$$M = T f_{in} \tag{10.14}$$

Combining the two equations we have:

$$M = N \frac{f_{in}}{f_S} \tag{10.15}$$

The requirement of an integer number of cycles of the input signal comes from the DFT properties. In fact, the sinusoidal signal, which is principle infinite, is truncated because the observation window is limited to a finite acquisition time T. If the non-integer number of periods is considered, when the DFT is calculated the information

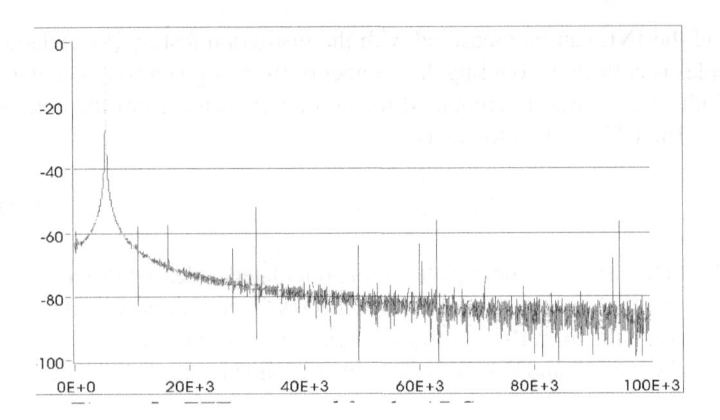

FIGURE 10.8 Spectrum obtained with a DFT algorithm.

"leaks" outside the fundamental harmonic, thus corrupting the spectrum. The leakage phenomenon can also be reduced by using suitable windows functions, that give more weight to the data in the center of the record and less to the ones at the borders [7, 8]. Fig. 10.8 shows the example of a spectrum obtained with an FFT. The "broadening" of the fundamental, that should ideally be a monochromatic peak, shows the effect of the leakage in the DFT. In the FFT test, the sampling and the input frequencies should not be related by an integer number, because otherwise the input signal is repetitively sampled in the same points and some codes of the ADC are more explored than others, introducing a bias in the measurement.

From the DFT data, the main ADC dynamic parameters can be calculated. In case of a quantizer with infinite resolution, only the fundamental should be present in the spectrum. However, the ADC approximates the perfectly smooth sinusoidal waveform with a succession of small steps, each corresponding to one LSB. This introduces other harmonics at higher frequencies. The Total Harmonic Distortion (THD) is thus defined as the ratio between the amplitude of the first six or ten harmonics present in the spectrum and the one of the fundamental. More precisely, we have:

$$THD = 20\log\sqrt{\frac{\sum_{k=2}^{6} A_k^2}{A_f^2}} \tag{10.16}$$

where A_k is the amplitude of the kth harmonic and A_f is the amplitude of the fundamental, corresponding to the frequency of the input signal. The Signal to Noise and Distortion (SINAD) evaluates the ratio between the amplitude of the fundamental and the one of all other components present in the DFT spectrum. If the fundamental has amplitude A_f and is located in the bin f, we can write the SINAD as:

$$SINAD = 20\log\sqrt{\frac{A_f^2}{\sum_{k=0}^{f-1} A_k^2 + \sum_{k=f+1}^{N/2} A_k^2}} \tag{10.17}$$

Applying now equation (10.6) and replacing the *SNR* with the *SINAD*, we obtain the Effective Number of Bits (ENOB):

$$ENOB = \frac{SINAD - 1.76}{6.02} \tag{10.18}$$

which is of course smaller than the one offered by an ideal quantizer. Finally, another parameter of interest that can be obtained from the DFT spectrum is the "Spurious Free Dynamic Range" (SFDR), which is simply the ratio between the fundamental and the highest extra components found in the spectrum (not necessarily located at a multiple of the input frequency):

$$SFDR = 20\log\sqrt{\frac{A_f^2}{A_{spurious}^2}} \tag{10.19}$$

The effective number of bits, the power consumption P and the sampling frequency f_S are combined together in a Figure of Merit (FoM), which is frequently used to quantify how good a converter is:

$$FoM = \frac{P}{f_S \times 2^{ENOB}} \tag{10.20}$$

The FoM is measured in unit of joules per conversion step (J/step) and for a state of the art ADC is below the pJ/step. For instance, an ADC with 7 effective bits, 50 MS/s and 1 mW of power has an FoM of 0.15 pJ/step. In the DFT test, the DNL errors and missing codes manifest themselves as an increased noise floor, while the INL increments the distortion. Gain and offset errors can not be detected by the DFT analysis, but they can be identified with the histogram testing. Before closing this session on ADC performance, it is important to observe that the sampling signal of the ADC must have a very low jitter, otherwise the converter performance can be easily compromised. To understand this point, consider an ADC with a full scale range V_{REF}, processing a sinusoidal signal with a peak to peak equal to the full scale, that means:

$$V_{in} = \frac{V_{REF}}{2}\sin(\omega_{in}t) \tag{10.21}$$

The maximum slope of the input signal is given by:

$$\left.\frac{dV}{dt}\right|_{MAX} = \left.\frac{V_{REF}}{2}\omega_{in}\cos(\omega_{in}t)\right|_{MAX} = \frac{V_{REF}}{2}\omega_{in} \tag{10.22}$$

At the point of maximum slope, a sampling signal with an rms jitter σ_t induces an rms uncertainty on the captured voltage given by:

$$\sigma_V = \frac{V_{REF}}{2}\omega_{in}\sigma_t \tag{10.23}$$

Called "aperture jitter", this effect determines an additional noise in the converter output data. In principle, this noise should not dominate the ADC performance and should be therefore kept below the quantization error. We can thus write:

$$\sigma_V = V_{REF}\pi f_{in}\sigma_t < \frac{V_{REF}}{2^N\sqrt{12}} \rightarrow \sigma_t < \frac{1}{2^N 2\pi f_{in}\sqrt{3}} \qquad (10.24)$$

where we have put $\omega_{in} = 2\pi f_{in}$. The constraint on the jitter comes from the maximum frequency that the converter is expected to digitize and for a Nyquist rate ADC this is one half the sampling frequency f_S. For instance, for a 10 bit ADC designed to acquire 50 MS/s, f_{in} is 25 MHz and the maximum tolerable jitter is 3.6 ps rms.

10.2 ADC ARCHITECTURES

Among the many possible ADC architectures, the most recurrent ones in front-end chips for radiation detectors are: Flash, Single ramp, Wilkinson and Successive Approximation Register (SAR). Our discussion on architectures will be therefore limited to these topologies.

10.2.1 FLASH ADC

The Flash ADC is the most straightforward implementation of an analog to digital converter and consists of three fundamental blocks: a resistor ladder, a comparator bank and an output encoder. Fig. 10.9 shows the concept of a 2 bit flash ADC. The reference voltage V_{REF} is partitioned into a series of taps by the voltage divider. The taps can be uniform (all resistors are equal) or can be scaled, so that the relative digitization error is approximately constant for small and large signals. The input signal is presented simultaneously to all comparators, and the ones for which the input is higher than the threshold respond with a logic one at the output, while all the other ones give a zero. The bit pattern generated by the comparator bank is thus a thermometric code, that is converted into a binary sequence by a digital encoder. The comparators usually employ some form of positive feedback and the circuit operation is regulated by a sampling clock. The clear advantage of the flash digitization is the conversion speed, as the output digital word is made available within a single clock cycle. The disadvantage is that the complexity increases exponentially with the number of bits and the area and power consumption quickly reach levels that prevent the integration of such a topology on a per-channel basis in a high density front-end. In fact, the number of comparators in a flash ADC is $2^N - 1$, so 255 comparators are necessary for an 8 bit system. Due to the high number of comparators, the input capacitance of the converter can be fairly high. The comparator offset affects the ADC linearity and the kick-back noise is also an issue. In practice, high resolution flash ADC are not implemented with a straightforward approach, but using the folding and interpolating method that allows us to reduce significantly the number of comparators [9–11]. However, the use of folding converters in radiation detection application is not very common. In multi-channel front-ends, standard flash ADC are typically limited to 3-4 bits resolution in contexts where a coarse information on the signal amplitude is adequate [12, 13].

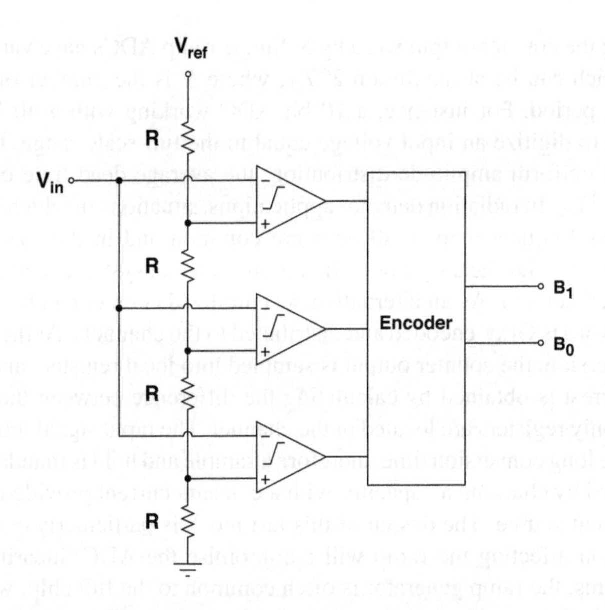

FIGURE 10.9 Flash ADC. A 2 bit converter is shown in the figure.

10.2.2 SINGLE RAMP AND WILKINSON ADC

The single ramp ADC, shown in Fig. 10.10 is widely used in front-end design because of its simplicity and low power consumption. The fundamental building blocks of a single ramp ADC are the comparator, the ramp generator and the counter. The analog voltage to be digitized is presented to one input of the comparator. An analog ramp is then started and a clock counter is enabled. A suitable digital logic, not shown in the figure, synchronizes the start of the ramp with the clock, so that the ramp begins simultaneously with a clock transition. When the analog ramp equalizes the unknown input voltage, the comparator flips and the counter is frozen. If S is the slope of the analog ramp measured in volt per clock cycle, the input voltage is simply

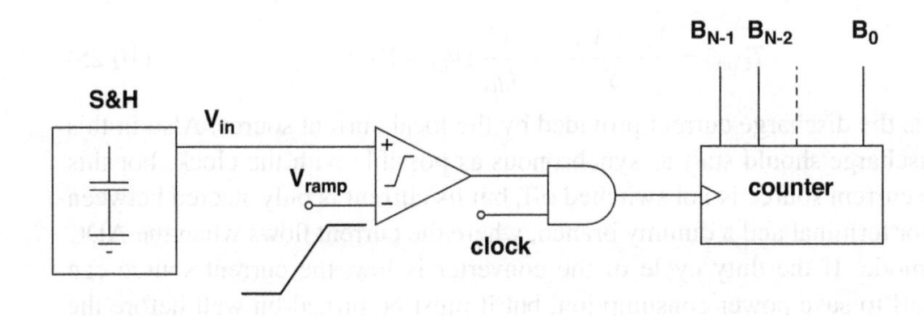

FIGURE 10.10 Single ramp ADC.

obtained multiplying the counter output word by S. Single ramp ADCs have variable conversion time, which can be at maximum $2^N T_{ck}$, where N is the number of bits and T_{ck} is the clock period. For instance, a 10 bit ADC working with a 40 MHz clock needs 25.6 μs to digitize an input voltage equal to the full-scale range. If the input signals have a uniform amplitude distribution, the average dead time of the converter will be $2^{N-1} T_{ck}$. In radiation detector applications, situations in which large signals are much less frequent than small ones are common and in this case the average dead-time can be significantly lower. In a multi-channel system, each ADC can have its individual counter. As an alternative, a centralized counter can be used. The counter output word is Gray-encoded and distributed to the channels. At the start and stop of the conversion, the counter output is sampled into local registers and the measurement of interest is obtained by calculating the difference between the two values. In this way, only registers are located in the channel. The input signal must be kept stable during the long conversion time, therefore a sample and hold is mandatory. The ramp is generated by charging a capacitor with a constant current provided by a high impedance current source. The design of this last block is particularly critical, because any distortion affecting the ramp will compromise the ADC linearity. In multi-channels systems, the ramp generator is often common to the full chip, which implies that the digitization occurs simultaneously in all channels. To avoid that the input capacitance of the comparators loads the ramp generator, reducing its slope, the former can be buffered. However, the buffer itself can introduce non-linearity in the ramp. As an alternative, one should employ in the ramp generator a capacitor large enough to make negligible the comparator loading effect.

In the detector community, single-ramp converters are often called "Wilkinson ADCs". Although similar, in the Wilkinson ADC [14] the signal is first used to charge a capacitor which is then discharged by a constant current, so it is basically a dual ramp converter. In most applications, the charging ramp is much faster than the discharging one. The latter thus defines the conversion speed, which is similar to that of the single ramp converter. A possible implementation of a Wilkinson ADC is shown in Fig. 10.11. Here the input voltage is first used to charge the capacitor C. The current source is then connected to the input, the counter is started and the capacitor is discharged. The conversion is ready when the voltage at the positive comparator input reaches again the baseline, freezing the counter. The digitization time is given by:

$$T_{Conv} = \frac{V_{in} - V_{BL}}{S} = \frac{C}{I_{dis}} (V_{in} - V_{BL}) \tag{10.25}$$

where I_{dis} is the discharge current provided by the local current source. Also in this case the discharge should start as synchronous as possible with the clock. For this reason, the current source is not switched off, but its current is only steered between the capacitor terminal and a dummy branch, where the current flows when the ADC is in idle mode. If the duty cycle of the converter is low, the current source can be turned-off to save power consumption, but it must be turned-on well before the discharge is started to make sure the switch-on transients do not compromise the performance. In multi-channel implementations, the value of the current source changes

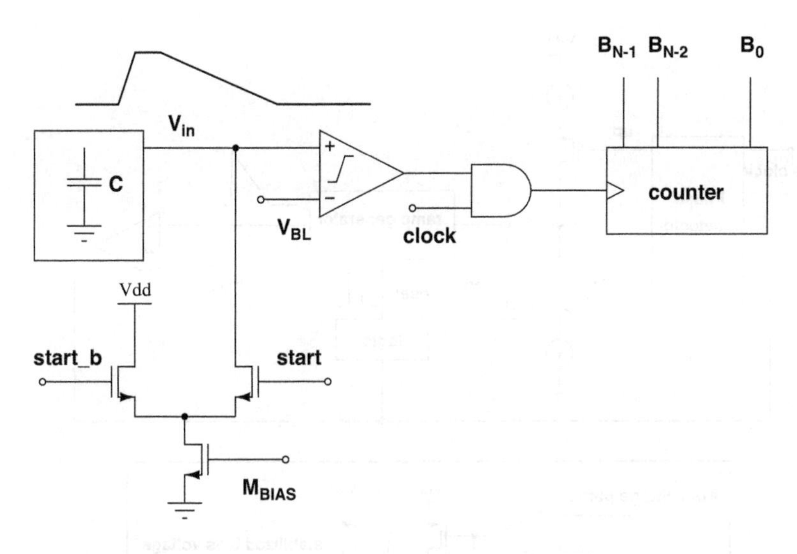

FIGURE 10.11 Wilkinson ADC.

from channel to channel due to the mismatch between transistors and capacitors. Systematic mismatch can however be easily calibrated and compensated by adjusting locally the value of the current source with a Digital to Analog Converter. Additional problems may come from the variations due to process spread, power supply voltage and temperature (PVT) which affect the circuit that is used to generate the reference bias voltage of the current source. This issue can be addressed with the method proposed in [15], which uses a feedback technique similar to the ones employed in Delay Locked Loops (DLL). The principle is sketched in Fig. 10.12. The building blocks are a phase detector, a charge pump and a ramp generator, formed by a voltage-controlled current source, which is shown in detail in the bottom part of the figure. The ramp is started and it is reset when it reaches the reference voltage. In this way at the output of the comparator a pulse is generated each time the ramp crosses the reference voltage. This time should correspond to the maximum conversion time of the ADC. The phase of the comparator pulses is compared with the one of a precise external clock. The feedback mechanism adjusts the value of the current source so that the duration of the ramp matches the period of the external clock, which is provided with a very stable oscillator, thus reducing the influence of the temperature and of the on-chip power supply variation on the reference current. More details on the circuit design can be found in [15]. The DNL of single ramp and Wilkinson ADCs is primarily defined by the uniformity of the clock pulses and it is thus very good. Some deterioration in the DNL figure can be observed if, to speed-up the conversion, both clock edges are used to trigger the counter. In fact, in this case, the circuit becomes sensitive to the duty cycle of the clock and a duty cycle different from 50% originates an odd-even pattern in the DNL profile. Interesting examples of the use of single ramp and Wilkinson ADCs can be found in [16–19].

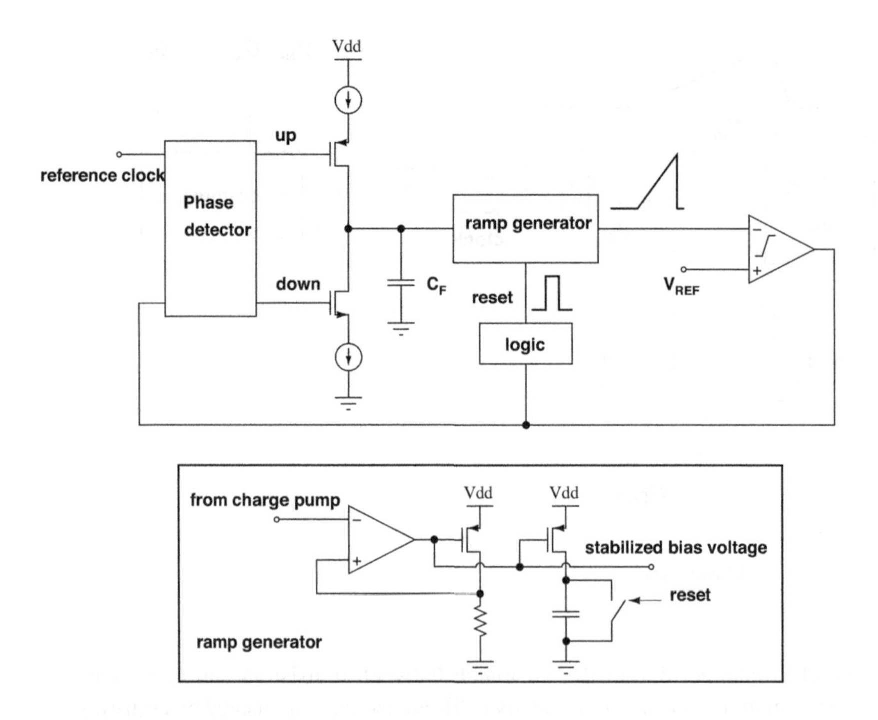

FIGURE 10.12 Correction of PVT variations in the ramp generator of a Wilkinson ADC.

10.3 SUCCESSIVE APPROXIMATION REGISTER ADC (SAR ADC)

The successive approximation technique is another of the oldest methods of analog to digital conversion, dating back to the times of vacuum tube technology. The first implementation exploiting MOS integrated circuits was described by Gray and McCreary in 1975 [20]. Since then and for about 30 years, SAR ADCs have been extensively employed in applications requiring medium to high resolution and moderate conversion speed, with maximum sampling rates in the order of a few MS/s. Due to the evolution of deep submicron CMOS technologies and the recent progress made in SAR design, the speed of this type of ADC has been substantially increased and their power consumption strongly reduced, making them one of the most attractive options to realize analog to digital converters for highly integrated radiation detectors front-ends. We examine first the conventional SAR ADC and we discuss afterwards the advantages of most recent implementations.

10.3.1 CONVENTIONAL SAR ADC

The most widely used SAR topology is based on the charge redistribution method. Fig. 10.13 shows an example of a 3 bit SAR ADC employing this technique. The circuit consists of an array of binary weighed capacitors, analog switches and a latched comparator. The top plate of the capacitors is common to the full array and can be

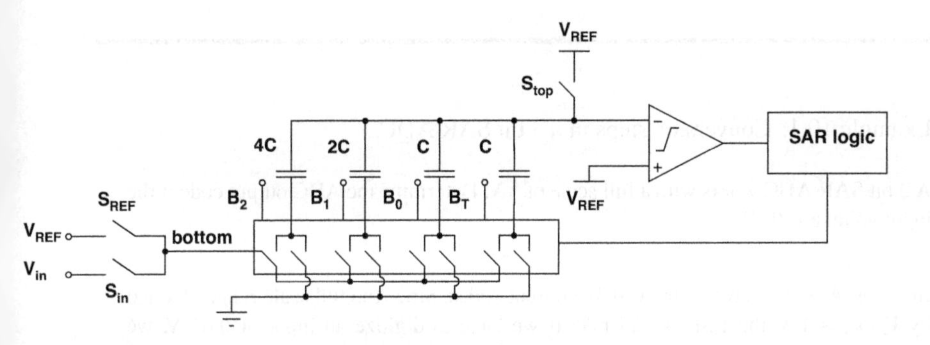

FIGURE 10.13 Example of a 3 bit SAR ADC.

connected to the reference voltage or left floating through the switch S_{top}. The bottom plate of each capacitor can be switched between ground and the bottom line. The latter can in turn be connected either to the input signal or to the reference voltage. The converter is controlled by a digital logic, partially shown in Fig. 10.14. The operation is divided in two phases: sampling and redistribution. During the sampling phase, the switch S_{top} is closed and the top plates of the capacitors are connected to V_{REF}. All the bottom plates are connected to the bottom line. The switch S_{REF} is open and S_{in} is closed, so that the input signal is sampled by the array. Conceptually, after the sampling is complete, the top plate is left floating and the bottom plates are switched from V_{in} to ground, bringing the voltage on the top plate to $V_{REF} - V_{in}$. At the same time the bottom line is switched from V_{in} to V_{REF}. The most significant bit is then tested by connecting the bottom plate of the largest capacitor (4C) to V_{REF}. The capacitor switched to V_{REF} is found in series with the sum of all the other capacitors in the array forming a voltage divider. When the MSB is tested, this voltage divider divides by two the reference voltage, adding $V_{REF}/2$ to the top plate. The top plate thus rises to $V_{REF} - V_{in} + V_{REF}/2$, which is compared against V_{REF} by the comparator. If the result is positive (i.e. $V_{REF} - V_{in} + V_{REF}/2$ is smaller than V_{REF}), the bottom plate of the tested bit is left connected to V_{REF}, otherwise it is switched back to ground. The other bits are tested with the same procedure till the final code is found. An additional capacitor with the same value of the one of the LSB must be added in order to have exact binary partitions of the reference voltage. This capacitor is used in the sampling phase, but is left always connected to ground during the conversion. From the above discussion, we see that during the sampling phase the capacitor array works as a sample and hold and during the conversion phase it performs in addition the function of Digital-to-Analog Converter that generates fractions of the reference voltage to be compared with the input signal. Although it is possible to implement the DAC functions with other methods, such resistor ladders or current division techniques, the approach of the capacitor array is the most widely employed and it is the only one considered here. A simple numerical example helps in understanding how the conversion proceeds.

Example 10.1: Conversion steps in a 3 bit SAR ADC.

A 3 bit SAR ADC works with a full scale of 1 V. Determine the ADC output code if the input voltage is 0.67 V.

* * *

In a 3 bit ADC we have eight available output codes. Since the full scale range, defined by V_{REF}, is 1 V the LSB is 125 mV. If we have to digitize an input of 0.67 V, we expect to find the code $0.67/0.125 = 5.36$, which, truncated to 5, yields 101 in binary code. With our numbers, we have that when the MSB is tested, the top plate goes to $V_{REF} - V_{in} + V_{REF}/2 = 1 - 0.67 + 0.5 = 0.83$ V, which is smaller than V_{REF}. Therefore the *MSB* is one and the bottom plate of capacitor 4C is left connected to V_{REF}. The second most significant bit is then tested by switching the bottom plate of 2C to V_{REF}. The resulting voltage divider divides the reference voltage by four. This brings the top plate voltage to $V_{REF} - V_{in} + V_{REF}/2 + V_{REF}/4 = 0.83 + 0.25 = 1.08$ V. Now the voltage on the top plate is greater than V_{REF} and the comparator output goes to zero. The bottom plate of 2C is thus brought back to ground, the voltage on the top plate is set back to 0.83 V and a zero is encoded. Finally, the last bit is tested. Switching the bottom plate of the third capacitor from ground to V_{REF} brings the top plate to $V_{REF} - V_{in} + V_{REF}/2 + V_{REF}/8 = 0.83 + 0.125 = 0.955$ V, which is smaller than V_{REF}. The comparator output goes to one and the conversion is completed, yielding as expected the pattern 101 in binary code.

A simplified version of the SAR logic is reported in Fig. 10.14. It consists of a shift register and a few flip flops to store the result of the conversion. The first flip flop in the shift register is of the set type, so its output goes to one when it is reset, while the others are of the reset type. The sample and reset signals are generated by a finite control state machine, not shown in the figure. When $B_2 - B_0$ are high, the bottom plates of the capacitors are connected to the common line, otherwise they are connected to ground. The sample signal thus shorts all the bottom plate to the common line. When "sample" is put to zero, the value of $B_2 - B_0$ depends on the preceding logic. If the Q output of the shift register flip-flop is one, the multiplexer select the path A and the comparator output is presented to the storage cell. At the same time, the bottom plate of the interested capacitor is connected to the common line and thus to V_{REF} to test the bit. The comparator is enabled on the falling edge of the clock, to allow enough time to give a valid output result. When the clock arrives, the result is stored in the storage flip flop and the output of the shift register flip-flop is brought back to zero, shifting the multiplexer in position B, so that the bit is stored. Depending on the result of the comparator, the output of the storage flip flop will be either zero or one, thus keeping the interested capacitor connected to V_{REF} or switching it back to ground. Note that, to save time, the possible switch back of the previous bit is done simultaneously with the test of the current bit.

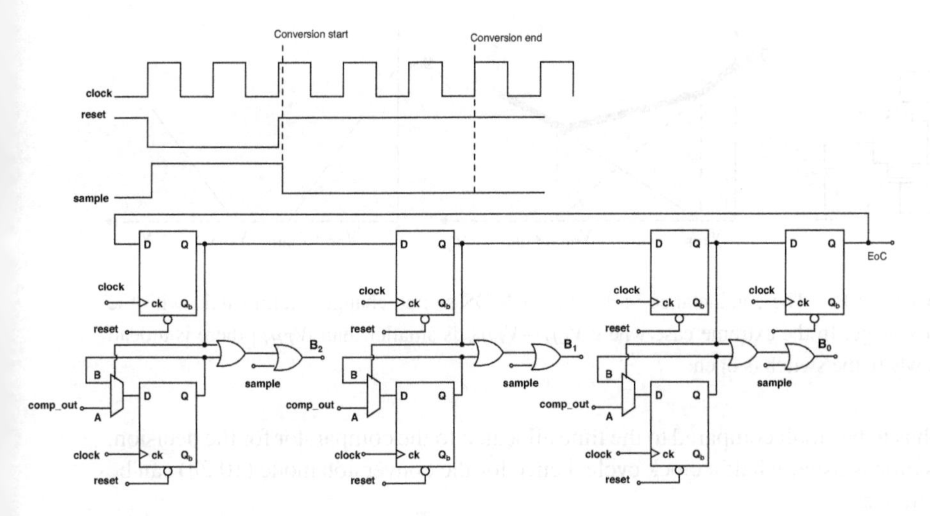

FIGURE 10.14 Simplified logic for a 3 bit SAR ADC.

10.3.2 KEY ISSUES IN SAR ADC DESIGN

The design of an SAR ADC entails several interesting issues. First, the switches must be properly sized in order to allow the settling time of the DAC to the desired accuracy. The switches of the bottom plate are usually sized to give equal time constants for every capacitor in the array, i.e. $R_iC_i = R_kC_k \forall i,k$. Under this hypothesis, when all capacitors are connected in parallel in the sampling phase, the total capacitance is $C_{TOT} = 2^N C$ and the total resistance is $R_{MSB}/2$, where R_{MSB} is the resistance of the switch driving the MSB capacitor. The time constant during the sampling phase is given by:

$$\tau_S = \left(R_{top} + R_{in} + \frac{R_{MSB}}{2}\right) C_{TOT} \tag{10.26}$$

where R_{top} is the resistance of the switch connecting the top plate to V_{REF} and R_{in} is the resistance of the switch connecting the bottom plate to the input signal. Within the sampling time, the converter must settle to the final value with an error smaller than 1/2 LSB. We therefore have:

$$\frac{V_{out}}{V_{in}} - 1 = -e^{-\frac{T_S}{\tau_S}} \leq \frac{1}{2^{N+1}} \tag{10.27}$$

where T_S is the time that must be allocated to the sampling phase. Taking the equality, we can derive the following upper limit for τ_S for a given sampling time:

$$T_S = \tau_S 0.693\,(N+1) \tag{10.28}$$

For instance, for a 10 bit ADC that needs to settle in 10 ns, τ_S must be smaller than 1.3 ns. During the conversion phase, the time constant can be approximated as:

$$\tau_C \approx \frac{R_{MSB}}{2} C_{TOT} \tag{10.29}$$

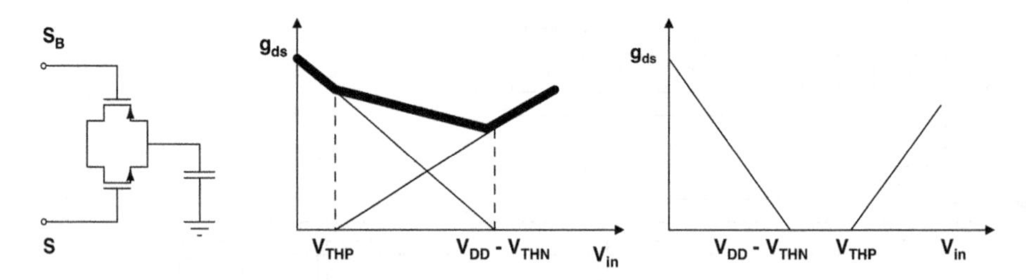

FIGURE 10.15 The conductance of a simple CMOS switch changes significantly with the input voltage. In the extreme case where $V_{DD} - V_{THN}$ is smaller than $|V_{THP}|$ there is a dead zone where the switch is open.

and has to be small compared to the time allocated to the comparator for the decision. This time is usually half a clock cycle, hence for the conversion mode (10.28) can be rewritten as:

$$T_C = 0.69\,(N+1)\,\tau_C = \frac{T_{ck}}{2} \tag{10.30}$$

In the sampling mode, the small signal bandwidth has a cut-off frequency defined by:

$$f_T = \frac{1}{2\pi\tau_S} \tag{10.31}$$

However, this is only an approximation, because the resistance of the sampling switch changes with the input signal, potentially leading to signal-dependent delay and harmonic distortion. This point can be explained with the help of Fig. 10.15, which sketches the typical profile of the conductance of a CMOS switch as a function of the input signal. When the input is close to zero, the conductance of the NMOS transistor is maximum, while the PMOS is cut-off. As the input signal rises, the conductance of the NMOS starts decreasing, because the source voltage is brought progressively closer to the gate voltage. The degradation is further exacerbated by the body effect, that increases the threshold voltage. The PMOS transistor, on the other hand, starts conducting only when V_{in} is greater than $|V_{THP}|$. In the middle of the dynamic range, the conductance reaches a minimum and afterwards the PMOS takes over, while the NMOS switches completely off when V_{in} becomes greater than $V_{DD} - V_{THN}$. In case $V_{DD} - V_{THN}$ is lower than $|V_{THP}|$, there is a dead zone where the switch stops conducting because the NMOS is cut-off before the PMOS is turned on, as shown in the rightmost part of Fig. 10.15. In a deep submicron CMOS technology powered at 1.2 V the threshold voltage around mid dynamic range (0.6 V) can be in the order of 0.5 V, so the switch resistance may increase significantly. A common way to cure the problem is to use in the sampling path a bootstrapped switch. The concept is shown schematically in Fig. 10.16. When the switch is open, the gate of the NMOS transistor is connected to ground. At the same time, a capacitor is charged to V_{DD}. When the switch is on, the capacitor is connected between the gate and source of the device, so that it behaves as a floating battery that keeps the $V_{GS} - V_{TH}$ of the transistor constant as the input voltage rises. Transistor-level examples of bootstrapped switches can be found in [21–24].

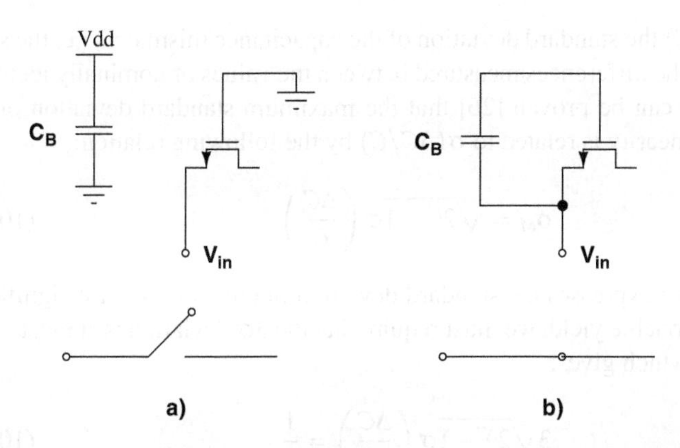

FIGURE 10.16 Concept of a bootstrap switch. The capacitor C_B is precharged to V_{DD} and during sampling it is connected between the gate and the source of the NMOS transistor, behaving as a floating battery.

Another critical point in the design of SAR ADCs is the optimization of the capacitor array. A first constraint comes from the noise introduced during the sampling action, when all capacitors are in parallel and form an RC filter with the resistance of the switches. Therefore, the noise is given by k_BT/C_{TOT}, where k_B is the Boltzmann constant and T the absolute temperature. The upper limit for the noise is obtained by requiring that it is at most equal to the rms of the quantization error. For an N bit quantizer, we can therefore write:

$$\frac{k_BT}{C_{TOT}} = \frac{V_{REF}^2}{2^{2N}12} \rightarrow C_{TOT} = \frac{12k_BT2^{2N}}{V_{REF}^2} \rightarrow C = \frac{12k_BT2^N}{V_{REF}^2} \tag{10.32}$$

where to derive the rightmost part of the equation we have used the fact that the total array capacitance is 2^N times the smallest capacitor used in the converter. Applying (10.32) we find that for a 10 bit ADC working with a full scale range of 1 V, the total DAC capacitance must be 52 fF and the smallest capacitor can be as small as 0.05 fF. For a 13 bit ADC, the limits become respectively 3.33 pF and 0.41 fF. These considerations show that, as far as noise in concerned, the capacitance in a charge redistribution SAR ADC can be very small. Twelve bit ADCs with a unit capacitor as small as 0.25 fF have been reported [25].

However, also the mismatch between capacitors must be taken into account, as it results in non-linearity that can easily deteriorate the converter performance, making it impossible to attain the nominal design resolution. To improve matching, the capacitive array is built using a single fundamental unit, corresponding to the LSB capacitor. All the other capacitors are obtained by connecting in parallel a suitable number of basic units. The MSB capacitor is therefore realized by connecting in parallel 2^{N-1} unit capacitors, and so on. It is very important to pay attention to the symmetry of the layout to avoid introducing systematic effects.

Let be $\sigma(\Delta C/C)$ the standard deviation of the capacitance mismatch, i.e. the standard deviation of the differences measured between the values of nominally identical unit capacitors. It can be proven [26] that the maximum standard deviation of the differential non-linearity is related to $\sigma(\Delta C/C)$ by the following relation:

$$\sigma_M = \sqrt{2^N - 1}\sigma\left(\frac{\Delta C}{C}\right) \tag{10.33}$$

The above equation expresses the standard deviation in units of the least significant bit. To have a reasonable yield, we must require that the 3σ deviation is at most equal to one half LSB, which gives:

$$3\sqrt{2^N - 1}\sigma\left(\frac{\Delta C}{C}\right) = \frac{1}{2} \tag{10.34}$$

The mismatch between capacitors is inversely proportional to the area A, so we can write:

$$\sigma\left(\frac{\Delta C}{C}\right) = \frac{K_\sigma}{\sqrt{A}} \tag{10.35}$$

where K_σ is a process-dependent coefficient and is expressed as $\% \cdot \mu$m. If K_C is the capacitance density, a capacitor of value C requires an area $A = C/K_C$. Inserting the value of A in (10.35) and combining (10.35) with (10.34), we can express the minimum unit capacitance which is necessary to fulfill the DNL constraint as a function of the number of bits of the ADC, the matching coefficient of the capacitors and the capacitor density offered by the chosen technology:

$$C = 36\left(2^N - 1\right) K_\sigma^2 K_C \tag{10.36}$$

The above equation shows that if K_C is 0.7 fF/μm^2 and $K_\sigma = 1\%\cdot\mu$m, the minimum capacitance required for a 10 bit ADC is 2.6 fF. Note that this is the smallest capacitance in the array, so the total capacitance would be $2^N C = 2.66$ pF. As it is seen from the previous considerations, the limit on capacitor size is primarily dictated by matching considerations and not by thermal noise limits.

Another important aspect to take into account is that each capacitor has parasitic capacitance associated to its top and bottom plates. Depending on the chosen capacitor structure, the parasitic capacitance can be symmetric or not. In chapter 8 we have seen that three different types of very linear capacitors are commonly available in modern CMOS technologies: planar Metal-Oxide-Metal capacitors (MOM), planar Metal-Insulator-Metal (MIM) and Vertical Parallel Plate Capacitors (VPPC). The structure of planar MIM and MOM capacitors is shown again in Fig. 10.17. It is important to remember that MIM capacitors require a dedicated oxide and therefore their production imposes additional fabrication steps. This increases the cost of ASIC production, but in general MIM capacitors offer the best matching performance and good capacitance density. Planar MOM capacitors, on the other hand, exploit the parasitic capacitance between the standard routing layers of the process located on different levels and do not require extra masks to be produced. This comes at the expense of a reduced

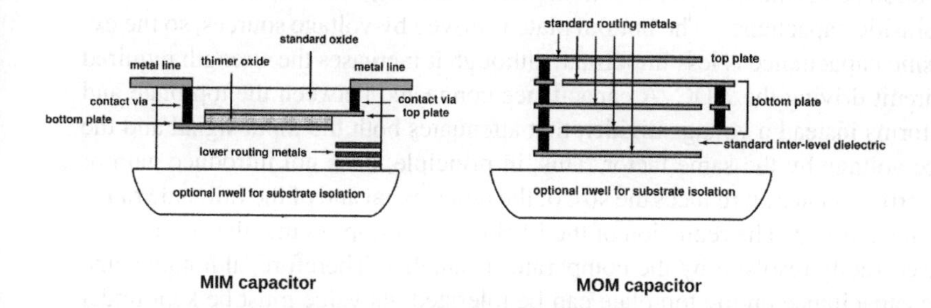

FIGURE 10.17 A Metal Insulated Metal capacitor (MIM) is formed by sandwiching a dedicate thin oxide between two metal layers and require additional masks and fabrication steps. A parallel Metal-Oxide-Metal capacitor exploits the parasitic capacitance between standard routing layers located on different levels and does not require extra processing steps.

density and a poorer matching. Often, planar MOM structures are not characterized, therefore simulation models and statistical data for Monte Carlo mismatch simulations are not available. Both MIM and planar MOM structures have in common that the top plate has less parasitic capacitance towards the circuit substrate, therefore the capacitor has a preferential arm to be connected to sensitive nodes. Vertical Parallel Plate capacitors, reported in Fig. 10.18, exploit the lateral capacitance between routing layers located on the same level, therefore they do not require additional masks. They have a good density and a fairly good matching. However the two plates are symmetric and thus they have the same amount of parasitic capacitance. In a charge

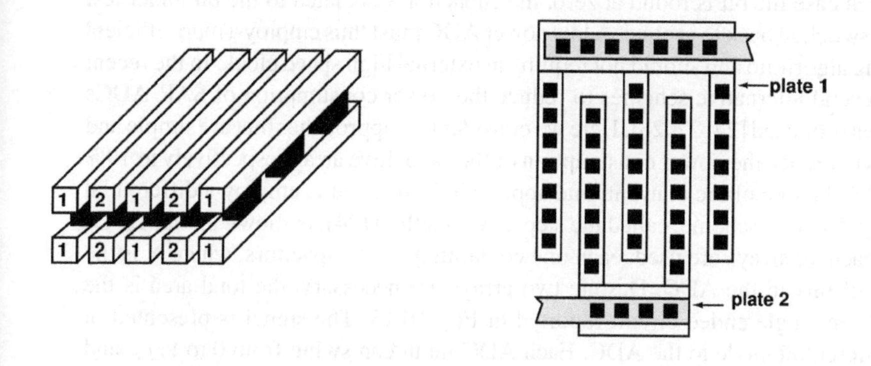

FIGURE 10.18 A vertical parallel plate capacitor exploits the natural capacitance present between closely spaced routing metals located on the same level. The parasitic capacitance is the same for both plates. Fabrication of VPPC does not require additional process steps.

redistribution ADC, the top and the bottom plate of the array have different sensitivity to the parasitic capacitance. The bottom plate is driven by voltage sources, so the extra parasitic capacitance is less important, although it increases the strength required to the circuit driving the ADC. A capacitance connected between the top plate and ground forms instead a voltage divider, that attenuates both the input signal and the reference voltage by the same factor. This, in principle, does not introduce gain or linearity error, because it reduces the size of the input signal and of the full scale range by the same amount. The reduction of the LSB however implies that the voltage that must be correctly resolved by the comparator is smaller. Therefore, although some parasitic capacitance on the top plate can be tolerated, its value must be kept under control. Finally, care must be paid to the junction capacitance of the switches that might be connected to the top plate. The junction capacitance changes in fact with the applied voltage and can thus degrade the converter linearity.

10.3.3 LOW POWER SAR ADC DESIGN

The conventional SAR ADC reported in Fig. 10.13 suffers from several drawbacks. First, the single ended structure makes it more susceptible to interference from external noise sources. Second, the control logic requires an external clock. In an SAR ADC, one bit is decided in one clock cycle, therefore the conversion time is $T_S + NT_{ck}$, where T_S is the time allocated for sampling the input signal and N is the number of bits. For a fast ADC with 50 MS/s conversion speed, a new sample must be captured every 20 ns. Assuming that 20% of the time is allocated to sampling, 16 ns are left for the conversion. For a 10 bit ADC, this implies that a single bit must be processed in 1.6 ns, requiring a clock frequency of 625 MHz. Distributing such high speed clock to a fairly large ASIC as a multichannel front-end typically is, entails a significant power consumption. Furthermore, the switching scheme described is not power efficient because in case the bit is found at zero, the capacitor associated to the bit under test must be switched back to ground. A low power ADC must thus employ a more efficient switching algorithm and should not require an external high speed clock. In the recent years, several alternative schemes to reduce the power consumption of SAR ADCs have been proposed [24, 27, 28]. Here we consider two approaches that are simple and allow us to reduce the power consumption of the capacitive array respectively to 19% and to 7% the one of the conventional approach. Both circuits employ a differential topology. The first scheme, called the step-down method [24], is shown in Fig. 10.19. Two capacitive arrays are used, each one containing 2^{N-1} capacitors, where N is the number of bits in the ADC. Despite two arrays are necessary, the total area is the same of the single ended circuit reported in Fig. 10.13. The signal is presented in fully differential mode to the ADC. Each ADC input can swing from 0 to V_{REF} and the difference between the V_p and V_n is measured and digitized. The full scale range of the ADC is thus $2V_{REF}$, while the amplitude of the LSB is $2V_{REF}/2^N$, so it is twice the one of the single-ended counterpart. However, also the input signal is doubled due to the fully differential operation, therefore the resolution of the circuit of Fig. 10.19 and of the one of Fig. 10.13 is the same. If $V_p = 0$ and $V_n = 1$, the ADC produces the binary code 000. As V_p starts increasing and V_n decreases, the output word increases. An

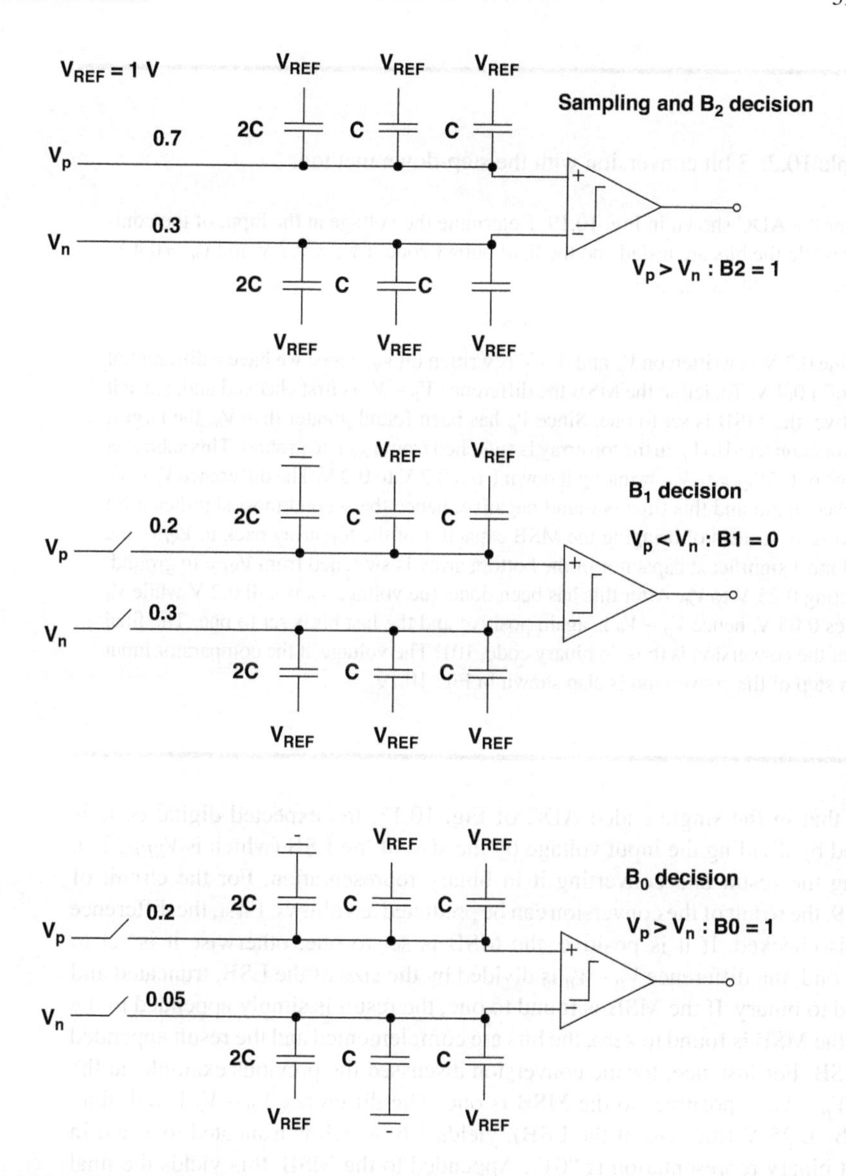

FIGURE 10.19 SAR ADC employing the step-down switching algorithm.

important point is that, during sampling, the input signal is written on the top plate of the array, while the capacitor bottom plates are held at V_{REF}. At each comparison, the sign of the different $V_p - V_n$ is checked. If it is positive the bit is set to one, otherwise it is set to zero. The MSB can therefore be decided just by checking the difference $V_p - V_n$ without the need of switching any capacitor. The following example shows in details the steps of a complete conversion.

Example 10.2: 3 bit conversion with the step-down method.

Consider the ADC shown in Fig. 10.19. Determine the voltage at the input of the comparator while the bits are tested and the final output code if $V_p = 0.7$ V and $V_n = 0.4$ V. The reference voltage V_{REF} is 1 V.

* * *

The value 0.7 V is written on V_p and 0.3 V is written on V_n, hence we have a differential signal of +0.4 V. To define the MSB the difference $V_p - V_n$ is first checked and, since it is positive, the MSB is set to one. Since V_p has been found greater than V_n, the largest capacitor connected to V_p in the top array is switched from V_{REF} to ground. This subtracts a voltage of $0.5V_{REF}$ to V_p, bringing it down from 0.7 V to 0.2 V. The difference $V_p - V_n$ is checked again and this time is found negative, hence the second most significant bit is set to zero. Instead of bringing the MSB capacitor of the top array back to V_{REF}, the second most significant capacitor of the bottom array is switched from V_{REF} to ground, subtracting 0.25 V to V_n. After this has been done, the voltage V_p is still 0.2 V while V_n becomes 0.05 V, hence $V_p - V_n$ is again positive and the last bit is set to one. The final result of the conversion is thus, in binary code, 101. The voltage at the comparator input in each step of the conversion is also shown in Fig. 10.19

Observe that in the single ended ADC of Fig. 10.13, the expected digital code is calculated by dividing the input voltage by the size of the LSB (which is $V_{REF}/2^N$), truncating the result and converting it in binary representation. For the circuit of Fig. 10.19, the result of the conversion can be predicted as follows. First, the difference $V_p - V_n$ is checked. If it is positive, the MSB is set to one, otherwise it is set to zero. Second, the difference $V_p - V_n$ is divided by the size of the LSB, truncated and converted to binary. If the MSB is found to one, the result is simply appended to the MSB. If the MSB is found to zero, the bits are complemented and the result appended to the MSB. For instance, for the conversion discussed the previous example, at the start-up $V_p - V_n$ is positive, so the MSB is one. The difference $V_p - V_n$ is 0.4, that, divided by 0.25 V (the size of the LSB), yields 1.6, which is truncated to 1 and in a two bit binary representation is "01". Appended to the MSB, this yields the final result 101. Consider now the case in which V_p is 0.1 V and V_n is 0.9 V. The difference $V_p - V_n$ is negative, so the MSB is zero. The absolute value of $V_p - V_n$ is divided by the LSB and yields 3.2, that truncated and converted to binary gives 11. Since the MSB is zero, the bits are inverted and appended to the MSB, so the result of the conversion is 000. Fig. 10.20 shows the conversion of $V_p - V_n = -0.8$ V done step by step. Note that the conversion results are in practice available at the output of the converter logic without further digital operations.

One drawback of the method just discussed is that the common mode at the input of the comparator is not constant, but it converges towards zero. The change of

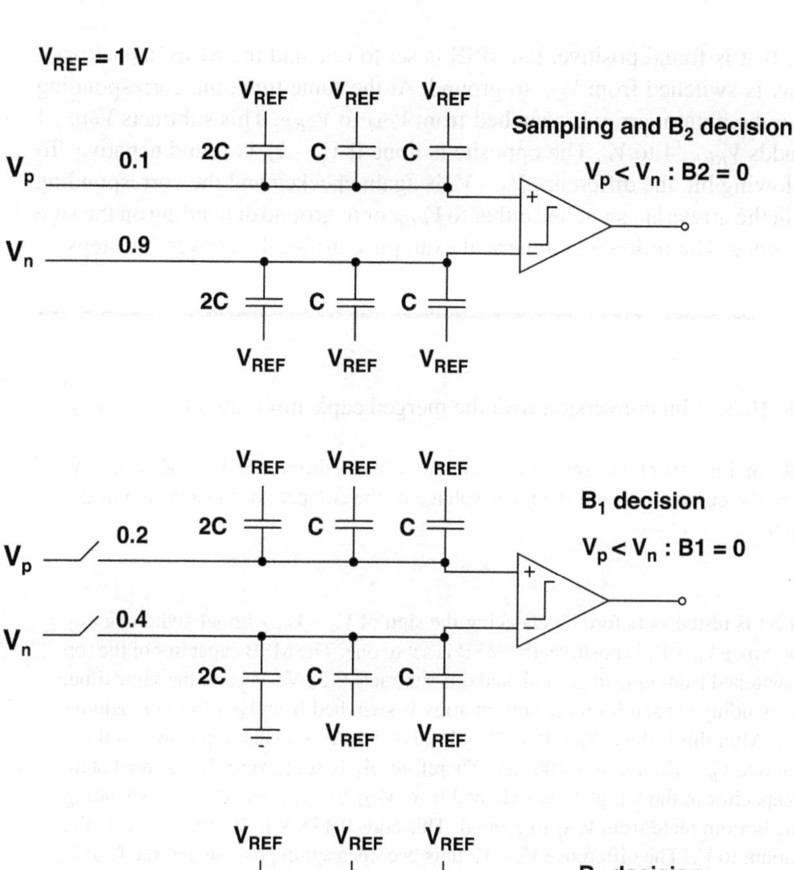

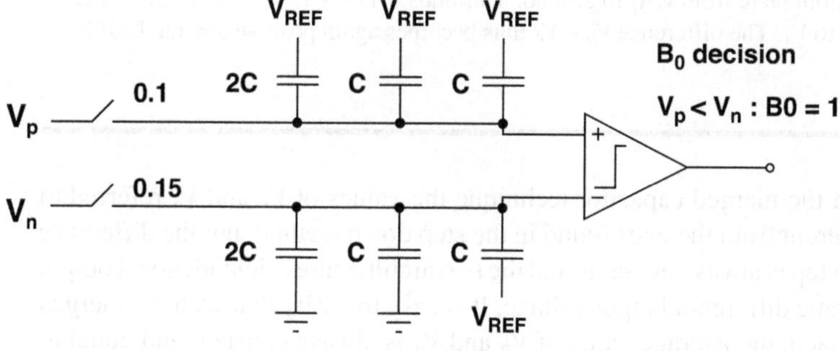

FIGURE 10.20 SAR ADC employing the step-down switching algorithm.The figure shows the conversion of $V_p - V_n = -0.8$ V.

common mode can affect the comparator operation and demands the use of PMOS input transistors, which are slower. This side-effect can be avoided by slightly modifying the circuit as shown in Fig. 10.21 [28], reporting the so called merged capacitor switching method. Here, a common mode reference located halfway between ground and V_{REF} is introduced. During sampling, the bottom plates of the capacitors are connected to V_{CM}. The conversion starts with the test of the MSB. The difference $V_p - V_n$

is checked. If it is found positive, the MSB is set to one and the MSB capacitor of the top array is switched from V_{CM} to ground. At the same time, the corresponding capacitor on the bottom array is switched from V_{CM} to V_{REF}. This subtracts $V_{REF}/4$ to V_p and adds $V_{REF}/4$ to V_n. The opposite is done if $V_p - V_n$ is found negative. To test the following bit, the difference $V_p - V_n$ is again checked and the corresponding capacitors in the arrays are switched either to V_{REF} or to ground depending on the sign of the difference. The following numerical example clarifies the conversion steps.

Example 10.3: 3 bit conversion with the merged capacitor method.

The ADC of Fig. 10.21 is used to convert the voltage difference $V_p - V_n = 0.4$ V. Determine the output code and the input voltage of the comparator in each step of the conversion.

* * *

The first bit is tested as before by checking the sign of $V_p - V_n$, without switching any capacitor. Since $V_p - V_n$ is positive, the MSB is set to one. The MSB capacitor of the top array is switched from V_{CM} to ground, and this subtracts 0.25 V to V_p. At the same time, the corresponding capacitor on the bottom array is switched from V_{CM} to V_{REF}, adding 0.25 to V_n. After this is done, V_p is 0.45 V and V_n is 0.55. As with the step-down method, the difference $V_p - V_n$ is now -100 mV. Therefore, B_1 is set to zero. In the final step, the last capacitor in the top plate is switched from V_{CM} to V_{REF} and the corresponding one in the bottom plate from V_{CM} to ground. This adds 0.125 V to V_p and subtracts the same amount to V_n. The difference $V_p - V_n$ thus becomes again positive and the LSB is set to one.

Observe that in the merged capacitor technique the values of V_p and V_n referred to ground are different from the ones found in the step down method, but the difference $V_p - V_n$ at each step is always the same and the two architectures yield identical output codes for the same differential input voltage. It is easy to verify that with the merged capacitor approach, the average value of V_p and V_n is always constant and equal to 0.5, therefore the input common mode of the comparator does not change. As for any differential circuit, in both methods, any common mode component present is the input signal is rejected and only the difference $V_p - V_n$ is digitized. Due to the straightforward switching schemes, the two described algorithms require a simple control logic and are thus very suitable for applications demanding a large number of channels. To further reduce the power consumption, the ADC should not need an external clock reference. Fig. 10.22 describes a method that can be used to generate a local clock for the converter. Observe that when the latch signal is low, the comparator is disabled and both outputs are forced to V_{DD}. When "latch" goes high, the comparator makes its decision and, due to the differential structure, the outputs takes opposite

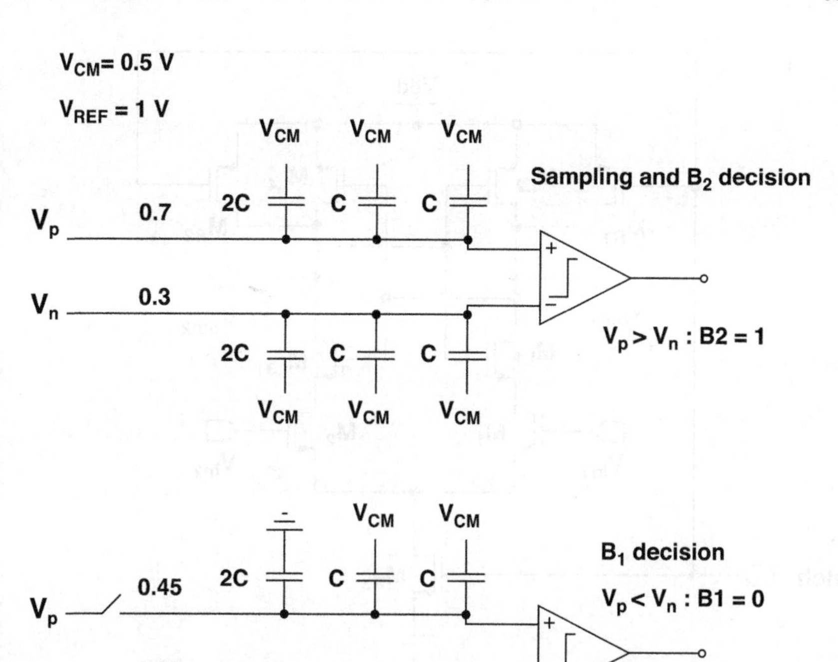

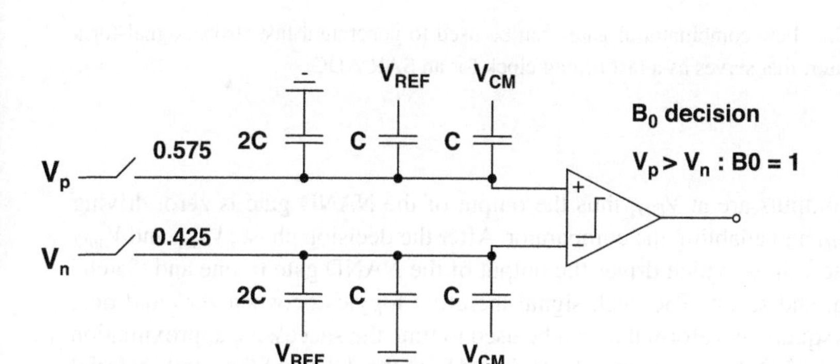

FIGURE 10.21 SAR ADC employing the merged capacitor switching algorithm.The figure shows the conversion of $V_p - V_n = +0.4$ V.

values. The simple combinatorial circuit shown in the right of the figure can be used to strobe the comparator. During the sampling phase, "sample" is high and "latch" is driven low, keeping the comparator in reset mode. When "sample" is released, both

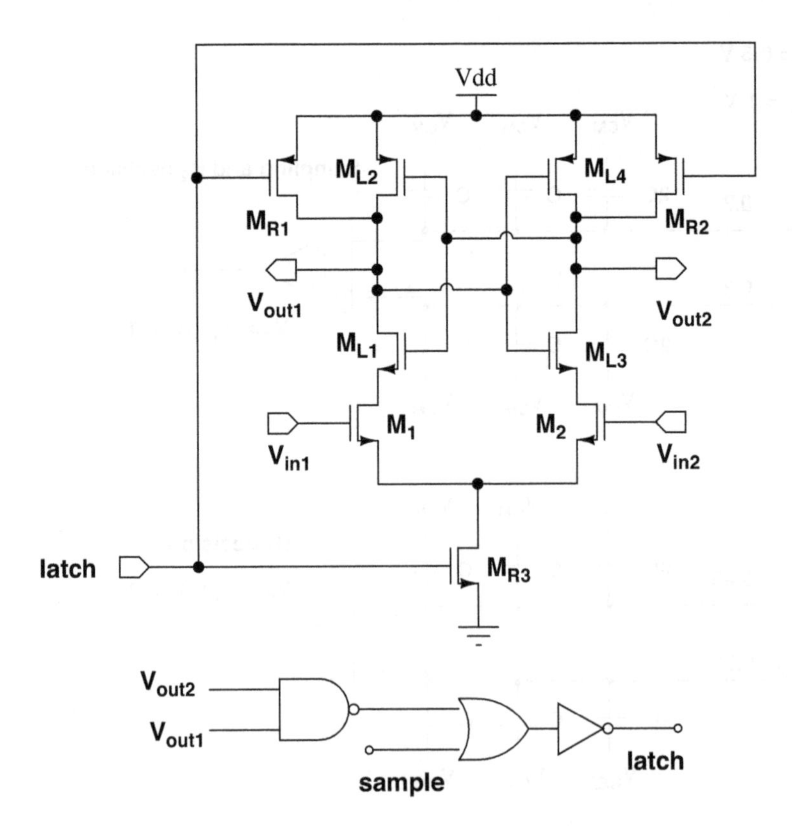

FIGURE 10.22 Few combinatorial gates can be used to generate a fast strobe signal for a latch comparator, that serves as a fast timing clock for an SAR ADC.

comparator outputs are at V_{DD}, thus the output of the NAND gate is zero, driving "latch" to V_{DD} and enabling the comparator. After the decision phase, V_{out1} and V_{out2} have opposite values, which drives the output of the NAND gate to one and "latch" back to zero, and so on. The latch signal therefore toggles between zero and one, generating a square waveform that can be used to time the successive approximation logic. The period of the square waveform depends on the delays of the combinatorial gates and of the comparator and can be controlled by limiting the maximum current available to the latter. Depending on the technology, this generates clock signals with a frequency in the 500 MHz-1 GHz range, resulting in sampling speeds that exceed 100 MS/s. A detailed example of the SAR logic can be found, for instance, in [24].

The area of the capacitor arrays employed in an SAR ADC doubles for every extra bit of resolution required. Beyond 10-11 bit resolution, the straightforward implementation becomes thus impractical, because it leads to structures occupying

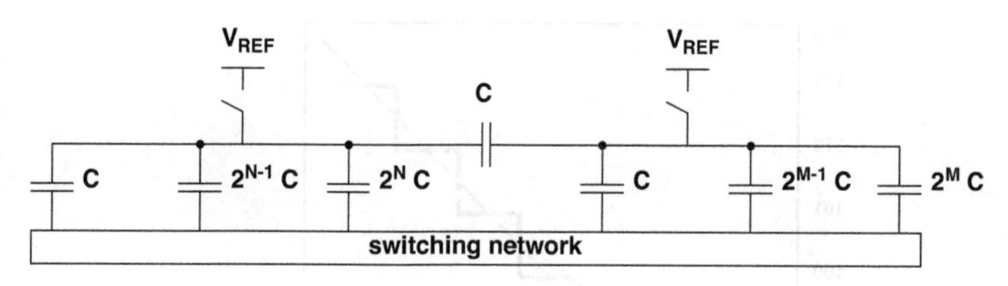

FIGURE 10.23 An M+N bit DAC can be obtained by coupling through a capacitor two sub-DACs with N and M bit resolution.

a very large area and presenting a huge load to the circuit driving the ADC. The alternative to reduce the DAC area is to use segmentation. The array is partitioned in two sub-arrays, which are connected by a coupling capacitor, as shown in Fig. 10.23. High resolution ADCs require also a matching between capacitors which is beyond the capability of the technology. Digital algorithms have thus been developed to correct for the DAC non-linearity. In general, up to 10 bits resolution can be conveniently obtained with a straightforward design without digital calibration. Above ten bits, DAC segmentation and digital error correction need to be combined to achieve the desired performance within an acceptable area and power consumption.

Two system-level issues which are particularly relevant for front-end electronics must finally be mentioned. Front-end chips for radiation detectors are multi-channel ASICs where the same building blocks are repeated along the chip. The capacitor array in a SAR ADC is fairly large and, if reproduced many times in multiple channels, it can generate violations of pattern density rules, as an anomalous high concentration of metal if found in a restricted area. Constraints coming from pattern density rules must therefore be already considered in an early design stage. In addition, the reference voltage to the converter is often provided from a precise off-chip circuit. During switching, the series combination of the DAC capacitance with the inductance of the bonding wires and of the PCB traces can result in ringing on the reference voltage that corrupts the ADC accuracy well beyond the LSB. The problem is exacerbated as more converters are integrated on chip and operated in parallel. An adequate number of bonding pads and bonding wires must be provided to lower as much as possible the contribution of the parasitic inductance and its effect must be carefully modeled with computer simulations.

Further interesting readings on SAR ADC can be found in [29–42].

10.4 BASIC TDC PROPERTIES

A Time to Digital Converter (TDC) associates a digital word with N bit to a time difference between two events. The maximum time difference the TDC is able to measure defines the full scale range T_{REF} and the smallest step that can be detected is the LSB and is equal to $T_{REF}/2^N$. Reported in Fig. 10.24, the ideal characteristics of a TDC is similar to the one of an ADC, with the difference that on the *x*-axis there is a time variable instead of a voltage. Since the continuous input time is approximated

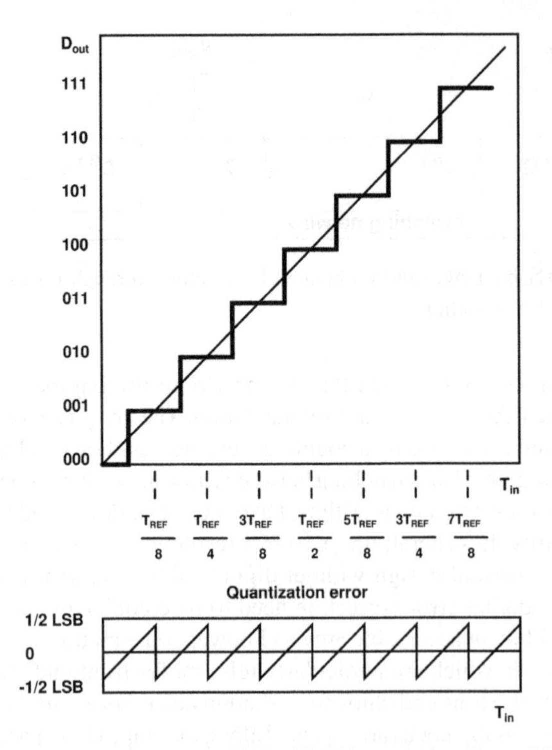

FIGURE 10.24 Ideal characteristics of a TDC with 3 bit resolution. For an N bit TDC, if T_{REF} is the full scale range, the LSB is $T_{REF}/2^N$ and the rms of the quantization error is LSB/$\sqrt{12}$.

with a discrete output word, also a TDC suffers from unavoidable quantization error, with an rms given by LSB/$\sqrt{12}$. In the same way as ADCs, also TDCs are subject to offset, gain and non-linearity errors. The same definitions of DNL and INL thus apply also to time to digital converters. Fig. 10.25 shows the characteristics of a 3 bit TDC affected by non-linearity. If the DNL exceeds one LSB, missing codes have to be expected. A convenient way of characterizing the TDC linearity is to measure the difference between a reference pulse and a test pulse that walks randomly through the converter input range. If a large number of events is collected, all the bins should be equally explored. The number of events falling in each bin is counted and it is compared to the theoretical one. Equation (10.9) can then be used to calculate the DNL, while the INL for a given bin is obtained by summing the DNL of the preceding bins. Other important TDC parameters are the conversion rate, expressed as for ADC in number of acquired samples per second, the latency (i.e. the time necessary to yield a valid output code) and the single shot accuracy, that is the precision in the measurement of an individual hit.

TDC can be broadly classified in two main categories: asynchronous and synchronous. An asynchronous TDC measures the time difference between two pulses, without referring the measurement to an external reference signal. In many

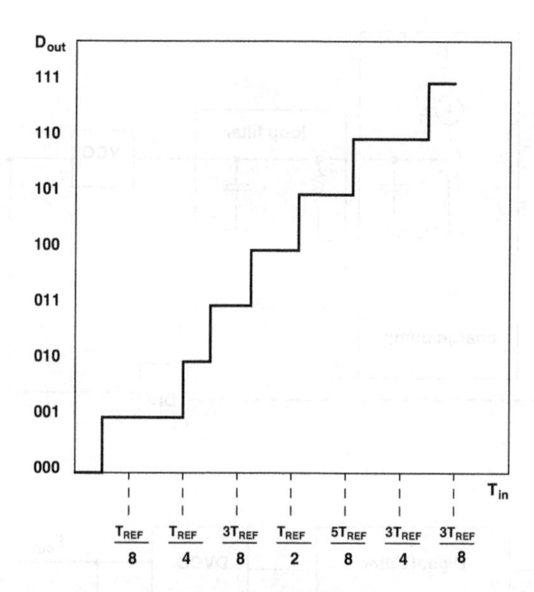

FIGURE 10.25 A real TDC's characteristics are affected by non-idealities similar to those found in ADCs. The figure shows a 3 bit TDC with non-uniform quantization steps.

applications, an external clock is normally used to extend the dynamic range. The clock provides a coarse time base and the difference between the event of interest and one clock edge is then interpolated with suitable techniques. In this case, one speaks of a synchronous TDC. Even though the clock is running at low speed, it provides nevertheless the reference for the accurate time measurement, therefore its jitter must be below the rms quantization error of the converter. Until a few years ego, TDC were employed only in high precision instrumentation, such as high energy physics detectors, range finders and testing equipment. Recently, TDC started being used as phase detectors in All Digital PLL (ADPPL). Phase locked loops are among the most ubiquitous building blocks in modern integrated circuits, because they provide the fundamental functions for frequency synthesis and clock generation. The traditional analog PLL is shown in the top part of Fig. 10.26. The frequency of a voltage controlled oscillator is divided and compared to a very stable external reference frequency, usually generated with a quartz oscillator. The phase detector measures the difference between the phases of the two signals. If the signal from the VCO comes too early, the current I_2 is selected to discharge the capacitor, reducing the VCO speed. If the VCO signal arrives too late, the current I_1 charges the capacitor to speed up the VCO. In the ADPLL, sketched in the bottom of the figure, the phase difference is measured by a TDC. The resulting digital signal is processed by a numeric filter whose output is used to control a digitally-programmable VCO. The widespread use of PLLs has thus triggered a strong interest in TDC design, which have now become mainstream components. In ADPLL design, the emphasis is more on measuring fairly

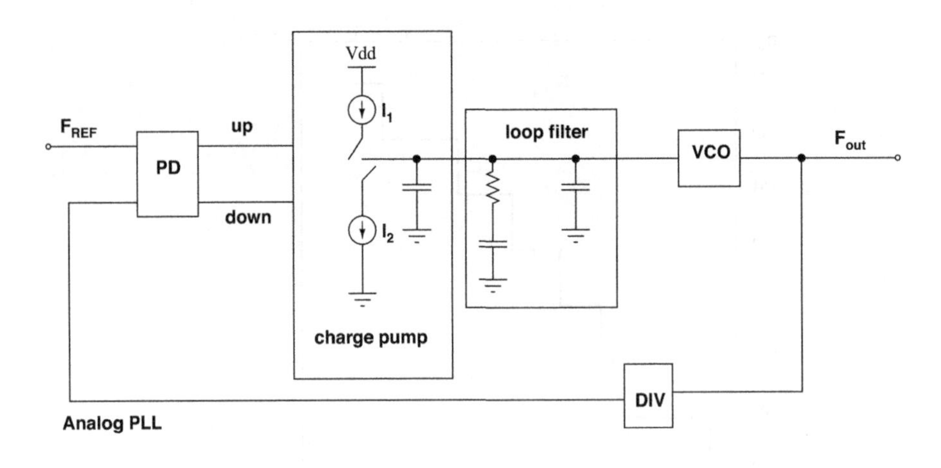

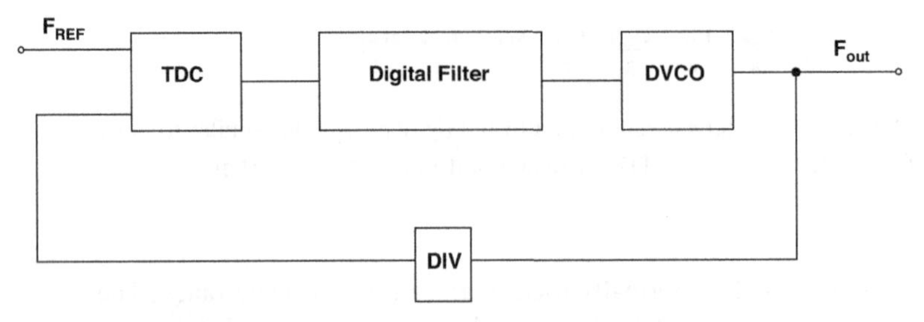

FIGURE 10.26 Comparison between analog (top) and All Digital PLL (ADPLL). In the ADPLL, the traditional PLL control loop based on a phase detector, a charge pump and an analog filter is replaced with a TDC to measure the phase error, followed by a digital filter.

short time intervals with a very good resolution, so the typical design shows a very small LSB (in some designs even below 1 ps) and a short dynamic range, often below the nanosecond. Interestingly, while SAR-based ADC have gained more and more popularity, a larger variety of architectures is employed to implement high resolution TDCs. In the following, we will focus our attention on the ones which are of greatest interest in radiation detector instrumentation.

10.5 TDC ARCHITECTURES

10.5.1 TDC WITH ANALOG INTERPOLATORS

The oldest method to measure time intervals is to charge a capacitor with a constant current source and digitize the resulting voltage, as shown in Fig. 10.27 [43, 44]. First, the capacitor C_1 is reset by closing the reset switch S_{R1}. In idle mode, S_{T1} is closed and S_{C1} is open and the current delivered by the source flows to ground. When the

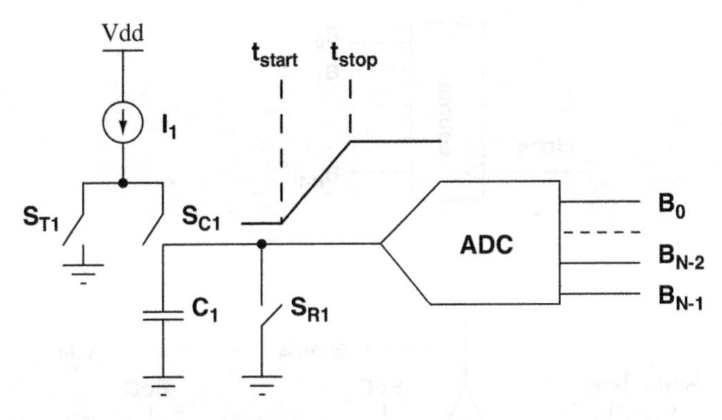

FIGURE 10.27 Time to Amplitude Converter (TAC). During the time interval to be measured, a capacitor is charged with a constant current source. The voltage is then digitized by an ADC.

start signal arrives, S_{T1} is open and S_{C2} is closed, steering the current from ground to the capacitor. The voltage on the top plate of C_1 rises linearly, and when the stop signal is asserted, the current is diverted back to ground and the capacitor is put in the hold mode. The final output voltage is given by:

$$V_{out} = \frac{I_1}{C_1}(t_{stop} - t_{start}) \tag{10.37}$$

and it can be digitized by an ADC. Observe that the method allows only to infer the difference between two generic signals. If the stop signal is the edge of a reference clock providing a coarse time base, the TDC becomes synchronous. The resolution of the ADC determines the number of bits in the TDC. In principle, any ADC can be employed, but the Wilkinson topology is the most frequently used with time to amplitude converters because it is simple to implement, requires little extra circuitry and can be operated at low power. Even more important, the Wilkinson technique offers an excellent differential non-linearity. Fig. 10.28 shows a possible digitization scheme. The capacitor on which the voltage is held is connected to one of the comparator inputs. The other input is connected to a second capacitor, which can be charged by a second current source I_2. When I_2 is connected to C_2, a counter is started and the clock pulses are counted. When the voltage on C_2 reaches the one stored on C_1, the counter is stopped and the conversion ends. To understand the method, suppose that the unknown time of arrival of an event must be measured with a resolution of 200 ps. A clock is used to provide a coarse time reference and the TAC measures the difference between the unknown signal and the next clock edge. If the clock runs at a frequency of 80 MHz, this difference can be up to 12.5 ns, which defines the full scale range that must be accommodated by the TAC. Digitizing 12.5 ns with a 200 ps binning requires 62.5 bit, so we need a resolution of 6 bits. A time difference between zero and one clock cycle is thus mapped into a time difference between 1 and 64 clock cycles. For this reason, a TDC employing the digitization with the Wilkinson method

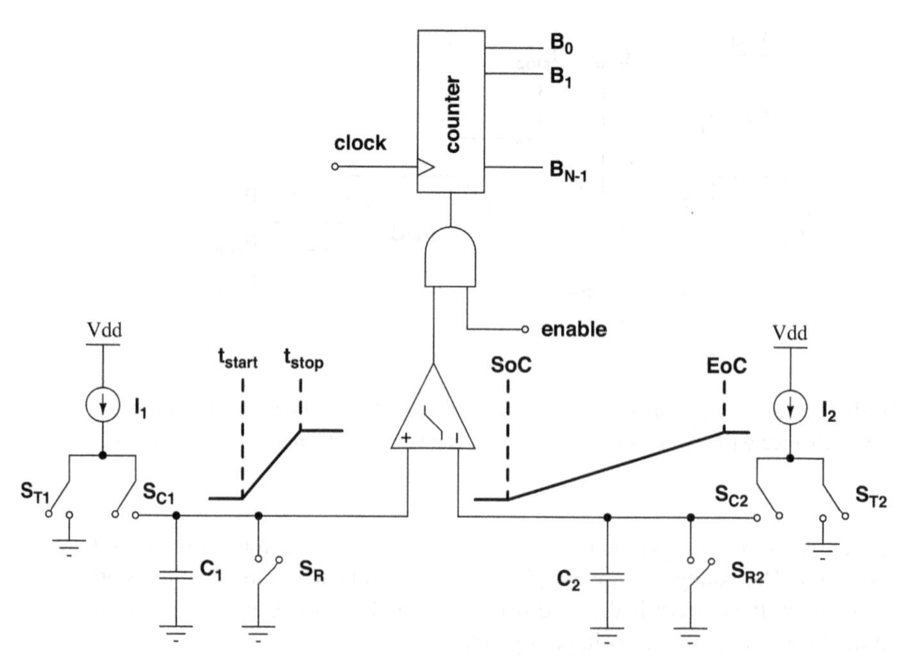

FIGURE 10.28 Time to amplitude conversion combined with a Wilkinson TDC. SoC and EoC indicate respectively the Start of Conversion and the End of Conversion. The slopes of the two ramps are not in scale.

is also called a "time stretcher". The current I_2 must therefore be 64 times smaller than I_2. A possible alternative to reduce the ratio between I_2 and I_1 is to increase the ratio between C_2 and C_1. For instance, if $C_2 = 4C_1$, then I_2 needs to be only 16 times smaller than C_1. The dead time introduced by the Wilkinson conversion is one of the major drawbacks of the circuit. In fact, for an N bit resolution, the maximum conversion time is $T_{conv} = 2^N T_{ck}$, where T_{ck} is the clock period. If the phenomenon under study is asynchronous with the system clock, the events will be uniformly distributed in the clock window and the average dead time is $2^{N-1}T_{ck}$. The impact of the long conversion time can be mitigated by using multi-buffering schemes [45, 46]. More capacitors are made available, so, if one capacitor is waiting for digitization, another one is ready to capture a new event. With a suitable number of buffers, events which are non-uniformly distributed in time can be acquired at a rate $1/T_{conv}$ without incurring in significant losses. For instance, under the hypothesis that we have used, the maximum conversion time is 812.5 ns, so a TDC with multi-buffer can process more than 1 million events per channel while maintaining a good efficiency. An important parameter of TDCs is the double pulse resolution, i.e. the capability of resolving two pulses closely spaced in time. In case of a time-to-amplitude converter, this is limited by the time needed to charge the capacitor, i.e. it can be up to one clock cycle. If this is not sufficient, completely independent circuits must be used in a time-interleaved configuration.

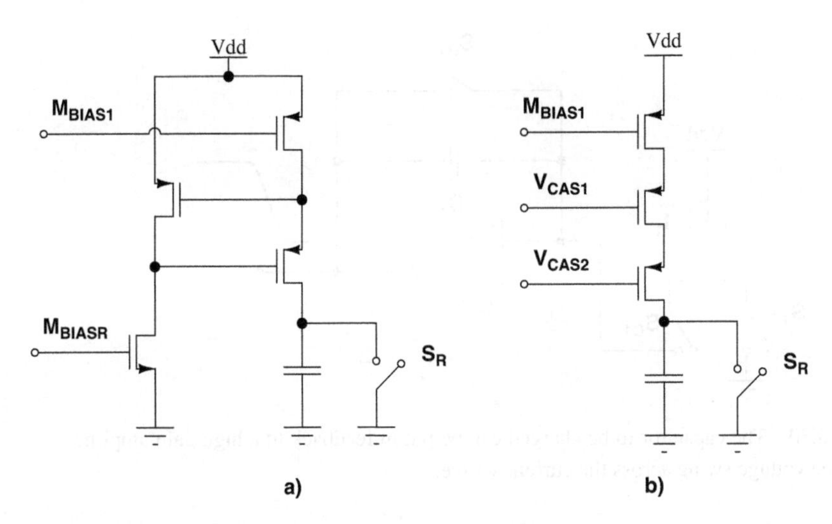

FIGURE 10.29 The current source used in the TAC must have very high impedance. An adequate impedance can be achieved with a regulated cascode or a triple cascode, but the former may suffer from bandwidth limitation in the regulating loop.

The ratio between I_1 and I_2 must be precisely controlled. Usually, a DAC is necessary to counteract the effect of mismatch between I_1 and I_2, that would introduce a gain error in the TDC. Since a gain error of 10% on the full scale range typically has a negligible effect on the quantization error, another possibility is also to correct externally with a look-up table. While the signal is being processed, the voltage across the current sources changes. Therefore, it is very important that I_1 and I_2 have a very high output impedance. If this is not the case, their currents will change during operation, introducting a non-linearity in the measurement. In deep submicron technologies, simple cascodes usually yield insufficient performance for high accuracy applications. To further boost the output impedance of the current source, two possibilities are available, shown in Fig. 10.29. The first option is to use a regulated cascode and the second one is to have a triple cascode. However, the regulated cascode implies a feedback loop, that must be properly stabilized. If the loop is underdamped, the ringing can corrupt the measurement. If the loop is properly compensated, it might become too slow and thus its effectiveness in increasing the output impedance on the signal can be strongly reduced. A triple cascode is thus faster and can provide better results. The drawback is that the extra cascode transistor requires headroom to stay in saturation. However, if the time to be measured is not very long, the voltage difference across C_1 can be kept small enough to allow a proper biasing of the circuit under all the operating conditions. Another approach is reported in Fig. 10.30. Here C_1 is connected in feedback to an integrator. The high gain voltage amplifier keeps the input at a virtual ground, greatly reducing the voltage excursion across the current source, thereby relaxing the requirement on its output impedance. Owing to its finite bandwidth, the voltage at the output of the integrator may not be a linear ramp. However, after the circuit has settled, the output voltage would be given by Q/C_1 and

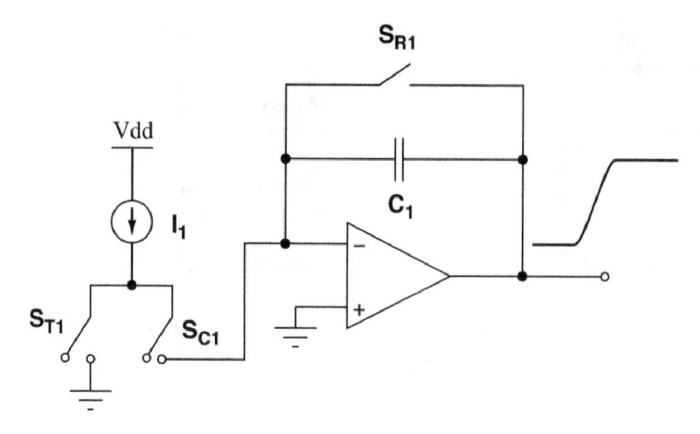

FIGURE 10.30 The capacitor to be charged can be put in feedback to a high gain amplifier to reduce the voltage swing across the current source.

thus still proportional to the time during which I_1 has been connected to the input. Since the current in I_2 is much smaller than the one in I_1, it is easier for I_2 to maintain an adequate output impedance. Recent example of TDCs based on time to amplitude converters can be found in [47–50].

10.5.2 TDC WITH DIGITAL DELAY LINES

The large majority of TDC architectures exploit the intrinsic delay of digital gates to measure the time interval. The most straightforward implementation is shown in Fig. 10.31. The start signal is sent to a buffer chain. Each element in the buffer chain is a non-inverting gate that can be obtained, for instance, by cascading two digital CMOS inverters. The cell delay is therefore twice the minimum gate delay of the technology and in a 130 nm process this is about 20-30 ps. The output of each buffer is sent to a sampling cell, represented in the figure with a D flip-flop. The stop signal is used to strobe the sampling of the buffer outputs into the register. The situation sketched in Fig. 10.32 is thus observed. Depending on when the stop signal arrives, a given number of buffers has already made the zero-to-one transition, therefore for these cells a logic one is captured by the associated flip-flops. The start signal has not yet reached the rest of the buffers, which still have a zero at their output. A thermometric pattern is thereby generated, and the number of "ones", multiplied by the delay of the single cell yields the sought time interval measurement. A thermometric to binary encoder is used to reduce the number of bits to be readout. The stop signal must reach simultaneously the flip-flops, hence it is distributed with a buffer tree, not shown in Fig. 10.31. If T is the maximum time interval to be measured and τ_1 is the delay of a

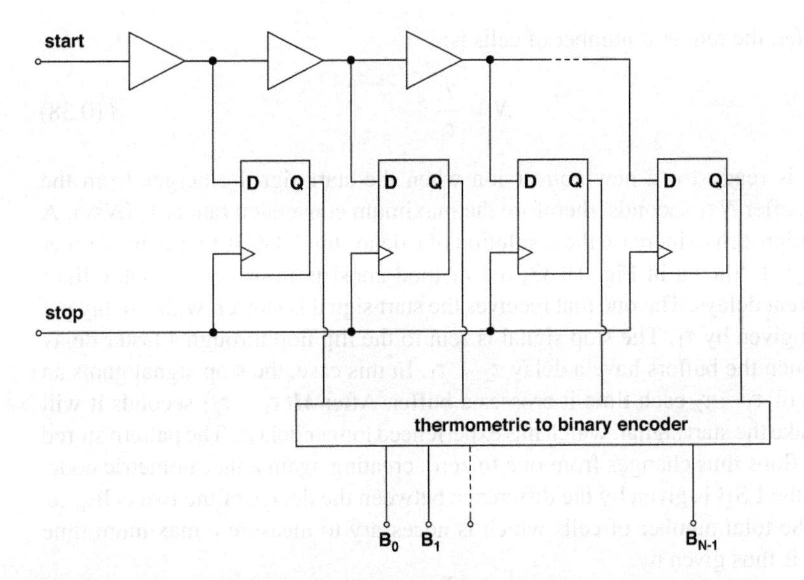

FIGURE 10.31 A time difference between two events can be measured by sampling the status of a digital delay line. The start signal is fed to the buffer chain, the stop signal is used to strobe the register. The buffers are non-inverting.

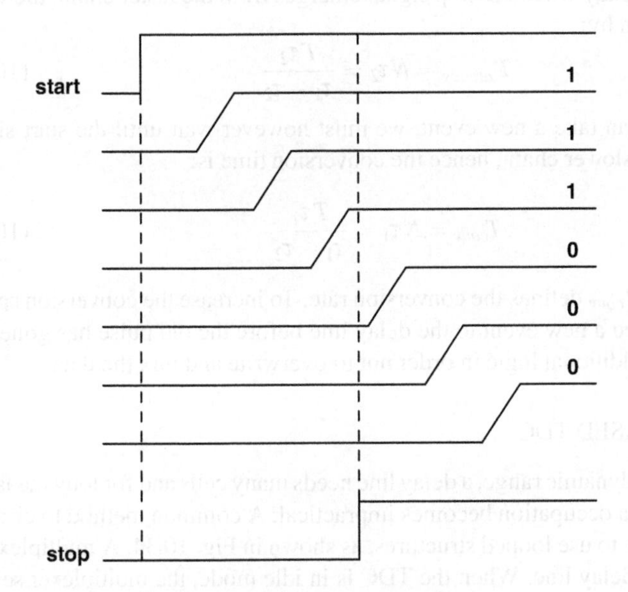

FIGURE 10.32 Thermometric bit pattern in a delay-line TDC.

single buffer, the required number of cells is:

$$N = \frac{T}{\tau_1} \tag{10.38}$$

The TDC is ready for a new conversion when the start signal emerges from the delay line, after $N\tau_1$ seconds, therefore the maximum conversion rate is $1/(N\tau_1)$. A common approach to increase the resolution of a delay-line TDC is to use the Vernier topology [51]. Shown in Fig. 10.33, the method consists in using two delay lines with different delays. The one that receives the start signal is slower, with a delay per buffer cell given by τ_1. The stop signal is sent to the flip flop through a faster delay line, in which the buffers have a delay $\tau_2 < \tau_1$. In this case, the stop signal gains an advantage of $\tau_1 - \tau_2$ each time it crosses a buffer. After $M(\tau_1 - \tau_2)$ seconds it will thus overtake the start signal, which has experienced longer delays. The pattern stored in the flip-flops thus changes from one to zero, creating again a thermometric code. However, the LSB is given by the difference between the delays of the two cells, i.e. $\tau_1 - \tau_2$. The total number of cells which is necessary to measure a maximum time interval T is thus given by:

$$N = \frac{T}{\tau_1 - \tau_2} \tag{10.39}$$

and it thus significantly larger than that of the single delay line TDC. We know that the conversion is ready when the stop signal emerges from the faster chain, therefore the latency is given by:

$$T_{latency} = N\tau_2 = \frac{T\tau_2}{\tau_1 - \tau_2} \tag{10.40}$$

Before the TDC can take a new event, we must however wait until the start signal emerges from the slower chain, hence the conversion time is:

$$T_{conv} = N\tau_1 = \frac{T\tau_1}{\tau_1 - \tau_2} \tag{10.41}$$

The reciprocal of T_{conv} defines the conversion rate. To increase the conversion speed, it is possible to feed a new event to the delay line before the old pulse has gone out, but this requires additional logic in order not to overwrite and mix the data.

10.5.3 LOOP-BASED TDC

To achieve a large dynamic range, a delay line needs many cells and for long measurement times the area occupation becomes impractical. A common method to circumvent the problem is to use looped structures, as shown in Fig. 10.34. A multiplexer is placed before the delay line. When the TDC is in idle mode, the multiplexer selects the path A, sending the start line to the delay line input. When a start pulse arrives, a suitable control logic shifts the multiplexer in position B before the pulse reaches the end of the line. The output of the last cell is folded back towards the input and resent into the fist cell. A counter is incremented each time the pulse emerges from the last buffer. The counter thereby provides a first coarse information, while the fine time

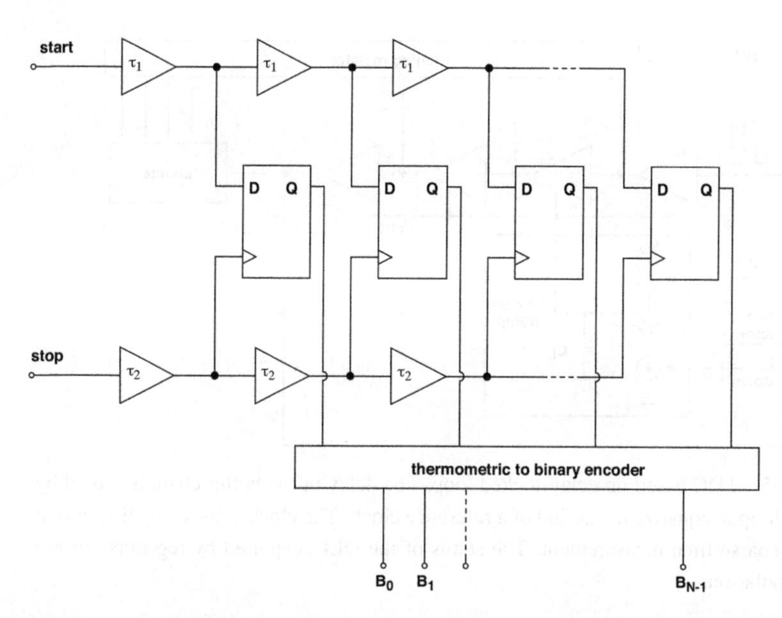

FIGURE 10.33 Vernier delay line TDC. The start and the stop signals propagate through two delay lines with different delays. The LSB is given by the delay difference.

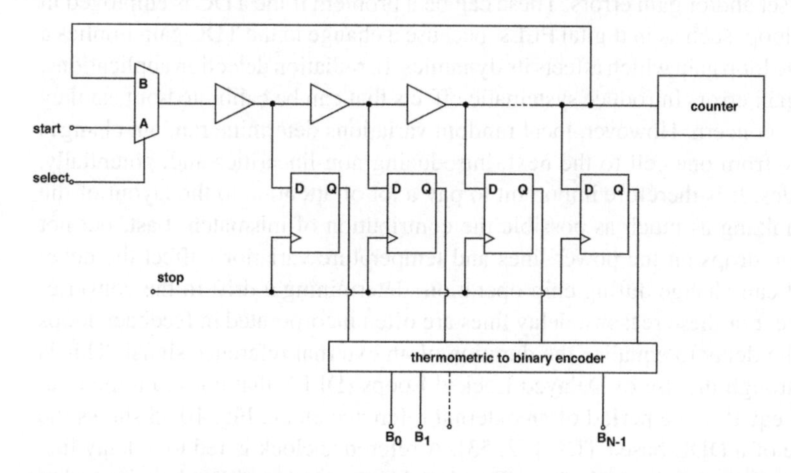

FIGURE 10.34 Example of looped delay line. The pulse is folded back to the input of the delay line, so that the same cells are re-used many times in the measurement. A counter counts how many times the pulse circulates in the loop and provides a coarse time information.

is obtained by capturing with the flip flops the status of the delay line. An important point of the techniques just described is that although they are realized in principle with digital components, these can not be taken as "black boxes" with granted performance. In fact, the effective delay is sensitive to process, temperature and voltage supply variations. As with any other circuits, process variations can induce global

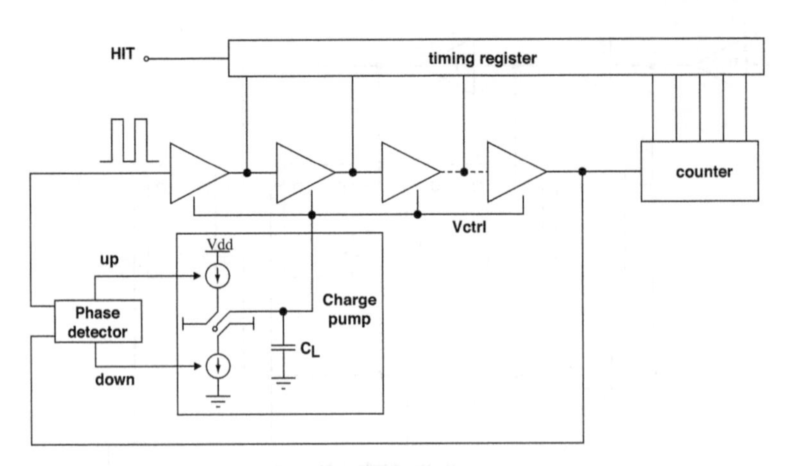

FIGURE 10.35 TDC based on delay-locked loop. The delay of the buffer chain is forced by the feedback loop to equalize the period of a reference clock. The clock at the delay line output provides the coarse time measurement. The status of the DLL, captured by registers, allows the fine interpolation.

or local parameter changes. Global changes affect all the cells in the same way and result in offset and/or gain errors. These can be a problem if the TDC is employed in a feedback loop, such as in digital PLLs, because a change in the TDC gain implies a change in the loop gain which affects its dynamics. In radiation detection applications, offset and gain errors introduce systematic effects that can be calibrated-out, so they are less of a concern. However, local random variations determine random changes of the delay from one cell to the next, introducing non-linearities and, potentially, missing codes. It is therefore important to pay a lot of attention to the layout of the TDC, minimizing as much as possible the contribution of mismatch. Last, but not least, voltage drops on the power lines and temperature variations affect the delay as well and can change during chip operation, determining a drift in the converter performance. For these reasons, delay lines are often incorporated in feedback loops that force the delay to equalize the duration of an external reference signal. This is achieved through the use of Delayed Locked Loops (DLL), that constrain the total delay to be equal to the period of an external reference clock. Fig. 10.35 shows the architecture of a DDL-based TDC [52, 53]. A reference clock is fed to a delay line containing a given number of buffers. The signal at the output of the chain is used to increment a counter, thereby providing the coarse time base. At the same time, the output signal is feedback to a phase detector that compares its phase to the one of the reference clock. If the reference signal comes before the delay line output, the up signal is asserted and the capacitor is charged by the top current source, increasing the control voltage. This increases the bias current in the delay cells making them faster. If, on the other hand, the DLL output leads the reference clock, the down signal is asserted, discharging the capacitor and making the cell slower by reducing their bias current. The concept of phase detector is shown in Fig. 10.36. Two D-type flip-flops

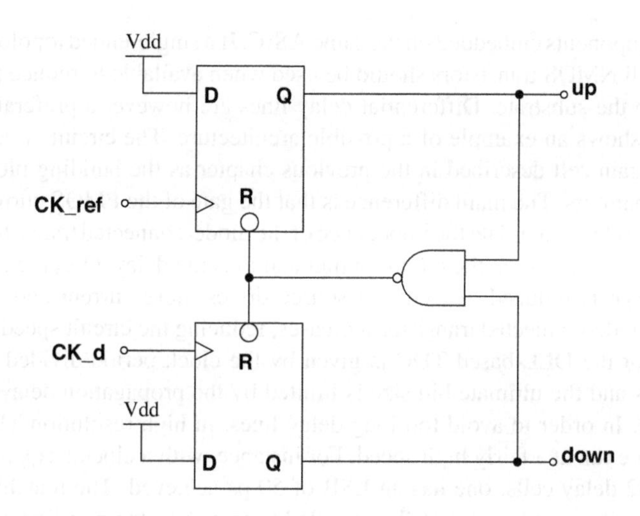

FIGURE 10.36 Principle of a phase detector that compares the phases of two digital signals.

receive on their clock input, respectively, the reference clock and the feedback signal. The D input of both of them is tied to V_{DD}. If the reference clock arrives first, the "up" signal goes to logic one, connecting the current source to the capacitor and charging it. When the feedback arrives, also the "down" signal goes at one, creating a logic zero at the output of the NAND gate that resets both flip flops. The cycle is then repeated in the next clock period. At each cycle, a pack of charge ΔQ is injected on the capacitor. In the loop filter, the capacitor voltage is low pass filtered to provide a smooth control signal. If the feedback signal is leading, the "down" signal is first asserted, subtracting ΔQ from the capacitor, thereby decreasing its voltage. When the reference and the feedback signal are too close in time, the phase detector may fail, creating a "dead-zone" is which the feedback does not control the delay, introducing an uncertainty on its value. This can be addressed by modifying the phase detector introducing a delay in the reset path, as shown in Fig. 10.37.

The buffers providing the delay unit in the DLL can be constructed with single-ended or differential topologies. Fig. 10.38 shows the possible building blocks of a single-ended delay line. A transistor acting as a voltage controlled resistor is put in series to an inverter gate. Three different configurations are possible: NMOS transistor in the ground path, PMOS transistor in the V_{DD} path or both. Observe that if the choice of the PMOS transistor is made, the control voltage must be decreased to increase the current in the inverter cell. The logic of the phase detector-charge pump is therefore reversed with respect to what was described above, i.e. the down signal must be asserted to speed-up the delay line and the up signal must be used to slow it down. Two inverters must be cascaded to form a non-inverting buffer. Single-ended delay lines have the advantage of requiring a small area. However, the inverters are more sensitive to power supply or ground noise that can deteriorate the TDC accuracy. Furthermore, the inverters generate at the output full swing CMOS signal spanning from 0 to V_{DD}, therefore they can inject noise into the chip substrate that can affect sensitive analog components embedded on the same ASIC. If a single-ended topology is chosen, triple well NMOS transistors should be used when available to reduce the noise injection into the substrate. Differential delay lines are however a preferable choice. Fig. 10.39 shows an example of a possible architecture. The circuit is very similar to the low-gain cell described in the previous chapter as the building block of high speed comparators. The main difference is that the gate of the PMOS current sources is now controlled to regulate the impedance of the diode-connected transistors and, with that, the RC constant at the circuit output and thus the delay. Observe that if the control voltage is reduced, the current source drives more current and the impedance of the diode-connected transistor increases, reducing the circuit speed.

The resolution of the DLL-based TDC is given by the clock period divided by the number of cells and the ultimate bin size is limited by the propagation delay of a single buffer cell. In order to avoid too long delay lines, in high resolution TDC the clock is therefore run at a fairly high speed. For instance, with a clock frequency of 640 MHz and 32 delay cells, one has an LSB of 50 ps achieved. The matching between the delay cells must be very well controlled to minimize the non-linearity. One of the biggest advantages of DLL-based TDC is the conversion speed. In fact, the hit signal just triggers the storage into registers of the status of both the coarse counter and the delay line. Therefore, the conversion result is immediately ready and the sampling rate can be very high. The down-side is that DLL-based TDC have fairly high power consumption. This comes both from the fact that the reference clock runs at high speed and from the use of current steering topology in the delay buffers. To mitigate the power consumption in multi-channel systems, the DLL can be shared among several channels. A global DLL generates the time-stamp and the counter and delay line outputs are propagated to a bank of registers, where one or more registers are associated to a given channel. The skew of the fine-time stamping signals must

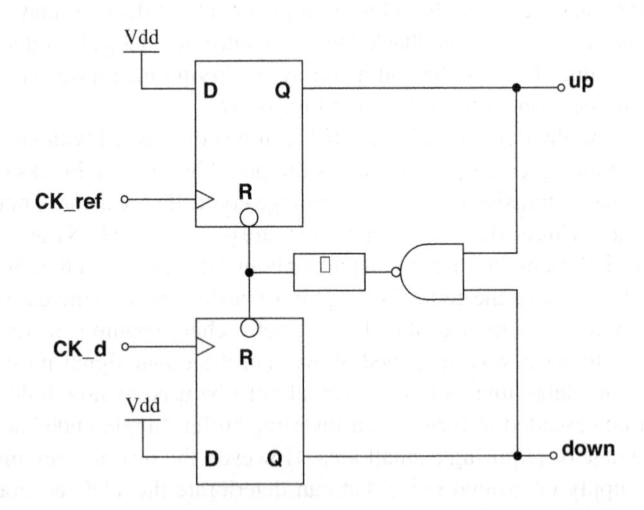

FIGURE 10.37 A delay inserted in the reset path of the phase detector reduces its dead-zone.

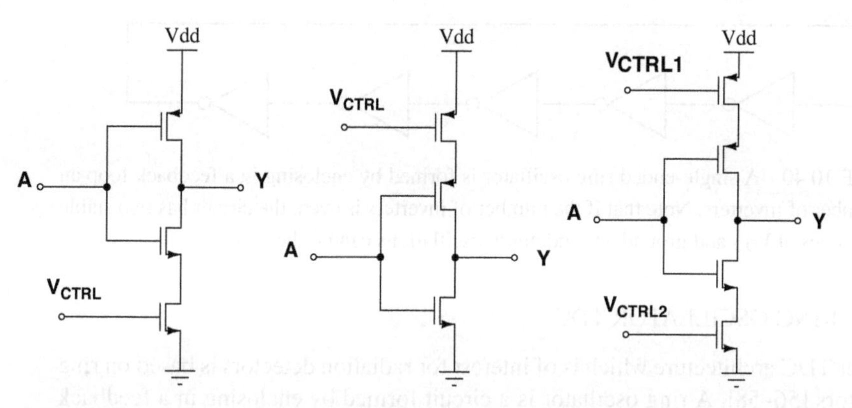

FIGURE 10.38 Starved inverters form the basis of single ended delay cells. The current in the inverter is limited by transistors connected in series to the V_{DD} and/or ground path that act as voltage-controlled resistor. Two inverters need to be cascade to obtain a non-inverting delay unit.

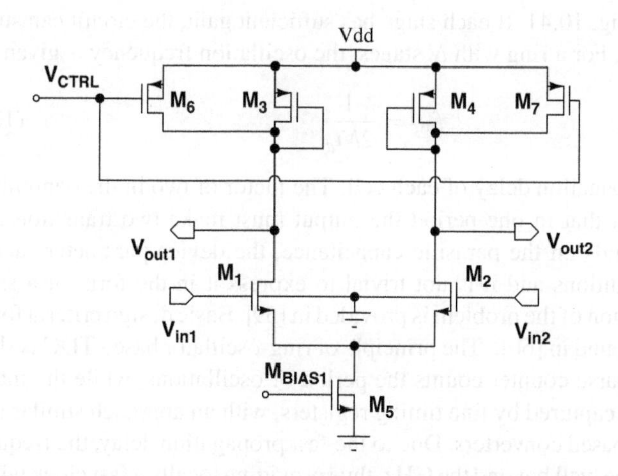

FIGURE 10.39 A current-steering differential delay cell allows to minimize both the circuit sensitivity to external interference noise and the injection of noise from the cell into the chip substrate.

however be well controlled. If the signals have to be propagated over a large area, a robust buffering scheme is needed, which again brings in extra power consumption. The use of local passive interpolation techniques reduces significantly the bin size of a DLL-based TDC [54] and converters with LSB in the order of 4-5 ps have been demonstrated [55].

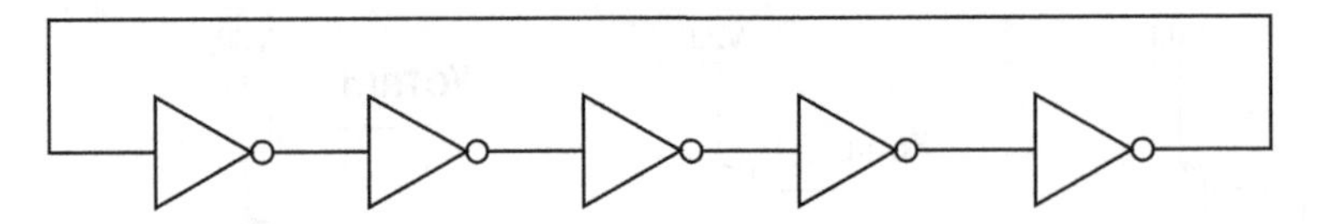

FIGURE 10.40 A single-ended ring oscillator is formed by enclosing in a feedback loop an odd number of inverters. Note that if the number of inverters is even, the circuit has two stable output values at V_{DD} and ground and indefinite oscillations can not be sustained.

10.5.4 RING OSCILLATOR TDC

Another TDC architecture which is of interest for radiation detectors is based on ring oscillators [56–58]. A ring oscillator is a circuit formed by enclosing in a feedback loop an odd number of inverting stages, as sketched in Fig. 10.40. Ring oscillators can be based on single-ended inverters or on differential cells. In the latter case, the oscillator is formed by an even number of stages, with a cross coupled connection inserted in one point to ensure that an odd number of inversions is encompassed by the loop, as shown in Fig. 10.41. If each stage has sufficient gain, the circuit can sustain a steady oscillation. For a ring with N stages, the oscillation frequency is given by:

$$f_{osc} = \frac{1}{2Nt_d} \tag{10.42}$$

where t_d is the propagation delay of each cell. The factor of two in the denominator stems from the fact that in one period the output must make two transitions. The delay time t_d depends on the parasitic capacitance, the device parameters and the environmental conditions and it is not trivial to express it in the form of a simple equation. A discussion of the problem is provided in [59]. Basic design criteria for ring oscillators can be found in [60]. The principle of ring-oscillator based TDC is shown in Fig. 10.42. A coarse counter counts the period of oscillations, while the internal status of the ring is captured by fine timing registers, with an approach similar to the one seen for DLL-based converters. Due to the fast propagation delay, the frequency of oscillations can be well beyond the GHz, thus providing locally a fast clock without the need of propagating it over long distances across the chip. Since fast oscillations are desired, a low number of gates is used, making ring oscillator TDCs one of the

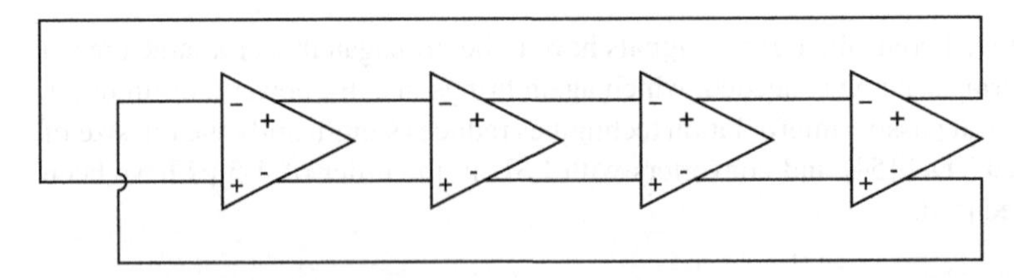

FIGURE 10.41 A fully differential ring oscillator is formed by an even number of fully differential gain stages. The requirement of having an odd number of inversions inside the loop is satisfied by introducing one cross-coupled connection between two stages.

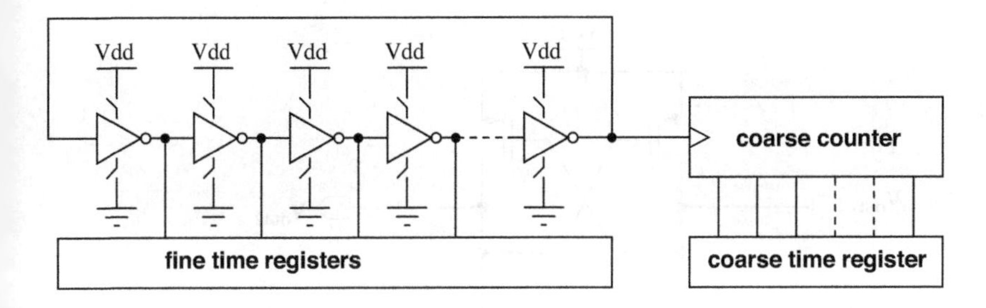

FIGURE 10.42 Principle of the gated ring oscillator TDC.

most compact solutions to implement time-to-digital converters. They are therefore a natural choice to design time measuring systems in high granularity front-ends, such as pixel detectors [61, 62]. Ring oscillators TDC are normally employed in a gated form, meaning that the oscillator works only during the time interval to be measured. Care must be taken to ensure that oscillations promptly begin when the start signal is issued to avoid introducing non-linearity in the measurement. It must be observed that the oscillation frequency depends on device parameters and environmental conditions, thus it is very sensitive to process, power supply and temperature variations. Actually, ring oscillators are one of the standard process monitoring structures which are put on wafer to have a first feedback of the quality of the transistors in a particular production lot. The sensitivity to ring oscillators to temperature is also exploited to make compact on-chip temperature sensors [63]. In a TDC, all these effects are of course undesirable, as one would like to keep the oscillation frequency as stable as possible to provide a reliable timing measurement. Fully differential oscillators help in mitigating the impact of power-supply variations. Process variations introduced a steady effect once the chip is fabricated and temperature variations are normally slow. One possibility to circumvent the problem thus consists in making a calibration immediately after each measurement by sending to the TDC reference signals of well known duration. A better alternative is to lock the oscillation frequency to a stable reference clock with a phase locked-loop. In this case, a bias voltage controls the rise time of the cells, transforming the ring oscillator into a Voltage Controlled Oscillator (VCO). Fig. 10.43 shows as an example a differential cell whose bias is defined by an external voltage. In a multi-channel system, controlling each oscillator individually is cumbersome. A replica oscillator is therefore provided in a point of the chip and its oscillation frequency is stabilized by the PLL. The generated control voltage is thus used to bias the individual ring oscillators. The correction method is however not perfect, because on a large chip process and temperature gradients create substantial mismatch between the reference and the in-channel oscillators. Nevertheless, circuits designed with this principle have shown a good uniformity among the channels [64].

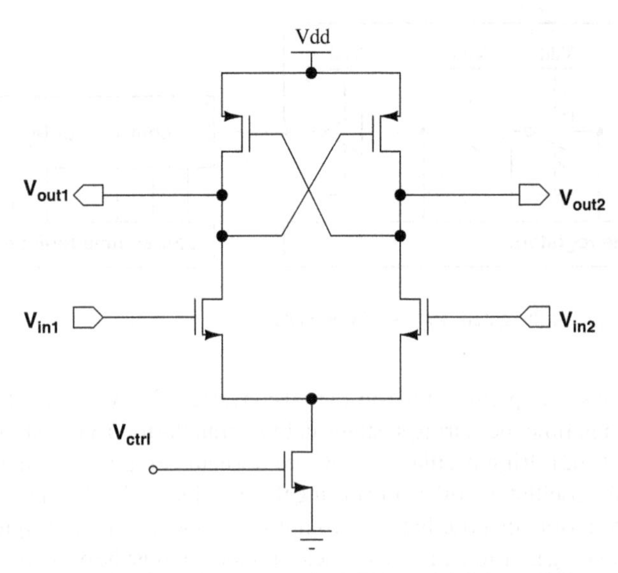

FIGURE 10.43 Example of a voltage-controlled differential delay cell that can be used in a VCO.

10.5.5 PULSE SHRINKING TDC

Pulse shrinking is another method that provides timing measurements with sub-gate accuracy. The basic principle is shown in Fig. 10.44 [65, 66]. The time interval to be measured is fed to a circular inverter chain that contains one asymmetric delay element with different rise and fall times. When the pulse goes through the asymmetric element it is shorten by a given amount ΔT corresponding to the difference between the rise and fall time of the cell. Each time the pulse emerges from the chain a counter is incremented and the number of turns necessary to make the pulse disappear provides the required output code. Fig. 10.45 shows the schematic of a typical pulse shrinking cell. The rise or the fall time of a digital inverter is limited by putting an additional transistor is series to the PMOS or NMOS device of the inverter. The cell is then buffered with a standard symmetric inverter to obtain a final pulse which is shorter, but still has steep rising and falling edges. With the described method, the value of the

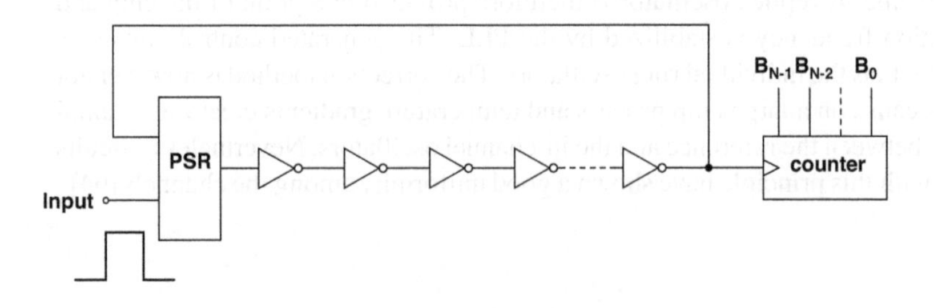

FIGURE 10.44 Possible implementation of a pulse shrinking TDC.

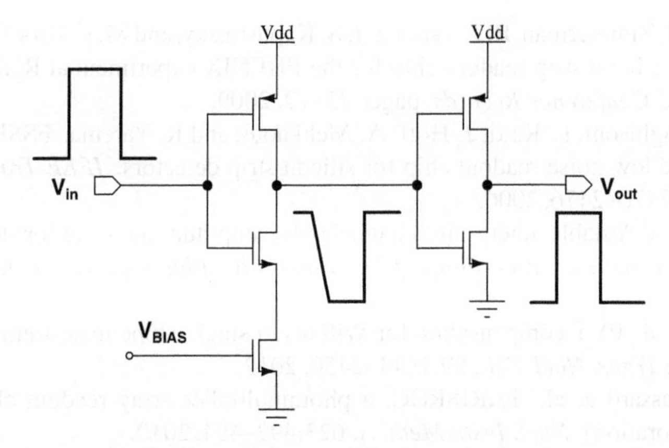

FIGURE 10.45 Example of a pulse shrinking cell. The fall time of the inverter is limited by starving the pull-down transistor. The signal is then buffered to get a shorter output pulse, but with fast rise and fall times.

LSB depends only on the parameter of one critical cell and the DNL is typically very good. Alternative implementations in which the pulse is shrunk by more elements in the chain have also been reported [67].

REFERENCES

1. M. Pelgrom. *Analog-to-Digital Conversion*. Springer, 2012.
2. F. Maloberti. *Data Conveters*. Springer, 2007.
3. R. van de Plassche. *CMOS Integrated Analog-to-Digital and Digital-to-Analog Converters*. Kluwer Academic Publishers, 2007.
4. B. Razavi. *Principles of Data Converter System Design*. IEEE Press, 1995.
5. S. Henzler. *Time-to-Digital Converters*. IEEE Press, 2010.
6. J. Doerenberg, H. S. Lee, and D. A. Hodges. Full-speed testing of A/D converters. *IEEE J. Solid-state Circ.*, 19:820–827, 1984.
7. F. J. Harris. On the use of windows for harmonic analysis with the discrete Fourier transform. *Proceedings of the IEEE*, 66:51–83, 1978.
8. A. Nuttall. Some windows with very good sidelobe behavior. *IEEE Transactions on Acoustics, Speech, and Signal Processing*, 29:84–91, 1981.
9. R. van de Grift, I. W. J. M. Rutten, and M. van der Veen. An 8-bit video ADC incorporating folding and interpolation techniques. *IEEE J. Solid-state Circ.*, 22:944–953, 1987.
10. Y. Nakajima et al. A background self-calibrated 6b 2.7GS/s ADC with cascade-calibrated folding-interpolating architecture. *IEEE J. Solid-state Circ.*, 45:707–718, 2010.
11. S. Dramico, G. Cocciolo, M. De Matteis, and A. Baschirotto. A 7.65mW 5bits 90nm 1Gs/s ADC folded-interpolated without calibration. In *Solid-state Circuits European Conference*, pages 151–154, 2011.

12. J. R. Hoff, T. N. Zimmerman, R. J. Yarema, J. S. Kapustinsky, and M. L. Brookes. FPHX: A new silicon strip readout chip for the PHENIX experiment at RHIC. *IEEE NSS-MIC Conference Records*, pages 75–79, 2009.
13. V. Re, M. Manghisoni, L. Ratti, J. Hoff, A. Mekkaoui, and R. Yarema. FSSR2, a self-triggered low noise readout chip for silicon strip detectors. *IEEE Trans. Nucl. Sci.*, 53:2470–2476, 2006.
14. D. H. Wilkinson. A stable ninety-nine channel pulse amplitude analyzer for slow counting. *Mathematical Proceedings of The Cambridge Philosophical Society*, 46, 1950.
15. K. V. Tham et al. PVT compensation for Wilkinson single-slope measurement systems. *IEEE Trans. Nucl. Sci.*, 59:2444–2450, 2012.
16. G. Martin-Chassard et al. PARISROC, a photomultiplier array readout chip (PMm2 collaboration). *Nucl. Instr. Meth. A*, 623:492–494, 2010.
17. V. Ferragina et al. Implementation of a novel read-out strategy based on a Wilkinson ADC for a 16×16 pixel X-ray detector array. In *IEEE International Symposium on Circuits and Systems*, pages 5569–5572, 2005.
18. W. Huang, S. Wood Chiang, and S. Kleinfelder. Waveform digitization with programmable windowed real-time trigger capability. *IEEE NSS-MIC Conference Records*, pages 422–427, 2009.
19. S. Tedja, J. Van der Speigel, and H. H. Williams. A CMOS low-noise and low-power charge sampling integrated circuit for capacitive detector/sensor interfaces. *IEEE J. of Solid-state Circ.*, 30:110–119, 1995.
20. J. L. McCreary, P. R. Gray, and D. A. Hodges. All-MOS charge redistribution analog-to-digital conversion techniques—Part I. *IEEE J. Solid-state Circ.*, 10:379–385, 1975.
21. A. M. Abo and P. R. Gray. A 1.5-V, 10-bit, 14.3MS/s CMOS pipeline analog-to-digital converter. *IEEE J. Solid-state Circ.*, 34:599–606, 1999.
22. J. Steensgaard. Bootstrapped low-voltage analog switches. In *IEEE International Symposium on Circuits and Systems*, pages 29–32, 1999.
23. M. Dessouky and A. Kaiser. Input switch configuration suitable for rail-to-rail operation of switched opamp circuits. *Electronics Letters*, 35, 1999.
24. C. Liu, S. Chang, G. Huang, and Y. Lin. A 10-bit 50MS/s SAR ADC with a monotonic capacitor switching procedure. *IEEE J. Solid-state Circ.*, 45:731–740, 2010.
25. P. Harpe, E. Cantatore, and A. van Roermund. A 10b/12b 40 kS/s SAR ADC with data-driven noise reduction achieving up to 10.1b ENOB at 2.2 fJ/conversion-step. *IEEE J. Solid-state Circ.*, 48:3011–3018, 2013.
26. D. Zhang, C. Svensson, and A. Alvandpour. Power consumption bounds for SAR ADCs. In *European Conference on Circuit Theory and Design*, 2011.
27. B. P. Ginsburg and A. P. Chandrakasan. An energy-efficient charge recycling approach for an SAR converter with capacitive DAC. In *IEEE International Symposium on Circuits and Systems*, pages 184–187, 2005.
28. V. Hariprasath, J. Guerber, S.-H. Lee, and U.-K. Moon. Merged capacitor switching based SAR ADC with highest switching energy-efficiency. *Electronics Letters*, 46, 2010.

29. P. Nuzzo, C. Nani, C. Armiento, A. Sangiovanni-Vincentelli, J. Craninckx, and G. Van der Plas. A 6Bit 50MS/s threshold configuring SAR ADC in 90-nm digital CMOS. *IEEE Trans. on Circ. and Syst. I-regular Papers*, 59:80–92, 2012.
30. A. Shikata, R. Sekimoto, T. Kuroda, and H. Ishikuro. A 0.5 V 1.1 MS/sec 6.3 fJ/conversion-step SAR-ADC with tri-level comparator in 40 nm CMOS. *IEEE J. Solid-state Circ.*, 47:1022–1030, 2012.
31. P. J. A. Harpe, B. Busze, K. Philips, and H. de Groot. A 0.47–1.6 mW 5-bit 0.5–1 GS/s time-interleaved SAR ADC for low-power UWB radios. *IEEE J. Solid-state Circ.*, 47:1594–1602, 2012.
32. B. Verbruggen, M. Iriguchi, and J. Craninckx. A 1.7mW 11b 250MS/s 2× interleaved fully dynamic pipelined SAR ADC in 40nm digital CMOS. In *Solid-state Circuits IEEE International Conference*, pages 466–468, 2012.
33. C. Zhang and H. Wang. Reduction of parasitic capacitance impact in low-power SAR ADC. *IEEE Transactions on Instrumentation and Measurement*, 61:587–594, 2012.
34. H. Wei, C-H. Chan, U-F. Chio, S-W. Sin, U. Seng-Pan, R. Martins, and F. Maloberti. A 0.024mm2 8b 400MS/s SAR ADC with 2b/cycle and resistive DAC in 65nm CMOS. In *Solid-state Circuits IEEE International Conference*, pages 188–190, 2011.
35. S-J. Cho, Y. Hong, T. Yoo, and K-H. Baek. A 10Bit, 50 MS/s, 55 fJ/conversion-step SAR ADC with split capacitor array. In *International Conference on ASIC*, pages 472–475, 2011.
36. Al. H. Chang, H-S. Lee, and D. S. Boning. Redundancy in SAR ADCs. In *ACM Great Lakes Symposium on VLSI*, pages 283–288, 2011.
37. R. Sekimoto, A. Shikata, T. Kuroda, and H. Ishikuro. A 40nm 50S/s–8MS/s ultra low voltage SAR ADC with timing optimized asynchronous clock generator. In *Solid-state Circuits European Conference*, pages 471–474, 2011.
38. M. Yoshioka, K. Ishikawa, T. Takayama, and S. Tsukamoto. A 10b 50MS/s 820μW SAR ADC with on-chip digital calibration. In *Solid-state Circuits IEEE International Conference*, pages 384–385, 2010.
39. Y. Zhu et al. A 10-bit 100MS/s reference-free SAR ADC in 90 nm CMOS. *IEEE J. Solid-state Circ.*, 45:1111–1121, 2010.
40. T. Ogawa et al. SAR ADC algorithm with redundancy and digital error correction. *Ieice Transactions*, 93-A:415–423, 2010.
41. C.-C. Liu et al. A 10b 100MS/s 1.13 mW SAR ADC with binary-scaled error compensation. In *Solid-state Circuits IEEE International Conference*, pages 386–387, 2010.
42. T. Jang et al. Single-channel, 1.25GS/s, 6-bit, loop-unrolled asynchronous SAR-ADC in 40nm-CMOS. In *Custom Integrated Circuits Conference*, pages 1–4, 2010.
43. R. E. Green and R. E. Bell. Notes on a fast time-to-amplitude converter. *Nuclear Instruments*, 3:127–132, 1958.
44. J. L. Rodda, J. E. Griffin, and M. G. Stewart. A transistorized time-amplitude converter for subnanosecond lifetime measurements. *Nuclear Instruments and Methods*, 23:137–140, 1963.

45. A. E. Stevens, R. P. Van Berg, J. Van der Spiegel, and H. H. Williams. A time-to-voltage converter and analog memory for colliding beam detectors. *IEEE J. Solid-state Circ.*, 24:1748–1752, 1989.
46. A. Rivetti et al. A pixel front-end ASIC in 0.13 μm CMOS for the NA62 experiment with on pixel 100 ps time-to-digital converter. *IEEE NSS-MIC Conference Records*, pages 55–60, 2009.
47. M. Crotti, I. Rech, and M. Ghioni. Four-channel, 40 ps resolution, fully integrated time-to-amplitude converter for time-resolved photon counting. *IEEE J. Solid-state Circ.*, 47:699–708, 2012.
48. P. Keranen, K. Maatta, and J. Kostamovaara. Wide-range time-to-digital converter with 1-ps single-shot precision. *IEEE Trans. on Instrum. and Meas.*, 60:3162–3172, 2011.
49. D. Stoppa et al. A 32x32-pixel array with in-pixel photon counting and arrival time measurement in the analog domain. In *Solid-state Circuits European Conference*, pages 204–207, 2009.
50. K. Koch, H. Hardel, R. Schulze, E. Badura, and J. Hoffmann. A new TAC-based multichannel front-end electronics for TOF experiments with very high time resolution. *IEEE Trans. Nucl. Sci.*, 52:745–747, 2005.
51. J. Yu, F. Foster Dai, and R. C. Jaeger. A 12bit Vernier ring time-to-digital converter in 0.13 CMOS technology. *IEEE J. Solid-state Circ.*, 45:830–842, 2010.
52. J. Christiansen. An integrated high resolution CMOS timing generator based on an array of delay locked loops. *IEEE J. Solid-state Circ.*, 31:952–957, 1996.
53. M. Mota and J. Christiansen. A high-resolution time interpolator based on a delay locked loop and an RC delay line. *IEEE J. Solid-state Circ.*, 34:1360–1366, 1999.
54. S. Henzler et al. A local passive time interpolation concept for variation-tolerant high-resolution time-to-digital conversion. *IEEE J. Solid-state Circ.*, 43:1666–1676, 2008.
55. L. Perktold and J. Christiansen. A flexible 5 ps bin-width timing core for next generation high-energy-physics time-to-digital converter applications. *Proceedings of the 8th Conference on Ph.D. Research in Microelectronics, PRIME 2012, Aachen, Germany*, pages 1–4.
56. M. Z. Straayer and M. H. Perrott. A multiPath gated ring oscillator TDC with first-order noise shaping. *IEEE J. Solid-state Circ.*, 44:1089–1098, 2009.
57. K. Hwang and L. Kim. An area efficient asynchronous gated ring oscillator TDC with minimum GRO stages. In *IEEE International Symposium on Circuits and Systems*, pages 3973–3976, 2010.
58. A. Elshazly, S. Rao, B. Young, and P. K. Hanumolu. A 13b 315fsrms 2mW 500MS/s 1MHz bandwidth highly digital time-to-digital converter using switched ring oscillators. In *Solid-state Circuits IEEE International Conference*, pages 464–466, 2012.
59. S. Docking and M. Sachdev. An analytical equation for the oscillation frequency of high-frequency ring oscillators. *IEEE J. Solid-state Circ.*, 39:533–537, 2004.
60. B. Razavi. *Design of Analog CMOS Integrated Circuits*. McGraw-Hill, 2001.

61. F. Zappon, M. van Beuzekom, V. Gromov, R. Kluit, X. Fang, and A. Kruth. GOSSIPO-4: an array of high resolution TDCs with a PLL control. *Journal of Instrumentation*, 7, 2012.
62. C. Veerappan et al. A 160×128 single-photon image sensor with on-pixel 55ps 10b time-to-digital converter. In *Solid-state Circuits IEEE International Conference*, pages 312–314, 2011.
63. S. Park, C. Min, and S. Cho. A 95 nW ring oscillator-based temperature sensor for RFID tags in 0.13 μm CMOS. In *IEEE International Symposium on Circuits and Systems*, 2009.
64. C. Veerappan et al. Characterization of large-scale non-uniformities in a 20k TDC/SPAD array integrated in a 130nm CMOS process. In *Solid-state Device Research European Conference*, pages 331–334, 2011.
65. C. Chen, S. Lin, and C. Hwang. An area-efficient CMOS time-to-digital converter based on a pulse-shrinking scheme. *IEEE Transaction on Circuits and Systems—II: Express Briefs*, 2014.
66. S. Tisa, A. Lotito, A. Giudice, and F. Zappa. Monolithic time-to-digital converter with 20ps resolution. In *Solid-state Circuits European Conference*, 2003.
67. E. Raisanen-Ruotsalainen, T. Rahkonen, and J. Kostamovaara. A low-power CMOS time-to-digital converter. *IEEE J. Solid-state Circ.*, 30:984–990, 1995.

A Differential and Operational Amplifiers

Due to the single-ended nature of most radiation sensors, differential configurations are seldom used in the input stage of front-end ASICs. However, they are extensively employed in the remaining circuitry which is found on the chip. In the first part of this Appendix, the properties of basic CMOS differential cells are reviewed. In the second part, the most common architectures of operational amplifiers are discussed. Before delving into circuit details, a few fundamental definitions are revised.

A.1 DIFFERENTIAL AND COMMON MODE SIGNALS

Fig. A.1 shows a few important differences between single-ended and differential signal transmission. In a single-ended system, reported in Fig. A.1 a), the information is encoded in the difference between the signal line and a reference (usually the ground) which is common to both the transmitter and the receiver. In differential architectures, the information is embedded in the difference between two closely matched lines. As a consequence, any disturbance which is common to both lines is rejected by the receiver, as sketched in the figure. Another key difference is that in the circuit of Fig. A.1 a), the current returns back to the transmitter via the common ground. This defines a loop whose amplitude is not well controlled and can thus favor electromagnetic pick-up. Additionally, the grounds of the two systems must be firmly connected as they provide the references against which the signal is measured. For practical reasons, power supplies and decoupling capacitors can not always be placed close to the circuits and the parasitic inductance of the interconnecting wires is not negligible. The situation is particularly critical if circuit A contains sensitive analog blocks whereas circuit B is a pure digital circuit. In B, the current towards ground will experience periodical surges that will determine bounces on the common line due to the parasitic inductance L_p. This "common impedance noise" can severely compromise the functionality of the driving circuit. If the connection between A and B is realized with a differential protocol, the return of the signal current towards the receiver is clearly defined by the path shown by the arrows and the loop area is kept under control. Furthermore, no strong common reference is required between the transmitting and the receiving ends because the receiver senses only the difference between the two signal lines, therefore the grounds can only be weakly coupled. This is shown by the inductance L (this time introduced on purpose) that connects the two grounds, decoupling them at high frequency. Differential signal processing must hence be used when possible on chip and becomes a must when the data need to be transmitted off-chip.

Fig. A.2 shows the elementary differential cell. Two *identical* transistors have the source in common and are biased by a current source, providing the tail current I_{BIAS}.

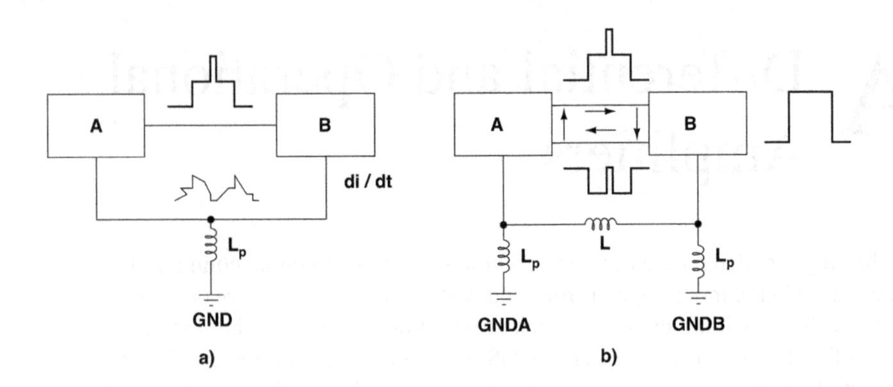

FIGURE A.1 Single-ended and differential systems.

The configuration based on NMOS input transistors is shown in Fig. A.2 a), while the one employing PMOS input devices is reported in b). The purpose of the circuit is to sense and amplify only the difference $V_{in1} - V_{in2}$ between two generic signals. Each input can be written as the sum of half the difference between the two input signals plus their average value, as shown by the following identity:

$$\begin{aligned} V_{in1} &= \frac{V_{in1}}{2} + \frac{V_{in1}}{2} + \frac{V_{in2}}{2} - \frac{V_{in2}}{2} = \frac{V_{in1}-V_{in2}}{2} + \frac{V_{in1}+V_{in2}}{2} \\ V_{in2} &= \frac{V_{in2}}{2} + \frac{V_{in2}}{2} + \frac{V_{in1}}{2} - \frac{V_{in1}}{2} = \frac{V_{in2}-V_{in1}}{2} + \frac{V_{in1}+V_{in2}}{2} \end{aligned} \tag{A.1}$$

Note that the first term has opposite sign for the two inputs. The quantity:

$$V_{CM,in} = \frac{V_{in1} + V_{in2}}{2} \tag{A.2}$$

is called the common mode component of the input signal and it should be rejected by the differential amplifier because it affects both inputs equally. Consider now

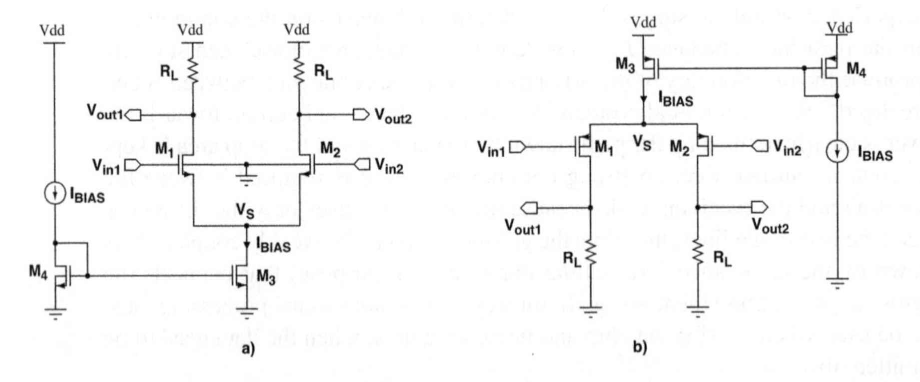

FIGURE A.2 Differential amplifiers with resistive loads. In a): configuration with NMOS input transistors. In b): complementary topology with PMOS input.

the situation in which $V_{in1} = V_{in}/2$ and $V_{in2} = -V_{in}/2$. Applying (A.1), we find that $V_{in1} - V_{in2} = V_{in}$, while $V_{CM,in} = 0$. In case the two input signals are equal in magnitude and opposite in phase, no common mode variation is applied and the circuit is said to be driven in "purely differential mode". Let's clarify the above points with some numerical examples. Suppose that we start from a common voltage of 0.6 V and apply to the inputs rectangular pulses with the same magnitude of 50 mV, but opposite in phase. As a consequence, V_{G1} will be 0.65 V and V_{G2} 0.55 V, where V_{G1} and V_{G2} are the gate voltages of M_1 and M_2 measured with respect to ground. In small signal terms (see equation A.1), we have $V_{in1} = 50$ mV and $V_{in2} = -50$ mV. The total differential signal is $V_{in1} - V_{in2} = 100$ mV and the common mode *variation* $V_{CM,in}$ is zero. In fact, when the signal is applied, $V_{G1} - V_{G2}$ is 100 mV, while $(V_{G1} + V_{G2})/2$ is 0.6 V. Therefore, if V_{in1} and V_{in2} are equal in magnitude and opposite in phase, we apply a pure differential signal that does not change the value of the common mode voltage. Consider now the case in which $V_{in1} = 50$ mV and $V_{in2} = -30$ mV. This means that V_{G1} will go to 0.65 V and V_{G1} to 0.57 V. In this case, the differential signal is 70 mV. The common mode variation is 10 mV and $(V_{G1} + V_{G2})/2$ rises to 0.61 V. Let's now apply to our circuit $V_{in1} = 50$ mV and $V_{in2} = 30$ mV, so that V_{G1} becomes 0.65 V and V_{G2} 0.63 V. Equation (A.1) predicts a common mode variation of 40 mV, in fact $(V_{G1} + V_{G2})/2$ becomes 0.64 V. Finally, if $V_{in1} = V_{in2} = 50$ mV the differential signal is zero and the common mode variation is equal to the full input signal. We can summarize the above results as follows:

- In a differential system, if the input signals are equal in magnitude and opposite in phase, a pure differential signal is applied without change of the input common mode.
- If the input signals are equal in magnitude and phase, a pure common mode signal is applied.
- In any other situation, the input signal has both a differential and a common mode component.

The ideal differential amplifier should amplify only the difference between its inputs, rejecting the common mode part.

A.2 DIFFERENTIAL CELL LARGE SIGNAL BEHAVIOR

We study first the behavior of the differential cell for large signals and then we perform the small signal analysis. We consider initially the case of the NMOS topology. In the following, we call V_{G1} and V_{G2} the gate voltage of M_1 and M_2 measured with respect to ground. We start from the situation shown in Fig. A.3 a), in which both inputs are connected to the same voltage source $V_{G1} = V_{G2} = V_{CM}$ and we explore what happens as V_{CM} is changed. In other words, we investigate the behavior of the circuit for large changes in the input common mode level. To understand how the

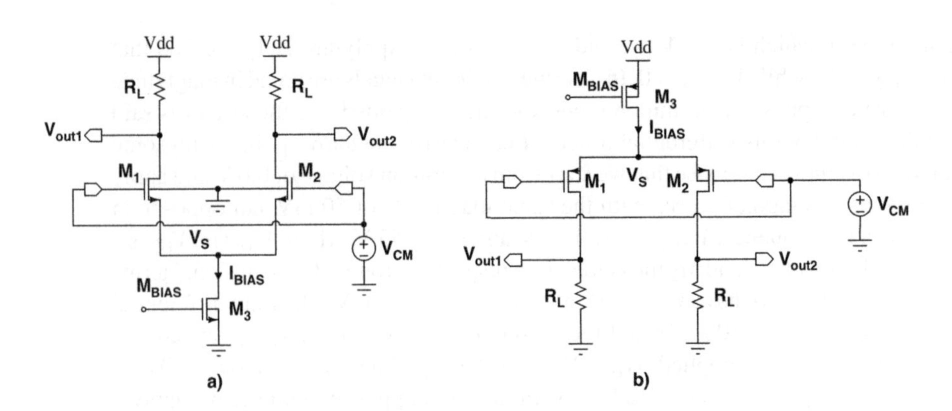

FIGURE A.3 Circuits to study the common mode behavior in differential amplifiers.

potential V_S at the source node of M_1 and M_2 is linked to V_{CM}, let's suppose that the current source M_3 is initially off. In this condition, no current can flow in the circuit and M_1 and M_2 must be off as well. Therefore, their source must be closer to V_{CM} than a threshold voltage, i.e $V_S > V_G - V_{THN}$. Now let's switch on the current source. Since V_S is relatively high, M_3 starts working in saturation. The charges needed to support the current in M_3 are initially provided by the parasitic capacitances present at the node V_S, which are discharged pulling V_S down. The gate voltage of M_1 and M_2 is fixed, hence lowering V_S increases the V_{GS} of the input transistors and drive them into conduction. When the gate-source voltage of M_1 and M_2 is such that all the current I_{BIAS} flows from V_{DD} to the current source through M_1 and M_2, the source node stops being discharged and the system reaches its equilibrium point. In general, we can write:

$$V_{G1} - V_{GS1} = V_S = V_{G2} - V_{GS2} \rightarrow V_{G1} - V_{G2} = V_{GS1} - V_{GS2} \tag{A.3}$$

where V_{GS1} and V_{GS2} are the differences between the gate and source voltages of M_1 and M_2, respectively. If we keep $V_{G1} = V_{G2} = V_{CM}$, equation (A.3) shows that $V_{GS1} = V_{GS2}$. If the two transistors are *identical*, imposing the same gate-source voltage makes their currents also equal:

$$\begin{gathered} I_{DS1} = I_{DS2} \\ I_{DS1} + I_{DS2} = I_{BIAS} \rightarrow I_{DS1} = I_{DS2} = \frac{I_{BIAS}}{2} \end{gathered} \tag{A.4}$$

Note that in a differential cell the relationship $I_{DS1} + I_{DS2} = I_{BIAS}$ is *always* true because the total bias current is imposed by the current source[1]. On the other hand, the condition $I_{DS1} = I_{DS2} = I_{BIAS}/2$ is verified *only* at the equilibrium, i.e. when the

[1]The statement holds if the transistor M_3 works in saturation and it is exactly true only if its output resistance is infinite. However it can be used for a simplified analysis of the differential cell.

gate voltages of M_1 and M_2 are equal. When $V_{G1} = V_{G2}$, the DC voltage at the output will be:

$$V_{out1,DC} = V_{out2,DC} = V_{DD} - \frac{I_{BIAS}}{2} R_L \tag{A.5}$$

Note that to guarantee that $V_{GS1} = V_{GS2}$ it is sufficient that $V_{G1} = V_{G2}$, since the transistors have the source is common by construction. Assuming that M_1 and M_2 are in strong inversion, their gate-source voltage at the equilibrium can be found as:

$$V_{GS1} = V_{GS2} = V_{GS,eq} = V_{THN} + \sqrt{\frac{2LI_{BIAS}/2}{\mu_n C_{ox} W}} \tag{A.6}$$

At the equilibrium, the voltage V_S is thus equal to the gate voltage applied to M_1 and M_2 minus the gate-source voltage needed by M_1 and M_2 to support the current $I_{BIAS}/2$. In formula, we can write:

$$V_S = V_{CM} - V_{GS,eq} \tag{A.7}$$

If we now increase or decrease V_{CM} by a given amount ΔV, while keeping $V_{G1} = V_{G2}$, equation (A.6) always holds, so V_{GS} of M_1 and M_2 can not change. To maintain the gate-source voltage of M_1 and M_2 constant, V_S must follow V_{CM}. In case there is no bulk effect, V_S follows V_{CM} by the same amount ΔV. If the body effect is present, like in the circuit of Fig. A.3 a), V_S will rise less than ΔV because the increased threshold voltage requires a greater V_{GS} to support the same current. In the following discussion, the modulation of V_{TH} with V_S is neglected for simplicity.

If we decrease V_{CM} and thus V_S, we reduce the drain-source voltage of M_3, which needs to work in saturation to perform well as a current source. The source voltage of M_1 and M_2 must thus be greater than saturation voltage of M_3:

$$V_S > V_{DS3,sat} \rightarrow V_{CM} - V_{GS,eq} > V_{DS3,sat} \rightarrow V_{CM} > V_{GS,eq} + V_{DS,sat3} \tag{A.8}$$

The above equation allows us to calculate the lower point that V_{CM} can reach without compromising the functionality of the differential cell.

If V_{CM} rises, also V_S rises. Since $I_{DS1} = I_{DS2} = I_{BIAS}/2$, the output DC levels defined by (A.5) do not change. This means that rising V_{CM} we increase the source voltage of the input transistors, keeping constant the drain voltage. As a consequence, we reduce the drain-source voltage of M_1 and M_2, eventually driving them in the linear region. We must hence require that:

$$V_{DD} - \frac{I_{BIAS}}{2} R_L - V_S = V_{DS1,2} > V_{DSsat1,2} \tag{A.9}$$

We can now introduce into (A.9) the explicit definition of $V_S = V_{CM} - V_{GS,eq}$ and $V_{DSsat1,2} = V_{GS,eq} - V_{TH}$ to get:

$$V_{DD} - \frac{I_{BIAS}}{2} R_L > V_{CM} - V_{GS,eq} + \left(V_{GS,eq} - V_{THN}\right) \rightarrow V_{CM} < V_{DD} + V_{THN} - \frac{I_{BIAS}}{2} R_L \tag{A.10}$$

Equation (A.10) shows that in case $(I_{BIAS}/2)R_L < V_{THN}$, V_{CM} can even be higher than V_{DD}, a condition in general not exploited to avoid the risk of damaging the gate oxide of the input transistors. From the above discussion, we see that changing the common mode voltage does not change in principle the circuit outputs, so V_{CM} is not measured by the differential cell. However, its value can not be fully arbitrary, as it determines the drain-source voltages of the current source and of the input devices, which must work in saturation.

A similar study can be conducted for the cell with PMOS input transistors, shown in Fig. A.3 b). Here, when V_{CM} rises driving V_S up, it reduces the margin that the biasing current source has to stay in saturation. We must therefore require that the following condition is met:

$$V_{DD} - V_S > V_{SD,sat3} \tag{A.11}$$

This can be rewritten as:

$$V_{DD} - (V_{CM} + V_{SG1,2}) > V_{SG3} - |V_{THP}| \tag{A.12}$$

which finally leads to:

$$V_{CM} < V_{DD} - V_{SG1,2} - V_{SG3} + |V_{THP}| \tag{A.13}$$

When V_{CM} decreases, the lowering of V_S reduces the drain-source voltage of the input pair, because the output voltage stays constant at $(I_{BIAS}/2)R_L$. We must therefore impose that V_S is still high enough to allow M_1 and M_2 to work in saturation, obtaining the following equations:

$$V_S - \frac{I_{BIAS}}{2}R_L > V_{SD,sat1,2} \tag{A.14}$$

$$V_{CM} + V_{SG1,2} - \frac{I_{BIAS}}{2}R_L > V_{SG1,2} - |V_{THP}| \rightarrow V_{CM} > \frac{I_{BIAS}}{2}R_L - |V_{THP}| \tag{A.15}$$

From (A.15) we see that if $|V_{THP}|$ is greater than the voltage drop across the load, V_{CM} can in principle be negative. The analysis shows that an NMOS differential pair is more suited to handle common mode input voltages above $V_{DD}/2$, while a PMOS one is preferable when the input common mode voltage is below $V_{DD}/2$. In the mid region around $V_{CM} = V_{DD}/2$ both configurations can be used. Additional insights in the response of the differential cell to the change in V_{CM} will be provided by the small signal analysis.

Let's now examine the response of the differential cell to large differential input signals. From equation (A.3), we see that for an NMOS pair the difference $V_{GS1} - V_{GS2}$ has the same sign of $V_{G1} - V_{G2}$. Therefore, if we start from the equilibrium point and make $V_{G1} > V_{G2}$, we have $V_{GS1} > V_{GS2}$, so M_1 drives more current than M_1. The total current is constrained to be equal to I_{BIAS}, hence if the current in M_1 changes from $I_{BIAS}/2$ to $I_{BIAS}/2 + \Delta I$, the one in M_2 must drop from $I_{BIAS}/2$ to $I_{BIAS}/2 - \Delta I$. Consequently, the output voltage at the drain of M_1 decreases by $R_L \Delta I$ and the one at the drain of M_2 increases by the same amount. If $V_{G1} - V_{G2}$ is large enough, all the current I_{BIAS} flows in M_1. In this case, $V_{out1} = V_{DD} - I_{BIAS}R_L$ and $V_{out2} = V_{DD}$. From this point onward, the outputs do not track anymore the inputs and the circuit saturates.

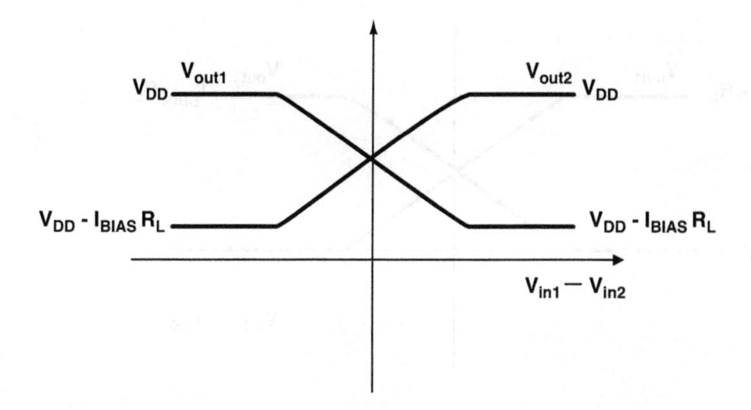

FIGURE A.4 Large signal behavior of an NMOS-input differential amplifier.

If we make now $V_{G1} < V_{G2}$, the role of M_1 and M_2 is interchanged. Hence, M_1 drives less current than M_2, V_{out1} rises and V_{out2} decreases. When the difference is negative enough, I_{BIAS} fully flows through M_2 and the circuit saturates with $V_{out1} = V_{DD}$ and $V_{out2} = V_{DD} - I_{BIAS}R_L$. Fig. A.4 schematically shows the circuit outputs as a function of $V_{G1} - V_{G2}$[2].
For a PMOS pair, we can write:

$$V_{in1} - V_{in2} = V_{SG2} - V_{SG1} \tag{A.16}$$

Therefore, if $V_{in1} \ll V_{in2}$, all the current flows in M_1, V_{out1} rises to $I_{BIAS}R_L$ and V_{out2} drops to zero. The vice versa is true if $V_{in1} \gg V_{in2}$, while for $V_{in1} = V_{in2}$, we have that $V_{out1} = V_{out2} = (I_{BIAS}/2)R_L$. The large signal behavior of the PMOS differential cell is sketched in Fig. A.5.

A.3 DIFFERENTIAL CELL SMALL SIGNAL ANALYSIS

In the small signal analysis we neglect for simplicity the bulk effect and the output resistance of the input transistors and we use the small signal equivalent circuit shown in Fig. A.6. We first write the nodal equations:

$$\begin{cases} \frac{V_{out1}}{R_L} + g_{m1}V_{gs1} = 0 \\ \frac{V_{out2}}{R_L} + g_{m2}V_{gs2} = 0 \\ \frac{V_s}{R_S} - g_{m1}V_{gs1} - g_{m2}V_{gs2} = 0 \end{cases} \tag{A.17}$$

[2]Observe that due to the intrinsic non-linearity of the I-V curve of the MOS transistor, the actual large signal characteristics of the differential cell is not fully linear.

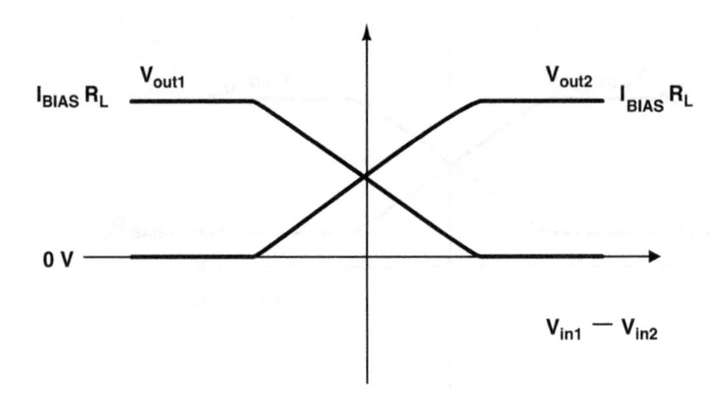

FIGURE A.5 Large signal behavior of a PMOS-input differential amplifier.

We now insert into the above equations the explicit definition of V_{gs1} and V_{gs2}, $V_{gs1} = V_{in1} - V_s$ and $V_{gs2} = V_{in2} - V_s$ to obtain:

$$\begin{cases} \frac{V_{out1}}{R_L} + g_{m1}V_{in1} - g_{m1}V_s = 0 \\ \frac{V_{out2}}{R_L} + g_{m2}V_{in2} - g_{m2}V_s = 0 \\ V_s\left(\frac{1}{R_S} + g_{m1} + g_{m2}\right) = g_{m1}V_{in1} + g_{m2}V_{in2} \end{cases} \tag{A.18}$$

To further simplify our study, we suppose that the circuit is fully symmetric and we take g_{m1}=g_{m2}=g_m. If the input signal has no common mode component, we have $V_{in1} = -V_{in2}$ and defining $V_{in} = V_{in1} - V_{in2}$, we have $V_{in1} = V_{in}/2$ and $V_{in2} = -V_{in}/2$.

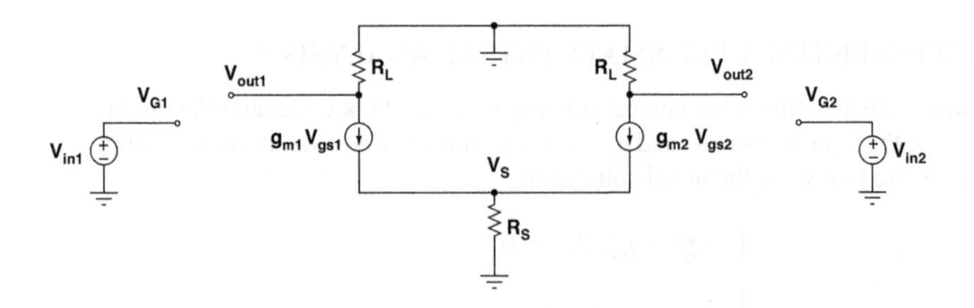

FIGURE A.6 Simplified small signal equivalent circuit of a differential amplifier.

Inserting these relationships in (A.18) we have:

$$\begin{cases} V_{out1} = -g_m \frac{V_{in}}{2} R_L \\ V_{out2} = g_m \frac{V_{in}}{2} R_L \\ V_{out2} - V_{out1} = g_m V_{in} R_L \\ V_s = 0 \end{cases} \tag{A.19}$$

We note that in case of a pure differential mode input signal, the source node of the differential pairs does not change, so it can be considered as a virtual signal ground. The circuit then reduces to two common source amplifiers, each driven with a signal which has half the amplitude of the total differential input. This simplifies significantly the study of differential amplifiers, because one can reuse the already known properties of the common source configuration. Note that the condition $V_s = 0$ is valid *only if* the input signal does not have any common mode component. The quantity:

$$A_{DM} = \frac{V_{out2} - V_{out1}}{V_{in1} - V_{in2}} \tag{A.20}$$

defines the differential gain of the circuit. Let's now consider what happens when we have a pure common mode input, i.e. when $V_{in1} = V_{in2} = V_{CM,in}$. In this case, we calculate first the value of V_s:

$$V_s = \frac{2g_m V_{CM,in}}{1 + 2g_m R_S} R_S \tag{A.21}$$

which means that a *variation* of the common mode at the input determines a change of the same phase at the source node of the differential pair, confirming that V_s "follows" the common mode signal. Note that if $R_S \to \infty$, $V_s = V_{CM,in}$, so V_s exactly tracks the common mode variation. In practice, R_S has a finite value. In this case, modulating the source voltage of the differential pair changes the current in R_S and hence the biasing point of the circuit. The variation of V_s has still the same phase of the common mode signal, while its amplitude is given by (A.21). Using again the relationships in (A.18) with the hypothesis $V_{in1} = V_{in2} = V_{CM,in}$ and $g_{m1} = g_{m2}$ we can calculate the response of the output nodes to the common mode input signal, which is:

$$V_{out1} = V_{out2} = -\frac{g_m V_{CM,in}}{1 + 2g_m R_S} R_L \tag{A.22}$$

From the above equation we see that, in case of full symmetry, both outputs move by the same amount. Therefore, a *common mode variation* at the input causes a *common mode variation* at the output but, due to the symmetry of the circuit, no *differential* output signal is generated. Equation (A.22) also shows that the change in the common mode output voltage becomes smaller and smaller as R_S gets bigger. Therefore, to maximize the common mode rejection, the biasing current source must

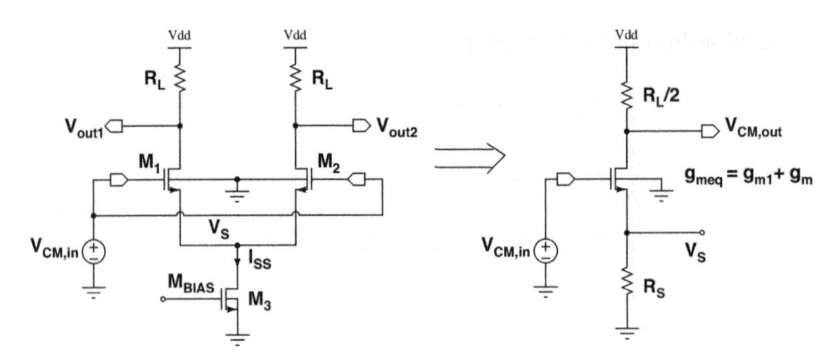

FIGURE A.7 Analogy between a differential amplifier driven by a pure common mode signal and a single-ended common source amplifier with source degeneration.

have an impedance as big as possible. To gain further insight into the circuit, we can calculate the current change in R_S, which follows immediately from equation (A.21):

$$\Delta I_s = \frac{V_s}{R_S} = \frac{2g_m}{1+2g_m R_S} V_{CM,in} \tag{A.23}$$

Remember now that the common source amplifier with source degeneration has an equivalent transconductance expressed by:

$$G_m = \frac{g_m}{1+g_m R_S} \tag{A.24}$$

We see that (A.24) and (A.23) coincide if in (A.24) we make the substitution $g_m \to 2g_m$. If the two transistors are identical, they equally share the current ΔI_S and both output voltages move by $-(\Delta I_s/2)R_L$. In other words, the response of the source node to the variation of the common mode voltage can be calculated considering the differential pair as a source degenerated transistor with a transconductance which is the sum of the g_m of the two input transistors. The output resistance of the bias current source acts as the degenerating resistor. The equivalent load resistor has half the value of the resistors loading each branch of the pair, as shown in Fig. A.7, because each load resistor receives only one half of the current generated by the common mode change. Applying this model, we can calculate again the common mode output voltages as:

$$V_{out1} = V_{out2} = -\frac{2g_m V_{CM,in}}{1+2g_m R_S}\frac{R_L}{2} = -\frac{g_m V_{CM,in}}{1+2g_m R_S} R_L \tag{A.25}$$

It is interesting to study what happens in differential pairs when there are asymmetries between the two branches of the circuit. Suppose, for instance, that the two load resistors have different values, R_{L1} and R_{L2}. If M_1 and M_2 are identical, they still share equally the current ΔI_s generated by a common mode input variation. The change in the output voltages is now:

$$V_{out1} = -\frac{g_m}{1+2g_m R_S} R_{L1} V_{CM,in} \tag{A.26a}$$

$$V_{out2} = -\frac{g_m}{1+2g_m R_S} R_{L2} V_{CM,in} \tag{A.26b}$$

Therefore, taking the difference $V_{out2} - V_{out1}$ we have:

$$V_{out2} - V_{out1} = \frac{g_m}{1 + 2g_m R_S} (R_{L1} - R_{L2}) V_{CM,in} \tag{A.27}$$

We can extend further this result to incorporate also the mismatch between the parameters of the input transistors, (threshold voltages, μC_{ox}, etc.), which result in different transconductances for M_1 and M_2. In this case, the model of Fig. A.7 can still be used to calculate V_s, substituting in (A.24) g_m with $g_{m1} + g_{m2}$. We can hence write:

$$V_s = \frac{(g_{m1} + g_{m2}) R_S}{1 + (g_{m1} + g_{m2}) R_S} V_{CM,in} \tag{A.28}$$

To calculate the output voltage, we first note that the small signal currents in M_1 and M_2 are respectively given by:

$$I_{ds1} = g_{m1} V_{gs1} = g_{m1} (V_{CM,in} - V_s) \tag{A.29a}$$

$$I_{ds2} = g_{m2} V_{gs2} = g_{m2} (V_{CM,in} - V_s) \tag{A.29b}$$

We then insert into (A.29a)-(A.29b) the value of V_s given by (A.28). Finally, we get the output voltages V_{out1} and V_{out2} as:

$$V_{out1} = -I_{ds1} R_{L1} = -\frac{g_{m1} R_{L1}}{1 + (g_{m1} + g_{m2}) R_S} V_{CM,in} \tag{A.30a}$$

$$V_{out2} = -I_{ds2} R_{L2} = -\frac{g_{m2} R_{L2}}{1 + (g_{m1} + g_{m2}) R_S} V_{CM,in} \tag{A.30b}$$

We can define a common mode to differential voltage, $A_{CM \to DM}$ as:

$$A_{CM \to DM} = \frac{V_{out2} - V_{out1}}{V_{CM,in}} = \frac{g_{m1} R_{L1} - g_{m2} R_{L2}}{1 + (g_{m1} + g_{m2}) R_S} \tag{A.31}$$

Equation (A.31) shows an extremely important result. The asymmetries introduced in the differential pair translate a common mode input variation into a *differential* output signal, which can not be rejected by a differential receiver and is hence amplified along the processing chain. This is of course an undesirable effect. The Common Mode Rejection Ratio (CMRR) is the performance metrics that measure the ability of a differential circuit to reject common mode input variations. It is defined as the ratio, usually expressed in dB, between the differential gain, A_{DM} and the common-mode to differential gain, i.e:

$$CMMR = 20 \log \left| \frac{A_{DM}}{A_{CM \to DM}} \right| \tag{A.32}$$

To improve as much as possible the $CMRR$ figure, the following actions must be taken:

- Increase the output impedance R_S of the biasing current source. To get a very high impedance, M_3 can be replaced by a cascode current mirror, which however requires a greater voltage headroom to stay in saturation. Therefore there is a trade-off between the common mode rejection ratio and the common mode input range.

- Minimize any asymmetry in the circuit through a very careful design of the layout.

In standard simulations, the finite output impedance of the transistors is taken into account, but not the asymmetries stemming from the unavoidable imperfections of the fabrication process. A thorough study of the *CMRR* of a differential circuit can be done with Monte Carlo simulations, which are the appropriate tool to investigate the impact on circuit performance of the mismatch between the parameters of nominally identical devices.

A.4 DIFFERENTIAL CELLS WITH ACTIVE LOAD

The load resistors of the differential pair can be replaced by transistors. Diode connected PMOS are used when the input transistors are NMOS and vice versa, with the exception of folded topologies in which the input stage and the load employ the same type of device. Fig. A.8 shows the direct configurations, while an example of a folded architecture is reported in Fig. A.9. In all these circuits the differential gain is given by the ratio between the transconductance of the input transistor and the one of the diode-connected load:

$$A_{DM} = \frac{g_{m,in}}{g_{mL}} \tag{A.33}$$

It must be observed that in the folded cascode topology, the current in M_3-M_4 must be greater than the one in M_1 and M_2 in order to provide sufficient bias current also to the cascode branches. If, for instance, $I_{DS}(M_9) = I_{DS}(M_{3,4}) = I_{BIAS}$, at the equilibrium M_1 and M_2 will sink only $I_{BIAS}/2$ and the additional $I_{BIAS}/2$ provided by M_3 and M_4 will bias the folding branches.

If the transistors work in weak inversion, the transconductance depends only on the bias current and the cell gain can be close to one. The gain can be increased by

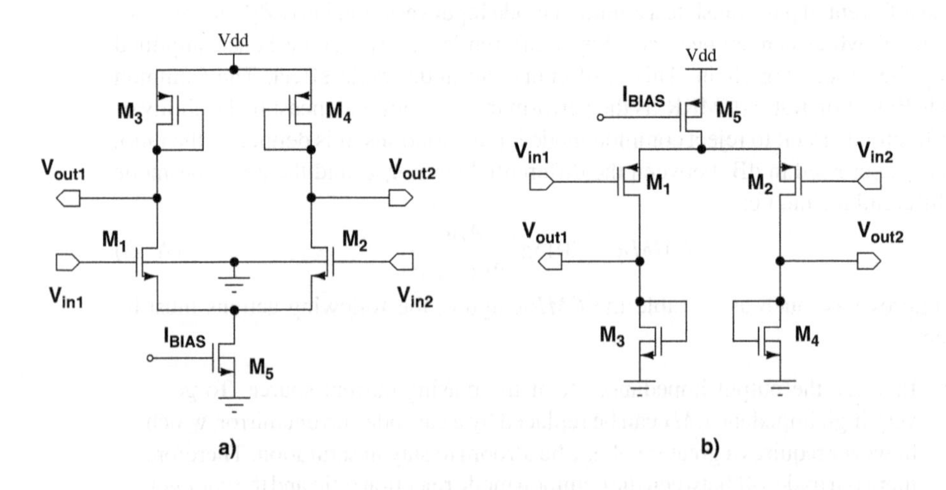

FIGURE A.8 Differential amplifiers with diode connected loads.

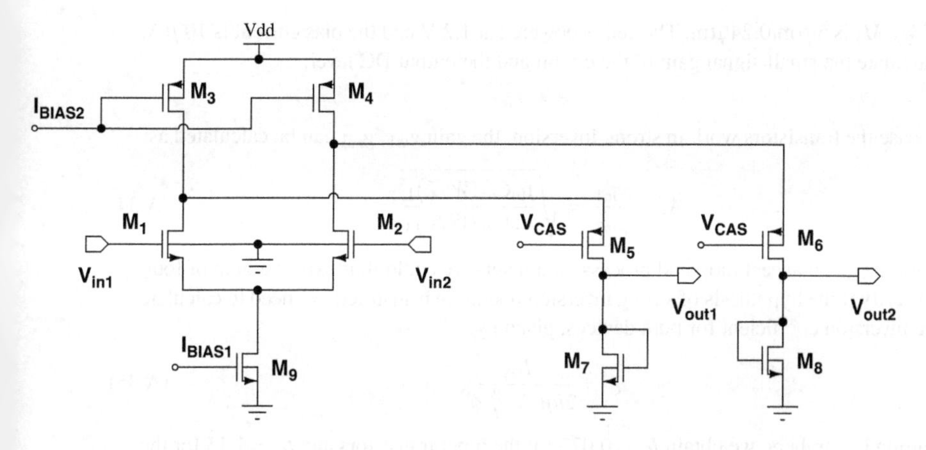

FIGURE A.9 Folded cascode differential amplifier with diode connected load.

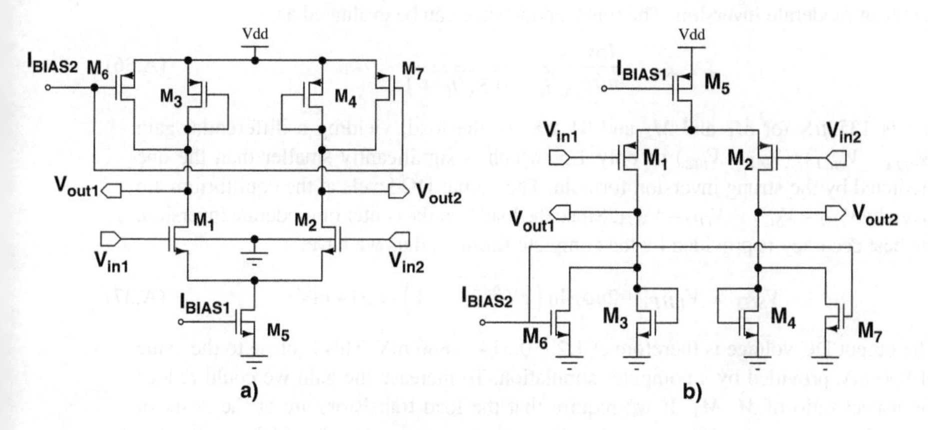

FIGURE A.10 Differential amplifiers with load current starving to increase the gain.

starving the load, reducing its current by connecting in parallel a current source that provides part of the bias current required by the input device, as shown in Fig. A.10. Differential cells with diode connected loads however typically provides a gain < 10 and are mostly used in the implementation of fast, multi-stage comparators discussed in chapter 9. The following example illustrates the design of this building block.

Example A.1: Differential cell with diode connected loads.

Consider the circuit of Fig. A.8 a). The amplifier is implemented in a 130 nm technology having the following parameters: $\mu_n C_{ox} = 480\ \mu\text{A/V}^2$, $\mu_p C_{ox} = 120\ \mu\text{A/V}^2$, $V_{THN} = 0.27\ V$, $|V_{THP}| = 0.27$ V. The aspect ratio of M_1 and M_2 is 20 μm/0.24 μm and the one of M_3-M_4 is 5 μm/0.24 μm. The cell is powered at 1.2 V and the bias current is 10 μA. Calculate the small signal gain of the circuit and the output DC level.

* * *

In case the transistors work in strong inversion, the gain g_{m1}/g_{m3} can be calculated as:

$$A_v = \frac{g_{m1}}{g_{m3}} = \sqrt{\frac{\mu_n C_{ox}(W/L)_1}{\mu_p C_{ox}(W/L)_3}} \tag{A.34}$$

With the given aspect ratio and process parameters we could thus expect a gain of four. To verify if the hypothesis of strong inversion operation is justified, we need to calculate the inversion coefficient for both devices, given by:

$$I_C = \frac{I_{DS}}{2n\mu C_{ox}\frac{W}{L}\phi_T^2} \tag{A.35}$$

Putting in numbers, we obtain $I_C = 0.072$ for the input transistors and $I_C = 1.15$ for the loads. The input transistors thus work deep in weak inversion, while the loads are at the center of moderate inversion. The transconductance can be evaluated as:

$$g_m = \frac{I_{DS}}{n\phi_T}\frac{1}{\sqrt{I_C + 0.5\sqrt{I_C} + 1}} \tag{A.36}$$

and is 135 μS for M_1 and M_2 and 91 μS for the load, yielding a differential gain $(V_{out2} - V_{out1})/(V_{in1} - V_{in2})$ of only 1.5, which is significantly smaller than the one predicted by the strong inversion formula. The output DC levels at the equilibrium are given by $V_{DD} - V_{SG3} = V_{DD} - V_{SG4}$. Since the load is at the center of moderate inversion, the best accuracy is provided by the complete formula, thus we have:

$$V_{SG3} = |V_{THP}| + 2n\phi_T \ln\left(e^{\sqrt{I_C(M_3)}} - 1\right) \approx 314 \text{ mV} \tag{A.37}$$

The output DC voltage is therefore at $1.2 - 0.314 = 886$ mV. This is close to the value of 890 mV provided by a computer simulation. To increase the gain we could reduce the aspect ratio of M_3-M_4. If we require that the load transistors are at the onset of strong inversion ($I_C = 10$), we find an aspect ratio of about $5/2$, which results in a transconductance of 41 μS and an overall gain of 3.25. Reducing the aspect ratio of the load we increase also its source-gate voltage. With the strong inversion formula, we can calculate that the V_{SG} of M_3 and M_4 is about 450 mV, which implies a reduction of the output DC voltage of 140 mV with respect to the previous case. As an alternative, load starving can be used to increase the gain, as shown in Fig. A.11. The current source delivers a current I_L smaller than $I_{BIAS}/2$, and only the difference between $I_{BIAS}/2$ and I_L flows in the load. For instance, if $I_L = 4$ μA, only 1 μA is left to M_3 and M_4. If we keep the aspect ratio of M_3 and M_4 unaltered, we have now an inversion coefficient of 0.22. However, owing to the smaller current, the transconductance is reduced to 24 μS, which yields a gain of 5.5. If, on the other hand, we want to keep the output DC level constant, we can change the aspect ratio of M_3 and M_4 to maintain I_C unaffected. Interestingly, with devices not far from minimum size, if the W or L are changed, differences in the gate-source voltages can be observed for the same inversion coefficient due to changes in the threshold voltages induced by short and narrow channel effects.

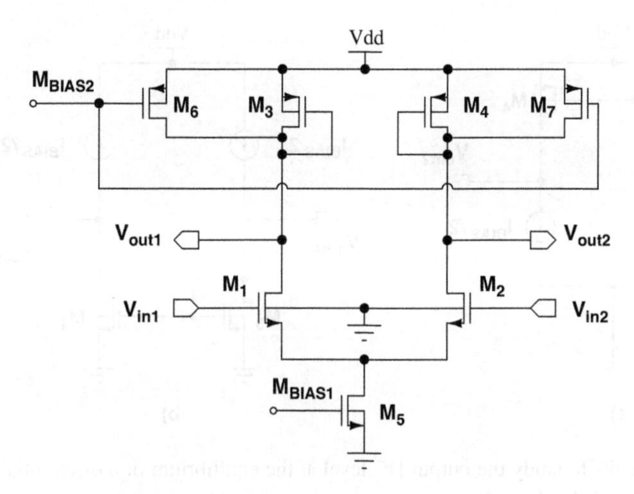

FIGURE A.11 Diode connected loads with current source in parallel. The current source reduces the current in the load, increasing the gain.

A.5 DIFFERENTIAL CELL WITH CURRENT MIRROR LOAD

Fig. A.12 shows differential pairs with current mirror load. This configuration has two differential inputs and one single ended output. To understand the behavior of these circuits, let's first consider the situation at the equilibrium, when $V_{in1} = V_{in2}$ and the two input transistors equally share the bias current, $I_{DS1} = I_{DS2} = I_{BIAS}/2$. We suppose for the moment that the circuit is not connected to any external load. The diode connected transistor M_3, having a relatively low impedance, easily accepts the current $I_{BIAS}/2$ and mirrors it towards M_4. The current of M_4 is absorbed (in case of NMOS input stage) or provided by (in case of PMOS input stage) M_2. All

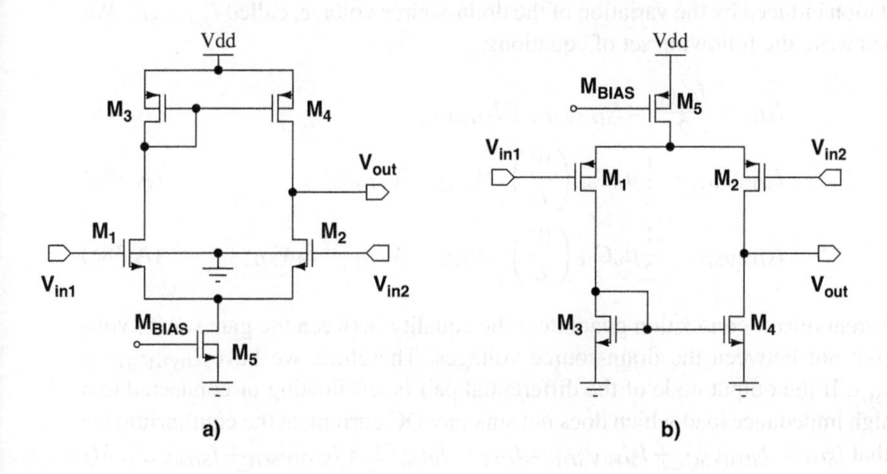

FIGURE A.12 Differential amplifiers with current mirror load.

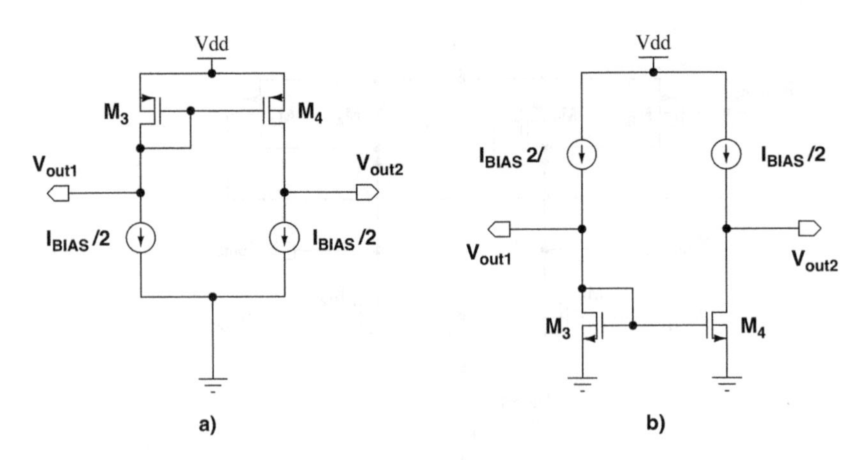

FIGURE A.13 Circuits to study the output DC level at the equilibrium of a differential amplifier with current mirror load.

transistors in the circuits must work in saturation. If the devices had infinite output impedance, in absence of an external load the condition $I_{DS4} = I_{DS2} = I_{BIAS}/2$ would be the only possible one and the output voltage would be undetermined. In practice, all the transistors in the circuit have a finite output resistance due to the channel length modulation effect. To study what happens at the equilibrium, we can redesign the circuit as shown in A.13. Here, the bias current source and the input transistors have been replaced by two independent generators, each providing $I_{BIAS}/2$. In the following, we limit the study to the case of the circuit with NMOS inputs and PMOS loads, as the extension to the complementary case is straightforward. The current flowing in M_3 can be seen as made of two components, one due to the control of the gate over the channel, indicated as $I_{SD3(VSG)}$ and one due to the channel length modulation induced by the variation of the drain-source voltage, called $I_{SD3(VSD)}$. We can then write the following set of equations:

$$I_{SD3} = \frac{I_{BIAS}}{2} = I_{SD3(VSG)} + I_{SD3(VSD)} \tag{A.38a}$$

$$I_{SD3(VSG)} = \frac{1}{2}\mu_p C_{ox}\left(\frac{W}{L}\right)_3 (V_{SG3} - |V_{THP}|)^2 \tag{A.38b}$$

$$I_{SD3(VSD)} = \frac{1}{2}\mu_p C_{ox}\left(\frac{W}{L}\right)_3 (V_{SG3} - |V_{THP}|)^2 \lambda_P V_{SD3} \tag{A.38c}$$

The current mirror connection guarantees the equality between the gate source voltages, but not between the drain-source voltages. Therefore, we have $I_{DS3(VSG)} = I_{DS4(VSG)}$. If the output node of the differential pair is left floating or connected to a very high impedance load which does not sink any DC current, at the equilibrium we have that $I_{SD4} = I_{SD4(VSG)} + I_{SD4(VSD)} = I_{DS2} = I_{BIAS}/2 = I_{SD3(VSG)} + I_{SD3(VSD)}$. M_3 and M_4 form a current mirror, therefore $V_{SG3} = V_{SG4}$ by construction, which imposes that $I_{SD3(VSG)} = I_{SD4(VSG)}$. This, combined with the constraint that the total current in

M_3 and M_4 is equal to $I_{BIAS}/2$ implies also that $I_{SD3(VSD)} = I_{SD4(VSD)}$. The last equality is satisfied only if $V_{SD3} = V_{SD4}$. In other words, forcing M_3 and M_4 to drive the *same* current, implies that *both* their gate-source voltages and drain-source voltages must be equal. Since M_3 is diode connected, we have that $V_{SD4} = V_{SD3} = V_{SG3}$. Observe that this condition is true only at the equilibrium, when $I_{SD3} = I_{SD4} = I_{BIAS/2}$. Most of the current in saturation is controlled by the gate action, therefore in hand calculations we can estimate the voltages as:

$$V_{SD4} = V_{SD3} = V_{SG3} \approx |V_{THP3}| + \sqrt{\frac{2L_3 I_{BIAS}/2}{\mu_p C_{ox} W_3}} \tag{A.39}$$

The analysis of the common mode input range proceeds in the same way as for the differential pair with resistive load. The inferior limit is defined by the requirement that M_5 is saturated, i.e. $V_S > V_{DSsat5}$, which leads to the following relationship:

$$V_{CM,in} - V_{GS1,2} > V_{DSsat5} \rightarrow V_{CM,in} > V_{GS1,2} + V_{GS5} - V_{TH5} \tag{A.40}$$

When the common mode voltage is increased, V_S follows $V_{CM,in}$ while the voltage at the drain node of M_1 and M_2 stays constant and the input transistors are eventually driven in the linear region. To prevent this, we must have:

$$V_{DD} - V_{SG3} - V_S > V_{DSsat1} = V_{GS1} - V_{THN1} \tag{A.41a}$$

$$V_{CM,in} < V_{DD} - V_{SG3} + V_{THN1} \tag{A.41b}$$

where, to get (A.41b), we have again used the fact that $V_S = V_{CM,in} - V_{GS1}$.

To understand the behavior of the active cell when a voltage difference is applied between its inputs, let's first study the circuits of Fig. A.14. The output node has been connected to a resistor. A voltage source keeps the other arm of the resistor at the same DC voltage that the amplifier output has when it is not loaded (i.e. $V_{DD} - V_{SD4} = V_{DD} - V_{SG3}$). In the study of the simple differential pair we have seen that $V_{in1} > V_{in2}$ implies that $V_{GS1} > V_{GS2}$. The current in M_1 is increased by ΔI and to maintain I_{BIAS} constant, the one in M_1 is decreased by the same amount. M_3 receives from M_1 the current $I_{BIAS}/2 + \Delta I$ and mirrors it in M_4, which pushes this current down towards M_2. However, the current in M_2 has been decreased to $I_{BIAS}/2 - \Delta I$. The difference between I_{SD4} and I_{DS2} is $2\Delta I$ and this excess current flows in R_L, determining a rise of the output voltage of $2\Delta I R_L$. The reasoning is specular for $V_{in2} < V_{in1}$. In this case, $I_{DS2} = I_{BIAS}/2 + \Delta I$ and $I_{SD4} = I_{BIAS}/2 - \Delta I$, so there is a difference of $2\Delta I$ between the current sunk by M_2 and the one provided by M_4. M_2 then sinks $2\Delta I$ from the resistor, lowering the output voltage. Note that when V_{in1} rises, also the output voltage rises and vice versa. V_{in1} is hence the non-inverting input of the cell and V_{in2} the inverting one. The current mirror thus copies the current variation generated by M_1 to the drain of M_2, where it is summed to the one determined by M_2. The signals generated by M_1 and M_2 are not sensed separately at the output but they are combined to form a single signal which is measured with respect to ground. The current mirror action thus converts a differential input signal into a single ended output one.

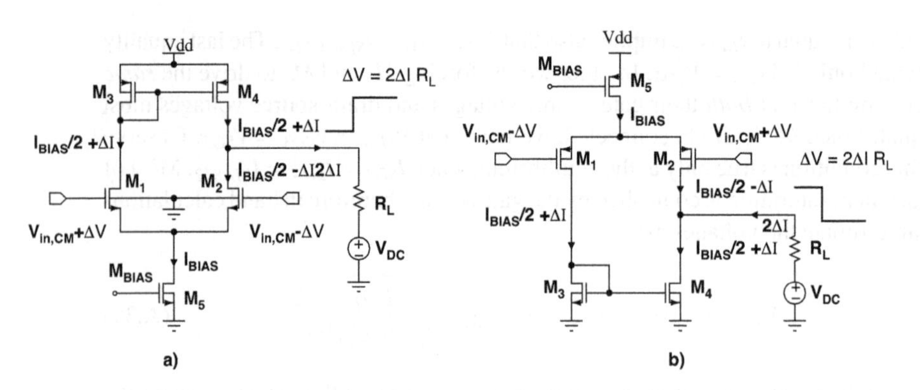

FIGURE A.14 Circuits to study the response of differential amplifiers with current mirror load to a differential input signal.

To calculate the small signal gain, we have to remember that when a pure differential-mode input signal is applied $V_s = 0$. The circuit can hence be redesigned as shown in Fig. A.15, where the output resistance of the transistors is explicitly shown. The small signal current in M_1 is $g_{m1}(V_{in}/2)$, which is read by M_3 and copied to M_4. The current in M_2 is $-g_{m2}(V_{in}/2)$. If $g_{m1} = g_{m2} = g_m$, at the output node we have a current imbalance given by:

$$I_{out} = g_m \frac{V_{in}}{2} - (-g_m \frac{V_{in}}{2}) = g_m V_{in} \tag{A.42}$$

This current flows in the output load. If the amplifier output is floating, the load is provided by the output resistance of M_2 and M_4, which on the small signal, are in parallel. Note that neither r_{02} nor r_{04} are connected to a DC ground. However, both V_{DD} and V_s, experiencing zero *variations*, become AC grounds in the small signal

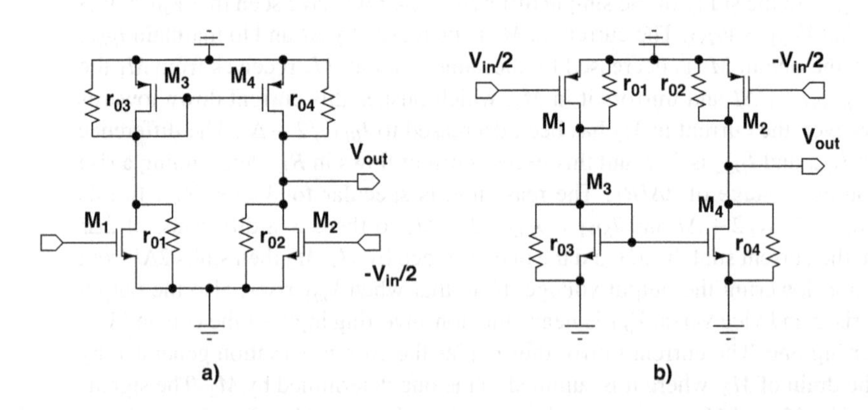

FIGURE A.15 Simplified small signal model of differential amplifiers with current-mirror load.

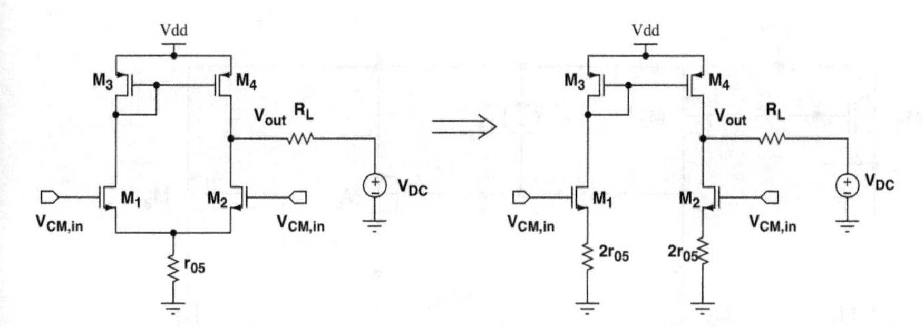

FIGURE A.16 Circuit to study the common mode response of the differential cell with current mirror load.

analysis. The gain of the cell is finally given as:

$$A_v = g_m (r_{02} // r_{04}) \quad \text{(A.43)}$$

In general, we can write $A_v = g_m R_L$, where R_L is the parallel combination of all the resistors that, in the small signal model, appear connected between the amplifier output and any small signal ground. We see from the above equation that the expression of the gain for a differential cell with current mirror load is identical to the one of the common source amplifier with active load. Observe that the impedance seen from the drain of M_1 is $\approx 1/g_{m3}$ and thus fairly low, while the one seen from the drain of M_2 is $r_{02}//r_{04}$ and hence much higher. Using the circuit of Fig. A.16, we can finally evaluate the common mode gain. Here the biasing transistor is replaced with its output resistance and the circuit is divided into two symmetrical halves. Under the effect of a common mode change, the current in M_1 becomes:

$$I_{CM1} = \frac{g_m V_{CM,in}}{1 + 2 g_m r_{05}} \quad \text{(A.44)}$$

The current is received by M_3 and mirrored downwards by M_4. However, also the current in M_2 has increased by the amount given by (A.44). The extra current coming from M_4 is thus fully absorbed by M_2 and no current flows into the load. In case of perfect symmetry, the common mode gain of the circuit is thus zero. In practice, the unavoidable mismatch reduces the amplifier common mode rejection.

The differential cell with current mirror load is suitable when a moderate gain is sufficient. In the design of front-end amplifiers, this block finds widespread use as the input stage of a baseline holder circuit. Other very common applications are the generation of reference currents and voltages. Fig. A.17 shows a scheme that can serve to provide a stable reference current with little sensitivity to the positive power supply variations. The method relies on the availability of a stable and clean reference voltage, usually obtained with bandgap voltage generators [1]. The open loop circuit can be seen as a differential cell with current mirror load, followed by a common source amplifier with resistive load. When the loop is closed, the feedback action minimizes the voltage difference between the inputs of the differential pair.

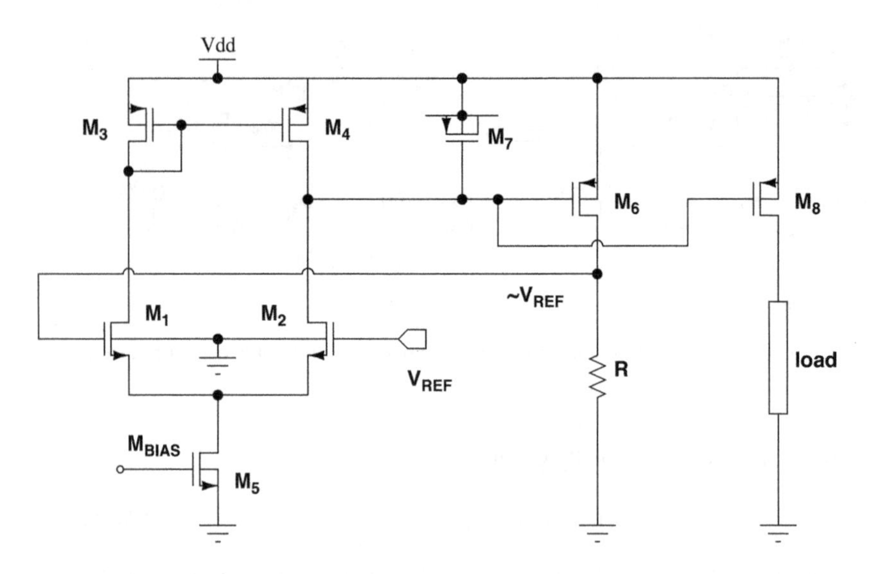

FIGURE A.17 Generation of a V_{DD} insensitive reference current.

Therefore, a voltage approximately equal to V_{REF} appears across R, determining in it a current $I_{REF} = V_{REF}/R$ that flows in M_6. The feedback loop regulates the gate voltage of M_6 to the appropriate level to deliver I_{REF}. Connecting other transistors in parallel to M_6 allows us to mirror and scale I_{REF}, using it as a bias current for other circuits. Note that these additional transistors will be outside the feedback loop that defines the reference current.

The positive power supply, V_{DD}, does not enter into the definition of I_{REF}. If a disturbance is superimposed on V_{DD}, two situations may arise. If the noise is at low frequency, the loop has time to react. Therefore, the gate voltage of M_6 replicates the disturbance on the power supply, so that V_{SG6} does not change, keeping I_{REF} constant. If the disturbance is fast, the loop may not react in time. For this reason, M_7 is added. The device has its drain-source and bulk connected to V_{DD} and serves as a decoupling capacitor, that shorts at a high frequency the source and gain of M_7, thus preventing fast transients on V_{DD} from modulating the current in M_6. A variant of the circuit to generate a reference voltage with good current drive capability is shown in Fig. A.18. Here V_{REF} is copied across R_1, generating a current V_{REF}/R_1. This current flows both in R_1 and R_2, establishing an output voltage given by:

$$V_{out} = \frac{V_{REF}}{R_1}(R_1 + R_2) = V_{REF}\left(1 + \frac{R_2}{R_1}\right) \tag{A.45}$$

The principle is at the basis of the "Low-Drop Out regulator (LDO)", which is used to provide clean power supplies to sensitive circuits. The output capacitors C_F works as a battery that supplies the load in case of abrupt current variations, giving time to the regulation loop to react and adapt to the new current value. The following terminology applies to low drop-out regulators:

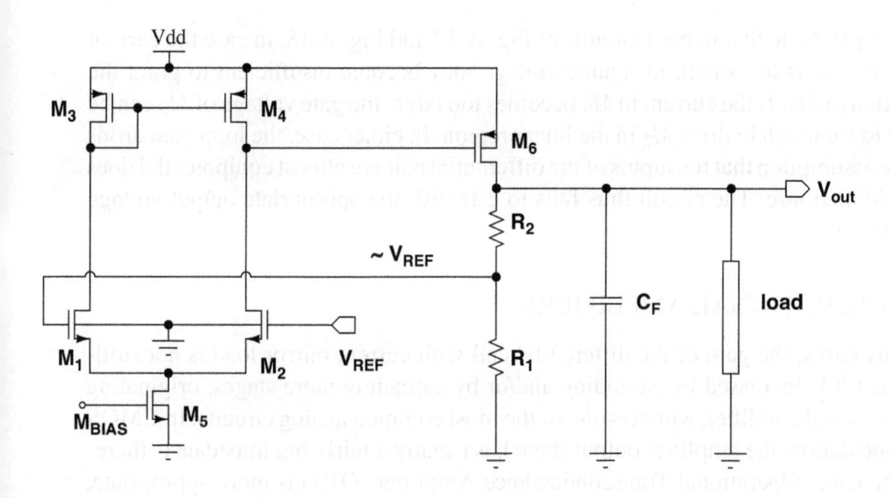

FIGURE A.18 Simplified scheme of a low drop-out regulator.

- Transistor M_6 is the pass-transistor and must be wide enough to deliver the maximum current requested by the load while requiring a small saturation voltage. Transistor width of several tens or even hundreds of millimeters are common in LDO design.
- The power supply voltage (V_{DD} in the figure) that powers the regulator is called the input voltage.
- The difference between the input and output voltage is the drop-out voltage and corresponds to the source-drain voltage of the pass transistor. Since the latter works in saturation, there is a minimum drop-out that can be accepted. All the current flowing in the load comes from the pass transistor and the higher the drop-out, the bigger is the additional power dissipated in the LDO.
- Changing the load current, the output voltage may change owing to the non-infinite speed and gain of the feedback loop. The load regulation in percentage is defined as following:

$$\text{Load Regulation} = 100\frac{V_{out,min} - V_{out,max}}{V_{out,nom}} \tag{A.46}$$

  where $V_{out,min}$ is the output voltage corresponding to the minimum load current, $V_{out,max}$ is the output voltage corresponding to the maximum load current and $V_{out,nom}$ the nominal output voltage.
- The line regulation is the ratio between the variations of the input and output voltages:

$$\text{Line Regulation} = \frac{\Delta V_{out}}{\Delta V_{in}} \tag{A.47}$$

LDOs are systematically employed as external components to power front-end ASICs. However, a number of designs incorporate the LDO also on board of the front-end chip. For more details and practical design of these circuits the reader can see, for instance [2]. Note that in both circuits of Fig. A.17 and Fig. A.18, in case the current driven by M_6 is too small, its source voltage may become insufficient to grant the saturation of M_4. If the current in M_6 becomes too large, the gate voltage of M_6 can be pulled low enough to drive M_2 in the linear region. In either case, the loop gain drops and the assumption that the inputs of the differential pair are almost equipotential does not hold anymore. The circuit thus fails to establish the appropriate output voltage and current.

A.6 OPERATIONAL AMPLIFIERS

In many cases, the gain of the differential cell with current mirror load is not sufficient and it is increased by cascoding and/or by cascading more stages, originating an operational amplifier, which is one of the most common analog circuits. In CMOS implementation, the amplifier output stage has usually a fairly big impedance, therefore the name Operational Transconductance Amplifier (OTA) is more appropriate. In front-end for radiation detectors, operational amplifiers are rarely employed in the input stage because, for the same power, they have higher intrinsic noise than a pure single-ended solution. However, they find widespread use in the other parts of the system. Operational amplifiers are often adopted in shapers, where the extra noise contribution is less critical and the direct access to the non-inverting terminal offers additional flexibility in mastering the circuit DC levels. In addition, the high driving capability of some configuration makes them indispensable when analog voltages need to be delivered off chip. In the following we examine a few configurations that are particularly useful in front-end design. A complete treatment on operational amplifiers can be found in [3].

A.6.1 SINGLE STAGE ARCHITECTURES

Single stage OTAs are obtained by cascoding both the differential input pair and the load. Their DC gain is thus given by the product between the transconductance of the input transistor and the parallel between the equivalent resistance of the input and output cascodes. The gain-bandwidth product is defined by the ratio between the transconductance of the input transistor and the load capacitor, $GBW = g_{m1}/C_L$, while the slew rate is the ratio between the tail current source of the differential pair and the output capacitance. Fig. A.19 shows the straightforward approach of the telescopic cascode, reporting two different flavors, one using a low-voltage current mirror and the other a standard cascode mirror. If the circuit is implemented in a deep submicron CMOS technology and biased at very low current, the configuration of Fig. A.19 a) can be preferable, because a single source-gate voltage drop might not be sufficient to keep two transistors in saturation with enough margin. The telescopic cascode OTA stacks five transistors. Assuming weak inversion operation and that each device can be kept are saturation with 0.1 V, the circuit has a maximum output dynamic range of $V_{DD} - 0.5$ V, which results in 0.7 V if V_{DD} is 1.2 V. Telescopic cascodes are thus useful in those applications where the linear output range required to the amplifier is limited. A very common OTA topology is the folded cascode,

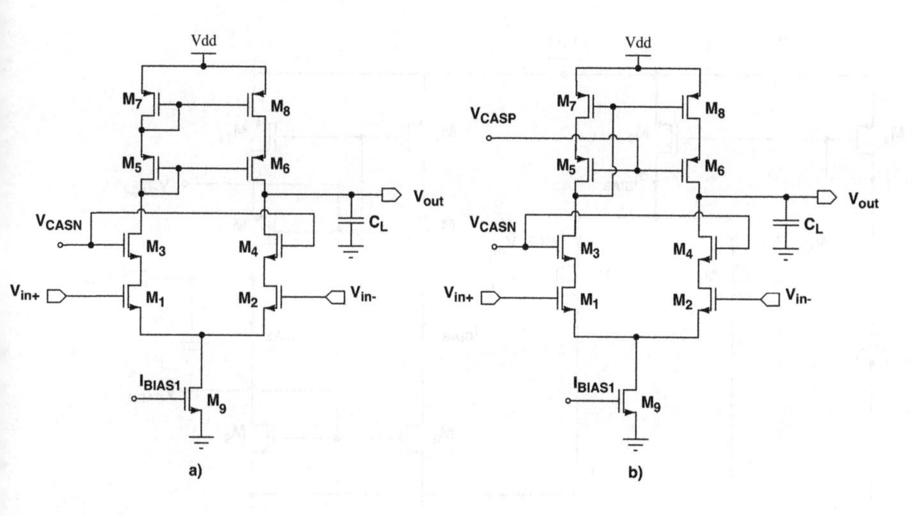

FIGURE A.19 Telescopic CMOS OTA.

shown in Fig. A.20. At the equilibrium, the current in M_6 and M_7 is twice the one in M_1 and M_2. Therefore, M_6 has an excess current that flows in the diode connected branch, is mirrored downwards by M_{13} and sunk by M_7. If the gate voltage of M_2 is lowered, M_2 drives more current than M_1. The current in M_9 decreases while the one in M_{13}-M_{11} increases and the difference $I(M_{11}) - I(M_9)$ is sourced to the output load. The opposite happens if the gate voltage of M_1 is lower than the one of M_2. Stacking four transistors in the output branch, the folded cascode OTA increases the output dynamic range by one saturation voltage with respect to the telescopic topology. In telescopic and folded architectures, the second non-dominant pole that limits the circuit bandwidth is given by C_{par}/g_{m7} (telescopic) or C_{par}/g_{m12} (folded), where g_{m7} and g_{m12} are the transconductance of the diode connected transistor and C_{par} the parasitic capacitance seen from its gate. In the folded cascode, the circuit bandwidth can be improved with the configuration of Fig. A.21. Here the mirroring action is transferred on the NMOS transistors. Since PMOS and NMOS devices are usually matched to have the same current drive capability, NMOS transistors can be made smaller due to their higher transconductance parameter $\mu_n C_{ox}$ and thus have less parasitic capacitance.

A.6.2 TWO STAGE OTA

To maximize the output dynamic range, only two transistors can be stacked in the output branch, originating the so called rail-to-rail output stage. However, the gain obtainable by a single stage without cascoding is in general insufficient, forcing the use of multi-stage configurations. One of the most common CMOS operational amplifiers is shown in Fig. A.22 and it consists of a common source amplifier with active load following a gain cell with current mirror load. The overall DC gain is given by the

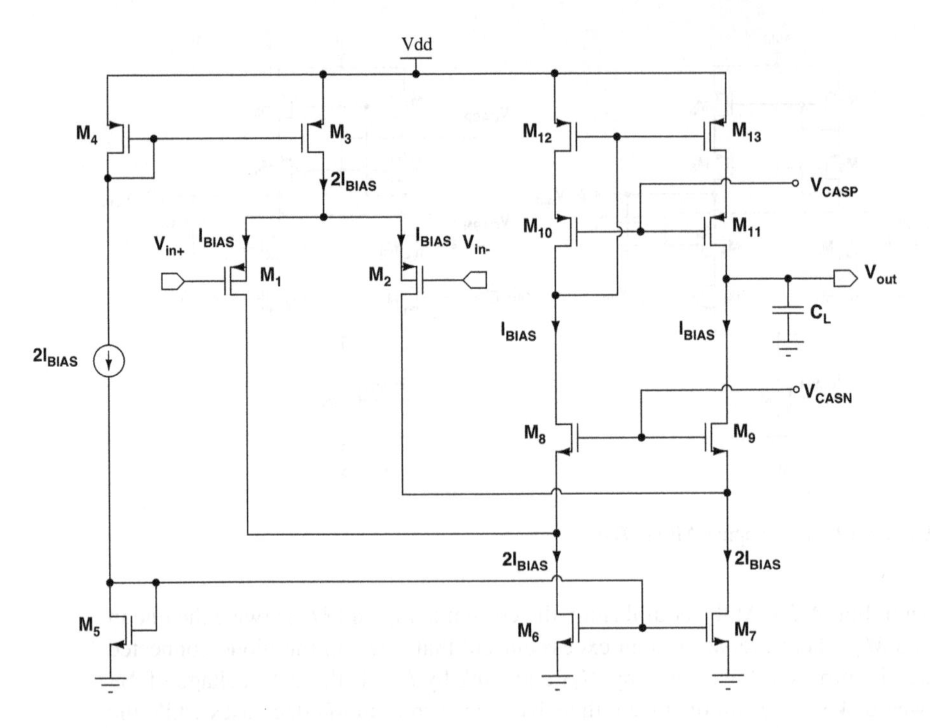

FIGURE A.20 Folded cascode OTA.

product of the gain of the two stages, that is:

$$A_{v0} = A_{v01}A_{v02} = g_{m2}\frac{r_{02}r_{04}}{r_{02}+r_{04}}g_{m6}\frac{r_{06}r_{07}}{r_{06}+r_{07}} \tag{A.48}$$

As discussed in chapter 4, a two stage OTA presents two closely spaced poles and needs to be compensated through the introduction of a capacitance between the gate and the drain of M_6. The other relevant parameters of the amplifier are the following:

- Gain-bandwidth product: $GBW = g_{m2}/C_M = g_{m1}/C_M$
- Slew rate: $SR = I(M_5)/C_M$. For a 60 ° phase margin, the following relationship holds between C_M and the load capacitance, C_L:

$$C_M > 0.22C_L. \tag{A.49}$$

An important point about two-stage OTAs concerns the proper DC balancing of the output stage to avoid introducing systematic offsets. In fact, at the equilibrium, the drain-source voltage of M_3 and M_4 are identical, thus the gate voltage of M_6 is equal to the one of M_3. Therefore, at the equilibrium and only at the equilibrium, M_6 mirrors the current of M_3, $I_{BIAS}/2$, so we can write:

$$I(M_6) = \frac{(W/L)_6}{(W/L)_3}\frac{IBIAS}{2} = k\frac{IBIAS}{2} \tag{A.50}$$

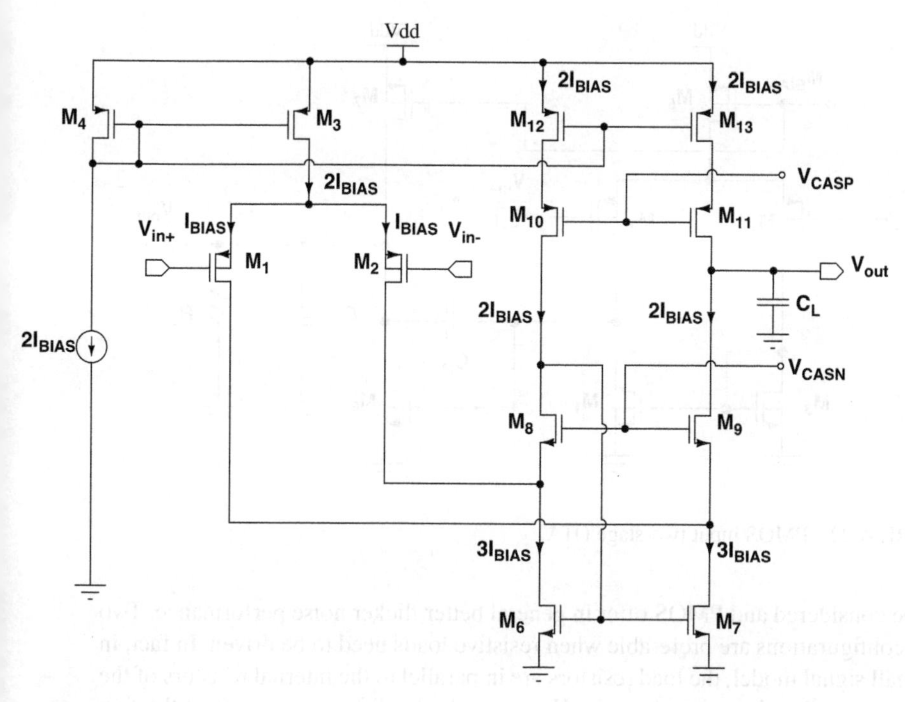

FIGURE A.21 Exploiting the NMOS transistors to implement the load current mirror improves the bandwidth of a folded cascode CMOS OTA.

The load M_7 must provide to M_6 the current given by (A.50), hence the aspect ratio of M_7 must be $k/2$ times the one of the tail current source of the differential pair, M_5. Each OTA configuration with input transistors of a given type has of course its complementary implementation. As an example, Fig. A.23 shows a two-stage OTA with NMOS input. PMOS input implementation are traditionally preferred because of two reasons:

- When triple-well NMOS were not available, only PMOS transistors could be put in a dedicated well, isolating them from the substrate. It is also important that the bulk is connected to the source, because if it is tied to V_{DD}, the power supply noise modulates the current in the differential pair through the bulk transconductance.
- A high transconductance in the second stage is necessary for the amplifier stability, therefore an NMOS is preferred in the second stage.

In modern technologies, the first argument is not so strong anymore, because also NMOS transistors can be put in dedicated pwell. The second always holds, even though in low power application where all transistors work in weak inversion, the transconductance depends only on the bias current, so a PMOS second stage working in weak inversion would have the same g_m of an NMOS one. However, noise must

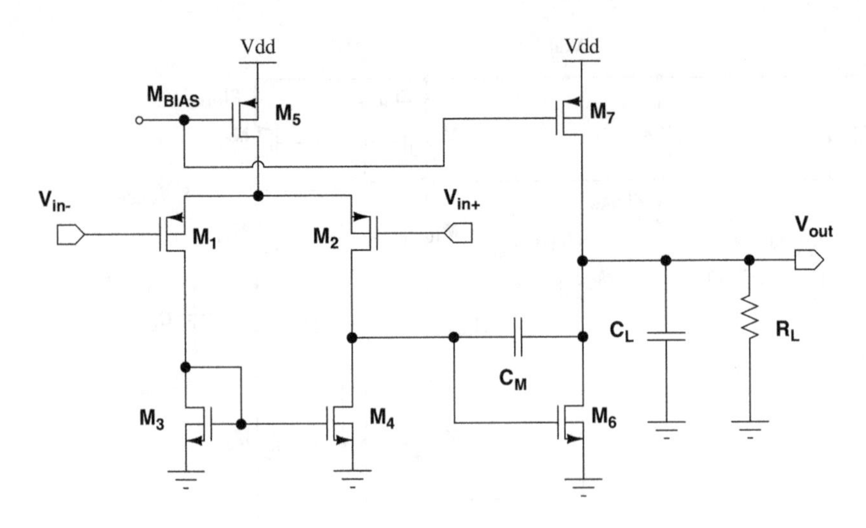

FIGURE A.22 PMOS input two-stage OTA.

also be considered and PMOS offer in general better flicker noise performance. Two stage configurations are preferable when resistive loads need to be driven. In fact, in the small signal model, the load resistors are in parallel to the internal resistors of the second stage, thus lowering the gain. However, the loading does not affect the first stage, so the amplifier can still provide sufficient gain even in case the output stage has to drive fairly small resistors. A detailed design plan for this type of amplifier is discussed in [4].

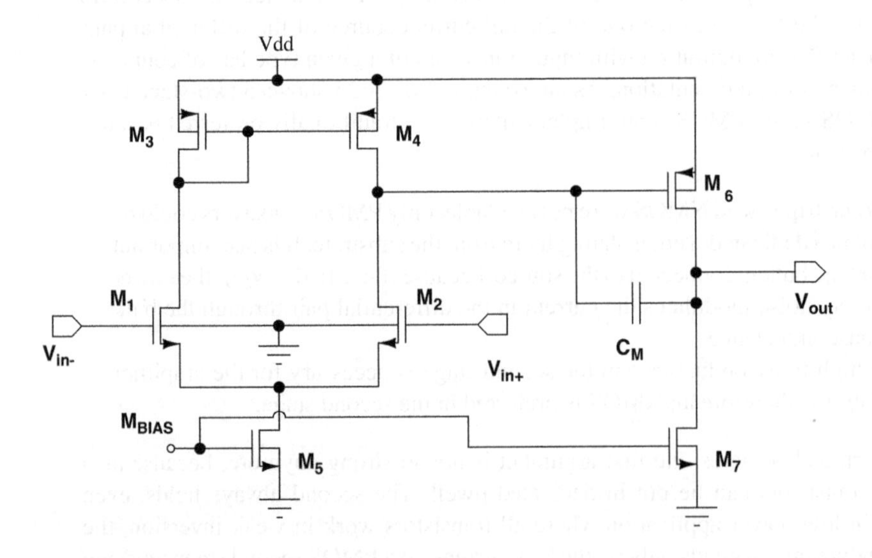

FIGURE A.23 NMOS input two-stage OTA.

A.6.3 OTA WITH CLASS AB OUTPUT

The standard CMOS OTA has an asymmetric current drive capability. Consider, for instance, the circuit of Fig. A.22. The maximum current that can be delivered to the load is equal to the current flowing in the load transistor M_7. Therefore, M_7 must be biased in class A with the maximum expected output current, resulting in high power consumption. The maximum current that can be sunk from the load is primarily defined by the aspect ratio of M_6. However, not all the current provided by M_6 is available for the load, but only the amount that exceeds the bias current of M_7. When high output current must be provided, it is convenient to design the output stage for class AB operation. In this way, at the equilibrium only a small portion of the maximum output current that the stage is able to deliver flows as a DC bias for the output transistors. When a signal is present, the current in the appropriate transistor is increased and the one in the complementary device is reduced, to allow the maximum efficiency in charging or discharging the load. For instance, if the output signal is negative, the current in the NMOS transistor is incremented and the one in the PMOS is minimized. The opposite happens if the output signal is positive. Fig. A.24 shows a common circuit to obtain a class AB control of the output stage. The gates of M_2 and M_3 are biased by a constant voltage and the transistors are sized so that at the equilibrium they equally share the current of M_1, which is equal to the one of M_4. M_2 forms a cascode for M_1 and M_3 is a cascode for M_4. The voltage at the gate of M_6, V_{G6} is given by $V_{B2} + V_{SG2}$ while V_{G5} is $V_{B3} - V_{GS3}$. The gate voltages V_{B2} and V_{B3} are thus chosen to bias the output transistors with a fairly small DC current to keep the power consumption low. When a signal comes, the current in M_1 is increased and the one in M_4 is decreased, or vice-versa. In the former case, both V_{G6} and V_{GS5} rise, because the current in M_2 increases and the one in M_3 decreases. The current in M_5 is thus increased and the one in M_6 is decreased, maximizing the current that can be sunk from the load. The opposite happens when the current in M_1 decreases and the one in M_4 increases. Observe that V_{G5} and V_{G6} are driven in phase, so that the current is

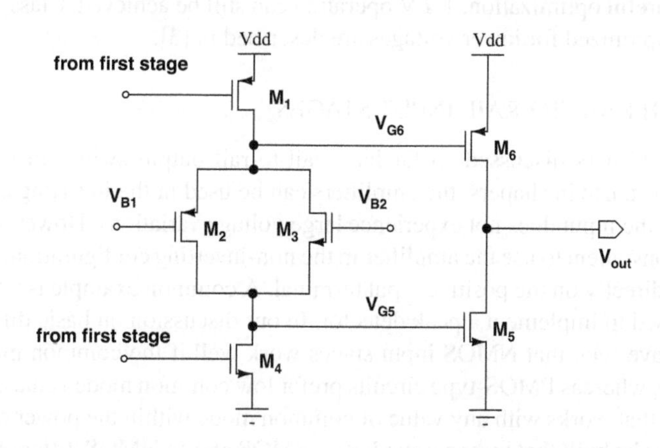

FIGURE A.24 Example of circuit to bias the OTA output stage in class AB.

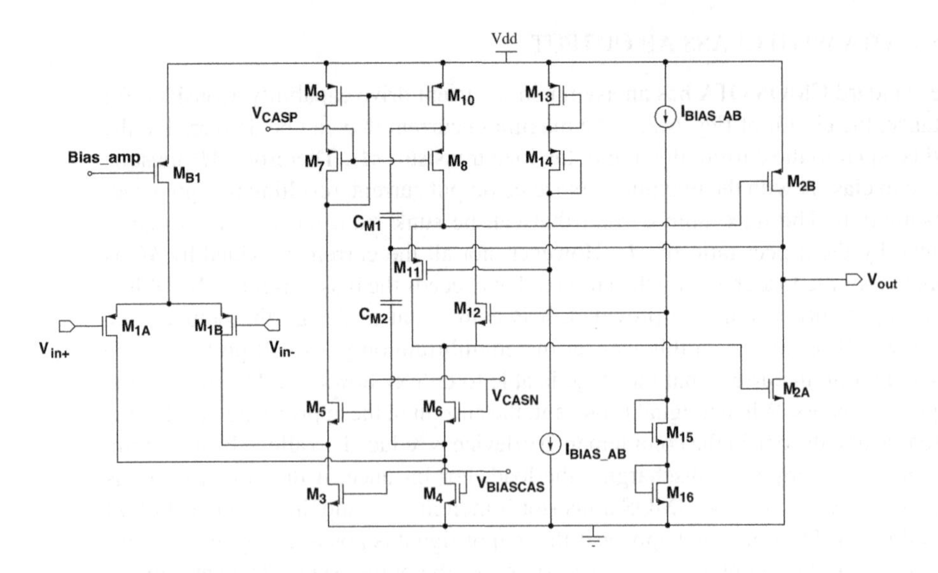

FIGURE A.25 Complete CMOS OTA with class AB output stage.

increased in one device and reduced in the other. This maximizes the current available to the load in either signal polarity. Balancing the aspect ratios of M_5 and M_6 gives to the stage symmetric drive capability. Fig. A.25 reports an example of the class AB control applied to an operational amplifier with folded-cascode input stage. This configuration can be very useful as a general purpose amplifier when large output loads need to be driven or a large slew rate is required. Note that five transistors are stacked in the branch containing the class AB control. However, owing to the fact that the output stage provides gain, the voltage swing in the internal nodes is small. The circuit of Fig. A.25 can thus be implemented also with a supply voltage of only 1.5 V and, with careful optimization, 1.2 V operation can still be achieved. Class AB control schemes optimized for lower voltages are described in [3].

A.6.4 OTA WITH RAIL-TO-RAIL INPUT STAGES

The two stage amplifiers discussed so far have rail-to-rail output swing. In most applications, for instance in shapers, the amplifiers can be used in the inverting configuration, so that the input does not experience large voltage variations. However in some cases it is convenient to use the amplifier in the non-inverting configuration, applying the signal directly on the positive input terminal. A common example is when the amplifier is used to implement a peak detector. In our discussion on basic differential cells, we have seen that NMOS input stages work well if the common mode input level is high, whereas PMOS-type circuits prefer low common mode values. To have an amplifier that works with any value of common mode within the power rails an input stage must be built that incorporates both a PMOS and an NMOS differential pair. One example of the concept is shown in Fig. A.26. Here, when the common

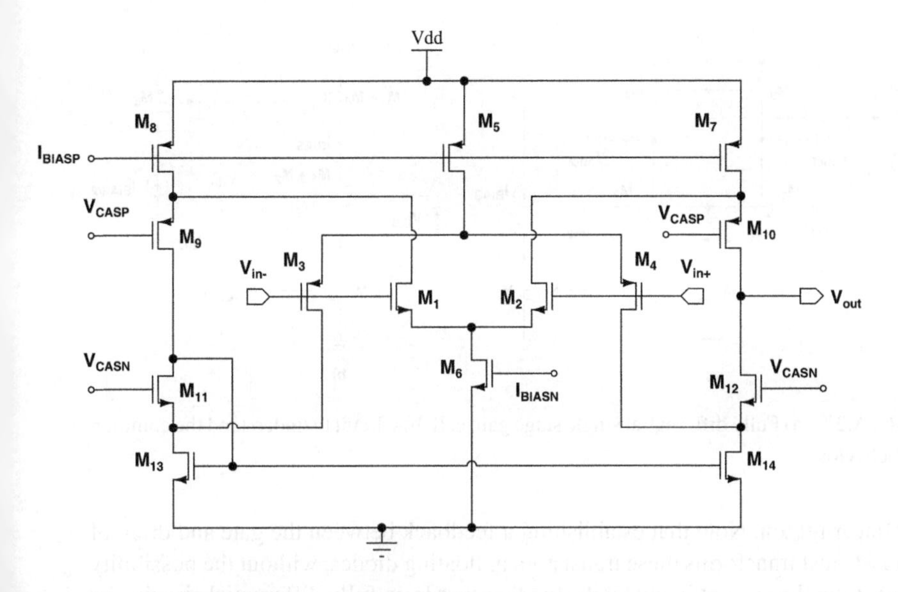

FIGURE A.26 CMOS OTA with rail-to-rail input.

mode voltage is low, the NMOS pair is off, but the PMOS one is active, while the opposite happens if the signal is close to the positive rail. In the intermediate region around $V_{DD}/2$ both pairs are active. In the simple implementation shown in the figure, the transconductance of the input stage is not constant, because it is halved when only one pair is on. This leads to variation in the DC gain and signal bandwidth of the amplifier that can cause signal distortion. To avoid that, techniques have been devised to increase the current in one active pair when the other is off, to keep the overall input transconductance as constant as possible. Such methods are discussed in [3].

A.7 FULLY DIFFERENTIAL AMPLIFIERS

The high gain amplifiers discussed so far have single-ended output. To keep the benefit of differential signal processing, architectures with differential outputs should be employed. Such fully differential structures can be obtained by symmetrizing the loads, breaking the diode connection that implements the differential to single-ended conversion. Fig. A.27 a) shows the concept applied to the single stage differential cell. Both outputs are loaded with a current mirror. Since at the equilibrium M_1 and M_2 share equally the bias current, the loads are design to deliver $I_{BIAS}/2$. To study the common-mode behavior of the circuit, the scheme of Fig. A.27 b) can be used. When both gates of M_1 and M_2 are moved in the same direction. the two transistors can be considered as a single one having the same length of M_1 and M_2 and double width. The same is true for the load transistors. From the figure, we see that on common mode, the circuit is equivalent to a cascode current mirror loaded by another current mirror. Even though the currents are designed to be same, in practice mismatch make a perfect balancing impossible. As usual, when two current mirrors are connected in series, only one works in saturation while the one that needs higher current is pushed

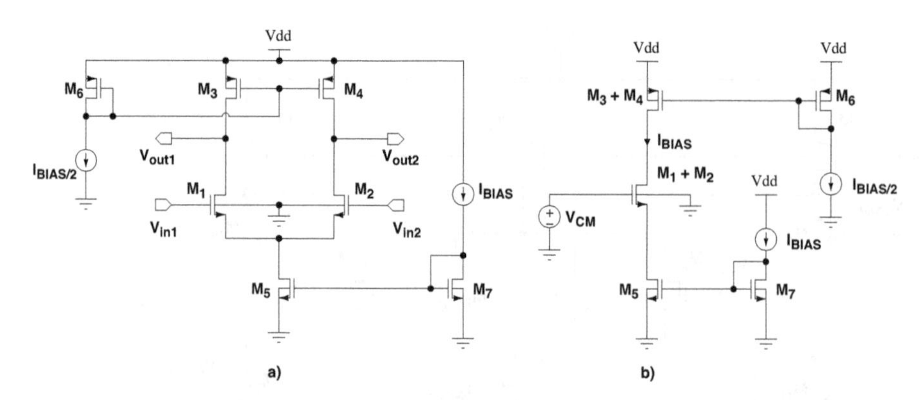

FIGURE A.27 a) Fully differential single stage gain cell. b) Circuit to understand the common mode behavior.

in the linear region. Note that establishing a feedback between the gate and drain of M_1 and M_2 just transforms these transistors in floating diodes, without the possibility of controlling the current in the loads. In other words, in fully differential circuits the signal feedback does not help in stabilizing the DC operating point and an additional network to define the output common mode level is necessary.

Fig. A.28 depicts the principle of the method. The first step consists in generating a signal that is only sensitive to common mode variations, while it ignores the differential ones. As shown in the figure, this in principle can be achieved by connecting two equal resistors ($R_{CM1} = R_{CM2}$) between the arms of the differential pair and taking the middle point. The voltage $V_{CM,sense}$, in fact, follows V_{out1} and V_{out2} when they both move in the same direction, while it behaves as a virtual ground when V_{out1} and V_{out2} swing by the same amount but with opposite phase. $V_{CM,sense}$ thus measures only the common mode variation at the output, ignoring the differential signals and is fed to a differential amplifier. The other terminal of the differential amplifier is connected to a DC reference voltage. The amplifier then drives the main tail source M_5. If the current provided by M_3 or M_4 (or both) is greater than the one arriving from the differential pair, $V_{CM,sense}$ rises with respect to $V_{CM,ref}$. This increases the output voltage of the servo amplifier and thus the gate voltage of M_5, adapting its current to the one provided by the PMOS loads. The opposite happens if the load current is initially smaller than the one requested by M_5. This shows that to stabilize the output common mode level in a fully differential amplifier only one current source can be kept independent, while the other is regulated through the common mode feedback. In our example we have kept independent the loads and controlled the tail current of the differential pair, but the opposite could also be done. It is important to observe that the common mode is controlled through an independent negative feedback network, whose stability must be ensured. The common mode feedback must also be fast. One drawback of the scheme of Fig. A.28 is that the resistors sensing the common mode should be large because they load the amplifier output and can thus reduce the gain. Big resistors occupy however a significant silicon area and may add non-negligible

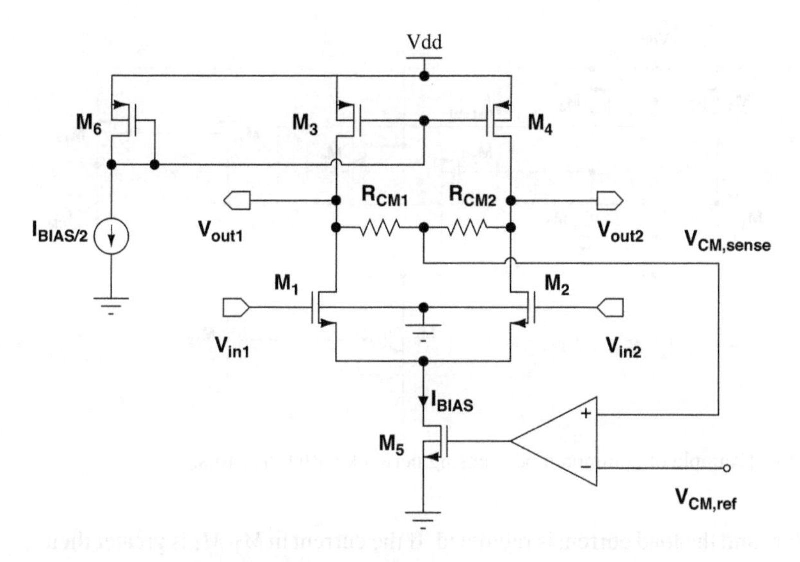

FIGURE A.28 Common mode feedback principle.

parasitic capacitance. The amplifier outputs must hence be buffered before connecting them to a resistive sensing network. Fig. A.29 shows as an example a fully differential folded cascode amplifier with its common mode feedback network. Capacitors C_1 and C_2 are added to stabilize the common mode feedback. It must be also noted that the differential amplifier regulating the common mode loop has an active load and thus it has a small gain. In fact, most of the loop gain is still provided by the active loads of the main amplifier and using too much gain in the auxiliary amplifier can favor the loop instability.

Fig. A.30 shows an alternative scheme in which the common mode feedback is sensed by transistors that do not load the amplifier. In this case, the tail current source

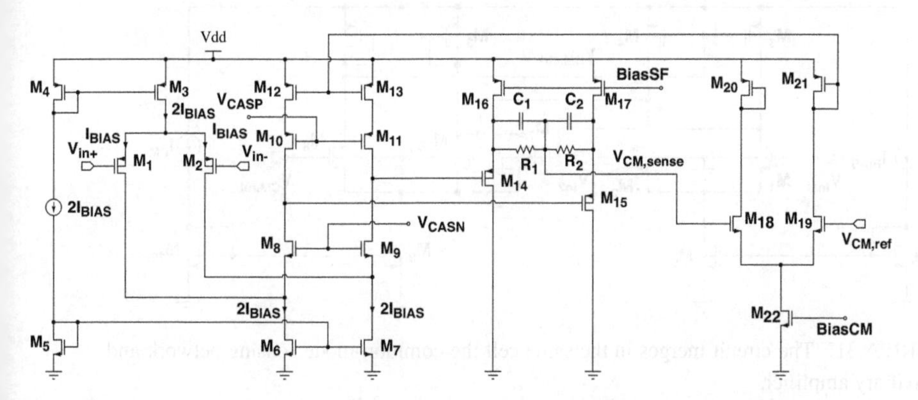

FIGURE A.29 Fully differential folded cascode amplifier with its common mode feedback network.

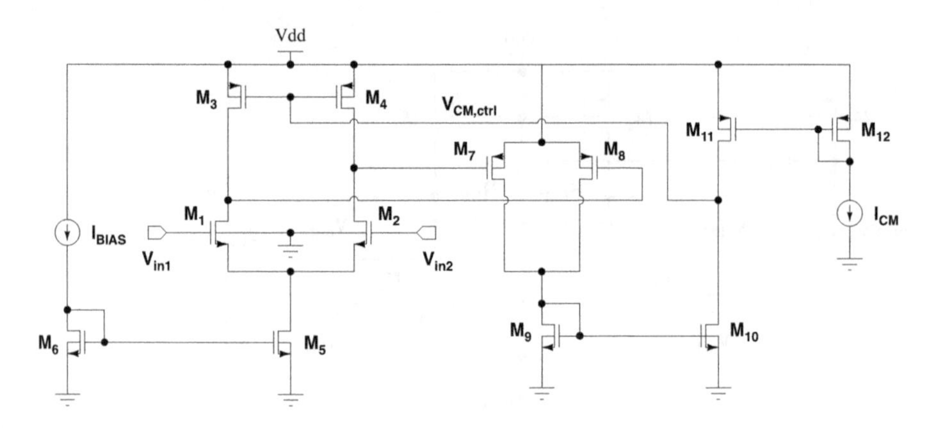

FIGURE A.30 Example of common mode sensing network with transistors.

is independent and the load current is regulated. If the current in M_3-M_4 is greater then necessary, the output voltage rises, reducing the current in M_7-M_8 and, consequently, in M_9. This decreases the gate voltage of M_{10}, thus rising $V_{CM,ctrl}$ and reducing the load currents. In this scheme, the common mode output voltage is determined by transistor size and by the value of I_{CM}. Fig. A.31 shows a better option that locks the common mode voltage to an external reference. Here the sensing network and the auxiliary amplifier are merged in the same circuit. In fact, one of the input transistors of the auxiliary amplifier is split in two equal devices. In case the common mode output voltage changes, so does the sum of the currents in M_{7A} and M_{7B}, thereby activating the common mode feedback. Observe that in both circuits the total current in the common mode sensing devices does not change when a pure differential mode signal is applied.

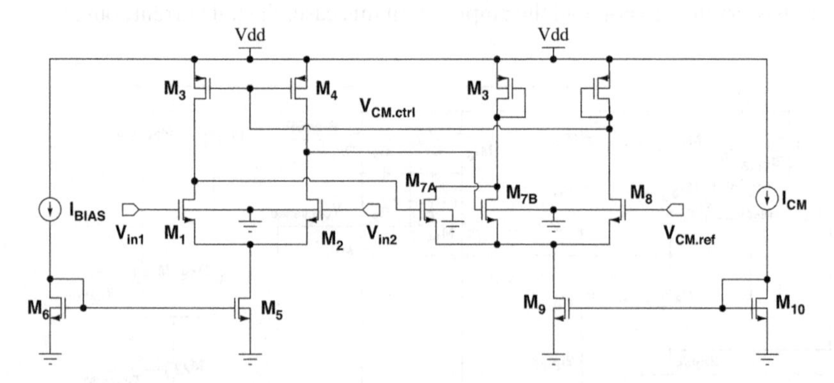

FIGURE A.31 The circuit merges in the same cell the common-mode sensing network and the auxiliary amplifier.

REFERENCES

1. B. Razavi. *Design of Analog CMOS Integrated Circuits.* McGraw-Hill, 2000.
2. G. Rincon-Mora. *Analog IC Design with Low-Dropout Regulators.* McGraw-Hill, 2nd edition, 2014.
3. J. Huijsing. *Operational Amplifiers: Theory and Design.* Springer, 2nd edition, 2011.
4. P. E. Allen and D. R. Holberg. *CMOS Analog Circuits Design.* Oxford University Press, 2nd edition, 2002.

B Practical Aspects in Front-end Design

In this appendix, a few important practical aspects of front-end design are introduced. First, the critical steps leading to a submission-ready ASIC are considered. Following this, the topics of substrate noise, off-chip data transmission, biasing distribution, system calibration and radiation hardness are briefly reviewed.

B.1 THE FRONT-END DESIGN CYCLE

Modern front-end ASICs are often complex mixed-signal chips, in which sensitive analog circuits are integrated together with many digital gates providing storage, control and signal processing functions. The design of such chips require several man-years of work which is completed by different designers. Therefore, the implementation of the analog and digital parts of the ASIC usually proceeds in parallel.

The design flow for analog and digital circuits is substantially different. The implementation of an analog circuit starts with high-level simulations, in which simplified components are used to achieve the desired specifications. A common example can be the study of a shaper in which suitable macromodels are initially employed to represent the core amplifiers. In this early phase, it is also useful to compute analytically the circuit transfer function, so that the interplay between different parameters can be further clarified. It must be pointed out that the specifications of a complex circuit defined by the end-user are often incomplete. For instance, in a front-end amplifier the peaking time, the power consumption, the event rate and the noise are usually given, but it is left to the designer to understand the DC gain, the bandwidth and the slew-rate that must be provided by the individual building blocks to meet the overall goals. Starting with highly idealized models and introducing progressively the performance limitations observed in real circuits is thus an effective way of defining the design targets for the different subcircuits. After this initial step, the transistor level implementation begins by drawing and simulating the circuit schematics. In the schematic editor, transistors and other electrical devices are represented with symbols similar to those used in textbooks. To each component, a model file specific of the chosen technology is attached in order to allow the electrical simulation of the circuit. Once the performance is satisfactory, the layout design is started. The layout is a simplified bidimensional representation of the patterns that will be transferred on the silicon wafer and the layout database is required to generate the production masks. Despite the improvements in CAD tools, analog layout is still a largely hand-crafted activity. When the layout is complete, additional effects due, for instance, to parasitic capacitors and resistors that were not accounted for in the schematics can be identified through post-layout simulations and, if necessary, fixed. Fabricating nanometer-scale devices in a reliable way is not an easy task and the layout design rules of modern deep submicron technologies are very complex. Although the compliance with the design rules is verified by dedicated computer programs, understanding the critical effects introduced in the circuit by the layout is of utmost importance. Very often, silicon foundries provide baseline design rules (usually intended to optimize the area usage) and recommended ones (finalized to achieve the best performance and reliability) and the choice between the two may not be obvious. Furthermore, curing problems emerging from the post-layout verification may require a deep revision of the circuit and can force the designer to go back to the schematics and also to the high level simulations. Sometimes, even the basic architectural choices must be reconsidered. It is therefore important that the layout-driven constraints are considered as early as possible in the design cycle. Several guidelines must be followed to layout analog circuits well performing on silicon. For a detailed treatment, the reader can consult, for instance, [1, 2].

The digital design flow is much more automatized and it begins with a functional description of the circuit written in a Hardware Description Language (HDL). The design is first synthesized, i.e. the gates necessary to perform the required functions are automatically picked-up from a library of standard cells and interconnected by the software. At this stage, key effects that will arise in silicon (such as the load a given cell has to drive) starts to be taken into account. The layout is then automatically generated and routed by a Place&Route program. Achieving a well-performing circuit requires however that the tools are properly driven by the designer. Verification of a digital design is a cumbersome task, because it can be difficult to foresee a priori all the different conditions that will be presented to the circuit in real-life operation.

When the digital and the analog blocks are completed, the final chip integration is performed. At this point, a clear floorplan of the overall ASIC should be already available together with the pads assignment. Integration of a mixed-signal ASIC can be done with two different methodologies. In the so called “analog on top approach” the ASIC is assembled using the same editor employed for analog layout and the routing between the digital and analog blocks is hand-crafted as well as the layout of the power grids and the interconnections between the chip core and the bonding pads. In the “digital on top” technique the analog blocks are processed as custom designed standard cells and the full chip is automatically assembled by the Place&Route software. Although the second method seems much more attractive, in a mixed-signal chip some layout customization can be at the end necessary and the “digital on top” can be particularly advantageous with large chips with a limited number of analog components, such as pixellated binary front-ends. Assembly and top level verification of a complex ASIC can easily take several weeks.

Once the design is completed, it can be sent to the foundry for manufacturing. Layout databases are exchanged through the GDSII format, which is a de facto standard to transfer layouts among different sites. The GDSII file is automatically generated by the CAD tool. However, to prevent possible errors in the format translation, it is important that the GDSII file is re-imported and all verifications are re-run on the imported design before a final green light for production can be given.

The fabrication of integrated circuits requires the use of dedicated masks that allow the selective processing of the different areas of the wafer (see [1] for details). A modern deep submicron process may require more than 50 masks and 200 fabrication steps. In particular, mask fabrication is a lengthy and expensive process and the mask cost increases significantly as the feature size is shrunk. To give a very qualitative reference, at the time of this writing the cost of a full set of masks can range from 40000 US dollars for a 0.35 μm technology to more than 1000000 dollars for a 45 nm process. The total mask area is smaller than the wafer size, therefore a fundamental unit, the reticle, is stepped across the wafer. Typical reticle sizes are between 2 cm$\times$2 cm and 2 cm$\times$3 cm and this in general corresponds to the largest ASIC that can be fabricated in a given process[1]. Very often, the size of a front-end ASIC is smaller than the reticle. This is particularly true when a new front-end is prototyped and "baby-chips" with a limited number of channels are fabricated to test new design concepts. In this case, more chips can be hosted in the same reticle in what is called a "Multi-Project Wafer" that shares the mask cost among different users, making affordable the access also to very advanced technologies, at least for prototyping purposes. Nevertheless, if a full size chip is expected to be produced in a dedicated run, the cost of the final production must be taken into consideration from the very beginning in the choice of the target technology. Fig. B.1 shows a processed silicon wafer. Observe the regular pattern (the reticle) which is repeated over the wafer surface.

B.2 SUBSTRATE NOISE IN MIXED-SIGNAL ASICS

Substrate noise arises when analog and digital circuits operate concurrently on the same chip. The phenomenon is therefore of great importance for all mixed-signal integrated circuits and a rich literature is available on the topic [3–8]. To illustrate the problem, we refer to Fig. B.2, where the action of the digital logic is represented by a simple inverter. When the gate changes its state, its internal and load capacitances must be either charged from ground to V_{DD} or discharged from V_{DD} to ground. This originates fast current spikes in correspondence of the transition points. The duration of the spikes is short (of the order of 50 ps or less), but the current peak can be substantial. During transitions, three main mechanisms can inject parasitic currents into the substrate. First, electrons accelerated by the electric field in the device can have sufficient energy to create electron-hole pairs (impact ionization), with the holes flowing into the substrate. Second, current can be injected via the capacitance associated with the reverse-biased junctions formed between the electrodes and the substrate. Third, the fast current transients create bounce on the supply rails due to the non-zero inductance presented by the bonding wires and the traces on the printed circuit board. The bounce on V_{DD} generates current into the substrate via the capacitive coupling between the nwells in which PMOS transistors are fabricated and the substrate itself. The bounce on the ground terminal couples to the substrate due to the finite resistance of the substrate contacts and of the on-chip ground distribution grid. This last

[1] A technique called stitching is however available to produce larger-than-reticle chips. This technique is of particular interest, for instance, to implement large CMOS imaging sensors.

FIGURE B.1 The figure shows an 8-inch silicon wafer processed in a 0.25 μm CMOS technology. The reticle which is stepped across the wafer contains two chips and is zoomed in the upper-right corner of the picture. Courtesy of the ALICE SDD group of Turin.

mechanism was found to be dominant in an experimental study presented in [5] and it is confirmed to be the key player in substrate noise injection in deep-submicron technologies [7]. Parasitic currents flowing into the substrate can affect a sensitive device, like the input transistor of a front-end amplifier, again in three possible ways: tweaking the local bulk potential and hence the threshold voltage, coupling AC currents via the junction capacitances and modulating the analog ground line which is usually also used to bias the substrate in the analog section of the chip.

Several techniques for substrate noise reduction have been proposed, some of which involves also rather elaborated active noise cancellation schemes [6]. We give hereafter a brief summary of the most common countermeasures:

- The analog section of the ASIC should have its own lines for power, ground and substrate biasing. In particular, it is important that the first stage of the analog front-end does not share any of these lines with other circuits. Up to now we have been concerned with noise injected by digital gates. However, for low-power applications, it can be helpful to use class AB analog circuits to reduce the power consumption. A common example is the output stage of a fast shaper that has to deliver relatively large signals to the rest of the chain.

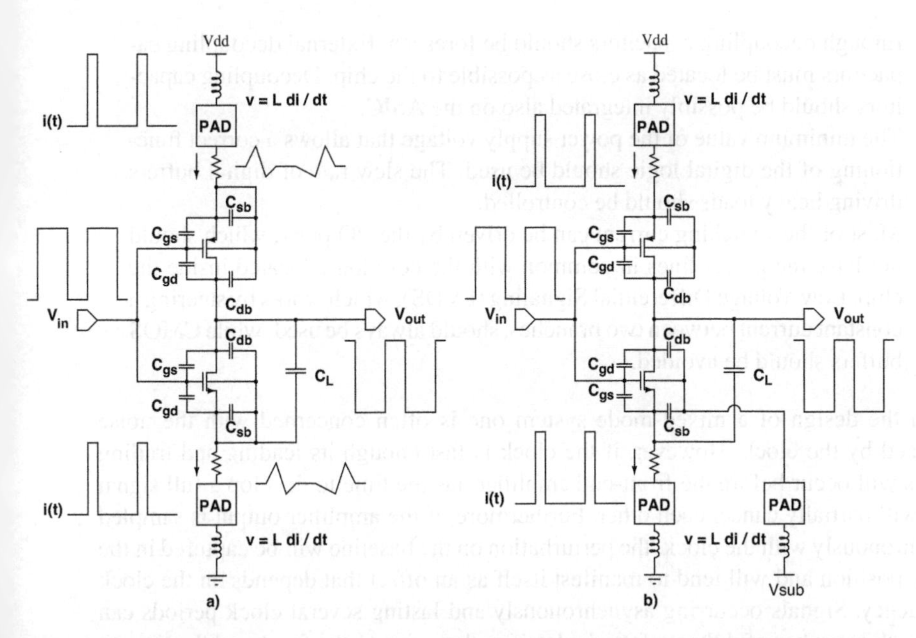

FIGURE B.2 Generation of substrate noise by digital gates switching between the power rails. See text for more details.

A Class AB amplifier can drive a signal current much larger than the one of the quiescent bias point, therefore the current change in its power supply lines can be significant. Sharing them with the first amplifying stage can be detrimental.

- Analog and digital blocks should be spaced as much as possible and shielded by guardrings of substrate contacts. Although the substrate potential is nominally the same in the whole ASIC, these guardrings should nevertheless be connected to different pads on the I/O ring to decrease the risk of contamination between the digital and the analog regions of the bulk. Dedicated isolation trenches that increase the impedance between different regional substrates must be used if available.
- A careful pad assignment should be made, alternating groups of V_{DD} and ground pads on the same domain. Enough pads should be reserved to the power and ground lines in order to reduce the series inductance presented by the wire bondings and the off-chip interconnection.
- An appropriate design of the on-chip power and ground distribution grids must be done to ensure that their resistance is as small as possible.
- If available, one should use a standard cell library in which the bulks of the transistors are not tied to their source. Independent lines should be used to bias the substrate of the cells, and should not be shared with their counterparts employed in the analog section.

- Enough decoupling capacitors should be foreseen. External decoupling capacitors must be located as close as possible to the chip. Decoupling capacitors should be possibly integrated also on the ASIC.
- The minimum value of the power supply voltage that allows a correct functioning of the digital logic should be used. The slew rate of digital buffers driving heavy loads should be controlled.
- Most of the switching current can be driven by the I/O ports, which should not have the power lines in common with the core logic located inside the chip. Low Voltage Differential Signaling (LVDS), which works by steering a constant current between two branches, should always be used, while CMOS buffers should be avoided.

In the design of a mixed-mode system one is often concerned with the noise induced by the clock. However, if the clock is fast enough its leading and trailing edges will occur before the front-end amplifier has the time to develop a full signal and will partially cancel each other. Furthermore, if the amplifier output is sampled synchronously with the clock, the perturbation on the baseline will be captured in the same position and will tend to manifest itself as an offset that depends on the clock frequency. Signals occurring asynchronously and lasting several clock periods can be much more harmful than a fast clock, since they give to the front-end the time to develop a full response.

Finally, it must be stressed that the way parasitic currents spread in a circuit is heavily influenced by the type of substrate. In epitaxial wafers the currents quickly reach the highly-doped part, which behaves as a lumped node. A disturbance in this node is easily picked-up by the analogue circuits through the substrate contacts. Therefore in this case protective guardrings around the sensitive circuits and physical separations between digital and analog blocks, although advisable, may not be very effective in minimizing digital noise pick-up. It has been suggested that a backside contact with very low impedance may help in mitigating the problem [9]. In lightly doped substrates, on the other hand, currents tend to propagate close to the wafer surface. In this case physical separation and guardrings are more helpful. In [5] its is found that the same test circuit is less noisy if fabricated on a lightly doped substrate. Modern very deep submicron processes offer the possibility of having triple well NMOS transistors. In these devices, a deep nwell is fabricated in the substrate. Inside the nwell a pwell is formed to host the NMOS transistor. This has the advantage that the transistor can be isolated from the substrate. Furthermore, the source and the bulk can always be connected together, thereby suppressing the body effect. Although they take significantly more layout area, triple well NMOS are natural candidates in implementing very sensitive devices like the input stage of front-end amplifiers. Triple-well isolation is particularly strong for low-to-medium frequencies, while for high frequencies the nwell-to-substrate capacitance allows the passage of parasitic currents, reducing the shielding effectiveness.

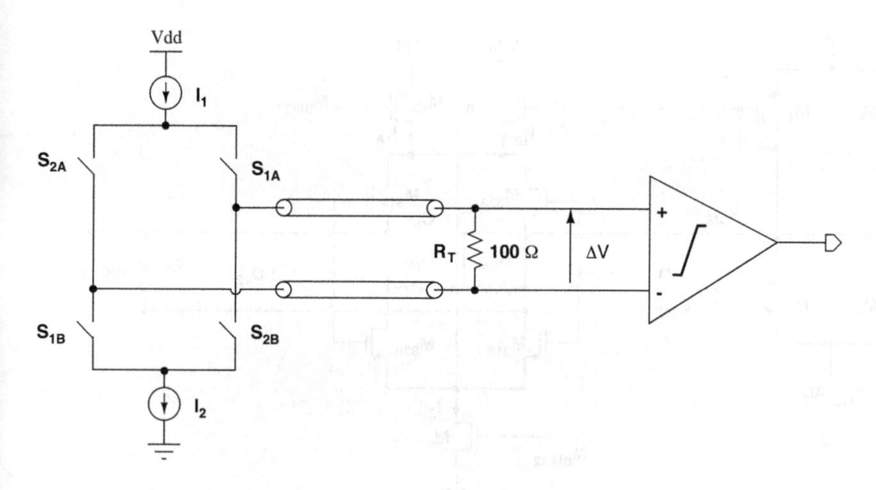

FIGURE B.3 Schematic representation of an LVDS digital link.

B.3 OFF-CHIP DATA TRANSMISSION AND LVDS I/O PORTS

It is advisable that analog data are digitized already on the front-end chip and that the communication between the front-end ASIC and the rest of the readout system occurs only in the digital domain. Several options are available to transmit digital data [10], among which LVDS is particularly suitable in front-end applications. Fig. B.3 shows the principle of an LVDS link [11]. The driver is based on two matched current sources and a four switches organized in pairs. If S_{1A} and S_{1B} are closed, S_{2A} and S_{2B} are open. In this case, the current flows through S_{1A} and reaches the output resistor R_T, which is located outside the chip, determining a voltage drop $+\Delta V$ across it. The current then returns back to the driver arriving at I_2 via S_{1B}. If S_{2A} and S_{2B} are closed, S_{1A} and S_{1B} must be open. The direction of the current flow is reversed and the voltage across R_T becomes $-\Delta V$. In the LVDS standard, the current $I_1 = I_2$ is 3.5 mA[2] and the termination resistance is 100 Ω to match the characteristic impedance of twisted pair cables. The two logic levels are hence separated by a voltage difference of 0.7 V peak-to-peak.

Fig. B.4 reports a simplified scheme of a typical LVDS driver. D_1 and D_{1B} are two complementary CMOS signals. If D_1 is logic high (i.e. equal to V_{DD}), D_{1B} is zero. In this case, the current from the top sources reaches R_T via M_{S1A} and returns back to the bottom NMOS current source through M_{S1B}, while M_{S2A} and M_{S2B} are off. The opposite happens if D_1 is low and D_{1B} is high. The current sources I_1 and I_2 are connected back to back, therefore common mode feedback is necessary to establish proper DC levels. In the circuit reported in the figure, transistors M_2 and M_{1A} are biased independently. The current in M_{1A} is smaller than the one in M_2. The difference between the two is provided by M_{1B} which is controlled by the common mode feedback network, made by the sensing resistors R_{CM1} and R_{CM2} and by the

[2]Current values between 2 mA and 4 mA are however acceptable.

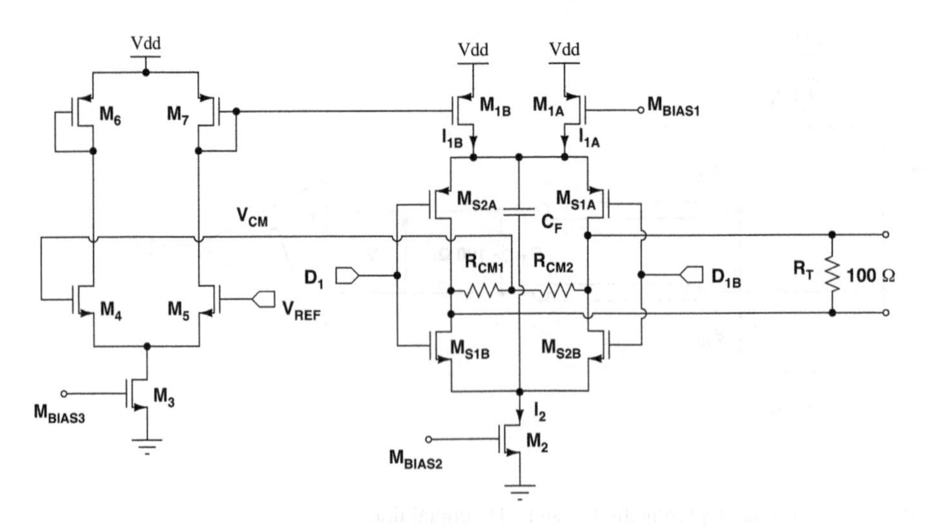

FIGURE B.4 Simplified schematic of an LVDS driver.

differential amplifier formed by transistors $M_3 - M_7$. If, for instance, the current in M_2 is greater than the sum of the ones provided by M_{1A} and M_{1B}, V_{CM} becomes eventually smaller than V_{REF}. This increases the current in $M_5 - M_7$ and in M_{1B}, which delivers the additional current necessary to balance the circuit. If, on the other hand, the current in M_2 is smaller, V_{CM} rises and the current in M_{1B} is reduced. The loop thus locks V_{CM} to V_{REF} by controlling the additional current that M_{1B} provides to the node. The circuit hence works only if the current in M_{1A} is smaller than the one in M_2. Care must be paid to verify that this condition is met also in presence of process variation and mismatch. In the LVDS standard, the common mode voltage is kept at 1.2 V.

A possible receiver design is shown in Fig. B.5. The circuit consists of a differential amplifier followed by a differential-to-CMOS converter and a CMOS buffer. To extend the input common mode range to rail-to-rail operation and increase the link robustness, two complementary circuits are connected in parallel, so that at least one works properly if the input common mode is close to ground or V_{DD}.

The power consumption of LVDS links is sometimes a source of concern. In fact, the driver works by steering between two branches a constant current of 2-4 mA and the need of having an output common mode of 1.2 V forces the use of a minimum power supply voltage of 1.8 V, leading to a power consumption close to 10 mW per driver. As a reference, this is the same power dissipated by a CMOS buffer working at 1.8 V and driving a 320 MHz signal into a 10 pF load. The use of CMOS output pads must nevertheless be avoided, especially if off-chip digital data transmission is concurrent with signal acquisition from the sensor. To understand better this point, consider the following example. A front-end ASIC is specified to provide an ENC of 500 electrons rms. When a channel is hit, the chip raises a flag to the data acquisition system trough a CMOS output port with a 1.8 V swing. A 0.5 fF parasitic capacitance on the PCB between the flag line and the input of the front-end amplifier determines

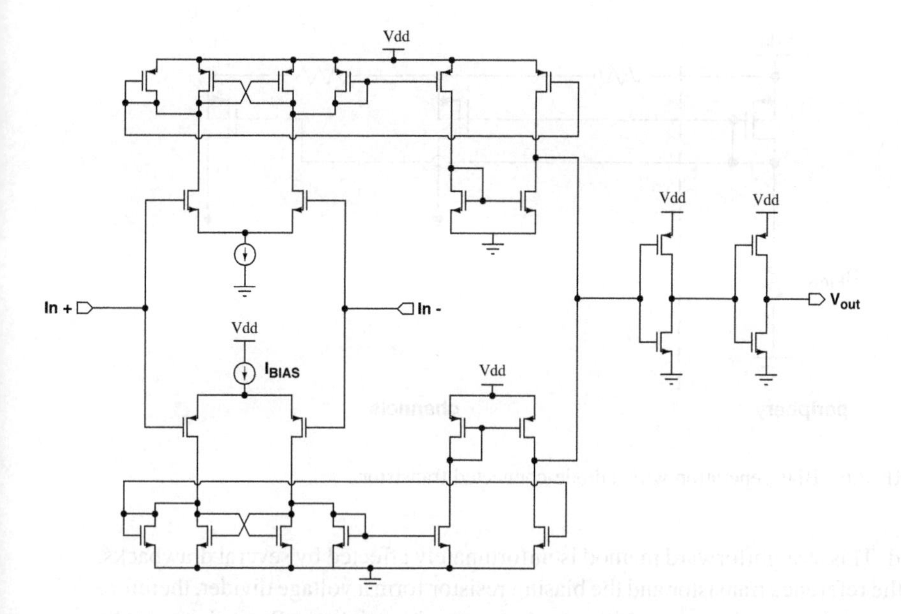

FIGURE B.5 Example of LVDS receiver.

the injection of a charge of almost 0.9 fC (5625 electrons), which is ten times the target noise limit. Even worse, if the spurious signal exceeds the specified threshold, it may trigger another switching event, leading to a disastrous positive feedback. The feedthrough is greatly suppressed by sending digital signals over a differential, low-swing protocol. Furthermore, LVDS links offer a transmission bandwidth in excess of 1Gbit/s, allowing the serialization of many data over a single port.

The use of LVDS greatly reduces also the injection of digital noise into the chip substrate because the current that flows between V_{DD} and ground is constant. As discussed in Appendix 1, differential communication allows moreover a better management of the ground loops at the system level, since the return path of the transmitted signal is clearly defined.

To reduce power consumption, differential standards similar to LVDS, but operating at lower supply voltages such as the Scalable Low-Voltage Signaling (SLVS) [12] can also be considered. An example of this approach can be found in [13].

B.4 BIAS DISTRIBUTION IN MULTI-CHANNEL FRONT-ENDS

An issue that should not be overlooked in multi-channel systems is the generation and distribution of the currents and voltages needed to bias the analog circuits. In principle, each front-end channel should have its own biasing cells. However, this is not always possible due to constraints on area and power consumption, which force to share the biasing blocks among several channels. Fig. B.6 shows the simplest approach. A resistor is put in series to a diode-connected transistor. The gate voltage of the latter is then distributed to the different channels, where a copy of the reference device is

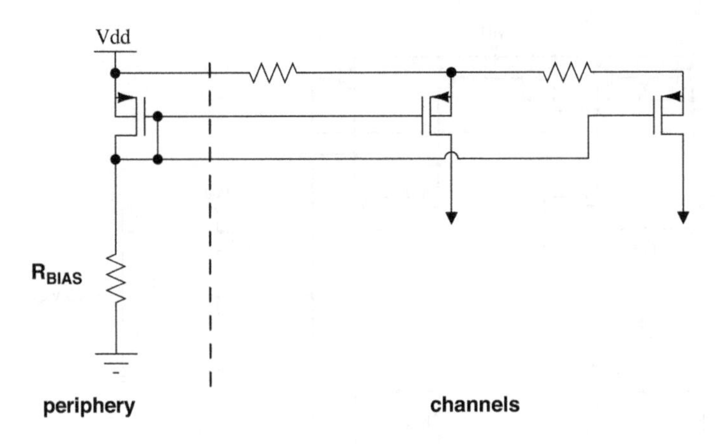

FIGURE B.6 Bias generation with a diode-connected transistor.

located. This straightforward method is unfortunately affected by several drawbacks. First, the reference transistor and the biasing resistor form a voltage divider, therefore the generated current is very sensitive to power supply variations. Second, it must be considered that the resistance of the power and ground distribution grid is not zero and the current circulating in the power rails originates voltage drops (known as IR drops) along its path. Gradients on the effective value of V_{DD} and "ground" inside the chip must thus be expected. As a consequence, the source of the receiving transistors is found on a different potential than the one of the reference device, originating systematic gradients in the bias current which is provided to the different channels. The problem becomes particularly severe when the power pads are located only on one side of the ASIC. An additional issue is that the gate line, which is common to all channels, can favor the cross-talk between them, especially when large signal swings are expected in the biased circuits. Fig. B.7 shows a better circuit that cures the first and the third problem. Assuming that a clean voltage reference is available in the system, a current-to-voltage converter is used to generate a current which is weakly sensitive to the value of V_{DD}. RC filters are also inserted along the voltage distribution line to dump spurious signals appearing on the common gate line and to allow a better decoupling between the channels. In the scheme of Fig. B.7 the current gradients caused by the IR drops however remain. These can be mitigated as shown in Fig. B.8. Here, the PMOS transistors are located close to the reference device so that their source is kept at the same potential. The currents are then distributed to the channels. The disadvantage of the method is that, in principle, at least one dedicated bias line per channel is necessary. In front-end circuits, however, it is not uncommon to have circuits whose bias currents differ by orders of magnitude. The baseline holder, for instance, needs only few nano-amperes, whereas the preamplifier input transistor may require hundreds of micro-amperes. In practice more independent lines per channel could hence be necessary, leading to a complicated routing. A compromise consists in generating a limited number of independent currents in the periphery. Locally, a

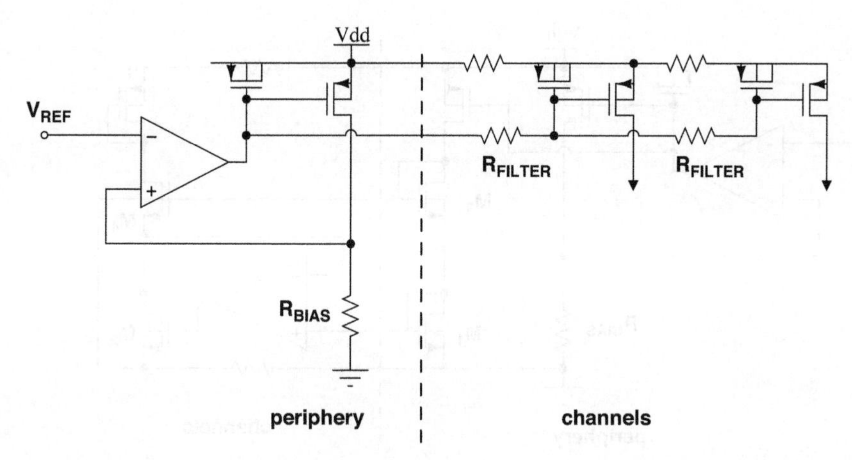

FIGURE B.7 A current-to-voltage converter generates a bias current independent of V_{DD}. Filters are used in the bias line to reduce the cross-talk between the different channels.

receiving diode-connected transistor provides a reference voltage which is fanned-out to nearby channels, where "nearby" means a distance over which the effect of the IR drops is not relevant [14]. Finally, a technique to distribute an IR-drop insensitive bias while using a limited number of lines is shown in Fig. B.9. Here the reference current is measured by two diode-connected transistors. An operational amplifier, located in the receiving channel, equalizes through the virtual ground principle the source potential of M_3 (in the periphery) and M_4 (in the channel). The gate of M_3 and M_4 are

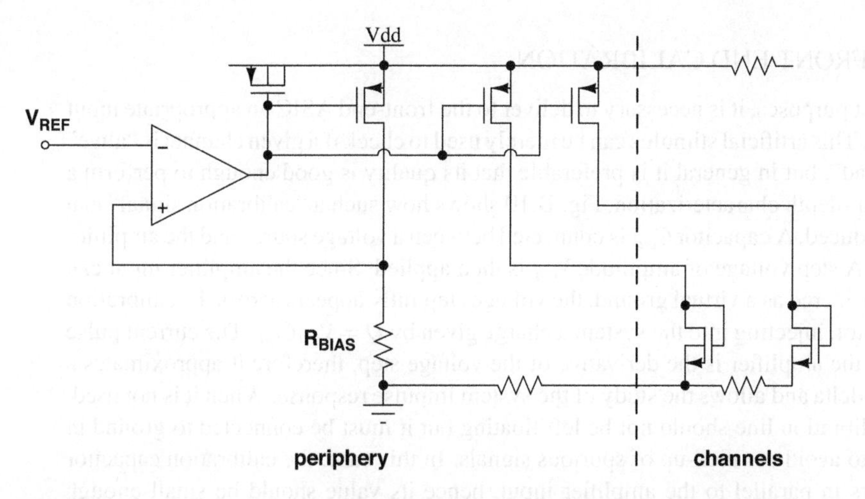

FIGURE B.8 Current mode biasing. The bias currents are generated in the chip periphery and then individually distributed to the channels. As a compromise, the receiving diode connected transistor can be shared by nearby channels.

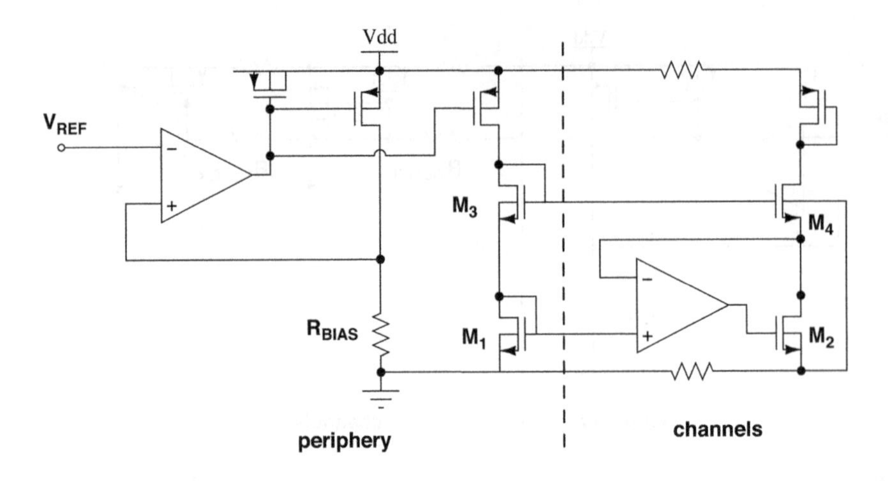

FIGURE B.9 IR drop insensitive biasing using op-amps.

directly tied together since the gate terminals do not absorb current and no IR drop is expected over this line. With this scheme, M_3 and M_4 thus work with the same gate-source voltage, regardless of the exact local value of the V_{DD} and "ground" nodes. An application of this principle is discussed in [15]. One downside of the method is that the op-amp takes additional space and its offset can introduce further random differences in the bias current between the channels. As for the circuit of Fig. B.8, as a compromise the receiving circuit can be shared between channels sufficiently close by.

B.5 FRONT-END CALIBRATION

For test purposes, it is necessary to deliver to the front-end ASIC an appropriate input signal. This artificial stimulus can be merely used to check if a given channel is "alive" or "dead", but in general it is preferable that its quality is good enough to perform a more in-depth characterization. Fig. B.10 shows how such a "calibration signal" can be produced. A capacitor C_{cal} is connected between a voltage source and the amplifier input. A step voltage of amplitude V_{cal} is then applied. Since the amplifier input can be considered as a virtual ground, the voltage step fully appears across the calibration capacitor, injecting into the system a charge given by $Q = V_{cal}C_{cal}$. The current pulse fed to the amplifier is the derivative of the voltage step, therefore it approximates a Dirac-delta and allows the study of the system impulse response. When it is not used, the calibration line should not be left floating but it must be connected to ground in order to avoid the pick-up of spurious signals. In this case, the calibration capacitor appears in parallel to the amplifier input, hence its value should be small enough to make negligible its impact on the amplifier noise. In the simplest approach, the calibration capacitors are connected to a line which is driven by an external signal. In general, more of these "calibration lines" are used, each one serving non-contiguous

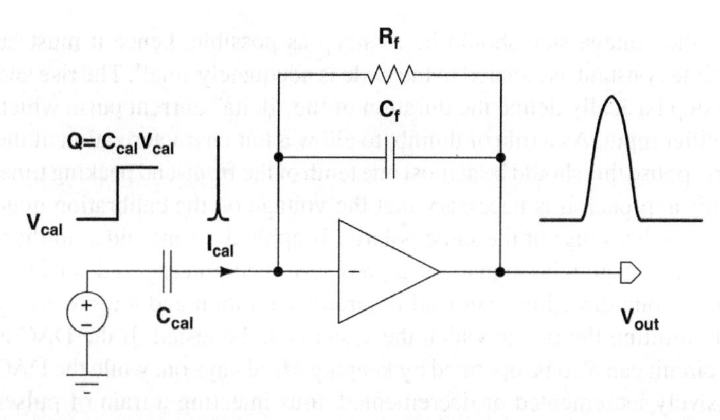

FIGURE B.10 A test signal can be injected by applying a voltage step to a capacitor connected in series to the system input.

channels to permit also the study of cross-talk. It is however more preferable that the test pulse is generated directly on chip. Fig. B.11 shows how this can be achieved. A current-mode DAC generates a programmable current which is fed to a resistor. Inside each channel, an appropriate switch connects the calibration capacitor either to the calibration line or to ground, in which case the channel does not respond to the injected stimulus. Transistors M_1 and M_2 act as a current steering switch and they are triggered by an external digital signal. If the DAC current is dumped from M_1 to M_2, the node connected to the calibration line is discharged, injecting a negative step into the circuit. The opposite is true when the current is steered from M_2 to M_1. Observe

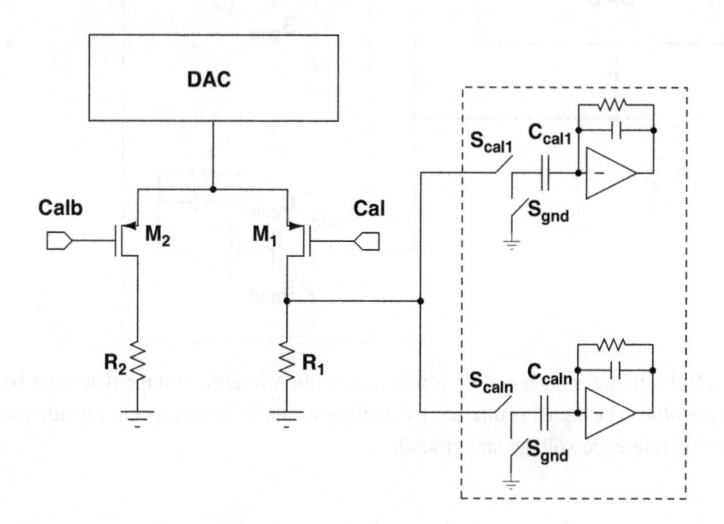

FIGURE B.11 Calibration circuit with current steering.

that the edges of the voltage step should be as steep as possible, hence it must be ensured that the time constant associated to the node is adequately small. The rise and fall times of the step basically define the duration of the "delta" current pulse which is fed to the amplifier input. As a rule of thumb, to allow a fair characterization of the system impulse response this should be at most one tenth of the front-end peaking time. Note that with this approach it is necessary that the voltage on the calibration node is restored before another pulse of the same polarity is applied. Front-end amplifiers are often optimized to cope with signals of a predetermined polarity and injecting a stimulus in the "wrong direction" can lead to circuit saturation and long recovery times, potentially limiting the rate at which the system can be tested. If the DAC is fast enough, the circuit can also be operated by keeping M_1 always on, while the DAC word is progressively incremented or decremented, thus injecting a train of pulses always with the same polarity. The potential of the node must however be restored once the DAC reaches the maximum or the minimum possible code. Fig. B.12 reports another method. Here, the calibration line is programmed to a given voltage and then it is held as stable as possible by filtering it with a sufficiently big capacitor. The signal is injected by steering the test capacitor inside the channel from the calibration line to ground or vice-versa. The described approaches just provide two examples and many alternatives are possible. It is nevertheless important to retain that an on-chip programmable pulse injector is an integral part of a high performance front-end ASIC.

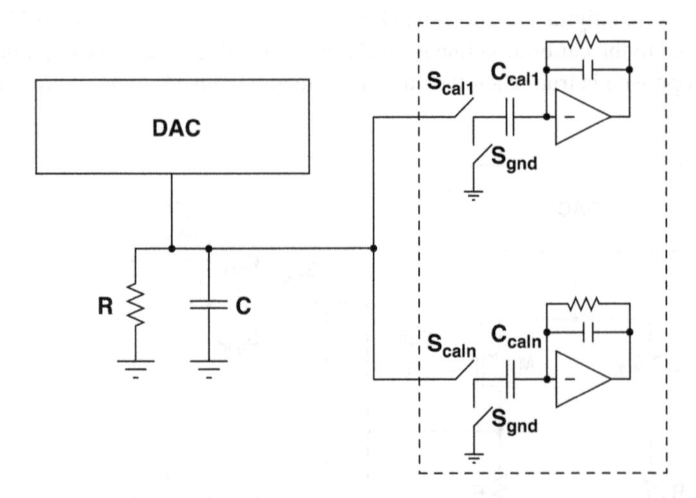

FIGURE B.12 Calibration circuit based on a programmable reference voltage that must be kept as stable as possible. The signal is injected by steering the calibration capacitor inside the channel between this reference voltage and ground.

B.6 RADIATION DAMAGE

In a number of applications the sensor and its front-end electronics are exposed to high doses of radiation that can cause severe damage to the system performance. Prominent examples are high energy physics detectors equipping particle accelerators (especially when hadron beams are involved), X-ray detectors employed at high intensity sources such as the free electron lasers and, although to a smaller extent, space instrumentation. Owing to relevance to the military field and to the avionics industry, the effects of radiation on MOS transistors and electronic devices in general are extremely well studied. For a broad coverage of the topic the reader can consult [16, 17] and references therein.

The effects of radiation on electronic circuits can be grouped in two categories: cumulative effects and Single Event Effects (SEE). Cumulative effects in turn can be caused by two different types of radiation. Total Ionizing Dose (TID) effects, as the name implies, are generated by charged particles or photons that can ionize the traversed medium. Displacement damage, on the other hand, is determined by particles such as neutrons that can knock an atom out of its intended position in the material. In silicon, the main result of the displacement damage is the reduction of the minority carrier lifetime. This is of scarce relevance for MOS transistors, in which the conduction is granted only by majority carriers flowing in a thin layer located very close to the device surface. Ionizing radiation instead can severely affect the device characteristics.

In CMOS technologies, the critical material is silicon dioxide (SiO_2) that is used for the gate oxide, for device isolation and for separating the routing metals. In the oxide, charged particles and sufficiently energetic photons produce electron-hole pairs. Electrons and holes in SiO_2 have very different mobility, hence if an electric field is applied, the two are quickly separated. In absence of electrons, the hole can not recombine and are trapped in the oxide, where they can originate two types of defects. In the oxide itself, the trapped holes behave as positive charges. At the interface between the oxide and the crystalline silicon they can generate the so called interface states. The build-up of interface states is slower than the direct hole trapping, therefore the device characteristics still evolve also after the irradiation has been stopped. Interface states are amphoterous, so they can behave as donors or acceptors. A donor trap releases an electron when it passes from below to above the Fermi level, whereas an acceptor trap captures an electron when it passes from above to below the Fermi level. A donor trap is neutral when full and positively charged when empty, whereas an acceptor trap is negatively charged when full and neutral if empty.

Fig. B.13 shows the top view of an MOS transistor. The intersection between the gate and the diffusion region, where the source and drain are fabricated, defines the area where the thin oxide that provides the capacitive coupling between the gate and the channel is grown. To assure a complete coverage of the channel, the gate extends also over the STI. Charge trapping and interface state build-up in the gate and in the STI oxide are therefore the mechanisms that alter the performance of the MOS device.

When a transistor is irradiated with the gate biased to a given potential, the resulting electric field separates the electrons and the holes. In an NMOS transistor, the gate

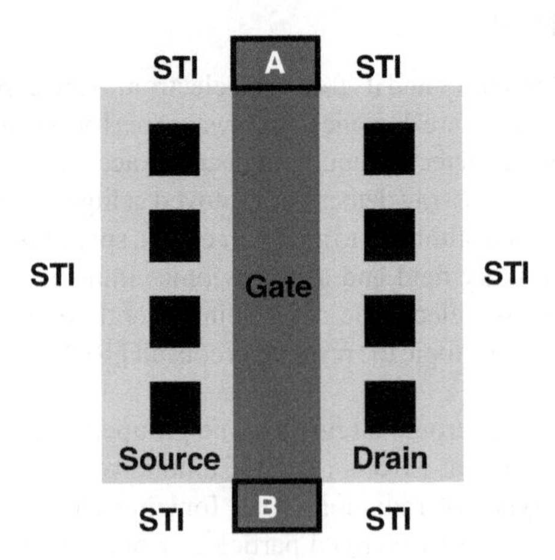

FIGURE B.13 Top view of an MOS transistor. Note that the gate electrode extends over the STI region, where the oxide is thicker.

is positively biased, so the holes migrate toward the gate-channel interface, while the electrons are attracted towards the oxide-gate boundary. In an PMOS transistor, the opposite is true. In an NMOS, the holes directly trapped in the gate oxide attract electrons in the channel and they favor the channel inversion, lowering the device threshold. The interface states, on the other hand, trap electrons becoming negatively charged, therefore they act in the opposite direction. This is at the origin of the "rebound" phenomenon observed in irradiated MOS devices, where the threshold voltage first decreases and then somewhat recovers. In PMOS transistors, the holes trapped in the oxide repel the ones in the channel, hence they hinder the channel inversion. The empty interface states are also positively charged and act in the same sense, therefore the effect of irradiation is to increase, in absolute value, the threshold of a PMOS device.

Modern CMOS technologies employ however very thin gate oxides, with a thickness of 2 nm or less. Tunneling of electrons from the silicon bulk into the oxide is thus an effective mechanism to remove the trapped charge. Therefore, only very modest threshold voltage shifts are observed that usually are not relevant for circuit performance. The situation is different in the regions where the gate and the STI overlap. The oxide in the STI areas is in fact much thicker, and a significant amount of trapped holes and interface states is formed here. In an NMOS device, the holes trapped in the STI can be sufficient to locally invert the channel even if the potential applied on the gate is meant to keep the transistor off. This creates parasitic conductive paths at the channel boundaries that put the source and drain in contact, creating an undesired leakage current. The action of the interface states counteracts this phenomenon, but with a delay due to their slower formation. It is thus observed that the leakage current first increases with irradiation and start decreasing again at higher doses, after a sufficient number of interface states has been created [18]. In PMOS transistors instead the threshold of the parasitic devices in increased, therefore radiation damage is a primary concern for NMOS devices.

In modern CMOS technologies the channel of minimum size devices can be so narrow that the gate regions overlapping the STI (indicated as "A" and "B" in Fig. B.13) becomes very close to each other. The charge trapped in these areas can therefore influence directly the channel formation underneath the gate, competing with the gate electrode itself in the control of the transistor current. It has in fact been observed that, for the same irradiation dose, the leakage current can significantly increase for very narrow channel device, originating the so called Radiation Induced Narrow Channel Effect (RINCE) [18].

In analog circuits, radiation-induced leakage currents can impair the functionality of analog switches and of circuits working deep in weak inversion, where the leakage current may become comparable with the intended bias current of the device, whereas they are less important for circuits that operate with bias currents above a few μA. In digital circuits, even a small leakage current can significantly increase the static power consumption of the digital gates. Finally, it must be mentioned that the presence of interface states, which are located at the boundary between the oxide and the conducting channel, reduces the carrier mobility and thus the transconductance of the device. This can affect the noise of analog circuits and the switching speed of digital ones. In irradiated MOS transistors, an increase of both flicker and thermal noise is also observed, but to an extent that does not prevent the design of high performance analog circuits [19].

Fig. B.14 shows another situation that can lead to radiation-induced leakage. This time, the parasitic conductive path is formed between the diffusions of two different devices. The figure shows a strip of polysilicon drawn over the STI that separates two transistors. The additional polysilicon could be used, for instance, to implement a passive resistor or it could be a dummy structure inserted to respect the pattern density constraint imposed by the technology design rules. If the strip is biased during irradiation it can favor the separation of electrons and holes in the STI. The charge trapped in the oxide can induce an inversion layer at the boundary between the STI and the silicon bulk located underneath. If the source and drain junctions of the two nearby transistors are held at different potentials, a parasitic current can thus flow, originating an inter-device leakage. A similar situation may occur between an n-type diffusion (source or drain of an NMOS device) and the nwell hosting a PMOS transistor. TID effects heavily depend on the detailed properties of the STI oxide, therefore the TID sensitivity strongly changes from one process to the other even in the same technology node.

In modern CMOS processes, radiation damage can be largely mitigated by applying Hardening By Design (HBD) methodologies. For TID effects, the key point is to cut the possible conductive paths between drain and source of the same transistor and between the diffusions belonging to different devices. This can be achieved employing Enclosed Layout Transistors (ELT) and guardrings [19]. The difference between a

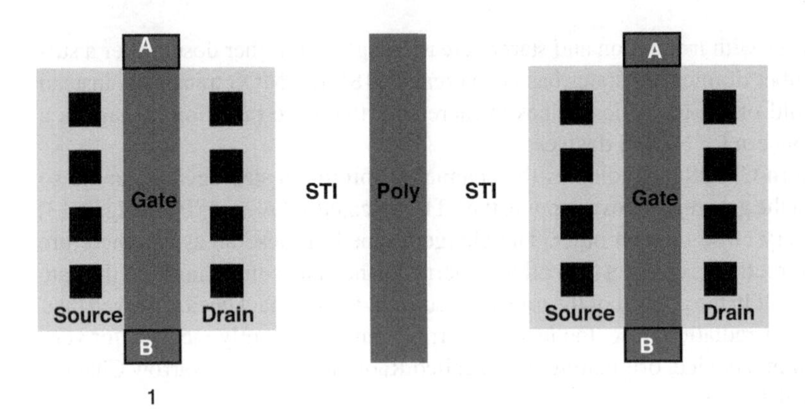

FIGURE B.14 Situation leading to radiation-induced leakage between different devices. A strip of polysilicon or metal can act as the gate of a parasitic device which has the STI as its gate oxide.

standard device and an ELT is shown in Fig. B.15. In the ELT the gate fully encloses one of the electrodes. The transistor thus becomes asymmetrical, because the external electrode is larger than the internal one. This asymmetry introduces a difference in the output resistance, which is higher if the external electrode is chosen as the drain [19]. The anular gate prevents the formation of any conductive path between drain and source, thereby solving the issue of device-level leakage. To cut possible parasitic paths towards neighboring devices, the transistor is surrounded by a ring of p^+ substrate contacts.

The use of ELTs allows us to implement circuits with extremely good radiation hardness in fully standard CMOS technologies, where designs tolerating more than 100 Mrd[3] of total ionizing dose have been reported[20]. Although it greatly enhances the TID tolerance, the adoption of ELTs in a design has also several drawbacks [19]:

- Enclosed layout geometries occupy more area;
- Calculating the effective aspect ratio of an ELT is not trivial and complicated formulas must be used;
- Matching between ELT devices is somewhat worse than for standard ones.

An additional, but very important, side effect is that foundry design kits and digital standard cell libraries commercially available consider only linear geometries. Therefore, the adoption of ELT implies that custom design kits and digital libraries must be developed, adding a huge burden to the design effort.

Starting from the 130 nm technology node, it has been observed that a good TID tolerance can however be achieved also with linear geometries, provided that some guidelines (e.g. avoiding very narrow devices) are followed [18].

[3]The SI unit of the absorbed dose is the Gray, which corresponds to the absorption of 1 J of energy by one kg of matter. The rad (rd) is however still used. One Gray corresponds to 100 rd.

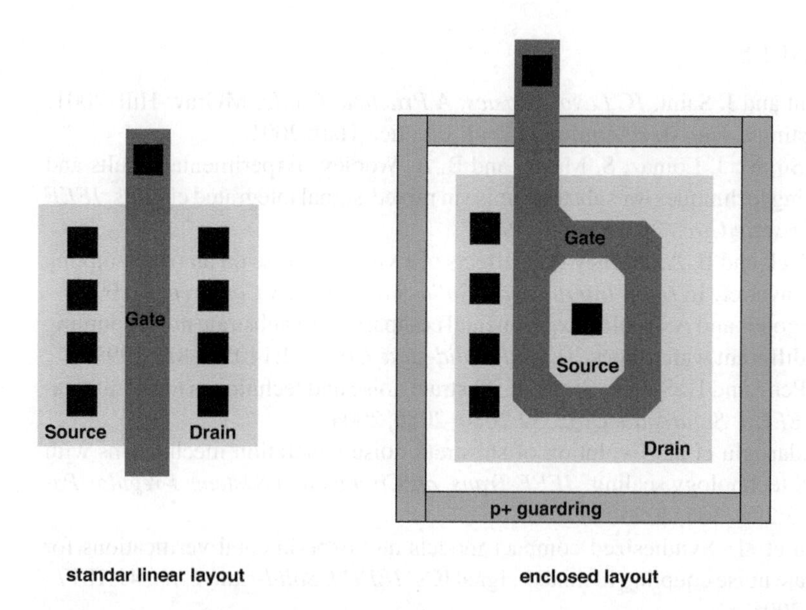

FIGURE B.15 Typical enclosed layout transistor. For comparison, a device with standard linear geometry is shown on the left.

Single event effects originate from the interaction of an individual particle with the circuit. Although many different types of SEEs exist, the one of utmost importance in mixed signal front-ends is the Single Event Upset (SEU) which consists in the flipping of one or more bits in a digital register. In general, this is a threshold phenomenon that depends on the Linear Energy Transfer (LET) of the impinging particle, because the deposited charge must be sufficient to change the status of a circuit node. The smaller the node capacitance, the more probable the upset is, therefore SEU sensitivity typically worsens by moving to more scaled technologies. The relevance of an SEU depends on the function of registers which is affected and it is a primary concern only when the register is used to store a configuration pattern or the states of a Finite State Machine (FSM) that controls critical operations on the chip. SEU effects are counteracted mainly by using appropriate architectures in the design of the digital logic. For instance, the states of an FSM can use Hamming encoding, so that two allowed states always differ by more than one bit and the flipping of a single bit is not sufficient to impair the circuit. Other methods use redundancy, for example by triplicating the registers and performing a majority voting on their contents, so that a corruption of one register can be tolerated. These methods do not require the design of dedicated standard cell libraries, but imply the use of more gates for performing the same functions, thereby increasing the circuit area and power consumption. The immunity to SEU can of course also be increased by designing custom storage cells, again at the expense of a larger area.

REFERENCES

1. C. Saint and J. Saint. *IC Layout Basics: A Practical Guide*. MGraw-Hill, 2001.
2. A. Hastings. *The Art of Analog Layout*. Prentice-Hall, 2001.
3. D. K. Su, M. J. Loinaz, S. Masui, and B. A. Wooley. Experimental results and modeling techniques for substrate noise in mixed-signal integrated circuits. *IEEE J. Solid-state Circ.*, 28:420–430, 1993.
4. T. Blalack and B. A. Wooley. The effects of switching noise on an oversampling A/D converter. In *IEEE International Solid-State Circuits Conference*, 1995.
5. X. Aragonés and A. Rubio. Experimental comparison of substrate noise coupling using different wafer types. *IEEE J. Solid-state Circ.*, 34:1405–1409, 1999.
6. M. S. Peng and H. S. Lee. Study of substrate noise and techniques for minimization. *IEEE J. Solid-state Circ.*, 39:2080–2086, 2004.
7. M. Badaroglu et al. Evolution of substrate noise generation mechanisms with CMOS technology scaling. *IEEE Trans. on Circuits and Systems I-regular Papers*, 53:296–305, 2006.
8. H. Lan et al. Synthesized compact models and experimental verifications for substrate noise coupling in mixed-signal ICs. *IEEE J. Solid-state Circ.*, 41:1817–1829, 2006.
9. T. Gabara. Reduced ground bounce and improved latch-up suppression through substrate conduction. *IEEE J. Solid-state Circ.*, 23:1224–1232, 1988.
10. J. M. Rabaey, A. Chandrakasan, and B. NiKolic. *Digital Integrated Circuits: A design perspective*. 2nd edition, Prentice Hall, 2003.
11. *LVDS Owner's Manual*. Texas Instruments, 4th edition, 2008.
12. JEDEC solid-state technology association. Scalable Low-Voltage Signaling for 400 mV (SLVS-400). *JEDEC8-13*, October 2001.
13. S Bonacini, K Kloukinas, and P Moreira. E-link: A radiation-hard low-power electrical link for chip-to-chip communication. In *Topical Workshop on Electronics for Particle Physics*, pages 422–425, 2009.
14. M. D. Rolo. *Integrated Circuit Design for Time-of-Flight PET*. PhD thesis, University of Turin, 2014.
15. M. Manghisoni, E. Quartieri, L. Ratti, and G. Traversi. High accuracy injection circuit for pixel-level calibration of readout electronics. In *IEEE NSS-MIC Conference Records*, pages 1312–1318, 2010.
16. D. M. Fleetwood and R. D. Schrimpf (editors). *Radiation Effects and Soft Errors in Integrated Circuits and Electronic Devices*. World Scientific, 2004.
17. K. Iniewski (editor). *Radiation Effects in Semiconductors*. CRC Press, 2011.
18. F. Faccio and G. Cervelli. Radiation-induced edge effects in deep submicron CMOS transistors. *IEEE Trans. Nucl. Sci*, 52:2413–2420, 2005.
19. G. Anelli et al. Radiation tolerant VLSI circuits in standard deep submicron CMOS technologies for the LHC experiments: practical design aspects. *IEEE Trans. Nucl. Sci.*, 46:1690–1696, 1999.
20. W. Snoeys et al. Layout techniques to enhance the radiation tolerance of standard CMOS technologies demonstrated on a pixel detector readout chip. *Nucl. Instr. Meth.*, A439:349–360, 2000.

Index

U

V